PATRICIA CUMMINGS

BENSON'S ANATOMY & PHYSIOLOGY LABORATORY MANUAL

Complete Version

WCB/McGraw-Hill

A Division of The **McGraw·Hill** *Companies*

BENSON'S ANATOMY AND PHYSIOLOGY Laboratory Manual, Complete Version

This book is printed on recycled paper.

1 2 3 4 5 6 7 8 9 QPD/QPD 9 3 2 1 0 9 8

ISBN 0–697–34201–8

Vice president and editorial director: *Kevin T. Kane*
Executive editor: *Colin H. Wheatley*
Sponsoring editor: *Kristine Tibbetts*
Developmental editor: *Brittany J. Rossman*
Marketing manager: *Heather K. Wagner*
Project manager: *Susan J. Brusch*
Production supervisor: *Sandy Ludovissy*
Coordinator of freelance design: *Michelle D. Whitaker*
Photo research coordinator: *John C. Leland*
Art editor: *Joyce Watters*
Supplement coordinator: *David A. Welsh*
Compositor: *Carlisle Communications Ltd.*
Typeface: *11/12 Times Roman*
Printer: *Quebecor Printing Book Group/Dubuque, IA*

Freelance cover designer: *Terri W. Ellerbach*
Cover images: *Inset: © Don W. Fawcett/PhotoResearches; Background: From THE HUMAN FIGURE by David K. Rubins. Copyright 1953 by The Studio Publishers. Copyright © renewed 1981 by David K. Rubins. Used by permission of Viking Penguin, a division of Penguin Books USA Inc.*
Photo Credits: *Chapter 48: 48.8-48.11: McGraw-Hill Companies/photographer Bob Coyle*
Chapter 53: 53.1(1): Courtesy Becton Dickinson; 53.1(2): McGraw-Hill Companies/photographer James Shaffer

www.mhhe.com

BENSON'S

ANATOMY & PHYSIOLOGY LABORATORY MANUAL

Complete Version

PATRICK GALLIART
North Iowa Area Community College

Boston Burr Ridge, IL Dubuque, IA Madison, WI New York San Francisco St. Louis
Bangkok Bogotá Caracas Lisbon London Madrid
Mexico City Milan New Delhi Seoul Singapore Sydney Taipei Toronto

Contents

Preface

This is the first edition of the new laboratory manual series based on Harold Benson's *Anatomy & Physiology Laboratory Textbook.* This alternate version of Benson's *Complete Version* differs from the standard version in several ways, while still maintaining the focus and depth of coverage of the standard version.

The most obvious difference between this version and the standard version involves the illustrations. Most have been revised substantially: all the unlabeled figures from the standard version are now labeled, and most pieces have been rendered in full color. It is our hope that these changes will facilitate both instruction and learning. The labeled diagrams should help those students whose teachers choose not to hold the students responsible for every structure described in the laboratory textbook. Another significant change is the replacement of many line drawings with new photographs of individuals performing laboratory procedures.

There has also been a concerted attempt during the development of this version to clarify what exactly students should be doing as they work through the laboratory exercises. Most of the exercises contain assignments that either relate to laboratory procedures or involve the students answering questions on the Laboratory Reports. In many exercises, students are directed to complete each part of a Laboratory Report as the material is presented in the exercise.

Key Features

1. **Labeled Art Program.** All figures are completely labeled for ease of student comprehension and identification of anatomical features on dissection specimens.
2. **Full Color.** The artwork and design in this manual brings a whole new look to the Benson series through the use of full-color art and design elements throughout the entire manual.
3. **Clarity in Writing.** This manual has been carefully written and edited to avoid wordy explanations and passive-voice constructions which may lead to student misunderstandings.
4. **Emphasis on Safety.** Safety instructions are provided in the Hematology section in order to help students protect themselves and others from infectious diseases while performing specific blood tests. Similar instructions are also provided at other points in this version where living tissues or body fluids are encountered.
5. **Multimedia Tie-In.** A CD-ROM icon appears in several of the exercises after the materials section. The icon represents *The Virtual Physiology Lab* CD-ROM and signals the reader that a supplemental laboratory exercise can be found on the CD-ROM. The CD-ROM can be packaged with this Laboratory Manual for a minimal fee, or can be bought separately.
6. **Unopette System.** The instructions for performing blood cell counts have been revised in this new edition, and the Unopette system for determining blood cell counts has been employed.
7. **Intellitool Exercises.** Four of the exercises in this version involve using Intellitool hardware and software. The instructions provided in these exercises have been updated to agree with the latest versions available in the Intellitool products line at the time this manual was prepared.

Supplemental Materials

1. *Instructor's Manual To Accompany Benson's Anatomy & Physiology Laboratory Textbook, Complete Version* by Patrick Galliart (0–697–38700–3) contains setup information, time allotments, answers to the questions in the Laboratory Reports, and bar codes for the *Slice of Life* videodisc series.
2. *The Dynamic Human CD-ROM,* Version 2.0 (0–697–38935–9), contains three-dimensional and other visualizations of relationships between human structure and function. This CD-ROM can be packaged with this Laboratory Manual for a minimal fee, or can be bought separately.
3. *The Dynamic Human Videodisc* (0–697–38937–5) contains all the animations (200+) from the first edition CD-ROM, with a bar code directory.
4. *Virtual Physiology Lab CD-ROM* (0–697–37994–9) has 10 simulations of animal-based experiments common in the physiology component of a laboratory course; allows students to repeat experiments for improved mastery.
5. *Laboratory Atlas of Anatomy and Physiology,* 2/e (0–697–39480–8) by Eder et al. is a full-color atlas containing histology, human skeletal anatomy, human muscular anatomy, dissections, and physiology reference tables.
6. *WCB Life Science Animations* Videotape Series contains 53 animations on five VHS videocassettes: Chemistry, The Cell, and Energetics (0–697–25068–7);

Cell Division, Heredity, Genetics, Reproduction, and Development (0–697–25069–5); Animal Biology I (0–697–25070–9); Animal Biology II (0–697–25071–7); and Plant Biology, Evolution, and Ecology (0–697–26600–1). Another videotape available is Physiological Concepts of Life Science (0–697–21512–1).

7. *Explorations in Human Biology CD-ROM* by George Johnson (0–697–37906–X IBM, 0–697–37907–8 Mac) consists of 16 interactive animations on human biology.
8. *Explorations in Cell Biology, and Genetics CD-ROM* by George Johnson (0–697–37908–6) contains 17 colorful animations that afford an engrossing way for students to delve into these often-challenging topics.
9. *Life Sciences Living Lexicon CD-ROM* by William Marchuk (0–697–37993–0) provides interactive vocabulary-building exercises, including meanings of word roots, prefixes, and suffixes, with illustrations and audio pronunciations.
10. *Anatomy and Physiology Videodisc* (0–697–27716–X) has more than 30 physiological animations, line art, and photomicrographs, with a bar code directory.
11. *WCB Anatomy and Physiology Video Series* consists of four videotapes, free to qualified adopters, including: Internal Organs and the Circulatory System of the Cat (0–697–13922–0); Blood Cell Counting, Identification, & Grouping (0–697–11629–8); Introduction to the Human Cadaver and Prosection (0–697–11177–6); and Introduction to Cat Dissection: Musculature (0–697–11630–1).
12. *Human Anatomy and Physiology Study Cards* (0–697–26447–5) by Van De Graaff et al. is a boxed set of 300 3-by-5 inch cards, each of which summarizes a concept in structure and function, defines terms, or provides a concise table of related information.
13. *Coloring Review Guide to Anatomy and Physiology* (0–697–17109–4) by Robert and Judith Stone emphasizes learning through the process of color association. The Coloring Guide provides a thorough review of anatomical and physiological concepts.
14. *Atlas of the Skeletal Muscles* (0–697–13790–2) by Robert and Judith Stone illustrates each skeletal muscle in a diagram that the student can color, and provides a concise table of the origin, insertion, action, and innervation of each muscle.
15. *Case Histories in Human Physiology,* 3 edition, by Van Wynsberghe and Cooley (Internet-based) stimulates analytical thinking through case studies and problem solving; includes an instructor's answer key.
16. *Survey of Infectious and Parasitic Diseases* (0–697–27535–3) by Kent M. Van De Graaff is a booklet of essential information on 100 of the most significant infectious diseases.

Acknowledgments

A number of people were of great help during the development of this Laboratory manual. I wish to thank the following persons at WCB/McGraw-Hill: Kris Noel-Tibbetts, Brittany Rossman, Susan Brusch, Michelle Whitaker, John Leland, and Joyce Watters. I would also like to thank Carol Schutte, my friend and colleague at North Iowa Area Community College for her input, and my wife Jo Ann for her encouragement. I owe much of the success this manual may achieve to these people, and to the following reviewers who kindly reviewed this manual during its development:

Joan I. Barber
Delaware Technical and Community College

Robert Bauman
Amarillo College

Ramesh Bhimani
New York University

Diana L. Curley
Catonsville Community College

Tom Dudley
Angelina College

Rev. Joseph C. Gregorek
Gannon University
Pennsylvania State University College of Medicine

Laura Hebert
Angelina College

Mary Katherine Lockwood
University of New Hampshire

Bradford Douglas Martin
LaSierra University

Holly J. Morris
Lehigh Carbon Community College

Margaret M. Nowack
Gwynedd Mercy College

Barry Palmerton
Kirtland Community College

John D. Pasto
Middle Georgia College

Brian K. Paulson
California University of Pennsylvania

John M. Ripper
Butler County Community College

Laura H. Ritt
Burlington County College

Geraldine Y. Ross
Highline Community College

Casey A. Shonis
Bloomsburg University

Patricia Turner
Howard Community College

Terry P. Wheeler
Halifax Community College

Peter J. Wilkin
Purdue University North Central

Introduction

These laboratory exercises have been developed to provide you with a basic understanding of the anatomical and physiological principles that underlie medicine, nursing, dentistry, and other related health professions. Laboratory procedures that reflect actual clinical practices are included whenever feasible. In each exercise you will find essential terminology that will become part of your working vocabulary. Mastery of all concepts, vocabulary, and techniques will provide you with a core of knowledge crucial to success in your chosen profession.

During the first week of this course your instructor will provide you with a schedule of laboratory exercises in the order of their performance. You will be expected to have familiarized yourself with each experiment prior to that week's session, thus ensuring that you will be properly prepared so as to minimize disorganization and mistakes.

The Laboratory Reports coinciding with each exercise are located at the back of the book. Remove each page by tearing it along the wire coil, and hand the report in to your instructor as you are directed.

The exercises of this laboratory guide consist of three types of activities: (1) anatomical dissections, (2) physiological experiments, and (3) microscopic studies. The following suggestions should be helpful:

Dissections

Although cadavers provide the ideal dissection specimens in human anatomy, they are expensive and thus may be impractical for large classes in introductory anatomy and physiology. Since much of cat anatomy is similar to human anatomy, the cat has been selected as the primary dissection specimen in this manual. The rat will also be used. Occasionally, sheep and cow organs will be studied. Frogs will frequently be used in physiological experiments.

When using live animals in experimental procedures, you must handle them with great care and minimize pain in all experiments. Inconsiderate or haphazard treatment of any animal (live or preserved) will not be tolerated.

Physiological Experiments

Before performing any physiological experiments, be sure that you understand the overall procedure. Reading the experiment prior to entering the laboratory will help a great deal.

Handle all instruments carefully. Most pieces of equipment are expensive, may be easily damaged, and are often irreplaceable. The best insurance against breakage or damage is to thoroughly understand how the equipment is expected to function.

Maintain astuteness in observation and record keeping. Record data immediately; postponement detracts from precision. Insightful data interpretation will also be expected.

Microscopic Studies

Cytology and histological studies will be made to lend meaning to text descriptions. Familiarize yourself with the contents of the Histology Atlas, which includes photomicrographs of most of the tissues you will study in this course. If drawings are required, execute them with care, and label those structures that are significant.

Laboratory Efficiency

Success in any science laboratory requires a few additional disciplines:

1. Always follow the instructor's verbal comments at the beginning of each laboratory session. It is at this time that difficulties will be pointed out, group assignments will be made, and procedural changes will be announced. Take careful notes on substitutions or changes in methods or materials.
2. Don't be late to class. Since so much takes place during the first ten minutes, tardiness can cause confusion.
3. Keep your work area tidy at all times. Books, bags, purses, and extraneous supplies should be located away from the work area. Tidiness should also extend to assembly of all apparatus.
4. Abstain from eating, drinking, or smoking in the laboratory.
5. Report immediately to the instructor any injuries that occur.
6. Be serious-minded and methodical. Horseplay, silliness, or flippancy will not be tolerated during experimental procedures.
7. Work independently, but cooperatively, in team experiments. Attend to your assigned responsibility, but be willing to lend a hand where necessary. Participation and development of laboratory techniques are an integral part of the course.

Laboratory Reports

When seeking answers to the questions and problems on the Laboratory Reports, work independently. The effort you expend to complete these reports is as essential as doing the experiment. The easier route of letting someone else solve the problems for you will handicap you at examination time. You are taking this course to learn anatomy and physiology. No one else can learn it for you.

Supplemental Materials

A variety of materials can be purchased separately to supplement this laboratory manual. Please see the preface for a list and description of these items, or call the WCB/McGraw-Hill Customer Service Department at 1-800-338-3987.

PART 1 Some Fundamentals

This part, which consists of only three exercises, has two objectives: (1) to equip you with some of the basic anatomical terminology that you will need to get started, and (2) to preview the systems and organs of the body that are studied in greater detail as you progress through the course.

1 Anatomical Terminology

Anatomical description would be extremely difficult without specific terminology. Many students believe that anatomists create multisyllabic words in a determined conspiracy to harass the beginner's already overburdened mind. Naturally, nothing could be further from the truth.

Scientific terminology is created out of necessity. It is a precise tool that allows us to say a great deal with a minimum of words. Conciseness in scientific discussion saves time and also promotes clarity of understanding.

Most of the exercises in this laboratory manual employ the terms defined in this exercise. The terms are used liberally to help you to locate structures on the illustrations, models, and specimens. Knowing the exact meanings of these words will enable you to complete the required assignments.

Relative Positions

The pairs of words that follow describe how one structure is positioned with respect to another. The Latin or Greek derivations of the terms are provided to help you understand their meanings.

Superior and Inferior The terms *superior* and *inferior* denote vertical levels of position. The Latin word *super* means *above;* thus, a structure that is located above another one is superior. Example: The nose is *superior* to the mouth.

The Latin word *inferus* means *below* or *low;* thus, an inferior structure is one that is below or under some other structure. Example: The abdomen is *inferior* to the chest.

Anterior and Posterior The terms *anterior* and *posterior* describe front and back positioning of structures. The word *anterior* is derived from the Latin, *ante,* meaning *before.* A structure that is anterior to another one is in front of it. Example: Bicuspids (premolars) are *anterior* to molars.

Anterior surfaces are the most forward surfaces of the body. The front portions of the face, chest, and abdomen are anterior surfaces.

Posterior is derived from the Latin *posterus,* which means *following.* The term is the opposite of anterior. Example: The heel is *posterior* to the toes.

When these two terms are applied to the surfaces of the hand and arm, the body is assumed to be in the *anatomical position,* which is shown in *figures 1.1* and *1.2.* In the anatomical position, the palms of the hands face forward.

Cranial and Caudal The terms *cranial* and *caudal* may be used in place of *anterior* and *posterior* to describe the location of structures of four-legged animals. Since the word *cranial* pertains to the skull (Greek: *kranion,* skull), it may be used in place of anterior. The word *caudal* (Latin: *cauda,* tail) may be used in place of *posterior.*

Dorsal and Ventral The terms *dorsal* and *ventral* are mainly used in discussing the comparative anatomy of animals. For an animal walking on all fours, dorsal (Latin: *dorsum,* back) surfaces are the *upper* surfaces (e.g., along the back), and ventral (Latin: *venter,* belly) surfaces are the *underneath* surfaces (e.g., along the belly).

When applied to humans, the word *dorsal* refers to the back of the trunk of the body and also may be used in describing the back of the head and the back of the hand. In the anatomical position, a human's dorsal surfaces are also posterior. A four-legged animal's back, on the other hand, occupies a superior position.

The word *ventral* generally pertains to the abdominal and chest surfaces. However, the underneath surfaces of the head and feet of four-legged animals are also often referred to as ventral surfaces. Likewise, the palm of the hand may be referred to as ventral.

Proximal and Distal The terms *proximal* and *distal* describe parts of a structure with respect to its point of attachment to some other structure. In the case of an arm or leg, the point of reference is where the limb attaches to the trunk of the body. In the case of a finger, the point of reference is where the finger attaches to the palm of the hand.

Proximal (Latin: *proximus,* nearest) refers to that part of the limb nearest the point of attachment. Example: The upper arm is the *proximal* portion of the arm.

Distal (Latin: *distare,* to stand apart) means the opposite of proximal. Anatomically, the distal

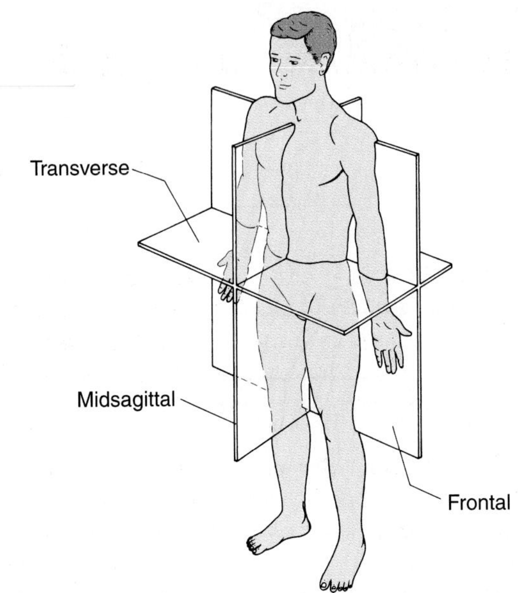

Figure 1.1 Body sections.

portion of a limb or other part of the body is that portion that is most remote from the point of reference (attachment). Example: The hand is *distal* to the arm.

Medial and Lateral The terms *medial* and *lateral* describe surface relationships with respect to the median line of the body. The *median line* is an imaginary line on a plane that divides the body into right and left halves.

The term *medial* (Latin: *medius,* middle) can be applied to surfaces of structures that are closest to the median line. The medial surface of the arm, for example, is the surface next to the body because it is closest to the median line. *Medial* can also be used to describe the location of one structure in relation to another. Example: The nose is *medial* to the eyes.

The term *lateral* is the opposite of medial. The Latin derivation of this word is *lateralis,* which pertains to *side.* The lateral surface of the arm is the outer surface, or that surface farthest away from the median line. The sides of the head are lateral surfaces, and the ears are lateral (and posterior) to the nose.

Assignment:
Before proceeding, use a skeleton or human torso model to practice using the six pairs of orientation terms just described.

Body Sections

Observing the structure and relative positions of internal organs requires viewing sections that have been cut through the body. The body as a whole has only three planes to identify. Figure 1.1 shows these sections.

Sagittal Sections A section that is parallel to the long axis of the body and that divides the body into right and left sides is a *sagittal section.* If such a section divides the body into equal halves, as in figure 1.1, it is a *midsagittal section.*

Frontal Section A longitudinal section that divides the body into front and back portions is a *frontal* or *coronal* section.

Transverse Sections Any section that cuts through the body in a direction that is perpendicular to the long axis is a *transverse* or *cross section.*

Although these sections have been described here only in relationship to the body as a whole, they can also be used on individual organs, such as the brain or kidneys.

Assignment:
Complete part A of the Laboratory Report for this exercise at this time.

Regional Terminology

Various terms, such as *flank, groin, brachium,* and *hypochondriac,* are applied to specific regions of the body to facilitate localization. Figures 1.2 and 1.3 show the locations and names of many body regions.

Trunk

The anterior surface of the trunk can be subdivided into two pectoral regions, an abdominal region, and the groin. The upper chest regions may be designated as **pectoral** or **mammary** regions. The anterior trunk region that the ribs do not cover is the **abdominal** region. The area where the thigh of the leg meets the abdomen is the **groin.**

The posterior surface of the trunk can be subdivided into the costal, lumbar, and buttocks regions. The **costal** (Latin: *costa,* rib) portion is the part of the dorsal surface of the trunk, or **dorsum,** that lies over the rib cage. The lower back region between the ribs and hips is the **lumbar** or **loin** region. The **buttocks** are the rounded eminences of the rump that the gluteal muscles form; this is also called the **gluteal** region.

The side of the trunk that adjoins the lumbar region is called the **flank.** The armpit region between the trunk and arm is the **axilla.**

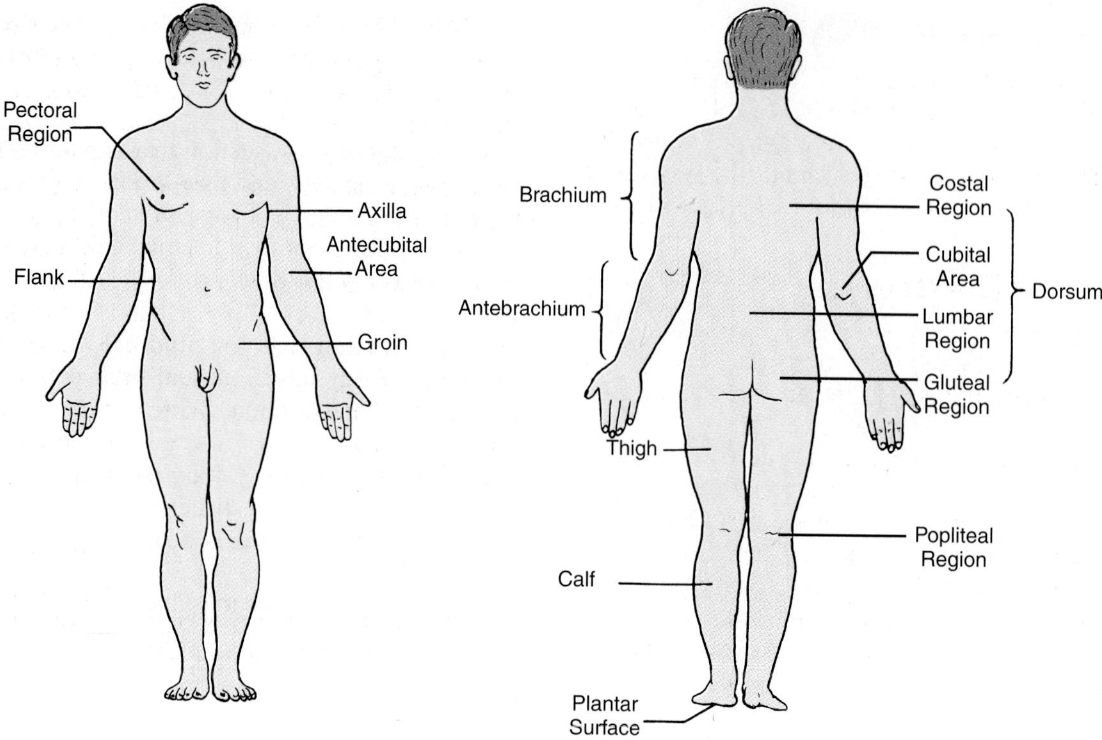

Figure 1.2 Reginal terminology.

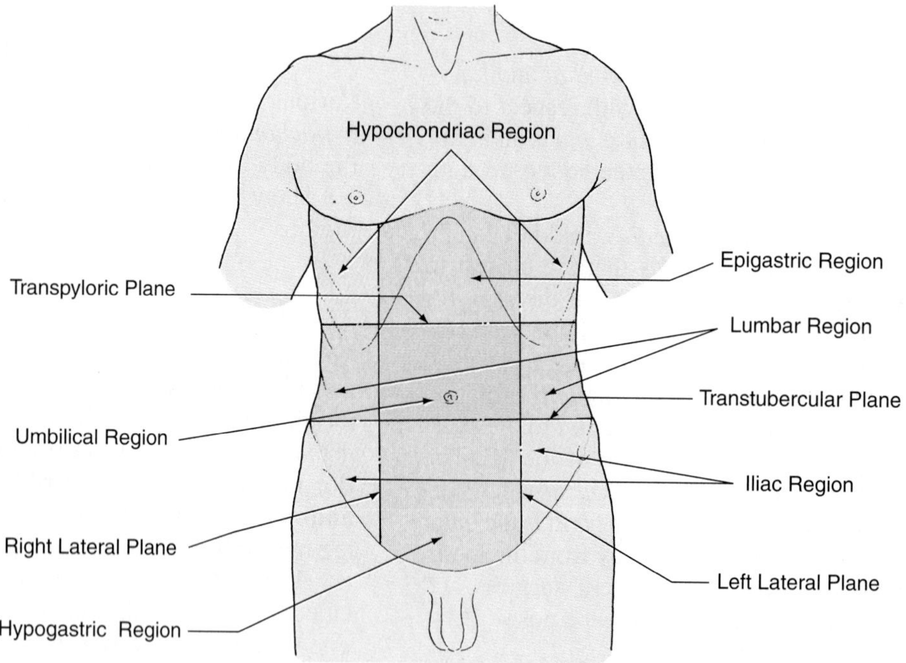

Figure 1.3 Abdominal regions and planes.

Upper Extremities

The term **brachium** refers to the upper arm, while the term **antebrachium** refers to the forearm (between the elbow and wrist). The elbow area on the posterior surface of the arm is the **cubital** area. That area on the opposite side of the elbow is the **antecubital area.** It is also correct to refer to the entire anterior surface of the antebrachium as antecubital.

Lower Extremities

The upper portion of the leg is the **thigh,** and the lower, fleshy, posterior portion is the **calf.** Between the thigh and calf on the posterior surface, opposite the knee, is a depression called the **popliteal region.** The sole of the foot is the **plantar surface.**

Abdominal Divisions

The abdominal surface can be divided into nine distinct areas. Dividing the abdomen into nine regions requires establishing four imaginary planes, two horizontal and two vertical (see figure 1.3). The **transpyloric plane** is the upper horizontal plane that passes through the lower portion of the stomach (pyloric portion). The **transtubercular plane** is the other horizontal plane; it touches the top surfaces of the hipbones (iliac crests). The two vertical planes, or **right** and **left lateral planes,** are approximately halfway between the midsagittal plane and the crests of the hips.

These four imaginary planes describe the umbilical, epigastric, hypogastric, hypochondriac, lumbar, and iliac regions. The **umbilical** area lies in the center, includes the navel, and is bordered by the two horizontal and two vertical planes. Immediately above the umbilical area is the **epigastric** area, which covers much of the stomach. Below the umbilical region is the **hypogastric,** or *pubic,* area. On the sides of the umbilical area are right and left **lumbar** areas. Above the lumbar areas are the right and left **hypochondriac** areas, and beneath the lumbar areas are the right and left **iliac** areas. (Note that the lumbar region, which is commonly thought of as only the lower back, extends to the anterior surface as well.)

Assignment:
Now complete part B of the Laboratory Report for this exercise. See how well you can match the regional terms to their descriptions without referring to the figures in this exercise.

2 Body Cavities and Membranes

Smooth membranes completely or partially line body cavities that contain the internal organs (*viscera*). This exercise explores the relationships of body cavities to each other, the organs they contain, and the membranes that line them.

Body Cavities

Figure 2.1 shows the principal cavities of the body. The two major cavities are the dorsal and ventral cavities. The **dorsal cavity,** which is nearest to the dorsal surface, includes the cranial and spinal cavities. The **cranial cavity** is the hollow portion of the skull that contains the brain. The **spinal cavity** is a long, tubular canal within the vertebrae that contains the spinal cord. The large **ventral cavity** encompasses the chest and abdominal regions.

A thin, dome-shaped muscle, the **diaphragm,** separates the superior and inferior portions of the ventral cavity. A membranous partition called the **mediastinum** separates the **thoracic cavity,** which is that part of the ventral cavity superior to the diaphragm, into right and left compartments. These compartments contain the lungs. The mediastinum encloses the heart, trachea, esophagus, and thymus gland.

Figure 2.2 shows the relationship of the lungs to the structures within the mediastinum. Note that within the thoracic cavity are a pair of right and left **pleural cavities** that contain the lungs and a **pericardial cavity** that contains the heart.

The **abdominopelvic cavity** is the portion of the ventral cavity that is inferior to the diaphragm. It consists of two portions: the abdominal and pelvic cavities. The **abdominal cavity** contains the stomach, liver, gallbladder, pancreas, spleen, and intestines. The **pelvic cavity** is the most inferior portion of the abdominopelvic cavity and contains the urinary bladder, sigmoid colon, rectum, uterus, and ovaries.

Assignment:
Before proceeding, identify the body cavities on a dissectible human torso model, and then complete part A of the Laboratory Report for this exercise.

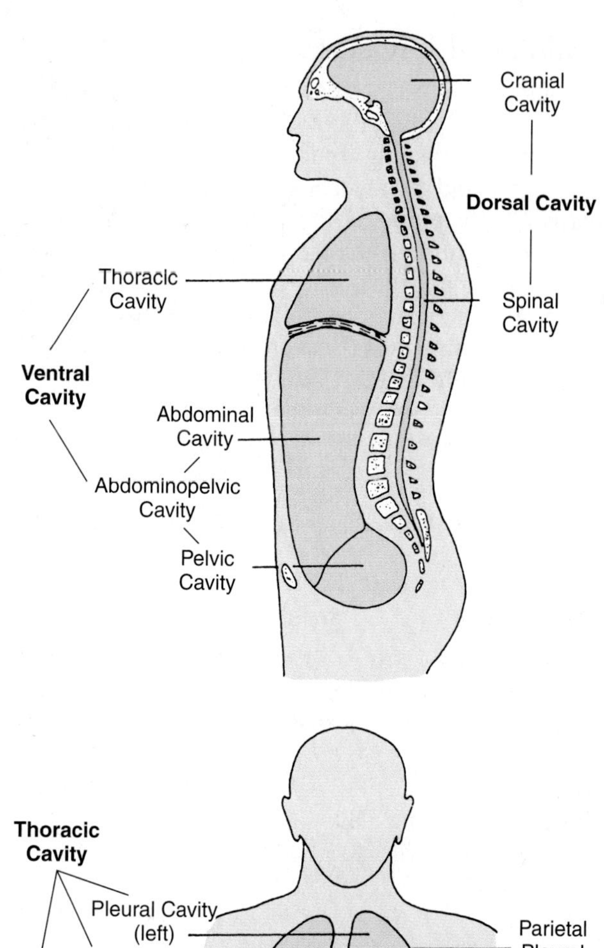

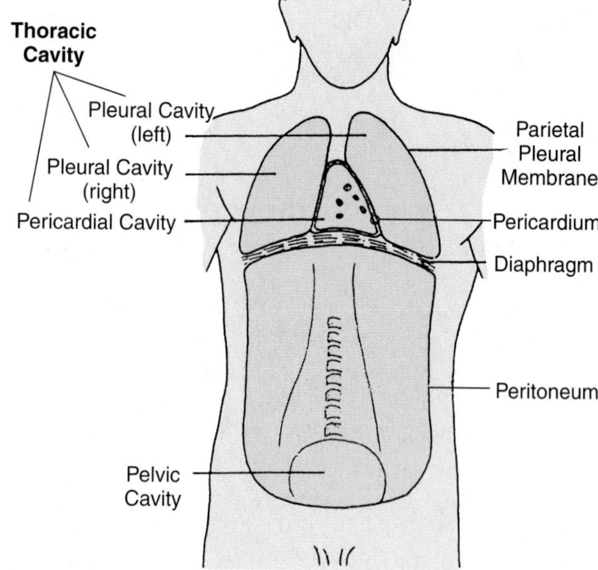

Figure 2.1 Body cavities.

Body Cavity Membranes

Serous membranes line the body cavities and provide a smooth surface for the enclosed internal organs. While thin, these membranes are strong and elastic. A self-secreted *serous fluid* moistens their surfaces and minimizes friction as viscera rub against cavity walls.

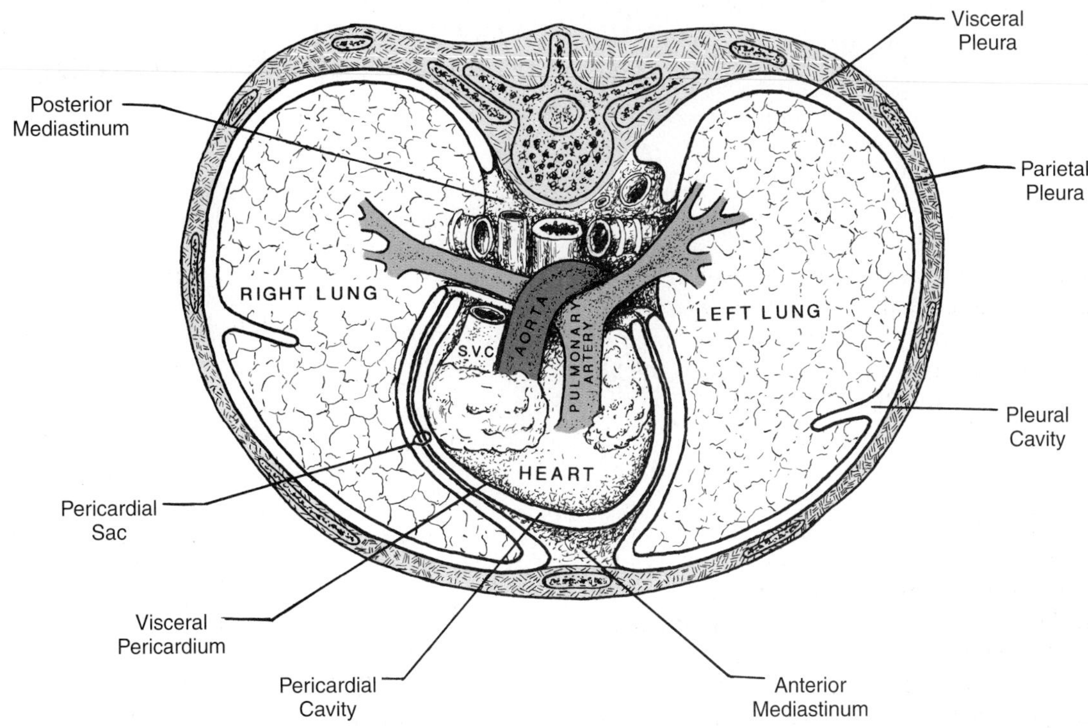

Figure 2.2 Transverse section through thoracic cavity. (S.V.C = Superior Vena Cava.)

Thoracic Cavity Membranes

The membranes that line the walls of the right and left thoracic compartments are the **parietal pleurae** (*pleura,* singular). **Visceral pleurae,** in turn, cover the lungs. Note in figure 2.2 that these pleurae are continuous with each other. Although these pleurae come into contact with each other, the potential cavity between the parietal and visceral pleurae is the **pleural cavity.** Inflammation of the pleural membranes results in *pleurisy,* a condition in which breathing becomes very painful.

The heart lies within the broadest portion of the mediastinum. A thin serous membrane, the **visceral pericardium,** or **epicardium,** covers the heart. Surrounding the heart is a double-layered fibroserous sac, the **pericardial sac.** The inner layer of this sac, or parietal pericardium is a serous membrane that is continuous with the epicardium of the heart. The fibrous outer layer strengthens the structure. The two serous membranes produce a small amount of serous fluid that lubricates the surface of the heart and minimizes friction as the heart contracts within the pericardial sac. The potential space between the visceral and parietal pericardia is the **pericardial cavity.**

Abdominal Cavity Membranes

The serous membrane of the abdominal cavity is the **peritoneum.** It does not extend deep into the pelvic cavity, however; instead, its most inferior boundary extends across the abdominal cavity at a level that is just superior to the pelvic cavity. The peritoneum covers the top portion of the urinary bladder.

The peritoneum both lines the abdominal cavity and covers the abdominal viscera. In addition, it has double-layered folds, called **mesenteries,** that extend from the dorsal body wall to the viscera and hold these organs in place. The mesenteries contain the blood vessels and nerves that supply the viscera.

The part of the peritoneum that attaches to the body wall is the **parietal peritoneum.** The peritoneum that covers the surfaces of the viscera is the **visceral peritoneum.** The potential cavity between the parietal and visceral peritoneums is the **peritoneal cavity.** Figure 2.3A shows the relationship between the peritoneums and the peritoneal cavity.

Extending downward from the inferior surface of the stomach is a large mesenteric fold called the **greater omentum.** This double-membrane structure passes downward from the stomach in front of the intestines, sometimes to the pelvis, and back up to the transverse colon, where it attaches. Because it folds upon itself, the greater omentum is essentially a double mesentery consisting of four layers. The potential cavity enclosed by the greater omentum is called the omental bursa. Fat accumulation in the greater omentum produces potbellies in some overweight people.

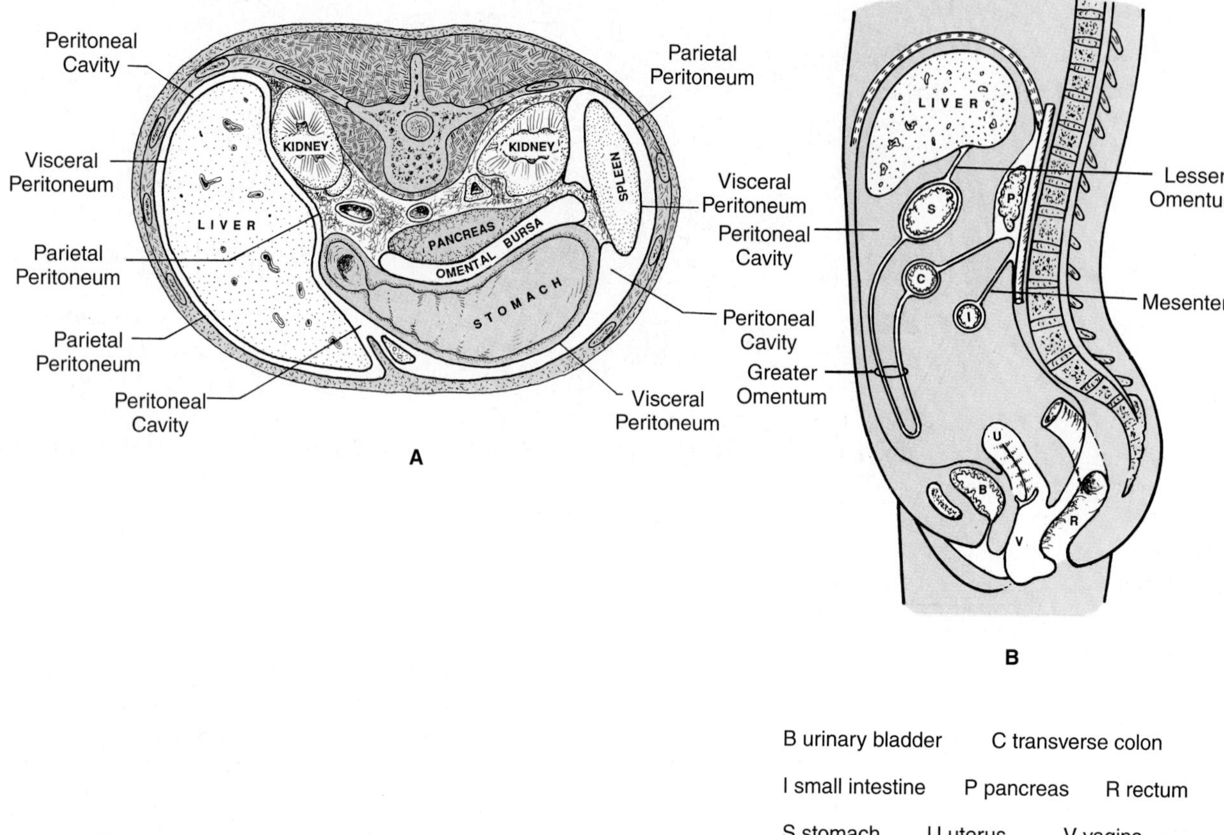

Figure 2.3 Transverse (A) and longitudinal (B) sections of the abdominal cavity.

A smaller mesenteric fold, the **lesser omentum,** extends between the liver and the superior surface of the stomach and a short portion of the duodenum. Figure 2.3B shows the relationship of these two omenta to the abdominal organs.

Assignment:
Complete the Laboratory Report for this exercise.

Organ Systems: Rat Dissection 3

In this exercise, a rat dissection provides a cursory study of the majority of organ systems. Since rats and humans have many anatomical and physiological similarities, much can be learned here about human anatomy.

A brief description of each of the 11 systems of the body is first. Keep in mind that an **organ** is composed of two or more tissues and performs one or more physiological functions. A **system,** on the other hand, is a group of organs that functionally relate to each other. To make the dissection more meaningful, answer the questions on the Laboratory Report for this exercise *prior to* doing the dissection.

The Integumentary System

The body surface covering that comprises the integumentum system includes the skin, hair, and nails.

The skin's principal function is to prevent harmful microorganisms from invading the body. In addition to being a mechanical barrier against invasion, the skin produces sweat and sebum (oil) that contain antimicrobial substances for further protection.

The skin also aids in temperature regulation and excretion. The evaporation of perspiration cools the body. Perspiration contains many of the same excretory products found in urine: thus, the skin aids the kidneys in eliminating water, salts, and some nitrogenous wastes from the blood.

To some extent, the skin also plays a role in nutrition. For example, vitamin D forms in the skin's deeper cell layers when a vitamin D precursor in those cells is exposed to ultraviolet rays of sunlight.

As long as the skin remains intact, the internal environment is protected. However, serious skin injury, such as deep burns, may result in excessive fluid loss and electrolytic imbalances.

The Skeletal System

The skeletal system, consisting of bones, cartilage, and ligaments, forms a solid framework for the body. It also supports and protects delicate organs, such as the lungs, heart, brain, and spinal cord. In addition to protection, the bones provide points of attachment for muscles and act as levers when the muscles contract. This makes movement possible.

Another important function of the skeletal system is mineral storage. Bones provide a pool of calcium, phosphorus, and other ions that can be used to stabilize the blood's mineral content. The skeletal system also produces blood cells. Both red and white blood cells form in the red marrow of certain bones of the body.

The Muscular System

Attached to the body's skeletal framework are muscles that make up nearly half the body's weight. Skeletal muscles consist primarily of long, multinucleated cells. Nerve impulses stimulate these cells to shorten, or contract, which enables the muscles to move parts of the body. This allows for walking, eating, breathing, and other activities. Although the purpose of most skeletal muscles is limb movement, some function in the reproductive, respiratory, and digestive systems.

In addition to skeletal muscle tissue, the body has two other kinds of muscle tissue: smooth and cardiac. Smooth muscle tissue is found in the walls of the stomach, intestines, arteries, veins, urinary bladder, and other organs. Cardiac muscle tissue is found in the walls of the heart. Like skeletal muscle fibers, these types of muscle tissue perform work by contracting. Skeletal muscles, however, are voluntarily controlled, while smooth and cardiac muscle tissues are involuntarily controlled.

The Nervous System

The nervous system consists of the brain, spinal cord, nerves, and receptors. While the nervous system has many functions, such as muscular control and regulation of circulation and breathing, its basic function is to help the body adjust to external and internal environmental changes.

Adapting to environmental changes involves, first of all, various kinds of stimuli, such as changes in temperature, pressure, chemicals, sound waves, and light. These stimuli activate many different types of **receptors** throughout the body. Nerve impulses then pass along **conduction pathways** (nerves and spinal cord) to **interpretation centers** in the brain, where stimuli are recognized and evaluated. The nervous system responds to the stimuli via outgoing conduction pathways and initiates adjustments, which may be muscular.

The Circulatory System

The circulatory system consists of the heart, blood vessels, and blood. The **heart** is a muscular pump that moves blood throughout the body. **Arteries** are thick-walled blood vessels that carry blood from the heart to the microscopic **capillaries** that permeate all body tissues. **Veins** are large blood vessels that convey blood from the capillaries back to the heart.

The circulatory system transports nutrients, oxygen, and carbon dioxide from one part of the body to another. It also transports hormones from glands, metabolic wastes from cells, and excess heat from muscles to the skin.

In addition to its transport function, the blood protects against microbial invasion. Phagocytic (cell-eating) white blood cells, antibodies, and special proteins in the blood prevent invading microorganisms from destroying the body.

The Lymphatic System

Tissue fluid containing nutrients and oxygen leaves the blood through capillary walls and passes into the spaces between cells. The lymphatic system is a network of lymphatic vessels that returns tissue fluid from the intercellular spaces of tissues to the blood. This system is also responsible for absorbing fats from the intestines. Although the simple sugars from carbohydrate digestion and the amino acids from protein digestion are absorbed directly into the blood through the intestinal wall, fats must pass first into the lymphatic system and then into the blood.

Once tissue fluid enters the lymphatic vessels, it is called **lymph.** As this fluid moves through the lymphatic vessels, it passes through specialized organs called **lymph nodes.** Phagocytic white blood cells called **macrophages** in these nodes remove bacteria and other foreign material, purifying the lymph before it returns to the blood. Lymphocytes are also produced in lymph nodes.

Macrophages are also found in the liver, lungs, spleen, tonsils, adenoids, appendix, and Peyer's patches (in the digestive tract). This diverse collection of phagocytic cells is collectively referred to as the **reticuloendothelial system** and is an important component of the immune system. The reticuloendothelial system is also called the **mononuclear phagocyte system** because macrophages have a single, unlobed nucleus.

The **spleen** is an organ of the lymphatic system that (with the aid of macrophages) removes foreign materials from the blood. The spleen also destroys worn-out red blood cells and is a reservoir for blood. In emergencies, smooth muscles in the spleen and the splenic blood vessels can contract, forcing the blood content of the spleen into the general circulation.

The Respiratory System

The respiratory system consists of two portions: the air passageways and the air sacs of the lungs. The air passageways consist of the **nasal cavity, nasopharynx, larynx, trachea, bronchi,** and the branches of the bronchi. The blood and the air exchange gases in the air sacs of the lungs, or **alveoli.** Each lung contains about 300 million alveoli. Together they provide a large surface area for oxygen and carbon dioxide exchange.

The Digestive System

The digestive system includes the **mouth, salivary glands, esophagus, stomach, small intestine, large intestine, rectum, pancreas, liver,** and **gallbladder.** The digestive system converts ingested food to molecules that are small enough to pass through the intestinal lining and into the capillaries and lymphatic vessels of the intestinal wall. This digestive process requires **digestive enzymes,** which the organs of the digestive system produce and which hydrolyze food molecules.

In addition to ingesting, digesting, and absorbing food, the digestive system eliminates undigested materials and protects against infection. Since the contents of the digestive system contain many foreign, potentially harmful microorganisms, the digestive tract lining, like the integument, is a barrier to microorganism invasion.

The Urinary System

Cellular metabolism produces waste materials, such as carbon dioxide, excess water, nitrogenous products, and excess products of metabolism. Although the skin, lungs, and large intestine assist in waste removal, the **kidneys** remove most waste products from the blood. During waste removal, the kidneys regulate the chemical composition of all body fluids to maintain homeostasis. The **ureters,** which drain the kidneys, the **urinary bladder,** which stores urine, and the **urethra,** which drains the bladder, assist the kidneys in excretion.

The Endocrine System

The endocrine system consists of a number of widely dispersed **endocrine glands** that dispense their secretions directly into the blood. These secretions, which capillaries in the glands absorb, are called **hormones.**

Hormones integrate various physiological activities (metabolism and growth), direct the differentiation and maturation of the ovaries and testes, and regulate specific enzymatic reactions.

The endocrine system includes the following glands: **pituitary, thyroid, parathyroid, thymus, adrenal, pancreas,** and **pineal.** The **ovaries, testes, stomach lining, placenta,** and **hypothalamus** also produce important hormones.

The Reproductive System

Continuity of the species is the function of the reproductive system. The testes of the male produce sperm, or spermatozoa, and the ovaries of the female produce eggs, or ova. Male reproductive organs include the **testes, penis, scrotum,** accessory glands, and various ducts. Female reproductive organs include the **ovaries, vagina, uterus, uterine** (fallopian) **tubes,** and accessory glands.

Assignment:
Complete the Laboratory Report for this exercise before beginning the rat dissection.

Rat Dissection

Work with a laboratory partner to perform this part of the exercise. Your principal objective is *to expose the organs for study, not to simply cut up the animal.* Make most cuts with scissors. Use a blunt probe for separating tissues.

Materials:
freshly killed or preserved rat
dissecting tray
dissecting scissors
scalpel
forceps
blunt probe
dissecting pins
disposable latex gloves

Skinning the Ventral Surface

1. Pin the four feet to the bottom of the dissecting pan, as figure 3.1 shows. Before making any incision, examine the oral cavity. Note the large **incisors** in the front of the mouth for biting off food particles. Force the mouth open sufficiently to examine the flattened **molars** at the back of the mouth. These teeth grind food into small particles.

   Note that the **tongue** is attached at its posterior end. Lightly scrape the surface of the tongue with a scalpel to determine its texture. The roof of the mouth consists of an anterior **hard palate** and a posterior **soft palate.** The throat is the **pharynx,** which is a component of both the digestive and respiratory systems.
2. Lift the skin along the midventral line with your forceps, and make a small incision with scissors, as figure 3.2 shows. Cut the skin upward to the lower jaw; then turn the pan around and complete the incision to the anus,

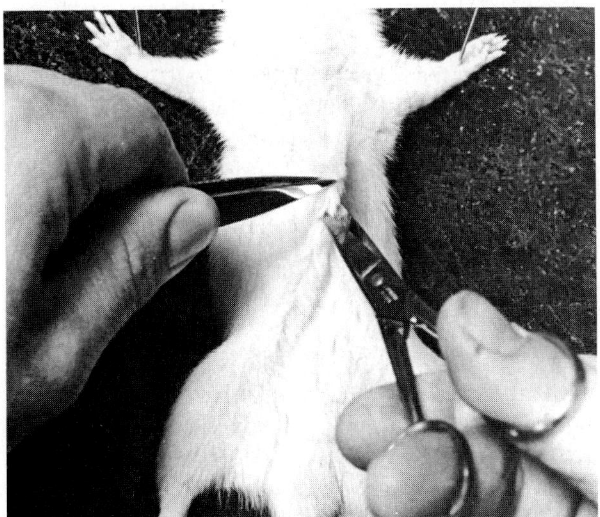

Figure 3.1 Start the incision on the median line with scissors.

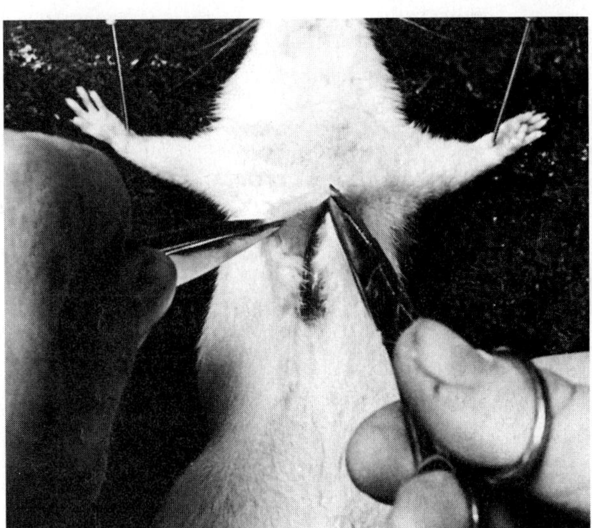

Figure 3.2 Extend the first cut to the lower jaw.

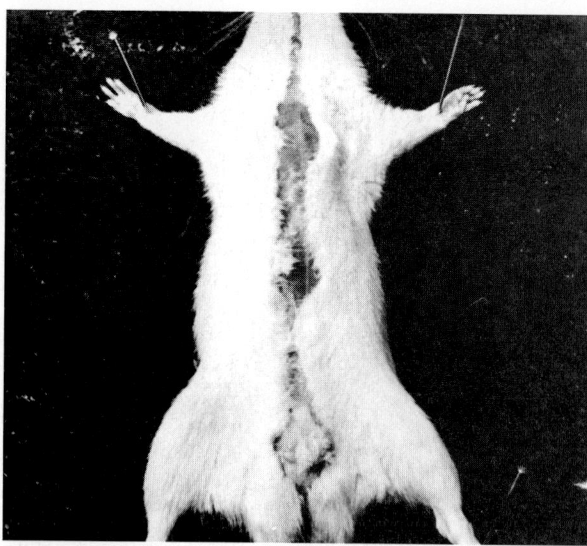

Figure 3.3 Completed incision from the lower jaw to the anus.

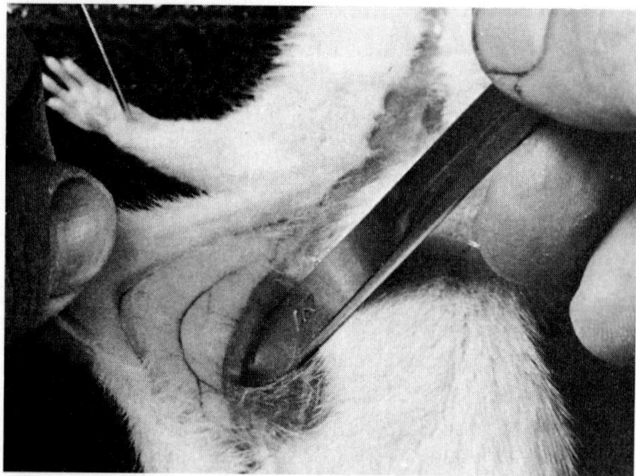

Figure 3.4 Separate the skin from the musculature with a blunt probe or a scalpel handle.

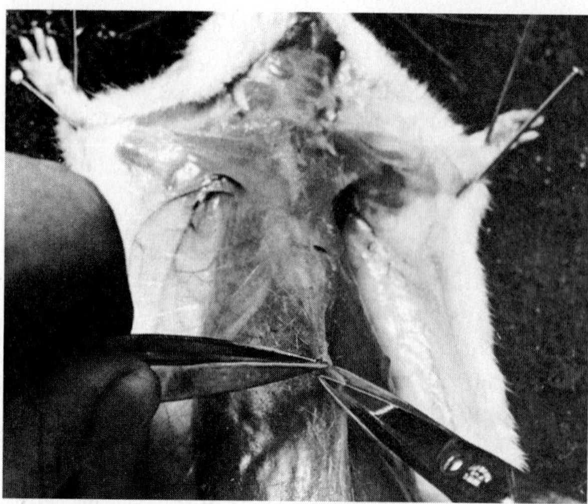

Figure 3.5 Begin the incision of musculature on the median line.

cutting around both sides of the genital openings. Figure 3.3 shows how the completed incision should appear.

3. With the blunt probe (or the scalpel handle), separate the skin from the musculature, as figure 3.4 shows. The fibrous connective tissue that lies between the skin and the musculature is the **superficial fascia** (Latin: *fascia,* band).
4. Skin the legs down to the "knees" or "elbows," and pin down the stretched-out skin. Examine the surfaces of the **muscles,** and note that **tendons,** which consist of tough, fibrous connective tissue, attach the muscles to the skeleton.

   Covering the surface of each muscle is a thin, gray, feltlike layer called the **deep fascia.** Fibers of the deep fascia are continuous with fibers of the superficial fascia, so separating the two membranes requires applying considerable force with a probe or scalpel handle.
5. At this stage, your specimen should appear as in figure 3.5. If your specimen is a female, the mammary glands will probably remain attached to the skin.

Opening the Abdominal Wall

1. As figure 3.5 shows, make an incision through the abdominal wall with scissors. Hold the muscle tissue with a pair of forceps while making the cut. **Caution:** Avoid damaging the underlying viscera as you cut.
2. Cut upward along the midline to the rib cage and downward along the midline to the genitalia.
3. To completely expose the abdominal organs, make two lateral cuts near the base of the rib cage—one to the left and the other to the right (see figure 3.6). The cuts should extend all the way to the pinned-back skin.
4. Fold out the flaps of the body wall, and pin them to the pan, as figure 3.7 shows. The abdominal organs are now well exposed.
5. Using figure 3.8 as a reference, identify all the labeled *abdominal* viscera without moving the organs out of place. Note in particular the position and structure of the **diaphragm.**

Examining the Thoracic Cavity

1. Using scissors, cut along the left side of the rib cage, as figure 3.9 shows. Cut through all of the ribs and connective tissue. Then cut along the right side of the rib cage in a similar manner.
2. Grasp the xiphoid process of the sternum with forceps, as figure 3.10 shows, and cut the

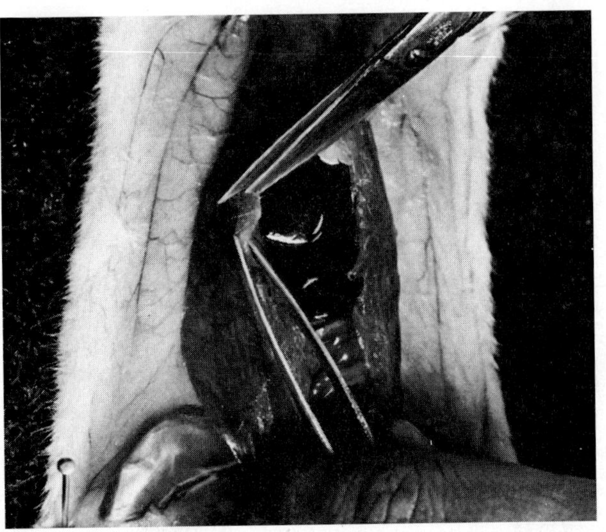

Figure 3.6 Make lateral cuts at the base of the rib cage in both directions.

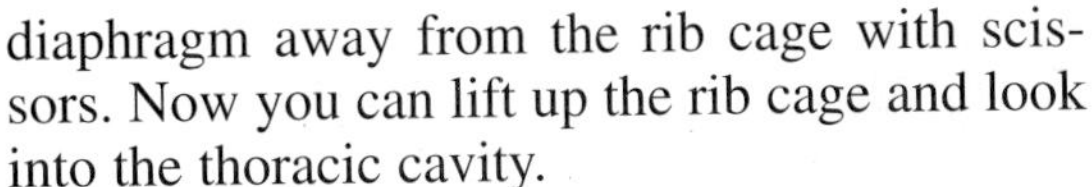

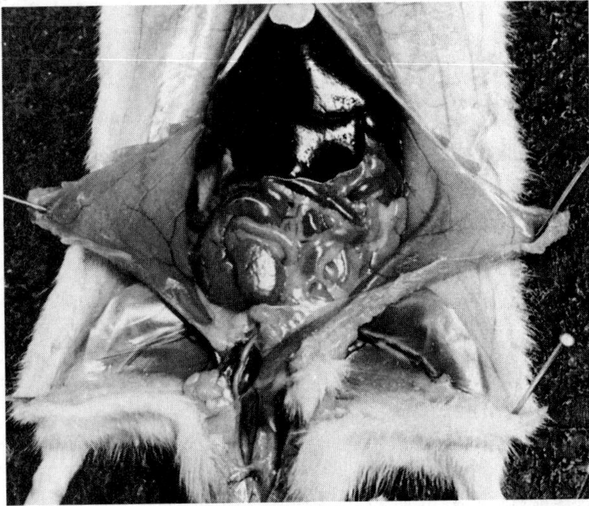

Figure 3.7 Pin the flaps of the abdominal wall back to expose the viscera.

diaphragm away from the rib cage with scissors. Now you can lift up the rib cage and look into the thoracic cavity.

3. With scissors, complete the removal of the rib cage by cutting off any remaining attachment tissue (see figure 3.11).
4. Now examine the structures exposed in the thoracic cavity. Refer to figure 3.8, and identify all the labeled structures.
5. Note the pale-colored **thymus gland** located just above the heart. Remove this gland.
6. Carefully remove the thin **pericardial sac** that encloses the **heart.**
7. Remove the heart by cutting through the major blood vessels attached to it. If working with a freshly killed rat, gently sponge away pools of blood with Kimwipes or other soft tissues.
8. Locate the **trachea** in the throat region. Can you see the **larynx** (voice box) at the anterior end of the trachea? Trace the trachea posteriorly to where it divides into two **bronchi** that enter the **lungs.** Squeeze the lungs with your fingers, noting their elasticity. Remove the lungs. See figure 3.12.
9. Probe under the trachea to locate the soft, tubular **esophagus** that runs from the oral cavity to the stomach. Remove a section of the trachea to reveal the esophagus, as figure 3.13 shows.

Deeper Examination of Abdominal Organs

1. Lift up the lobes of the reddish brown liver, and examine them. Note that rats lack a **gallbladder.** Carefully remove the liver, and wash out the abdominal cavity. The stomach and intestines are now clearly visible.
2. Lift out a portion of the intestines, and identify the membranous **mesentery,** which holds the intestines in place. Its blood vessels and nerves supply the digestive tract. If your specimen was a mature, healthy animal, the mesenteries will contain considerable fat.
3. Now lift the intestines out of the abdominal cavity, cutting the mesenteries, as necessary, for a better view of the organs. Note the great length of the **small intestine.** Obviously, its name refers to its diameter, not its length.

   The first portion of the small intestine, which is connected to the stomach, is the **duodenum.** At its distal end, the small intestine is connected to a large, saclike structure, the **cecum.** The *appendix* in humans is a vestigial portion of the cecum.

   The cecum communicates with the **large intestine,** which has **ascending, transverse, descending,** and **sigmoid** divisions. The sigmoid colon empties into the **rectum.**
4. Try to locate the **pancreas,** which is embedded in the mesentery alongside the duodenum. It is often difficult to see. Pancreatic enzymes enter the duodenum via the **pancreatic duct.** Try to locate this minute tube.
5. Locate the **spleen** on the left side of the abdomen near the stomach. It is reddish brown and held in place with mesentery. Do you recall the functions of this organ?
6. Remove the stomach, small intestine, and large intestine after severing the esophagus near the stomach and the end of the sigmoid colon.

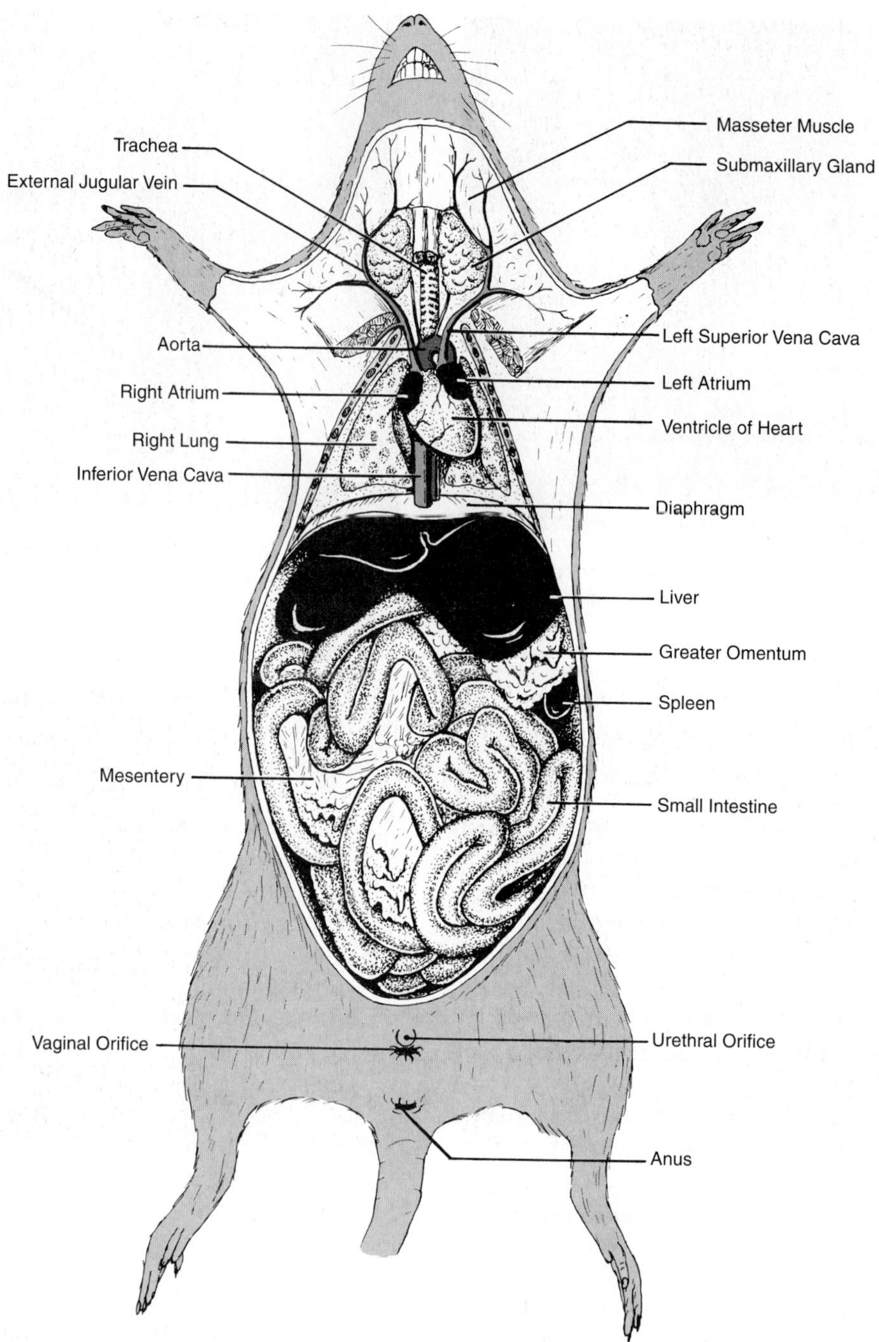

Figure 3.8 Viscera of a female rat.

Removal of these organs enables you to see the **descending aorta** and the **inferior vena cava.** The aorta carries blood posteriorly to body tissues. The inferior vena cava is a vein that returns blood from posterior regions to the heart.

7. Peel away the peritoneum and fat from the posterior wall of the abdominal cavity. *Removal of this fat requires special care to avoid damaging important structures.* This makes the kidneys, blood vessels, and reproductive structures more visible.

   Locate the two **kidneys** and the **urinary bladder.** Trace the two **ureters,** which extend from the kidneys to the bladder. Examine the anterior surfaces of the kidneys, and locate the **adrenal glands,** which are important components of the endocrine system.

8. **If your specimen is a female:** Compare it with figure 3.14. Locate the two **ovaries,** which lie lateral to the kidneys. From each ovary, a **uterine tube** leads posteriorly to join the **uterus.** Note that the uterus is a Y-shaped structure that joins the vagina.

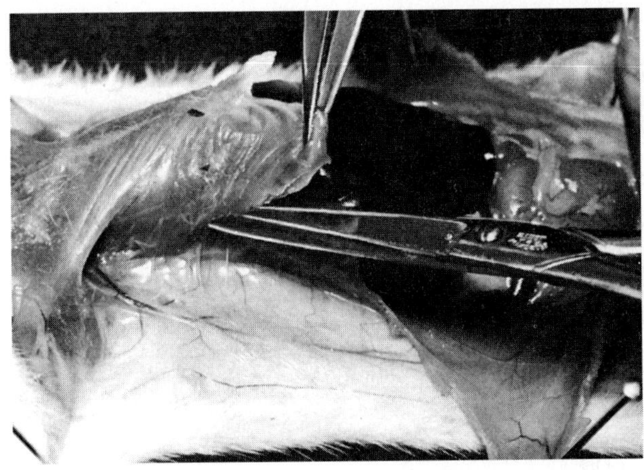

Figure 3.9 Sever the rib cage on each side with scissors.

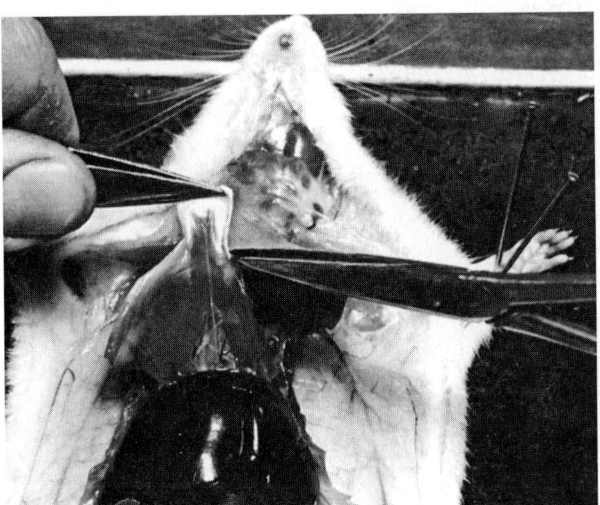

Figure 3.10 Cut the diaphragm free from the edge of the rib cage as it is held up by the xiphoid process of the sternum.

If your specimen appears to be pregnant, open up the uterus and examine the developing embryos. Note how they attach to the uterine wall.

9. **If your specimen is a male:** Compare your specimen to figure 3.15. The **urethra** is in the **penis.** Apply pressure to one of the testes through the wall of the **scrotum** to see if it can be forced up into the **inguinal canal.**

   Open up the scrotum on one side to expose a **testis, epididymis,** and **vas deferens.** Trace the vas deferens over the urinary bladder to where it penetrates the **prostate gland** to join the urethra.

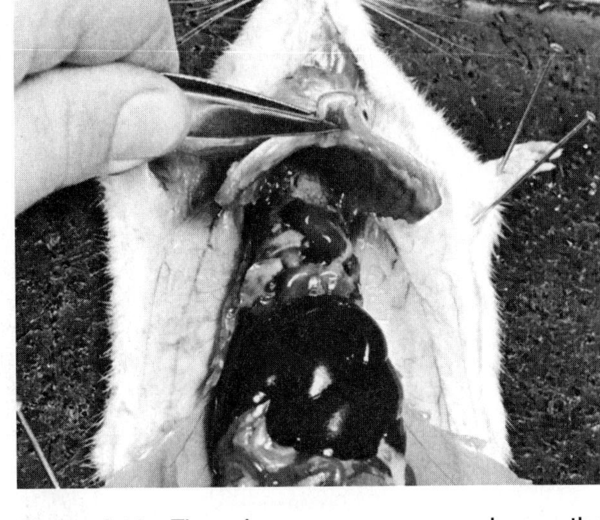

Figure 3.11 Thoracic organs are exposed once the rib cage has been cut loose.

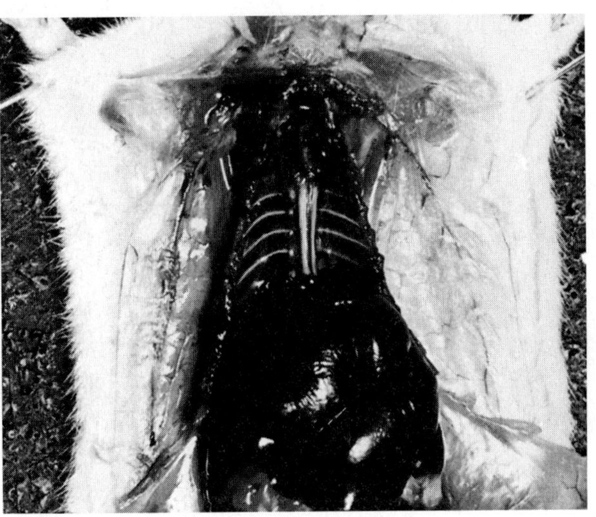

Figure 3.12 Specimen with heart, lungs, and thymus gland removed.

Summary This cursory dissection has acquainted you with the respiratory, circulatory, digestive, urinary, and reproductive systems. You have also seen portions of the endocrine and lymphatic systems. Four systems (integumentary, skeletal, muscular, and nervous) have been, for the most part, omitted at this time. These are studied later.

If you have done a careful and thoughtful rat dissection, you should have a good general understanding of the basic structural organization of the human body. Much that you have seen here has a human counterpart.

Cleanup Dispose of the specimen as your instructor directs. Scrub your instruments with soap and water, rinse, and dry them.

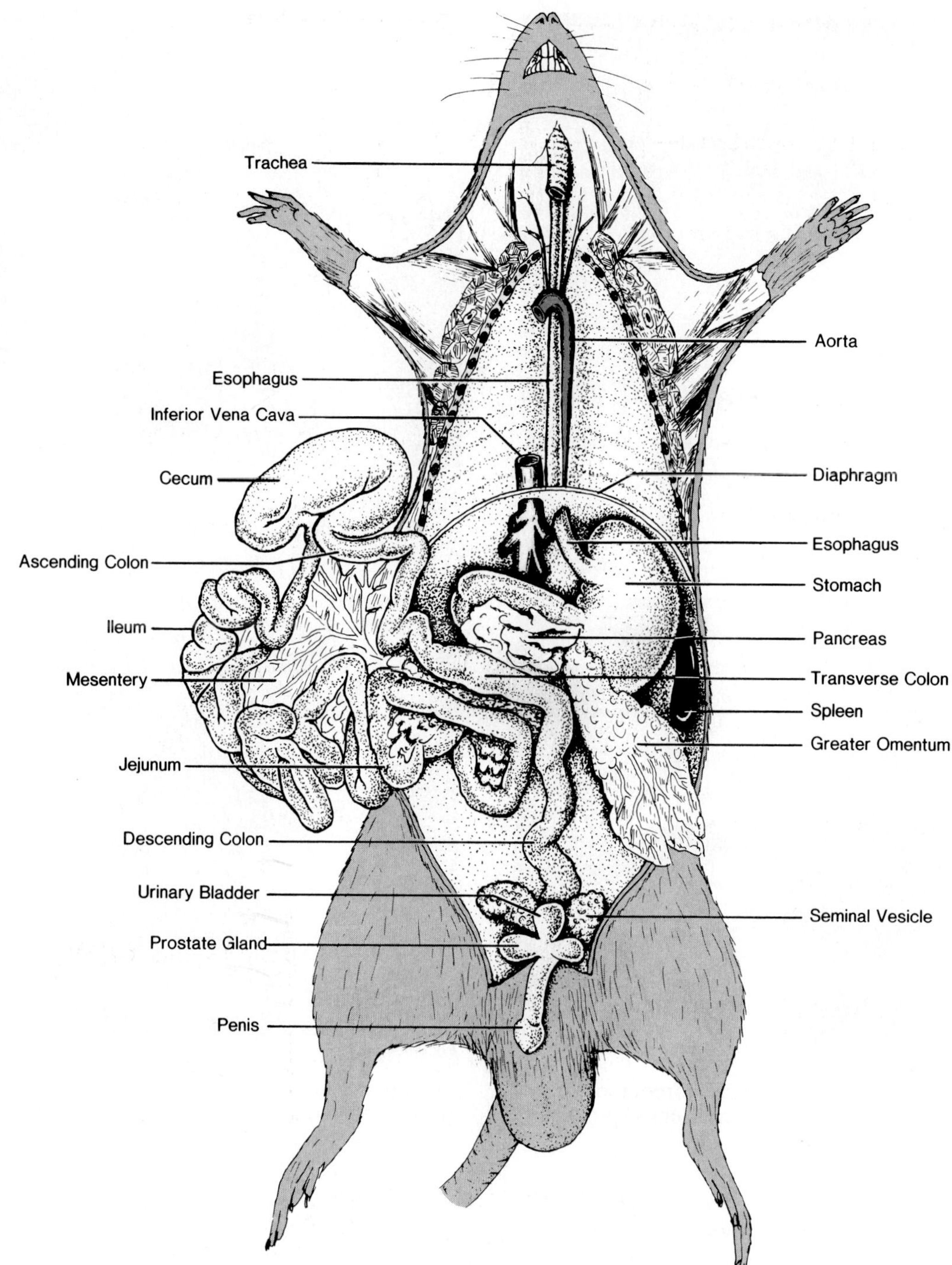

Figure 3.13 Viscera of a male rat (heart, lungs, and thymus removed).

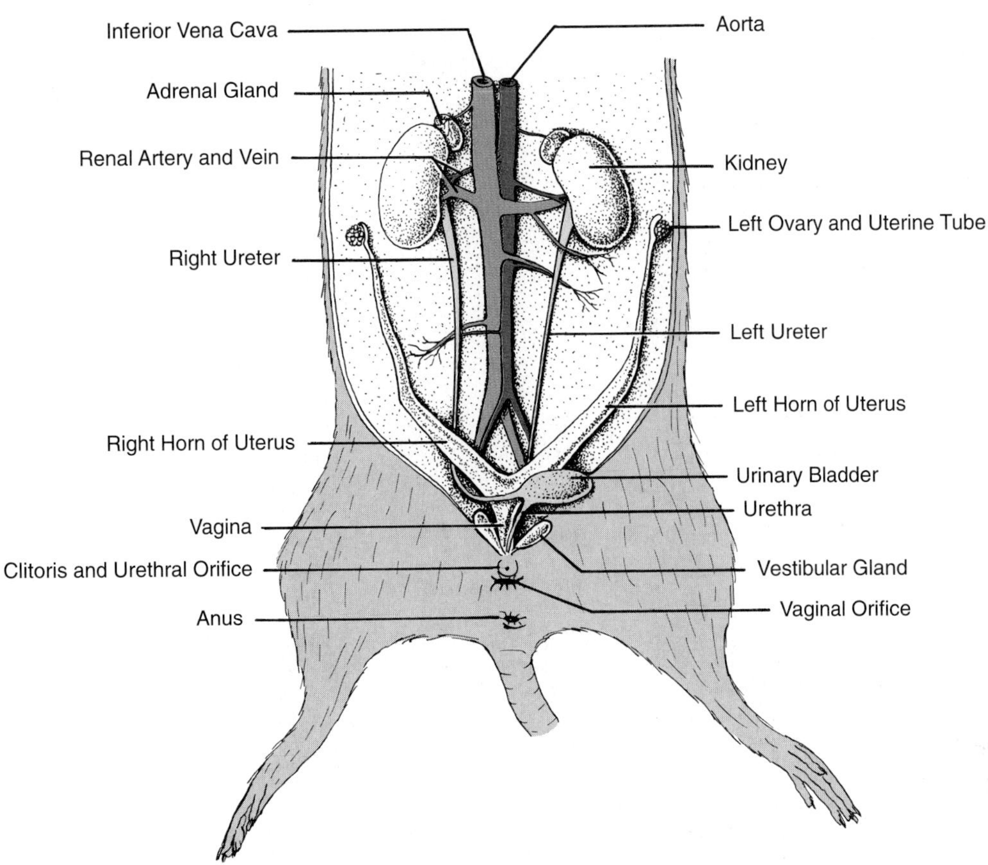

Figure 3.14 Abdominal cavity of a female rat with intestines and liver removed.

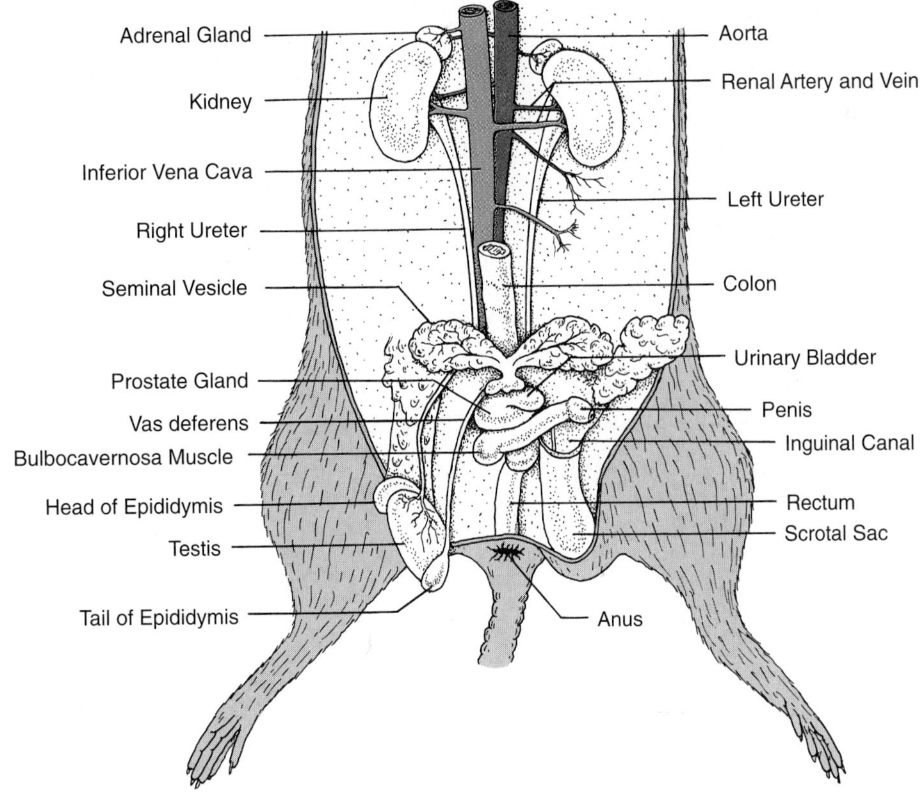

Figure 3.15 Abdominal cavity of a male rat (intestines and liver removed).

PART 2 Cells and Tissues

This part consists of nine exercises on cells and tissues. Since histological studies are emphasized throughout the course, this part, which is basic to anatomy and physiology, is extremely important.

Study Exercise 4 on microscopy before going to the laboratory. It provides some of the physical and mathematical reasons why certain techniques in microscopy work and others do not.

The next five exercises pertain to cytological studies. Exercises 5 and 6 examine cellular structure, physiology, and reproduction. Exercises 7, 8, and 9 focus on the immediate environment of cells with regard to the role of osmosis, pH, and buffers on cellular physiology.

Exercises 10 and 11 survey epithelial and connective tissues. Microscope slides are examined in the laboratory to note how the various types of tissues differ. Exercise 12, which pertains to the skin, is included here because the skin contains both epithelial and connective tissues.

4 Microscopy

In this exercise, you will use the compound light microscope for your studies. The term *compound* refers to the set of lenses used to magnify an image, and the term *light* refers to the light rays used to produce an image. Electron microscopes magnify and then focus a beam of electrons, using electromagnets. Electron microscopes have far greater magnifying and resolving abilities than light microscopes; however, they are expensive, and their operation is complex. The compound light microscope, in spite of its limitations, is still the most widely used type of microscope in hospitals and clinical laboratories. This exercise explains accepted procedures for using this instrument.

Microscopy can be fascinating or frustrating. Success in this endeavor depends to a large extent on how well you understand the mechanics and limitations of your microscope. This exercise outlines procedures that should minimize difficulties and enable you to get the most out of your subsequent attempts at microscopy.

Assignment:
Before using the microscope, answer all of the questions in the Laboratory Report. Your preparation prior to working through this laboratory exercise will greatly facilitate your understanding. Your instructor may wish to collect this report at the beginning of the period on the first day that the microscope is to be used in class.

Care of the Instrument

Microscopes are expensive, and they can be damaged rather easily if you do not observe certain precautions.

Transport Use two hands when carrying your microscope from one part of the room to another (see figure 4.1). If you carry it with only one hand and allow it to dangle at your side, the microscope may collide with furniture or some other object. Never try to carry two microscopes at once. Also, avoid jarring your microscope when setting it down on your workbench.

Clutter Keep your work area uncluttered while you use the microscope. Coats, purses, nonessential books, and so on should be stashed in a drawer or some other out-of-the-way place. A clear work area promotes efficiency and results in fewer accidents.

Electric Cord Dangling microscope electric cords can result in catastrophic accidents. In crowded quarters, feet can become entangled in dangling cords, resulting in a damaged microscope on a concrete floor. Keep your microscope cord out of harm's way.

Lens Care At the beginning of each laboratory period, use special lens paper to clean the lenses. If your lenses are too soiled to be cleaned with lens paper alone, ask your instructor for lens-cleaning solution. At the end of each lab session, wipe any immersion oil off the immersion lens if it has been used.

Dust Protection In most laboratories, dustcovers protect the microscopes during storage. If a dustcover is available, place it over the microscope at the end of the period.

Microscope Components

Figure 4.2 shows the principal microscope components.

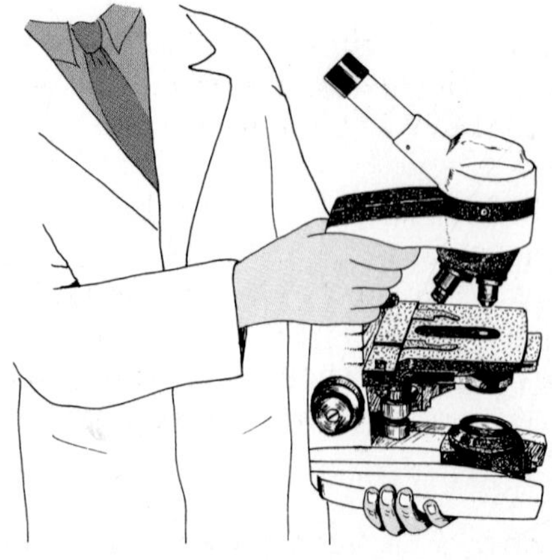

Figure 4.1 Hold the microscope firmly with both hands when carrying it.

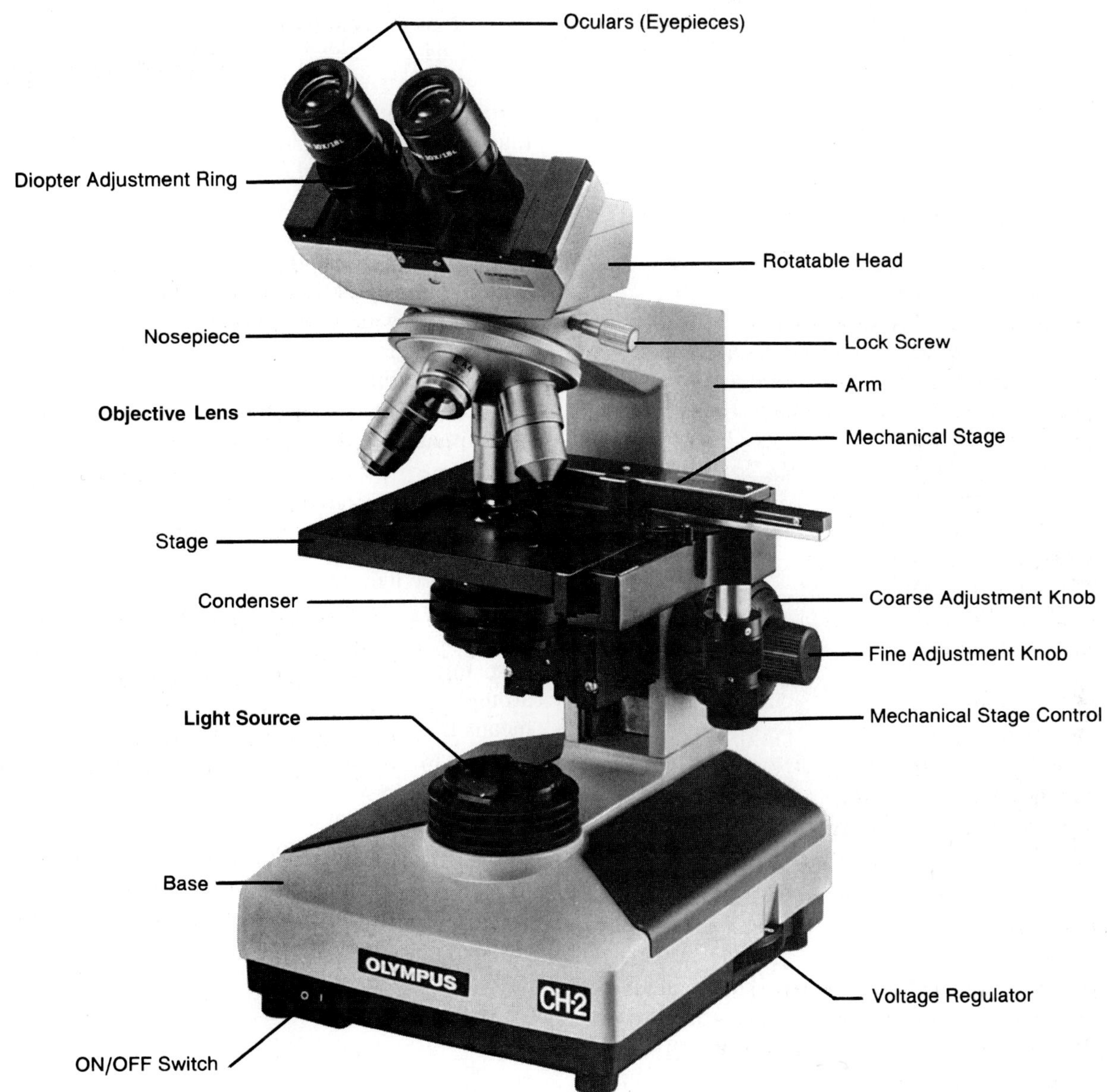

Figure 4.2 The compound microscope.

Framework All microscopes have a basic frame structure that includes the **arm** and the **base.** All other parts attach to this framework.

Stage The horizontal platform that supports the microscope slide is called the *stage.* Your microscope may be equipped with a **mechanical stage**—that is, one with a device for clamping onto a slide and **stage control knobs** for moving the slide around. Note the location of the stage control knobs in figure 4.2.

Light Source A light source is positioned in the base of most microscopes. Ideally, the lamp should have a **voltage regulator** or **rheostat** to vary the light intensity. The microscope in figure 4.2 has a knurled wheel on the right side of its base to regulate the voltage supplied to the lightbulb.

Lens Systems All compound microscopes have three lens systems: the ocular(s), the objectives, and the condenser. Figure 4.3 illustrates the light path through these three systems.

1. The **ocular,** or eyepiece, which is at the top of the instrument, consists of two or more internal lenses and usually has a magnification of 10×. Although the microscope in figure 4.2 has two oculars (binocular), a microscope often has only one.

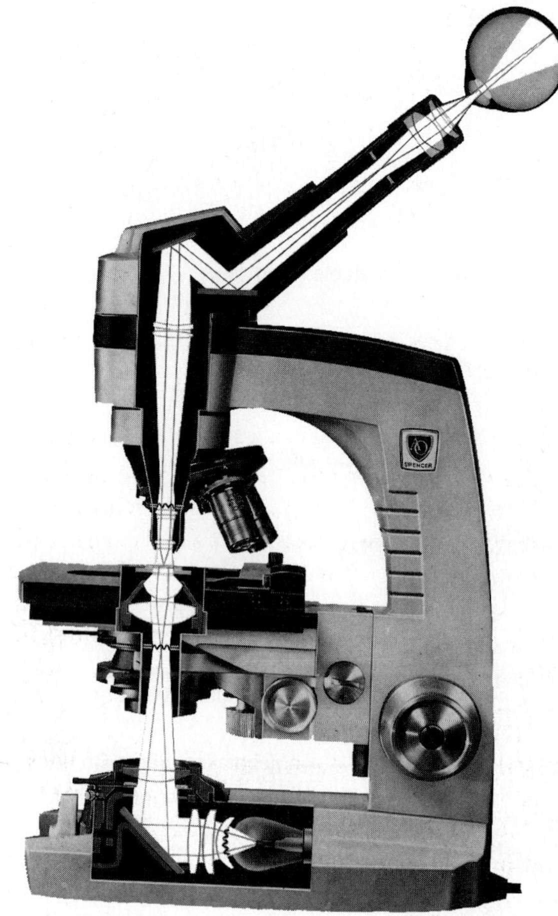

Figure 4.3 The light pathways of a microscope.

2. Three or more **objectives** are usually attached to a rotatable **nosepiece** that allows them to be moved into position over a slide. Objectives on most laboratory microscopes have magnifications of 4, 10, 45, and 100, designated as **scanning, low-power, high-dry,** and **oil immersion (high-wet),** respectively.
3. The third lens system is the **condenser** under the stage. It collects and directs the light from the lamp to the slide being studied. The condenser is either stationary or adjustable. A knob under the stage moves adjustable condensers up and down. A **diaphragm** within the condenser is simply an opening through which light passes, and the diameter of the opening can be varied to regulate the amount of light that reaches a slide. The diameter of the diaphragm opening is adjusted by either sliding a diaphragm lever or by turning a knurled ring, either of which is found on the condenser.

Focusing Knobs The concentrically arranged **coarse adjustment** and **fine adjustment knobs** on the side of the microscope bring objects on a slide into focus.

Ocular Adjustments Binocular microscopes must have ways to change the distance between the oculars and to make diopter changes for eye differences. Simply pulling apart or pushing together the oculars changes the interocular distance on most microscopes.

Diopter adjustments require focusing first with the right eye only. Diopter adjustments are then made on the left eye by turning the knurled **diopter adjustment ring** (see figure 4.2) on the left ocular until the image is sharp. Both eyes should now see sharp images.

Resolution

The resolution limit, or **resolving power,** of a microscope lens system is a function of its numerical aperture, the wavelength of light, and the design of the condenser. The maximum resolution of the best microscopes with oil immersion lenses is around 0.2μm. (**Note:** 1 mm = 1000μm). This means that two small objects that are 0.2 μm apart are seen as separate entities; objects closer than that are seen as a single object. The human eye can resolve objects that are about 100 μm apart.

The following factors maximize the resolution of a lens system:

- Place a **blue filter** over the light source. This maximizes resolution because of blue light's short wavelength.
- Keep the **condenser** at its highest position so that the maximum amount of light enters the objective.
- Open the **diaphragm** as much as possible. Although partially closing the diaphragm improves contrast, it reduces resolution when using high power or oil immersion.
- Use **immersion oil** between the slide and the 100× objective.

Lens Care

Keeping your microscope lenses clean is a constant concern. Unless all lenses are free of dust, oil, and other contaminants, they are unable to achieve the desired degree of resolution. The following are suggestions for cleaning the various lens components:

Lens Paper Use only lint-free, optically safe lens papers to clean lenses. Booklets of lens tissues are most widely used for this purpose. Although several types of boxed tissues are also

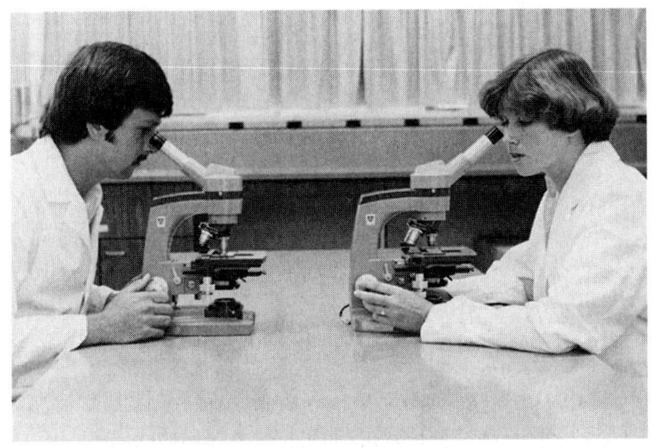

Figure 4.4 The microscope position on the right has the advantage of stage accessibility.

Figure 4.5 Properly position the slide while moving the retainer lever to the right.

safe, *use only the type of tissue your instructor recommends.* If your lenses are too soiled to be cleaned with lens paper alone, ask your instructor for lens-cleaning solution.

Oculars To determine if your eyepiece is clean, rotate it between your thumb and forefinger as you look through the microscope. A rotating pattern is evidence of dirt.

If cleaning the top lens of the ocular with lens tissue fails to remove the debris and the ocular is removable, try cleaning the lower lens. *Whenever the ocular is removed from the microscope, place a piece of lens paper over the open end of the microscope to prevent dust from entering the framework.*

Objectives Materials from slides or fingers often soil objective lenses. Lens tissue moistened with lens-cleaning solution usually removes whatever is on the lens. Sometimes a cotton swab dipped in the cleaning solution works better than lens tissue. Whenever the image on the slide is unclear or cloudy, assume at once that your objective is soiled.

Condenser Wipe off the top surface of the condenser with lens paper to remove the dust that often accumulates there.

Procedures

If your microscope has four objectives, you have four magnification options: (1) scanning power, or 40×, magnification; (2) low-power, or 100×, magnification; (3) high-dry magnification, which is 450× with a 45× objective; and (4) 1000× magnification with a 100× oil immersion objective. The total magnification seen through an objective is calculated by simply multiplying the power of the ocular (normally 10×) by the power of the objective.

Which objective you use depends on how much magnification is necessary. Generally speaking, however, starting with the scanning or low-power objectives and progressing to the higher magnifications is best. Consider the following suggestions for setting up your microscope and making microscopic observations:

Viewing Setup If your microscope has a rotatable head, such as the one the two students in figure 4.4 are using, you can use the instrument in two ways. Note that the student on the left has the arm of the microscope *near* him, while the student on the right has the arm *away from* her. With this type of microscope, the student on the right has the advantage in that the stage is easier to observe. Note also that, when focusing the instrument, the student on the right is able to rest her arm on the table. If the microscope head is not rotatable, you will need to use the position of the student on the left.

Low-power Examination Starting with either the scanning or low-power objective enables you to explore the slide and locate the object you are planning to study. Then you can proceed to higher magnifications. Use the following steps when exploring a slide with either the scanning or low-power objective:

Notes: If you are unfamiliar with microscope usage, always begin your observations with the scanning lens. This will make it easier for you to find the object of interest and to focus clearly. You need not look through the ocular(s) until you reach step 6 of the procedure that follows. Steps 1–5 are accomplished while observing the stage from the side or front.

1. Position the slide on the stage so that the material to be studied is on the slide's *upper* surface. Figure 4.5 illustrates how one common type of mechanical stage mechanism (a retainer lever) clamps the slide in place. Other types of

mechanical stages have different clamps for holding slides in place.

2. Turn on the light source, using an *intermediate* amount of voltage. If necessary, reposition the slide so that the stained material on the slide is in the *exact center* of the light source.
3. If the condenser is adjustable, check to see that it has been raised to its highest point.
4. Rotate either the scanning or low-power objective into position over the center of the stage. Be sure that the objective clicks into its locked position.
5. Turn the coarse adjustment knob to lower the objective *until it stops,* or as is the case with some microscopes, raise the stage *until it stops.* A built-in stop prevents the objective from touching the slide.
6. Look through the ocular (or oculars), and turn the fine adjustment knob to bring the object into focus. If this does not bring the object into focus, readjust the coarse adjustment knob. If you are using a binocular microscope, you will also need to adjust the interocular distance and make diopter changes to match your eyes (as described previously on p. 22).
7. Manipulate the diaphragm control to reduce or increase the light intensity to produce the clearest, sharpest image. Note that as you close the diaphragm to reduce the light intensity, contrast improves. Closing the diaphragm when using the scanning or low-power objectives does not decrease resolution.
8. Once an image is visible, move the slide to find the object you plan to study. Turning the knobs that move the mechanical stage moves the slide.
9. Once you have located the object to be studied, you may want to increase the magnification by proceeding to either high-dry or oil immersion magnification. However, before changing objectives, *be sure to center the object you wish to observe.*

High-dry Examination To change from low-power to high-dry magnification, rotate the high-dry objective into position and open up the diaphragm somewhat. Sharpen the image by slightly adjusting the fine adjustment knob if necessary, but *do not touch the coarse adjustment knob.*

A good-quality microscope requires only minor focusing adjustments when changing from low-power to high-dry because all the objectives are **parfocalized.** Parfocal microscopes do not require considerable refocusing when changing from one objective to another.

Use high-dry objectives only on slides with cover glasses; without cover glasses, images are usually unclear. When increasing the lighting, be sure to open up the diaphragm first instead of increasing the voltage on your lamp, because *lamp life is greatly extended when lamps are used at low voltage.* If the field is not bright enough after you open the diaphragm, increase the voltage. A final point: Keep the condenser at its highest point.

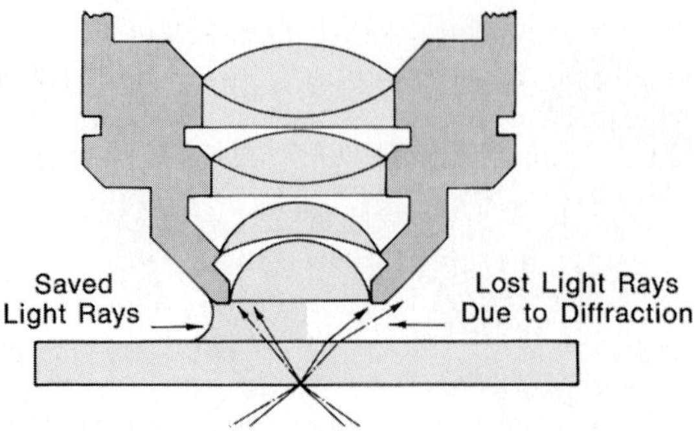

Figure 4.6 Immersion oil has the same refractive index as glass and prevents light loss due to diffraction.

Oil Immersion Techniques With an oil immersion lens, you place a drop of a special mineral oil between the lens and the microscope slide. The oil has the same refractive index as glass, which prevents the loss of light due to the bending of light rays as they pass from one medium to another. The use of oil in this way enhances the microscope's resolving power. Figure 4.6 illustrates this phenomenon.

Parfocal microscopes allow switching to oil immersion from either low-power or high-dry. Once you have focused the microscope at one magnification, you can rotate the oil immersion lens into position without fear of striking the slide. Before rotating the oil immersion lens into position, however, place a drop of immersion oil on the slide. If the oil appears cloudy, discard it.

When using the oil immersion lens, open the diaphragm as much as possible. Closing the diaphragm tends to limit the resolving power. In addition, keep the condenser at its highest point. If different-colored filters are available for the lamp housing, use blue or greenish filters to enhance the resolving power.

Using an oil immersion lens takes practice. The manipulation of lighting is critical. Before returning the microscope to the cabinet, remove all oil from the objective and stage, and use lens paper to wipe slides with oil clean.

Putting It Away

When you take a microscope from the cabinet at the beginning of the period, you expect it to be clean and in proper working condition. People coming after you expect the same consideration. A few moments of care at the end of the period ensures these conditions. Go through the following check-list at the end of each period before returning the microscope to the cabinet:

1. Remove the slide from the stage.
2. If immersion oil has been used, wipe it off the lens and stage with lens paper. If the slide has a cover glass on it, remove the oil with tissue. Slides that lack a cover glass (such as blood smears) should not be wiped dry. Simply place these slides in a slide box and let the oil drain off.
3. Rotate the low-power objective into position.
4. Wrap the electric cord around the microscope base.
5. Adjust the mechanical stage so that it does not project too far on either side.
6. Replace the dustcover.
7. Return the microscope to its correct place in the cabinet.

5 Basic Cell Structure

A review of cellular anatomy is prerequisite to the study of cellular physiology. Visualizing the structures of the plasma membrane, nucleus, and such organelles as the endoplasmic reticulum, centrosome, and lysosomes goes hand in hand with understanding the cell's physiological activities as a whole.

In this exercise, current knowledge of cellular structure and function is summarized first. Laboratory studies of epithelial cells and living microorganisms follow.

Since much of the discussion in this exercise pertains to cellular structure as seen with an electron microscope, and since the laboratory observations are made with light microscopes, you will not be able to see some cellular structures in great detail in the laboratory.

Assignment:
To get the most out of this exercise, answer the questions on the Laboratory Report before performing any laboratory activities.

The Basic Design

Electron microscopy and special techniques in the study of cellular physiology have rapidly advanced the study of cellular structure (**cytology**) over the past decades. A much more dynamic model has replaced the old notion of the cell as a "sac of protoplasm."

A cell is a highly complex miniature machine with an integrated, compartmentalized ultrastructure capable of communicating with its environment, altering its shape, processing information, and synthesizing a great variety of substances. This new viewpoint of cells has emerged largely from advancing knowledge of cellular organelles.

Figure 5.1 is a diagrammatic version of a pancreatic cell that shows the various organelles that an electron microscope reveals. A pancreatic cell has been chosen for study here because it lacks the degree of specialization seen in many other cells, and it contains a majority of the organelles present in most body cells.

Although you cannot see the fine structure of many organelles with an ordinary light microscope, the organelles' descriptions and current theories of their function are included here. As the various cell structures are discussed in the following text, locate them in figure 5.1.

The Plasma Membrane

The outer surface of every cell consists of an extremely thin, delicate cell membrane called the *plasma membrane.* Plasma membranes are primarily composed of **lipids** (e.g., phospholipids, glycolipids, and cholesterol) and **proteins** (including glycoproteins). Lipids account for half the mass of plasma membranes; proteins make up the other half.

Electron microscopes reveal that all plasma membranes have a similar basic structure. The current interpretation of the molecular organization of this membrane is that protein molecules are embedded in a fluid double layer of lipid molecules. Figure 5.1 shows an enlarged portion of the plasma membrane that illustrates how these components relate to each other. Note that some protein molecules protrude externally, others protrude internally, and some extend through both sides of the membrane. This asymmetrical architecture appears to account for the external and internal differences in receptor sites that affect the membrane's permeability. Protein molecules that extend completely through the membrane provide tubular passageways or channels for certain types of ions and molecules.

Plasma membranes are *selectively permeable* in that they allow certain molecules to pass through easily while preventing other molecules from gaining entrance to the cell. Plasma membranes play both active and passive roles in molecular movements into and out of a cell. Exercise 7 examines some of the forces involved in permeability.

The Nucleus

The largest and most conspicuous organelle in a cell is the large, spherical *nucleus* (shown in figure 5.1 with a portion cut out of it). The nucleus contains *nucleoplasm.* A double-layered **nuclear envelope** that is perforated by **nuclear pores** of significant size surrounds the nucleoplasm. The pore openings provide a passageway between the nucleoplasm and the rest of the cell contents (*cytoplasm*).

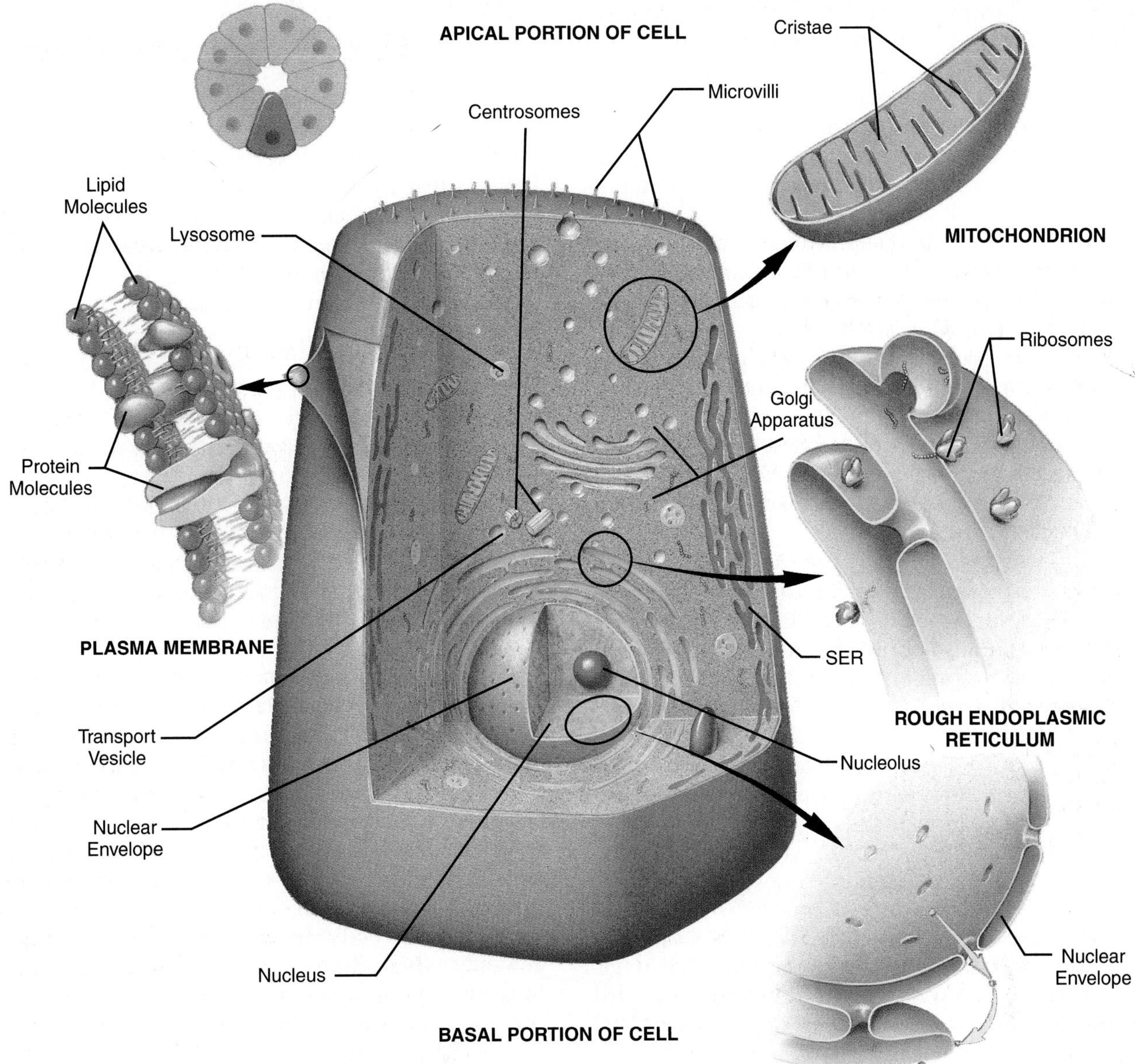

Figure 5.1 The microstructure of a cell.

A cell that has been stained with certain dyes exhibits darkly stained nucleic regions that are composed of *chromatin.* These regions are visible even with a light microscope. They represent highly condensed DNA molecules that comprise parts of the chromosomes. Current theories suggest that increased condensed chromosomal material indicates a metabolically less active cell.

The most conspicuous spherical region within the nucleus is the **nucleolus.** Although nucleoli of different cells vary considerably in size, number, and structure, they all consist primarily of RNA molecules and function chiefly in ribosome production. The arrow in the magnified view in the lower right-hand corner of figure 5.1 illustrates the passage of a ribosome from the nucleoplasm through a pore of the nuclear envelope.

The Cytoplasm

Between the plasma membrane and nucleus is the *cytoplasm.* This region is a heterogenous aggregation of many components involved in cellular metabolism. The numerous organelles, many of which are hollow and enclosed by membranes, structurally and functionally compartmentalize the cytoplasm. A description of each organelle follows.

Endoplasmic Reticulum The most extensive structure within the cytoplasm is a complex system of tubules, vesicles, and sacs called the *endoplasmic reticulum (ER)*. It is composed of membranes that are somewhat similar to the nuclear and plasma membranes; in some instances, it is continuous with these membranes. The ER functions, in part, as a microcirculatory system for the cell, providing a passageway for intracellular transport of molecules.

Detailed cell studies show that ER can have smooth or rough surfaces. **Rough endoplasmic reticulum (RER)** is a fluid-filled canalicular system studded with ribosomes. RER is the site of protein synthesis and storage. It is more developed in cells that are primarily secretory in function.

Smooth endoplasmic reticulum (SER) lacks ribosomes. Some layers of SER are shown in the cytoplasm near the plasma membrane in figure 5.1. SER is active in various types of metabolic processes, including lipid synthesis and carbohydrate metabolism.

Ribosomes As stated previously, ribosomes are small bodies attached to the surface of RER. *Free ribosomes* are also scattered throughout the cytoplasm. Since ribosomes are only 170 angstrom units (17 nm) in diameter, only electron microscopy can resolve them.

The two components for each ribosome originate in the nucleolus of the nucleus. They pass from the nucleolus through the pores of the nuclear membrane into the cytoplasm, where they unite to form the completed or intact ribosomal particle. Each ribosome consists of about 60% RNA and 40% protein. The enlarged view of RER in figure 5.1 shows ribosomes' proposed structure.

Ribosomes are sites of protein synthesis and are particularly numerous in cells that are actively synthesizing proteins. Free ribosomes appear to be involved in the synthesis of proteins for use within the cell; attached ribosomes on RER, on the other hand, are implicated in the synthesis of protein to be transported extracellularly (i.e., secreted).

Mitochondria Mitochondria are large (1×2–$3\ \mu m$), bean-shaped organelles that have double-layered membranous walls with inward-protruding partial partitions called **cristae** that increase the total membrane surface area. Mitochondria contain DNA and can replicate themselves. They also contain their own ribosomes. They are sometimes referred to as the "power plants" of the cell, since the energy-yielding reactions of aerobic cellular respiration occur in them. Simply stated, these reactions oxidize glucose ($C_6H_{12}O_6$) to yield energy in the form of ATP:

$$C_6H_{12}O_6 + 6\,O_2 \rightarrow 6\,CO_2 + 6\,H_2O + ATP + Heat$$

Golgi Apparatus The Golgi apparatus is a membranous organelle that is quite similar in basic structure to SER. In figure 5.1 it appears as a layered stack of vesicles, called cisternae.

The golgi apparatus modifies, packages, and distributes products that secretory cells will release. Proteins that the RER synthesizes are pinched off into membranous sacs, or **transport vesicles,** that coalesce with the cisternae within the Golgi apparatus. Inside the Golgi apparatus, the secretory products undergo further processing. Completely processed materials accumulate at the end of the Golgi apparatus that is closest to the plasma membrane. Additional transport vesicles bud off from the cisternae and carry the processed materials away. These vesicles move to the apex of the cell, where they fuse with the plasma membrane and release their contents to the extracellular environment via exocytosis. Figure 5.1 illustrates this process and how the vesicles fuse with the plasma membrane. The fate of the exocytosed product depends on the cell type.

Lysosomes *Lysosomes* are saclike structures in the cytoplasm that contain digestive enzymes. These sacs form by pinching off from the Golgi apparatus. The contents of lysosomes vary from cell to cell. Some lysosomal enzymes can hydrolyze all major categories of macromolecules.

Lysosomes function in *intracellular digestion.* For example, some white blood cells engulf foreign microbes, a process called *phagocytosis,* to defend the body against disease. Lysosomal enzymes then help to digest these foreign invaders. Lysosomes also function in the cycling of a cell's own organic materials. The observation that lysosomes often contain fragments of mitochondria as well as other cell organelles supports this. When a lysosome has performed its function, it is expelled from the cell through the plasma membrane.

When a cell dies, the membranes surrounding its lysosomes disintegrate, releasing enzymes into the cytoplasm. The enzymes' hydrolytic action hastens the cell's death. For this reason, lysosomes have been called "suicide bags."

The Centrosome Near the upper surface of the nucleus in figure 5.1 is a pair of microtubular bundles arranged at right angles to each other. Each bundle, called a *centriole,* consists of nine triplets of microtubules arranged in a circle. The two centrioles are collectively called the *centrosome.* Because of their small size, centrosomes appear as very tiny dots when observed with a light microscope.

During cell division, the centrioles play a role in the formation of *spindle fibers* that aid in the distribution of genetic material to daughter cells. They also help form cilia and flagella in certain types of cells.

Cytoskeletal Elements

Living cells have extensive permeating networks of fine fibers that contribute to such cellular properties as contractility and shape. These fibers also function in cellular support, movement, and the translocation of materials within a cell. While at least three categories of cytoskeletal elements have been identified, only two—microtubules and microfilaments—are described here.

Microtubules *Microtubules* are molecules of protein (*tubulin*), arranged in submicroscopic, cylindrical, hollow bundles. They are dispersed throughout the cytoplasm of most cells and converge especially around the centrioles to form *aster fibers* and *spindles* during cell division (Exercise 6). Microtubules also play a significant role in transporting vesicles from the Golgi apparatus to other parts of the cell and in maintaining cell shape.

Nerve cells with long processes, called axons, utilize microtubules to transport vesicles and mitochondria from the cell body to the axon terminal, as well as from the terminal back to the cell body.

Figure 5.2 shows how microtubules might look within the axon of a nerve cell. Note that the microtubules are propelling various-sized vesicles in both directions *simultaneously*. Mitochondria supply the energy for vesicle transport. Figure 5.2 shows a portion of a mitochondrion in the lower right quadrant; another is barely discernible in the upper left quadrant.

Figure 5.2 A computer-generated visualization of microtubules, vesicles, and mitochondria in the axon of a nerve cell. The microtubules are propelling the vesicles and mitochondria.

Microfilaments All cells show some degree of contractility. The cytoskeletal elements responsible for this phenomenon are the *microfilaments*. Microfilaments are composed of protein molecules arranged in solid, parallel bundles. They are most highly developed in muscle cells adapted specifically for contraction. In other types of cells, they are interwoven networks within the cytoplasm and are relatively inconspicuous, even in electron photomicrographs. Besides aiding in contractility, microfilaments also appear to be involved in cell movement, cell division, and cell support.

Cilia and Flagella

Cilia or *flagella* are hairlike extensions from cells that produce movement. In humans, cilia move substances past a stationary cell, and flagella propel a motile cell. While cilia are usually less than 20 μm long and are numerous, flagella may be thousands of micrometers in length and are few in number. Cells that line the respiratory tract and uterine tubes have cilia (see figures HA-2D and HA-31D, Histology Atlas). The beating of the cilia in the respiratory tract moves inhaled substances like pollen and dust upward to the pharynx so that they can be swallowed, while the beating of the cilia lining the uterine tubes propels an egg or zygote toward the uterus. The only human cells with flagella are sperm cells or spermatozoa. A sperm cell's flagellum beats to propel the sperm cell toward an egg. Both cilia and flagella originate from centrioles and are composed of microtubules.

Microvilli

Certain cells, such as the one in figure 5.1, have tiny protuberances known as *microvilli* on their apical or free surfaces. They may appear as a fine **brush border,** as in figure HA-2C (Histology Atlas), or as **stereocilia,** as in figure HA-34B (Histology Atlas). Both of these may be mistaken for cilia with a light microscope.

Each microvillus is an extension of cytoplasm that the plasma membrane encloses. On cells lining the intestine and kidney tubules, microvilli are common and increase cell surface area for water and nutrient absorption. Microvilli are not involved in any form of movement.

Assignment:
After answering the questions on the Laboratory Report pertaining to cell structure and function, do the following two cellular studies in the laboratory.

Laboratory Assignment

In the first laboratory study, an epithelial cell is removed from the inner surface of the cheek and examined with high-dry optics. Then microorganisms (primarily protozoa) are examined to observe the activity of living cells. In the second laboratory study, ciliary and flagellar action, amoeboid movement, secretion and excretion of cellular products, and cell division may be observed in viable cultures containing a mixture of protozoa.

Materials:
microscope slides
cover glasses
flat toothpicks
methylene blue stain
mixed cultures of protozoa
medicine droppers
microscope

Epithelial Cell Prepare a stained, wet-mount slide of some cells from the inside surface of your cheek as follows:

1. Obtain a microscope slide and cover glass.
2. Gently scrape the inside surface of your cheek with a clean toothpick. A light scraping is sufficient!
3. Wipe the toothpick onto the surface of your slide.
4. Place a drop of methylene blue stain on the cells, and mix the cells and stain with your toothpick.
5. Cover your specimen with a cover glass by placing one end of the cover glass down on your slide just to one side of your specimen and then carefully dropping the cover glass in place.
6. After locating a cell under scanning or low-power magnification, carefully examine the cell under high-dry magnification. Identify the *nucleus, cytoplasm,* and *plasma membrane.*
7. Draw a few cells on the Laboratory Report, labeling the structures noted in step 6.

Microorganisms Prepare a wet mount of protozoa by placing a drop of the culture on a slide and covering it with a cover glass. Be sure to obtain the drop from the very bottom of the protozoan culture, since most of the protozoa are found there.

Examine the slide first under scanning and/or low-power magnification, then under high-dry magnification. Study individual cells carefully, looking for nuclei, cilia, flagella, vacuoles of ingested food, excretory vacuoles, and so on. If you wish to identify the organisms you observe, consult Appendix E for representative types. Since algae are often present in protozoan cultures, the appendix also includes some illustrations of algae.

Assignment:
Answer the questions on the Laboratory Report that pertain to this exercise.

Mitosis

6

Cell division is the equal division of all cellular components to form two daughter cells that are identical in genetic composition to the original cell. This process allows for growth of body structures, tissue repair, and the replacement cells. While many cells in the body are able to divide, many do not after a certain age.

Division of the cytoplasm is called *cytokinesis;* nuclear division is called *karyokinesis,* or *mitosis.* Cells that are not undergoing mitosis are said to be in *interphase.*

During this exercise, you will study the various phases of mitosis on a microscope slide of hematoxylin-stained embryonic cells of *Ascaris,* a roundworm. The advantage of using *Ascaris* over other organisms is that it has few chromosomes.

Interphase

Although a cell in interphase does not appear to be active, it is carrying on the physiological activities characteristic of its particular specialization. The cell may be in one of three different phases: the G_1 phase, the S phase, or the G_2 phase (G stands for "gap," S for "synthesis").

After a round of cell division, cells enter the *G_1 phase.* During this and the other two phases of interphase, the cell grows. The cell then enters the *S phase*, characterized by *DNA replication.* During DNA replication, each double-stranded DNA molecule replicates itself to produce another identical molecule. All of the cell's genetic information is copied during this stage, which most cells complete within 20 hours. When DNA replication is complete, the cell enters the *G_2 phase.*

Note in part 1 of figure 6.1 that, during interphase, cells exhibit a more or less translucent nucleus with an intact nuclear membrane.

Stages of Mitosis

Mitotic cell division is a continuous process once the cell begins to divide; however, for discussion purposes, the process is divided into a series of four recognizable stages: prophase, metaphase, anaphase, and telophase.

Prophase In prophase, the first stage of mitosis, the nuclear envelope and nucleolus disappear, **chromosomes** become visible, and the two **centrioles** of

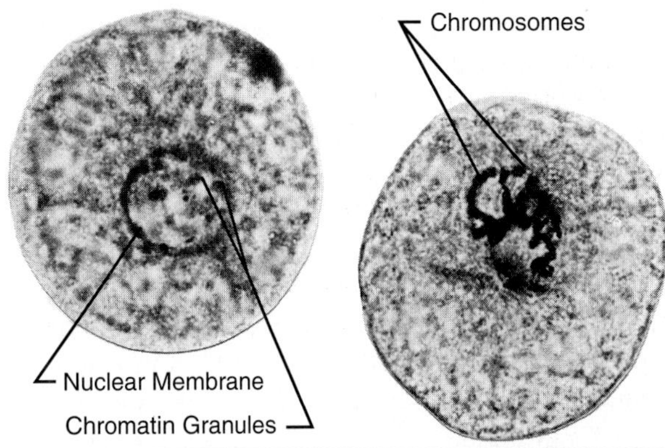

1 INTERPHASE and PROPHASE Note the distinct nuclear membrane and chromatin granules during interphase. Chromosomes form during prophase as the membrane disappears.

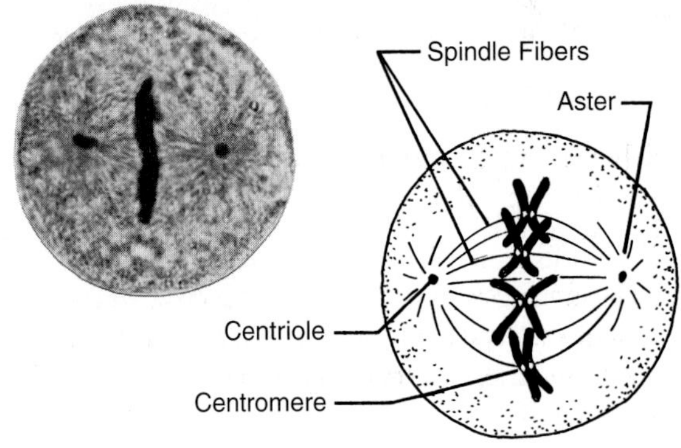

2 METAPHASE Chromosomes become arranged in the equatorial plane of the cell. Note the relationship of centromeres to spindle fibers.

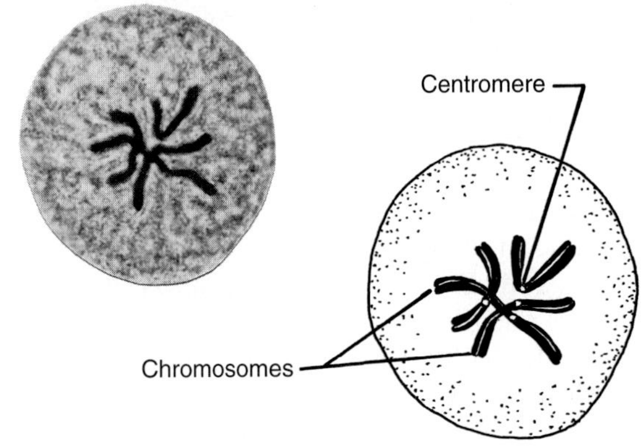

3 METAPHASE (Polar View) A polar view of chromosomes during metaphase shows their distinct shapes.

Figure 6.1 Early stages of mitosis.

the centrosome move to opposite poles of the cell to form **asters.** The dark-staining chromosomes are made up of DNA and protein. Each aster consists of a centriole and astral rays (microtubules). By the end of prophase, the microtubule assemblage radiating from each aster is quite extensive. This entire collection of microtubules is the **spindle apparatus,** and the individual fibers are called **spindle fibers.**

As prophase progresses, the chromosomes condense (become thicker and shorter). During this time, each chromosome has a double nature, consisting of a pair of identical **sister chromatids.** This dual structure is due to the replication of DNA molecules during interphase. The sister chromatids are connected in a region of the chromosome called the **centromere**. By the end of prophase, spindle fibers attached to each chromosome's centromere connect the chromosome to both asters (this is best seen in part 2 of figure 6.1).

Metaphase During metaphase, the spindle apparatus coordinates the migration of all of the chromosomes to the equatorial plane of the cell. Chromosomes are most condensed during metaphase.

Anaphase During anaphase, the individual chromatids of each chromosome separate from each other and are drawn to opposite poles of the cell by the spindle fibers, as indicated in part 1 of figure 6.2. Cytokinesis also begins during anaphase with the appearance of a **cleavage furrow** that results from the contraction of cytoskeletal elements that run along the equator of the cell. The cleavage furrow continues to deepen until the daughter cells are completely formed.

Telophase Telophase is the opposite of prophase. During telophase, nuclear envelopes form around the daughter nuclei, nucleoli reappear, chromosomes begin to decondense, the spindle apparatus is disassembled, and cytokinesis continues. Cytokinesis concludes shortly after telophase, and the resulting cells are called **daughter cells.**

Assignment:
Examine a prepared slide of *Ascaris* mitosis (Turtox slide E6.24). You may need to use both the high-dry and oil immersion objectives. Identify and draw cells that are in interphase and all four stages of mitosis in the spaces provided in the Laboratory Report. Label all identifiable structures. Also answer the questions in the Laboratory Report for this exercise.

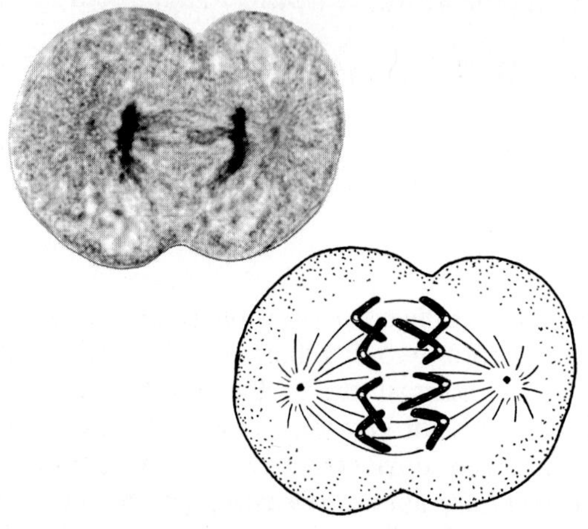

1 ANAPHASE Chromatids of each chromosome separate and move along spindle fibers to opposite poles of the cell.

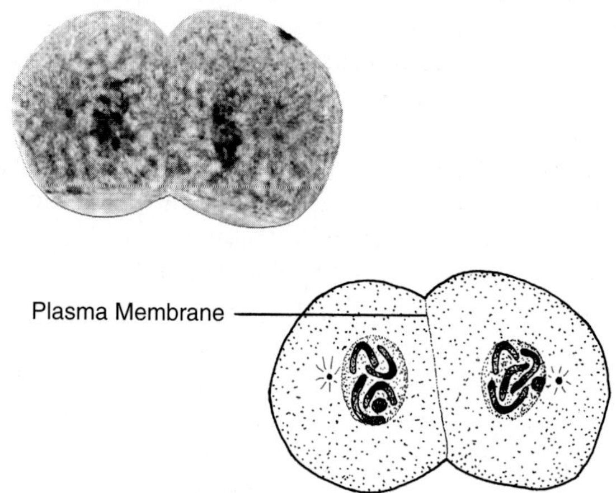

2 TELOPHASE Chromosomes lengthen, becoming less distinct. Nuclear membranes reappear, and a new plasma membrane forms.

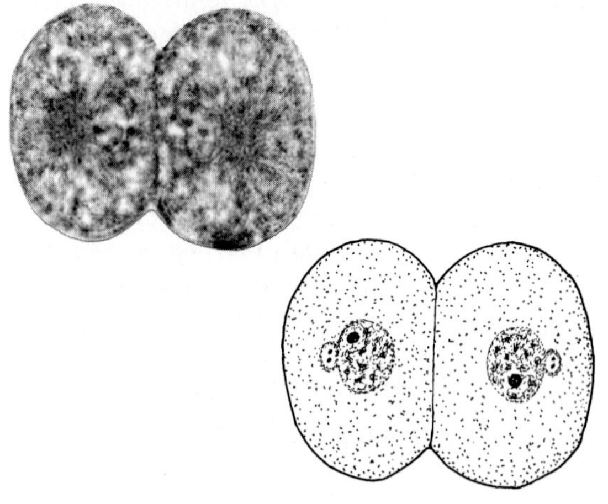

3 DAUGHTER CELLS Due to the preciseness of the mitotic process, these two cells are genetically identical to the original cell.

Figure 6.2 Later stages of mitosis.

Osmosis and Plasma Membrane Integrity 7

Every body cell is bathed in a watery mixture of molecules essential to the cell's survival. This fluid may be the plasma of blood or the tissue fluid in interstitial spaces. In either case, these molecules, whether water, nutrients, gases, or ions, pass in and out of the cell across the plasma membrane.

Some molecules, usually small ones, diffuse *passively* across the plasma membrane from areas of high concentration to areas of low concentration. Larger organic molecules, such as glucose and amino acids, and certain ions can only cross the plasma membrane by traveling through transport proteins embedded in it. When these molecules move from areas of high concentration to areas of low concentration (moving *with* the concentration gradient), it is *passive transport.* When these molecules move from areas of low concentration to areas of high concentration (moving *against* the concentration gradient), it is *active transport.* Active transport requires the cell to expend energy.

The integrity of plasma membranes of cells throughout the body is due largely to body fluids being *isotonic* to the intracellular fluid or cytoplasm. This means that the body fluids have the same concentration of dissolved molecules as the cytoplasm of the body cells. To see what happens to plasma membranes when isotonic conditions do not exist, you will study in this exercise the mechanics of **diffusion** and **osmosis.**

Molecular Movement

All molecules, whether in a gaseous or liquid state, move constantly. As the molecules bump into each other, they change direction and disperse randomly. Although molecular movement cannot be seen due to molecules' small size, both microscopic and macroscopic techniques allow indirect observation of this activity. Two such observations relate to Brownian movement and diffusion.

Brownian Movement

In 1827, the Scottish botanist Robert Brown observed that extremely small particles suspended in the protoplasm of plant cells are constantly vibrating. At first, he erroneously assumed that this movement was peculiar to living cells. Subsequent studies revealed, however, that invisible water molecules bombarding the small visible particles caused this vibratory movement, which became known as *Brownian movement.*

The simplest way to observe this phenomenon is to examine a wet-mount slide of India ink under the oil immersion objective of a microscope. India ink is composed of carbon particles in water, alcohol, and acetone.

Examine the demonstration setup provided in the laboratory. Brownian movement is of particular interest to microbiologists, since it must be differentiated from true motility in bacteria. Nonmotile bacteria are small enough that moving water molecules can displace them.

Diffusion

If a crystal of a soluble compound is placed on the bottom of a container of water, molecules from the crystal spread throughout the water to all parts of the container by *diffusion.* The molecules move away from the original site of high concentration to an area of lower concentration because the presence of fewer molecules in the lower-concentration areas means less obstruction and favors movement of molecules into those areas. The rate of diffusion varies and depends on ambient temperature and molecular size. Given sufficient time, diffusion eventually results in even dispersal of all molecules throughout the container.

You can demonstrate this type of molecular activity with crystals of colored compounds, such as potassium permanganate (purple) and methylene blue. Figure 7.1 illustrates how to set up such a demonstration.

Methylene blue has a molecular weight of 320. Potassium permanganate has a molecular weight of 158. A petri plate with 1.5% agar is used. This agar medium of 98.5% water allows molecules to move freely. Prepare such a plate as follows, and record your observations on the Laboratory Report.

Materials:
crystals of potassium permanganate and methylene blue
petri plate with about 12 ml of 1.5% agar
plastic ruler with metric scale

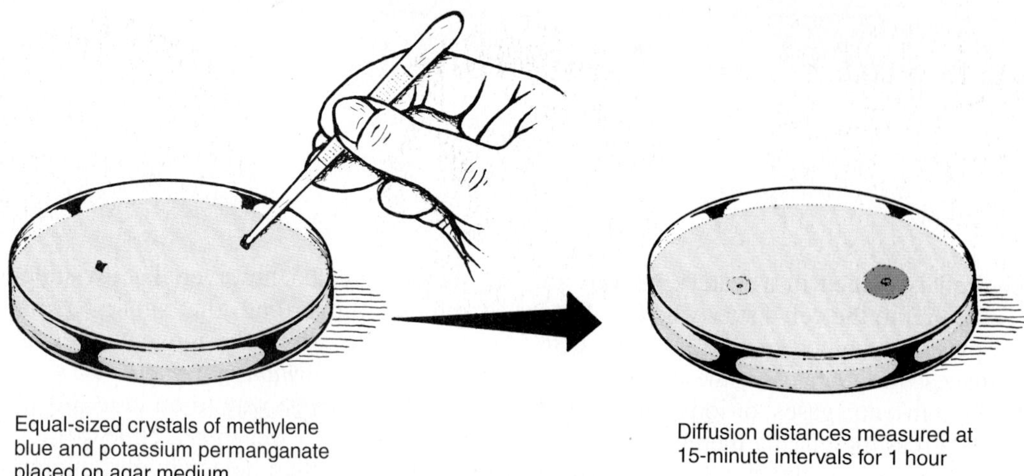

Figure 7.1 Comparing areas of diffusion.

1. Select one crystal of each of the two different chemicals, and place them about 5 cm apart on the surface of the agar medium. Try to select crystals of similar size.
2. Every 15 minutes, for 1 hour, use a small plastic metric ruler to measure the distance of diffusion (in millimeters) from the center of each crystal.
3. Record the measurements on the Laboratory Report.

Osmotic Effects

Because of their small size, water molecules move freely across a plasma membrane both into and out of a cell. Water molecules exhibit net directional movement into or out of a cell in response to the concentrations of dissolved substances (solutes) in the cytoplasm and extracellular fluid. The movement of water molecules through a selectively permeable membrane in response to solute concentrations is called *osmosis.*

Although water molecules are always moving in both directions across a plasma membrane, the concentrations of solutes on each side of the membrane determine the predominant direction of movement. If a funnel containing a sugar solution is immersed in a beaker of water and separated from the water with a semipermeable membrane, as in figure 7.2, more water molecules will enter the funnel through the membrane than will leave the sugar solution to enter the beaker of water.

As was observed in the diffusion experiment, molecules tend to move from areas of high concentration to areas of low concentration. In this case, the water molecules are more concentrated in the beaker than in the sugar solution and, therefore, move into the sugar solution. This example demonstrates why osmosis is often defined as *the diffusion of water across a selectively permeable membrane.* In the experiment, the inward flow of water molecules moves the sugar solution upward in the tube. The force required to restrain the upward flow is called the *osmotic pressure.*

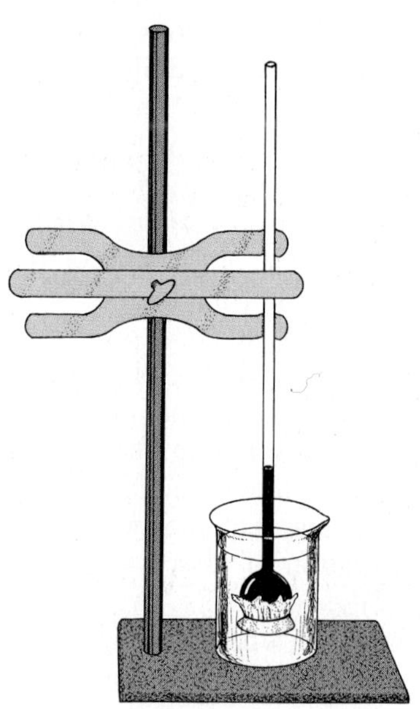

Figure 7.2 Osmosis setup.

Solutions that contain the same concentration of solutes as cells contain are **isotonic solutions.** Blood cells immersed in such solutions gain and lose water molecules at the same rate, establishing an *osmotic equilibrium,* as shown in figure 7.3.

If a solution contains a higher concentration of solutes than is present in cells, water leaves the cells faster than it enters, causing the cells to shrink and develop cupped (crenated) edges. This shrinkage is called *plasmolysis,* or *crenation.* Such solutions of high solute concentration are **hypertonic**

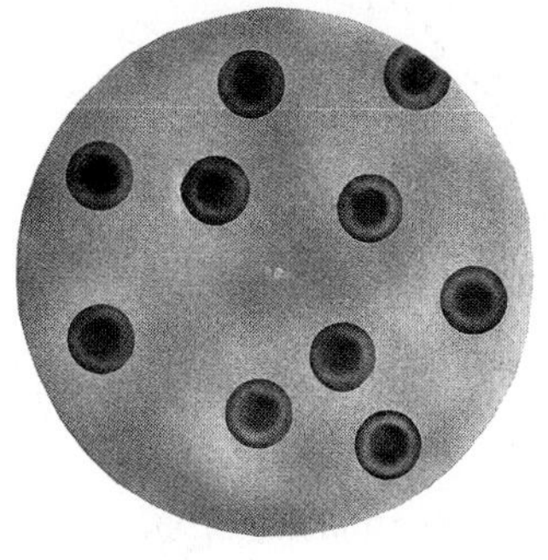

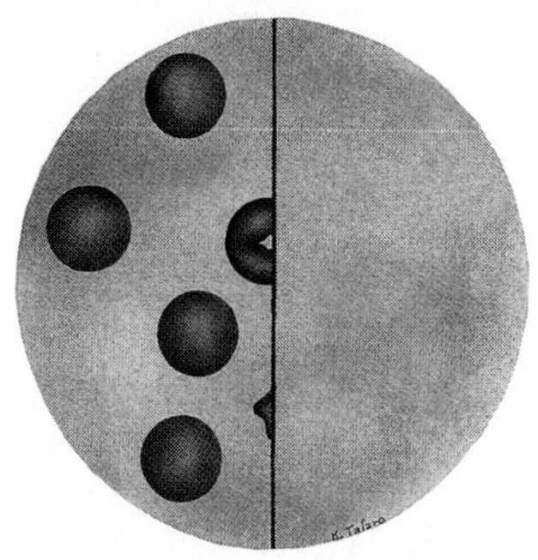

HYPERTONIC (Crenation)	ISOTONIC (Osmotic Equilibrium)	HYPOTONIC (Plasmoptysis-Left, Following Hemolysis-Right)

Figure 7.3 Effects of solutions on red blood cells.

solutions and have a higher osmotic potential than isotonic solutions.

Solutions with a lower solute concentration than cells are **hypotonic solutions.** In such solutions, water flows rapidly into the cells, causing them to swell *(plasmoptysis)* and eventually burst *(lyse)* as the plasma membrane disintegrates. Lysis of red blood cells is called *hemolysis.*

To observe the effects of the various types of solutions on red blood cells, follow the procedures outlined in figure 7.4. Blood cells are added to various concentrations of solutions. The effects of the solutions on the cells are determined macroscopically and microscopically. Read the precautions on page 283 relating to the use of fresh blood.

Materials:
five serological test tubes and tube rack
small beaker of distilled water (50 ml size)
two depression microscope slides, cover glasses
wax pencil, Vaseline, and toothpicks
serological pipette (5 ml size)
canister for used pipettes
syringes, needles, Pasteur pipettes
mechanical pipetting device
fresh blood
solutions of: .15M NaCl, .30M NaCl, .28M glucose, .30M glycerine, .30M urea

 **Lab 9:** Diffusion, osmosis, and tonicity.

1. Label five clean serological tubes 1 to 5, and arrange them sequentially in a test tube rack.
2. With a 5 ml pipette, deliver 2 ml of each solution to the appropriate tube, as shown in figure 7.4, using the delivery method shown in figure 7.5. Rinse the pipette with distilled water after each delivery.
3. Dispense two drops of blood to each tube, using a syringe. Shake each tube from side to side to mix; then let the tubes stand for 5 minutes.
4. Hold the rack of tubes up to the light, and compare them. A transparent solution indicates hemolysis. If you are unable to see through the tube, hemolysis has not occurred, and the cells are intact. Crenation is indicated if the solution

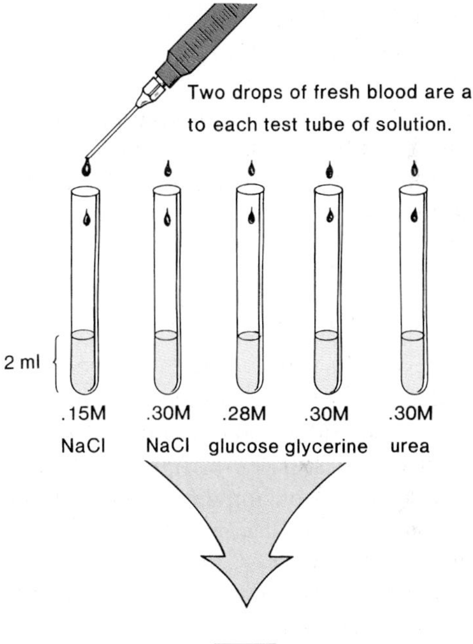

Figure 7.4 Routine for making cell suspensions.

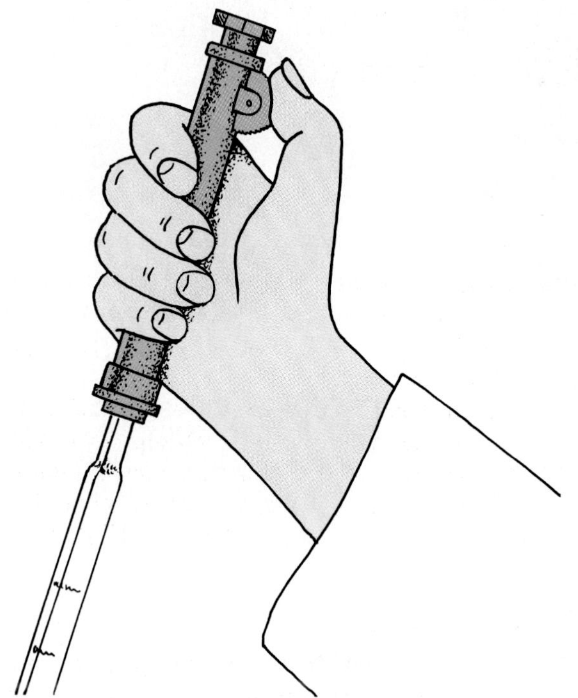

Figure 7.5 Use a mechanical pipetting device for all pipetting.

is somewhere between the clarity of hemolysis and the opacity of osmotic equilibrium.

5. Record your results on the Laboratory Report.
6. Make a hanging drop slide from each tube as follows:
   a. Use a toothpick to place a tiny speck of Vaseline near each corner of a cover glass, as shown in part 1 of figure 7.6.
   b. With a Pasteur pipette, transfer a drop of the cell suspension to the center of the cover glass.
   c. Place a depression slide on the cover glass, with the depression facing downward.
   d. With the cover glass held to the slide by the Vaseline, quickly invert.

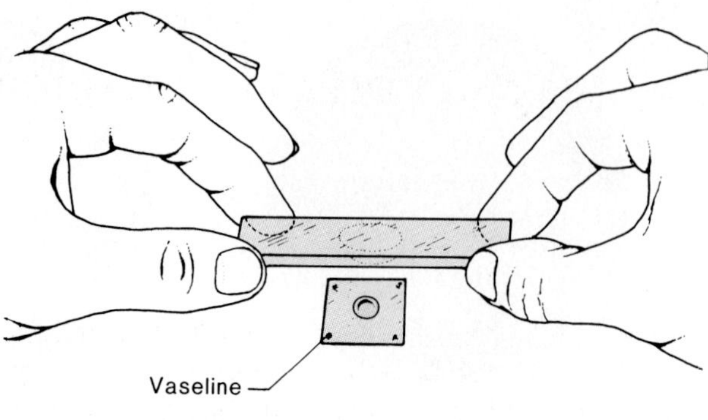

1 Depression slide is pressed against Vaseline on cover glass and quickly inverted.

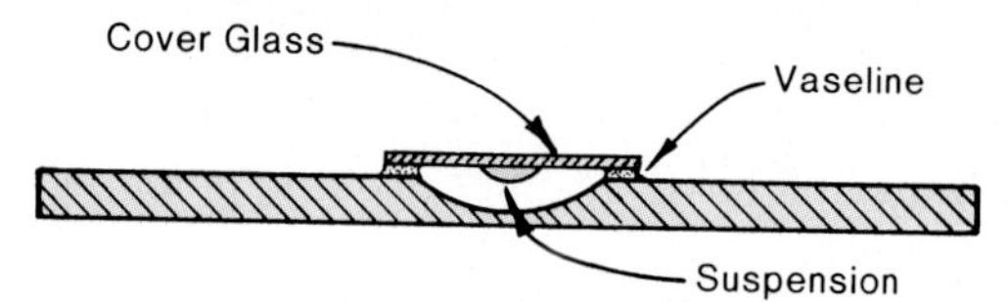

2 The completed slide is examined under the high-dry objective.

Figure 7.6 Procedure for making a hanging drop slide.

7. Examine each slide under the high-dry objective, and record your results on the Laboratory Report.

Assignment:

After recording your results on the Laboratory Report, complete the report by answering all the questions.

Hydrogen Ion (H^+) Concentration and pH

8

In addition to being isotonic, extracellular fluid must have a hydrogen ion concentration that is compatible with the cell's physiology. Hydrogen ions (protons) readily diffuse from extracellular fluid into the cell across the plasma membrane and profoundly affect the cell's enzymatic reactions. If the concentration of these ions is too great or too low, the cell's physiology is adversely affected.

The hydrogen ion concentration of extracellular fluid throughout the body is very close to pH 7.36. Arterial blood has a pH of 7.41. Venous blood is close to the pH of extracellular fluid. The pH of body fluids rarely falls below 7.0 or rises above 7.8, even under extremely abnormal conditions. As the hydrogen ion concentration of extracellular fluid deviates from the norm of pH 7.36, certain metabolic reactions in cells are accelerated or depressed.

Due to the significance of hydrogen ion concentration in physiological processes, this exercise focuses on two aspects of pH: (1) its mathematical derivation, and (2) the relative merits of colorimetric versus electronic methods of pH measurement. Unless you understand both aspects, you may find subsequent experiments difficult to comprehend.

A Review of pH

The Danish chemist Sørensen developed the pH scale to provide a simplified numerical system for expressing the concentration of hydrogen (H^+) and hydroxyl (OH^-) ions. The concentration of these ions is a function of the degree of ionization. For example, pure distilled water dissociates (comes apart) to produce these two ions as follows:

$$HOH \rightleftharpoons H^+ + OH^-$$

Water dissociates at a very low rate: In pure water at 25°C, only 1×10^{-7}, or one ten-millionth, of the water molecules have dissociated. This means that one out of 10 million molecules is in the ionized state at a given time and that the numbers of hydrogen and hydroxyl ions present at the same time are equal.

Strength of Acids and Bases

When acids are added to water, they dissociate a variable amount. *Strong acids,* such as hydrochloric acid (HCl), dissociate almost completely and maximize the number of free hydrogen ions. *Weak acids,* on the other hand, may dissociate only about 1% and release very few ions.

The same is true of bases. Most of the molecules of *strong bases,* such as sodium hydroxide (NaOH), ionize to produce an alkaline solution of high OH^- content. *Weak bases,* such as ammonium hydroxide (NH_4OH), produce fewer OH^- ions.

Definition of pH

To eliminate the cumbersome way of expressing ion concentrations in terms of 1×10^{-7}, or one part in 10 million, Sørensen devised the pH or *potential hydrogen* scale. He defined pH as *the logarithm of the reciprocal of the hydrogen ion concentration expressed as grams per liter:*

$$pH = \log \frac{1}{(H^+)}.$$

Given that the H^+ concentration of distilled water is 1×10^{-7}, pH 7 is calculated as follows:

$$pH = \log \frac{1}{1 \times 10^{-7}} = \log \frac{10^7}{1.0}$$

since $\log 10^7 = 7.0000$ and

$\log 1.0 = 0.0000$

$pH = 7.0000 - 0.0000 = 7$

To apply this equation to another problem, consider this example: If the hydrogen ion concentration of a sample of arterial blood is 0.0000000389 g H^+ per liter, is this concentration near the expected value of pH 7.41?

$$pH = \log \frac{1}{3.89 \times 10^{-8}} = \log \frac{10^8}{3.89}$$

since $\log 10^8 = 8.0000$ and

$\log 3.89 = 0.5899$

$pH = 8.0000 - 0.5899 = 7.41$

The answer to the previous question is, obviously, yes.

Refer to the logarithm table (table II) in Appendix A to confirm the log values for the previous two examples.

Table 8.1 Ion Concentrations as Related to pH.

HYDROGEN ION CONCENTRATION (Grams H^+ per Liter)		pH		HYDROXYL ION CONCENTRATION (Grams OH^- per Liter)
1.0	$= 10^{0}$	0	Acid	0.00000000000001
0.1	$= 10^{-1}$	1		0.0000000000001
0.01	$= 10^{-2}$	2		0.000000000001
0.001	$= 10^{-3}$	3		0.00000000001
0.0001	$= 10^{-4}$	4		0.0000000001
0.00001	$= 10^{-5}$	5		0.000000001
0.000001	$= 10^{-6}$	6		0.00000001
0.0000001	$= 10^{-7}$	7	Neutral	0.0000001
0.00000001	$= 10^{-8}$	8		0.000001
0.000000001	$= 10^{-9}$	9		0.00001
0.0000000001	$= 10^{-10}$	10		0.0001
0.00000000001	$= 10^{-11}$	11		0.001
0.000000000001	$= 10^{-12}$	12		0.01
0.0000000000001	$= 10^{-13}$	13		0.1
0.00000000000001	$= 10^{-14}$	14	Alkaline	1.0

The pH Scale and Normality

Table 8.1 shows that the pH scale extends from 0 to 14. Note that a solution of pH 0 contains 1 g of hydrogen ions per liter. Such a solution is also designated as being a 1 N solution (N = normal), since, by definition, a normal solution contains 1 g equivalent of hydrogen ions per liter of water.

The upper limit of the pH scale, which is 14, represents a hydrogen ion concentration of 10^{-14}, or 0.000,000,000,000,01 N. Since the hydroxyl ions vary inversely with the number of hydrogen ions, a solution of pH 14 has 1 g of hydroxyl ions.

pH Measurements

Two methods measure pH: one is colorimetric, and the other is electronic. In this portion of the exercise, both methods are compared by measuring the pH of several phosphate solutions and certain body fluids.

The **colorimetric method** utilizes color indicators, either in solution or in papers. To cover a broad pH range, you must use several indicators, as shown in table 8.2, where the range is from 1.2 through 8.8. To determine pH with any degree of accuracy by this method, however, you must use color standards. Instead of using indicator solutions in this experiment, you will use papers, which are more convenient.

The **electronic method** utilizes a pH meter that a glass electrode placed in the solution activates. When using a pH meter, you must standardize (calibrate) the unit for the pH range in which the meter is to be used. You must also make an adjustment for the temperature of the solution. When determining the pH of several different solutions, clean the electrode with a spray of distilled water between measurements. These instruments are extremely sensitive to pH changes.

Table 8.2 pH Indicators (Range of 1.2–8.8).

INDICATOR	COLOR		pH RANGE
	Acid	Alkaline	
Thymol blue	Red	Yellow	1.2–2.8
Bromphenol blue	Yellow	Blue	3.0–4.6
Methyl red	Red	Yellow	4.4–6.4
Bromcresol purple	Yellow	Purple	5.2–6.8
Bromthymol blue	Yellow	Blue	6.0–7.6
Phenol red	Yellow	Red	6.8–8.4
Cresol red	Yellow	Red	7.2–8.8

The hydrogen ion concentrations of 12 different phosphate solutions are determined here. The pH of each solution is measured first with pH papers and then with a pH meter. If only a small number of pH meters are available, the class may be divided into teams that are assigned a certain number of solutions. If the team approach is used, results should be posted on the chalkboard so that all class members have a complete set of results. If teams are unnecessary, students can work in pairs to perform the entire experiment.

Materials:

per pair of students:
four small beakers (30–50 ml size)
pH indicator paper
pH meter

onion
wax marking pencil
wash bottle of distilled water
large beaker (250 ml size)
Kimwipes

on demonstration table:
12 phosphate solutions in squeeze bottles (labeled A through L)

1. Label four small beakers A through D, and dispense solutions A through D into the appropriate beakers. Each beaker should receive about a half inch of solution (approximately 15–20 ml).
2. Determine the pH of each solution with pH paper, recording the pH in the table on the Laboratory Report.
3. Measure the pH of each solution with the pH meter. Be sure to rinse the electrode with distilled water from the wash bottle and dry it with Kimwipes between tests. See figure 8.1. Record your results in the table on the Laboratory Report.
4. Discard solutions A through D. Rinse the beakers, and relabel them E through H.
5. Dispense solutions E, F, G, and H to their respective containers.
6. Determine the pH of each solution by both methods, and record the results on the Laboratory Report.
7. Rinse the four beakers again, and repeat the procedure for solutions I, J, K, and L.

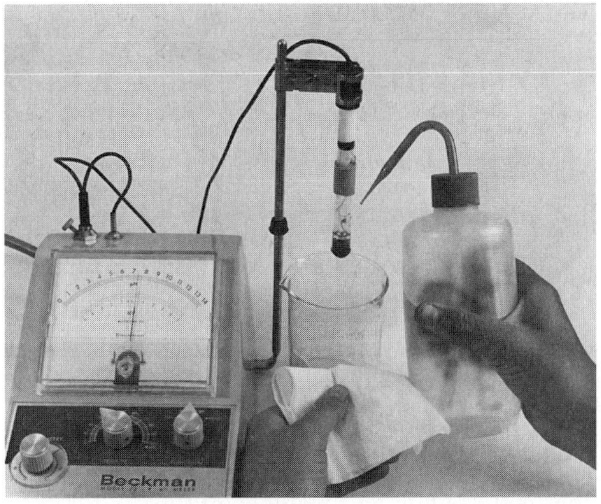

Figure 8.1 **Rinse the electrode tip with distilled water, and blot it with tissue between solutions.**

8. Use the pH meter to determine the pH of urine and blood, and pH paper to determine the pH of saliva and tears. When testing tears, use the onion to stimulate tear production, and do not place pH indicator paper into the eye. Instead, try to capture tear flow onto a slide for testing. Record your results.

Assignment:
After recording your results on the Laboratory Report, do the pH conversions in part C of the Laboratory Report.

9 Buffering Systems

As stated in Exercise 8, body cells function best in a very narrow pH range. When the pH of arterial blood becomes less than 7.40, **acidosis** occurs. Severe acidosis (below pH 7.0) results in a coma state and then death within minutes. When the pH of arterial blood exceeds 7.41, **alkalosis** occurs. Severe alkalosis (above pH 7.8) results in convulsions and death in an equally short time.

The energy-yielding processes in all body cells produce large quantities of carbon dioxide and organic acids. The carbon dioxide combines with water to form carbonic acid, which, in turn, dissociates to produce hydrogen and bicarbonate ions. Although carbonic acid is a weak acid, so much of it is produced from carbon dioxide that it is a major source of H^+ for the body fluids. These reactions are reversible and appear as follows:

$$CO_2 + H_2O \rightleftharpoons H_2CO_3 \rightleftharpoons H^+ + HCO_3^-$$

Three mechanisms—the kidneys, the respiratory system, and buffering systems—regulate the removal of excess hydrogen ions. Of the three, the kidneys are most effective in maintaining acid-base balance; however, they work very slowly, requiring *up to 24 hours* to make adjustments.

The respiratory system regulates H^+ concentration by controlling carbon dioxide levels in the blood. This adjustment requires approximately *1–3 minutes.*

The **buffering systems** adjust to pH change within *fractions of a second.* All buffering systems consist of a weak acid and its conjugate base. Each of these compounds has a common anion (negatively charged ion), but different cations (positively charged ions). The anion of the base reacts with excess H^+ to form the weak acid.

The most important buffering system in the blood is the bicarbonate system. In addition to this system, the blood contains proteins and inorganic phosphates that act as buffers. Even the erythrocytes act as buffers because they contain bicarbonate, phosphates, and hemoglobin. Figure 9.1 summarizes the blood's buffering systems.

The **bicarbonate system** utilizes carbonic acid as the weak acid (a base neutralizer) and sodium bicarbonate as the conjugate base (an acid neutralizer). These two compounds dissociate to produce the same anion (HCO_3^-):

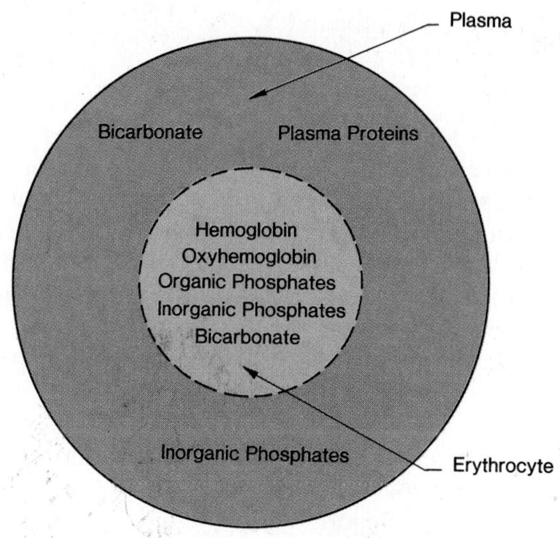

Figure 9.1 The total buffering capacity of the blood includes the buffers within erythrocytes as well as those in the plasma.

carbonic acid dissociation

$$H_2CO_3 \rightleftharpoons H^+ + HCO_3^-$$

sodium bicarbonate dissociation

$$NaHCO_3 \rightleftharpoons Na^+ + HCO_3^-$$

When hydrogen ions are added to this system, the anion from sodium bicarbonate combines with hydrogen ions to form carbonic acid, thus preventing the pH from falling precipitously.

When a strong base, such as NaOH, is added to this system, sodium bicarbonate and water form to prevent the pH from rising too much:

$$NaOH + H_2CO_3 \rightleftharpoons NaHCO_3 + H_2O$$

The ratio of H_2CO_3 to $NaHCO_3$ in extracellular fluid is approximately 1:20. The extra HCO_3^- provides an alkali reserve, or base excess, to react with H^+ that most normal activity produces.

In this exercise, you compare the relative effectiveness of five different buffering systems. Hydrochloric acid is added drop by drop to each solution to determine how much H^+ each buffer can take up in lowering the pH from 7.4 to 7.0. The solutions to be tested are (1) blood, (2) plasma, (3) albumin suspension, (4) a phosphate system, and (5) a bicarbonate system. The blood and plasma are diluted 1:10, so the values determined for these two solutions must be multiplied by 10 for an accurate comparison. A sample of distilled water is also compared.

Materials:

per pair of students:
pH meter or pH papers
six graduated beakers (30–50 ml size)
wax pencil
0.0001 N HCl in plastic squeeze bottle

test solutions (all at pH 7.4):
A—distilled water
B—bicarbonate buffer
C—phosphate buffer
D—albumin suspension
E—blood (1:10 saline dilution)
F—plasma (1:10 dilution)

1. Label six clean beakers A–F, and dispense 15 ml of the appropriate solution into each beaker. If the beakers are graduated, fill to the 15 ml mark. If the beakers are not graduated, use a graduated cylinder.
2. With the pH electrode (or pH paper) in the beaker of test solution A (distilled water), add 0.0001 N HCl slowly, drop by drop, counting drops until the pH reaches 7.0. Shake the beaker frequently for dispersal. Record the number of drops on the Laboratory Report.
3. Repeat the procedure for each beaker of solution. Rinse the electrode with distilled water after acidifying each solution.

Assignment:

Complete the Laboratory Report by answering the questions that pertain to this exercise.

10 Epithelial Tissues

Although all body cells share common structures, such as nuclei, centrosomes, Golgi apparatus, and so on, the cells differ considerably in size, shape, and structure according to their specialized functions. An aggregate of cells with similar structure and function is called a **tissue.** The study of tissues is called **histology.**

In this exercise, you study the different types of epithelial tissues with prepared microscope slides from portions of different organs. Learning how to identify specific epithelial tissues on slides that have several types of tissue is part of the challenge of this exercise.

Epithelial tissues, or **epithelia,** are aggregations of cells that perform specific protective, absorptive, secretory, transport, and excretory functions. They are often coverings for internal and external surfaces, and they rest on a bed of connective tissue. Characteristics common to all epithelial tissues are:

- Individual cells closely attach to each other at their margins to form tight sheets that lack any extracellular matrix or vascularization.
- Cell groupings are oriented so that they have an *apical* (free) surface and a *basal* (bound) region. The basal portion is closely anchored to underlying connective tissue. This thin adhesive margin between epithelial cells and connective tissue is the **basement lamina.** Although formerly referred to as the "basement membrane," it is not a true membrane. Unlike true membranes, the basement lamina is acellular. In reality, it is a colloidal complex of protein, polysaccharide, and reticular fibers.

Epithelial types are outlined in figure 10.1. The basic criteria for classification are stratification (layering), cell shape, and the presence or absence of cilia. The three divisions of cell shape are (1) *squamous* (flattened shape), (2) *cuboidal,* and (3) *columnar.* The three divisions of stratification are (1) *simple,* (2) *pseudostratified,* and (3) *stratified.* A discussion of each type of epithelium follows.

Simple Epithelia

Simple epithelial tissues are composed of a single cell layer that extends from the basement lamina to the free surface. Figure 10.2 shows three basic kinds of simple epithelia.

Simple Squamous Epithelia Simple squamous epithelial cells are very thin, flat, and irregular in outline, as part 1 of figure 10.2 shows. They form a pavementlike sheet in various organs that perform filtering or exchange functions. Capillary walls, alveolar walls in the lungs, the peritoneum, the pleurae, and blood vessel linings consist of simple squamous epithelia.

Cuboidal Epithelia Cuboidal epithelial cells (see part 2 in figure 10.2) are stout and blocklike in cross section and replace hexagonal with polygonal from a surface view. Many glands, such as the thyroid gland, salivary glands, and pancreas, as well as

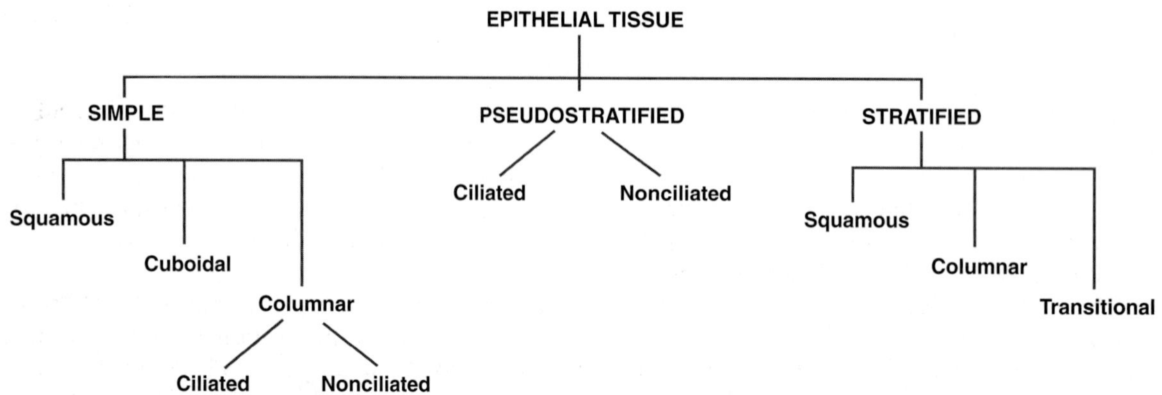

Figure 10.1 A morphologic classification of epithelial types.

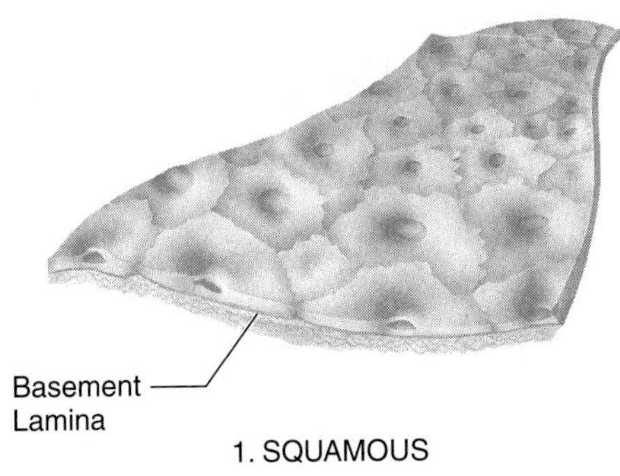

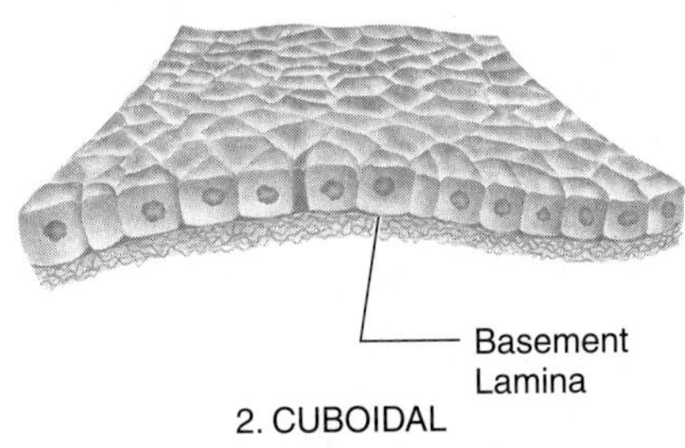

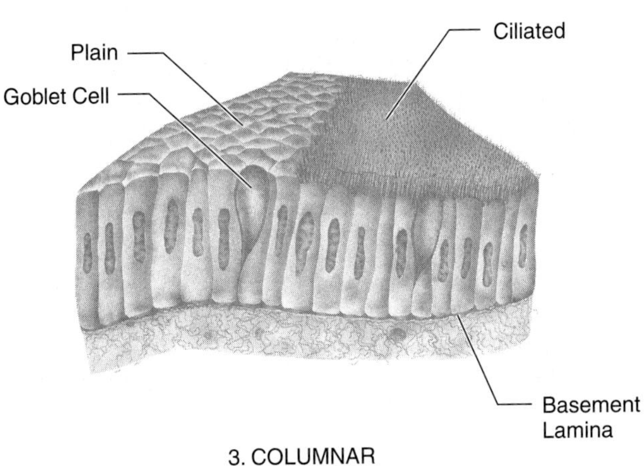

Figure 10.2 Simple epithelia.

the ovaries, kidneys and the capsule surrounding the lens of the eye have cuboidal epithelia.

Simple Columnar Epithelia Simple columnar epithelial cells are elongated between their apical and basal surfaces as part 3 of figure 10.2 shows. Their free surfaces give the cells a polygonal appearance.

Functionally, simple columnar epithelia may be specialized for protection, secretion, or absorption. As shown in the figure, certain secretory cells, called **goblet cells,** may be present and produce **mucus,** a protective glycoprotein. Also, note that the cells may be *ciliated* or not ciliated (*plain*).

Plain columnar epithelia line the stomach, intestines, and kidney collecting tubules. Ciliated columnar epithelia are found in the lining of the respiratory tract, uterine tubes, and portions of the uterus. Goblet cells can be present in both the plain and ciliated types.

The free surfaces of columnar cells may also have **microvilli** that are seen as a **brush border** when observed with a light microscope under oil immersion. Columnar cells that line the small intestine and make up the walls of the kidney collecting tubules exhibit distinct brush borders.

Stratified Epithelia

Stratified epithelia are squamous, columnar, and cuboidal epithelia that exist in layers. Figure 10.3 shows four layered epithelia that have distinct differences.

Stratified Squamous As shown in part 1 of figure 10.3, stratified squamous epithelia have superficial cells that are distinctly squamous, while the deepest layer of cells is columnar or, in some cases, cuboidal. Between the basal cell layer and the squamous cells are successive layers of irregular and polyhedral cells.

The chief function of stratified squamous epithelia is protection. Exposed inner and outer body surfaces, such as the skin, oral cavity, esophagus, vagina, and cornea, consist of this tissue type.

Stratified Columnar As shown in part 2 of figure 10.3, the superficial cells of stratified columnar epithelia are distinctly columnar and vary in height, while the deeper cells are irregular or polyhedral. Protection and secretion are the chief functions of stratified columnar tissue. This tissue type is present in some glands, the conjunctiva of the eye, the pharynx, a portion of the urethra, and the lining of the anus.

Stratified Cuboidal Although stratified cuboidal tissue is not listed in figure 10.1, this tissue type is present in the ducts of the skin's sweat glands. The ducts in figure HA-9C (Histology Atlas) show what stratified cuboidal epithelium looks like.

Transitional A unique characteristic of the transitional type of stratified epithelia is that the surface layer consists of large, round, dome-shaped cells that may be binucleate. The deeper strata are cuboidal, columnar, and polyhedral.

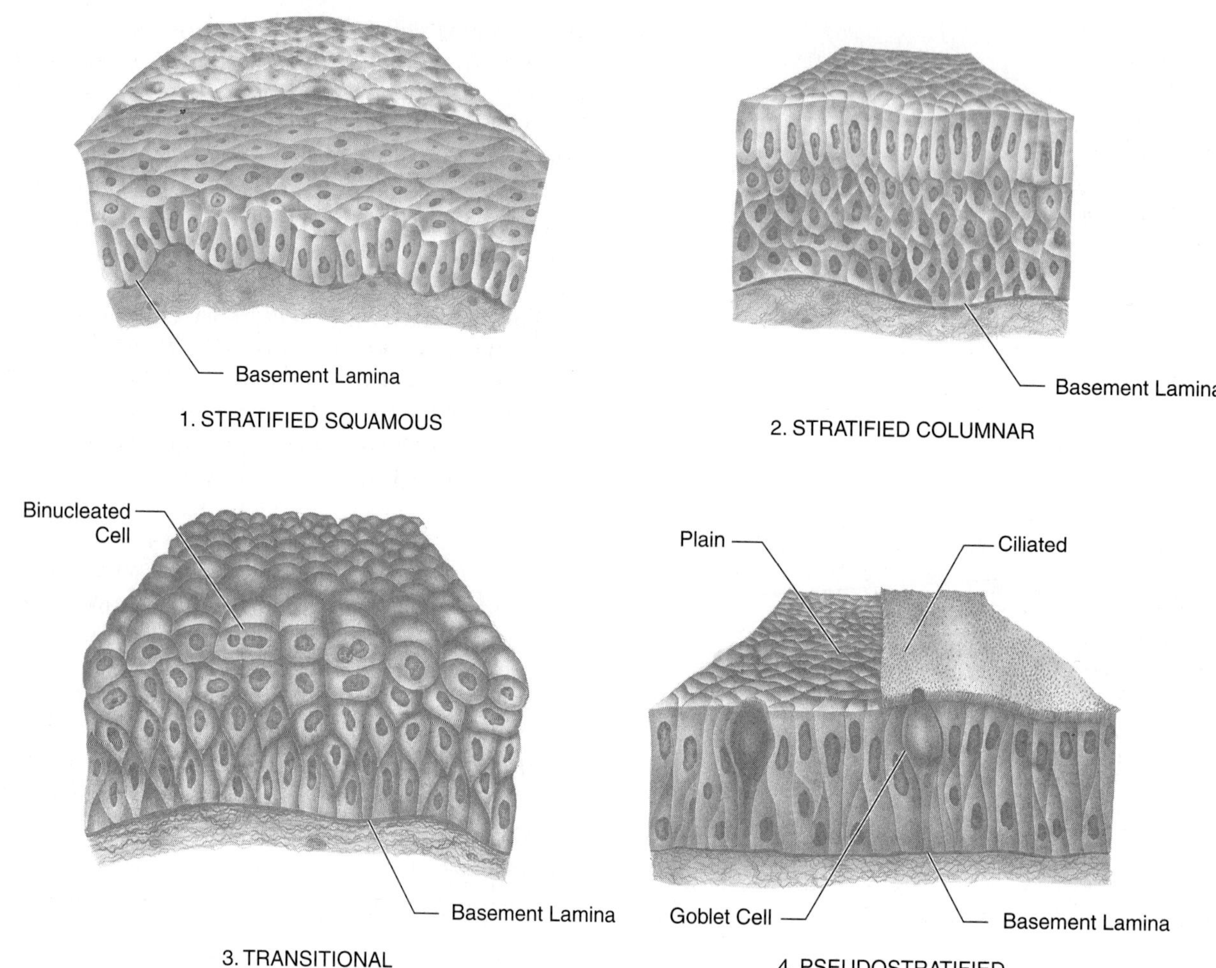

Figure 10.3 Stratified, transitional, and pseudostratified epithelia.

Note in part 3 of figure 10.3 that the deeper cells are not as closely packed as other stratified epithelia. The cells have distinct spaces between them, which imparts some elasticity to the tissue. The urinary bladder, ureters, and kidneys (the calyces) contain transitional epithelium that enables distension due to urine accumulation.

Pseudostratified Epithelia

Pseudostratified epithelia are not truly stratified but appear to be for two primary reasons: (1) Some nuclei are near the apical surface, while others are near the basal surface; and (2) some cells extend from the basement lamina to the free surface, while others do not. Close examination of pseudostratified epithelia reveals that while every cell is in contact with the basal lamina, only columnar cells extend to the free surface. The smaller cells wedged between them have no free surfaces.

Part 4 of figure 10.3 shows two types of pseudostratified epithelia: plain and ciliated. The *plain,* or nonciliated, type is present in the male urethra and parotid gland. The *ciliated* type lines the trachea, bronchi, auditory tubes, and part of the middle ear. Since both types possess goblet cells, they frequently produce mucus.

Laboratory Assignment

Examine prepared slides of each kind of epithelial tissue. Remember that epithelial cells have a free surface and a basement lamina, so always look for the tissue near the edge of a structure.

Note that preferred and optional slides are listed under "Materials." The *preferred* slides are limited to the type of tissue being studied. If the preferred slides are not available, use the *optional* slides. These are good but are more difficult to use because several kinds of tissue appear on each slide.

Materials:
preferred slides:
squamous
stratified squamous
cuboidal
pseudostratified ciliated
transitional

optional slides:
skin
trachea
stomach
ileum
kidney
thyroid gland

Squamous Epithelium Look for the flattened cells that are representative of squamous epithelium. Use figure HA-1 in the Histology Atlas for reference. The exfoliated cells are the same ones you studied in Exercise 5.

Explore each slide first with the low-power objective before using the high-dry or oil immersion objectives. Make drawings if required.

Columnar Epithelium Consult figure HA-2 in the Histology Atlas for representatives of columnar epithelial cells. To see the **brush border** or **cilia** on these cells, study the cells with the high-dry or oil immersion objectives. Can you recall how cilia and brush borders differ? Can you identify the **basement lamina** on each tissue?

The **lamina propria** consists of connective tissue, vascular and lymphatic channels, lymphocytes, and mast cells. All of these structures are studied later.

Cuboidal Epithelium Figures HA-3A and B in the Histology Atlas show cuboidal tissue. The preferred slide is usually made from the uterine lining of a pregnant guinea pig. If the preferred slide is unavailable, examine a slide of the thyroid gland for cuboidal tissue.

Transitional Tissue Figures HA-3C, D and HA-30C in the Histology Atlas provide good examples of transitional tissue. Note the loose arrangement of the cells. Look for **binucleate cells.**

Ciliated Pseudostratified Columnar Epithelium The best place to look for ciliated pseudostratified columnar epithelium is in a cross section of the trachea. Figure HA-2D in the Histology Atlas is the epithelium of the trachea.

Assignment:
Complete the Laboratory Report for this exercise.

11 Connective Tissues

Connective tissues bind, support, transport, and nourish organs and organ systems. All connective tissues have nonliving extracellular material that holds and surrounds various specialized cells.

The extracellular material, or **matrix,** is composed of fibers, fluid, organic ground substance, and/or inorganic components intimately associated with the cells. The matrix is generally a product of the cells in the tissue. The relative proportion of cells to matrix varies from one tissue to another.

Connective tissues are classified in several ways. Figure 11.1 shows the classification system used here. This system is based on the nature of the extracellular material; in some cases, categories overlap.

Connective Tissue Proper

Histologically, *connective tissue proper* is composed primarily of protein fibers, a ground substance that varies among the several types, and special cells. The fibers may differ in protein density and in whether they consist of either *collagen* or *elastin* protein.

The three basic fiber types are collagenous, reticular, and elastic. **Collagenous** (white) **fibers** are relatively long, thick bundles that appear in most ordinary connective tissues in varying amounts. **Reticular fibers** constitute minute networks of very fine threads. Although both collagenous and reticular fibers are composed of collagen, reticular fibers stain more readily with silver dyes (i.e., they are *argyrophilic*).

Elastic (yellow) **fibers** are common in connective tissue of organs that change shape. These are the only fibers to contain elastin protein. All three types of fibers are seen in loose (areolar) tissue, as figure 11.2 shows.

The ground substance of connective tissue proper usually consists of complex peptidoglycans that form an amorphous solution or gel around the cells and fibers. Chondroitin sulfate and hyaluronic acid are frequent components.

The matrix of connective tissue proper contains various types of cells, including fibroblasts, adipose cells, mast cells, plasma cells, macrophages, and other types of blood cells.

Loose Fibrous (Areolar) Connective Tissue

Loose fibrous connective tissue is interwoven into the framework of many organs throughout the body. Its cells and extracellular substances are loosely organized. Because of its preponderance of spaces, this tissue is often designated as *areolar* (a small space). It is highly flexible and can distend to accommodate excess extracellular fluid. Figure 11.2 shows its generalized structure. Note the presence of **mast cells, macrophages,** and **fibroblasts.**

Loose connective tissue has three functions: (1) it provides flexible support and a continuous network within organs, (2) its capillary network furnishes nutrition to cells in adjacent areas, and (3) it provides an arena for the activities of the immune system.

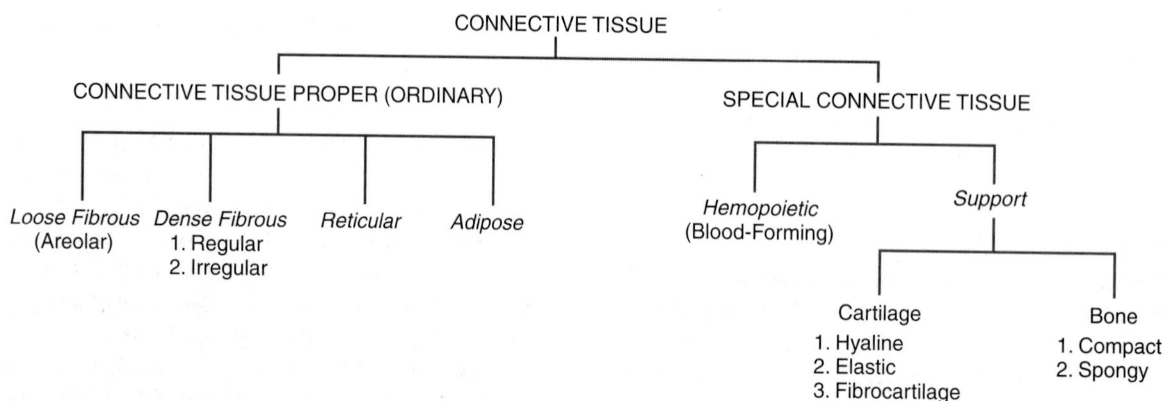

Figure 11.1 Types of connective tissue.

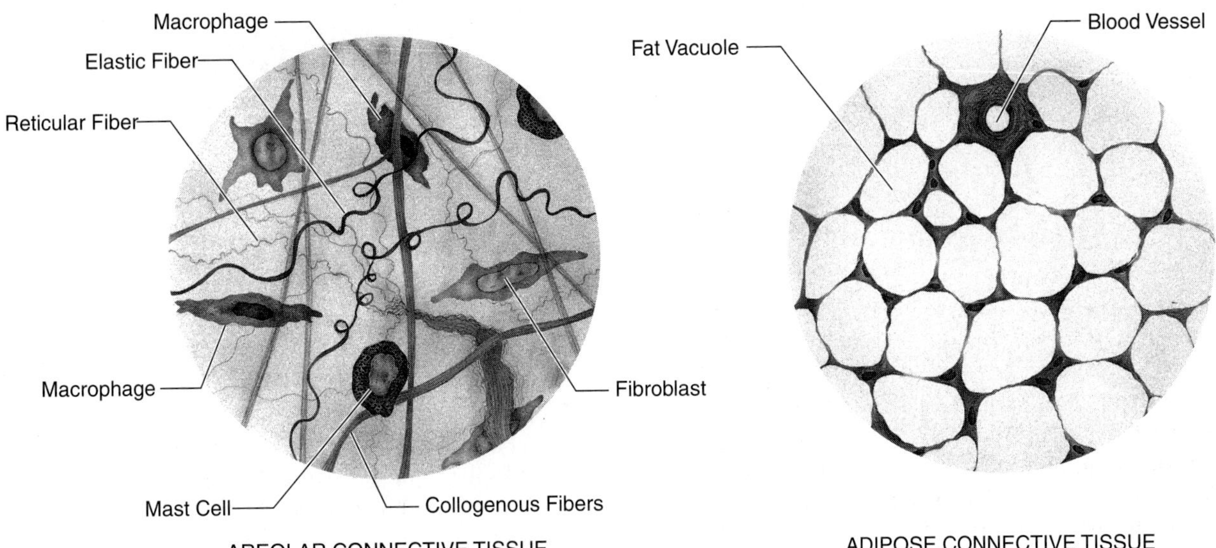

Figure 11.2 **Loose fibrous and adipose tissues.**

Loose connective tissue is found beneath epithelia, around and within muscles and nerves, and as part of the serous membranes. The deep and superficial *fasciae* encountered in the rat dissection in Exercise 3 are of this type.

Adipose Tissue

Fat or adipose cells collect in small groupings throughout the body. Because they are a storage depot for fat, however, they frequently accumulate in large areas to constitute the bulk of body fat. There are various types of adipose tissue, but ordinary adipose is the most common in humans (see figure 11.2).

Fat cells are very large and have a spherical or polygonal shape. As much as 95% of their mass may be stored fat. As fat is deposited, the cell cytoplasm becomes reduced and thin, and the vacuole, filled with lipids, appears as a large, open space with a thin periphery of cytoplasm. The nucleus is displaced to one side, often producing the appearance of a signet ring. A fine network of reticular fibers exists between the cells.

Fat tissue cushions and insulates the body and is also a potential energy source and heat generator. Its rich vascular supply points to its relatively high rate of metabolism and turnover.

Reticular Tissue

Reticular tissue contains a dense network of reticular fibers. The fibers' unique pattern and staining reaction identify them as the supporting framework of many vascular organs, such as the liver, lymphatic structures, hemopoietic tissue, and basement lamina. Figure 11.3 shows the reticular tissue of a lymph node.

Dense Fibrous (White) Connective Tissue

Dense fibrous connective tissue differs from loose fibrous connective tissue by its high content of fibrous elements and low content of cells, ground substance, and capillaries. The fiber arrangement separates this tissue into two groups: *dense regular* and *dense irregular* tissues. Figure 11.3 shows both types.

Dense regular connective tissue is found in tendons, ligaments, fascia, and aponeuroses. It consists of thick groupings of longitudinally organized collagenous fibers and some elastic fibers. These tough strands have enormous tensile strength and can withstand strong pulling forces without stretching. Fibroblasts are the principal cells, although they are scarce.

Dense irregular connective tissue consists of elastic fibers woven into flat sheets to form capsules around certain organs. The sheaths that encircle nerves and tendons are of this type. Much of the dermis of the skin also consists of the dense irregular type.

Special Connective Tissue

As indicated in figure 11.1, special connective tissue consists of two divisions: hemopoietic tissue and support tissue. *Hemopoietic tissue* consists of red bone marrow connective tissue that produces the formed elements of blood (red blood cells, white blood cells, and platelets). Exercises 52 and 53 examine the products of this tissue.

The *support tissue* includes cartilage and bone, which are adapted to bear weight. These tissues share the following structural characteristics: (1) a

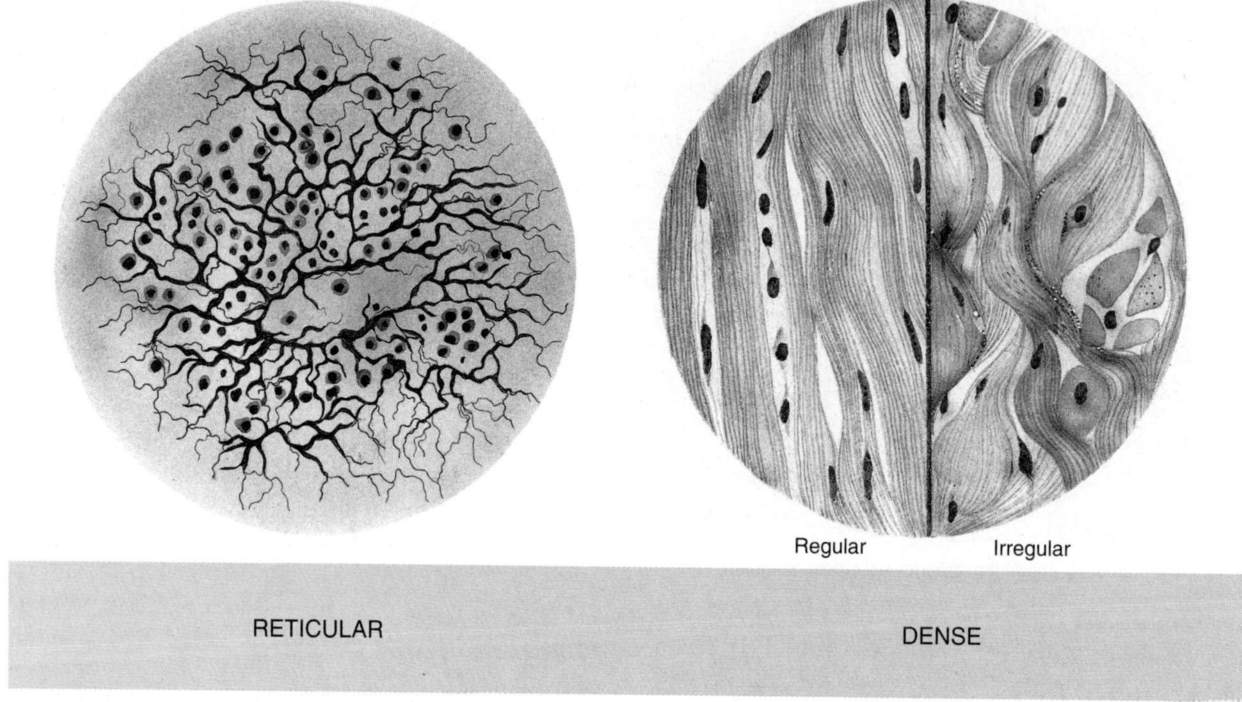

Figure 11.3 Reticular and dense connective tissue.

solid, flexible, yet strong *extracellular matrix;* (2) cells in matrix cavities called *lacunae;* and (3) an external covering (*periosteum* or *perichondrium*) capable of generating new tissue.

Cartilage

In general, cartilage consists of a stiff yet pliable matrix that lubricates and also bears weight. For this reason, cartilage is found in areas that require support and movement (skeleton and joints).

The three types of human cartilage consist of cartilage cells **(chondrocytes)** embedded within a matrix that contains ground substance and fibers. A much larger proportion of cartilage tissue is devoted to matrix than to chondrocytes. Unlike other connective tissues, cartilage is devoid of a vascular supply and receives all nutrients through diffusion. Figure 11.4 shows the three classes of cartilage.

Hyaline cartilage (part 1 in figure 11.4) looks like white plastic and makes up a very large part of the fetal skeleton. During fetal development, bone gradually replaces the cartilage, except for certain areas in the joints, ears, larynx, trachea, and ribs. The matrix is a firm, homogenous gel of chondroitin sulfate and collagen. Cells occur singly or in "nests" of several cells, the sides of which may be flattened. Often, the areas directly around lacunae appear denser in sections.

Elastic cartilage (part 2 in figure 11.4) is similar in basic structure to hyaline cartilage but differs in that the matrix contains many elastic fibers. The resulting tissue is very flexible and elastic. Elastic cartilage reinforces the pinna of the external ear, the epiglottis, and the auditory tube.

Fibrocartilage (part 3 in figure 11.4) differs from the other two types in that its chondrocytes are grouped between bundles of collagenous fibers. Fibrocartilage cushions strategic joint ligaments and tendons. It is the major component, for instance, of the vertebral column's intervertebral disks and of the pubic symphysis.

Bone Tissue

The tissue that makes up the bones of the skeleton meets all the criteria of connective tissue, yet it has many unique features. It is hard, rigid, strong, and relatively lightweight. It is found in those parts of the anatomy that require maximum weight-bearing capacity, protection, and storage of certain materials.

In a sense, bone is a living organic cement. It consists of specialized cells, blood vessels, and nerves that are reinforced within a hard ground substance made up of organic secretions impregnated with mineral salts. It arises during embryonic and fetal development from cartilage and/or fibrous connective tissue precursors. The initial organic matrix consists of collagenous fibers that provide a

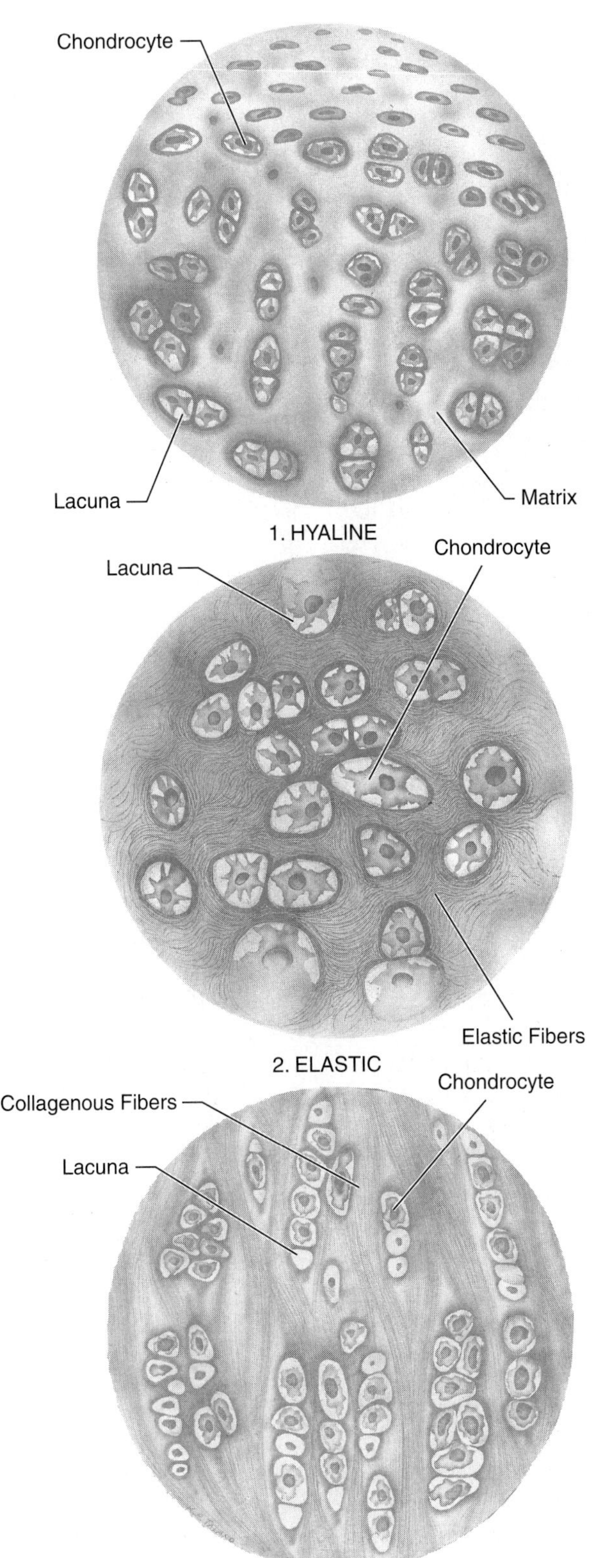

Figure 11.4 Types of cartilage.

framework on which special bone cells, the *osteoblasts,* deposit calcium and phosphate salts. Bone is far from being an inactive tissue; both metabolic and external influences continuously modify and reconstruct it.

Macroscopic sections of bones show two frameworks: the solid, dense **compact bone** that makes up the outermost layer and the more porous **spongy** (*cancellous*) **bone** located internally. Each has a distinctive histological character.

Compact Bone Figure 11.5 A shows that a hard matrix surrounds a microscopic framework of tunnels, channels, and interconnecting networks that permeate compact bone. Within this hollow network are the living cells that nourish and maintain bone.

The functional and structural unit of compact bone is a cylindrical component called the **osteon,** or Haversian system. In the center of each osteon is a hollow space, the **central** (Haversian) **canal,** which contains one or two blood vessels. Surrounding each central canal are several concentric rings of matrix called the **lamellae. Osteocytes,** which secrete the lamellae, are within small, hollow cavities, the **lacunae.** Note that these cavities, which are oriented between the lamellae, have many minute, hollow tunnels, called **canaliculi,** that radiate outward, imparting a spiderlike appearance. The canaliculi contain processes (cellular extensions) of the osteocytes.

The various levels of structure in figure 11.5 show a continuous communication system—from the central canal to the lacunae and between adjacent lacunae via the canaliculi. The entrapped osteocytes can receive nourishment and exchange materials within the hard space of the matrix because of intimate contact between the processes of adjacent osteocytes through the canaliculi.

Groups of osteons lie in vertical array, with adjacent lamellae separated at lines of demarcation called **cement lines.** The central canals of adjacent osteons are continuous because of **perforating** (Volkmann's) **canals** that penetrate the bone obliquely or at right angles.

Spongy Bone Spongy bone lies adjacent to compact bone and is continuous with it; the line of demarcation between the two regions is not distinct. Histologically, spongy bone has a lesser degree of organization than compact bone. Figure 11.5 A shows that spongy bone's outstanding feature is a series of branching, overlapping plates of matrix called **trabeculae.** These plates are oriented to produce large, interconnecting, cavelike spaces. These spaces function well in storage and as pockets to hold the

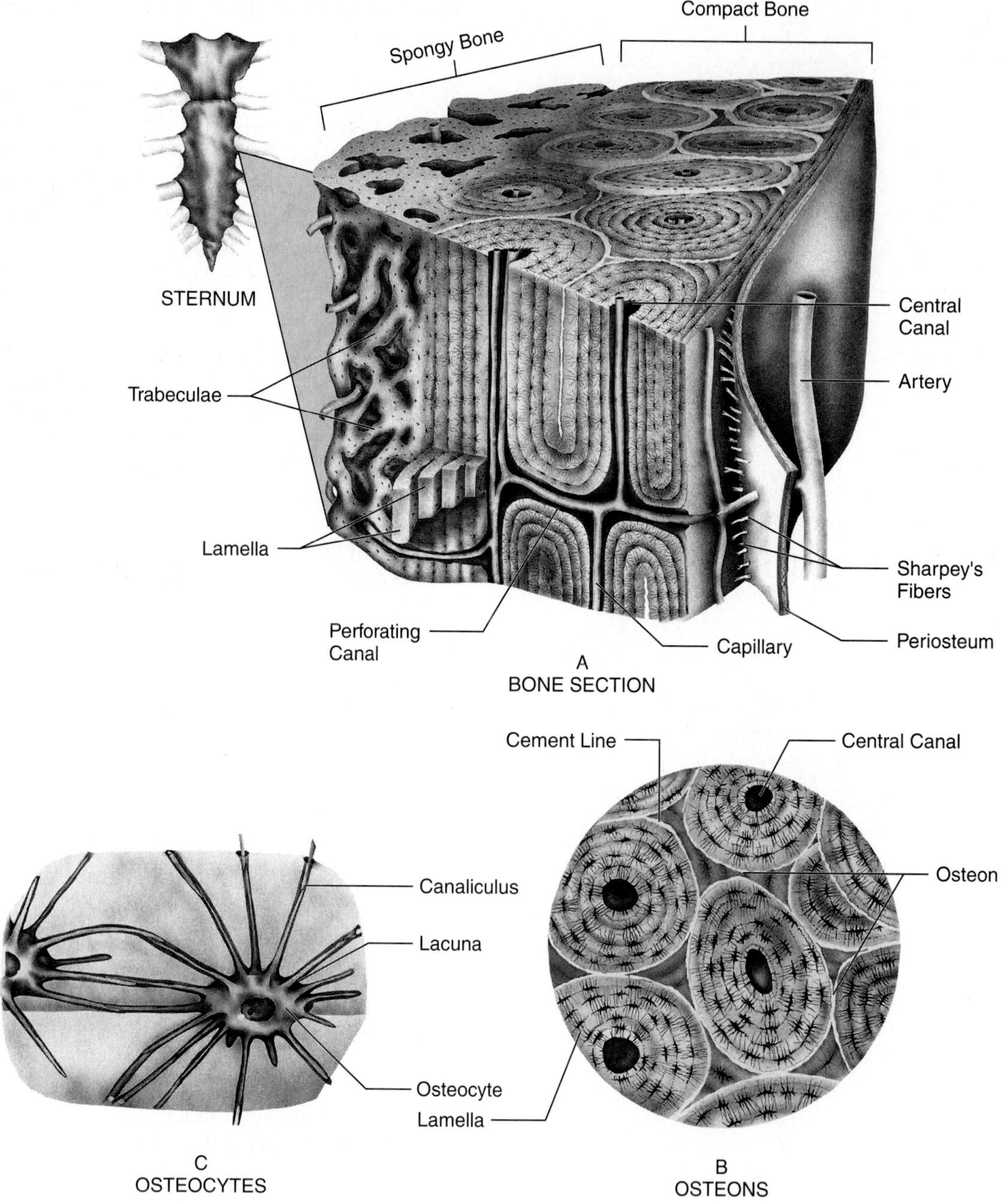

Figure 11.5 Bone tissue.

blood-forming cells (hemopoietic tissue) of the bone marrow. They also function in weight reduction.

The osteocyte-holding spaces, or lacunae, randomly punctuate the trabeculae. Blood vessels meander through the large spaces between the trabeculae, nourishing nearby osteocytes.

The Periosteum Contiguous with the outer layer of compact bone and tightly adherent to it is a thick, tough membrane called the *periosteum.* It is composed of an outer layer of fibrous connective tissue and an inner *osteogenic layer* that is a source of new bone-forming cells and provides access for blood vessels. Bundles of collagenous fibers that perforate and become firmly embedded within the outer lamellae tightly anchor the periosteum to compact bone. These minute attachments are called **Sharpey's fibers.**

Histological Study

Study prepared slides of each type of connective tissue. Note the Histology Atlas references indicated for each type of tissue.

Materials:
prepared slides of:
loose (areolar) connective tissue
adipose tissue
white (dense) fibrous connective tissue
yellow (elastic) fibrous connective tissue
reticular tissue
hyaline cartilage
fibrocartilage
elastic cartilage
bone, x.s.
developing bone

Connective Tissue Proper Examine prepared slides of areolar, adipose, fibrous, and reticular connective tissues. On your prepared slides, identify all the structures shown in figure HA-4 (Histology Atlas). Note the comments in the legend for figure HA-4 about the function of mast cells seen in loose connective tissue.

Cartilage Examine prepared slides of the three different types of cartilage (hyaline cartilage, fibrocartilage, and elastic cartilage), differentiating their distinct characteristics. On your prepared slides, identify the structures labeled in figure HA-5 (Histology Atlas).

Examine prepared slides of bone and developing bone.

Bone When studying the prepared slide of developing membrane bone, try to differentiate the osteoblasts from the osteoclasts (see figure HA-6, Histology Atlas). Note that an osteoclast is a large, multinucleate cell with a clear area between it and the bony matrix.

Assignment:
Complete the Laboratory Report for this exercise.

12 The Integument

Skin is constructed of epithelial and connective tissues. Thus, a study of the skin provides a review of Exercises 10 and 11.

During this exercise, you examine prepared slides of skin. The skin consists of two layers: an outer multilayered **epidermis** and a deeper **dermis.**

The Epidermis

The skin's outermost layer, or epidermis, is about 0.1 mm thick across most of the body and up to 1.5 mm thick on the palms of the hands and soles of the feet. The enlargement of the epidermis on the right side of figure 12.1 shows that the epidermis consists of four distinct layers: (1) an outer **stratum corneum;** (2) a thin, translucent **stratum lucidum;** (3) a darkly staining **stratum granulosum;** and (4) a multilayered **stratum spinosum** (stratum mucosum).

All four layers of the epidermis originate from the deepest cell layer of the stratum spinosum. This deep layer, which lies adjacent to the dermis, is the **stratum basale** (stratum germinativum). The columnar cells of this deep layer constantly divide to produce new cells that move outward to undergo metamorphosis at different levels. The brown skin pigment *melanin,* which is produced by stellate *melanocytes* of the stratum basale, is responsible for skin color. Skin color differences among individuals are due to the amount of melanin present rather than to the abundance of melanocytes.

The stratum corneum of the epidermis has many layers of dead, scalelike epithelial cells filled with a water-repellent protein called *keratin.* As the cells of the stratum spinosum push outward, they move away from the nourishment of the capillaries, die, and undergo *keratinization. Eleidin* granules in the cells of the stratum basale are believed to be an intermediate product of keratinization. The translucent stratum lucidum consists of closely packed cells with traces of flattened nuclei.

The Dermis

The dermis is often referred to as the "true skin." It varies in thickness from less than 1 mm to over 3 mm. It is highly vascular and provides nourishment for the epidermis.

The outer portion of the dermis, which lies next to the epidermis, is the **papillary layer.** It derives its name from numerous projections, or **dermal papillae,** that extend into the upper layers of the epidermis. In most regions of the body, these papillae form no pattern; however, on the fingertips, palms, and soles of the feet, they form regularly arranged patterns of parallel ridges that improve frictional characteristics in these areas. The "prints" that result from these papillae are unique to each individual.

The deeper portion, of the dermis, or **reticular layer,** contains more collagenous fibers than the papillary layer. These fibers greatly enhance the skin's strength. The surface texture of suede leather is, essentially, the reticular layer of animal hides.

Subcutaneous Tissue

Beneath the dermis lies the **subcutaneous layer,** or *hypodermis;* it is also called the *superficial fascia.* It consists of loose connective tissue, nerves, and blood vessels. One of its prime functions is attaching the skin to underlying structures.

Hair Structure

Hair consists of keratinized cells that are compactly cemented together. A tube of epithelial cells, the **hair follicle,** surrounds each hair shaft. The terminal end of the hair shaft, or **root,** forms an enlarged onion-shaped region called the **bulb.** Within the bulb is an involution of loose connective tissue called the **follicular** (*hair*) **papilla,** through which nourishment enters the shaft. An **internal root sheath** and an **external root sheath** encase the hair root.

Extending diagonally from the wall of the hair follicle to the epidermis is a band of smooth muscle fibers, the **arrector pili muscle.** Contraction of these muscle fibers moves the hair to a more perpendicular position, causing elevations on the skin surface commonly referred to as "goose bumps."

Glands

The skin has two kinds of glands: sebaceous and sweat.

Sebaceous Glands Sebaceous glands are formed from the epithelial tissues that surround each hair

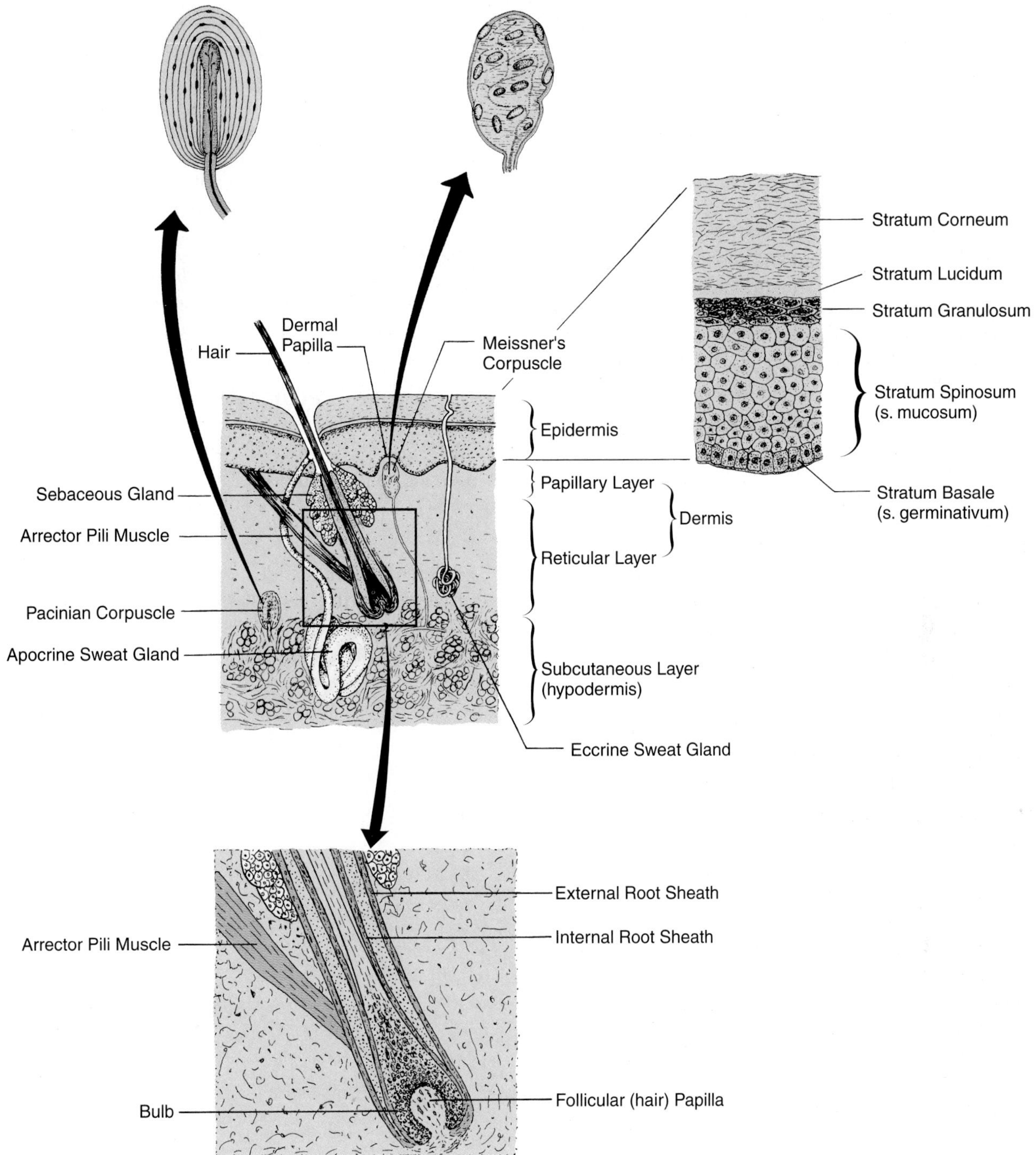

Figure 12.1 **Skin structure.**

follicle. They secrete an oily substance called *sebum* into the hair follicles and out onto the skin surface. Contraction of the arrector pili muscles facilitates this secretion to some extent. Sebum keeps hair pliable and helps to waterproof the skin.

Sweat Glands Sweat glands are of two types: eccrine and apocrine. Small **eccrine glands** empty directly through the surface of the skin. These glands are simple tubular structures whose coiled basal portions are deep in the dermis. Except for the lips, glans penis, and clitoris, eccrine glands are widely distributed throughout the body.

Although all eccrine gland secretions are similar in composition, there are two different controlling stimuli. Emotional factors trigger the sweat glands of some parts of the body, such as the palms and axillae. Thermal stimuli regulate glands in other regions, such as the forehead, neck, and back.

Apocrine sweat glands are much larger than the eccrine type and have their secretory coiled portions in the hypodermis. Instead of emptying onto the surface of the epidermis, all apocrine glands empty directly into a hair follicle canal.

Apocrine glands are found in the axillae, the male scrotum, the female perigenital region, the external ear canal, and nasal passages. While eccrine sweat is watery, the apocrine glands secrete a thick white, gray, or yellowish substance. Malodorous substances in apocrine sweat are the principal contributors to body odors. Emotional factors (e.g., fear, stress) and sexual arousal, rather than temperature changes, contribute mostly to apocrine secretions.

Receptors

The receptors shown in figure 12.1 are Meissner's corpuscles, and Pacinian corpuscles. **Meissner's corpuscles** are in the papillary layer of the dermis and project up into papillae of the epidermis. They are receptors of touch.

Pacinian corpuscles are spherical receptors with onionlike laminations. They lie deep in the reticular layer of the dermis. Pacinian corpuscles are sensitive to variations in sustained pressure.

Free nerve endings are also present in the skin. These project up into the epidermis and fall into three categories. Some free nerve endings sense pain (tissue damage), and some are *heat receptors* or *cold receptors.*

Assignment:
Answer the questions on the Laboratory Report for this exercise, and then proceed with the Laboratory Assignment.

Laboratory Assignment

Materials:
prepared slide of the skin

Examine prepared slides of sections through the skin, and identify the structures seen in figure 12.1 and figures HA-9, HA-10, and HA-11 in the Histology Atlas. Make drawings if required.

PART 3 The Skeletal System

Many phases of medical science demand a thorough understanding of the skeletal system. Since most of the muscles of the body are anchored to specific loci on bones, you need to learn the names of many processes, ridges, and grooves on individual bones.

Your comprehension of the skeletal system also affects your understanding of the nervous and circulatory systems. This is particularly true when studying how nerves and blood vessels pass through openings in bones of the skull. Each passageway (foramen or meatus) has a name. A complete comprehension of the skeletal system includes knowing the names of all these openings.

X-ray technology is an important branch of medical practice that relies heavily on osteology. In dentistry, the structure of the mandible and surrounding facial bones is important in taking X rays of the teeth. In medicine, a thorough understanding of all parts of the skeleton is critical.

13 The Skeletal Plan

This exercise examines the structure of the skeleton as a whole and the anatomy of a typical long bone. Individual parts of the skeleton are examined in detail in subsequent exercises.

> ***Materials:***
> fresh beef bones, sawed longitudinally
> articulated human skeleton

The adult skeleton is made up of 206 named bones and many smaller unnamed ones. They are classified as long, short, flat, irregular, or sesamoid.

The *long* bones include the bones of the arms and legs, and the metacarpals, metatarsals, and phalanges. The *short* bones are in the wrists and ankles. In addition to being shorter, the short bones differ from the long ones in being filled with spongy (cancellous) bone instead of having a medullary cavity.

The *flat* bones are the protective bones of the skull. Bones that are neither long, short, nor flat are *irregular.* The vertebrae and the bones of the middle ear fall into this category.

Round bones embedded in tendons are called *sesamoid* bones (shape resembles sesame seeds). The kneecap is the most prominent bone of this type.

Bone Structure

Before studying the structures of a long bone, become familiar with the following terms that apply to depressions and cavities on all bones of the skeleton:

1. ***Foramen:*** An opening in a bone that provides a passageway for nerves and blood vessels.
2. ***Fossa:*** A shallow depression in a bone. In some instances, the fossa is a socket into which another bone fits.
3. ***Sulcus:*** A groove or furrow.
4. ***Meatus:*** A canal or tubelike passageway.
5. ***Fissure:*** A narrow slit.
6. ***Sinus*** *(antrum):* A cavity in a bone.

The following terms apply to different types of *processes* on bones:

1. ***Condyle:*** A rounded, knucklelike eminence on a bone that articulates with another bone.
2. ***Tuberosity:*** A large, roughened process on a bone.
3. ***Tubercle:*** A small, rounded process.
4. ***Trochanter:*** A very large process on a bone.
5. ***Head:*** A portion of a bone supported by a constricted part, or *neck.*
6. ***Crest:*** A narrow ridge of bone.
7. ***Spine:*** A sharp, slender process.

A long bone, such as the femur in figure 13.1, is a complex structure made up of many different tissues. In addition to supporting the body, it performs many other functions. The femur in figure 13.1 is

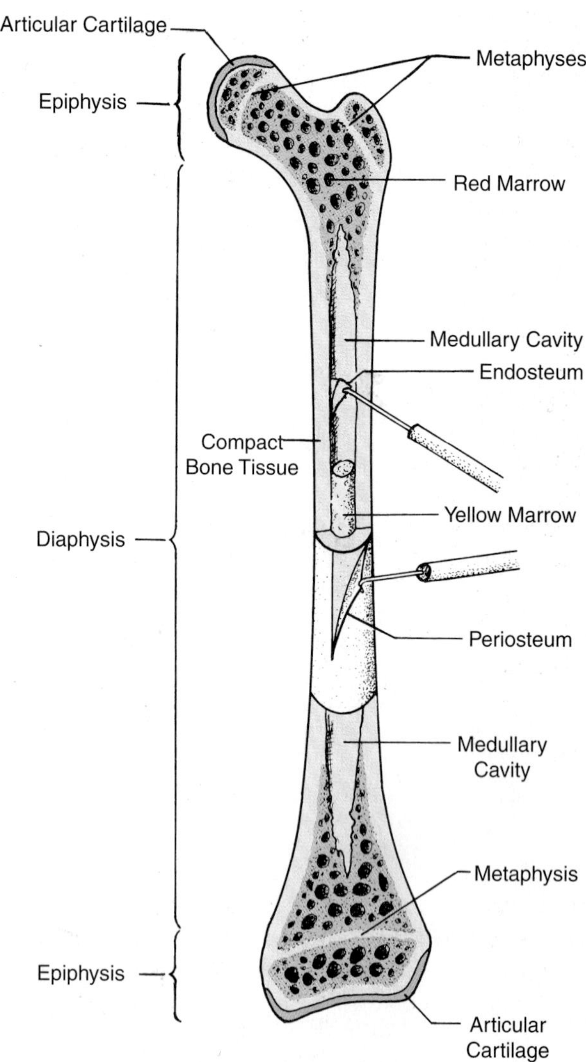

Figure 13.1 Long bone structure.

cut open longitudinally to reveal its internal structure. Note that it has a long shaft called the **diaphysis** and two enlarged ends, the **epiphyses.** Where the epiphyses meet the diaphysis there are growth zones called **metaphyses.**

During the growing years, a plate of hyaline cartilage, the *epiphyseal disk,* exists in each metaphysis. As new cartilage forms on the epiphyseal side, it is destroyed and replaced by bone on the diaphyseal side. During the growing years, the metaphysis consists of the epiphyseal disk and calcified cartilage; at maturity, however, the area becomes completely ossified, and linear growth ceases.

Note that the central portion of the diaphysis is a hollow chamber, the **medullary cavity.** The strength of the shaft's compact bone tissue eliminates the need for central bone tissue. A membrane called the **endosteum** lines the medullary cavity. This membrane is continuous with the linings of the central canals of the osteons.

The entire medullary cavity and much of the spongy bone extremities contain **yellow marrow,** a fatty substance in most long bones. The spongy bone of the epiphyses of the femur (and humerus) contains **red marrow** in the adult. Red marrow in adults is also found in the ribs, sternum, and vertebrae.

A tough covering, the **periosteum,** envelops the entire bone except for the areas of articulation. This covering consists of fibrous connective tissue that is quite vascular. Smooth **articular cartilage** of the hyaline type covers the surfaces of epiphyses that contact adjacent bones.

Beef Bone Study Examine a freshly cut section of a beef bone. Identify all of the structures that figure 13.1 shows. Probe into the periosteum near a torn ligament or tendon; note the continuity of fibers between the periosteum and these structures. Probe into the marrow, and note its texture.

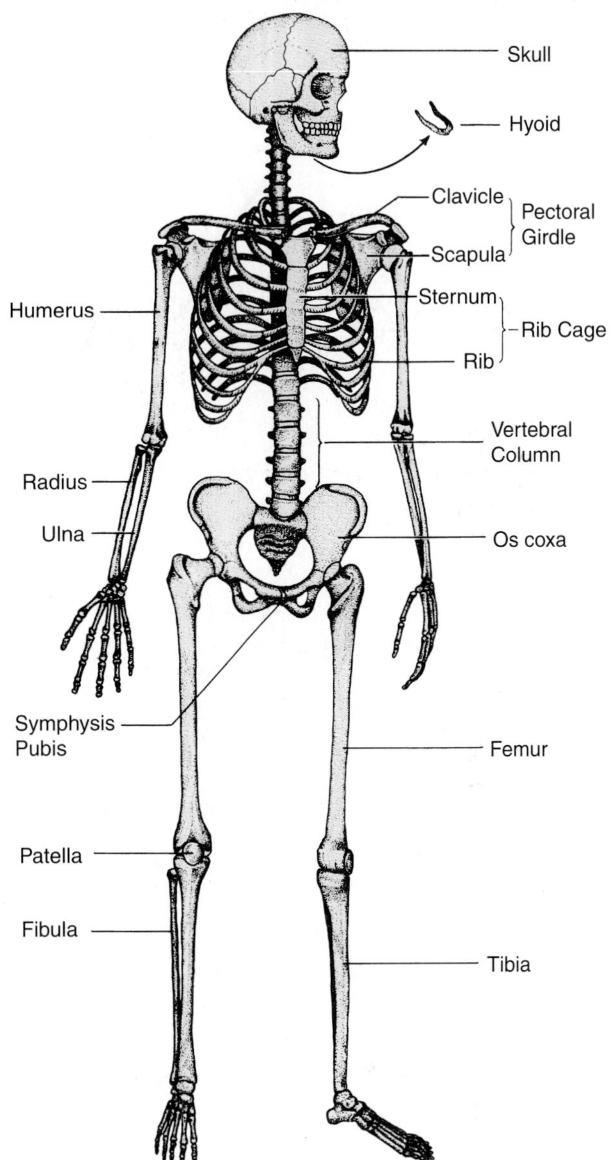

Figure 13.2 The human skeleton.

Bone Names

The bones of the skeleton fall into two main groups: those forming the axial skeleton and those forming the appendicular skeleton. Refer to figure 13.2 as you learn the names of some of the major bones.

The Axial Skeleton The axial skeleton includes the **skull, hyoid bone, vertebral column** (spine), and **rib cage.** The hyoid bone is a horseshoe-shaped bone in the neck under the lower jaw. The rib cage consists of 12 pairs of **ribs** and a **sternum** (breastbone).

The Appendicular Skeleton The appendicular skeleton includes the upper and lower extremities.

Each upper extremity consists of a pectoral girdle, arm, and hand. The **pectoral girdle** consists of a **scapula** (shoulder blade) and **clavicle** (collarbone). Each arm consists of an upper portion (the **humerus**) and two forearm bones (the **radius** and **ulna).** The radius is lateral to the ulna. The hand includes the **carpals** of the wrist, the **metacarpals** of the palm, and the **phalanges** of the fingers.

The lower extremities consist of the pelvic girdle, legs, and feet. The **pelvic girdle** is formed by two bones, the **ossa coxae,** which attach posteriorly to the sacrum of the vertebral column and anteriorly to each other. The anterior joint where the ossa coxae unite on the median line is the **symphysis pubis.**

Each leg consists of four bones: (1) the **femur** in the upper leg; (2) the **tibia** (shinbone); (3) the

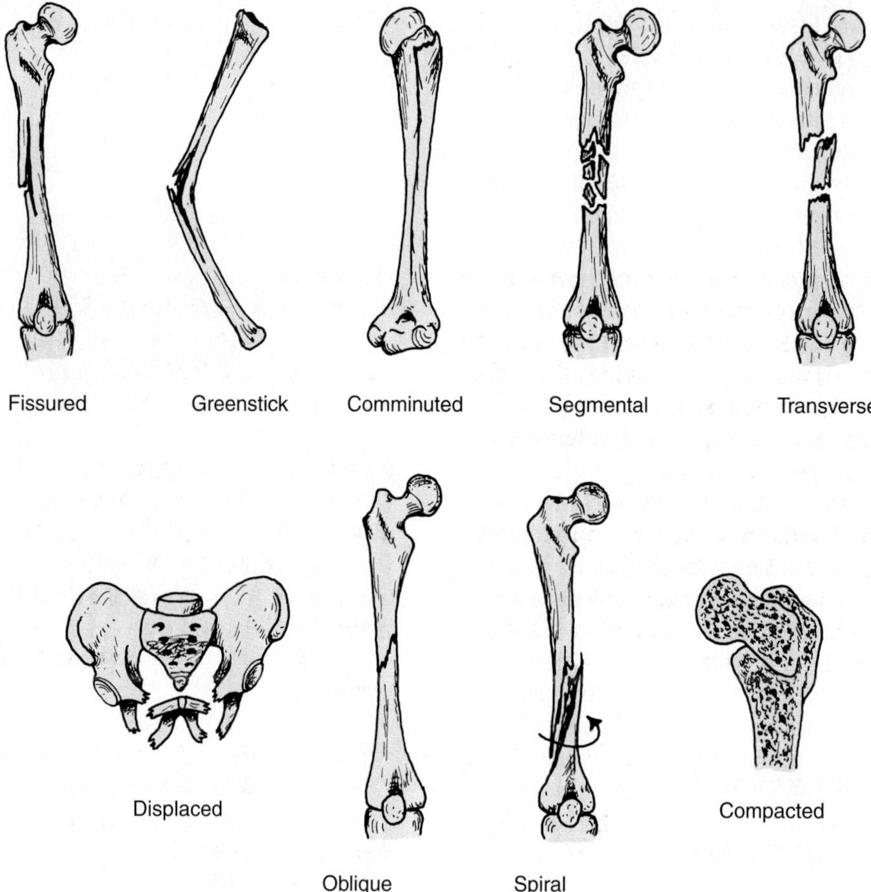

Figure 13.3 Types of bone fractures.

long, thin **fibula** parallel to the tibia; and (4) the **patella,** or kneecap. The feet include the **tarsals** near the ankle, the **metatarsals** that form the instep, and the **phalanges** of the toes.

Bone Fractures

When subjected to excessive stress, bones can fracture in various ways. The type of fracture depends on the nature and direction of applied forces. Figure 13.3 shows some of the more common fracture types.

A **closed,** or **simple,** fracture is contained in the soft tissues and does not in any way affect the skin or mucous membranes. Most of the fractures in figure 13.3 probably fall in this category. An **open,** or **compound,** fracture, on the other hand, does affect external surfaces. Fractures of this type can be considerably more difficult to treat because of potential bone marrow infections (osteomyelitis).

Fractures may be complete or incomplete. With **incomplete** fractures, the bone is split, splintered, or only partially broken. A **greenstick** fracture occurs when a bone breaks through on only one side as a result of bending. These fractures are most common among the young. Linear splitting of a long bone is a **fissured** fracture.

Complete fractures are those in which the bone is broken clear through. A **transverse** fracture is a break at right angles to the long axis. Breaks that are at an angle to the long axis are **oblique** fractures. Breaks that result from torsional forces are **spiral** fractures.

A **segmental** fracture occurs when a piece of bone breaks out of the shaft. More extensive fractures involving two or more fragments are **comminuted** fractures. When bone fragments move out of alignment, the fracture is **displaced.**

Severe vertical forces can result in compacted or compression bone fractures. A broken portion of bone driven into another portion of the same bone results in a **compacted** fracture. This is often seen in femur fractures. **Compression** fractures (not shown in figure 13.3) often occur in the vertebral column when falls from excessive heights crush vertebrae.

Assignment:
Complete the Laboratory Report for this exercise.

The Skull

14

The cranium, (which protects the brain), the facial bones, and the lower jawbone or mandible comprise the bones of the skull. As you read through the discussion of the various bones, identify them first in the illustrations and then on specimens in the laboratory. Compare the specimens with the illustrations to note the degree of variance.

Handle laboratory skulls carefully to avoid damaging them. **Never use pencils as pointers** because they leave marks on the bones. Use metal probes or pipe cleaners instead. If using a metal probe, **touch the bones very gently** to avoid damaging areas where bone is thin.

Materials:
whole and disarticulated skulls
fetal skulls
metal probes or pipe cleaners

The Cranium

The *cranium* encases the brain and consists of the following bones: a single frontal, two parietals, two temporals, one sphenoid, one ethmoid, and one occipital. Irregular, interlocking joints called *sutures* join these bones at their margins. Figures 14.1 and 14.2 show the lateral and inferior aspects of the cranium. Figure 14.6 shows a sagittal section.

Frontal The anterior superior portion of the skull consists of the *frontal* bone. It forms the eyebrow ridges and the ridge above the nose. The most inferior edge of this bone extends well into the orbit of the eye to form the **orbital plates** of the frontal bone. On the superior ridges of the eye orbits is a pair of foramina, the **supraorbital foramina** (see figure 14.3).

Parietals Directly posterior to the frontal bone on the sides of the skull are the *parietal* bones. The lateral view of the skull in figure 14.1 shows only the left parietal bone. The right parietal is on the other side of the skull.

The right and left parietals meet on the midline of the skull to form the **sagittal suture.** Between the frontal and each parietal bone is another suture, the **coronal suture.** Two semicircular bony ridges that extend from the forehead (frontal bone) and

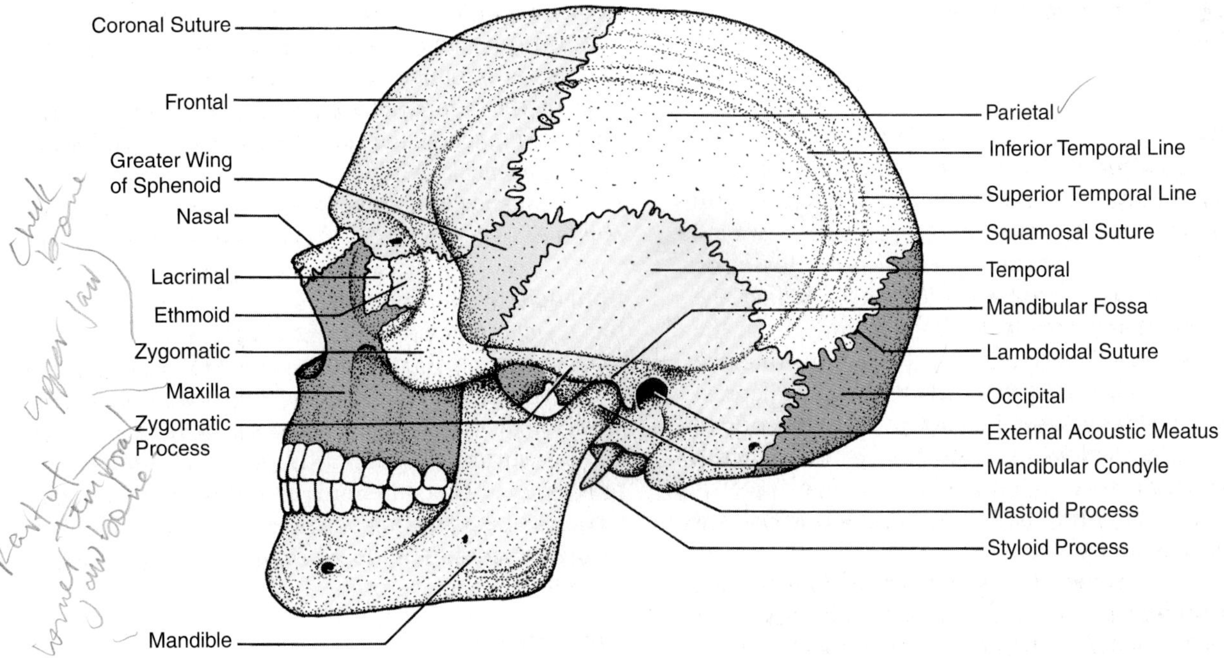

Figure 14.1 Lateral view of skull.

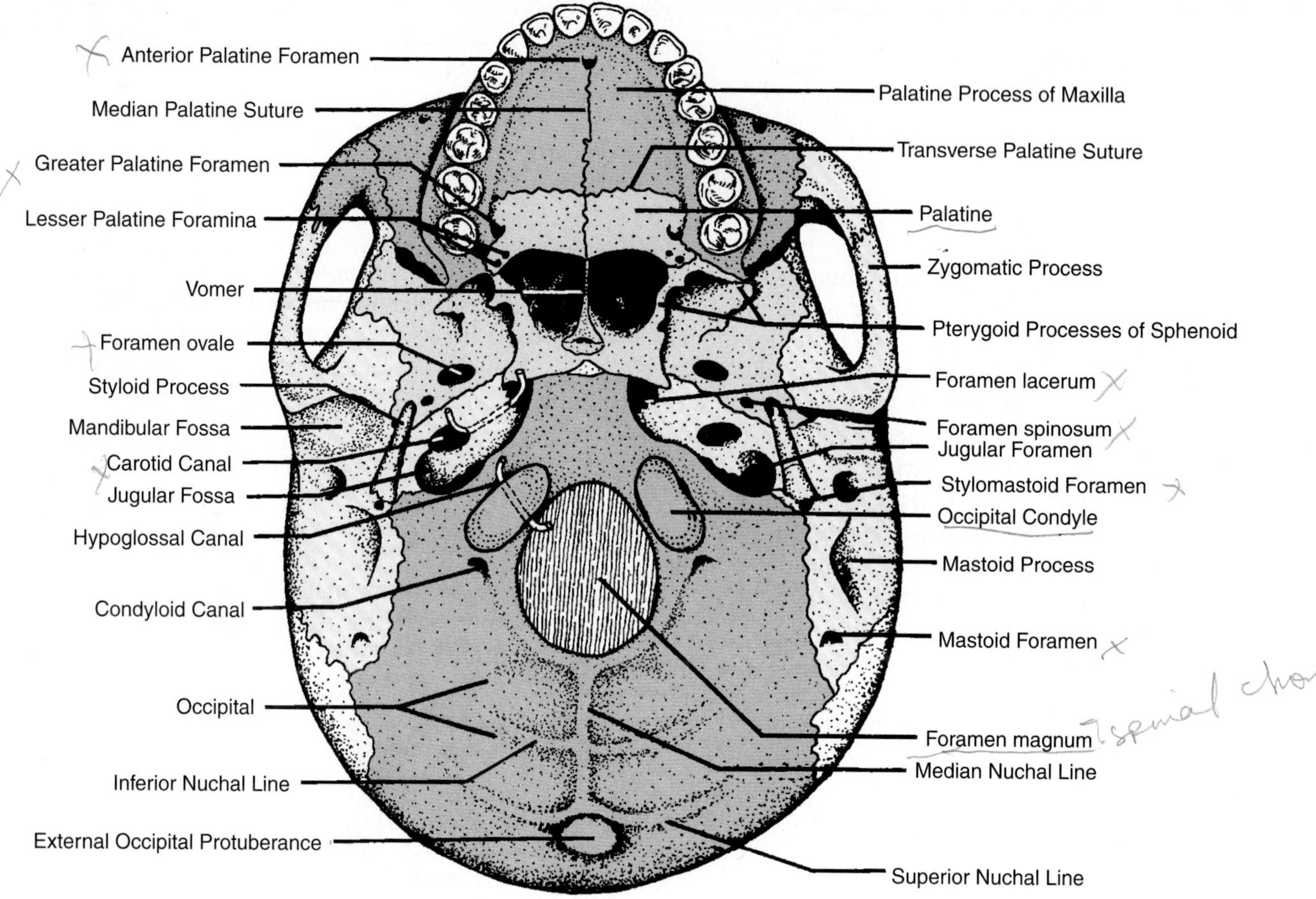

Figure 14.2 Inferior surface of skull.

over the parietal bone are the **superior temporal line** and **inferior temporal line.** These ridges form the points of attachment for the longest muscle fibers of the *temporalis* muscle. Figure 36.1, page 149, shows the position of this muscle. The upper extremity of the muscle falls on the superior temporal line.

Temporals Inferior to the parietal bones on each side of the skull are the *temporals.* The **squamosal suture** joins each temporal to its adjacent parietal. A depression, the **mandibular** (*glenoid*) **fossa,** on each temporal provides a recess into which the lower jaw articulates. Pull the jaw away from the skull to note the shape of this fossa. The rounded eminence of the mandible that fits into this depression is the **mandibular condyle.** Just posterior to the mandibular fossa is the ear canal, or **external acoustic meatus.**

The temporal bone has three significant processes: the zygomatic, styloid, and mastoid. The **zygomatic process** is a long, slender process that extends forward to articulate with the zygomatic bone. The zygomatic process and a portion of the zygomatic bone constitute the *zygomatic arch.* The **styloid process** is a slender, spinelike process that extends downward from the bottom of the temporal bone to form a point of attachment for some muscles of the tongue and pharynx. This process is often broken off on laboratory specimens. The **mastoid process** is a rounded eminence on the inferior surface of the temporal just posterior to the styloid process. It anchors the *sternocleidomastoideus* muscle of the neck. Middle ear infections that spread into the cancellous bone of this process are called *mastoiditis.*

Sphenoid The light pink bone in figure 14.1 is the *sphenoid* bone. Note in figure 14.2 that this bone extends from one side of the skull to the other.

The **greater wings of the sphenoid** are on the sides of the skull in the "temple" region. On the ventral surface (figure 14.2) are the **pterygoid processes of the sphenoid,** to which the pterygoid muscles of mastication attach. The **orbital surfaces of the sphenoid** (figure 14.3) make up the posterior walls of each eye orbit.

Ethmoid The *ethmoid* bone is on the medial surface of each orbit of the eye. The ethmoid forms

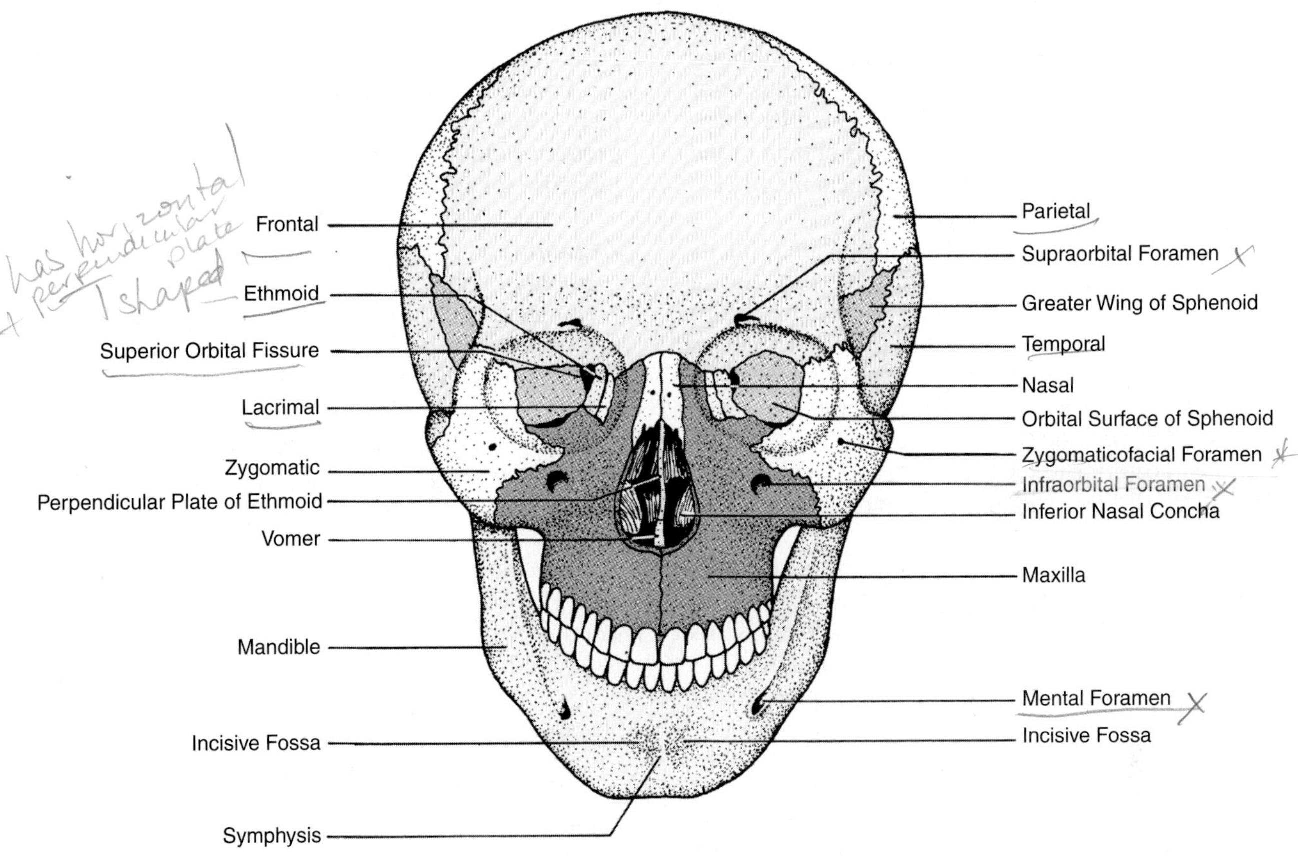

Figure 14.3 Anterior aspect of skull.

part of the roof of the nasal cavity and closes the anterior portion of the cranium.

Examine the upper portion of the nasal cavity of a laboratory skull specimen. Note that the inferior portion of the ethmoid has a downward-extending **perpendicular plate** on the median line (see figure 14.6). This portion articulates anteriorly with the nasal and frontal bones. Posteriorly, it articulates with the sphenoid and vomer. On each side of the perpendicular plate are irregular curved plates, the **superior** and **middle nasal conchae.** They provide bony reinforcement for the fleshy **upper nasal conchae** of the nasal cavity.

Occipital The posterior inferior portion of the skull consists primarily of the *occipital* bone. The **lambdoidal suture** (see figure 14.1) joins the occipital to the parietal and temporal bones.

Compare the inferior surface of your laboratory skull with figure 14.2. Note the large **foramen magnum** that surrounds the brain stem in life. On each side of this opening is a pair of **occipital condyles.** These two condyles rest on fossae of the *atlas,* the first cervical vertebra of the spinal column.

Near the base of each occipital condyle are two passageways: the condyloid and hypoglossal canals. The **condyloid canal** opens posterior to the occipital condyle. The **hypoglossal canal** in figure 14.2 has a piece of wire passing through it. The hypoglossal canal provides a passageway for the hypoglossal (twelfth) cranial nerve.

Three prominent ridges form a distinctive pattern posterior to the foramen magnum on the occipital bone. The **median nuchal line** is a ridge that extends posteriorly from the foramen magnum. It provides a point of attachment for the *ligamentum nuchae.* The **inferior** and **superior nuchal lines** lie parallel to each other. The inferior one lies between the foramen magnum and the superior nuchal line. These ridges form points of attachment for various neck muscles. Note that where the superior nuchal and medial nuchal lines meet is a distinct prominence, the **external occipital protuberance.**

Foramina on Inferior Surface Compare the inferior surface of your laboratory skull to figure 14.2 to identify the foramina discussed next. The significance of these openings will become more apparent when you study the circulatory and nervous systems.

Two foramina, the foramen ovale and foramen spinosum, are on each half of the sphenoid bone.

The **foramen ovale** is a large, elliptical foramen that provides a passageway for the mandibular branch of the trigeminal nerve. Slightly posterior and lateral to the foramen ovale is a smaller **foramen spinosum,** which allows a small branch of the mandibular nerve and middle meningeal blood vessels to pass through the skull.

Each temporal bone has five openings on its inferior surface. The **foramen lacerum** has a jagged margin. A piece of wire (bent paper clip) carefully inserted into the foramen lacerum as shown in figure 14.2 reveals a passageway, the **carotid canal.** The internal carotid artery supplies the brain with blood through this canal.

Just posterior to the carotid canal opening is an irregular, slitlike opening, the **jugular foramen,** which drains blood from the cranial cavity via the inferior petrosal sinus. A depression, the **jugular fossa,** is adjacent to the jugular foramen. The **stylomastoid foramen** is a small opening at the base of the styloid process. The **mastoid foramen** is the most posterior foramen on the temporal bone and is sometimes absent.

The Face

The face consists of 13 fused bones, plus a movable mandible. Of the 13 fused bones, only one bone—the vomer—is not paired. Figure 14.3 shows most of the facial bones.

Maxillae The upper jaw consists of two maxillary bones (maxillae) that a suture on the median line joins. Remove the mandible from your laboratory skull, and compare the hard palate with figure 14.2. Note that the anterior portion of the hard palate consists of two **palatine processes of the maxillae.** A **median palatine suture** joins the two bones on the median line.

The maxillae of an adult support 16 permanent teeth. Each tooth is contained in a socket, or *dental alveolus.* That portion of the maxillae that contains the teeth is called the *alveolar process.* The combined alveolar processes of the maxillae are the *alveolar arch,* or dental arch.

The maxillae have three significant foramina: two infraorbital and one anterior palatine. The **infraorbital foramina** are on the front of the face under each eye orbit. Nerves and blood vessels emerge from each of these foramina to supply the nose. The **anterior palatine** (incisive) **foramen** is in the anterior region of the hard palate, just posterior to the central incisors.

Palatines In addition to the horizontal palatine processes of the maxillae, the hard palate also consists of two **palatine** bones. The palatines form the posterior third of the palate.

Locate the palatines on your laboratory specimen. Note that each palatine bone has a large **greater palatine foramen** and two smaller **lesser palatine foramina.**

Zygomatics On each side of the face are two *zygomatic* bones. They form the prominence of each cheek and the inferior, lateral surface of each eye orbit. Each zygomatic has a small foramen, the **zygomaticofacial foramen.**

Lacrimals Between the ethmoid and upper portion of the maxillary bones is a pair of lacrimal (Latin: *lacrima,* tear) bones—one in each orbit.

Locate a groove on the surface of each *lacrimal* that is continuous with a groove on the maxilla. This groove provides a recess for the lacrimal duct, through which tears flow from the eye into the nasal cavity.

Nasals A pair of thin, rectangular *nasal* bones form the bridge of the nose.

Vomer The thin vomer bone is in the nasal cavity on the median line. Its posterior upper edge articulates with the back portion of the perpendicular plate of the ethmoid and the rostrum of the sphenoid. The lower border of the vomer joins the maxillae and palatines. The *septal cartilage* of the nose extends between the anterior margin of the vomer and the perpendicular plate of the ethmoid. Locate the vomer on figures 14.2, 14.3, and 14.6, as well as on your laboratory skull.

Inferior Nasal Conchae The inferior nasal conchae are curved bones attached to the walls of the nasal fossa. They are beneath the superior and middle nasal conchae, which are part of the ethmoid bone.

Floor of the Cranium

Now that all of the bones of the skull have been identified, examine the inside of the cranium to note the significant structural details. Remove the top of your laboratory skull, and compare the floor of the cranium with figure 14.4 to identify the structures that follow.

Cranial Fossae The floor of the cranium is divided into three large depressions, called *cranial fossae.* The orbital plates of the frontal form the **anterior cranial fossa.** The large depression formed within the occipital bone is the **posterior cranial fossa.** It is the deepest fossa. Between these two

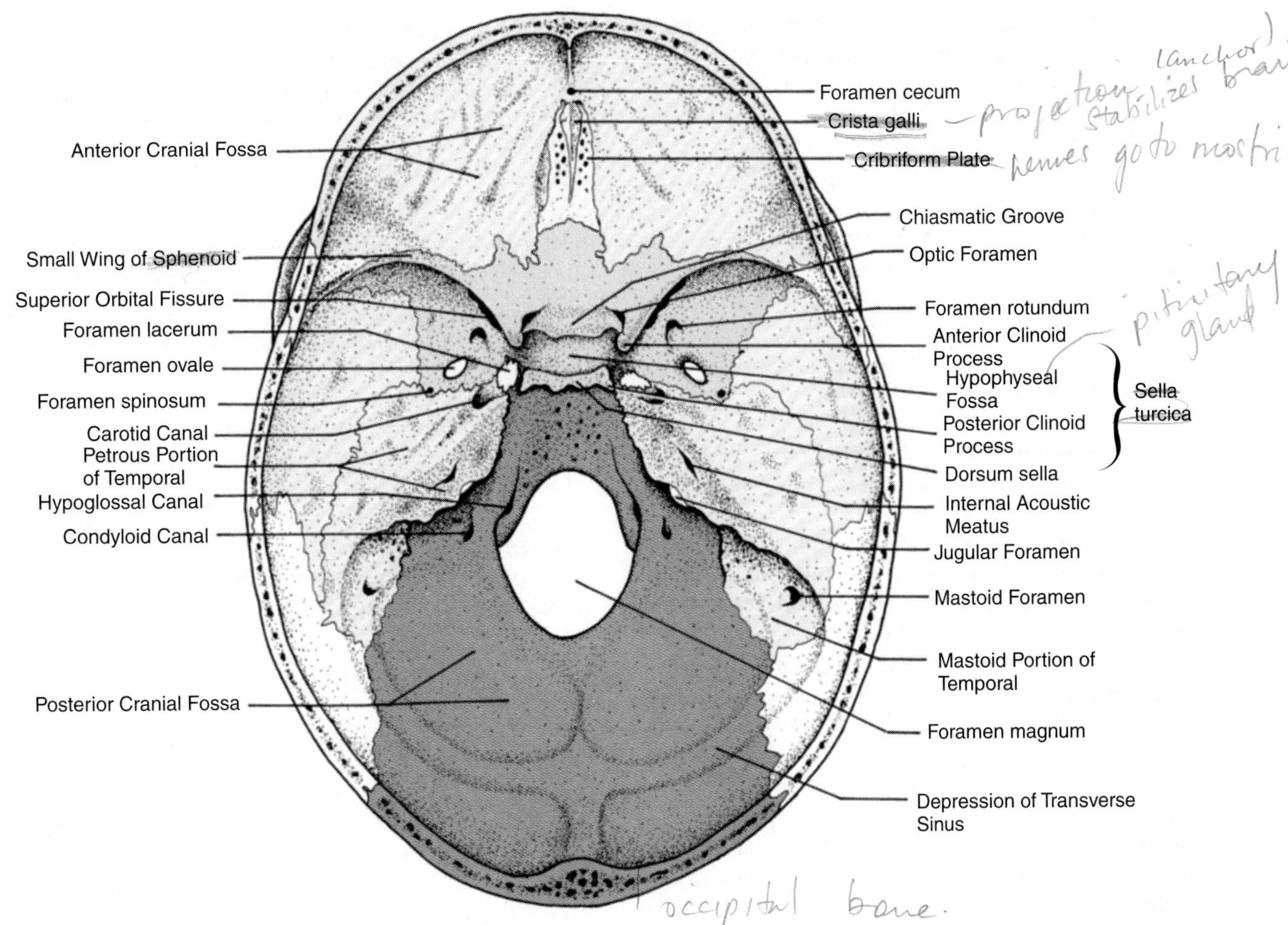

Figure 14.4 Floor of cranium.

fossae at an intermediate level is the **middle cranial fossa,** which involves the sphenoid and temporal bones.

Ethmoid The *ethmoid* bone in the anterior cranial fossa is between the orbital plates of the frontal bone. It consists of a perforated horizontal portion, the **cribriform plate,** and an upward projecting process, the **crista galli** (cock's comb). The holes in the cribriform plate allow branches of the olfactory nerve to pass from the brain into the nasal cavity.

Locate the small **foramen cecum,** which perforates the frontal bone just anterior to the ethmoid bone. This foramen provides a passageway for a small vein.

Sphenoid The sphenoid bone has a batlike configuration, with the greater wings extending out on each side. The anterior leading edges of the wings are called the **small wings of the sphenoid.**

On the median line of the sphenoid is a deep depression called the **hypophyseal fossa.** This depression contains the pituitary gland *(hypophysis)* in life. Posterior to this fossa is an elevated ridge called the **dorsum sella.** The two spinelike processes anterior and lateral to the hypophyseal fossa that project backward are the **anterior clinoid processes.** The outer spiny processes of the dorsum sella are the **posterior clinoid processes.** The hypophyseal fossa, dorsum sella, and clinoid processes, collectively, make up the **sella turcica,** or *Turkish saddle.*

Just anterior to the sella turcica is a pair of openings called the **optic foramina,** which lead into a pair of short **optic canals.** Locate the optic canals on your laboratory skull. The optic nerves pass through these canals from the eyes to the brain.

Extending from one optic foramen to the other is a narrow, bony shelf called the **chiasmatic groove.** Locate it on your specimen. Fibers of the right and left optic nerves cross over the chiasmatic groove in a formation called the optic chiasma.

Lying under each anterior clinoid process of the sphenoid is a **superior orbital fissure.** Figure 14.4 shows very little of it, but it is also seen in figure 14.3. Locate the superior orbital fissure on your

Table 14.1 Locations and Functions of Skull Foramina.

FORAMEN	LOCATION	PASSAGEWAY FOR
Anterior palatine foramen	Maxillae	Nasopalatine nerves and descending palatine vessels
Carotid canal	Temporal bone	Internal carotid artery
Foramen lacerum	Between sphenoid and temporal bone	Internal carotid artery and plexus of sympathetic nerves
Foramen magnum	Occipital bone	Brain stem
Foramen ovale	Sphenoid bone	Mandibular nerve and accessory meningeal artery
Foramen spinosum	Sphenoid bone	Branch of mandibular nerve
Greater palatine foramen	Palatine bone	Anterior palatine nerve and descending palatine vessels
Hypoglossal canal	Occipital bone	Twelfth cranial nerve (hypoglossal)
Jugular foramen	Between temporal and occipital bones	Inferior petrosal sinus, vagus nerve, glossopharyngeal nerve, accessory nerve, etc.
Mandibular foramen	Mandible	Inferior alveolar nerve (branch of fifth cranial nerve) and blood vessels
Mental foramen	Mandible	Mental nerve and blood vessels

laboratory specimen. It enables the third, fourth, fifth, and sixth cranial nerves to enter the eye orbit from the cranial cavity. Certain blood vessels also pass through it.

Just posterior to the superior orbital fissure is the **foramen rotundum,** which provides a passageway for the maxillary branch of the fifth cranial (trigeminal) nerve. Just posterior and slightly lateral to the foramen rotundum is the large **foramen ovale.** The small foramen posterior and lateral to the foramen ovale is the **foramen spinosum.** The **foramen lacerum** is between the margins of the sphenoid and temporal bones. Note the proximity of the **carotid canal** (on the temporal bones) to the foramen lacerum.

Temporals The thin **squamous portion of the temporal** forms a part of the side of the skull. The **petrous portion of the temporal** is probably the hardest portion of the skull. Its medial sloping surface has an opening to the **internal acoustic meatus.** This canal contains the facial and vestibulocochlear cranial nerves. The latter nerve passes from the inner ear to the brain. The **mastoid portion of the temporal** is the most posterior part. It has an enlargement, the **mastoid process,** and a **mastoid foramen,** which is posterior to the process.

Occipital The large occipital bone has a semicircular groove called the **depression of the transverse sinus.** Blood from the brain collects in a large vessel in this groove. Locate the **jugular foramen,** where the internal jugular vein originates. It appears as an irregular slit between the anterolateral margin of the occipital bone and the petrous portion of the temporal bone. The ninth, tenth, and eleventh cranial nerves also pass through this foramen.

Identify the openings to the **condyloid** and **hypoglossal canals.** Figure 14.2 shows that the hypoglossal canal passes through the base of the occipital condyle. Explore these foramina with a slender probe or pipe cleaner to differentiate them.

Foramina Summary

Table 14.1 summarizes the more important foramina of the skull. Use the table for quick referencing.

The Paranasal Sinuses

Some bones of the skull contain cavities, the *paranasal sinuses,* that reduce the weight of the skull without appreciably weakening it. All of the sinuses have passageways leading into the nasal cavity and are lined with a mucous membrane similar to the type that lines the nasal cavities.

The paranasal sinuses are named after the bones in which they are situated, as shown in figure 14.5. Above the eyes in the forehead are the **frontal sinuses.** The largest sinuses are the **maxillary sinuses,** which are in the maxillary bones. The **sphenoidal sinus** is the most posterior sinus in figure 14.5. Between the frontal and sphenoidal sinuses is a group of small spaces called the **ethmoid air cells.**

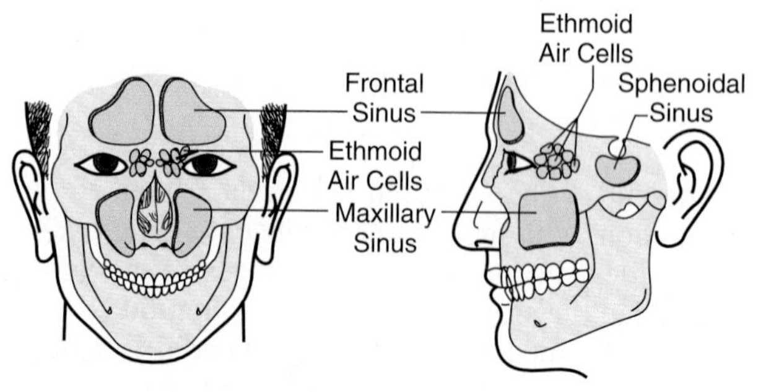

Figure 14.5 The paranasal sinuses.

Sagittal Section

A sagittal section of the skull, as shown in figure 14.6, is helpful for visualizing the sinuses and other structures in the center of the skull. Note how the frontal bone is hollowed out in the forehead region to form the **frontal sinus** and how the **sphenoidal sinus** is quite large. Immediately above the sphenoidal sinus is a distinct, saddlelike structure, the **sella turcica.**

A sagittal section is also best for showing the relationship of the **ethmoid** to the **vomer.** Extending upward from the superior margin of the vomer is the **perpendicular plate of the ethmoid.** These two bones, combined, form a bony septum in the nasal cavity. The uppermost projecting structure of the ethmoid is the **crista galli.** The air cells of the ethmoid are not shown in a sagittal section because they are not on the median line.

The vomer fuses with the two bones of the hard palate. The anterior portion of the hard palate is the **palatine process of the maxilla.** Posterior to it is the **palatine bone.**

The Mandible

The only bone of the skull that is not fused as an integral part of the skull is the lower jaw, or **mandible** (see figure 14.7). The mandible consists of a horizontal U-shaped portion, the **body,** and two vertical portions, the **rami** (*ramus,* singular). Embryologically, the mandible forms from two centers of ossification, one on each side of the face. As the bone develops toward the median line, the two halves meet and fuse to form a solid ridge. This point of fusion on the midline is called the **symphysis** (see figure 14.3). On each side of the symphysis are two depressions, the **incisive fossae.**

The superior portion of each ramus has a condyle, a coronoid process, and a notch. The **mandibular condyle** occupies the posterior superior terminus of the ramus. The **coronoid process** is on the superior anterior portion of the ramus. This process provides attachment for the *temporalis* muscle (see figure 36.1*A*). Between the mandibular condyle and the coronoid process is the **mandibular notch.** At the posterior inferior corners of the mandible, where the body and rami meet, are two protuberances, the **angles.** The angles provide attachment for the *masseter* and *internal pterygoid* muscles (see figure 36.1).

A ridge of bone, the **oblique line,** extends at an angle from the ramus down the lateral surface of the body to a point near the mental foramen. This bony elevation is strong and prominent in its upper part, but gradually flattens out and disappears, as a rule, just below the first molar.

On the internal (medial) surface of the mandible is another diagonal line, the **mylohyoid**

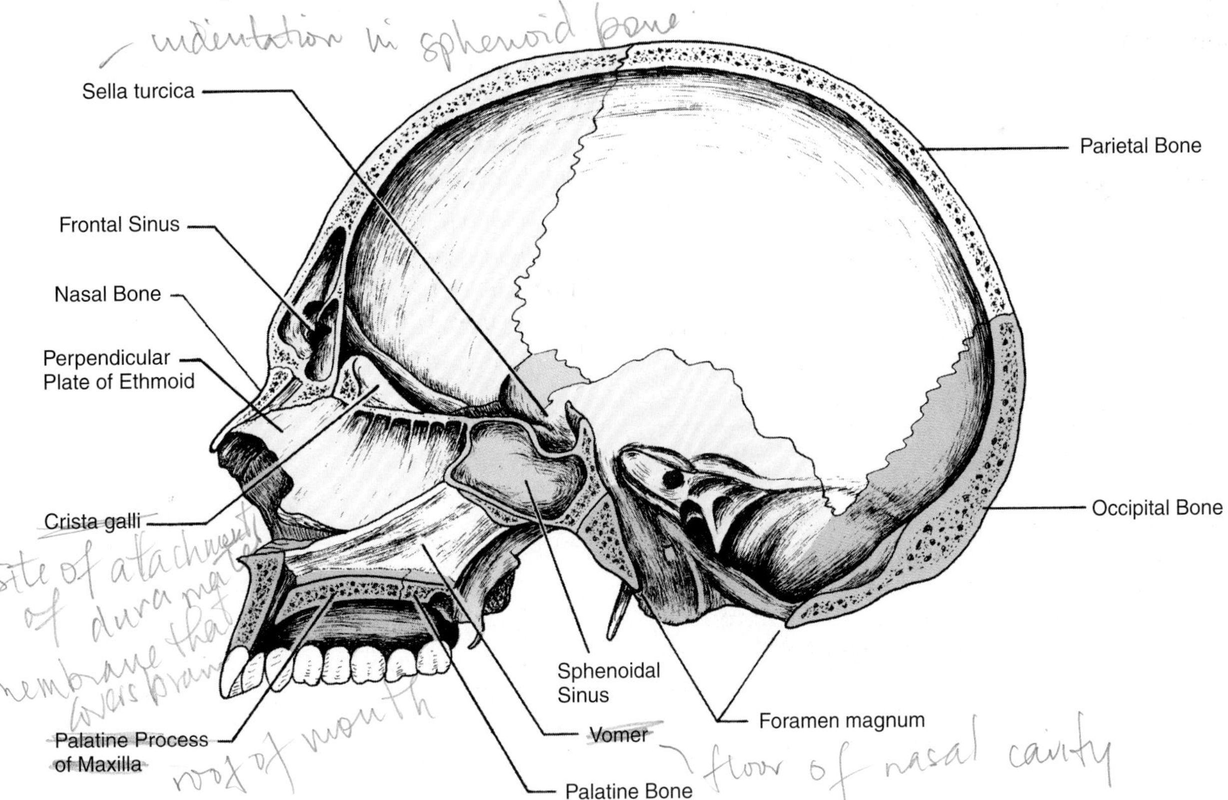

Figure 14.6 Sagittal section of skull.

line. that extends from the ramus down to the body. To this crest is attached a muscle, the *mylohyoid,* which forms the floor of the oral cavity. The bony portion of the body above this line makes up a portion of the sides of the oral cavity proper.

Each tooth lies in a socket of bone called a *dental alveolus.* As in the case of the maxillae, the portion of this bone that contains the teeth is called the **alveolar process.** The alveolar process consists of two compact tissue bony plates, the *external* and *internal alveolar plates.* Partitions, or *septae,* that lie between the teeth and make up the transverse walls of the alveoli join the external and internal alveolar plates.

On the medial surfaces of the rami are the two **mandibular foramina.** On the external surface of the body are two prominent openings, the **mental foramina.**

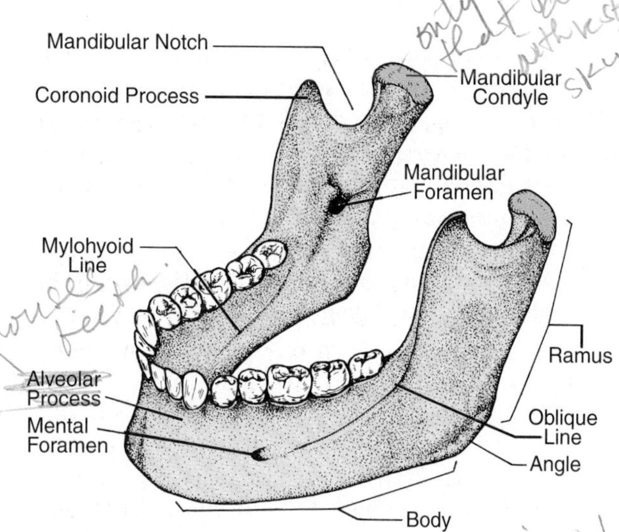

Figure 14.7 The mandible.

The Fetal Skull

The human skull at birth is incompletely ossified. Figure 14.8 shows its structure. The unossified membranous areas, called *fontanels,* allow compression of the skull at childbirth. During labor, the bones of the skull are able to overlap without causing injury to the brain as the infant passes down the birth canal.

The six fontanels are joined by five areas where future sutures of the skull will form. The largest fontanel is the **anterior fontanel,** a somewhat diamond-shaped membrane that lies on the median line at the juncture of the frontal and parietal bones. The **posterior fontanel** is somewhat smaller and lies on the median line at the juncture of the parietal and occipital bones. Between these two fontanels on the median line is a membranous area where the **future sagittal suture** of the skull will form.

On each side of the skull, where the frontal, parietal, sphenoid, and temporal bones come together behind the eye orbit, is an **anterolateral fontanel.** Between the anterior and anterolateral fontanels is a membranous area where the **future coronal suture** will develop.

The most posterior fontanel on the side of the skull is the **posterolateral fontanel** that lies at the juncture of the parietal, temporal, and occipital bones. Between the anterolateral and posterolateral fontanels is a membranous line that will develop into the **future squamosal suture.** The fontanels and membranous future sutures are usually completely ossified by age two.

Assignment:
Complete the Laboratory Report for this exercise.

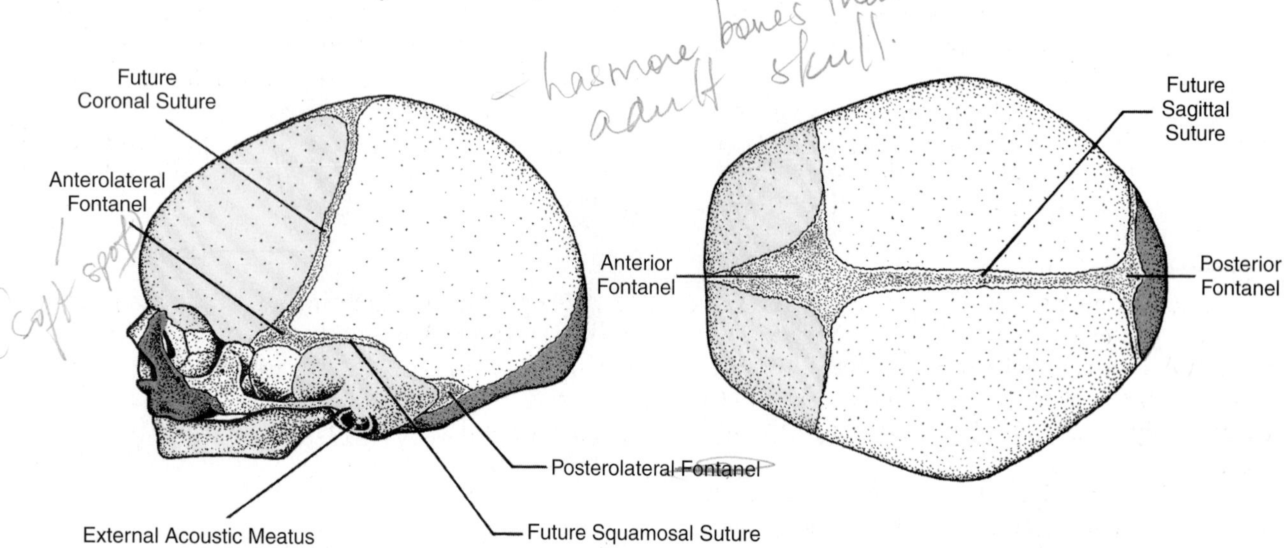

Figure 14.8 The fetal skull.

The Vertebral Column and Thorax 15

In this exercise, you study the skeletal structure of the trunk of the body, which includes the vertebral column, ribs, sternum, and hyoid bone. As in Exercise 14, locate the described features first on the illustrations and then on the skeletons in the laboratory.

Materials:
skeleton, articulated
skeleton, disarticulated
vertebral column

The Vertebral Column

The vertebral column consists of 33 bones, 24 of which are individual movable vertebrae (see figure 15.1). Note that the individual vertebrae are numbered from the top.

The Vertebrae

Although the vertebrae in different regions of the vertebral column vary considerably in size and configuration, they do have certain features in common. Each one has a structural mass, the **body,** which is the principal load-bearing contact area between adjacent vertebrae. A fibrocartilaginous **intervertebral disk** occupies the space between the surfaces of adjacent vertebral bodies. The collective action of these 24 disks cushions the spinal column.

Adjacent vertebrae also contact each other on articulating surfaces located on the **transverse processes.** The **superior articular surfaces** of each vertebra contact the **inferior articular surfaces** of the vertebra above it. At these three points, ligaments secure adjacent vertebrae, uniting the entire column into a single functioning unit.

In the center of each vertebra is an opening, the **spinal** (vertebral) **foramen,** which contains the spinal cord. Note that this foramen becomes progressively smaller further down the vertebral column.

In addition to the two transverse processes, each vertebra has a **spinous process** that projects out on its posterior surface. Spinous processes provide points of attachment for various muscles of the neck and back.

Extending backward from the body of each vertebra are two processes, the **pedicles,** which form a portion of the bony arch around the spinal foramen (see figures 15.1 C, D). The openings between the pedicles of adjacent vertebrae allow spinal nerves to exit from the spinal cord. These openings are the **intervertebral foramina** of the vertebral column. Between the transverse processes and the spinous process are two broad plates, called **laminae.** The two pedicles and two laminae constitute the *neural arch* of each vertebra.

Cervical Vertebrae The upper seven bones are the *cervical vertebrae* of the neck. The first of these seven is the **atlas** (see figure 15.1 A). The spinal foramen is much larger here to accommodate a short portion of the brain stem that extends down into the vertebral column.

Note in figures 15.1 A, B, and C that the cervical vertebrae have a small **transverse foramen** in each transverse process. Collectively, these foramina form a passageway on each side of the spinal column for the vertebral artery and vein.

The second cervical vertebra, or **axis** (see figure 15.1 B), is unique in having a vertical protrusion, the **odontoid process** (dens). The atlas pivots around this process when the head turns from side to side.

Thoracic Vertebrae Below the seven cervical vertebrae are twelve *thoracic vertebrae.* Figure 15.1 D shows the structure of a typical thoracic vertebra. Note that these bones are larger and thicker than those in the neck. A distinguishing feature of thoracic vertebrae is that all twelve of them have **articular facets for ribs** on their transverse processes.

Lumbar Vertebrae Inferior to the thoracic vertebrae lie five *lumbar vertebrae.* As figure 15.1 E shows, the bodies of lumbar vertebrae are much thicker than those of the other vertebrae, due to the greater stress that this region of the vertebral column endures.

The Sacrum

Inferior to the fifth lumbar vertebra lies the *sacrum,* consisting of five fused vertebrae. Two oval **superior**

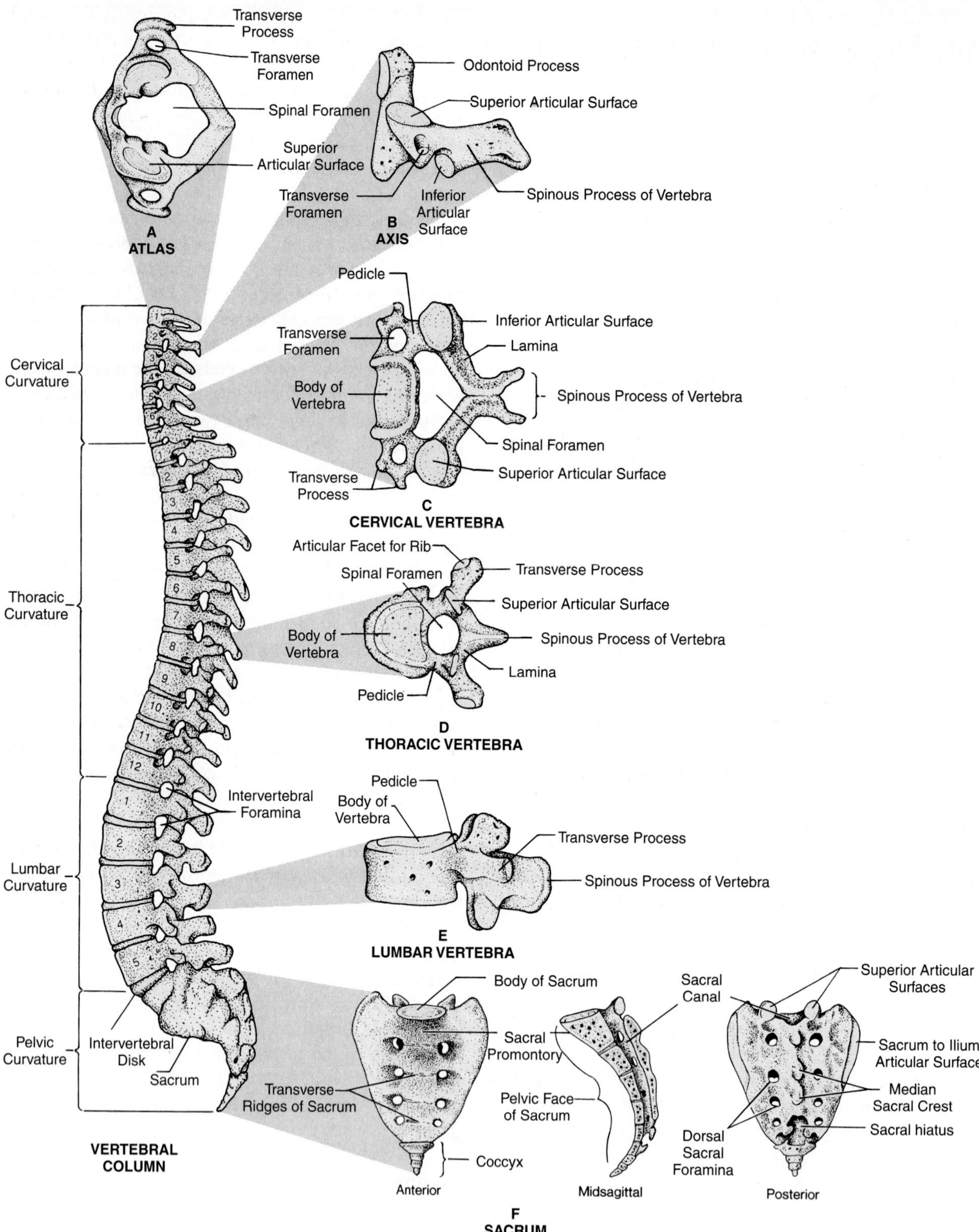

Figure 15.1 The vertebral column.

articular surfaces (facets) articulate with the inferior articular surfaces of the fifth lumbar vertebra. On each lateral surface of the sacrum is a **sacrum to ilium articular surface.** The **body of the sacrum** articulates with the intervertebral disk superior to it.

The anterior surface of the sacrum, or **pelvic face,** is concave. The body of the first sacral vertebra forms a protrusion called the **sacral promontory.** Four **transverse ridges** on the pelvic face reveal where the five vertebrae fuse.

Locate the **median sacral crest** and the **dorsal sacral foramina** on the posterior surface. The neural arches of the fused sacral vertebrae form the **sacral canal,** which exits at the lower end as the **sacral hiatus.**

The Coccyx

The "tailbone" of the vertebral column is the triangular **coccyx,** consisting of three to five small, irregularly shaped vertebrae. Ligaments attach the coccyx to the sacrum.

Spinal Curvatures

Four spinal curvatures, together with the intervertebral disks, impart considerable springiness to the vertebral column. Three of these curves are identified by the type of vertebrae in each region: the **cervical, thoracic,** and **lumbar curves.** The sacrum and coccyx form the fourth curvature, the **pelvic curve.**

The Thorax

The sternum, ribs, costal cartilages, and thoracic vertebrae form an enclosure called the *thorax.* Figures 15.2 and 15.3 show its components.

The Sternum The sternum, or breastbone, consists of three separate bones: an upper **manubrium,** a middle **body** (*gladiolus*), and a lower **xiphoid** (*ensiform*) **process.** A **sternal angle** forms where the inferior border of the manubrium articulates with the body.

On both sides of the sternum are notches (facets) where the sternal ends of the costal cartilages attach. Note that the second rib fits into a pair of *demifacets* (*demi,* half) at the sternal angle.

The Ribs There are twelve pairs of ribs. Costal cartilages attach the first seven pairs, called **vertebrosternal** or **true ribs,** directly to the sternum. The remaining pairs are called **false ribs.** The upper three pairs of false ribs, the **vertebrochondral ribs,** have cartilaginous attachments on their anterior ends but do not attach directly to the sternum. The lowest false ribs, the **vertebral** or **floating ribs,** are unattached anteriorly.

Figure 15.3 illustrates the structure of a central rib, a lateral view of a thoracic vertebra, and articulation details. Although sizes and configurations of ribs vary considerably, the central rib has structures common to most ribs.

The principal parts of each rib are a head, neck, tubercle, and body. The **head** is the enlarged end of the rib that articulates with the vertebral column. The **tubercle** consists of two portions: an **articular portion** and a **nonarticular portion.** The **neck** is a flattened portion, about 2.5 cm long, between the head and tubercle. The **body** is the flattened, curved remainder of the rib.

Note that the head of the rib in figure 15.3 C has two **articular facets** on its medial surface that contact two adjacent vertebrae. Between these two facets is a roughened **interarticular crest** that anchors ligamentous tissue. Figure 15.3 B shows how the rib articulates with two points on a thoracic vertebra. Although the rib shown in figure 15.3 C has two articular facets on its head, the first, second, tenth, eleventh, and twelfth ribs have only one.

The body of the rib in figure 15.3 C has two landmarks: an angle and a costal groove. The **angle** is an external ridge that anchors the *iliocostalis* muscle of the back. Another back muscle, the *longissimus dorsi,* attaches on the rough, irregular

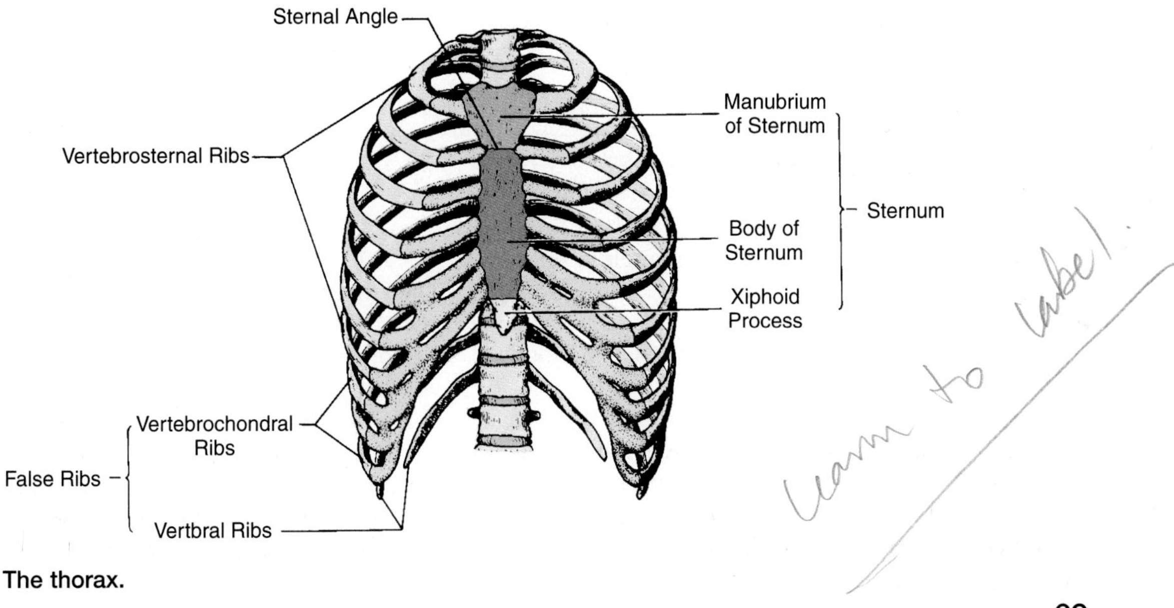

Figure 15.2 The thorax.

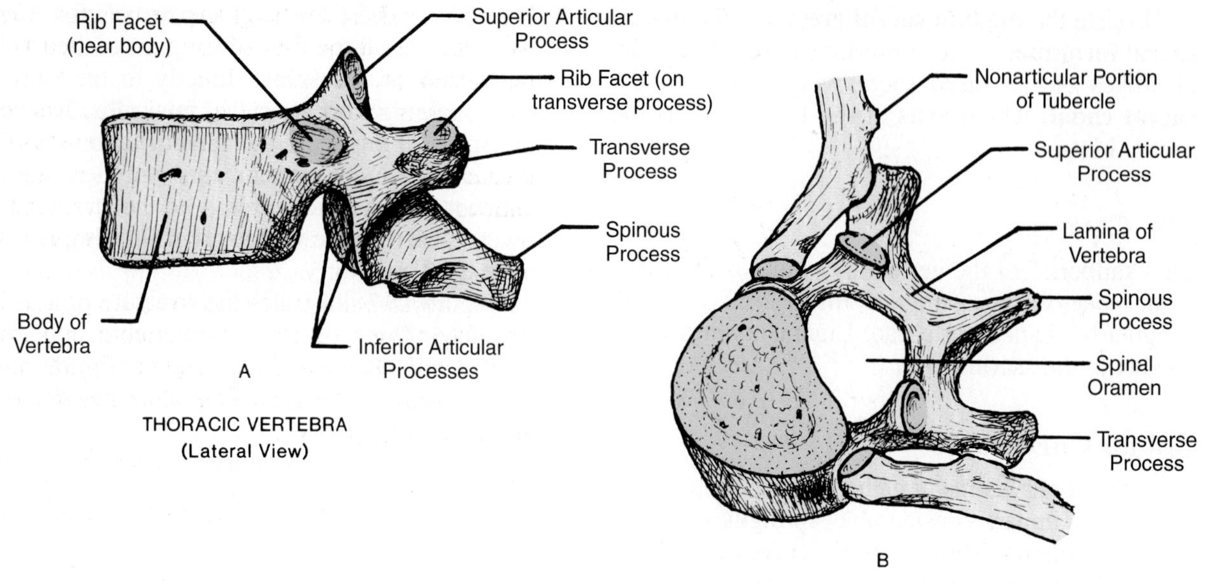

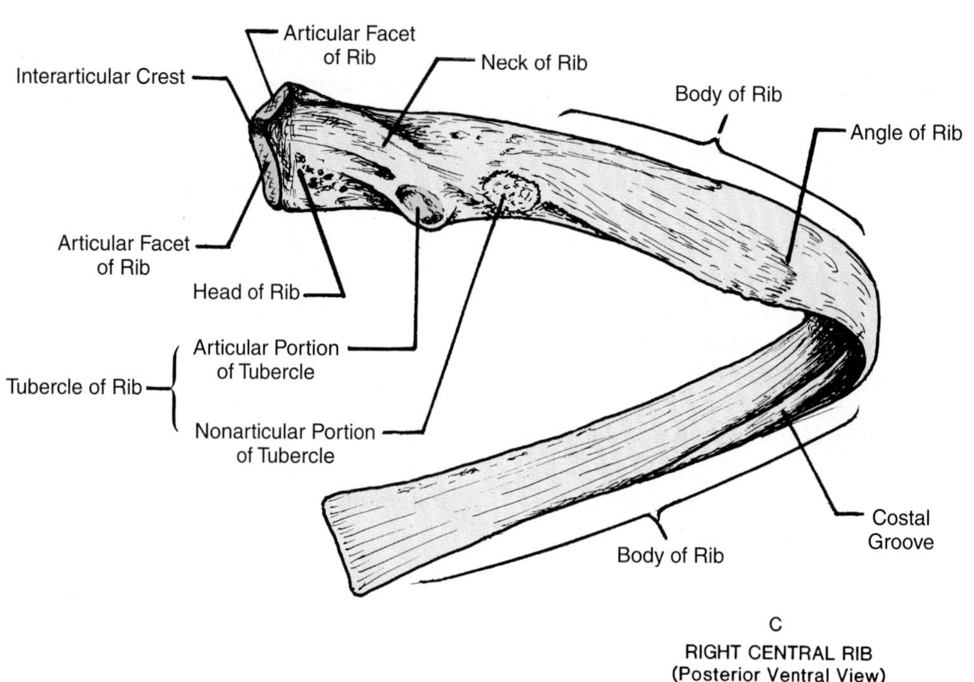

Figure 15.3 Rib anatomy and its articulation.

surface between the angle and the tubercle. Figure 37.2 shows both of these muscles. The **costal groove,** a depression on the ventral side of the rib, provides a recess for the intercostal nerve and blood vessels.

In addition to showing the location of the two rib facets, figure 15.3A shows the **superior articular process,** which contacts the inferior articular process of the vertebra above it. On its underside are the two **inferior articular processes.**

The Hyoid Bone

The *hyoid* bone is a horseshoe-shaped bone in the neck region between the mandible and larynx. Although the hyoid bone does not articulate directly with any other bone, various ligaments and muscles hold it in place. Figure 15.4 illustrates its structure.

The hyoid bone consists of five segments: a body, two greater cornua (*cornu,* singular), and two lesser cornua. The massive central portion of the bone is the **body.** The two long arms that extend

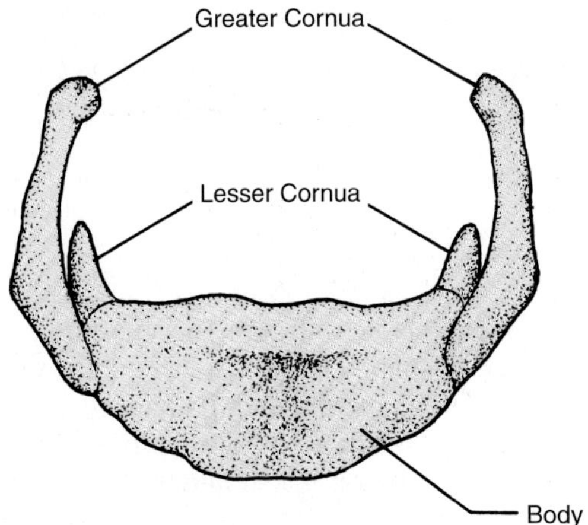

Figure 15.4 The hyoid bone.

from each side of the body are the **greater cornua.** Note that the distal end of each greater cornu terminates in a tubercle.

The **lesser cornua** are the conical eminences on the sides of the body, superior to the bases of the greater cornua. The lesser cornua are separate bone entities that fibrous connective tissue connects to the body, and occasionally, to the greater cornua. These joints are usually diarthrotic (freely movable), but occasionally become fused (ankylosed) in later life.

The following muscles have points of attachment on the hyoid bone: *mylohyoid, sternohyoid, omohyoid, thyrohyoid, digastric, hyoglossus, genioglossus,* and *constrictor pharyngis medius.*

Assignment:
Complete the Laboratory Report for this exercise.

16 The Appendicular Skeleton

In this exercise, you study the individual bones of the upper and lower extremities. As before, locate all described structures in the illustrations first and then on the laboratory specimens.

Materials:
skeleton, articulated
skeleton, disarticulated
male pelvis and female pelvis

The Upper Extremities

A pair of *pectoral* (shoulder) *girdles,* each of which includes a scapula and clavicle, holds the upper extremities in place. Figure 16.1 illustrates the upper arm and its pectoral girdle.

The Clavicle Note in figure 16.1 that the *clavicle* is a slender, S-shaped bone that articulates with the manubrium on its medial end. Its lateral end articulates with the scapula to form a portion of the shoulder joint. Important shoulder muscles attach to this bone.

The Scapula As shown in figure 16.2, the *scapula* is a flattened, triangular bone with a cartilage-lined socket, the **glenoid cavity,** on its upper lateral edge. Instead of firmly attaching to the axial skeleton, the scapula is loosely held in place by muscles attached to the upper arm, back, and chest.

Observe in figure 16.2 that the scapula's posterior surface has a pronounced elongated diagonal ridge called the **scapular spine.** This ridge divides the back surface of the scapula into two depressions:

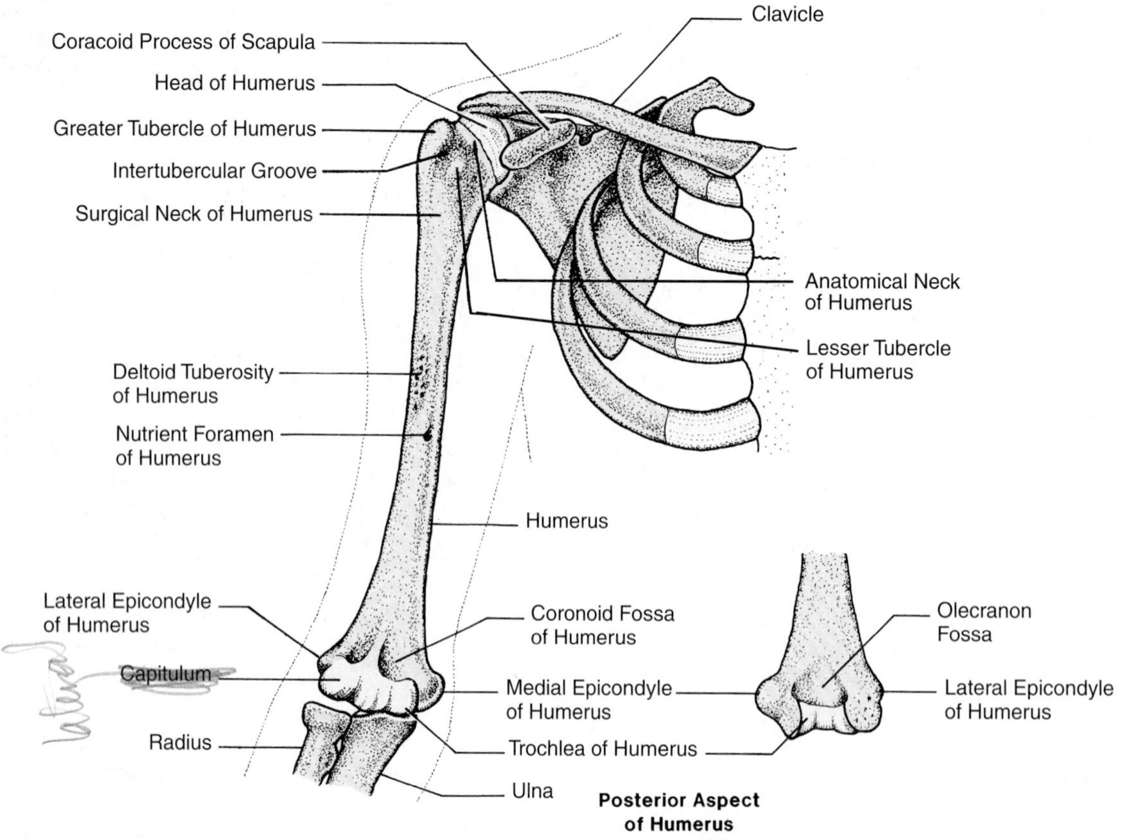

Figure 16.1 The upper arm.

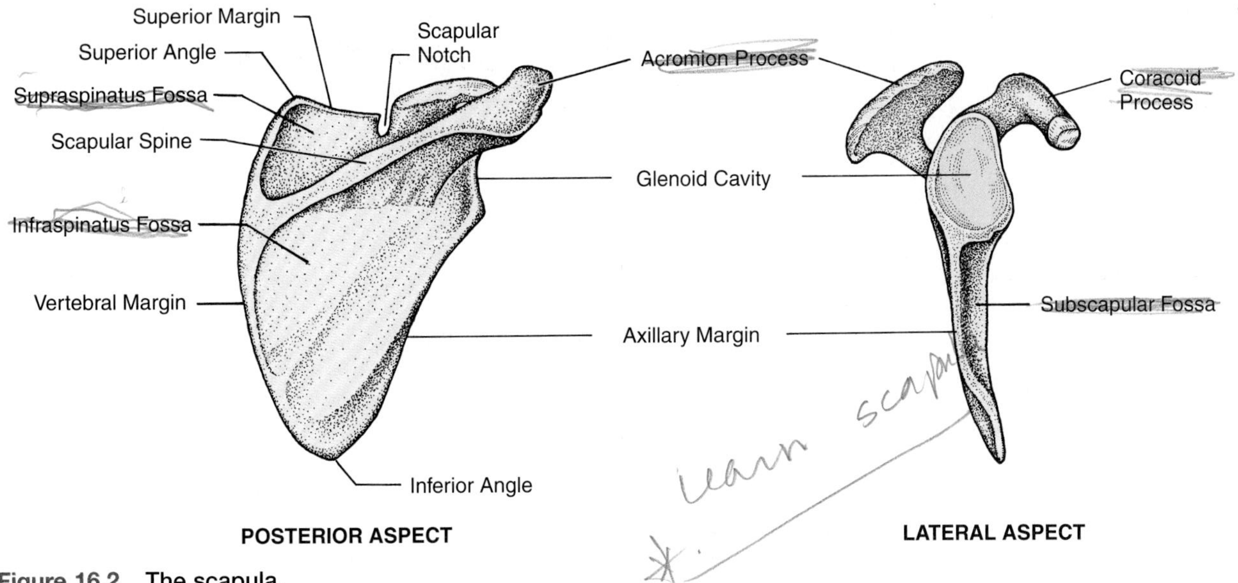

Figure 16.2 The scapula.

an upper **supraspinatus fossa** and a larger **infraspinatus fossa** below it. The anterior surface of the scapula has a single large concavity, the **subscapular fossa.** These three fossae anchor various shoulder muscles.

Two processes are prominent on the lateral aspect of figure 16.2. The large **acromion process** forms at the lateral terminus of the scapular spine. The lateral end of the clavicle attaches to this process. The **coracoid process** is anterior to the acromion process and glenoid cavity.

The perimeter of the scapula consists of three margins. The convex edge on the medial side of the posterior aspect (figure 16.2) is the **vertebral** (medial) **margin.** This border extends from the **superior angle** down to the **inferior angle** at the bottom. The **axillary** (lateral) **margin** makes up the border opposite the vertebral border and extends from the inferior angle to the glenoid cavity. The third border is the **superior margin,** which extends from the superior angle to a deep depression called the **scapular notch.**

Assignment:
Answer the questions on the Laboratory Report that pertain to the shoulder.

The Upper Arm The skeletal structure of the upper arm consists of a single bone, the **humerus.** Its smooth, rounded, proximal end, the **head,** fits into the glenoid cavity of the scapula. Just below the head is a narrowing section called the **anatomical neck.**

Inferior to the head and anatomical neck are two prominent eminences, the greater and lesser tubercles. The **greater tubercle** is the larger process and is lateral to the **lesser tubercle.** Between these two processes is a depression called the **intertubercular groove.** The narrowing section just distal to the two tubercles is the **surgical neck,** so named because of the frequency of bone fractures in this area.

The anterior surface of the diaphysis of the humerus has a roughened, raised area near its midregion called the **deltoid tuberosity.** A small opening, the **nutrient foramen,** is also in this same general area.

The distal terminus of the humerus has two condyles, the capitulum and trochlea, which contact the bones of the forearm. The **capitulum** is the lateral condyle that articulates with the radius. The **trochlea** is the medial condyle that articulates with the ulna. Superior and lateral to the capitulum is an eminence, the **lateral epicondyle.** On the opposite side is a larger tuberosity, the **medial epicondyle.** Above the trochlea on the anterior surface is a depression, the **coronoid fossa.** A depression on the posterior surface of this end of the humerus is the **olecranon fossa.**

The Forearm The radius and ulna constitute the skeletal components of the forearm. Figure 16.3 shows how these two bones relate to each other and to the hand.

The **radius** is the lateral bone of the forearm. The proximal end of this bone has a slightly concave, disk-shaped **head** that articulates with the capitulum of the humerus. The head's disklike shape allows the radius to rotate at the upper end when the palm of the hand is changed from palm down to palm up (pronation to supination).

A few centimeters below the head of the radius on its medial surface is an eminence, the **radial**

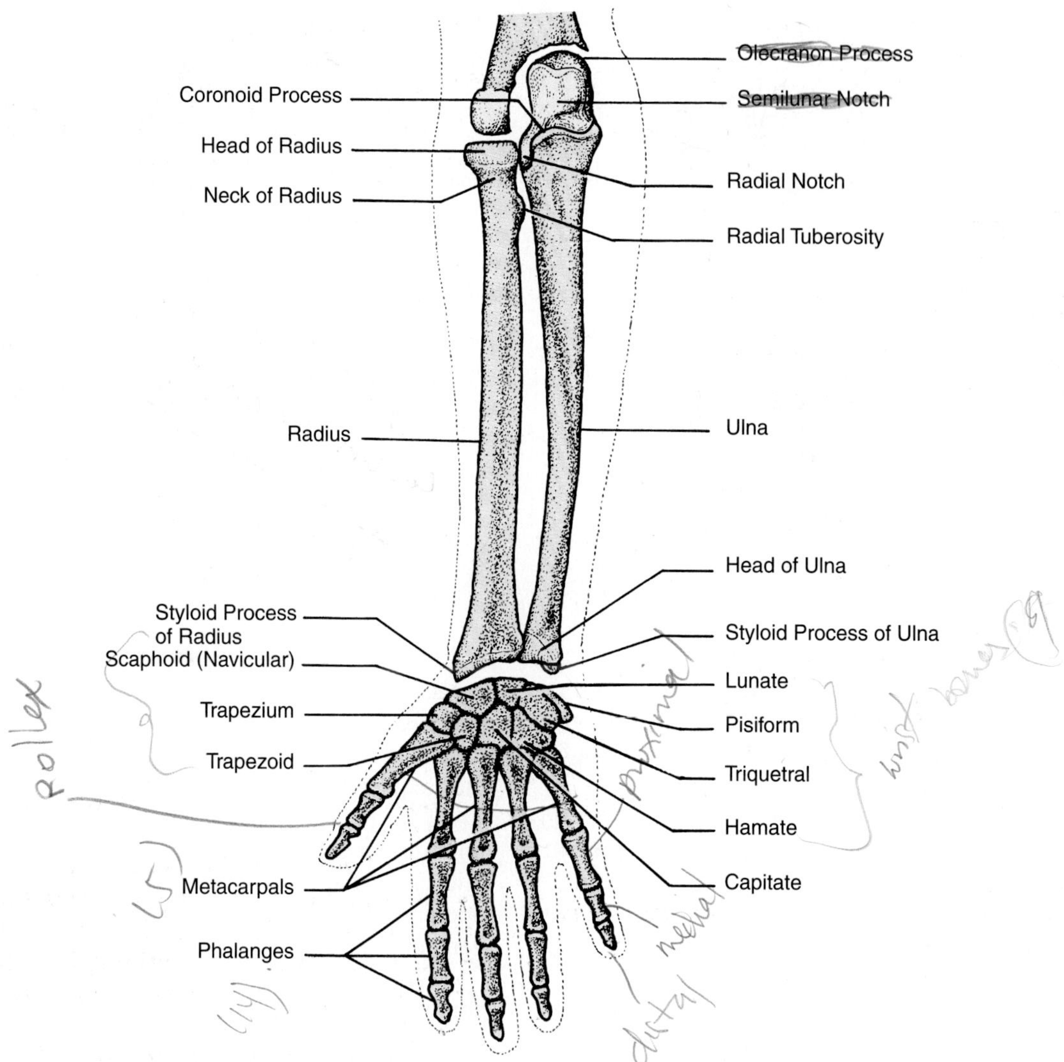

Figure 16.3 The forearm and hand.

tuberosity. This process is the point of attachment for the *biceps brachii,* a flexor muscle of the arm. The region between the head and the radial tuberosity is the **neck** of the radius. The distal lateral prominence of the radius that articulates with the wrist is the **styloid process.** Locate this process on your own wrist.

The **ulna,** or elbow bone, is the largest bone in the forearm. A portion of the humerus in figure 16.3 is cut away to reveal the prominence of the elbow, which is called the **olecranon process.** Within this process is a depression, the **semilunar notch,** that articulates with the trochlea of the humerus. The eminence just below the semilunar notch on the anterior surface of the ulna is the **coronoid process.** Where the head of the radius contacts the ulna is a depression, the **radial notch.** The lower end of the ulna is small and terminates in two eminences: a large portion, the **head,** and a small **styloid process.** The head articulates with a fibrocartilaginous disk, which separates the head from the wrist. The styloid process is a point of attachment for a ligament of the wrist joint.

Assignment:
Answer the questions on the Laboratory Report that pertain to the arm.

The Hand Each hand consists of a **carpus** (wrist), a **metacarpus** (palm), and **phalanges** (fingers). The carpus consists of eight small bones arranged in two rows of four bones each. Named from lateral to medial, the proximal carpal bones are the **scaphoid** (navicular), **lunate, triquetral,** and **pisiform.** The distal carpal bones from lateral to medial are **trapezium** (greater multangular), **trapezoid** (lesser multangular), **capitate,** and **hamate** (see figure 16.3). The metacarpus consists of five bones that are numbered one to five, starting at the thumb.

The phalanges are the skeletal elements of the fingers distal to the metacarpal bones. Each finger

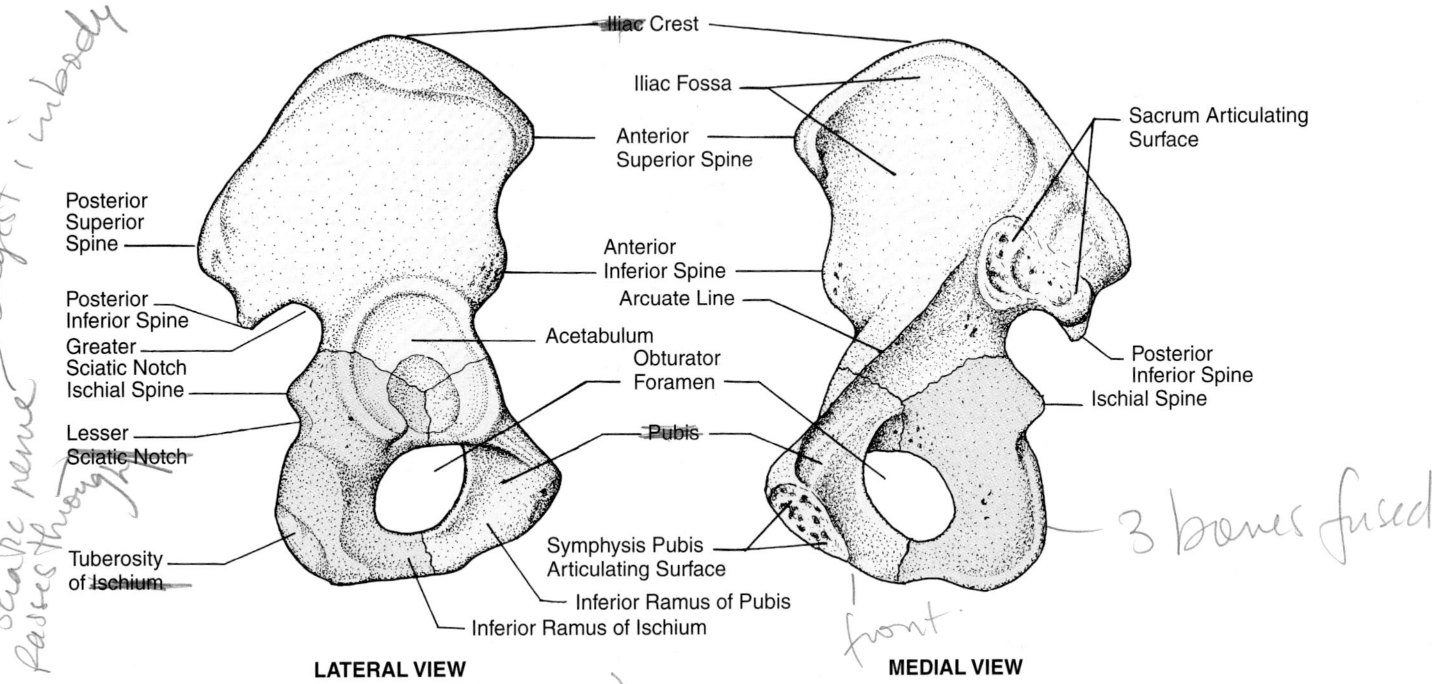

Figure 16.4 The os coxa.

has three phalanges (proximal, middle, and distal), and the thumb has two.

Assignment:
Answer the questions on the Laboratory Report that pertain to the hand.

The Lower Extremities

The Pelvic Girdle The two hipbones (*ossa coxae*) articulate in front to form a bony arch called the *pelvic girdle.* The union of the ossa coxae with the sacrum forms the back of this arch. The enclosure formed within this bony ring is the **pelvis**.

Figure 16.4 presents lateral and medial views of the right os coxa. The large, circular depression into which the head of the femur fits is the **acetabulum.** Note that each os coxa consists of three fused bones (ilium, ischium, and pubis) that join in the center of the acetabulum. Although these ossification lines are easy to discern in young children, they are generally obliterated in the adult.

Two articulating surfaces are visible on the medial aspect of the os coxa in figure 16.4. The **sacrum articulating surface** interfaces with the sacrum; the **symphysis pubis articulating surface** joins with the other os coxa to form the **symphysis pubis.** The line of juncture between the ilium and sacrum is the **sacroiliac joint.**

The **ilium** is the yellow portion of the bone in figure 16.4. On the greater portion of its medial surface is a large concavity called the **iliac fossa.** The **arcuate line** is a ridge of demarcation below the iliac fossa.

The **iliac crest** is the upper margin of the ilium. It extends from the **anterior superior spine** to the **posterior superior spine.** Just below the latter process is the **posterior inferior spine.** The **anterior inferior spine** is on the opposite margin below the anterior superior spine.

The pink portion of the os coxa in figure 16.4 is the **ischium.** It has two distinct processes: a small eminence, the **ischial spine,** and a larger one, the **tuberosity of the ischium,** which makes up the lower bulk of the bone. When you sit, your body weight rests on this tuberosity. Just below the ischial spine is a depression, the **lesser sciatic notch.** The larger depression superior to it is the **greater sciatic notch.**

The blue portion of the bone in figure 16.4 is the **pubis.** The opening that the pubis and ischium surround is the **obturator foramen.** Note that the **inferior ramus of the ischium** and the **inferior ramus of the pubis** form the lower bony arch around this foramen.

Assignment:
Answer the questions in the Laboratory Report that pertain to the pelvis. Then compare a male pelvis to a female pelvis, and answer questions about their anatomical differences on the Laboratory Report.

The Upper Leg One bone, the **femur,** supports the thigh. The femur's upper end consists of a

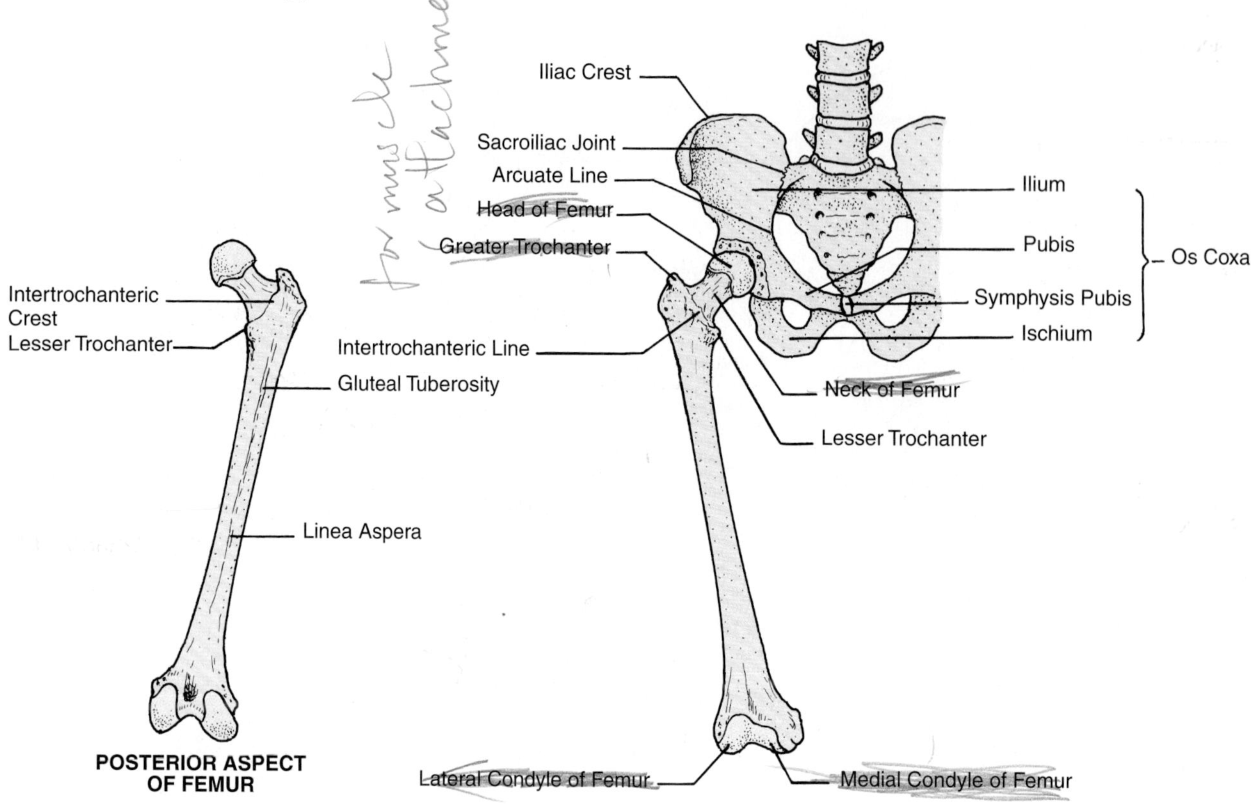

Figure 16.5 The upper leg.

hemispherical **head,** a **neck,** and two processes, the greater and lesser trochanters (see figure 16.5). The **greater trochanter** is the large process on the femur's lateral surface. The **lesser trochanter** is farther down on the femur's medial surface. A ridge, the **intertrochanteric line,** lies obliquely between these two trochanters on the anterior surface. On the posterior surface, the ridge between these two trochanters is the **intertrochanteric crest.**

Along the posterior surface of the diaphysis of the femur is a pronounced ridge. The upper portion of this ridge is called the **gluteal tuberosity;** the lower portion is the **linea aspera.** Several muscles of the thigh attach to this prominence.

The lower extremity of the femur is larger than the upper end and divides into two condyles: the **lateral** and **medial condyles.** Superior to these are two prominences: the **lateral** and **medial epicondyles.** The **patellar surface** is on the anterior distal portion of the femur. The **patella,** or kneebone, is a sesamoid bone that attaches to the quadriceps femoris tendon on its superior surface and to the patellar ligament on its inferior surface. Examine the articulated skeleton for this bone.

The Lower Leg Figure 16.6 shows the skeletal structure of the lower leg, which consists of the tibia and fibula. The **tibia** is the stronger of the two bones. Its upper portion expands to form a **medial condyle,** a **lateral condyle** on the opposite side, and a **tibial tuberosity** inferior to the two condyles on the anterior surface that provides an attachment site for the patellar ligament (which extends from the patellar tendon). Projecting upward between the condyles are two small processes that constitute the **intercondylar eminence.**

The distal extremity of the tibia is smaller than the upper portion. A strong process, the **medial malleolus,** forms the inner prominence of the ankle. Locate this process on your leg. Note that a triangular cross section of the tibia would result in a sharp ridge, the **anterior crest,** being present on its anterior surface.

The **fibula** is lateral and parallel to the tibia. Its upper extremity, or **head,** articulates with the tibia, but it does not form a part of the knee joint. Below the head is the **neck** of the fibula. The lower extremity of the fibula terminates in a pointed process, the **lateral malleolus,** which lies just under the skin and forms the outer prominance of the ankle. Locate this process on your leg. Like the tibia, the fibula has an **anterior crest** extending down its anterior surface.

Assignment:

Answer the questions on the Laboratory Report that pertain to the leg.

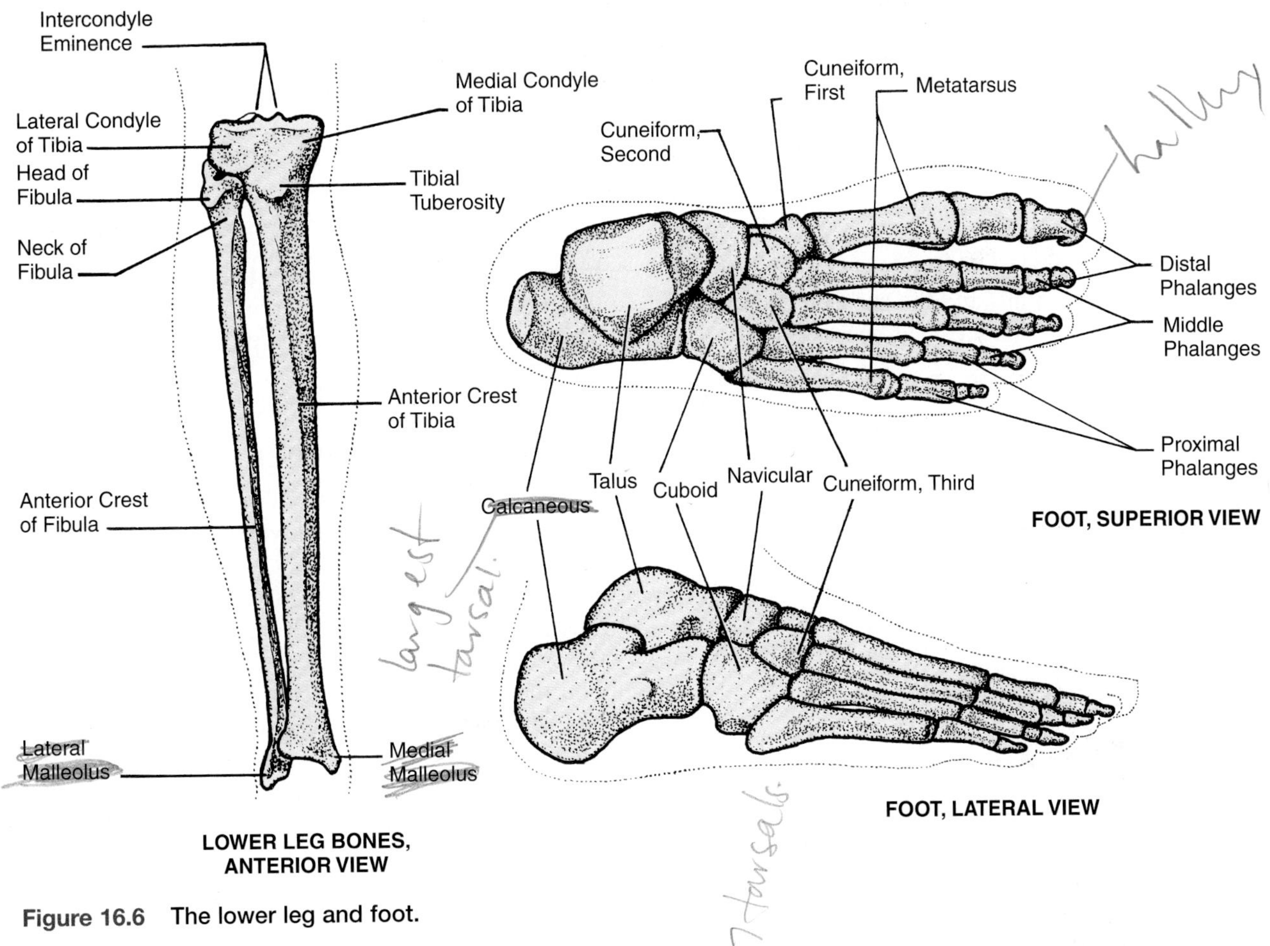

Figure 16.6 The lower leg and foot.

The Foot Figure 16.6 also shows superior and lateral views of the skeleton of the foot, consisting of part of the ankle, the instep, and toes.

The **tarsus,** or ankle portion of the foot, consists of seven tarsal bones: the calcaneous, talus, navicular, cuboid, and three cuneiforms. The **calcaneous,** or heel bone, is the largest tarsal bone. It forms a strong lever for muscles of the calf of the leg. The **talus** occupies the uppermost central position of the tarsus and articulates with the tibia. In front of the talus on the medial side of the foot is the **navicular.** The **cuboid** is on the lateral side of the foot in front of the calcaneous. The three **cuneiform** bones are anterior to the navicular. They are numbered one, two, and three, with number one being on the medial side of the foot.

The **metatarsus** forms the anterior portion of the instep of the foot and consists of five elongated **metatarsal** bones. They are numbered one through five, with number one being on the medial side of the foot.

The **phalanges** that make up the toes resemble the phalanges of the hand in general shape and number: two in the great toe and three (proximal, middle, and distal) in each of the other toes.

Assignment:

Complete on the Laboratory Report for this exercise. Answer the questions on the Laboratory Report that pertain to the foot.

17 Articulations

In this exercise, you study the basic characteristics of three types of joints, and the specific anatomical details of the shoulder, hip, and knee joints.

> *Materials:*
> fresh knee joint of cow or lamb, sawed through longitudinally

Three Basic Types of Joints

All body joints fall into one of the three categories illustrated in figure 17.1. Note that classification is based on degree of movement.

Immovable Joints (Synarthroses)

Immovable joints lack mobility because fibrous connective tissue or cartilage bonds adjacent bones in the joint. The two kinds of immovable joints are sutures and synchondroses.

Sutures Figure 17.1A shows the structure of a *suture,* an irregular joint between the flat bones of the skull. The "glue" between these bones is **fibrous connective tissue,** which is continuous with the **periosteum** on the outside of the skull and the **dura mater** on the inside of the skull.

Synchondroses The bonding agent in synchondroses is cartilage instead of fibrous tissue. The **metaphyses** of long bones in children are joints of this type. As noted in Exercise 13, these growth areas in children consist of hyaline cartilage during the growing years.

Slightly Movable Joints (Amphiarthroses)

Joints that exhibit slight movement are either symphyses or syndesmoses.

Symphyses Intervertebral joints of the spine and the symphysis pubis are representative symphyses. The most characteristic feature of symphyses is a pad of **fibrocartilage** that cushions the joint (see figure 17.1B). **Articular cartilage** (hyaline type) on the bone surfaces provides a frictionless surface adjacent to the fibrocartilage. A fibroelastic **capsule** holds the joints together.

Syndesmoses Syndesmoses are slightly movable joints that lack fibrocartilage and are held together by an interosseous ligament (a ligament connecting two bones). A good example is the attachment of the fibula to the tibia, as figure 17.5 shows.

Freely Movable Joints (Diarthroses)

Articulations that move easily are *diarthrotic,* or *synovial,* joints. Note in figure 17.1C that smooth **articular cartilage** covers bone ends. A fibrous

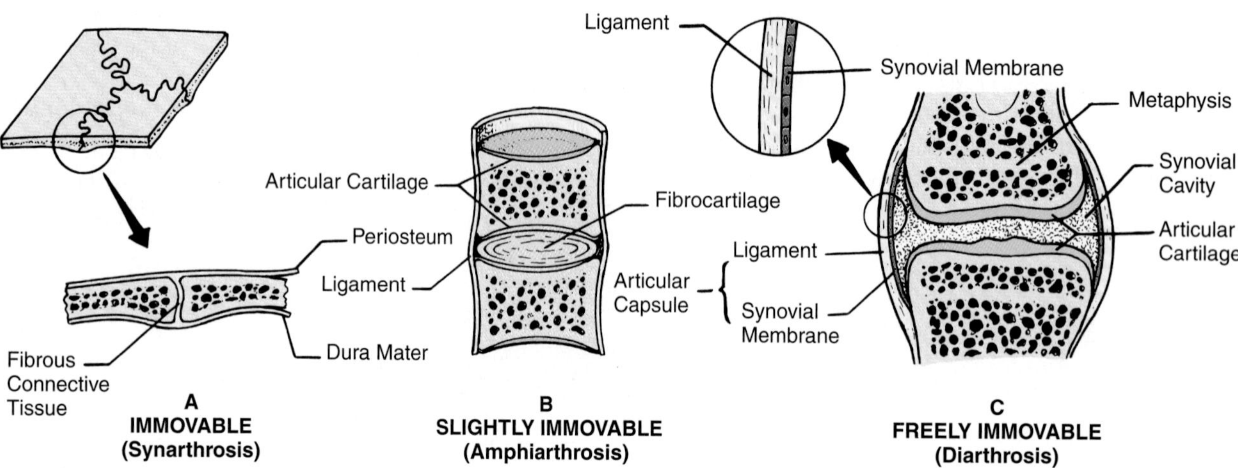

Figure 17.1 Types of articulations.

articular capsule that consists of an outer layer of **ligaments** and an inner lining of **synovial membrane** holds the joint together and encloses the **synovial cavity.** Synovial membrane produces a viscous fluid called *synovium* that lubricates the joint. Freely movable joints also contain sacs, or **bursae,** of synovial tissue and may have fibrocartilaginous pads.

Types of Freely Movable Joints

Figure 17.2 shows the six categories of diarthrotic, or synovial, joints.

Gliding Joints The articular surfaces in gliding joints are nearly flat or are slightly convex. Bones that exhibit gliding action are present between carpal bones of the wrist and between tarsal bones of the ankle.

Hinge Joints Ankle, elbow, and knee joints are hingelike, moving primarily in one plane. Movement between the occipital condyles of the skull and the atlas is also hingelike.

Condyloid Joints Joints that have bones with oval-shaped heads, or condyles, that move in elliptical cavities can move in two directions. The wrist is a typical condyloid joint.

Saddle Joints Like the condyloid joints, saddle joints allow movement in two directions. The ends of the bones differ, however, in that each bone end is convex in one direction and concave in the other

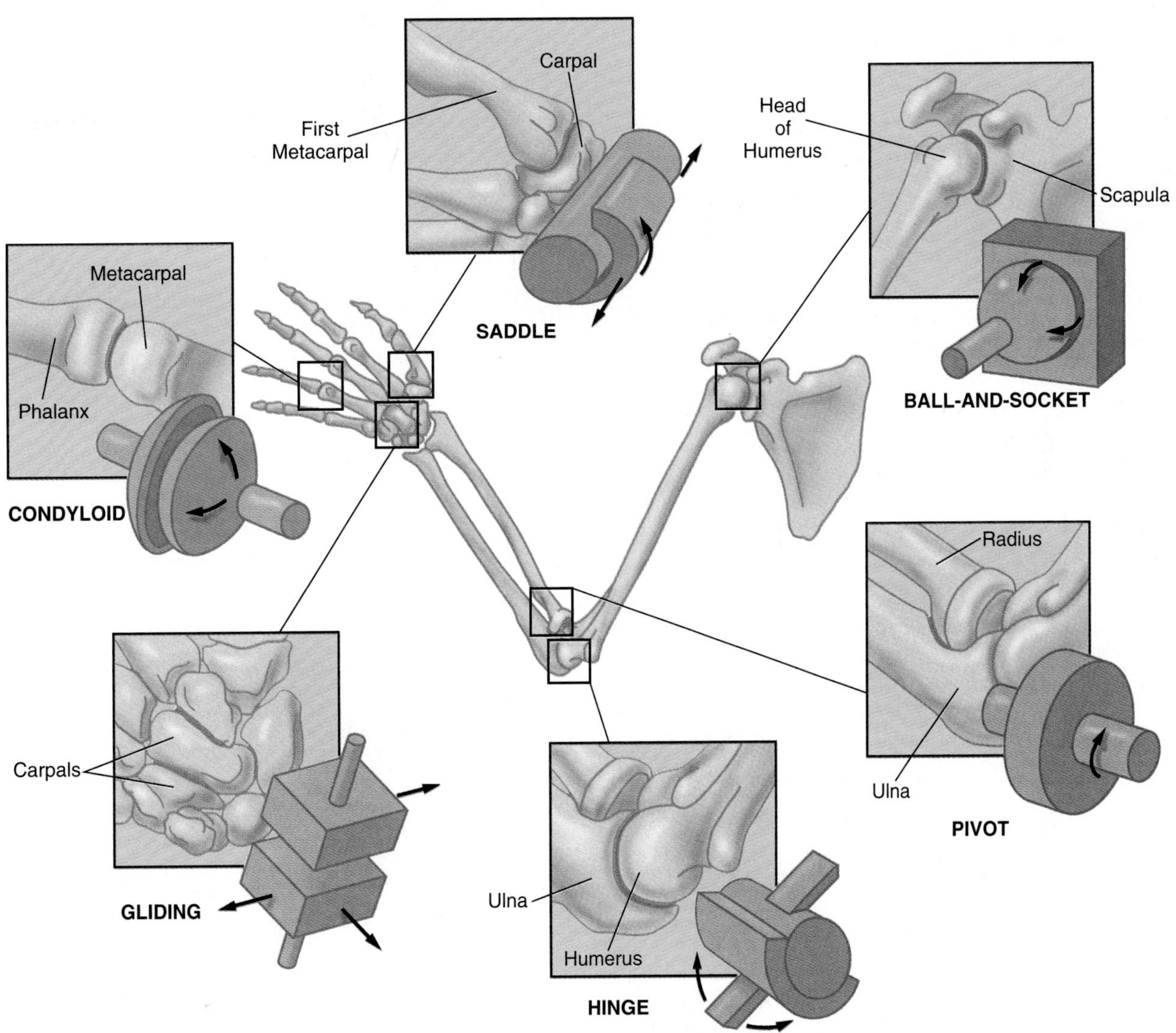

Figure 17.2 Types of synovial joints.

direction. The joint between the thumb metacarpal and trapezium is a saddle joint.

Pivot Joints Movement rotates around an axis in pivot joints. Rotation of the atlas around the odontoid process of the axis is a good example.

Ball-and-Socket Joints Ball-and-socket joints have angular movement in all directions. The shoulder and hip joints are representative.

The Shoulder Joint

The shoulder joint is the most freely movable articulation of the body. Note in figure 17.3 that **articular cartilage** covers the articulating surfaces of the humerus and glenoid cavity.

Although six or seven bursae are usually in the neighborhood of this joint, figure 17.3 shows only one, the **subdeltoid bursa.** Note that the **tendon of the supraspinatus muscle** inserts on the **greater tubercle of the humerus.** Bony sections in the upper portion of the joint are the **acromion process** and a portion of the **clavicle.** The membranous fold beneath the head of the humerus is a portion of the **articular capsule.**

The Hip Joint

Figure 17.4 shows the hip joint, another ball-and-socket joint. The acetabular labrum and the transverse acetabular ligament confine the rounded head of the femur in the acetabulum of the os coxa. The **acetabular labrum** is a fibrocartilaginous rim attached to the margin of the acetabulum. The **transverse acetabular ligament** is an extension of the acetabular labrum.

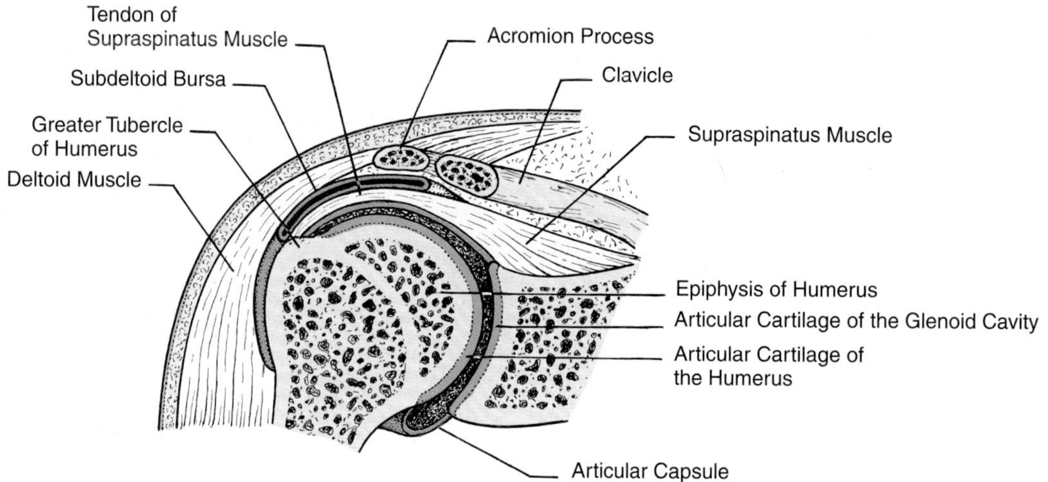

Figure 17.3 The shoulder joint.

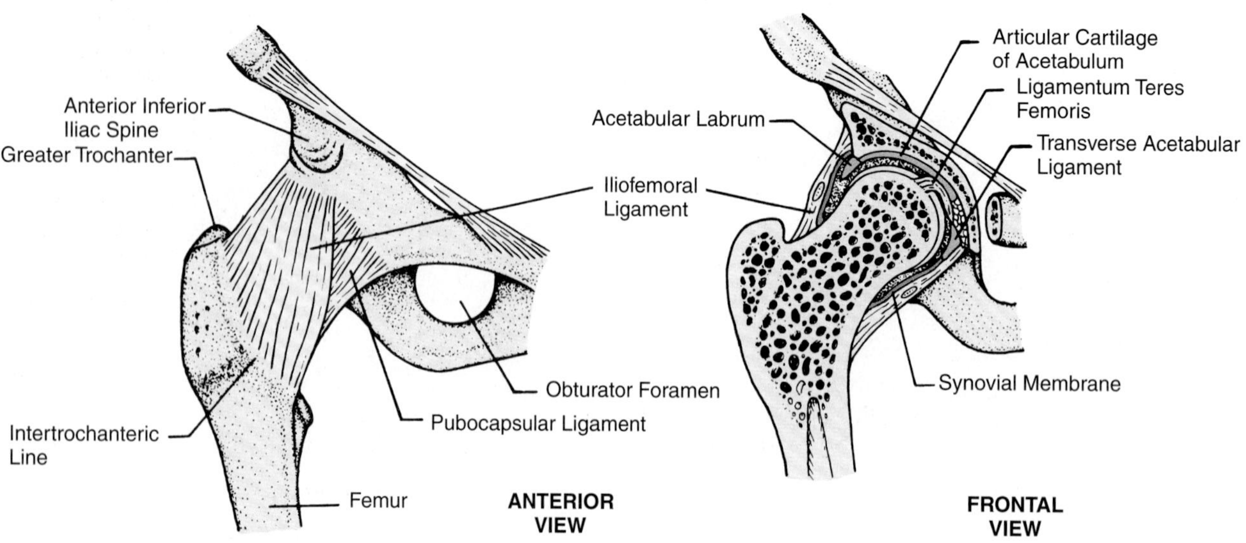

Figure 17.4 The hip joint.

Note in the sectional view in figure 17.4 that the **ligamentum teres femoris** attaches to the middle of the curved condylar surface of the femur. The other end of this ligament attaches to the surface of the acetabulum. This ligament adds nothing to the strength of the joint; instead, it helps to nourish the head of the femur and supplies synovial fluid to the joint.

An articular capsule encloses the entire joint, including the three structures just discussed. This capsule consists of longitudinal and circular fibers surrounded by three external accessory ligaments, the iliofemoral, pubocapsular, and ischiocapsular ligaments.

The **iliofemoral ligament** is a broad band on the joint's anterior surface. It attaches to the **anterior inferior iliac spine** at its upper margin and to the **intertrochanteric line** on its lower margin. Adjacent and medial to the iliofemoral ligament lies the **pubocapsular ligament.** The **ischiocapsular ligament** reinforces the posterior surface of the capsule. As with the knee joint, a **synovial membrane** lines the capsule and lubricates the joint.

The hip joint allows flexion, extension, abduction, adduction, rotation, and circumduction of the thigh. The unique angle of the neck of the femur and the relationship of the condyle to the acetabulum make this possible. To some extent, the iliofemoral ligament limits excessive backward movement of the body at the joint, which takes some of the strain off certain muscles.

The Knee Joint

Figure 17.5 shows two views and a sagittal section of the knee. Although the action of this joint was described earlier as being essentially hingelike, it is by no means a simple hinge. The curved surfaces of the condyles of the femur allow rolling and gliding movements within the joint. The nature of the hip and foot alignment also allows some rotary movement.

The knee joint is probably the most highly stressed joint in the body. Two *semilunar cartilages,* or **menisci,** in each joint absorb some of this stress. As shown in the sagittal section in figure 17.5, these fibrocartilaginous pads are thick at the periphery and thin in the center of the joint, providing a deep recess for the condyles.

The **lateral meniscus** lies between the lateral condyle of the femur and the tibia; the **medial meniscus** is between the medial condyle and the tibia. These menisci are best seen in the anterior and posterior views of figure 17.5. A **transverse ligament** connects them anteriorly.

Several layers of ligaments hold the entire joint together. Innermost are two cruciate and two

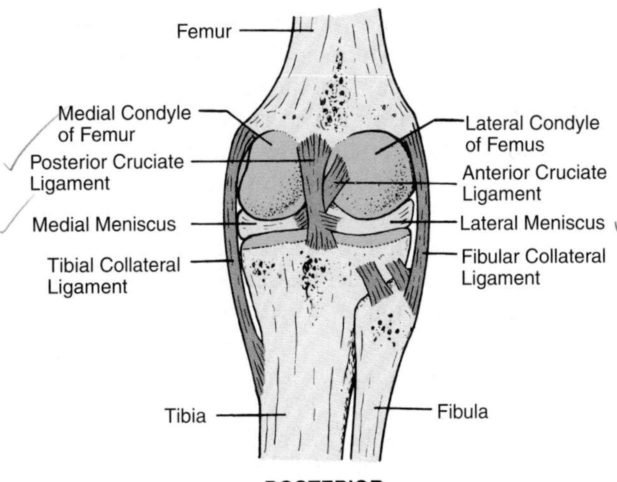

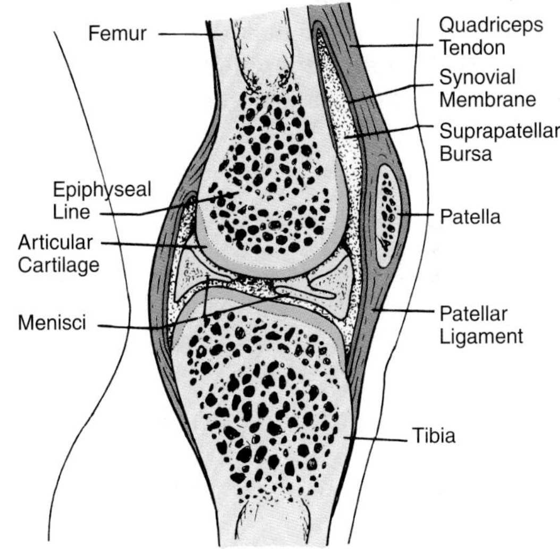

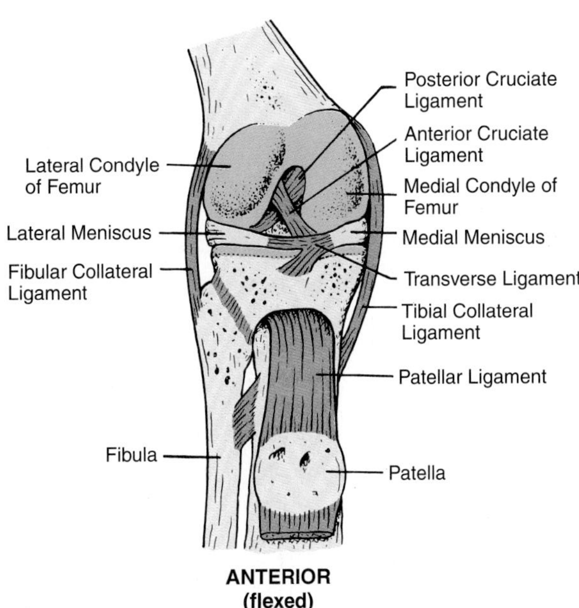

Figure 17.5 The knee joint.

collateral ligaments. The **posterior** and **anterior cruciate ligaments** form an X on the median line of the posterior surface, with the posterior cruciate ligament being outermost. The anterior cruciate ligament is outermost on the anterior surface when the leg is flexed. The **fibular collateral ligament** is on the lateral surface, extending from the lateral epicondyle of the femur to the head of the fibula. The **tibial collateral ligament** extends from the medial epicondyle of the femur to the upper medial surface of the tibia. In addition to these four ligaments are the *oblique* and *arcuate popliteal ligaments* on the posterior surface. Figure 17.5 does not show these.

Encompassing the entire joint is the *fibrous capsule,* a complicated structure of special ligaments, united with muscle tendons that pass over the entire joint.

Observe in the sagittal section in figure 17.5 that an upper **quadriceps tendon** and a lower **patellar ligament** hold the kneecap in place. **Synovial membrane** lines the innermost surfaces of these structures. The space between this synovial membrane and the femur is the **suprapatellar bursa.**

Although the knee joint can sustain considerable stress, it lacks bony reinforcement to prevent dislocations in almost any direction. It relies almost entirely on soft tissues to hold the bones in place. This is why knee injuries are so common in vigorous sport activities. The knee joint is particularly vulnerable to lateral and rotational forces. An understanding of its anatomy should alert you to its limitations.

Animal Joint Study

Examine a cow or lamb knee joint that has been sawed through longitudinally. Identify as many of the structures in figure 17.5 as possible.

Assignment:
Complete the Laboratory Report for this exercise.

PART 

The Nerve-Muscle Relationship

The three exercises in Part 4 and the six exercises in Part 5 are grouped here to provide sufficient background information for skeletal muscle physiology experiments performed in Part 6. Since skeletal muscle cannot function independent of nerve stimulation, studying the two together makes sense.

Although nerve and muscle cells are morphologically different, they have one property in common that sets them apart from all other body cells: their ability to transmit electrochemical impulses along their membranes. This property is called *excitability*.

Excitability is a function of plasma membrane electrical potential. Although virtually all body cells exhibit measurable plasma membrane electrical potential, only nerve and muscle cells demonstrate the ability to generate *action potentials* (impulses) because of this electrical characteristic.

In Exercises 18 and 19, you study the structural differences of the various kinds of nerve and muscle cells. Exercise 20 focuses on action potential transmission between nerve and muscle cells.

18 The Nerve Cells

The nervous system has two kinds of cells: neurons and neuroglia. **Neurons** are functional units of the nervous system in that they transmit nerve impulses. **Neuroglia** are connective, supportive, and nutritive; excitability is not characteristic. Although the prime concern here is with neurons, both kinds of nerve tissue are studied.

Study the first portion of this exercise, and answer the questions on the Laboratory Report before examining prepared microscope slides in the laboratory. As you study the neurons in figure 18.1, keep in mind that not all the structures shown are visible with your laboratory microscope. Once you start studying microscope slides of the spinal cord and brain tissues, however, you will be able to see most of the structures to some extent.

Neurons

Since neurons perform different roles in different parts of the nervous system, they exist in a variety of sizes and configurations. In spite of this diversity, all neurons have much in common, as figure 18.1 shows.

The Cell Body and Its Processes

The sensory neurons, motor neurons, and interneurons in figure 18.1 are the principal neurons associated with skeletal muscle physiology. Sensory and motor neurons are peripheral neurons because they are found outside of the central nervous system (brain and spinal cord). The cross section through the spinal cord in figure 18.1 illustrates how these neurons relate to the spinal cord and to each other. These three types of neurons, together with the receptor and effector, make up a **reflex arc,** described in more detail in Exercise 41.

A neuron cell body, called a **perikaryon,** usually contains a rather large nucleus surrounded by Nissl bodies and neurofibrils in the cytoplasm. **Nissl bodies** are dense aggregations of rough endoplasmic reticulum and are sites of protein synthesis. They stain readily with basic aniline dyes such as toluidine blue or cresyl violet. Some Nissl bodies are shown in the perikaryon of the motor neuron in figure 18.1.

Neurofibrils are slender protein filaments that extend throughout the cytoplasm parallel to the long axis. They resemble roadways that circumvent the Nissl bodies. These minute fibers converge into the axon to form the core of the nerve fiber.

The most remarkable features of neurons are their cytoplasmic processes: axons and dendrites. Traditional terminology once categorized these processes purely from a morphological sense; however, numerous exceptions require more accurate and universal definitions that emphasize functional attributes.

A **dendrite,** or **dendritic zone,** receives signals from receptors or other neurons and helps to integrate information. An **axon,** on the other hand, is a single, elongated extension of the cytoplasm that transmits impulses away from the dendritic zone.

The perikaryon can be at any position along the conduction pathway. In the motor neuron of figure 18.1, the dendrites are the short, branching processes of the perikaryon; the axon is the long single process. A neuron may have several dendrites, but only one axon. Some neurons, such as the *amacrine cells* of the retina, have no axon.

Fiber Construction

A covering of **Schwann cells** (neurolemmocytes) encloses the neural fibers of all peripheral neurons and extends the entire length of the axon. In the larger peripheral neurons, such as those seen in figure 18.1, the neurolemmocytes wrap around the axon in a unique manner to form a **myelin sheath** of lipoprotein. The outer portions of the neurolemmocytes that contain nuclei and cytoplasm make up the **neurolemma** (*neurolemmal sheath*).

Axons that have a thick myelin sheath with a neurolemma are *myelinated.* Neurons that lack a myelin sheath have *unmyelinated* fibers. An important function of neurolemmocytes is to help regenerate damaged fibers. When a fiber is damaged, regeneration occurs along the pathway the neurolemmocytes form.

The myelin sheath and neurolemma form during embryological development when neurolemmocytes fold around the axons. The axon of the motor neuron in figure 18.1 shows an enlarged cross section of the myelin sheath and neurolemma.

In figure 18.1, note that the axons are exposed at points between adjacent Schwann cells. These

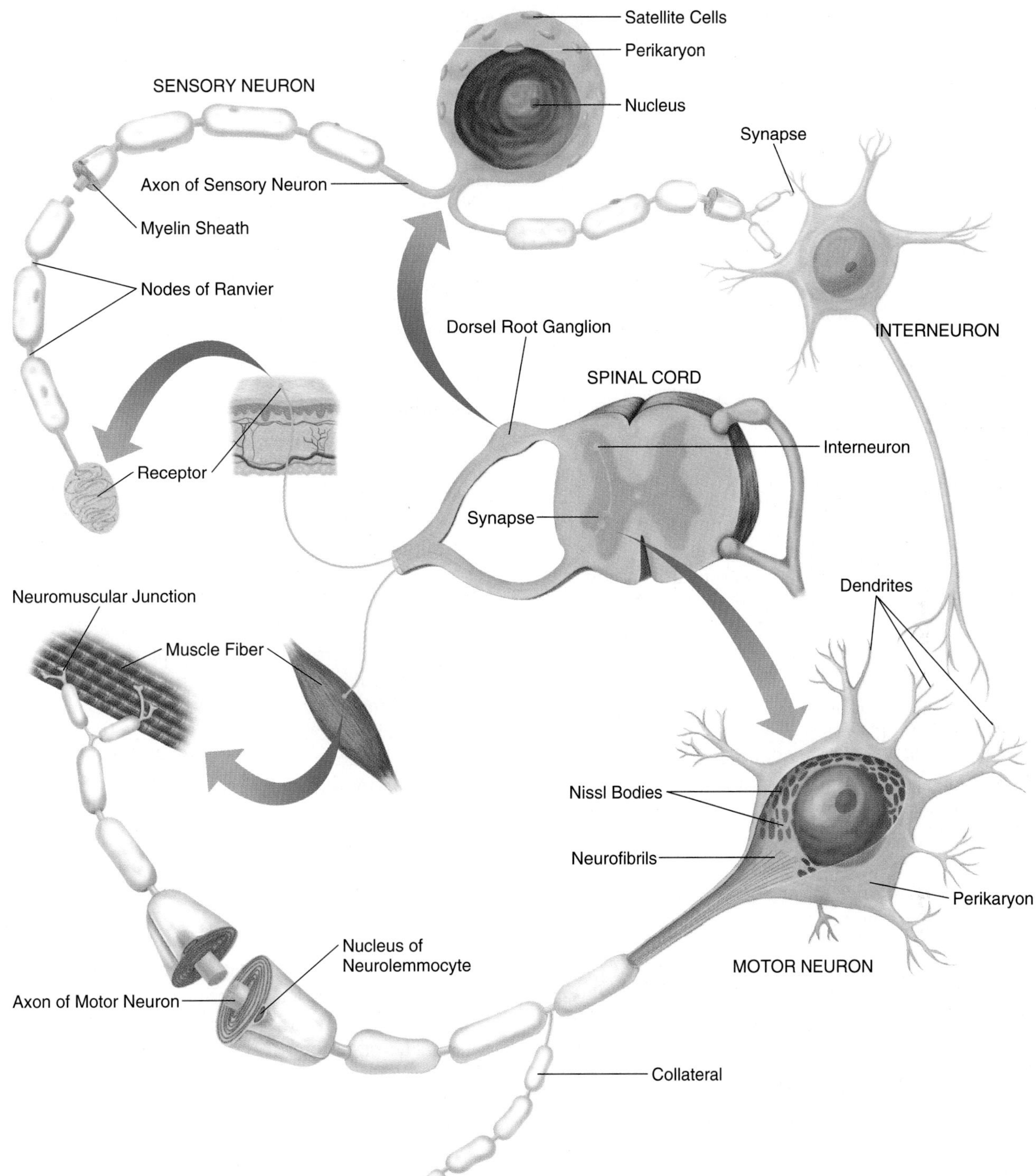

Figure 18.1 Sensory and motor neurons.

points are called **nodes of Ranvier.** Each internode along the length of an axon consists of a single neurolemmocyte, and each node of Ranvier represents a place where the exposed axon lacks the myelin and neurolemma. The nodes speed up nerve impulse transmission by *saltatory conduction,* a phenomenon studied in Exercise 44.

Types of Neurons

Neurons vary extensively with regard to size, shape, and the branching of axons and dendrites. Neurons may be multipolar like the motor neuron in figure 18.1, unipolar, like the sensory neuron in figure 18.1, or bipolar with two nerve fibers arising from the perikaryon (not pictured). Some have perikarya as small as 4 μm; others may have perikarya as large as 150 μm. The perikarya of some neurons are encapsulated with satellite cells; others are not encapsulated. Many have myelinated fibers; others do not. Detailed structural features of spinal cord and brain neurons follow.

Neurons of the Spinal Cord

The principal neurons of the spinal cord are motor neurons, sensory neurons, and interneurons. Each type has unique characteristics.

Motor Neurons As indicated in figure 18.1, motor neurons are multipolar, and their perikarya are in the gray matter of the central nervous system (CNS). They are also called *efferent neurons* because they carry nerve impulses away from the CNS. The axons of these neurons are myelinated and terminate in skeletal muscle fibers. Motor neurons help maintain voluntary control over skeletal muscles.

The perikarya of motor neurons are somewhat angular or star-shaped rather than oval. The cytoplasm in a properly stained perikaryon reveals a nucleus, many Nissl bodies, and neurofibrils. Many short, branching dendrites make synaptic connections with axons of interneurons or sensory neurons within the spinal cord. Figure HA-12C and D in the Histology Atlas shows motor neurons when observed with an oil immersion lens.

The myelinated axons of motor neurons often have **collaterals** emanating at right angles from a node of Ranvier. Collaterals are branches that innervate additional muscle fibers, other neurons, and even the same neuron in some instances.

Sensory Neurons Sensory neurons are also called *ganglion cells* because their perikarya are in dorsal root ganglia outside of the CNS (see the spinal cord diagram in figure 18.1). Because sensory neurons carry nerve impulses from a receptor toward the CNS, they are also often called *afferent neurons.*

As illustrated in figure 18.1, a capsule of **satellite cells** covers the ovoid-shaped perikaryon of a sensory neuron. Satellite cells have the same embryological origin as Schwann cells. Figure HA-12E of the Histology Atlas shows satellite cells in a sectioned sensory ganglion.

Although sensory neurons appear to be unipolar, histologists classify them as *pseudounipolar* on the basis of their embryological development. Note that the entire fiber is myelinated from the receptor in the skin to the synaptic connection (synapse) in the spinal cord. The entire myelinated fiber is called an **axon,** even though part of it functions like a dendrite in carrying nerve impulses toward the cell body. A **synapse** is the point where the axon terminal of one neuron contacts the dendritic zone of another neuron.

Interneurons Interneurons, also called *association neurons,* are multipolar unmyelinated neurons that exist in large numbers in the gray matter of the CNS. Essentially, they are the integrator cells between sensory and motor neurons. Most reflexes involve one or more interneurons between the sensory and motor neurons; however, some automatic responses, such as the knee-jerk reflex, are monosynaptic in that they lack interneurons.

Neurons of the Brain

Figure 18.2 shows four neuron types found in various parts of the brain. They are pyramidal, Purkinje, stellate, and basket cells.

Pyramidal Cells Pyramidal cells (figure 18.2A) are multipolar neurons and the principal cells of the cerebral cortex. Figure HA-13 of the Histology Atlas shows three photomicrographs of these cells that were made from three different slide preparations. The perikarya of these neurons are somewhat triangular, imparting a pyramid-like shape.

A unique characteristic of pyramidal cells is their two separate sets of dendrites: One long, trunk-like structure with many branches ascends vertically in the cortex, while several basal branching dendrites extend from the basal portion of the perikaryon. The smoother surface of the single axon distinguishes it from the basal dendrites.

Purkinje Cells Figure 18.2B shows a typical Purkinje cell. The perikarya of these multipolar neurons are located deep within the molecular layer of the cerebellum (see figure HA-14, Histology Atlas). The large, branching dendritic processes of these cells fill most of the space in the molecular layer of the cerebellum. The axons of these neurons extend into the granule cell layer of the cerebellum.

Stellate Cells Stellate cells are small neurons found in both the cerebrum and cerebellum, where they act as association cells for the pyramidal cells of the cerebrum and the Purkinje cells of the cerebellum. Both multipolar and bipolar types exist. See figure 18.2C and figures HA-13 and HA-14 (Histology Atlas).

Basket Cells Basket cells are multipolar neurons and are variants of large stellate cells located deep in the molecular layer of the cerebellum in close association with the Purkinje cells (see figures 18.2D and HA-14, Histology Atlas). Basket cells

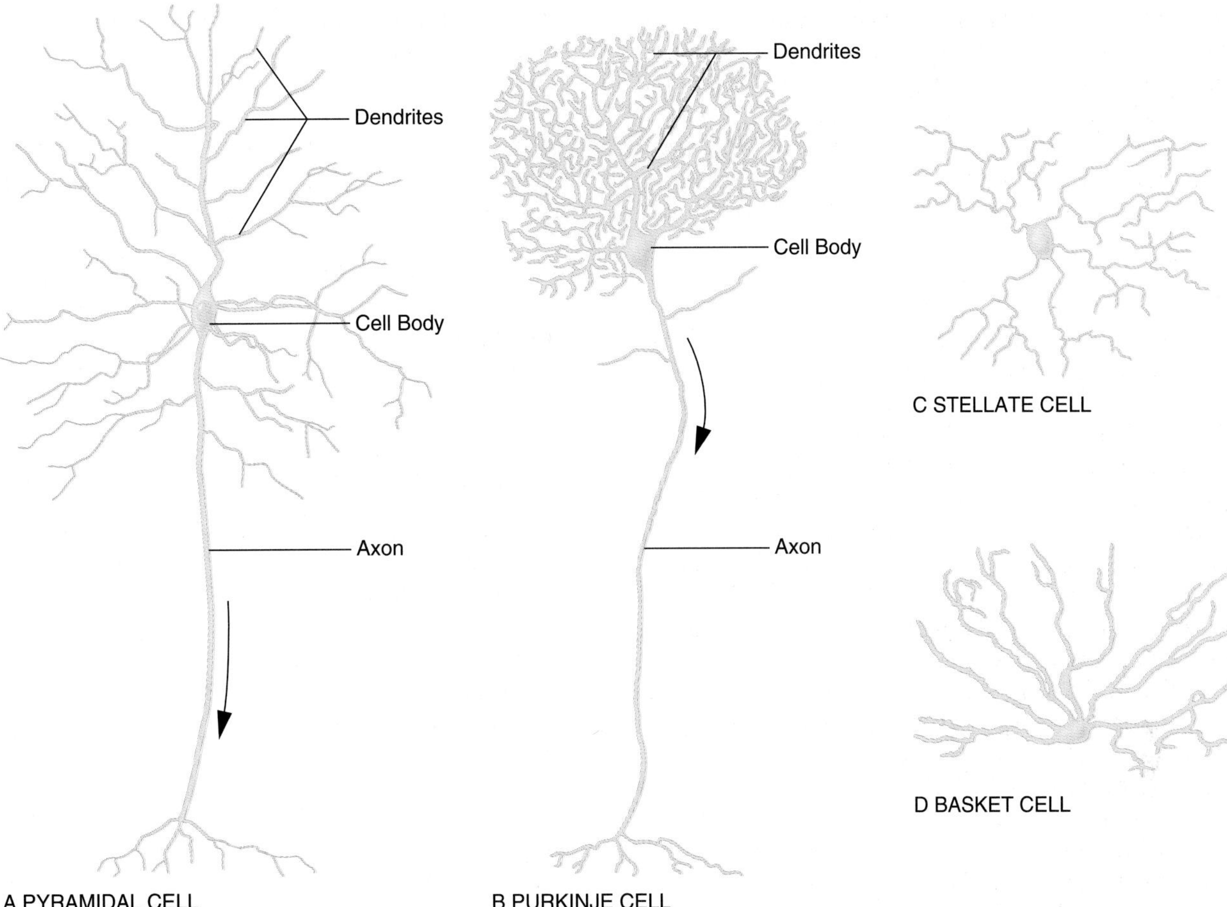

Figure 18.2 Neurons of the brain.

have short, thick, branching dendrites and a long axon. The axons of basket cells usually have five or six collaterals that make synaptic connections with dendrites of the Purkinje cells. Like the stellate cells, basket cells function as association neurons.

Neuroglia

Numerically, the neuroglia of the nervous system exceed neurons by a ratio of approximately 5:1. The term *neuroglia* (nerve glue) applies to astrocytes, microglia, ependymal cells, oligodendrites, Schwann cells, and satellite cells.

Schwann cells and satellite cells, described earlier, are found only in the peripheral nervous system and have a different embryonic origin than the other types of neuroglia. For these reasons, they are often called *peripheral neuroglia.*

Although the neuroglia of the CNS do not carry nerve impulses, they perform many important functions. During embryological development, they provide a framework in the CNS for young, developing neurons. As the nervous system develops, the neuroglia seem to regulate the formation of synaptic connections to control the development of functional reflex patterns in the CNS. They also appear to mediate the normal metabolism of neurons, and they are involved in the degeneration and regeneration of nerve fibers. Unfortunately, the neuroglia are also the chief source of tumors of the CNS.

The neuroglia of the CNS are divided into four distinct morphological categories: astrocytes, microglia, ependymal cells, and oligodendrocytes. Figure 18.3 shows representatives of each group.

Astrocytes Astrocytes have large cell bodies with numerous processes radiating in all directions. At the ends of some processes are "feet" that form connections with blood vessels, as shown in figure 18.3, or with the pia mater or neurons. Astrocytes help regulate which substances from the blood reach the neurons.

Microglia Microglia are much smaller than other types of neuroglia, and their cytoplasmic processes are relatively short. Microglia are modified macrophages

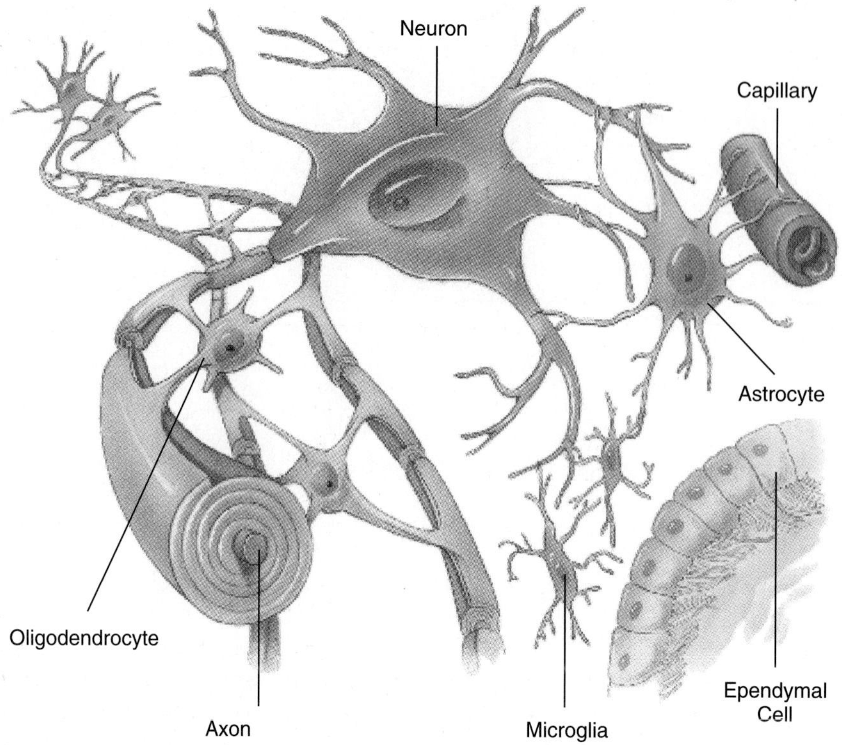

Figure 18.3 Neuroglial cells in the central nervous system.

scattered throughout the brain and spinal cord. They are CNS phagocytes, engulfing dead tissue, microorganisms, and foreign substances.

Ependymal Cells Ependymal cells line the ventricles of the brain and the central canal of the spinal cord. Some ependymal cells secrete the cerebrospinal fluid that fills the cavities of the CNS; other ependymal cells are covered with cilia. The beating of these cilia is responsible for the circulation of cerebrospinal fluid.

Oligodendrocytes Oligodendrocytes have much in common with Schwann cells, since they develop in rows along nerve fibers and form myelin sheaths within the CNS. Oligodendrocytes are large like astrocytes; together with astrocytes, they are often called *macroglia.* In tissue culture, oligodendrocytes pulsate rhythmically, a phenomenon that is not well understood.

Assignment:
Complete the Laboratory Report for this exercise.

Microscopic Studies

After answering the questions on the Laboratory Report about cells of the nervous system, examine prepared slides of the nervous system using oil immersion optics whenever necessary. The following procedure will involve most of the cells discussed in this exercise:

Materials:
prepared slides of:
x.s. spinal cord
spinal cord smear
spinal ganglion
cerebral cortex
cerebellum

Spinal Cord Examine the cross section of a rat spinal cord under low power, and identify the structures labeled in figure HA-12 (Histology Atlas). Study several of the large motor neurons in the gray matter with the high-dry objective. Also study a smear of spinal cord tissue, and compare the cells with the photomicrograph in figure HA-12 D.

Ganglion Cells Study a slide of a section through a spinal ganglion, and look for *satellite cells* that surround the sensory neurons. Refer to figure HA-12 E (Histology Atlas).

Cerebral Cortex Examine a slide of the cerebral cortex, and look for *pyramidal* and *stellate cells.* Refer to figure HA-13 (Histology Atlas). Try to differentiate dendrites from axons on the pyramidal cells.

Note the presence of spiny processes called *gemmules* on dendrite branches. These processes greatly increase dendrites' effective surface area, allowing large neurons to receive as many as 100,000 separate axon terminals or synapses.

Cerebellum Examine a slide of the cerebellum, and look for *Purkinje cells, stellate cells,* and *basket cells.* Refer to figure HA-14 (Histology Atlas).

Assignment:
Sketch representative cells of the various neuron types that you observed on the available slides.

19 The Muscle Cells

Excited muscle cells **contract** due to contractile protein fibers within the cells. This shortening of muscle cells is responsible for the various movements of body appendages and organs. Because of their elongated structure, muscle cells are often called **muscle fibers.**

On the basis of cytoplasmic differences, muscle fibers are categorized as striated or smooth. **Striated muscle** fibers have regularly spaced transverse bands, or striations. **Smooth muscle** fibers lack these transverse bands.

Striated muscle cells subdivide into skeletal and cardiac muscle cells. **Skeletal muscle** fibers are multinucleated, or syncytial. Attached primarily to the skeleton, they are responsible for voluntary body movements. **Cardiac muscle** cells, which make up the muscle of the heart wall, are not syncytial and have less prominent striations than skeletal muscle.

Follow the same procedure for this exercise as for Exercise 18—that is, read the text material and answer the questions on the Laboratory Report before examining the various kinds of muscle tissue slides in the laboratory.

Smooth Muscle

Smooth muscle fibers are long, spindle-shaped cells, as figure 19.1 shows. Each cell has a single, elongated nucleus with two or more nucleoli. When studied with electron microscopy, the cytoplasm reveals fine linear filaments of the same protein composition that exists in striated muscle. Because both smooth and striated fibers contain actin, myosin, troponin, and tropomyosin, the chemistry of contraction is probably similar or identical in all types of muscle fibers.

Smooth muscle cells are in the walls of blood vessels, the walls of organs of the digestive tract, the urinary bladder, and other internal organs. They are innervated by the autonomic nervous system and, thus, are under involuntary nervous control.

Skeletal Muscle

Skeletal muscle fibers are long, cylindrical, and multinucleated. Large numbers of these fibers grouped into bundles are called **fasciculi.** A thin layer of connective tissue called **endomysium** separates individual cells within each fascicle (see figure 19.1). This enables each cell to respond independently to nerve stimuli.

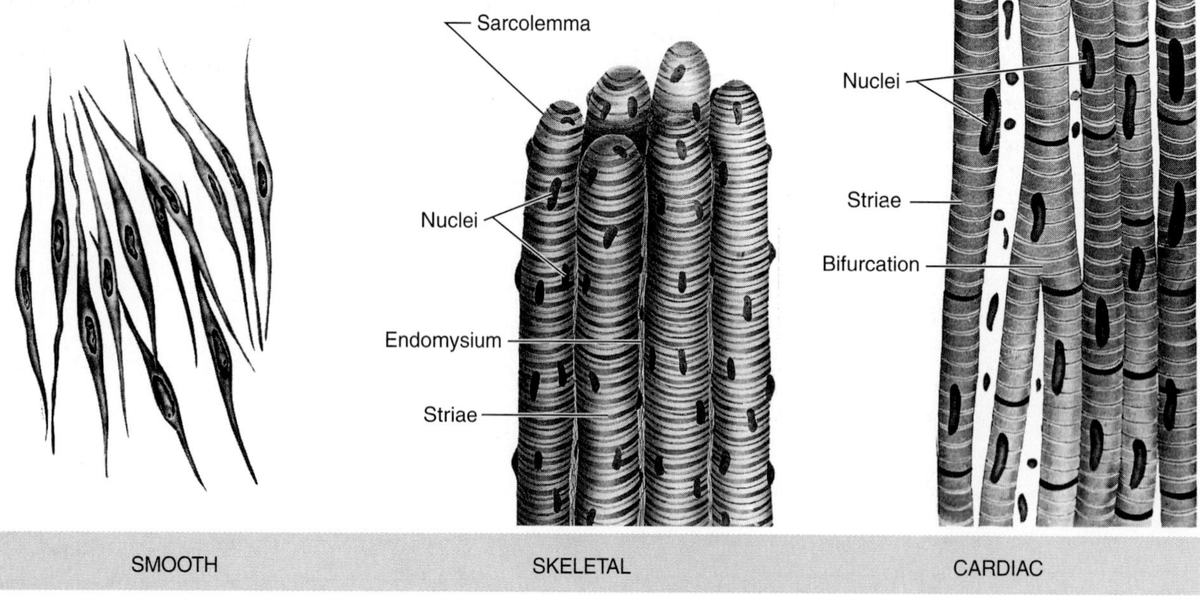

Figure 19.1 Smooth muscle fibers, teased apart. A fascicle of skeletal muscle fibers. Cardiac muscle.

Examination of the cytoplasm, or **sarcoplasm,** of a skeletal muscle fiber with a light microscope reveals a series of distinct dark and light striations. The detailed structure of these bands is explored in the chemistry of muscle contraction in Exercise 27. Surrounding each cell is a plasma membrane called the **sarcolemma.** Note in figure 19.1 that the nuclei of each cell lie directly below the sarcolemma.

Although striated muscle is called "skeletal" and "voluntary," certain striated muscles are not attached to bone or subject to voluntary control. For example, striated muscles of the tongue and external anal sphincter have no skeletal attachment, and striated muscles of the pharynx and upper esophagus have no skeletal attachment and are not voluntarily controlled. In spite of these differences, these examples are structurally indistinguishable from other skeletal muscle.

Cardiac Muscle

Cardiac muscle cells are in the wall of the heart and in the walls of the venae cavae where they enter the right atrium of the heart (see figure 19.1). A unique characteristic of cardiac muscle is its ability to contract rhythmically and continuously without external stimulation.

Morphologically, cardiac muscle is readily distinguished from skeletal and smooth fibers, yet it shares some characteristics of each type. The principal distinguishing features of cardiac muscle fibers are as follows:

- Cardiac muscle consists of separate cellular units that **intercalated disks** separate from each other. These disks stain somewhat darker than transverse striations.
- Some fibers of cardiac cells are bifurcated to form a branched, three-dimensional network instead of long, straight cylinders.
- The elongated single nucleus of each cardiac muscle cell lies deep within the cell instead of near the surface, as in skeletal muscle.
- Cardiac tissue is not normally subject to voluntary control.

Assignment:
Answer the questions on the Laboratory Report that pertain to the characteristics of the various types of muscle tissue. Then proceed to the Laboratory Assignment that follows.

Laboratory Assignment

Materials:
prepared slides of:
striated muscle, teased
striated muscle, x.s and l.s.
muscle to tendon
cardiac muscle, teased
cardiac muscle, l.s.
smooth muscle, teased
smooth muscle, sectioned

Skeletal Muscle Scan a slide of striated muscle tissue with low power to find a muscle fiber with readily visible striations. Increase the magnification to 1000× (oil immersion), and see if you can differentiate the A and I bands (refer to figure HA-7, Histology Atlas). Note the position of the nuclei in these fibers. Also examine a cross section of this tissue to identify the endomysium.

If a slide of muscle to tendon structure is available, examine it under high-dry magnification, and refer to figure HA-7C (Histology Atlas).

Cardiac Muscle Scan a slide of cardiac muscle with low power, looking for pronounced striations and intercalated disks. Increase the magnification to 1000×, and identify the structures labeled on figure HA-8A (Histology Atlas). Can you see bifurcation of the fibers on the slide?

Smooth Muscle Study slides of smooth muscle tissue in the same manner as for cardiac muscle tissue. Finding individual fibers, separated from others, is not always as easy as it might seem. Compare your observations to figures HA-8B and C (Histology Atlas).

Assignment:
Sketch representative muscle cells that you observed on the slides.

20 The Neuromuscular Junction

As noted in Exercise 18, motor neurons emerge from the spinal cord through the anterior roots of spinal nerves to innervate skeletal muscle fibers throughout the body. Each large, myelinated nerve fiber branches many times through collaterals to stimulate from three to two thousand skeletal muscle fibers. At the end of each branch is a **neuromuscular junction,** where the nerve fiber contacts the muscle fiber. This junction is usually at the approximate midpoint of the fiber so that the action potential generated in the fiber reaches the opposite ends of the fiber at about the same time. Normally, there is only one junction per muscle fiber.

Figure 20.1 shows the structure of a neuromuscular junction as determined by electron microscopy. Chemical studies over the past several decades have revealed a series of events that occur here. The present understanding of these events is somewhat different than what it was only a few years ago, and future research will surely reveal a slightly different picture as present questions are answered.

After you have answered the questions on the Laboratory Report, you will have the opportunity to study microscope slides of the neuromuscular junction. Since figure 20.1 is the result of electron microscopy, do not expect to see this degree of detail on your slides.

Note in figure 20.1 that the end of the neuron consists of a cluster of knoblike structures that lie within invaginations of the **sarcolemma** of the muscle fiber. These knoblike projections are **axon terminals.**

Between the axon terminal and sarcolemma is a gap, the **synaptic cleft,** approximately 20 μm wide and filled with a gelatinous ground substance. The **presynaptic membrane** and the **postsynaptic membrane** form the surfaces on each side of this cleft. Note that further infoldings to form **subneural clefts** (junctional folds) considerably increase the surface area of the postsynaptic membrane.

Whereas impulse transmission along the length of an axon is an electrical event, impulse transmission across the synaptic cleft is not. Instead, synaptic transmission involves the release of a chemical neurotransmitter that diffuses across the synaptic cleft and stimulates the postsynaptic membrane. The sequence of events involved in action potential transfer and muscle fiber contraction are as follows:

1. The neurolemma depolarizes as sodium ions flow into the neuron and potassium ions flow out. This ionic interchange produces an action potential in the nerve fiber.
2. When the action potential reaches the end of the neuron, the permeability of the neurolemma changes, and calcium ions flow inward.
3. **Synaptic vesicles** in the cytoplasm of the neuron diffuse toward the neurolemma of the axon terminal (presynaptic membrane) and release the neurotransmitter acetylcholine (ACh) into the synaptic cleft via exocytosis.
4. ACh molecules diffuse across the synaptic cleft and bind with receptors on the sarcolemma (postsynaptic membrane).
5. The binding of ACh to the receptors on the sarcolemma changes the permeability of the sarcolemma, thereby initiating an action potential.

Once the action potential is initiated in the sarcolemma of the muscle fiber, the following events occur within the fiber:

1. The action potential moves into the interior of the muscle fiber via the **T-tubule system.**
2. The **sarcoplasmic reticulum** releases calcium ions, which permit the muscle fiber's bundles of contractile filaments, or **myofibrils,** to contract.
3. The enzyme acetylcholinesterase inactivates ACh within 1/500 of a second of ACh binding with receptors on the sarcolemma.
4. Mitochondria in the area supply ATP for muscle fiber contraction.

The enzyme acetylcholinesterase in step 3 is found on the surfaces of the presynaptic and postsynaptic membranes, as well as in the ground substance of the synaptic cleft. It breaks ACh down into acetate and choline. Once the choline and acetate diffuse back into the axon terminal, the enzyme choline acetyltransferase synthesizes them back into ACh. Choline acetyltransferase is produced in the perikaryon of the neuron and travels down microtubules of the axon to the terminal, where it is stored in the cytoplasm.

The preceding description of synaptic transmission across a neuromuscular junction has many

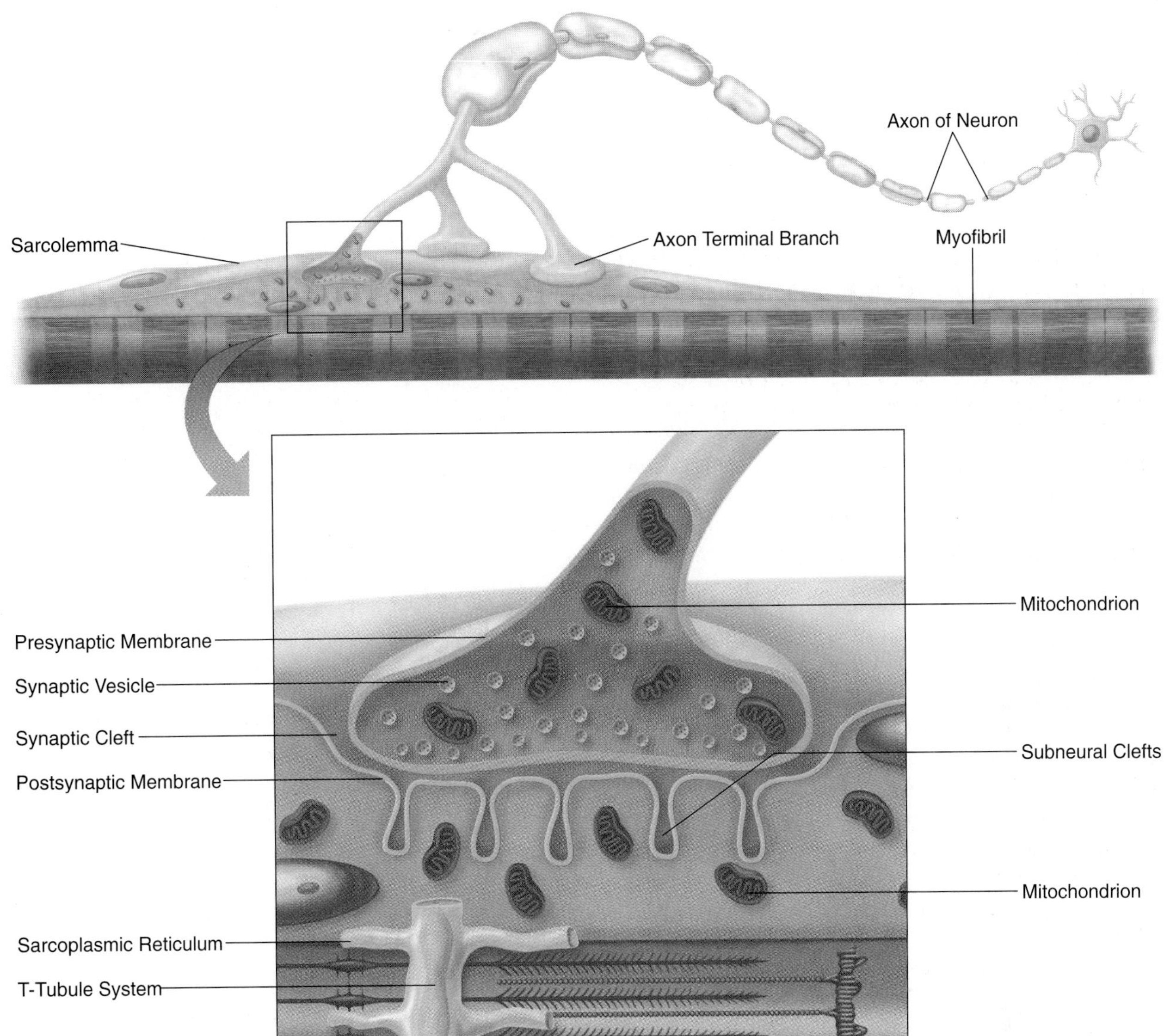

Figure 20.1 The neuromuscular junction.

points in common with synaptic transmission between two neurons and between a neuron and a secretory cell. However, keep in mind that acetylcholine is only one type of neurotransmitter. Other types of chemical synapses rely on different neurotransmitters to bridge the gap between presynaptic and postsynaptic membranes.

Assignment:
Complete the Laboratory Report for this exercise before proceeding to the microscopic study that follows.

Microscopic Study

After reviewing figure 20.1, examine a microscope slide of motor nerve endings under low-power and high-dry magnification. Consult figure HA-7D (Histology Atlas) for reference.

PART  Instrumentation

The study of physiological phenomena, such as muscle contraction, heart action, blood pressure, and cerebral activity, requires the use of electronic instruments, and you need to understand the nature and function of this type of equipment. This part introduces the principal types of equipment available. Complete the Laboratory Report for the exercises in this unit before performing any of the experiments in Part 6.

The illustration shows a typical equipment arrangement. First, a device, or combination of devices, picks up the biological signal being studied. This device will be an **electrode** if the signal is electrical, as in neural activity, or an **input transducer** if the signal is nonelectrical, as in temperature changes. Since the electrical signals that biological phenomena produce are often weak, the signals are then fed into an **amplifier.** Finally, an **output transducer** converts the amplified signals to some form of visual or audio display that the experimenter can observe and quantify. Output transducers can be electrical meters, oscilloscopes, loudspeakers, chart recorders, or computer monitors.

Each of the six exercises in Part 5 focuses on a particular type of equipment to be used in different experiments. You will not be able to assimilate all the information in these exercises in one reading. Study the material carefully, however, so that you can answer the questions on the Laboratory Report for Exercises 21–26. These questions are of a general nature; they do not pertain to all the specific minutiae of all of the equipment. Use the exercises in Part 5 as a reference when working with the various types of laboratory equipment.

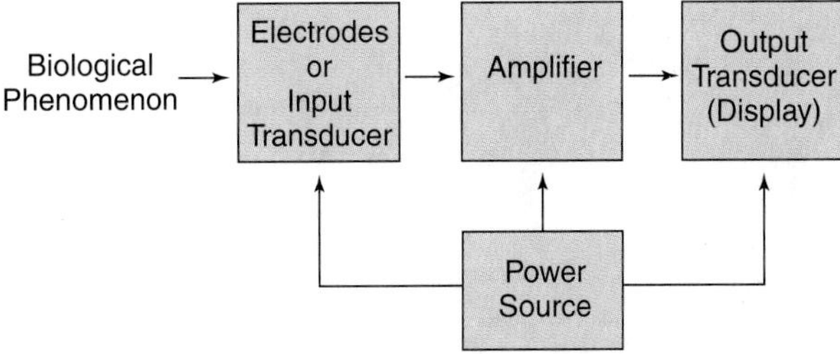

21 Electrodes and Input Transducers

Since electrodes and input transducers serve the same function in an experimental setup, we study them together here. **Electrodes** pick up electrical signals, and **input transducers** convert nonelectrical stimuli, such as heat, light, sound, or force, to an electrical signal that electronic equipment can manage. Another type of electrode, the **stimulating electrode,** is also studied here.

A review of some of the characteristics of electrical current is helpful at this point. Electricity flowing through a wire is the movement of free electrons from atom to atom of the conductor. The direction of movement is from the negative end of the wire, where there is a surplus of electrons, to the positive end. The negative terminal of a battery, called the cathode, supplies the electrons. The other terminal, which is positive, is the anode.

Electrical current was once believed to flow from the positive terminal of a power source to the negative terminal. However, scientists now know that electrons move from the negative terminal to the positive one—opposite to the conventional concept of current direction. Thus, the two directions must be distinguished by name. It is customary to speak of "electron flow" as from negative to positive, and of "current flow" as from positive to negative.

Another concept to keep in mind is the difference between direct current (DC) and alternating current (AC). With DC, the electrons flow in only one direction (fixed polarity). This type of electrical energy is preferred for stimulating muscle and nerve preparations. With AC, the electrons flow in one direction for a brief interval and then reverse their direction in the next interval. These direction reversals occur many times per second. The undulating nature of AC produces undesirable effects, such as heating and accommodation. A potential problem with DC current is that prolonged tissue stimulation can result in tissue damage due to hydrolysis. Exercise 25 on the electronic stimulator presents a more in-depth study of AC and DC currents.

Stimulating Electrodes

A stimulating electrode can initiate a nerve impulse in a nerve fiber or cause a muscle to contract. Figures 21.1 and 21.2 show two types of electrodes. In addition, there are "needle," "pin," and "sleeve" electrodes, which have different applications in nerve and muscle physiology experiments.

The type of electrode used depends on such factors as electrical conductivity, toxicity to tissues, size, chemical properties, ease of handling, and cost. Metals such as platinum, palladium, gold, and nickel-silver alloys are most frequently used because of their relative inertness and good conductivity. All stimulating electrodes have two contacts: a negative cathode and a positive anode. The wire ends of these electrodes usually have phone jack terminals that connect easily to an electronic stimulator.

Figure 21.1 This type of electrode stimulates muscles or nerves with electrical pulses to elicit desired responses.

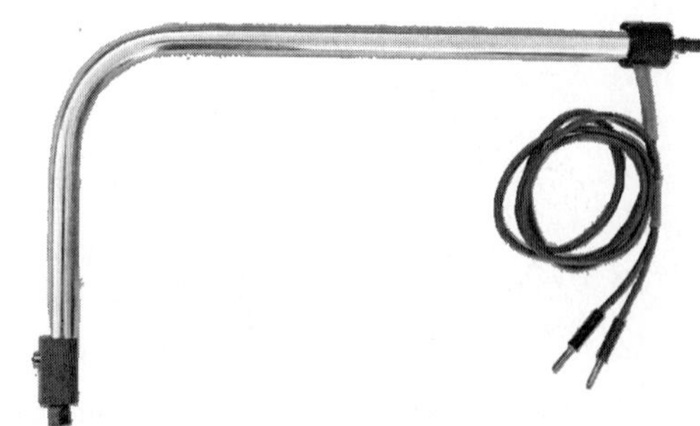

Figure 21.2 This type of electrode, called a Combelectrode, stimulates small portions of muscle tissue in certain types of experiments.

Pickup-Type Electrodes

Electrodes that pick up electrical signals to be amplified and displayed exist in many configurations. Figures 21.3 and 21.4 show their diversity. Note the small size of the electroencephalogram (EEG) disk electrode (figure 21.3) compared to the large, plate-type electrode for electrocardiograms (ECGs) (figure 21.4).

The primary problem with electrodes used for picking up electrical signals through the skin is achieving maximum skin contact. The cornified outer skin layer may have considerable resistance to low-voltage electrical transmission. Overcoming this resistance often requires using an abrasive material to rub off some of the outer dead skin cells and placing a high-conductance solution between the skin and electrode. Such a solution may be 0.05N saline or a commercially prepared paste or jelly.

Input Transducers

As previously stated, input transducers convert nonelectrical stimuli such as light, heat, pressure, sound, or movement to electrical signals that an electronic setup can amplify and study. Figures 21.5, 21.7, and 21.8 show examples of three transducers.

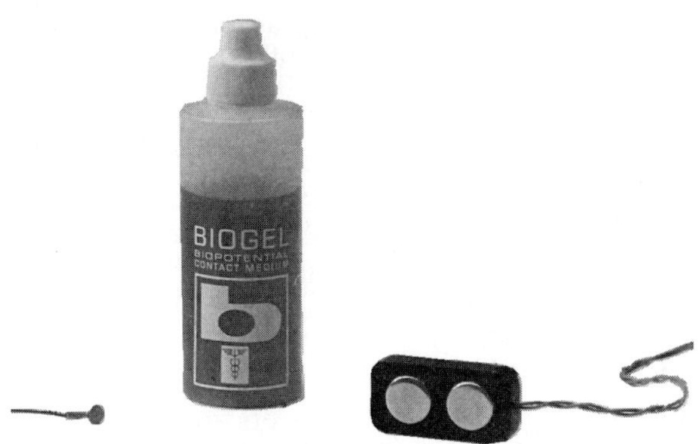

Figure 21.3 The small disk-type electrode on the left is for EEG measurements. The one on the right is a skin conduction electrode for measuring galvanic skin response. The electrode jelly in the middle reduces skin electrical resistance.

Figure 21.4 Two types of electrodes that can be used for ECG recording. The one on the left is a suction type that establishes good skin contact on the chest. The plate type on the right is held in place with rubber straps.

Force Transducers

Various kinds of electromechanical (force) transducers can convert muscular movements to electrical signals. The Trans-Med/Biocom Model 1030 force transducer shown in figure 21.5 is used with Gilson chart recorders, as figure 29.1 illustrates.

The Model 1030 is essentially a strain gauge consisting of five steel leaf springs. It can measure muscle contraction or similar forces from 10 mg to 10 kg, which gives it a sensitivity range of an incredible 1:1,000,000. Only the fixed single leaf is used to measure small forces. For loads approaching 10 kg, all five leaves are used. Obviously, the number of leaves used depends on the size of the load within this range.

This type of transducer functions through a series of resistors arranged as a Wheatstone bridge (see figure 21.6). A Wheatstone bridge has four resistors (R_1, R_2, R_3, and R_4) connected to an electrical source (E) and a galvanometer (G). Electrical current flows through two parallel circuits or half bridges ($R_1 + R_2$, and $R_3 + R_4$). If the voltage (potential) at A and B is equal, no current will flow through the galvanometer, and the bridge is balanced:

$$\frac{R_1}{R_2} = \frac{R_3}{R_4}$$

If all the resistors are constant except for R_1, which is variable, a change in R_1 upsets the balance and allows current to flow through the galvanometer.

Figure 21.5 A strain gauge force transducer, such as this one, utilizes a tubular silicon element that functions as the R_1 component in the Wheatstone bridge shown in figure 21.6. Bending changes its electrical resistance.

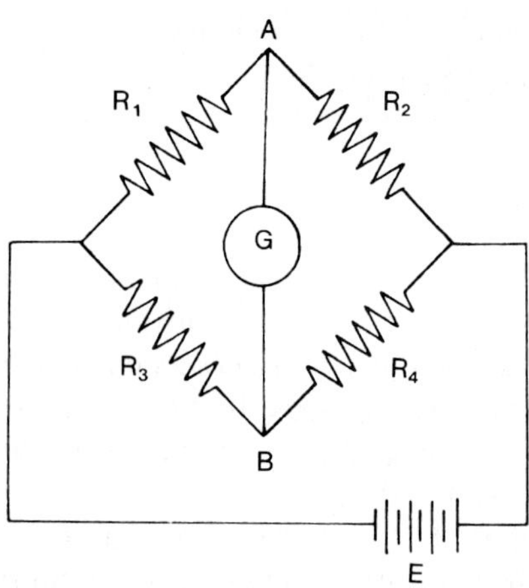

Figure 21.6 When this Wheatstone bridge is balanced, the electrical potential at points A and B are equal, and no current flows through the circuit. Bending in the leaves at R_1 causes unbalancing, producing an electrical signal.

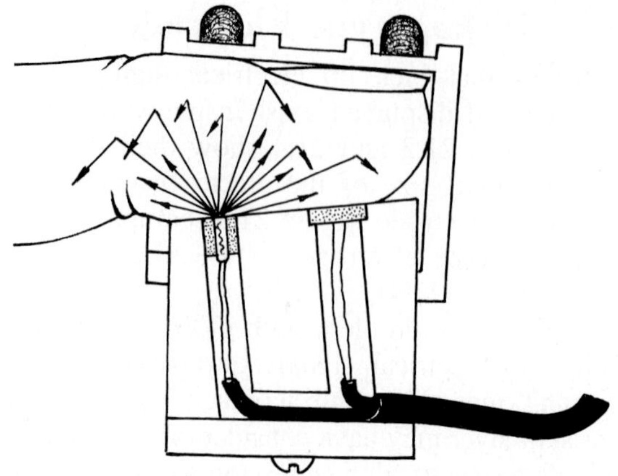

Figure 21.7 A pulse transducer uses light rays and a photoresistor to produce a signal.

Figure 21.8 This transducer can produce electrical signals from air pressure differentials.

In this type of strain gauge transducer, a silicon element acts as a variable resistor (R_1). It is a tubular component connected to the spring leaves. As the leaves are deformed due to loading, the silicon resistor bends proportionately. This induced bending increases the electrical resistance in the element, causing the Wheatstone bridge to become unbalanced. When the bridge is unbalanced, electrical current flows from the transducer to produce a signal. The signal is proportional to the applied stress. Other factors, such as temperature adjustments, also are considered in the design and manufacture of these transducers.

A strain gauge transducer usually must be balanced before use by making certain control adjustments on the output transducer (chart recorder). Appendix C outlines specific instructions for balancing.

Pulse Transducers

Several different kinds of transducers monitor the human pulse. Figure 21.7 illustrates, diagrammatically, how a photoelectric plethysmograph functions. A small light source produces a beam of light that is directed upward into the tissues of the finger. A few millimeters away from the light source is a photosensitive resistor (photoresistor). A surge of blood passing through the pad of the finger disrupts and scatters the light beam, and reflects rays to the photoresistor. Light striking the photoresistor produces a signal that is fed to the amplifier and output transducer.

In addition to determining the pulse rate, this setup can detect pulse pressure differences. This type of transducer lacks a Wheatstone bridge and thus does not require balancing.

Other Types of Transducers

Several other types of transducers are available for physiological work. For example, figure 21.8 shows a transducer that converts air pressure differentials into electrical signals. Other transducers can detect fluid pressure differentials. A transducer that monitors temperature changes is called a **thermistor.** Thermistors work because temperature changes alter the flow of electricity through a wire. The **microphone** is a transducer that converts sounds to electrical signals. You will encounter several of these various types of transducers in different experiments in this manual.

Assignment:
Answer the questions about electrodes in the Laboratory Report for Exercises 21–26.

Amplifiers, Panel Meters, and Oscilloscopes 22

This exercise pertains to three items—amplifiers, panel meters, and oscilloscopes—not used as independent components in any of the experiments in this book, yet you must understand them in the event you encounter them in your laboratory work. No attempt is made here to outline procedural protocol, however. If any of these components is used in an experiment, your instructor will provide supplemental instructions.

Amplifiers

Electronic amplifiers boost the weak electrical signals of biological systems many times (often a million or more) for monitoring. Although none of the experiments in this book use separate amplifier units, such as the ones shown in figure 22.1, all chart recorders have amplifier units incorporated into them.

Amplifiers usually function within a limited range. The input signal an amplifier receives must have sufficient intensity to overcome the "electrical noise" of the amplifier's electronic circuitry, yet not be so great that it overloads and distorts the amplifier output. An amplifier should increase the amplitude of a signal, not distort it.

Sensitivity Range of Amplifiers

Amplifiers vary in degree of sensitivity to match particular applications. While an electrocardiogram (ECG) amplifier has to be very sensitive to detect the weak electrical signals that cardiac activity generates, an amplifier that drives a chart pen need not be nearly as sensitive.

The sensitivity range of an amplifier determines what size of input signal the amplifier can accept. If the input signal is so weak that it does not exceed the inherent noise of the amplifier's circuitry, it is not within the amplifier's minimum range. On the other hand, the input signal must not be so great that it distorts amplifier output. The range of frequency in which an amplifier handles input without distortion is called the amplifier's linearity. Linearity information is important to the amplifier operator because it determines the fidelity of readout.

Power Supply

Amplifiers need direct current (DC) to function. Since their external source of electricity is usually conventional 110–120V alternating current (AC), they must have a power supply unit that converts AC to DC. In addition to making this conversion, the power supply unit either increases or decreases the voltage the amplifier requires. Since present-day amplifiers are completely transistorized, voltage requirements are quite low. This results in physically smaller power supply units that are usually incorporated into the amplifier unit.

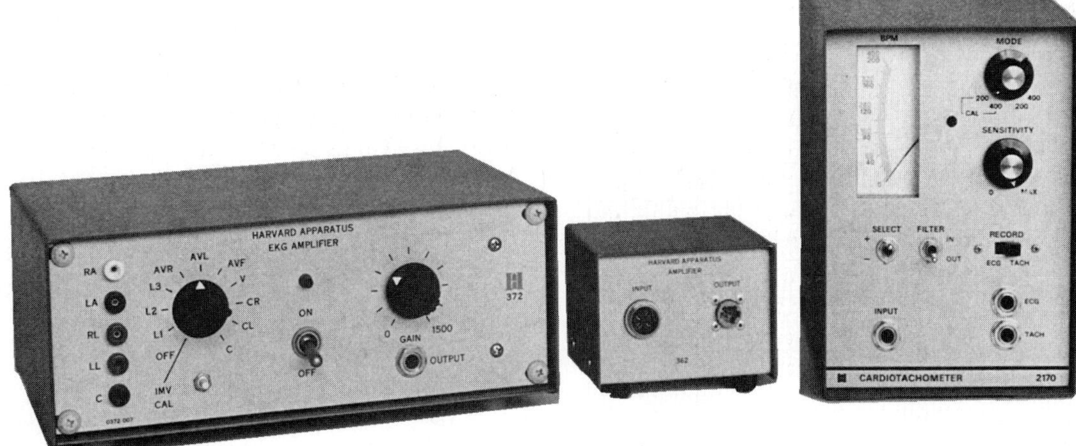

Figure 22.1 Amplifiers and amplifier-meter components.

Controls

An amplifier's function determines the types of controls needed. If the incoming signal is constant, the amplifier may be a very simple unit without controls, such as the central unit in figure 22.1. Most amplifiers, however, have mode and gain controls. If an amplifier is used in conjunction with a chart recorder, it may also have zero offset (centering) and calibration controls.

Mode A mode switch allows an amplifier to accommodate either AC or DC input signals. When the mode switch is set at the AC position, a capacitor in the circuitry prevents DC signals from entering the amplifier; only AC signals pass through. In the DC position, both alternating and direct current signals are amplified. A ground position on a mode switch grounds the input signal during connection adjustments at the input terminals. Grounding the amplifier input prevents severe deflection of recording instruments during these manipulations.

Gain A gain control adjusts the sensitivity of the amplifier to the strength of the incoming signal. If the input signal is strong, the gain control is turned to a low value. Weak signals require higher values. When the intensity, or level, of the input signal is unknown, setting the gain control at the least sensitive position is best. The final sensitivity setting is determined by advancing the control until a usable output signal is obtained.

Zero Offset A zero offset or centering control moves the zero line on the chart in either a positive or negative direction. This control permits the expansion of phenomena that are predominantly all positive or all negative by moving the pen away from the limits of its travel (see figure 22.2).

Electrical Interference

Electrical currents flowing through power cables, electrical appliances, and lighting systems generate magnetic and electrostatic fields that can cause unwanted interference in amplifier units. To prevent this type of electrical interference in electronic systems, all interconnecting cables are shielded. This shielding usually consists of braided metallic coverings between the outer and inner insulation coatings. These are grounded to carry away unwanted currents or voltages. Additional methods for minimizing electrical interference are outlined in various experiments where the problem becomes more acute.

Panel Meters

Electrical meters that monitor signals are output transducers based on the principle of the galvanometer (see figure 22.3). Such a meter is essentially a coil of wire mounted on a pair of pivots within the field of a permanent magnet. A current passing through the coil generates a magnetic field. Since like poles repel and unlike poles attract, the coil deflects against a spring in proportion to the current flowing through it. A pointer needle attached to the moving coil records the degree of angular displacement on a calibrated dial card. The meter on the cardiotachometer (right-hand unit, figure 22.1) is based on this principle.

This type of meter is readily converted to an ammeter or voltmeter. In an **ammeter,** which measures currents in amperes, an added electrical bypass, or shunt, diminishes the amount of current that passes through the coil. In the **voltmeter,** which measures potential differences in volts, a resistor is added in series with the coil to allow only a small current to pass through the coil.

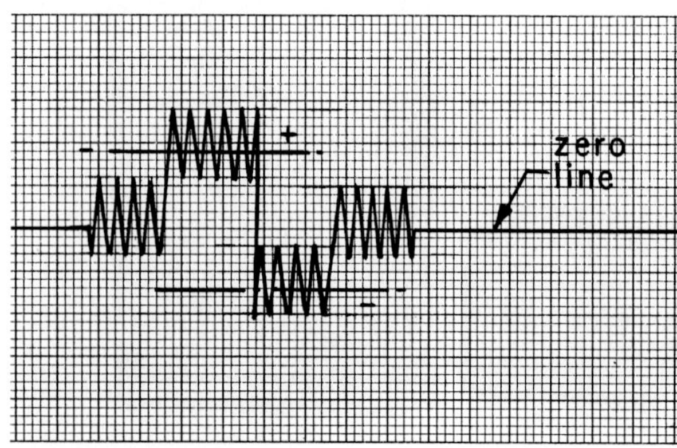

Figure 22.2 Using zero offset or centering control shifts the position of the tracing.

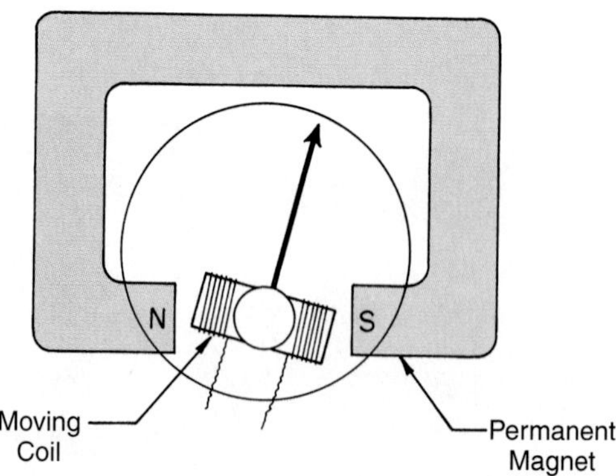

Figure 22.3 Current flowing through the magnet of the coil deflects the needle.

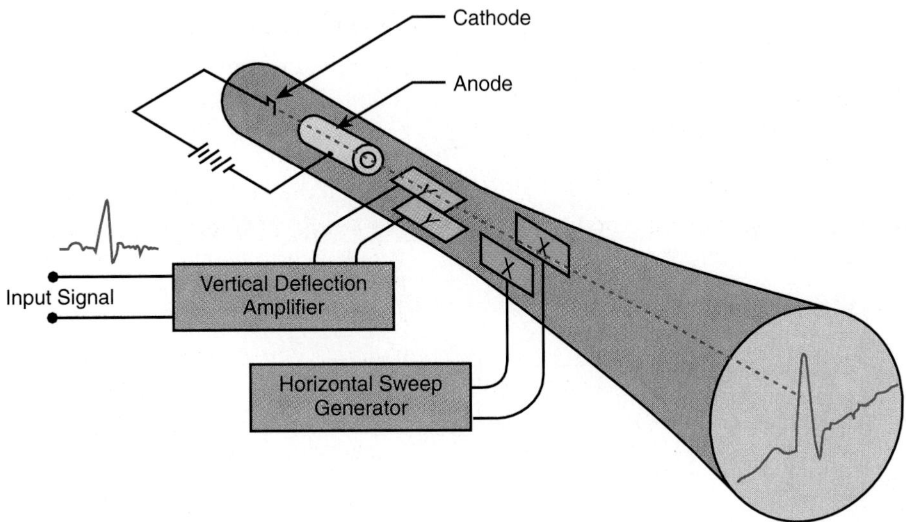

Figure 22.4 Schematic of an oscilloscope.

The Oscilloscope

The cathode ray oscilloscope is an output transducer that visually displays voltage signals on a fluorescent screen. The schematic diagram of an oscilloscope in figure 22.4 shows how an oscilloscope works. Figure 22.5 shows a Tektronix dual-beam oscilloscope.

The electrical potential difference (voltage) of the input signal is expressed vertically, and the time base is shown horizontally on the screen. This is achieved by an electron gun (cathode) at the back end of the tube generating a stream of electrons. The electrons pass through a hole in an anode on their way to the fluorescent screen, where they produce a tiny, luminous spot.

The electrostatic influence of two vertical (X) plates and two horizontal (Y) plates that lie alongside the beam pathway determines the position of the spot on the screen. The Y plates control vertical movement of the beam (and spot). These plates are energized by the input signal and represent the **voltage** of the signal. The X plates control the horizontal movement, or **sweep,** of the beam to establish the time base.

Vertical deflection of the beam of electrons depends on the electrical charge on the Y plates. Since the signal potential from the amplifier is fed to these plates, one of the plates becomes positive and the other negative. The electron beam, being negative, is attracted to the positively charged plates. When the positive charge reverses to the opposing plate, the beam is attracted to the other plate. The result is that the luminous spot moves up and down on the surface of the screen. The speed at which this occurs, in combination with the influence of the sweep circuit, produces a visual image of the characteristics of the biological phenomenon being studied.

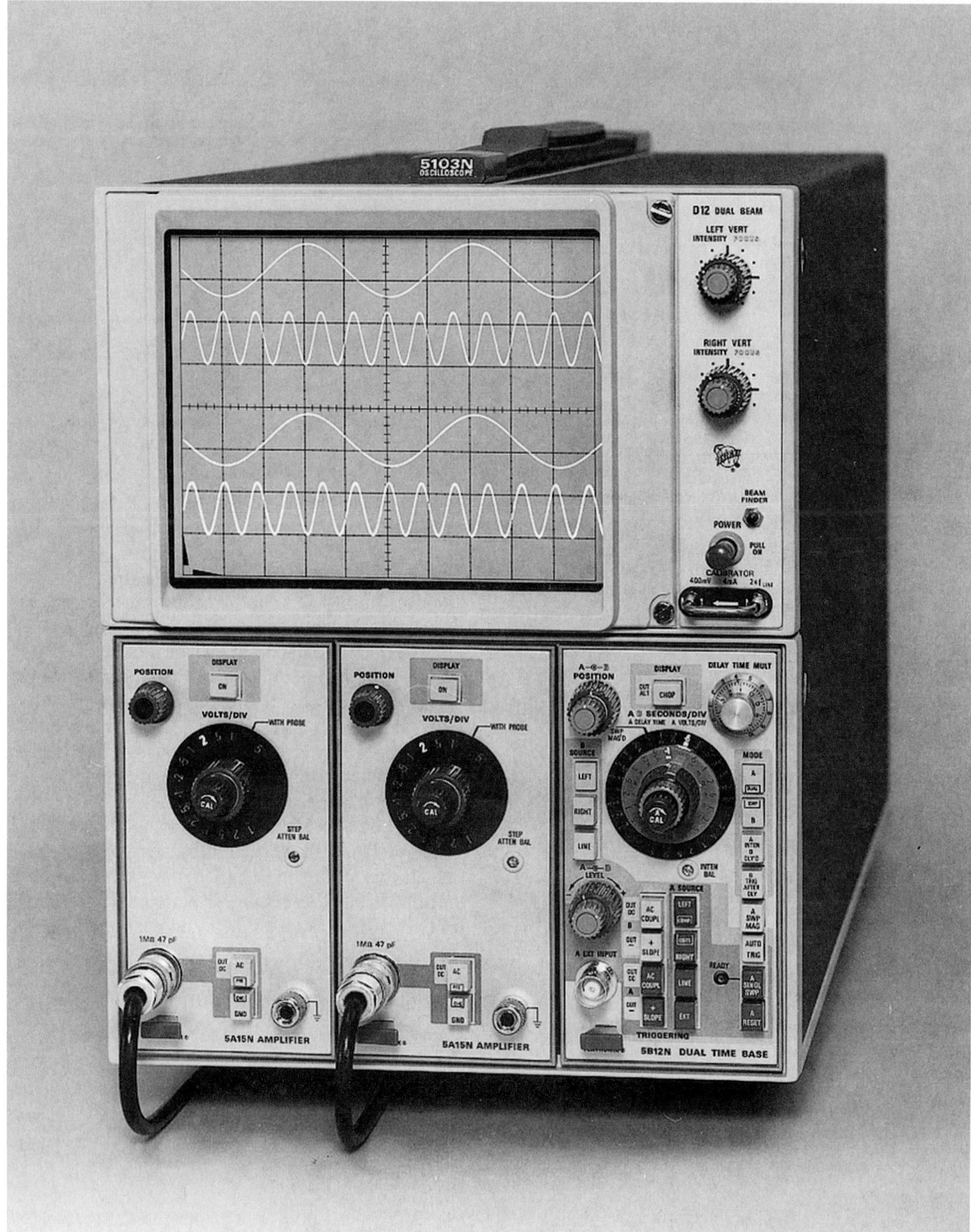

Figure 22.5 A duel-beam oscilloscope manufactured by Tektronix.

Gilson Chart Recorders

23

Of all the output transducers available in the physiology laboratory, the chart recorder is by far the one most widely used. Gilson Medical Electronics of Middleton, Wisconsin, manufactures and distributes three recorders described here. They are the Unigraph, Duograph, and 5\6H polygraph.

The Unigraph

Most single-event physiological experiments performed in this manual utilize the Gilson Unigraph (see figure 23.1). This unit is compact, rugged, and versatile, and learning how to use it is relatively easy.

The Unigraph can be used for studying skeletal muscle contraction, electrocardiograms (ECGs), electroencephalograms (EEGs), blood pressure variations, and many other activities. A heated stylus records the physiological event on calibrated, heat-sensitive paper. The paper moves at either 2.5 mm or 25 mm per second.

Controls

Before attempting to use the Unigraph, you should be familiar with the following controls:

Mode Selection Control The mode selection control knob can be set at one of the six following settings: EEG, ECG, CC–CAL, DC–CAL, DC, and TRANS. Refer to figure 23.2 to locate this knob.

The ECG setting is used to make an electrocardiogram. The EEG setting is selected for an electroencephalogram. The TRANS setting is chosen when a transducer is used. The CAL settings are used when calibrating a recording is necessary, as it is for ECGs.

Styluses The Unigraph has two styluses: a **recording stylus** and an **event marker stylus.** Figure 23.1 shows where the styluses are located, and figure 23.2 illustrates them more clearly.

Since stylus temperature is critical to produce a desired "trace" on the chart, a **stylus heat control** regulates it. This control is usually set at the two

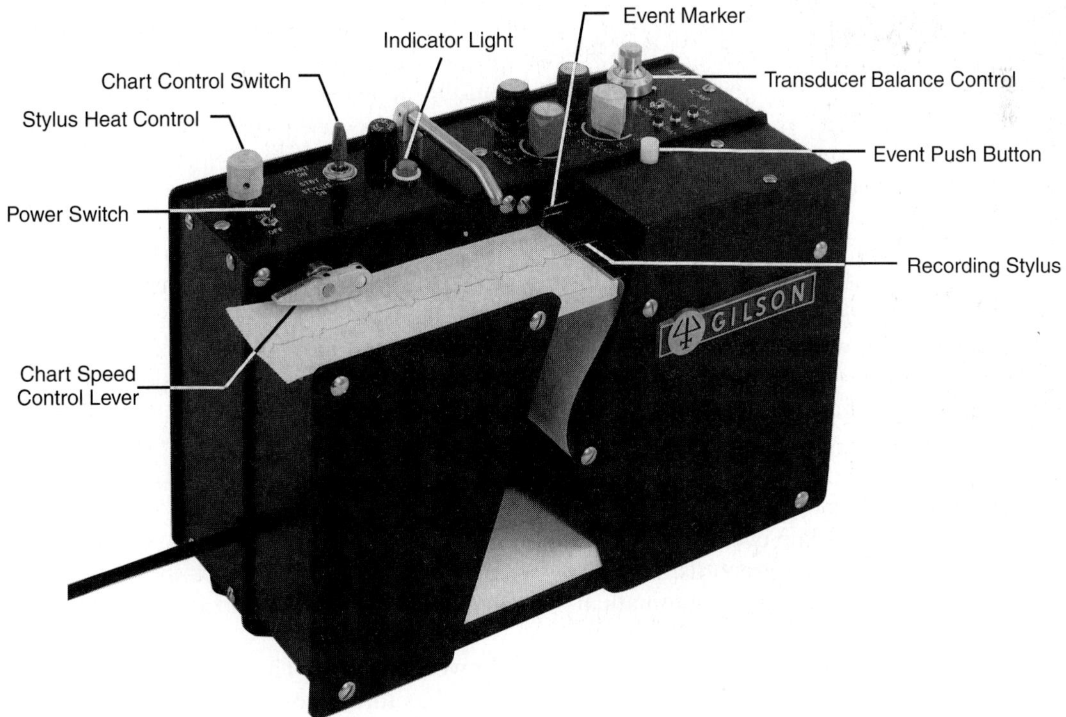

Figure 23.1 The Gilson Unigraph.

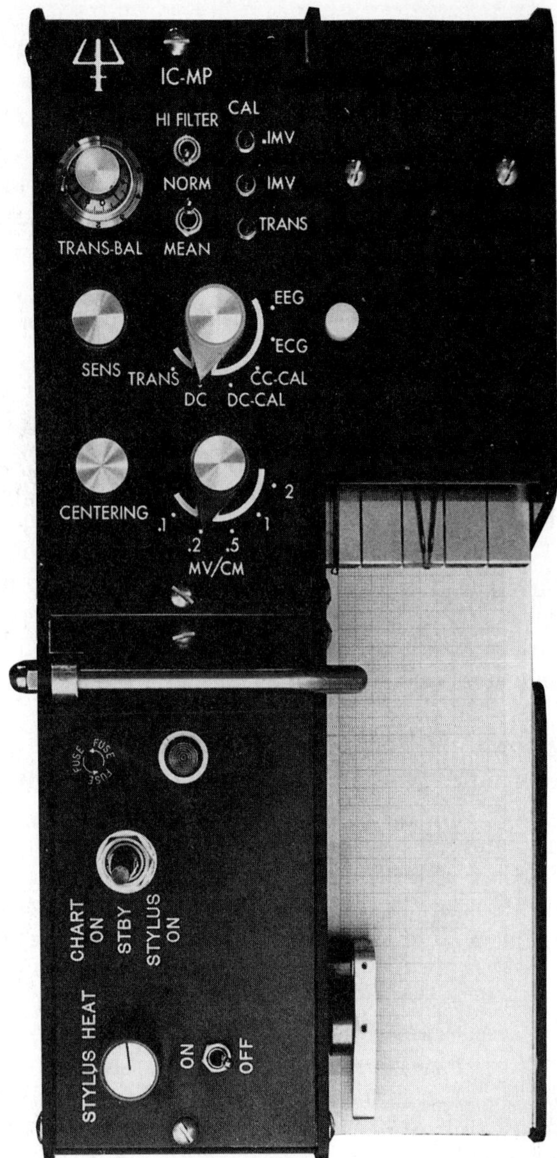

Figure 23.2 Unigraph controls.

o'clock position at the start of a recording. If the recording stylus is not warm enough, the tracing will be too faint. Turning the knob clockwise increases the temperature; counterclockwise movement reduces the temperature.

While the recording stylus records the amplified input signal, the event marker records when an event begins or when some new pertinent influence affects the event. Depressing the event push button (see figure 23.1) at the beginning records the event manually, or the event can be recorded automatically if an event synchronization cable connects the electronic stimulator and the Unigraph.

Centering Control (Zero Offset Control) The centering control, on the left side of figure 23.2, moves the baseline of the recording stylus to the most desirable position, which is not always the center of the paper. The most desirable position may be near one of the edges.

Chart Speed The **speed control lever** (see figure 23.1) controls the speed at which the paper moves. When the lever points to the left, as in figure 23.1, the paper moves at the faster speed of 25 mm per second. When the lever is positioned as shown in figure 23.2 (rotated backward 180°), the paper moves at the slower speed of 2.5 mm per second. For the chart to move at all, however, the chart control switch has to be in the Chart On position.

Toggle Switches The Unigraph has four toggle switches, two of which are extensively used in its operation. The main **power switch** (see figure 23.1) has two settings: On and Off. The **chart control switch** (see figure 23.1) has three settings: STBY (Standby), Chart On, and Stylus On (see figure 23.2). In the Stylus On position, the stylus heats up, but the chart does not move. At the Chart On position, the chart moves, and the recording stylus is activated (heated).

The other two small toggle switches at the other end of the unit come into play only occasionally. They should be set at the NORM positions for most experiments.

Sensitivity Controls The gain control and the sensitivity knob control Unigraph sensitivity. The **gain control** is calibrated in mV/cm (millivolts per centimeter) and has five settings: 2, 1, 0.5, 0.2, and 0.1. Look for this control in figure 23.2; note that it is adjacent to the centering control. The **sensitivity knob,** adjacent to the mode selection control, increases the sensitivity when it is turned clockwise.

Unigraph Operational Procedures

When setting up the Unigraph in an experiment, keep the following points in mind:

Input Insert the cable end of electrodes or the transducer into an input receptacle in the end of the Unigraph. The cable end must be compatible with the Unigraph. Some experiments will require using an adapter unit between the cable end and the Unigraph.

Synchronization Cable If you are using an electronic stimulator, connect the stimulator and the Unigraph with an event synchronization cable.

Power Source The end of the Unigraph power cord has three prongs, which indicates that a 110V grounded outlet is required. Before plugging in the Unigraph, however, turn the power switch to Off and the chart control switch to STBY.

Figure 23.3 The Duograph.

Stylus Heat Once the unit is plugged in, turn the power to On, and set the stylus heat control at the two o'clock position.

After allowing a few minutes for the stylus to heat up, test the trace by setting the chart control switch to the Chart On mode. If the trace line on the chart is too light, increase the darkness by turning the stylus heat control knob clockwise. Do this testing with the chart speed set at the slow rate.

Balancing If you are using a force transducer, balance the Wheatstone bridge in the circuitry, using the procedure outlined in Appendix C.

Calibration When monitoring EEGs, ECGs, or EMGs (electromyograms), follow calibration instructions provided in Appendix C.

Mode Control Before running the experiment, select the proper mode.

Sensitivity Control Start all experiments at the lowest sensitivity setting, unless instructed otherwise. Increase the sensitivity level gradually to arrive at the best level.

End of Experiment When finished with the experiment, always return all controls to their least sensitive settings, or to the Off position. Leave the chart control switch at STBY.

The Duograph

The Gilson Duograph is a two-channel recorder that consists of two multipurpose amplifier modules (see figure 23.3). The modules (IC-MPs) are identical to the unit used in the Unigraph. The Duograph can monitor the interaction of two parameters simultaneously. Although heavier than the Unigraph (25 versus 18 pounds), the Duograph is still extremely portable and can be carried with one hand. The chart is 127 mm wide and has two speeds (2.5 and 25 mm/sec) on the standard unit. An option of 10 speeds is also available. Since Unigraph and Duograph controls are essentially the same, learning to use one recorder makes it easy to use the other.

The Gilson Polygraph

Polygraphs are recorders that can monitor three or more events simultaneously. Gilson's polygraph is the 5\6H (see figure 23.4). To facilitate accessibility, the five amplifier modules are mounted on a cantilevered unit above the chart. Each amplifier module can be easily removed and replaced. This feature greatly enhances the recorder's versatility.

The chart on the 5\6H is 300 mm wide and can move at 10 different speeds: 0.1, 0.25, 0.5, 1.0, 2.5, 5, 10, 25, 50, and 100 mm per second. Five heated styluses provide individual tracings, and each one can be deactivated with a stylus lifter. A sixth stylus is also available if a servo unit is installed in the sixth space on the control panel. The stylus of a servo unit traces a pattern that overlaps the tracing of the other styluses. The Gilson polygraph also has a time marker that produces marks on the chart at 1- and 10-second intervals.

Types of Amplifier Modules

Six galvo-type amplifier modules and three servo modules are available for the Gilson 5\6H polygraph. Each **galvo module** has a rapid-response stylus that has a 2-inch excursion on the chart. These units occupy the five spaces shown in figure 23.4 on the cantilevered control panel. The **servo channels** have a stylus that moves slowly over the full width of the chart. Only one servo unit can be used on the Gilson polygraph, and it is mounted on the right-hand sixth space of the control panel. No servo unit is shown in figure 23.4. A brief description of most of these modules follows.

IC-MP Multipurpose Module The IC-MP is the same amplifier module that the Unigraph and Duograph use. It can monitor ECGs, EEGs, EMGs, strain gauge measurements, temperature changes, respiration, pulse, blood pressure, etc.

IC-UM Multipurpose Module The IC-UM is similar in application to the IC-MP but has a greatly increased sensitivity range. This module is a much more sophisticated unit than the IC-MP and a finer research tool, particularly for recording EEGs.

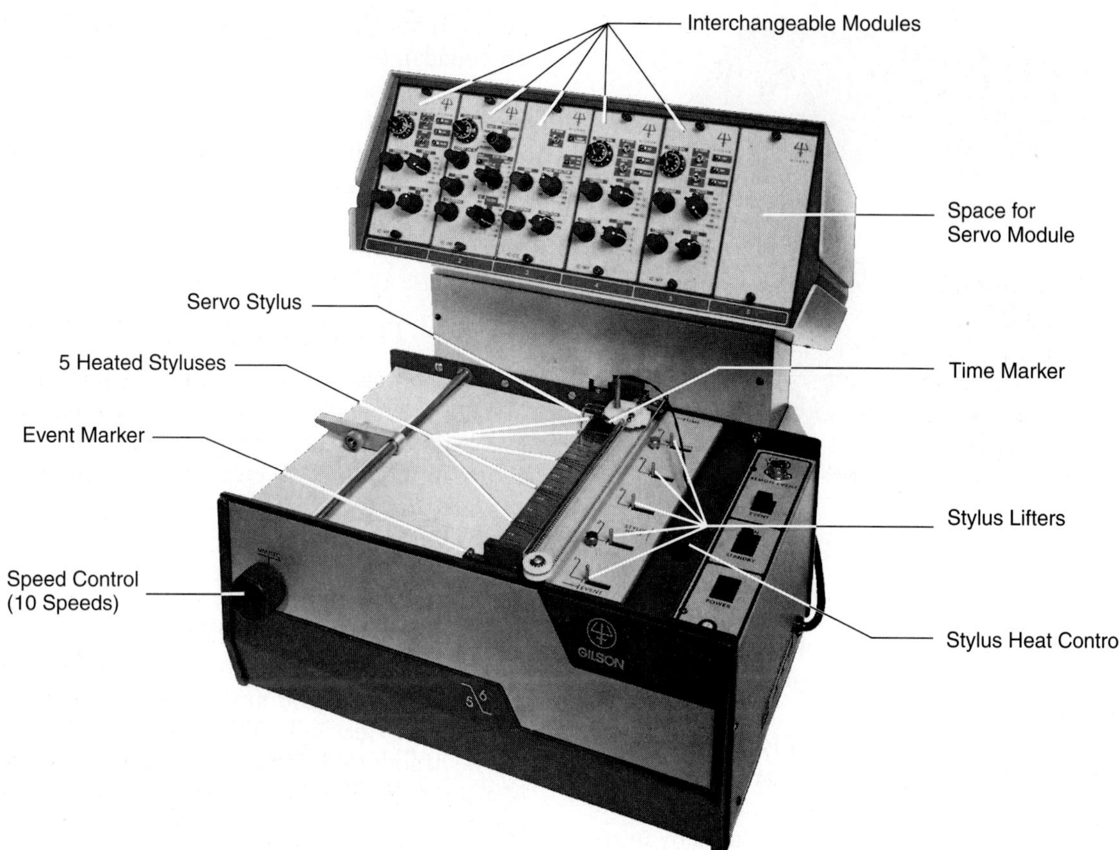

Figure 23.4 The Gilson polygraph.

IC-CC Electrocardiograph Module The IC-CC is intended primarily for EEGs and ECGs. It has a conventional 10-position lead switch that allows monitoring of ECGs from a standard five-lead patient cable.

IC-CT Cardiotachometer Module The IC-CT is an amplifier module that monitors the heart or pulse rate with a photoelectric finger pulse pickup like the one in figure 21.7.

IC-EMG Electromyograph Module The IC-EMG is designed specifically to monitor bioelectrical signals that muscle activity produces. It is used with a three-lead patient cable. Two red leads attach to electrodes on the belly of the muscle, and the black lead attaches to an electrode that acts as a ground. The electrode on the ground lead attaches to a point some distance away from the two red leads.

IC-S DC Servo Channel The IC-S DC servo module records phenomena that change relatively slowly, such as temperature (via a thermistor probe), skin resistance, and other low-frequency DC data. It can also be used with pH, CO_2, and O_2 analyzers. Stylus excursion is 200 mm.

BG-5 Blood Gas Servo Channel The BG-5 records pCO_2, pO_2, pH, or H_2 directly from conventional electrodes that require very small samples. Stylus excursion is 200 mm.

Operation Setting up a multichannel experiment on the polygraph requires attaching the electrodes or transducers to the subject and connecting the cables to the proper modules. Certain types of measurements require calibration. Before you perform any polygraph experiments, however, you should have considerable experience working with the Unigraph. The calibration procedures for the Unigraph are the same as for the IC-MP module on the polygraph.

Assignment:
Answer the questions that pertain to this exercise in the Laboratory Report for Exercises 21–26.

The Narco Physiograph

24

If your laboratory is equipped with Narco Physiographs instead of Gilson recorders, focus on this exercise instead of Exercise 23. The Narco Physiograph, manufactured by Narco Bio-Systems of Houston, Texas, is a four-channel polygraph with five ink-fed pens. Like the Gilson polygraph, it has various interchangeable modules.

Components

Refer to figure 24.1 to identify the following parts of the Narco Physiograph:

Input Couplers

On the upper portion of the sloping control panel of the Narco Physiograph are four units called input couplers (see figure 24.1). The transducers or electrodes feed the input signals to be monitored into these input couplers. Couplers are, essentially, preamplifier units that modify the signals before the signals enter the amplifier units.

Narco manufactures various types of couplers. The universal coupler is probably the most widely used. It is equivalent to the Gilson IC-MP module used in the Unigraph, Duograph, and polygraph. It can monitor electrocardiograms (ECGs), electroencephalograms (EEGs), electromyograms (EMGs), and other DC and AC potentials. Other couplers available are the strain gauge coupler, GSR coupler, transducer coupler, and temperature coupler.

Amplifier Controls

Note in figure 24.1 that below each input coupler is its individual cluster of amplifier controls, consisting of two knobs and one push button. The knob on the left has 10 settings for adjusting the channel's sensitivity. Its outer portion is for selecting the desired millivolt setting (1, 2, 5, 10, 20, 50, 100, 200, 500, or 1000 mV). At setting 1, it takes only 1 mV to produce 1 cm of pen movement; when set at 1000, it takes 1000 mV to produce the same amount of pen travel. Thus, the 1 mV setting is the most sensitive setting available here.

Adjusting the sensitivity between any of these settings requires rotating the inner portion of this control. When the inner knob is completely clock-

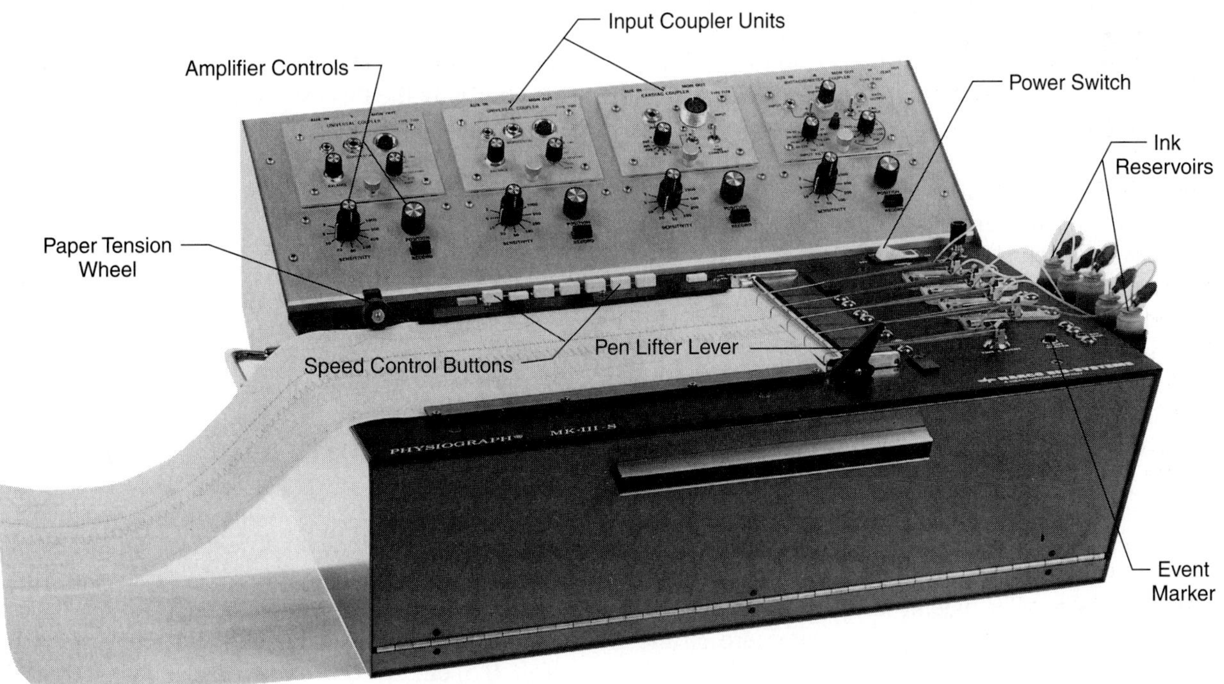

Figure 24.1 The Narco Physiograph Mark IIIS.

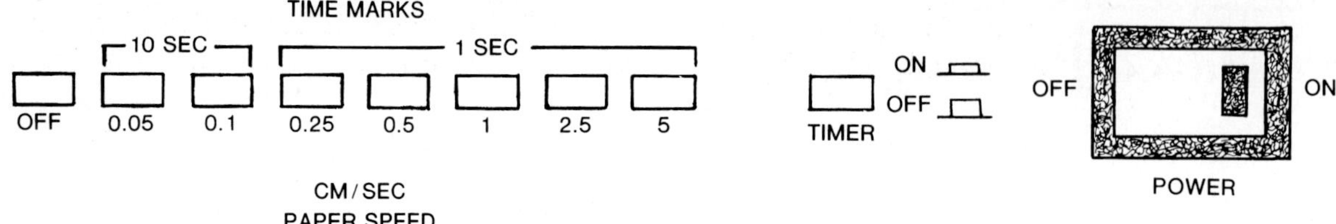

Figure 24.2 Push-button control panel of the Physiograph Mark IIIS.

wise, the sensitivity is identical to the reading on the dial. Counterclockwise rotation of the inner knob decreases the sensitivity.

The knob on the right (with "Position" under it) repositions the pen to locate the baseline of the recording in the most desirable spot. The degree of pen travel determines the location of the baseline. For example, if the pen deflection is to be as much as 6 cm, the baseline should be set substantially below the center of the arc of the pen. The total range of pen deflection on the Physiograph is approximately 8 cm.

The push button below the positioning knob starts the recording process. When depressed, this button is in the On mode, which activates the pen. If pressed again, the button returns to the Off position, which stops the recording process.

Ink Pens

The Physiograph Mark IIIS has five ink pens: one for each of the four couplers and one to record 1-second time intervals at the bottom of the chart. Five separate **ink reservoirs** supply the pens with ink (see figure 24.1). The **pen lifter lever** actuates a metal bar that simultaneously lifts all the pens off the paper. The pens should be in the raised position if not recording.

Ink reaches each pen through a small tube that connects with the pen's ink reservoir. Getting the ink to flow to the pen requires raising the reservoir 2–3 cm and then squeezing the rubber bulb at the top of the reservoir as the index finger is held over the small hole at the tip of the rubber bulb. Once the tube is full of ink, the fingertip is released before the bulb is released. If the flow of ink has to be increased or decreased during recording, the reservoirs are simply raised or lowered slightly.

To prevent clogging, all ink should be removed from the pens at the end of each laboratory period. With the pens in the raised position, the ink is drawn back into the reservoir by squeezing the reservoir bulb, then placing the index finger over the hole on the bulb and releasing pressure on the bulb. The vacuum thus created draws the ink back into the bulb.

Chart Speed

The Physiograph Mark IIIS has seven push buttons that allow the following chart speeds: 0.05, 0.1, 0.25, 0.5, 1.0, 2.5, and 5.0 cm/sec (see figure 24.2). It also has an Off push button to the left and a Timer push button to the right of the series.

Setting the desired speed requires depressing first the Timer button (the up position is Off), then the speed selection button, and finally, the Off button (which puts the button in the On position).

Changing the speed simply involves pushing a different speed button. Pressing the Off button stops the paper movement. As soon as the paper stops moving, the pens should be raised with the pen lifter lever. At the end of the period, both the Off and Timer buttons should be in the Off position.

Operational Procedures

Although each experiment performed on the Physiograph will differ in certain respects, the following general procedures should always be practiced:

Dustcover Remove the unit's dustcover, neatly fold it, and place it out of the way. Laboratory dust affects some electronic components over time, which is why the Physiograph must be covered during storage.

Recording Paper If the paper compartment is empty, access the compartment through the Physiograph's hinged front panel. Place a stack of paper in the compartment, and then feed the paper's leading edge up through the opening in the compartment's top right margin. Slide the paper under the paper guides on each side.

Draw the paper across the top of the recording surface, and slide it under the paper tension wheel (see figure 24.1) after lifting the wheel slightly by pulling the black lever straight up. When the paper is in place, release the lever. The paper should now be in firm contact with the paper drive mechanism and will move when the proper controls are activated.

Because each small block on the paper measures 0.5 cm by 0.5 cm, you can determine the exact speed of the paper by the time marks.

Power Cord The power cord is a loose component. Insert it first into the right side of the Physiograph and then into a grounded electrical outlet.

Power Switch When you turn on the power switch (see figure 24.1), a small indicator in the switch lights up. Leave the power switch On during the entire laboratory period; turn it Off at the end of the period.

Ink Pens Fill the ink pens, observing the procedure described earlier.

Paper Speed Set the paper speed by first putting the Timer push button in the On mode and depressing the speed push button the experiment calls for. Do not press the Off button until you are ready to start recording.

Coupler Attachment Connect the cable from the experimental setup to the receptacle of the proper input coupler.

Amplifier Setting Set the amplifier's sensitivity control at the sensitivity level the experiment recommends. Most experiments begin with a low sensitivity.

Transducer Balancing If you are using a transducer, balance it, as per the instructions in Appendix C.

Calibration In certain experiments, such as in Exercise 29, you will need to calibrate the transducer, as per the instructions in Appendix C.

Monitoring Once all the hookups are complete, lower the pens to the paper and press the Off button, which will start the paper moving. Then press the Record buttons on each amplifier cluster involved in the experiment. Experiment events will now record.

Assignment:
Answer the questions that pertain to the Narco Physiograph in the Laboratory Report for Exercises 21–26.

25 The Electronic Stimulator

Electronic stimulators are used to stimulate nerve and muscle cells in the laboratory. Although the focus here is on the Narco and Grass stimulators, the primary objective is learning the general principles of stimulator operation.

The electronic stimulator is essential to controlled experiments. First, it converts alternating current (AC) to direct current (DC). As stated previously, DC moves continuously in one direction only, while AC reverses its direction repeatedly and continuously. The advantage of DC for applying a pulse of electricity to a nerve or muscle is that voltage rises abruptly from zero to maximum and then suddenly returns to zero. This type of voltage control produces a **square wave** instead of a sine wave (see figure 25.1).

Second, the electronic stimulator, through a variety of controls and switches, can vary the voltage, duration, and frequency of pulses that a specimen receives. When **voltage** increases, the **amplitude,** or height, of the wave form increases. The length of time from the start to end of a single pulse of electrical stimulus is the **duration,** and the number of pulses delivered per second is the **frequency.** Some electronic stimulators, such as the Grass SD9, can also vary the time between twin pulses; this time interval is called the **delay.** Figure 25.2 illustrates these various characteristics.

Control Functions

Figures 25.3 and 25.4 show the Grass SD9 and Narco SM-1 electronic stimulators, respectively. Although the control panels of these two instruments differ in configuration and terminology, both stimulators accomplish essentially the same function.

Power Switch

The power switch for the Grass SD9 is at the bottom of the front panel near the middle. Immediately above it is a red indicator that lights up when the power is turned on (see figure 25.3).

The power switch for the Narco SM-1 is on the back panel. Like the Grass SD9, the Narco SM-1 has a red indicator light on the horizontal front panel (see figure 25.4).

Voltage Settings

Increasing or decreasing the voltage varies the intensity of any stimulus. The Grass SD9 has two voltage settings: a round knob in the upper right-hand corner and a decade (multiplier) switch below it. The voltage ranges from 0.1–100 V (see figure 25.3).

On the Narco SM-1, the voltage range switch and the variable control knob adjust voltage (see figure 25.4). The voltage range switch has three settings: 0–10, 0–100, and Off. When the voltage range switch is set at 0–100, each digit on the variable dial is multiplied by 10; thus, the range on this scale is from 0–100 V. When the voltage range switch is set at 0–10, the voltage output is from 0–10 V. In the Off mode, no electrical stimulus can occur. The lowest voltage output for this stimulator is 50 mV.

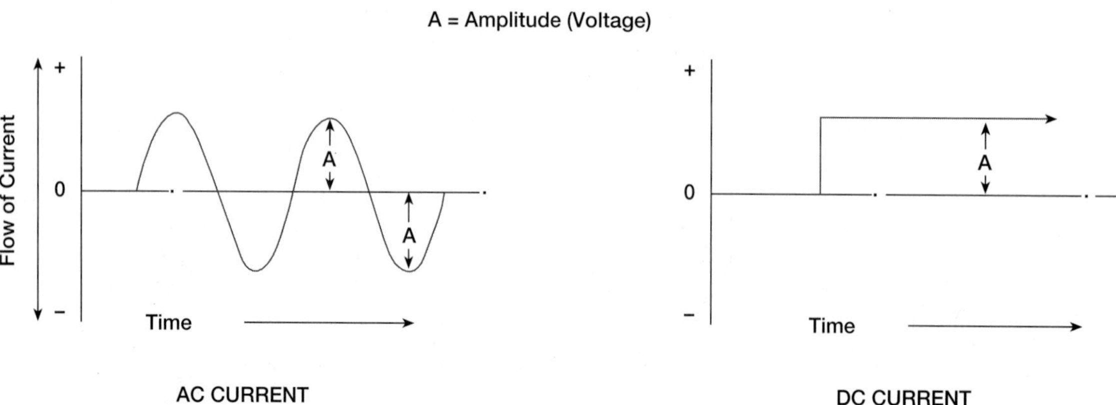

Figure 25.1 Comparison of waves produced with AC and DC current.

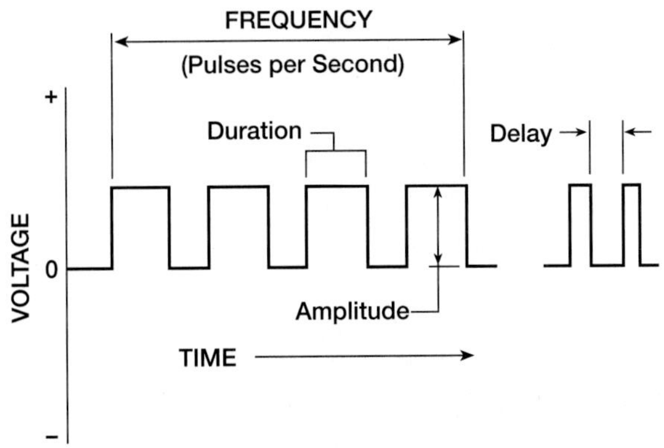

Figure 25.2 Square wave characteristics.

Duration Settings

The duration of each electrical pulse is measured in milliseconds (msec). On the Grass SD9, the duration knob and the four-position decade switch below it control this stimulus characteristic (see figure 25.3). The combination of these two controls yields a range of 0.02–200 msec. The duration setting should not exceed 50% of the interval between pulses.

On the Narco SM-1, the width control (the left-hand knob on the horizontal panel) regulates duration (see figure 25.4). The range of the Narco SM-1 is 0.1–2.0 msec.

Mode Settings

Since stimulators can produce single stimuli, continuous pulses, and even twin pulses, the mode preferred for a particular experiment must be selected.

The two mode selection switches on the Grass SD9 are under the "Stimulus" label (see figure 25.3). The right-hand mode switch has three positions: Repeat, Single, and Off. In the Repeat position, repetitive impulses are produced according to the setting on the frequency control. To produce single pulses, the switch lever is depressed manually to the Single position and released. One pulse is administered for each downward press of the lever; a spring returns the switch lever to the Off position.

The other three-position switch under the "Stimulus" label on the Grass SD9 is used for hooking up with another stimulator and for twin pulses. The switch is set to the MOD position when another stimulator is used in tandem with this one to modulate output. When an experiment calls for the stimulator to produce two pulses close together, the Twin Pulses setting is used. If neither a second stimulator nor twin pulses are required, the setting should be on Regular, as it is for most of the experiments in this manual.

The mode switch on the Narco SM-1 has three positions: CONT (continuous), Single, and EXT TRIG (external triggering) (see figure 25.4). In the CONT position, the stimulator produces repetitive pulses according to the setting on the frequency control. When the switch lever is pressed to Single, the stimulator delivers a single pulse. The EXT TRIG position is used when an external triggering device is plugged into a receptacle in the back of the stimulator.

Figure 25.3 Control panel of the Grass SD9 electronic stimulator.

Figure 25.4 The Narco Bio-Systems SM-1 stimulator.

Frequency Settings

The frequency control on a stimulator is set only when repetitive pulses are administered. In other words, the mode must be set at Repeat on the Grass SD9 or at CONT on the Narco SM-1.

On the Grass SD9, the frequency control with its decade switch has a range of 0.20–200 pulses per second (PPS) (see figure 25.3).

On the Narco SM-1, the frequency ranges from 0.1–100 Hertz (Hz). (The Hertz unit means essentially the same thing as PPS: 1 Hz represents 1 PPS.) A Rate switch, which is calibrated ×1, ×.1, and ×.01, must be used in conjunction with the round knob to get the proper setting (see figure 25.4).

Delay Settings

Stimulators that can produce twin pulses have a delay control to regulate the time interval between the pairs of pulses (see the left-hand illustration in figure 25.5). The delay on the Grass SD9 can vary from 0.02–200 msec on the control knob and four-place decade switch (see figure 25.3). The delay control can be used in either a single or repetitive mode. With the Repeat mode, the delay time should not exceed 50% of the period between each set of pulses to avoid overlap.

Since the Narco SM-1 stimulator does not produce twin pulses, it does not have a delay control.

Output Switches on the Grass SD9

In addition to the aforementioned controls, the Grass SD9 has two switches under the "Output" label that can modify the waveform (see figure 25.3). The three-position slide switch on the right is for selecting monophasic (MONO), biphasic (BI), or DC. The middle illustration in figure 25.5 shows the differences between these three types of current. The proper setting for most experiments is monophasic. However, when tissues are stimulated for a long period of time with the monophasic waveform, hydrolysis occurs, causing gas bubbles

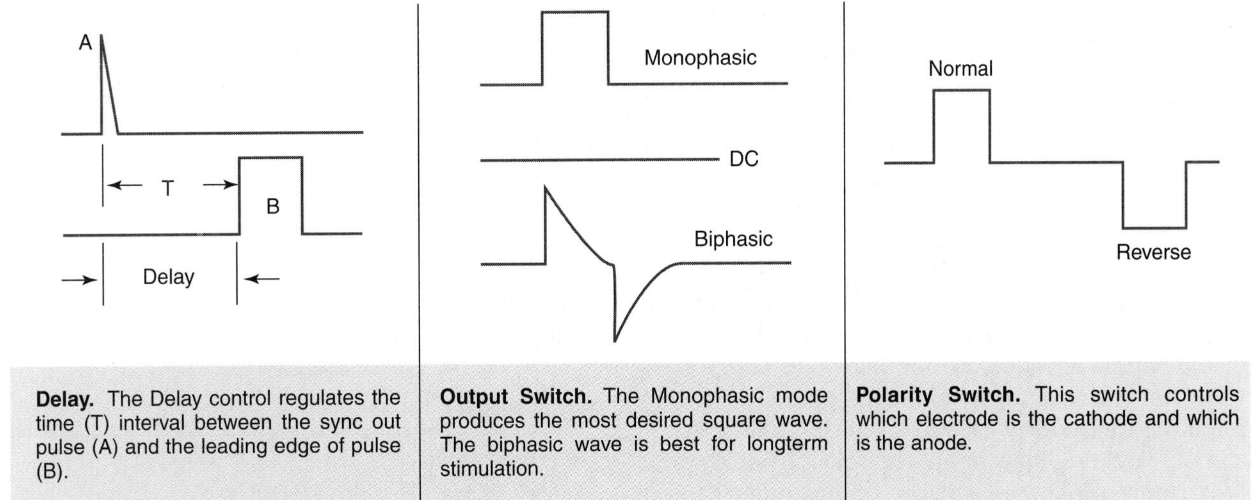

Figure 25.5 Waveform modifications (Grass SD9 stimulator).

to form around the electrodes. For longer experiments, the biphasic waveform is preferable.

The DC position is seldom used because electrolytes and gas bubbles form in the preparation. Activation of the DC position overrides all other controls except voltage and mode.

The polarity switch to the left of the other switch reverses the direction of electron flow. In the Normal position, the red output binding post is positive (+) with respect to the black binding post. In the Reverse position, the red binding post is negative (−) with respect to the black binding post (see the right-hand illustration in figure 25.5).

Operational Procedures

Operational procedures for the electronic stimulator are provided in the setup for each experiment. The primary concern here is that you understand how to use the various stimulator controls. When an electronic stimulator is first used in an experiment, review this exercise on stimulator controls.

Assignment:

Answer the questions that pertain to electronic stimulators in the Laboratory Report for Exercises 21–26.

26 The Intelitool System

Four experiments in this laboratory manual utilize transducers with a computer to monitor physiological activities. Intelitool, Inc, of Batavia, Illinois, manufactures and markets the hardware and software for performing these experiments. The four Intelitool transducers used in Exercises 30, 43, 62, and 77 are the Physiogrip (muscle physiology), Flexicomp (reflex monitoring), Cardiocomp (ECG monitoring), and Spirocomp (spirometry).

Each of these systems comes with computer software for Apple, Macintosh, and IBM-compatible computers and is relatively easy to set up. If a printer with graphics capability is available, a permanent record of experiment results can be printed out.

The Intelitool approach to these experiments has the following advantages: (1) data are accumulated automatically, (2) calculations are made instantly, and (3) transducer calibration is very easy. In addition, no animals (such as frogs) are needed. All experiments are performed on the human body.

The Intelitool systems, however, cannot do multichannel experimentation. When two or more parameters, such as ECG (electrocardiogram), blood pressure, and pulse, are monitored simultaneously, a Duograph or polygraph is required.

Assembling the various components for each experiment consists of connecting the cable from the Intelitool transducer unit to the computer after attaching the transducer to the subject. Only one system, the Physiogrip, uses an electronic stimulator, which is coupled to the subject's arm via a fitting on the front of the Physiogrip (see figure 30.1). All other systems have a single cable that leads to the computer from the transducer.

Once the equipment is hooked up, the program disk is inserted into the computer, and the computer is turned on. If the computer is an Apple IIe, IIc, or IIGS, the Caps Lock key must be in the down position. To get started, you simply press any key, and the program takes over. If the transducer has to be calibrated, the program immediately explains how to do it. When calibration is complete, the Main Menu appears on the screen.

The Main Menus of the different systems vary in the number of selections, but all of them have the following items in common:

MAIN MENU

( ) Experiment Mode
( ) Review/Analyze
( ) Save Current Data
( ) Load Old Data File
( ) Disk Commands

No Data in Memory

In place of "Experiment Mode," the selection may be "Run Spirocomp" or "Electrocardiogram," depending on which experiment you are performing. The experiment begins when you select "Experiment Mode" or its equivalent first. You can select "Review/Analyze" and "Save Current Data" only if experimental data are in the memory. If you try to select these items before putting data into the system, the "No Data in Memory" flashes to tell you that data must be in the system before selection can be made.

Menu selections are made by moving the selector arrow (cursor) up or down until it is in front of the item desired and then pressing the Return key. The command keys for moving the selector arrow vary to some extent with the type of computer used.

Once you have performed the experiment or loaded a file from a disk, you can select "Review/Analyze." In this mode, the data put into the system are reviewed. The number of selections possible in this mode vary with the system being used.

The "Disk Commands" selection provides instructions for saving and opening files.

Assignment:

Answer the questions that pertain to this exercise in the Laboratory Report for Exercises 21–26.

PART 

Skeletal Muscle Physiology

The six exercises in this part offer an in-depth study of most aspects of skeletal muscle contraction. Exercise 27 compares the ultrastructure of myofibrils to what is observed under the light microscope during muscle contraction. The roles of calcium ions and ATP in muscle fiber contraction are also examined. Exercise 28 provides experience in using an electronic stimulator. By using the instrument to map motor points on a subject's arm, you learn how to adjust the frequency and duration with voltage changes to produce optimum electrical stimuli.

Exercise 29 is a classic experiment involving the simple twitch, spatial summation (recruitment), and temporal (wave) summation. You electronically stimulate frog muscle to observe the speed and strength of skeletal muscle contraction. Exercise 30 explores the same physiological phenomena as Exercise 29, except with a computer and Intelitool software. Exercise 31 examines the effects of epinephrine and acetylcholine on muscle contraction. This experiment is open-ended in that a multiplicity of drugs can be tested. Exercise 32 focuses on electromyography.

27 The Physicochemical Properties of Muscle Contraction

This exercise explores the complex phenomenon of muscle contraction. You view the microscopic changes of skeletal muscle contraction in a controlled experimental setup.

Muscle Contraction

Skeletal muscle contraction involves many components within the muscle cell. An action potential that reaches the neuromuscular junction initiates the process. As noted in Exercise 20, acetylcholine initiates depolarization of the muscle fiber. The depolarization proceeds along the surface (sarcolemma) of the muscle fiber and deep into the sarcoplasm via the T-tubule system of the sarcoplasmic reticulum.

Figure 27.1 illustrates the ultrastructure of a single muscle fiber. Note that the sarcoplasm of the fiber contains many long, tubular **myofibrils,** the **sarcoplasmic reticulum,** and **mitochondria. Nuclei** are visible beneath the **sarcolemma.** The abundance of mitochondria and their close association to the myofibrils indicate that ATP is available for contraction.

Figures 27.1C and D reveal the molecular structure of the contractile **myofilaments** of the myofibrils. While myofibrils are visible with a good light microscope, electron microscopy is required for seeing myofilaments.

The two types of myofilaments are myosin and actin. **Myosin** filaments are the thickest filaments and are the principal constituents of the dark **A band.** They are slightly thicker in the middle and taper toward both ends. The thinner **actin** filaments extend about 1 μm in either direction from the **Z line** and make up the lighter **I band.** See figure 27.1C.

Myofibrils shorten and thereby contract muscle fibers due to the interaction of actin and myosin in the presence of calcium, magnesium, and ATP. Two other protein molecules—troponin and tropomyosin—are also involved.

As the depolarization wave (action potential) moves along the sarcolemma and T-tubule system, the sarcoplasmic reticulum releases large quantities of calcium ions. Attractive forces that then develop between the actin and myosin filaments pull the ends of the actin fibers toward each other (see figure 27.1D). The exact mechanism that causes myofibril shortening is unknown, but the following explanation is the most acceptable current theory:

The myosin molecule is made up of **heavy meromyosin** and **light meromyosin** (see figure 27.2). The heavy meromyosin forms a head that is flexible and able to pivot. The actin filament consists of three components: a double-stranded helix of **F-actin protein, tropomyosin,** and **troponin** (see figure 27.2). Tropomyosin is a long protein strand that lies within the groove of the F-actin protein. Troponin occurs at various active sites along the helix. The cross-bridges of heavy meromyosin bridge the gap between myosin and actin molecules.

The *ratchet theory of contraction* proposes the following series of events:

1. Calcium ions combine with troponin.
2. Calcium-bound troponin pulls the tropomyosin deeper into the groove of the two actin strands, exposing the active sites on the actin filaments.
3. ATP energizes the heads of the myosin molecules.
4. Cross-bridges form between the myosin heads and the actin filaments as the ATP molecules split.
5. As the myosin heads release the products of ATP hydrolysis, the heads pull on the actin filaments with a ratchetlike power stroke.
6. After ATP reenergizes them, the myosin heads release the actin filaments (cross-bridge dissociation).
7. The cycle of cross-bridge formation, power stroke, and dissociation continues.

Microscopic Examination

You can microscopically observe muscle contraction by inducing contraction in glycerinated rabbit muscle fibers with ATP and ions of magnesium and potassium. Measuring the muscle fibers before and after contraction determines the percentage of contraction. Microscopic comparisons of relaxed and contracted fibers show the changes that occur.

Materials:

test tube of glycerinated rabbit psoas muscle (muscle is tied to a small stick)
dropping bottle of ATP in triple-distilled water
dropping bottle of 0.25% ATP, 0.05M KCl, and 0.001M $MgCl_2$

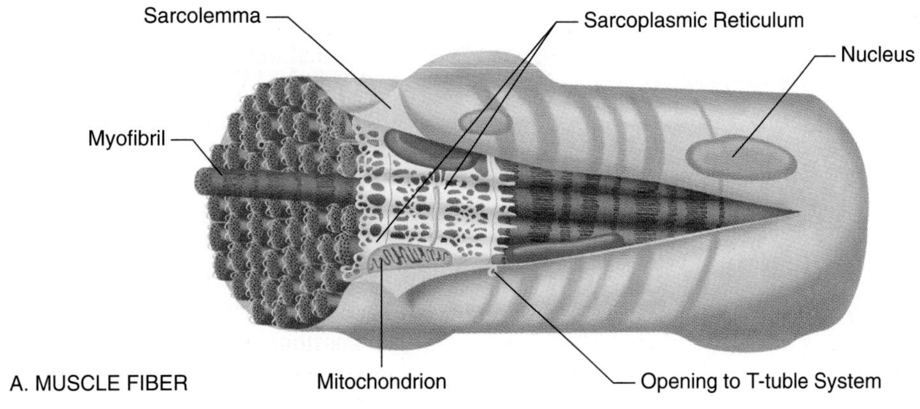

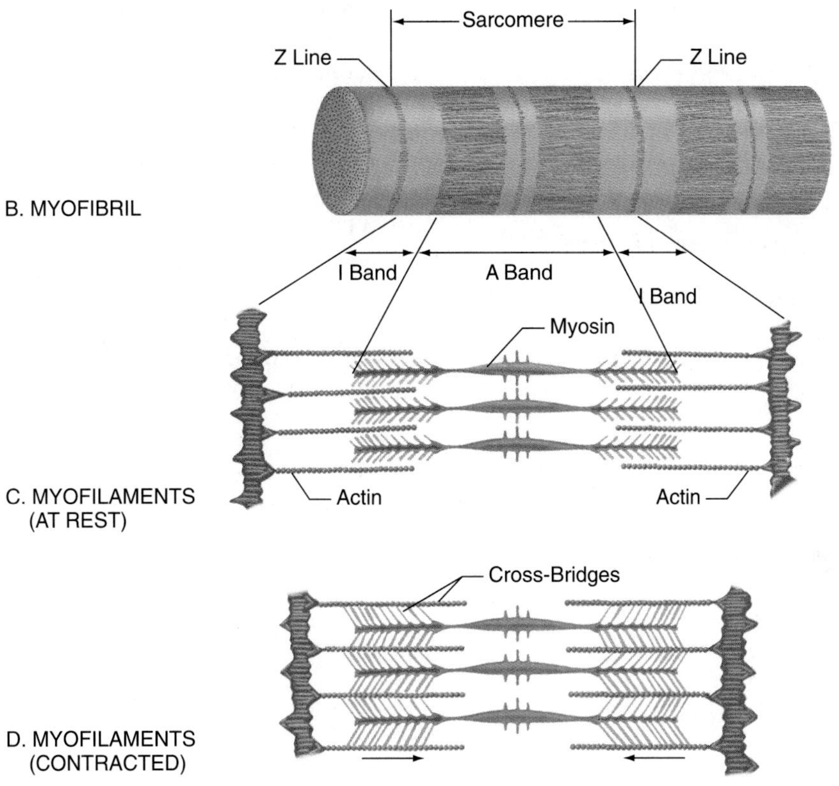

Figure 27.1 The ultrastructure of a muscle fiber.

four microscope slides and cover glasses
sharp scissors
forceps
dissecting needle
Petri dish
plastic ruler, metric scale
microscopes, dissecting and compound

1. Remove the stick of glycerinated muscle tissue from the test tube, and pour some of the glycerol from the tube into a Petri dish.
2. With scissors, cut the muscle bundle into 2 cm segments and allow them to fall into the Petri dish of glycerol.
3. With forceps and dissecting needle, tease one of the segments into very thin groups of muscle fibers; single fibers demonstrate the greatest amount of contraction. Do not use strands of muscle fibers exceeding 0.2 mm in cross-sectional diameter.
4. Place one strand on a clean microscope slide, and cover with a cover glass. Examine under a high-dry or oil immersion objective. Sketch the appearance of striations on the Laboratory Report.

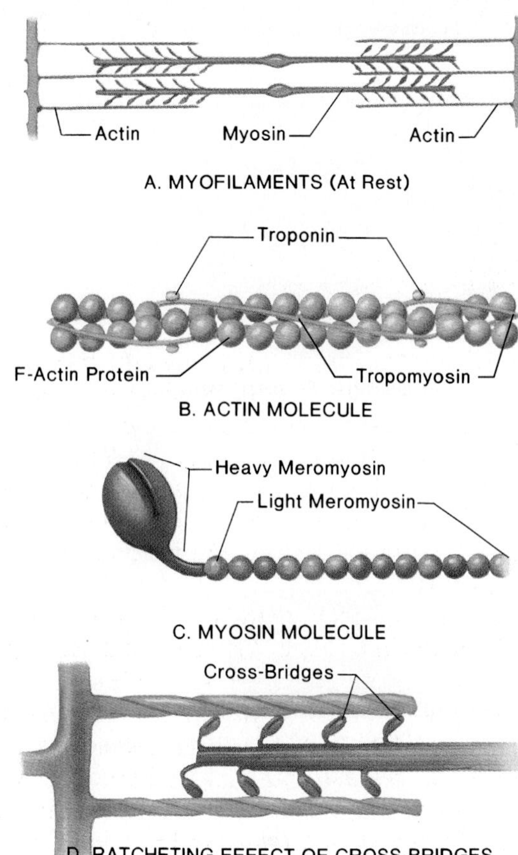

Figure 27.2 Myofilament structure.

5. Transfer three or more of the thinnest strands to a minimal amount of glycerol on a second microscope slide. Orient the strands straight and parallel to each other.
6. Measure the lengths of the fibers in millimeters by placing a plastic ruler underneath the slide. Use a dissecting microscope, if available. Record lengths on the Laboratory Report.
7. With the pipette from the dropping bottle, cover the fibers with a solution of ATP and ions of potassium and magnesium. Observe the reaction.
8. After 30 seconds or more, remeasure the fibers, and calculate the percentage of contraction. Has the width of the fibers changed? Record the results on the Laboratory Report.
9. Remove one of the contracted strands to another slide. Examine it under a compound microscope, and compare the fibers with those seen in step 4. Record the differences on the Laboratory Report.
10. Place some fresh fibers on another clean microscope slide, and cover them with a solution of ATP and distilled water (no potassium or magnesium ions). Do the fibers contract?

Assignment:

Record your data on the Laboratory Report for this exercise, and answer all of the questions.

Mapping of Motor Points

28

A muscle fiber's sensitivity to electrical stimulation through the skin surface is a function of the proximity of a neuromuscular junction, skin thickness, localized variability in skin conductivity, and other factors. An electrical stimulus that causes a muscle twitch at one spot may have no effect on muscle fibers only a centimeter away. **Motor points** are specific spots on the skin where minimal electrical stimuli cause muscle twitching. In this exercise, you attempt to identify different motor points.

An electrical stimulus applied close enough to naked nerve endings in the skin commonly produces a slight pricking sensation. A muscle twitch, however, may or may not be produced simultaneously with the pricking sensation. In this experiment, the pricking sensations are ignored, only the motor points are recorded.

This simple experiment on motor points also familiarizes you with the electronic stimulator, on which the next few experiments in this manual rely heavily. By applying electrical stimuli to your skin, you acquire a greater appreciation for the flexibility and safety of the stimulator. Before performing this experiment, review Exercise 25, "The Electronic Stimulator."

Experimental Procedure

In this experiment, you work in four-member teams. One member is the subject, another is the prober, a third is the stimulator operator, and the fourth is the recorder. If teams need to be larger due to lack of equipment, additional members can act as assistants to the stimulator operator or prober. If ample time is available, two or more members of the group can be subjects and stimulator operators to diversify team members' experiences and to enable more students to develop skills with the stimulator.

Materials:
electronic stimulator (a grounded electrical outlet is mandatory)
stimulator probe (Katech #105)
ruler
fine-point felt-tip pen
electrode paste
alcohol swabs
abrasive soap
toothpicks

Stimulator Settings

The electrical stimuli to be applied to the skin are minimal, or **threshold**—just enough to elicit a response. *Duration* (in milliseconds) and *voltage* govern the threshold stimulus. If the duration is too short, more voltage is required. You should start with a prolonged duration to find the minimum voltage and then gradually decrease the duration to where the minimum voltage still elicits a response.

Set the stimulator controls as follows to begin the stimulation procedure: frequency at 2 pulses per second (pps), delay at 200 msec (although delay does not really come into play in this experiment), duration at 100 msec, volts at 1.0, stimulus mode switches at Regular Pulses and Repeat, power switch On, polarity at Normal, and output switch at Biphasic. Connect the terminal jacks of the probe to the output posts of the stimulator.

Subject Preparation

The subject scrubs the back of one forearm with a bar of abrasive soap to dislodge superficial layers of the *stratum corneum* and to dissolve excess skin oil. After the subject's skin is completely dry, the prober lays out a grid of 18 inked dots 1 inch apart on the subject's forearm surface, as figure 28.1 shows. A fine-point felt-tip pen works well. (The dots wash off with soap and water at the end of the experiment.) Figure 28.2 illustrates the nature of the grid. Numbers shown here indicate the sequence of stimulation. *Do not put numbers on the skin.*

The prober places a small amount of electrode paste on each dot and positions the subject's arm in a relaxed manner, palm down, on the table. The subject should not look at the prober and stimulator operator while stimuli are applied.

The subject will signal with his or her free hand to indicate sensations:

- **No sensation**—no signal.
- **Tingle, pricking sensation, or vague, uncertain feeling**—hold up one finger.
- **Distinct pain**—hold up two fingers.

The recorder records voltages and spots where muscle twitches are observed. Your team is now ready to test each spot for motor points.

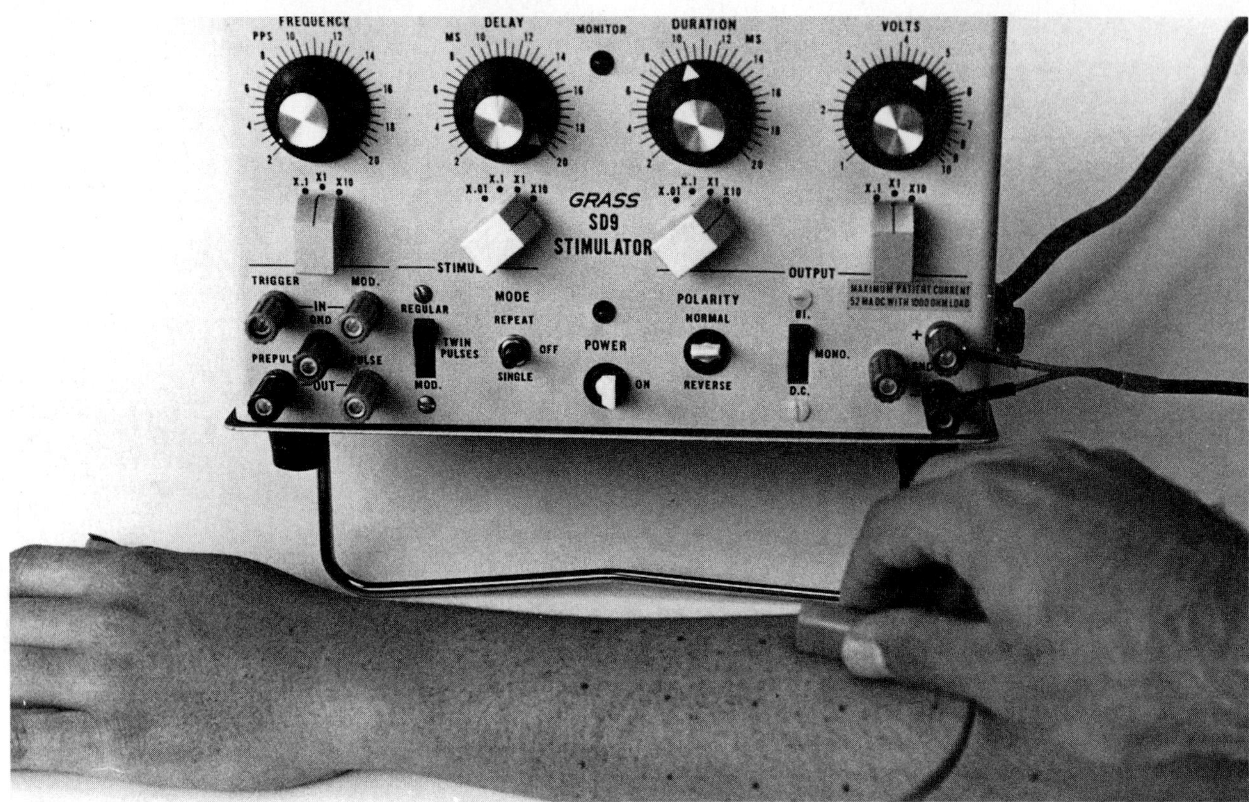

Figure 28.1 Experimental setup.

1	2	3	4	5	6
12	11	10	9	8	7
13	14	15	16	17	18

Figure 28.2 Grid pattern.

Testing with Voltage Increases

The stimulator settings are set for 1 V, 100 msec duration, and twin pulses. Proceed as follows:

1. Apply the stimulator probe to spot 1, and direct the stimulator operator to place the mode switch on Repeat. You are now applying two pulses per second of 1 V each for 100 msec. *The subject should not experience any sensation.*
2. Test each of the 18 grid points with this 1 V stimulation, recording in part A of the Laboratory Report any reaction that occurs at a particular point.
3. Clean the probe contacts by using a toothpick to remove excess electrode paste from the space between the inner and outer contacts.
4. Now increase the stimulus to 5 V, and stimulate each grid point for 3 seconds. *Try to maintain equal pressure on the probe.*
5. Repeat step 3.
6. Increase the voltage by 5 V, and stimulate each grid point again for 3 seconds.
7. Continue this process of testing at increased increments of 5 V until the prober and recorder observe a twitch, or the subject signals a sensation of pain. Be sure to repeat step 3 before each new test at an increased voltage. **Important:** Once you record a reaction at a particular grid point, do not stimulate that point again.
8. Continue testing all nonreactive points until you reach 100 V. **At all times, prevent the subject from observing stimulator settings.**

Determining Minimum Duration

In the previous series of tests, the duration was arbitrarily set at 100 msec and kept constant as the voltage increased. Now you try to determine the minimum duration that elicits a response at a few sensitive spots.

1. Place the probe at the first spot where twitching occurred, and set the voltage at the predetermined level.
2. Change the duration to 1 msec (large knob on 10, decade switch on 0.1×). Stimulate for 3 seconds, and record the reaction.

3. Move on to five other grid points where a contraction was previously noted, skipping grid points that showed no response.
4. Now increase the duration to 10 msec (move decade switch to 1×), and stimulate each of the same six grid points again.
5. Continue increasing the duration by increments of 10 msec until you reach 150 msec. Do not exceed 150 msec duration.
6. Record results on part B of the Laboratory Report.

Reversing Polarity

To observe whether reversing the polarity affects sensitivity to twitching, place the polarity switch on Reverse, and administer a threshold stimulus of minimum duration to the last reactive spot tested. Record your observations on part C of the Laboratory Report.

Assignment:

After recording all data on the Laboratory Report, answer all the questions.

29 Muscle Contraction Experiments: Using Chart Recorders

This exercise explores several phenomena of skeletal muscle contraction, including the **simple twitch, spatial summation** or **recruitment,** and **temporal summation.** All of these physiological activities are observed in muscle tissue from a freshly killed frog. Although the frog is dead, the tissues are still viable and respond to electrical stimulation. Frog tissue is used here in place of mammalian tissue because it tolerates temperature change and adverse handling, and because the results are similar to what a more carefully controlled mammalian experiment would produce.

Skeletal muscle physiology studies require considerable amounts of equipment and careful handling of tissue. First, the frog must be prepared carefully to keep the tissues viable. Second, several pieces of equipment must be properly arranged and hooked up. Third, skeletal muscle must be stimulated in a manner that elicits the desired responses.

The overall procedure is complex, and all tests must be completed within this laboratory period. For these reasons, the class will be divided into four or five teams of five or six students each. Team members can decide among themselves which roles, outlined in table 29.1, they prefer to perform on the team.

Figures 29.1 and 29.2 illustrate two different setups for this experiment. One setup utilizes Gilson and Grass equipment, while the other is essentially a Narco Bio-Systems setup. The type of equipment available in your laboratory determines which setup you use. Some of the compatible components may be interchanged so that your particular setup is not exactly like either illustration. Your instructor will indicate the setup you will use.

Materials:

For dissection:

small frog (one per team)
decapitation scissors
scalpel, small scissors, dissecting needle, and forceps
squeeze bottle of frog Ringer's solution
spool of cotton thread

For Gilson setup:

Unigraph, Duograph, or polygraph
electronic stimulator
event synchronization cable
electrode and electrode holder
T/M Biocom #1030 force transducer
ring stand
two double clamps
isometric clamp (Harvard Apparatus, Inc.)
galvanized iron wire and pliers

For Narco setup:

Physiograph IIIS with transducer coupler
SM-1 electronic stimulator

Table 29.1 Team Member Responsibilities.

EQUIPMENT ENGINEER
Hooks up the various components (recorder, stimulator, transducer, etc.). Balances the transducer and calibrates the recorder. Operates the stimulator. Responsible for correct use of equipment to prevent damage. Dismantles equipment and returns all components to proper storage places at the end of the period.

ENGINEER'S ASSISTANT
Works with engineer in setting up equipment to ensure that procedures are correct. Makes necessary tension adjustments before and during experiment. Applies electrode to nerve and muscle for stimulation.

RECORDER
Labels events produced on tracings. Keeps a log of the sequence of events as experiment proceeds. Sees that each member of team is provided with chart records for attachment to Laboratory Report.

SURGEON
Decapitates and piths frog. Removes skin on leg and exposes sciatic nerve. Responsible for maintaining viability of tissue. Attaches string to nerve and muscle end.

HEAD NURSE
Assists surgeon as needed during dissection. Keeps exposed tissues moist with Ringer's solution. Helps to keep work area clean.

COORDINATOR
Oversees entire operation. Communicates among team members. Reports to instructor any developing problems that seem insurmountable. Double-checks the setup and cleanup procedures. Keeps the process moving.

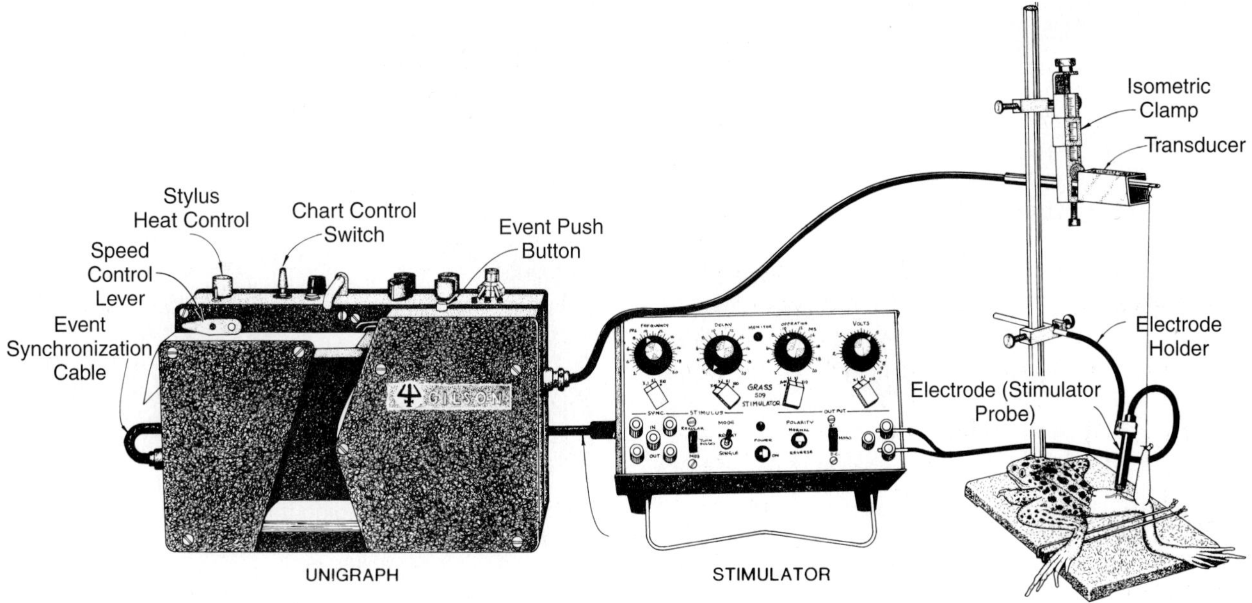

Figure 29.1 Equipment setup with Gilson recorder.

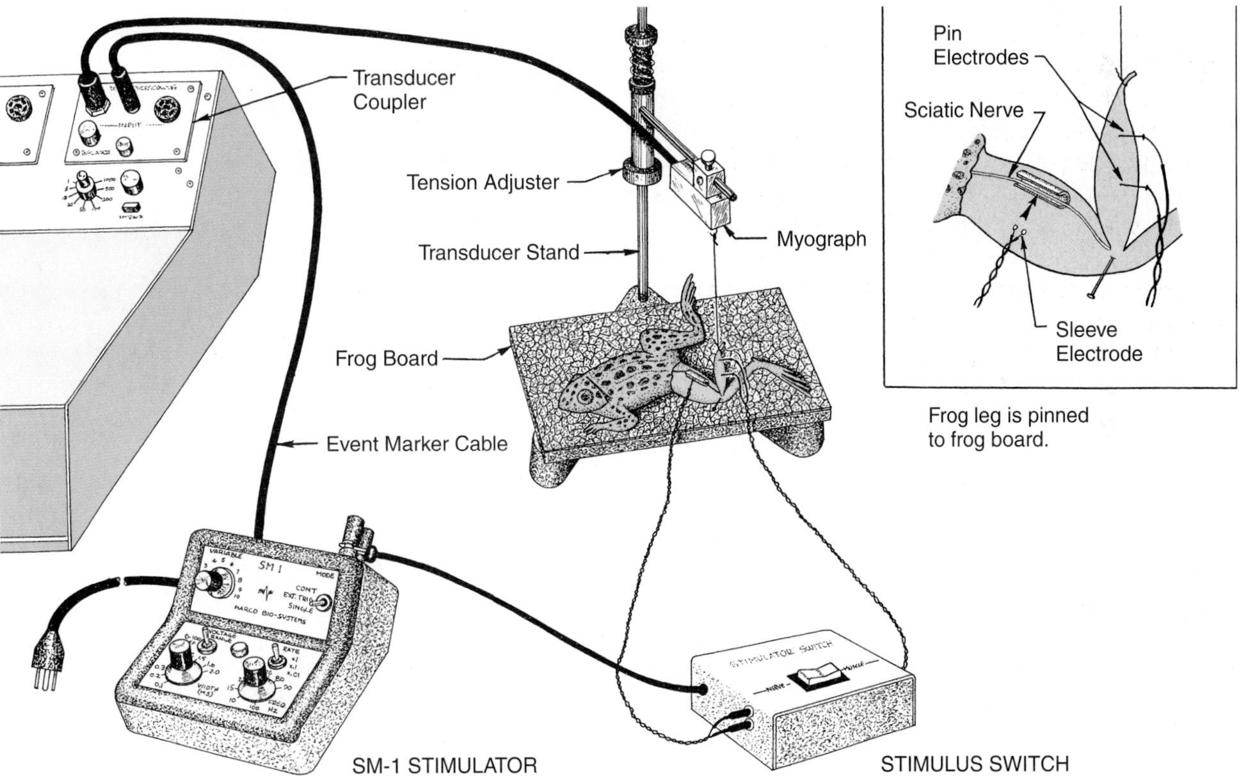

stimulus switch and myograph
transducer stand and frog board
myograph tension adjuster
event marker cable
sleeve electrode and pin electrodes
dissecting pins

Lab #1: Action Potential
Lab #3: Frog Muscle

Preparations

While some team members set up equipment, other team members decapitate the frog and perform the dissection.

Figure 29.3 Make the first incision through the skin at the base of the thigh.

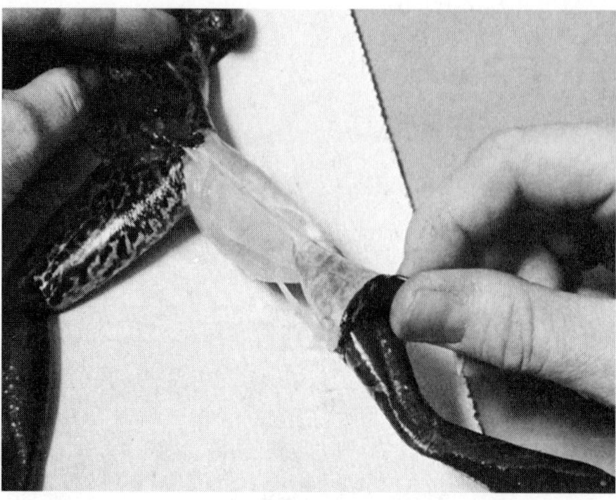

Figure 29.5 While holding the body firmly with the left hand, strip the skin off the entire leg.

Figure 29.4 After the skin is cut around the leg, pull the skin away from the muscles.

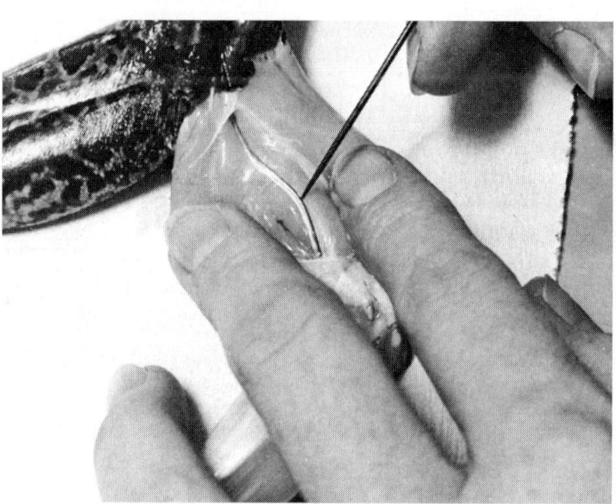

Figure 29.6 Separate the thigh muscles with a dissecting needle to expose the sciatic nerve.

Frog Dissection

Grip the frog with the left hand so that the forefinger is under the neck and the thumb is over the neck. Hold the frog under a water tap, allowing cool water to flow onto its head and body. This calms the animal and helps to wash away the blood.

Insert one blade of a sharp, heavy-duty scissors into the mouth, well into the corner; poise the other scissors blade over a line joining the tympanic membranes (eardrums). With a swift, clean action, snip off the head. Immediately after this, force a probe down into the spinal canal to destroy the spinal cord. The frog should become limp and show no signs of reflexes.

Strip the skin from the right or left hind leg, following the instructions in figures 29.3 through 29.5. *If you are using the Gilson setup, strip the right hind leg. If you are using the Narco setup, strip the left hind leg.*

With a sharp dissecting needle, tear through the fascia and muscle tissue of the thigh to expose the sciatic nerve, and insert a short (10-inch) length of cotton thread under it. You will use this thread to manipulate the nerve. Figures 29.6 through 29.8 outline the desired procedure. **Be careful in the way you handle the nerve.** The less it is traumatized or comes in contact with metal, the better. Keep the exposed tissues moist with frog Ringer's solution.

Gilson Equipment Hookup

Set up the equipment as figure 29.1 shows. Plug the power cords of both the Unigraph and stimulator

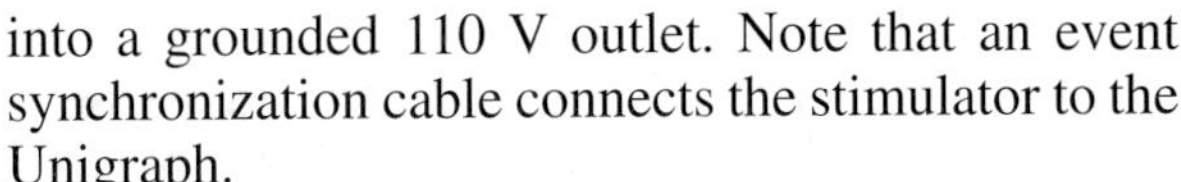

Figure 29.7 Insert forceps under the nerve to grasp the end of the string on the other side.

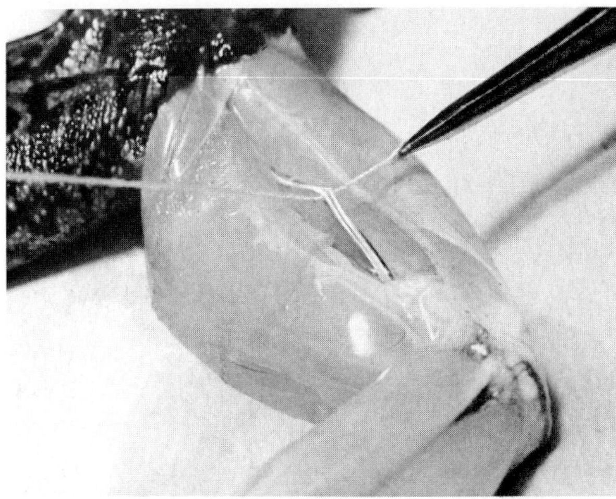

Figure 29.8 Pull through the thread, which is used for manipulating the nerve.

into a grounded 110 V outlet. Note that an event synchronization cable connects the stimulator to the Unigraph.

Attach the Harvard isometric clamp to the ring stand with a double clamp, keeping the adjustment screw of the isometric clamp upward. Attach the shaft of the transducer to the lower part of the isometric clamp, and plug the transducer cord into the Unigraph.

Put another double clamp on the ring stand shaft at the lower level, and clamp the shaft of the electrode holder in place.

Plug the jacks of the electrode into the stimulator at the position that figure 29.1 shows, but do not fix the electrode to the clamp at this time, since the electrode will be maneuvered about in the first tests.

After making all the necessary connections, balance the transducer and calibrate the Unigraph according to the instructions in Appendix C.

Narco Equipment Hookup

Set up the equipment as figure 29.2 shows. This setup uses a stimulus switch between the stimulator and the frog. This switch allows you to switch from direct muscle stimulation to nerve stimulation without disturbing the setup.

Note that an event marker cable connects the stimulator with the Physiograph. The receptacle for this cable is on the stimulator back and is labeled Marker Output. Insert the other end of the cable into the middle receptacle of the transducer coupler.

After affixing the myograph tension adjuster to the transducer stand, attach the myograph to the support rod, and tighten the thumbscrew to hold it in place. Insert the free end of the myograph cable into the transducer coupler.

Plug the power cords for the Physiograph and SM-1 stimulators into grounded 110 V outlets. Insert the plugs for the pin electrodes into the right side of the stimulus switch and the plugs for the sleeve electrode into the left side of the stimulus switch. Attach the cable from the stimulus switch to the output posts of the SM-1 stimulator.

After completing all the hookups, balance the channel and calibrate the myograph according to the instructions in Appendix C.

The Threshold Stimulus

A **threshold stimulus** is the minimum voltage that elicits a muscle twitch. A **subliminal stimulus** is less than threshold strength. Proceed as follows to compare threshold differences between muscle and nerve stimulation. *Be sure to keep the muscle and nerve moist with frog Ringer's solution.*

Gilson Procedure

1. Set the duration control at 15 msec, the voltage control at 0.1 V, the stimulus switches on Regular and Off, the polarity switch on Normal, and the output switch on Mono.
2. Turn on the power switch, and depress the mode switch to Single, while holding the electrode against the belly of the gastrocnemius muscle, as figure 29.9 illustrates. You are now administering a single pulse of 0.1 V of 15 msec duration to the muscle. This is the minimum voltage possible with the Grass stimulator.

   If the muscle twitches at this minimum voltage, reduce the duration to the lowest value that produces a twitch, and then record

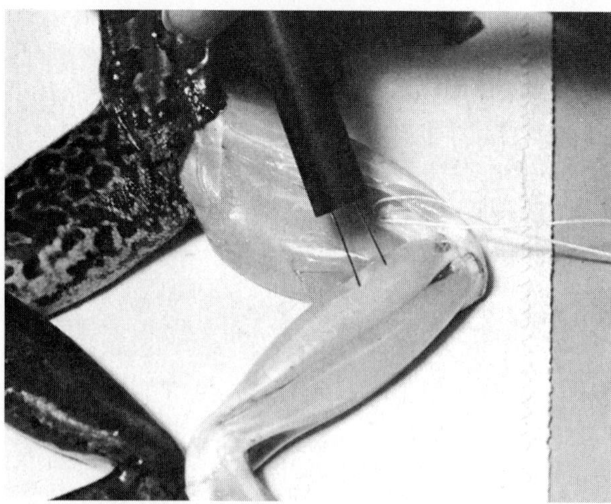

Figure 29.9 To determine direct muscle response to stimulation, stimulate the belly of the gastrocnemius muscle with a handheld electrode (Gilson setup).

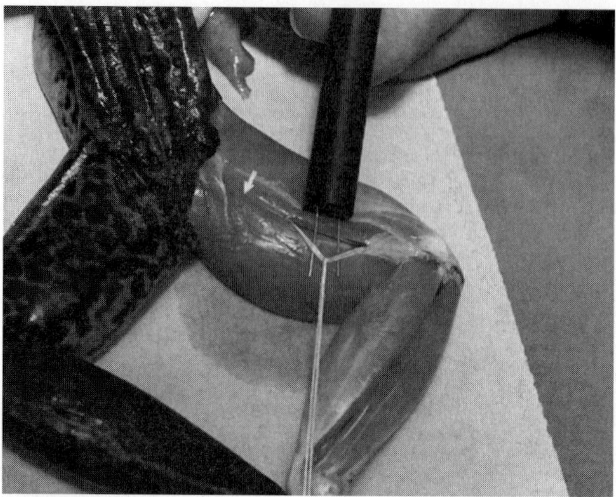

Figure 29.10 Cut the nerve at the point the white arrow indicates in both setups. Note that the probe is under the nerve for the Gilson setup.

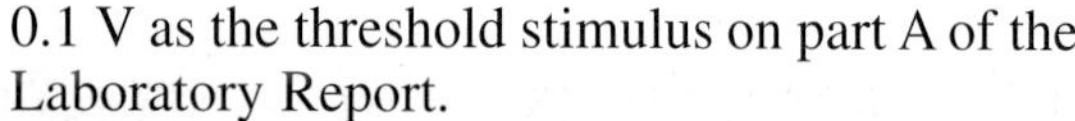

0.1 V as the threshold stimulus on part A of the Laboratory Report.

3. If the muscle does not twitch at 0.1 V, increase the voltage in increments of 0.1 V (duration at 15 msec) until you see a twitch. Record this voltage as the threshold stimulus on the Laboratory Report.
4. Cut the nerve where the white arrow in figure 29.10 indicates, and position the stimulator probe under the nerve as the illustration shows. The nerve is cut to prevent extraneous reflexes.
5. Place the foot in a flexed position.
6. Return the voltage control to 0.1 V, and stimulate by pressing the mode lever to Single. If you do not see a twitch, stimulate the nerve at increasing increments of 0.1 V until a twitch is visible. Record this voltage as the threshold stimulus for a visible twitch by nerve stimulation on part A of the Laboratory Report.
7. Now determine the threshold stimulus (minimum voltage of nerve stimulation) that extends the foot. Record this voltage on the Laboratory Report. If extension occurs with a stimulus of 0.1 V, try reducing the duration in the same manner as for direct muscle stimulation.

Narco Procedure

1. Set the width control at 2 msec, the voltage at 0.1 V, the range switch at 0–10, and the variable control at its lowest value.
2. Insert the pin electrodes into the muscle about 1 cm apart.
3. Press the Muscle side of the button on the stimulus switch to ensure that the muscle receives stimuli from the stimulator.
4. Now, deliver a single stimulus to the muscle by depressing the mode switch on the stimulator to Single. If the muscle twitches, reduce the width to the lowest value that produces a twitch, and then record 0.1 V as the threshold stimulus on part A of the Laboratory Report.

   If the muscle does not twitch at 0.1 V, increase the voltage in increments of 0.1 V until you see a twitch. Record this voltage as the threshold stimulus on the Laboratory Report.
5. Now, place the foot in a flexed position, and stimulate the muscle by gradually increasing the voltage until the foot extends completely due to muscle contraction. Record the voltage on the Laboratory Report.
6. Carefully position the rubber tubing sleeve around the nerve, and insert the sleeve electrode within the tubing next to the nerve. Use the string to manipulate the nerve, and *be sure to bathe the nerve in frog Ringer's solution.*
7. Press the Nerve side of the button on the stimulus switch.
8. Cut the nerve where the white arrow in figure 29.10 indicates to prevent extraneous reflexes.
9. Place the foot in a flexed position.
10. Return the voltage to 0.1 V, and stimulate by pressing the mode lever to Single. If you do not see a twitch, stimulate the nerve at increasing increments of 0.1 V until a twitch is visible. Record this voltage as the threshold stimulus by nerve stimulation on part A of the Laboratory Report.
11. Now determine the minimum voltage that extends the foot by stimulating the nerve. Refer to figure SR–1B in Appendix D for a Physiograph sample record. Record this voltage on the Laboratory Report.

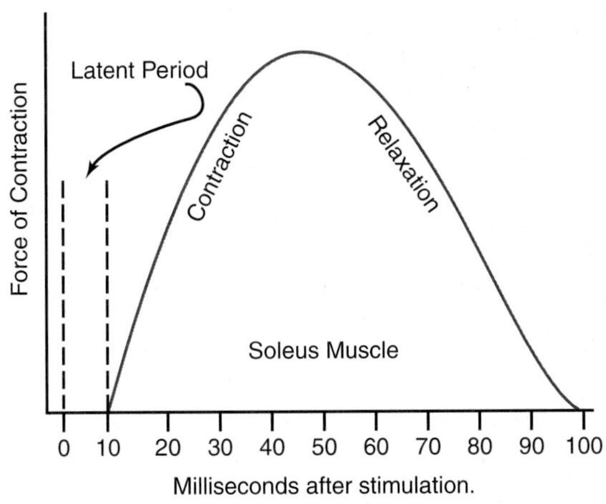

Figure 29.11 A simple twitch of the human soleus muscle.

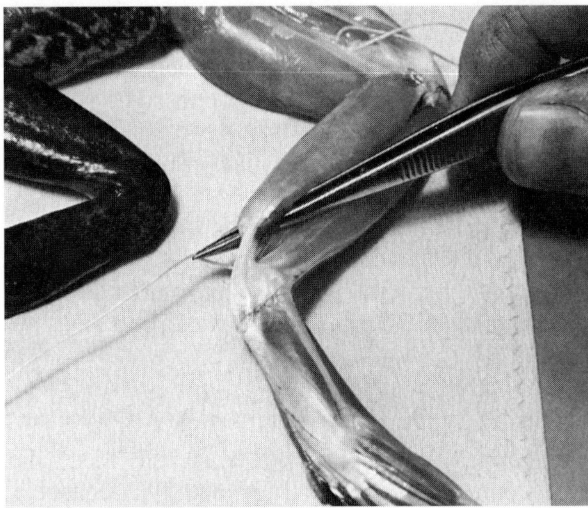

Figure 29.12 Draw the thread through the space behind the Achilles tendon by inserting forceps to grasp the thread.

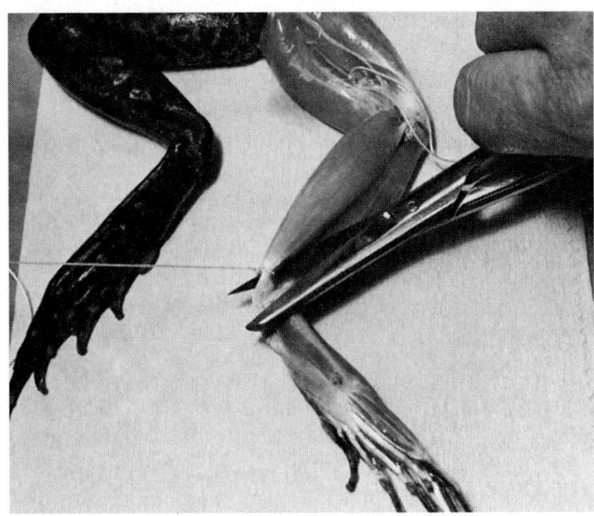

Figure 29.13 After securely tying the thread to the Achilles tendon, sever the tendon between the thread and joint.

Myogram of Muscle Twitch

Now that you have determined the threshold stimulus that induces a visible twitch in the muscle, you can produce a *myogram,* or tracing, of the twitch, using a transducer and recorder. Figure 29.11 shows a myogram of the human soleus muscle. The frog muscle twitch will have the same general configuration, but a different amplitude (force) and spread (milliseconds).

Note in figure 29.11 that the myogram of a muscle twitch has three distinct phases: a latent period, a contraction phase, and a relaxation phase.

1. The **latent period** is a very short time lapse between the time of stimulation and the start of contraction. In most muscles of the body, it lasts about 10 msec.
2. During the **contraction phase,** the muscle shortens due to chemical changes within the muscle fibers. Different muscles have different durations in this phase.
3. The **relaxation phase** follows the contraction phase. During this phase, the muscle returns to its former length.

Proceed as follows to attach the frog's leg to the transducer and to record a myogram:

Thread Hookup

Use a dissecting needle to free the connective tissue that holds the gastrocnemius muscle to adjacent muscles. Insert forceps under the muscle near the Achilles tendon (see figure 29.12) to grasp one end of an 18-inch length of cotton thread. Pull the thread through with forceps, and tie it to the tendon a short distance away from where the tendon will be severed. Now, with a pair of scissors, sever the tendon distal to where it attaches to the string (see figure 29.13).

If you are using the Gilson setup, tie the free end of the thread to two leaves (the stationary one and the one next to it) of the transducer. If you are using the Narco myograph, tie a loop in the thread, and slip the loop onto the little hook.

To prevent the leg from moving during contraction, secure the leg to the base with wire (Gilson setup; see figure 29.1) or pin the leg to the corkboard with a pin (Narco setup; see figure 29.2). *Handle these transducers gently. Placing undue stress on the leaves or hook can internally damage them.*

Use the fine adjustment knob on the tension adjuster to just take the slack off the thread.

Electrode Positioning

For the Gilson setup, position the electrode prongs so that the nerve rests on the prongs. The stem of the electrode holder consists of a soft metal that can be bent into any position. Maneuver this to hold the electrode firmly in place.

For the Narco setup, check the sleeve electrode to make sure the electrode is properly positioned.

Although figure 29.2 shows pin electrodes, they are not used in the remainder of the experiment.

Recorder Settings

Unigraph Turn on the power switch. Place the chart control (c.c.) lever at STBY, the stylus heat control knob at the two o'clock position, the speed selector lever at the slow position (opposite of position shown in figure 29.1), the gain control on 2 mV/cM, and the mode control on TRANS (transducer).

Now place the c.c. lever at Chart On, and note the position of the line produced on the chart. The line should be about 1 cm from the nearest margin of the paper. If it is not at this position, relocate the stylus with the centering knob. Now stop the chart by placing the c.c. lever at STBY. Proceed to the "Stimulator Settings" section.

Physiograph Turn on the power switch. Make sure that the myograph is calibrated so that 100 g of tension is equivalent to 4 cm of pen deflection (see Appendix C).

After the initial attachment of the muscle to the myograph and with minimal tension on the myograph, set the baseline to the centerline with the position control knob. Place the record button in the On position, and gradually increase the tension by moving the myograph upward with the myograph tension adjuster. Increase the tension until the pen has moved upward 0.5 cm (you have applied 12.5 g of tension).

By using the balance control, return the pen to centerline. Now, by using the position control, move the pen to any desired baseline; normally, this would be 2 cm below the centerline.

Be sure to periodically bathe the muscle and nerve preparation with frog Ringer's solution.

Stimulator Settings

Grass Stimulator Set the duration control at 15 msec and the voltage at the level that flexed the leg by nerve stimulation in the previous experiment. Check the switches to make sure that the mode switch is Off, the pulse setting is at Regular, polarity is at Normal, and the right-hand slide switch is at MONO.

SM-1 Stimulator Set the width control at 2 msec and the voltage at the level that flexed the leg by nerve stimulation in the previous experiment.

Recording

You are now ready to produce a simple muscle twitch myogram on the chart of your recorder and to determine the durations of each of the three phases in the contraction cycle. Proceed as follows:

Unigraph

1. Place the c.c. lever at Chart On. The paper should be moving at the slow rate (2.5 mm/sec).
2. Depress the mode switch to Single, and observe the tracing on the chart. The stylus travel should be approximately 1.5 cm. Adjust the sensitivity control to produce the desired stylus displacement.
3. If the stylus travel remains insufficient even after adjusting the sensitivity control, increase the sensitivity by changing the gain control to 1 mV/cm and readjusting the sensitivity control again.
4. Change the speed of the paper to 25 mm/sec by repositioning the speed control lever 180° to the left.
5. Administer 25 to 30 stimuli to provide enough chart material so that each team member has at least 5 inches of chart to attach to part B of the Laboratory Report.
   Note: If your setup does not have an event synchronization cable, you must simultaneously depress the event marker button on the Unigraph and the mode lever on stimulation.
6. Calculate the duration of each of the three phases of the contraction cycle, knowing that 1 mm on the chart equals 0.04 seconds (40 msec). Record your results on part C of the Laboratory Report.

Physiograph

1. Start the paper moving at 0.1 cm/sec, and lower the pens onto the recording paper. You do not need to turn on the timer.
2. Press the mode switch to Single, and observe the tracing on the chart. The pen travel should be approximately 1.5 cm. Adjust the sensitivity control to produce the desired pen travel.
3. Change the chart speed to 2.5 cm/sec.
4. Administer 25 to 30 stimuli to provide enough chart material so that each team member has at least 5 inches of chart to attach to part B of the Laboratory Report. Refer to figure SR–1A in Appendix D for a Physiograph sample recording.

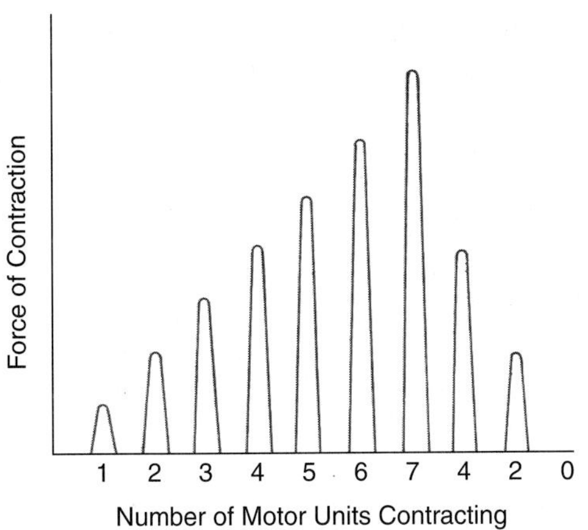

Figure 29.14 Spatial summation.

5. Calculate the duration of each of the three phases of the contraction cycle, and report your results on part C of the Laboratory Report.

Summation

Summation allows whole muscles to exert different degrees of pull. The two types of summation are spatial summation and temporal summation.

Spatial summation is an increased force caused by the recruitment of additional motor units. It is also known as *recruitment* or *multiple motor unit summation.* **Temporal summation** is due to an increase in the frequency of nerve impulses. It is also known as wave summation. In this portion of the experiment, you attempt to demonstrate both types of summation.

Spatial Summation

Increasing the strength of a stimulus (by voltage increase) beyond threshold causes increasingly greater degrees of contraction. This phenomenon is called *spatial summation.* As figure 29.14 schematically illustrates, the force of contraction is a function of the number of motor units stimulated.

A **motor unit** consists of a group of muscle fibers that a single motor neuron innervates. Small muscles that react quickly and precisely may have as few as two or three muscle fibers per motor unit. Large muscles that lack a fine degree of control, such as the gastrocnemius, may have as many as a thousand muscle fibers per motor unit. These large muscles are well adapted to a posture-sustaining function.

The muscle fibers of adjacent motor units overlap so that each unit supports its neighbors. When some nerve fibers to a muscle are destroyed, as in poliomyelitis, **macromotor units** form by extensive arborization of the ends of surviving motor neurons to provide neural connections to all muscle fibers. These large motor units may be four or five times as large as normal units.

As the voltage of a stimulus increases, the muscle reaches a degree of contraction that further voltage increase cannot exceed. This maximum contraction, called the **maximal response,** is due to all motor units being activated and additional stimuli not being able to produce further contraction. The lowest voltage that produces a maximal response is called the **maximal stimulus.**

Proceed as follows to demonstrate spatial summation, maximal stimulus, and maximal response:

1. **Unigraph:** Set the speed control at the slow speed (2.5 mm/sec) and the c.c. switch at Stylus On. Leave all other settings as they were for the previous experiment.
   Physiograph: Set the paper speed at 0.25 cm/sec, and lower the pens onto the paper, but do not start the paper at this time.
2. **Stimulator:** Turn the voltage to 0.1 V. Leave all other settings as they were in the previous experiment.
3. **Electrodes:** Check the electrodes on the frog's nerve to see that they have good contact with the nerve. *Remember to keep the tissues moist with frog Ringer's solution.*
4. With the chart still, stimulate the nerve by depressing the mode lever to Single. Look for stylus movement on the chart.
5. Advance the chart approximately 5 mm, and stop it again.
6. Raise the voltage by 0.2 V, and administer another single stimulus.
7. Continue this process of advancing the chart 5 mm and increasing the voltage by 0.2 V until you reach a maximal response. *Be sure to record all voltages on the chart.*
8. Repeat this overall process five or more times to provide enough chart material for all team members.

Temporal Summation

If spatial summation were the only way for muscles to achieve maximum contraction, muscles would have to be much larger than they are to do what they do. Fortunately, temporal summation augments spatial summation.

Temporal summation in skeletal muscle occurs when the muscle receives a series of stimuli in very rapid succession, as figure 29.15 illustrates. If the rate of stimulation is very slow, only single twitches occur. However, as figure 29.15 shows, the single twitches produced at 35 pps (pulses per second) produce some summation. As frequency increases

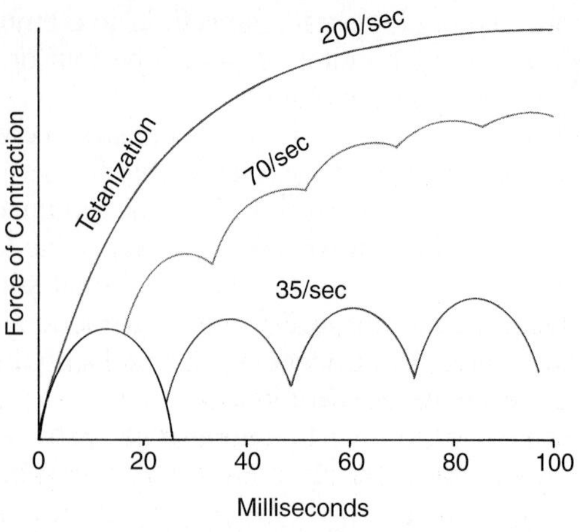

Figure 29.15 Temporal (wave) summation.

to 70 pps, considerable summation occurs, and at 200 pps, the muscle does not relax; instead, complete tetanization (sustained contraction) occurs.

In this portion of the experiment, use the frequency control to regulate the pulses per second or Hertz (Hz) units.

Gilson Setup

1. Set the voltage at the previously determined maximal stimulus, the frequency at 1 pps, and the stimulator mode switch at Off.
2. Put the c.c. switch at Chart On, with chart speed at slow setting (2.5 mm/sec).
3. Move the mode switch on the stimulator to Repeat. Note that the monitor lamp on the stimulator is blinking.
4. Record the tracing for 10 seconds; then return the stimulator mode switch to Off and the c.c. switch to STBY. Let the muscle rest for 2 minutes.
5. Increase the frequency to 2 pps, start the paper moving, and record for another 10 seconds. Stop the chart (STBY), and rest the muscle for another 2 minutes.
6. Now double the frequency to 4 pps for another 10 seconds. Rest again for 2 minutes.
7. Continue to double the frequency, record for 10 seconds, and rest the muscle for 2 minutes until the muscle goes into complete tetany.
8. After resting the muscle for a few minutes, repeat the experiment enough times to provide a record for each team member.

Narco Setup

1. Set the voltage at the previously determined maximal stimulus and the frequency at 1 Hz.
2. Lower the pens, and start the chart paper moving at a speed of 0.25 cm/sec.
3. Place the stimulator mode switch on CONT. Note that the monitor lamp on the stimulator is blinking.
4. Record the tracing for 10 seconds, and return the stimulator mode switch to Off. Stop the chart movement, and raise the pens.
5. Let the muscle rest for 2 minutes.
6. Increase the frequency to 2 Hz, set the stimulator on CONT, lower the pens, and start the paper moving again. Record for another 10 seconds, return the stimulator mode switch to Off, stop the paper, and raise the pens.
7. Allow the muscle to rest another 2 minutes.
8. Now double the frequency to 4 Hz, turn on the stimulator to CONT again, lower the pens, and start the paper moving. After recording for another 10 seconds, shut down the system again, and rest the muscle for another 2 minutes.
9. Continue to double the frequency, record for 10 seconds, and rest the muscle for 2 minutes until the muscle goes into complete tetany.
10. Refer to figure SR–1C in Appendix D for sample Physiograph records.
11. After resting the muscle for a few minutes, repeat the experiment enough times to provide a record for each team member.

Assignment:

Complete the Laboratory Report, and attach the charts from the experiments to it.

Muscle Contraction Experiments: Using Computerized Hardware

30

This exercise examines muscular contraction phenomena with an Intelitool Physiogrip controlled by computer software. Figure 30.1 illustrates the setup. Note that electrical stimuli are applied to nerves in the right arm that flex the third and fourth fingers. With either the third or fourth finger on the trigger of the Physiogrip, flexor response is fed into the computer for data accumulation and evaluation. The following muscle contraction phenomena are studied:

- threshold stimulus
- spatial summation (recruitment)
- temporal (wave) summation
- tetany
- single muscle twitch

Although you can perform this experiment alone, it is much easier if you work in pairs, with one person the subject while the other person reads the instructions and operates the stimulator and computer.

Note in the "Materials" list the two Physiogrip manuals that come with the equipment. Although you may not need them, these manuals are available for reference.

Materials:
Physiogrip transducer on wooden base
Physiogrip program disk
Physiogrip Lab Manual
Physiogrip User Manual
continued on next page

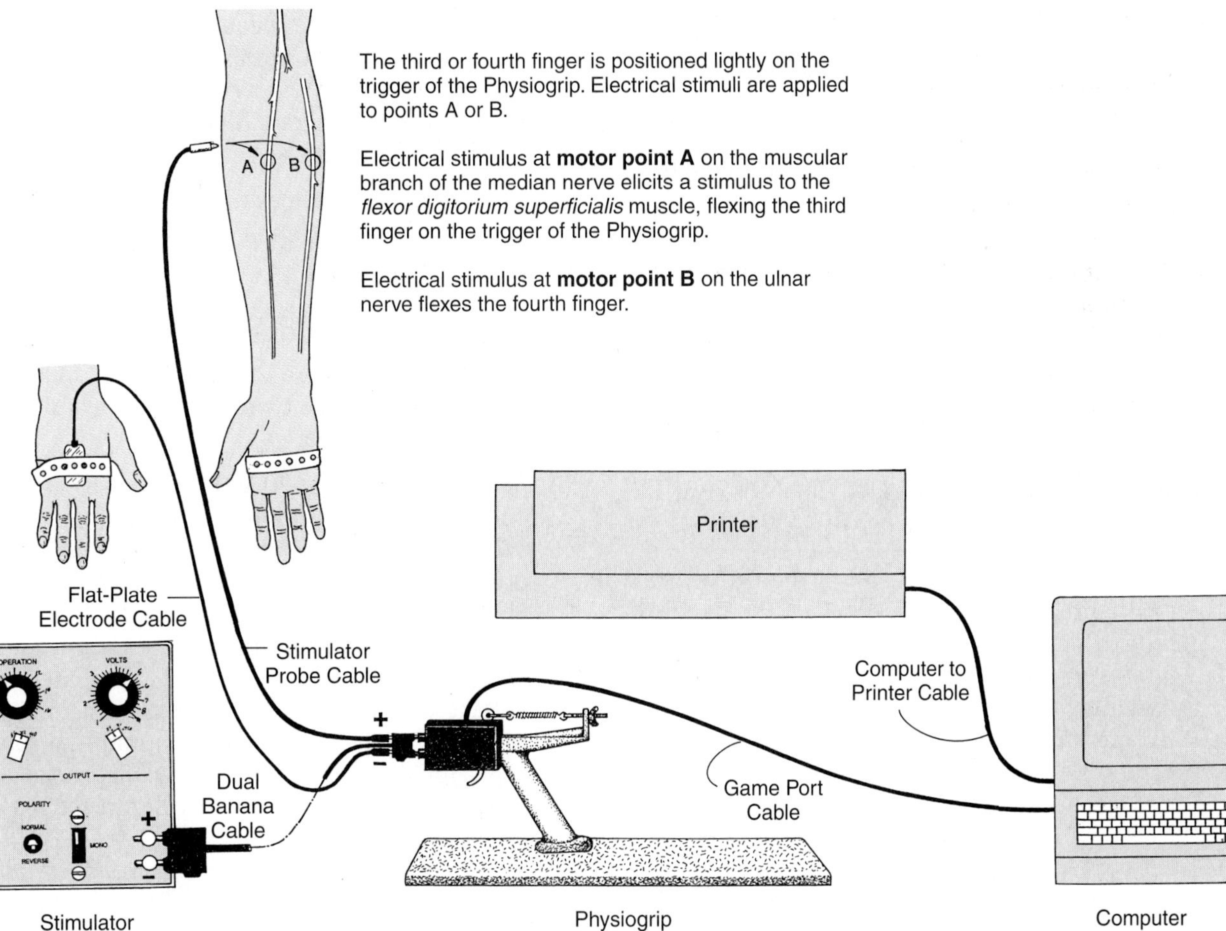

Figure 30.1 Equipment hookup for muscle contraction experiments.

blank disk for saving data
electronic stimulator, computer, and printer
flat-plate electrode cable and rubber strap
stimulator probe cable
dual banana cable, game port cable
electrode gel

Equipment Hookup

Arrange the various components in figure 30.1 in a manner that makes them convenient to use. Connect the electronic components with the appropriate cables according to the following sequence:

1. Plug the cables for the computer, printer, and stimulator into *grounded* 110 V electrical outlets.
2. *With the computer turned off,* connect the **game port cable** between the black box of the pistol grip and the computer. Be sure that the connector pins on the transducer end of the cable match *exactly* the holes in the socket of the black box.

   The location of the game port on the computer varies with the type of computer. Usually, it is on the back of the computer. On some of the early Apple models, it is inside the cabinet, which necessitates lifting the cover of the unit. Consult the *Physiogrip User Manual* if you are unsure of how to connect the game port cable.
3. Install the **dual banana cable** between the pistol grip and the output posts on the stimulator. Note that both ends of this cable have identical, dual-pronged banana plugs. One of the prongs on each plug has a bump near it. *Insert the prong with the bump into the negative stimulator post.* The other prong fits into the positive (red) post on the stimulator. Since the pistol grip has no polarity, it makes no difference how the cable plug is inserted into it.
4. Insert the prong on the end of the **flat-plate electrode cable** into the back of the banana plug at the pistol grip. *The cable prong must be inserted into the negative side of the banana plug, which has the bump near it.* Before strapping the plate electrode, which is the ground, to the back surface of the right hand, apply an ample amount of electrode gel to the plate to ensure good electrical contact.
5. Insert the prong of the **stimulator probe cable** into the other hole in the back of the banana plug at the pistol grip.

Getting Started

Now that all components are hooked up, proceed as follows:

1. Insert the program diskette into drive A and the data diskette into drive B.
2. Turn on the computer. If you are using an Apple IIe, IIc, or IIgs, be sure to depress the Caps Lock key, since all command keys are functional only in the caps-locked mode.
3. Press any key, and the Physiogrip program will take over and ask questions about the configuration. Press the Y key to respond "Yes" and the N key to respond "No."
4. Calibration procedures are next. Calibrating the Physiogrip requires only two steps: (a) holding the trigger on the Physiogrip all the way back and pressing any key, and (b) releasing the trigger to its foremost position and pressing any key. If the transducer is out of adjustment, you will be informed on the screen and told how to make adjustments.
5. When calibration is complete, the Main Menu appears on the screen:

   Experiment Menu
   Review/Analyze Menu
   Data File/Disk Commands
   Preferences
   ESCape to Program Menu
   ESCape to DOS

   Note: The last two lines of the Main Menu may not appear on the computer screen.

Threshold Stimulus

As explained in Exercise 29, a *threshold stimulus* barely induces a muscle twitch. Proceed as follows to determine its voltage:

1. Set the stimulator at 20 V, 1 pps (pulse per second), and 1 msec (millisecond) duration.
2. With the subject completely relaxed, apply the probe to the skin of the right arm in the area of the motor point (figure 30.1, label A). Use considerable pressure for good contact. If the subject does not feel any tingling, increase the voltage until the subject detects a sensation.
3. Move the probe around the area until the subject's middle finger flexes. You may need to increase the voltage to 70–80 V before the muscle reacts.
4. Select "Experiment Menu" from the Main Menu; then select "Alter Sweep Speed" from the Experiment Menu, and choose "15 seconds per frame."

5. Select "Continuous Experiment Mode" from the Experiment Menu.
6. Set the stimulator on single mode, and place the middle finger on the trigger of the Physiogrip.
7. Reduce the voltage to a point where the screen shows no response; then increase the voltage to a level where you see a slight response. Record this voltage as the threshold limit.
8. If you need further help, consult pages 9–11 of the *Physiogrip Lab Manual* (*PLM*). If you are having trouble locating the motor point for the flexor digitorum muscle, consult pages 6–8 of the PLM.

Spatial Summation (Recruitment)

With your finger still on the trigger, demonstrate *spatial summation* with the following procedure:

1. Increase the voltage to 10 V above the threshold stimulus, and administer a single stimulus. Note the increased contraction as more motor units respond.
2. Repeatedly stimulate the motor point, increasing the voltage by small increments each time, until all motor units are recruited (**maximal contraction**).
   Note the "stair-stepping" in the strength of responses. Do not increase the voltage beyond maximal contraction. **Do not exceed 100 V.**
3. Return to the Main Menu (use the ESC [escape] key), and select "Review/Analyze Menu."
4. Measure the height of the contraction the threshold stimulus caused and the height of the maximal contraction. (Reference: *PLM,* p. 11.)

Temporal Summation and Tetany

Temporal summation occurs when stimulation frequency increases to the extent that motor units do not have time to relax. *Tetany* occurs when the contractions fuse to produce a steady state of contraction. Proceed as follows to demonstrate these two phenomena:

1. Select "5 seconds per sweep" from "Alter Sweep Speed" from the Experiment Menu.
2. Select "Continuous Experiment Mode" from the Experiment Menu.
3. Position the middle finger on the Physiogrip trigger and the probe on motor point A in figure 30.1.
4. Set the stimulator on continuous (multiple) mode, 1 msec duration, 1 pps frequency, and sufficient voltage to produce a response that is 25% of the screen.
5. Gradually increase the stimulus frequency to a tetanic contraction within two screens (10 seconds).
   Do this several times, sustaining tetanic contractions for 1 to 2 seconds.
6. Go back to the Main Menu, and select "Review/Analyze Menu."
7. Determine and record the stimulation frequency at the point of tetanic contraction. (Reference: *PLM,* pp. 13–14.)

Single Muscle Twitch

Determine the durations of each of the phases in a muscle twitch as follows:

1. Select "Single Twitch Mode" from the Experiment Menu.
2. Set the stimulator as follows: continuous mode, lowest possible frequency, voltage of maximal contraction (determined previously), and duration of 1 msec.
3. Place the middle finger on the trigger and the probe on the motor point of the arm.
4. Increase the voltage until the response is 75% of the screen.
5. Press the B key and, instead of the screen registering the response, data are recorded into the computer's memory.
6. Select "Velocity Plot Review" from the Review/Analyze Menu of the Main Menu, and study the double plot.
7. Return to the Review/Analyze Menu. By moving the vertical line from left to right with the right arrow cursor key, determine the millisecond durations of the *latent period, contraction phase, relaxation phase,* and *entire twitch.* (Reference: *PLM,* pp. 14–18.)

Assignment:
Since no Laboratory Report is provided for this experiment, write up this report in the manner your instructor directs.

31 The Effects of Drugs on Muscle Contraction

Temperature changes, hormones, drugs, and certain ions profoundly affect the chemistry of muscle contraction. While many of these factors act through the nervous system to stimulate or depress excitability, some act directly on the muscle tissue itself.

As explained in Exercise 27, the most important ions involved in contractility are sodium, potassium, and calcium. In addition, epinephrine and acetylcholine produce known effects on muscle fibers. The semiquantitative experiment in this exercise studies the degree to which various substances affect muscle contraction.

Figure 31.1 shows the experiment setup. Instructions are provided for both Unigraph and Physiograph setups. If you are using a Physiograph, you will need a transducer input coupler.

In this experiment, a small strip of frog skeletal muscle is mounted on a Combelectrode and then immersed in a beaker of frog Ringer's solution (see figure 31.1). A thread attaches the muscle strip to the transducer. Drops of chemical agents are then added to the Ringer's solution as the stimulator stimulates the tissue. The Unigraph or Physiograph records contractions of the muscle strip. Oxygen supplied to the beaker by rubber tubing connected to the end of the Combelectrode prolongs tissue viability. The oxygen, which an air pump or tank of oxygen may produce, bubbles out into the Ringer's solution through the end of the Combelectrode.

Substances such as aspirin, Valium, epinephrine, acetylcholine, and EDTA (ethylenediamine tetraacetaic acid, disodium salt) are available for testing. You may also bring to class any drugs from home that you wish to test (e.g., over-the-counter drugs, such as NoDoz, antihistamines, etc.). Agents in tablet form will have to be pulverized with a mortar and pestle and dissolved in Ringer's solution so that they can be added with a dropper to the beaker.

When the testing of an agent is finished, the fluid is siphoned out of the beaker with a large syringe, as figure 31.2 shows. Fresh Ringer's solution is then added to the beaker for the next test reagent. During the experiment, the strip of muscle tissue may need

Isometric Clamp
Transducer
O2
Combelectrode
Skeletal Muscle Strip

Figure 31.1 Instrumentation setup.

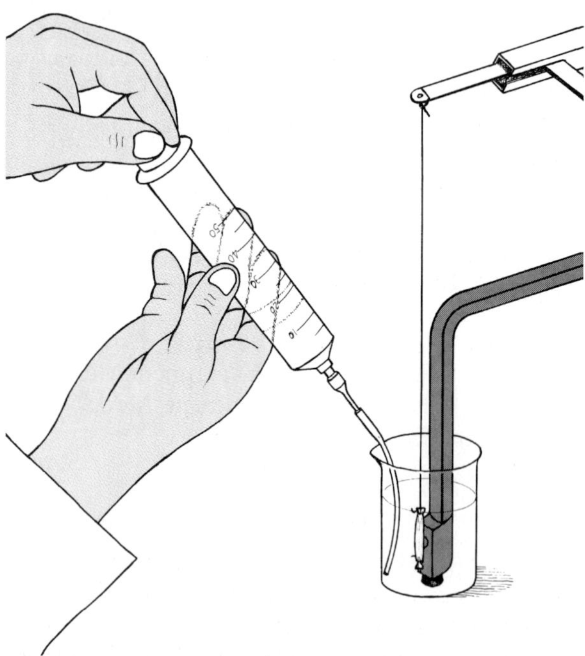

Figure 31.2 Remove test solutions with a large syringe, and replenish with fresh Ringer's solution.

to be replaced because of slow recovery or permanent damage that results from some agents.

The class will be divided into several teams, as in Exercise 29. Each team will have one or two individuals to set up the equipment, another pair to perform the dissection, a recorder, and a coordinator. This experiment may be expanded to include tests on cardiac and smooth muscle for comparison purposes.

Materials:

For dissection:
small frog
decapitation scissors, scalpel, small scissors, dissecting needle, and forceps
frog Ringer's solution
Petri dishes, medicine droppers, thread

Chemicals:
aspirin tablets (325 mg/100 ml Ringer's solution)
Valium (2 mg/100 ml Ringer's solution)
epinephrine (1:1000)
acetylcholine (1:1000)
EDTA (2%)

For Unigraph setup:
Unigraph and stimulator
event synchronization cable
force transducer (Trans-Med #1030)

For Physiograph setup:
Physiograph with transducer coupler
stimulator and event marker cable
myograph (Narco) and steel rod (1/4-inch diameter)

Other equipment:
Combelectrode (Katech)
isometric clamp (Harvard)
ring stand and two double clamps
oxygen supply and rubber tubing
metering valve for oxygen supply
hypodermic syringe (50 ml size) and needle
tubing to fit syringe
mortar and pestle
beakers and graduates

Preparations

Equipment Hookup

Set up a ring stand, isometric clamp, transducer and two double clamps, as figure 31.1 shows. If you are using the Physiograph, refer to figure 31.3 to see how the Narco myograph attaches to the Harvard isometric clamp. Note that the myograph attaches to a steel rod, which, in turn, clamps to the isometric clamp.

Insert the transducer cord into the proper receptacle of the recorder, and connect the stimulator to the recorder with the event synchronization (marker) cable. Do not attach the Combelectrode to the ring stand until the tissue has been mounted on the Combelectrode.

After completing the assembly, balance and calibrate the recorder, following the instructions in Appendix C for the type of recorder you are using.

Tissue Preparation

Decapitate a frog and pith its spine, using the procedure outlined in Exercise 29. Remove the skin from the hind legs, and excise several strips of muscle tissue (see figure 31.4). Attach one strip to the Combelectrode, and store the remaining strips in a beaker of oxygenated Ringer's solution. Figure 31.4 shows all steps for attaching the muscle strip to the Combelectrode and transducer.

Hook up the oxygen hose to the Combelectrode, and adjust the metering valve to produce only a few bubbles per second. A "T" fitting in the rubber tubing that supplies the Combelectrode will furnish oxygen to the beaker of spare tissue strips.

Preparation of Drugs

The solutions of epinephrine, acetylcholine, and EDTA are used as prepared. To prepare solutions from tablets (aspirin, Valium, etc.), pulverize each tablet with a pestle in a mortar, and add 10 ml of

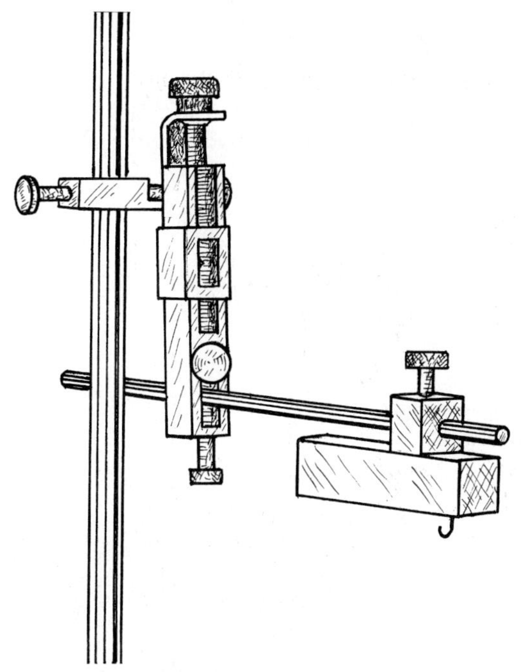

Figure 31.3 For the Physiograph setup, join the myograph and Harvard clamp with a metal rod.

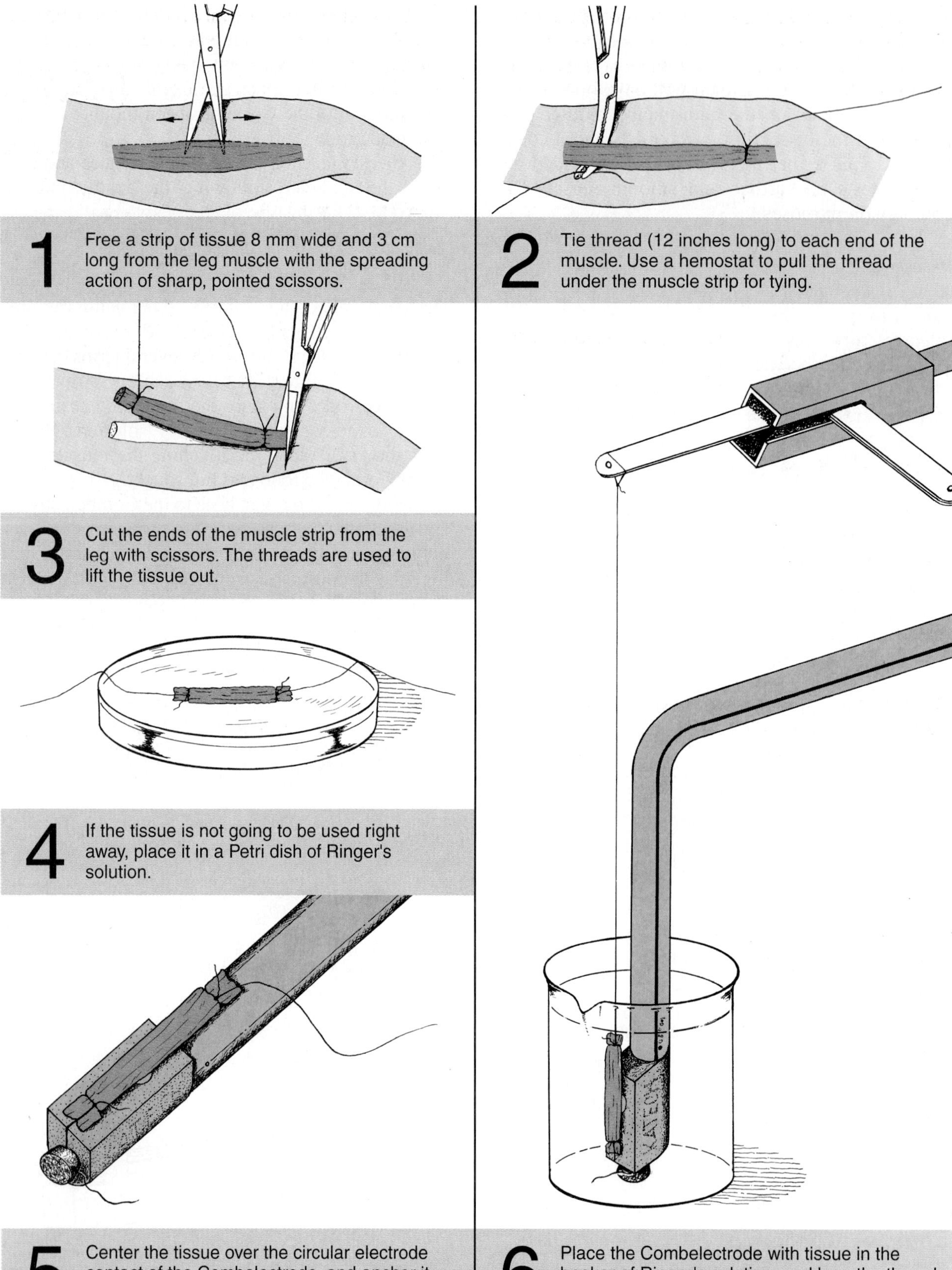

Figure 31.4 Tissue removal and attachment procedures.

Ringer's solution to the mortar. When the tablet is completely dissolved, pour the solution into a beaker, wash the mortar and pestle, and repeat for the other tablets.

Instrumentation Settings

Unigraph Turn on the power switch, place the c.c. switch at Stylus On, the stylus heat control at the two o'clock position, the speed selector lever at the slow position, the gain control at 2 mV/cm, and the mode control at TRANS.

Physiograph Turn on the power switch. Make sure the myograph is balanced and calibrated in the same manner as in Exercise 29. Calibration should produce 4 cm deflection with 100 g. Place the record button in the On position. Set the speed at 0.25 cm/sec, and lower the pens to the paper.

Stimulator After turning on the power switch, set the duration (width) at 15 msec on the Grass stimulator or 2 msec on the Narco SM-1. Set the voltage at a level slightly higher than the threshold stimulus, as determined in Exercise 29.

Test Procedure

The five drugs used in this experiment were chosen for their different effects on muscle tissue. Aspirin is a well-known analgesic. Valium is a widely used muscle relaxant and anxiety reducer. Acetylcholine transmits nerve impulses across synapses at motor end plates. Epinephrine affects muscle strength. EDTA ties up calcium. Proceed as follows:

1. With the stimulator mode at Repeat or CONT, adjust the sensitivity and gain controls to produce 1.5 cm stylus movement on the chart. Allow the muscle strip to equilibrate under these conditions for 2 minutes.
2. Add two drops of the aspirin solution to the Ringer's solution in the beaker, and mark "2 drops aspirin" on the chart with a pen. If you are using a Unigraph, depress the event button at the same time.
3. Wait 30 seconds, mark the chart, and add two more drops. *Do not change the gain or sensitivity settings during the test.*
4. Continue adding two drops every 2 minutes for 12 to 15 minutes. Mark the chart each time you add drops.
   Any changes will show on the chart within 15 minutes.
5. Remove the contaminated Ringer's solution with a large syringe (see figure 31.2), rinse out the beaker, and refill it with fresh Ringer's solution.
6. Repeat the previous procedure for all test drugs.

Assignment:
Complete the Laboratory Report for this exercise.

32 Electromyography

Early experiments during the 1920s proved that bioelectricity accompanies muscle contraction. Since then, electromyography (EMG) has become firmly established in physiology. Electrical activity associated with muscle contraction arises from nerve and muscle action potentials of the motor units. Depolarizations initiate contraction, while repolarizations accompany relaxation. Together, these activities produce voltage changes that skin and needle electrodes can detect.

The type of EMG recording depends on the electrode selection. Recordings from needle electrodes inserted into muscle tissue yield precise data about individual motor unit action potentials, allowing diagnosticians to distinguish between myopathic and neurogenic disorders. Recordings from skin electrodes give a general impression of whole muscle groups, although single muscles are sometimes detected. For various reasons, this exercise uses only skin electrodes. Skin electrodes cannot readily distinguish individual action potentials because the region beneath the electrode often consists of several motor units.

Figure 32.1 shows a typical EMG. The high-frequency, irregular, overlapping spikes of an EMG of this type are called an **interference pattern.** The voltage spread of these potentials may be as great as 50 mV.

With sufficient amplification, an interference pattern may be detectable even in completely relaxed muscles. This spontaneous activity is associated with maintaining **normal muscle tone.** Activation of the muscle, however, exaggerates the interference pattern.

An oscilloscope, audio monitor, or chart recorder can display an EMG. Although the Unigraph is the recorder of choice in this experiment because of its portability, a polygraph with an EMG module is excellent. A Unigraph in which an IC-EMG module has replaced the IC-MP module can also be used. In fact, the IC-EMG module is better adapted to this type of experiment.

The class will be divided into teams of three or four students. While one individual prepares the Unigraph, another attaches the electrodes to the subject and assists in recording information on the chart.

Materials:
Unigraph, Duograph, or polygraph
handgrip dynamometer (Stoelting #19117)
three-lead patient cable for Unigraph
three skin electrodes (Gilson self-adhering #E1081K)
adhesive pads for electrodes
electrode gel (EKG or other)
Scotchbrite pads (grade #7447)
alcohol swabs

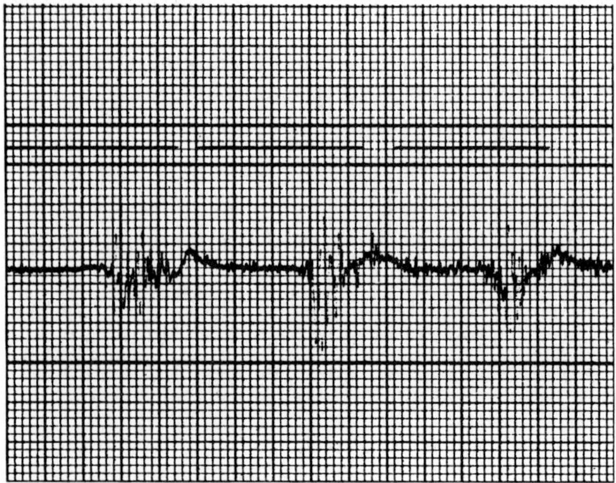

Figure 32.1 An electromyogram.

Preparations

Equipment Hookup

While the subject is being prepped for EMG recording, connect the three-lead cable to the Unigraph. Plug the power cord into an outlet. Turn on the power switch and the stylus heat control, and check the intensity of the stylus line on the paper at slow speed.

If the Unigraph does not have a special EMG module, set the mode control on the EEG setting. To quantify the EMG in terms of millivolts, you will need to calibrate the instrument according to instructions in Appendix C.

If you are using a Gilson polygraph, the following modules can be used for EMG monitoring:

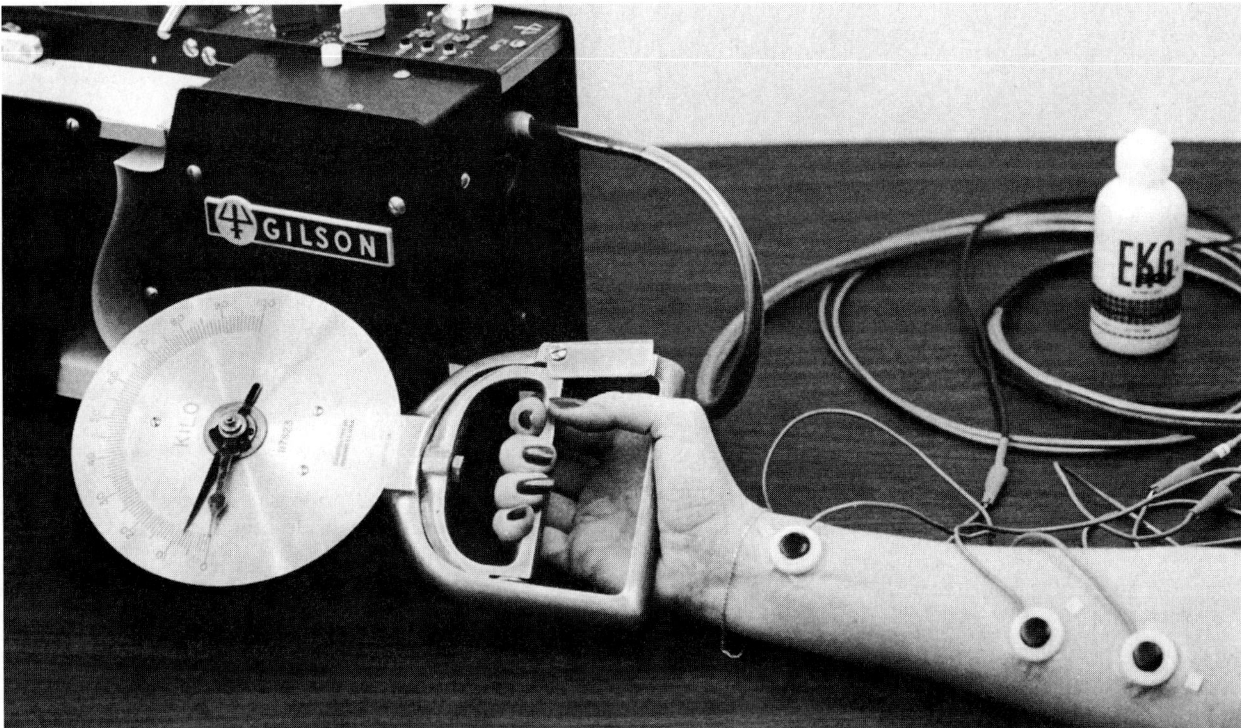

Figure 32.2 EMG monitoring setup.

IC-MP, IC-UM, and IC-EMG. If all modules are available, use the IC-EMG module.

Subject Preparation

Figure 32.2 shows the placement of three electrodes on the arm. Before attaching the electrodes, scrub the skin areas gently with a Scotchbrite pad, and disinfect the skin with alcohol swabs (see figure 32.3). Removing some of the dead cells from the skin surface greatly facilitates conductivity. Place the two recording red-wired electrodes within an inch of each other on the belly of the major flexor muscle of the forearm (*flexor digitorum superficialis*). Place the third electrode—a ground—some distance away over a bony area.

Figures 32.4 through 32.6 show how to place the self-adhesive pads and electrode gel on the electrodes before attaching the electrodes to the skin. Several other kinds of electrodes also can be used. Figure 32.7 shows how to substitute three ECG plate electrodes on the arm. This alternate method also requires scrubbing first with Scotchbrite and applying electrode gel before strapping on the electrodes.

When the electrodes are firmly attached to the skin, connect the wires to the appropriate wire of the three-lead cable. These cables usually have alligator clips and are color-coded to accept the correct electrodes. Be sure to attach the ground electrode to the black ground clip of the three-lead cable.

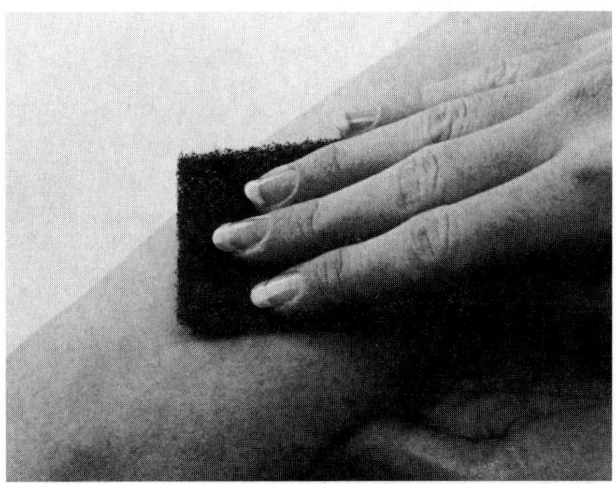

Figure 32.3 Rub the skin surface gently with a Scotchbrite pad to improve skin conductivity.

Monitoring

This setup allows you to study three muscle activities: spontaneous activity, recruitment, and fatigue. If a Duograph, dual-beam oscilloscope, or polygraph is available, you will also be able to monitor EMGs of agonist and antagonist muscles, simultaneously.

Spontaneous Activity

With the subject's arm completely relaxed and fully supported on the tabletop, turn on the Unigraph

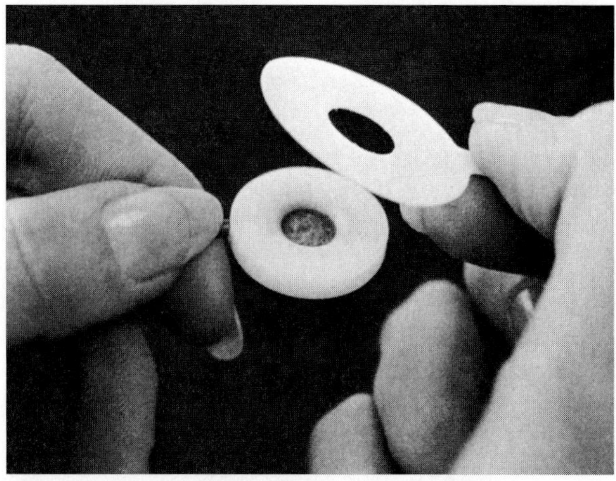

Figure 32.4 Apply an adhesive disk to the clean, dry undersurface of the skin electrode.

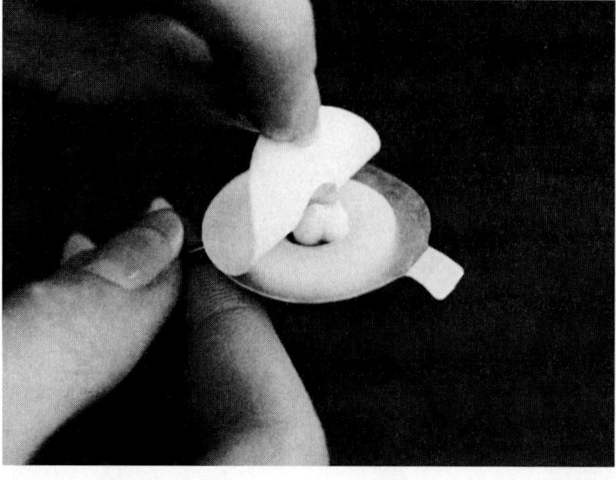

Figure 32.6 Remove the top covering of the adhesive disk, and place the electrode on the skin.

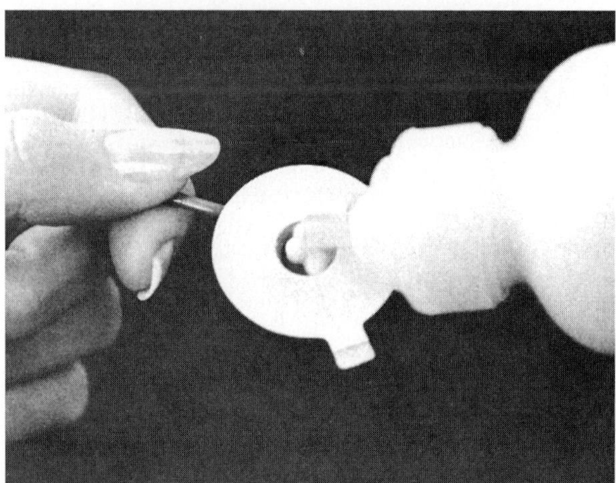

Figure 32.5 Apply electrode gel to the depression of the skin electrode.

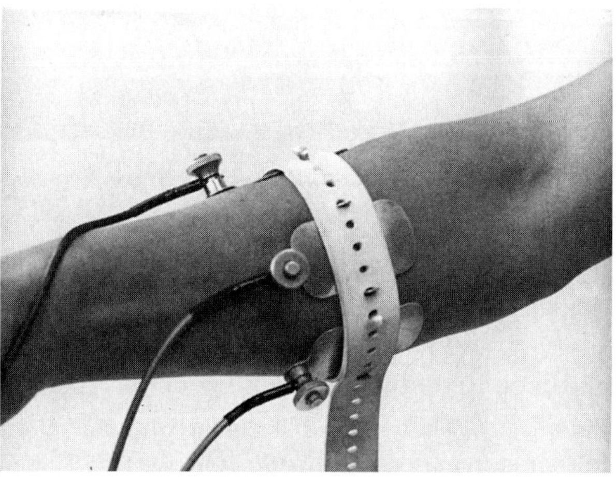

Figure 32.7 An alternate method of electrode application for an EMG demonstration.

(high speed), and observe if any evidence of motor unit activity is discernible at the lowest gain. Increase the sensitivity by changing the gain settings and turning the sensitivity control. Determine the magnitude and frequency of peaks.

Recruitment

Instruct the subject to clench the fingers into a fist, and note the burst of activity that forms a typical EMG interference pattern on the chart.

To demonstrate recruitment, have the subject grip a hand dynamometer, as in figure 32.2. First, record a few bursts at 5 kg. Then double the force to 10 kg. Continue to increase the force by 5 or 10 kg increments to maximum, recording each force magnitude on the chart with a pen or pencil.

Do you see an increase in amplitude on the chart as evidence of recruitment of motor units? Produce enough tracings for all team members.

Fatigue

With the Unigraph turned off, direct the subject to perform some hand-gripping work on the dynamometer until the hand muscles are fatigued. When the subject cannot continue, take the dynamometer away, and allow the subject's arm to recover on the tabletop. Start the recorder, and monitor any electrical activity. How does the EMG of fatigue compare with that of spontaneous activity?

Agonist Versus Antagonist Muscles

Agonist and *antagonist* muscle pairs oppose each other's movements. For example, contraction of the *biceps brachii,* the agonist, flexes the forearm against the upper arm. Contraction of the *triceps brachii,* the antagonist, extends the forearm. The action of these two muscles must be coordinated so that when one contracts, the other relaxes. Without such coordination, rigidity occurs.

To demonstrate coordination, the activity of both muscles can be recorded simultaneously with a Duograph, polygraph, or dual-beam oscilloscope. Hook up three electrodes to each muscle of one arm through separate channels of whatever type of equipment is available. Have the subject execute the following maneuvers:

1. Flexion of arm against resistance. Assistant restrains subject's fist while subject attempts flexion.
2. Extension of arm against resistance.
3. Isometric tension (contraction) of both agonist and antagonist, simultaneously.

Assignment:
Complete the Laboratory Report for this exercise.

Part 7 The Major Skeletal Muscles

The eight exercises of Part 7 focus on around 105 skeletal muscles of the human body. This muscle coverage, although not complete, includes almost all the surface muscles and most of the deeper ones. If time does not permit the study of all these muscles, your instructor will indicate which ones you should know.

This is the first unit in which the cat is used for dissection. Study human musculature first, however, by viewing illustrations and models. Although cat musculature differs somewhat from human musculature, the similarities far outweigh the differences.

The similarities of human and cat musculature, not the differences, should be your focus here.

That the cat has certain muscles that humans lack, or that muscle origins or insertions are slightly different in the cat, is unimportant. By studying the similarities rather than the differences, you reinforce your knowledge of human muscles.

33 Muscle Structure

The ability of skeletal muscles to move limbs and other structures is a function of the contractility of muscle fibers and of how muscles attach to the skeleton. This exercise focuses on the detailed structure of whole muscles and their mode of attachment to the skeleton.

Figure 33.1 shows two muscles of the arm and shoulder girdle. Several other upper arm muscles that would obscure the points of attachment of the muscles shown have been omitted. The longer muscle attached to the scapula and humerus at its upper end is the **triceps brachii.** The shorter muscle on the anterior aspect of the humerus is the **brachialis.**

Muscle Attachment

Each muscle of the body has an origin and insertion. The **origin** of the muscle is the immovable end. The **insertion** is the other end, which moves during contraction. Contraction of the muscle shortens the distance between the origin and insertion, causing the insertion end to move.

Muscles attach to bone in three different ways: (1) directly to the periosteum, (2) by means of a tendon, or (3) with an aponeurosis. The upper end of the brachialis (the origin) attaches by the first method—that is, directly to the periosteum. A short tendon, however, attaches the insertion end of this muscle.

A **tendon** is a band or cord of white fibrous connective tissue that provides a durable connection to the skeleton. The tendon of the triceps insertion in figure 33.1 is much longer and larger than that of the brachialis because the muscle exerts a greater force in moving the forearm.

Figure 39.1 shows an example of the aponeurosis type of attachment. An **aponeurosis** is a broad, flat sheet of fibrous connective tissue that attaches a muscle to the skeleton or another muscle.

Although most muscles attach directly to the skeleton, some attach to other muscles or to soft structures (skin) such as the lips and eyelids. Exercises 36 through 40 examine the precise origins and insertions of various muscles.

Microscopic Structure

Figures 33.1B and C reveal the microscopic structure of a portion of the brachialis muscle. Note that the muscle is made up of bundles, or **fasciculi,** of skeletal muscle cells. Figure 33.1C shows a single enlarged fasciculus.

Protruding upward from the upper cut surface of the fasciculus in figure 33.1C is a portion of an individual muscle cell, or **muscle fiber.** Between the muscle fibers of each fasciculus is a thin layer of connective tissue, the **endomysium.** This connective tissue enables each muscle fiber to react independently of the others when nerve impulses stimulate it. Surrounding each fasciculus is a tough sheath of fibrous connective tissue, the **perimysium.**

The **epimysium** is another layer of coarser connective tissue surrounding the fasciculi of the muscle. Exterior to the epimysium is the **deep fascia,** which covers the entire muscle. Since this muscle lies close to the skin, the deep fascia has fibers that intermesh with the **superficial fascia** (not shown) that lies between the muscle and the skin. The connective tissue fibers of the deep fascia are also continuous with the tissue in the tendons, ligaments, and periosteum.

Figure 33.1A shows a longitudinal section through the distal end of the brachialis and its tendon. Notice that the connective tissue of the tendon is continuous with the endomysium, perimysium, and epimysium.

Assignment:
Complete the Laboratory Report for this exercise. Then examine some muscles on a freshly killed or embalmed animal, identifying as many of the aforementioned structures as possible.

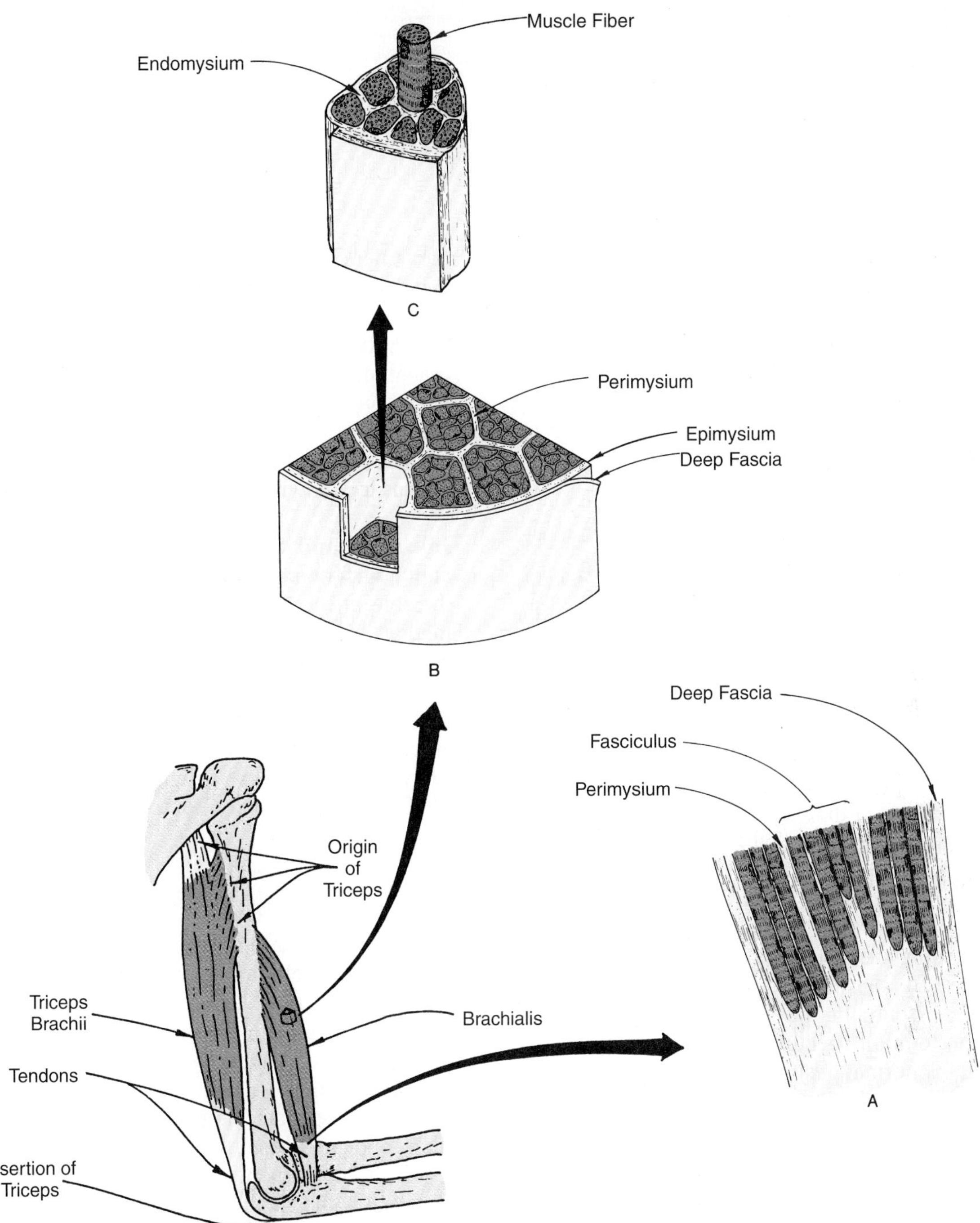

Figure 33.1 Muscle anatomy and attachments.

34 Body Movements

Muscles act with diarthrotic joints to produce a variety of body movements. The nature of the movements depends on muscle position and individual joint construction. Muscles working through hinge joints produce movements that are primarily in one plane. Ball-and-socket joints, on the other hand, have many axes through which movement can occur; thus, movement through these joints is in many planes. The different types of movement follow.

Types of Body Movements

Flexion When the angle between two parts of a limb decreases, the limb is said to be "flexed." The arm flexes at the elbow when the antebrachium moves toward the brachium. The term *flexion* also applies to movement of the head against the chest, and the thigh against the abdomen.

Upward flexion of the foot is called **dorsiflexion.** Movement of the sole of the foot downward, or flexion of the toes, is called **plantar flexion.**

Extension Increasing the angle between two portions of a limb or two parts of the body is *extension.* Straightening the arm from a flexed position is an example of extension. Extension of the foot at the ankle is essentially the same as plantar flexion. **Hyperextension** occurs when extension goes beyond the normal posture, as in leaning backward.

Abduction and Adduction Movement of a limb *away* from the body's median line is *abduction.* Movement of a limb *toward* the body's median line is *adduction.* These terms also apply to parts of a limb, such as the fingers and toes, when the longitudinal axes of the limbs are points of reference.

Rotation and Circumduction The movement of a bone or limb around its longitudinal axis without lateral displacement is *rotation.* The arm rotates through the shoulder joint. The head rotates through movement in the cervical vertebrae. If the rotational movement of a limb through a freely movable joint describes a circle at its terminus, the movement is called *circumduction.* The arms, legs, and fingers can be circumducted.

Supination and Pronation Rotation of the antebrachium at the elbow affects the position of the palm. *Supination* occurs when the palm is raised upward from a downward-facing position. *Pronation* occurs when the rotation is reversed so that the palm returns to a downward position.

Inversion and Eversion Inversion and eversion apply to foot movements. *Inversion* occurs when the sole of the foot is turned inward or toward the median line. *Eversion* occurs when the sole is turned outward.

Sphincter Action Circular muscles such as those around the lips (*orbicularis oris*) and the eye (*orbicularis oculi*) are *sphincter muscles.* When the fibers of these muscles contract, the openings they encircle close.

Muscle Grouping

For reasons of simplicity, muscles are often studied individually. Seldom, if ever, however, do they act alone. Rather, muscles are arranged in groups with specific functions to perform—for example, flexion and extension, abduction and adduction, supination and pronation.

The flexors are the prime movers, or **agonists.** The opposing muscles, or **antagonists,** contribute to smooth movements by maintaining tone and giving way to the flexor group's movement. Variance in the tension of flexor muscles results in a reverse reaction in the extensor muscles.

Muscles that help the agonists to reduce undesired action or unnecessary movement are **synergists. Fixation muscles** are muscle groups that hold structures in position for action.

Assignment:
Identify the types of movement in the figure on the Laboratory Report, and complete the matching exercise.

Cat Dissection: Skin Removal

35

Removing the skin from your laboratory cat is preparatory to the muscle studies in the next few exercises. Your instructor may have you retain the skin as a wrap for the carcass to prevent dehydration.

Laboratory cats are preserved with a mixture of alcohol and formaldehyde that is unpleasant to the nose, eyes, and skin; yet, this preservative is essential in preventing bacterial decomposition of the specimen. Protect your hands with rubber gloves. If gloves are not available, rubbing a protective cream, such as Protek, into the pores of the skin affords some protection.

You will work in pairs to dissect your cat. At the end of each laboratory period, the specimen will be sealed in a plastic bag with a name tag. Never use another student's cat for dissection.

When the skin removal dissection is complete, your specimen should look like the one in figure 35.1. Proceed as follows:

Materials:
cat (preserved and latex injected)
dissecting board or tray
dissecting instruments
plastic bag for cat
rubber surgical gloves or Protek hand lotion

1. Place the cat ventral side down on a dissecting board. Make a short incision on the midline of the neck region with a sharp scalpel, as figure 35.2 shows. Cut only through the skin, which is about 1/4 inch thick here. Avoid cutting into the muscles beneath the skin.
2. With scissors, continue the incision along the back midline all the way to the tail (see figure 35.3).
3. As figures 35.4 and 35.5 show, separate the skin from the body muscles by pulling on the skin and severing the superficial fascia with a scalpel.

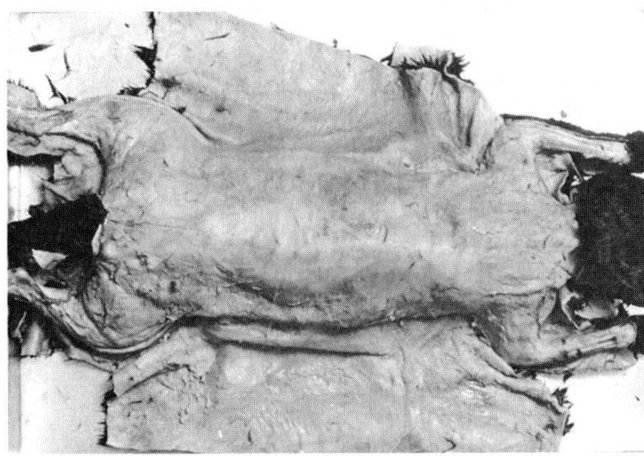

Figure 35.1 Cat with skin removed.

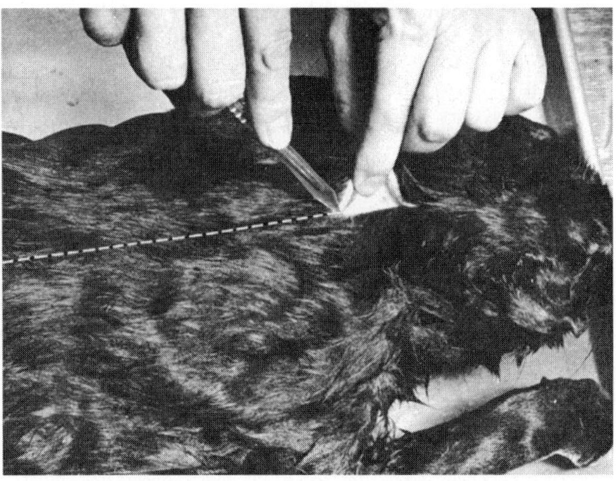

Figure 35.2 Make the first cut in the neck region with a sharp scalpel. Be careful to cut only through the skin, not into the muscles.

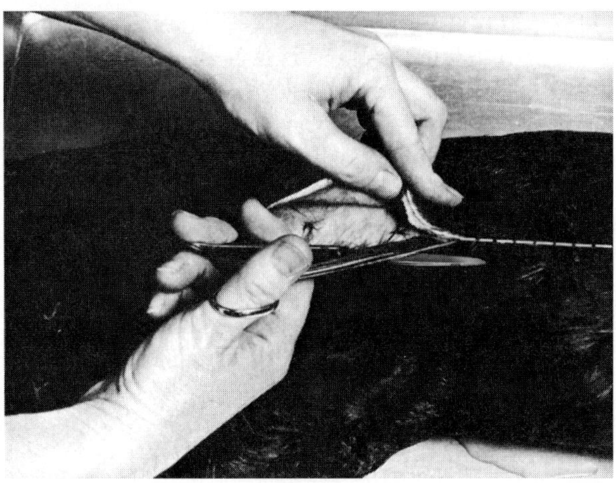

Figure 35.3 With sharp scissors (or a very sharp scalpel), cut the skin along the back from the neck to the tail region.

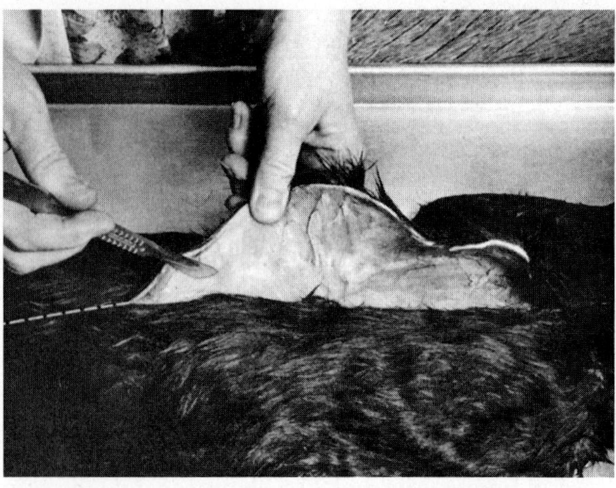

Figure 35.4 Pull the skin away from the muscles, using the scalpel to sever the superficial fascia.

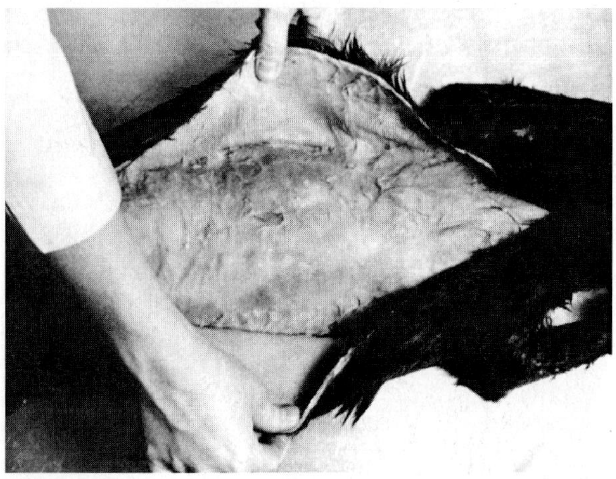

Figure 35.5 Separate as much of the skin as possible from the muscles of the body before proceeding to the next step.

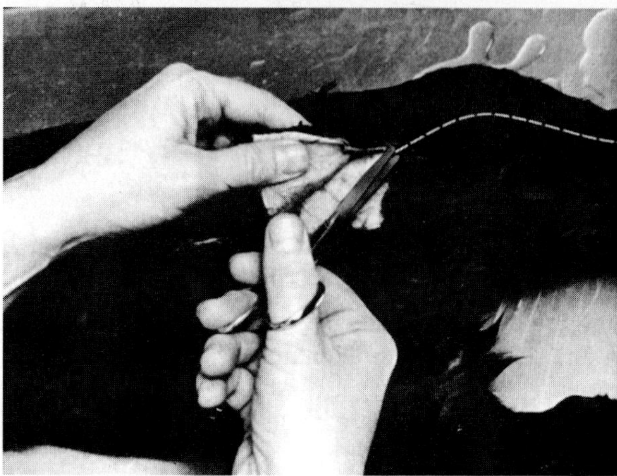

Figure 35.6 With scissors, cut the skin around the tail and down each of the hind legs to the ankles. Cut around each ankle.

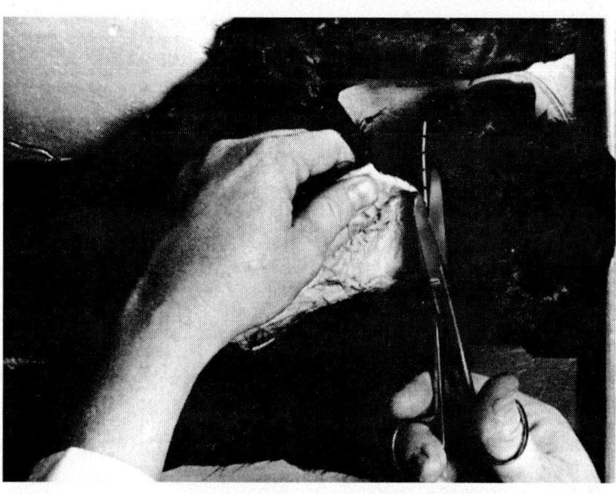

Figure 35.7 With scissors, cut the skin around the neck and down each side of the front legs to the wrists. Cut around each wrist.

4. From the tail, make incisions down each hind leg to the ankles, and cut all around each ankle (see figure 35.6).
5. As figure 35.7 shows, cut the skin around the neck and down each foreleg to the wrists. Cut the skin around each wrist.
6. Now, gripping the skin with your fingers, gently pull it away from the body. If your specimen is a female, look for the **mammary glands** on the underside of the abdomen and thorax. Remove the glands and discard them.
7. If laboratory time is still available, proceed to Exercise 36 to study the neck muscles of your cat.
8. If no more time is available, wash your instruments and tray with soap and water, and scrub your desktop. Leave your table clean and dry.
9. Wrap the specimen in the skin, and seal it in the plastic bag. Attach a label to the bag with your names on it. Place the cat in the assigned storage bin.
10. After washing your hands with soap and water, you may wish to apply hand lotion.

Head and Neck Muscles

36

This exercise focuses on the principal muscles of the head and neck. Laboratory studies may involve cadavers, muscle models, or cats, depending on availability. The head muscles are grouped here according to their functions.

Scalp Movements

Removal of the scalp reveals an underlying muscle known as the **epicranial.** It consists of the **frontalis** in the forehead region, an **occipitalis** in the occipital region, and an aponeurosis extending between these two over the top of the skull (see figure 36.1).

The occipitalis originates on the mastoid process and occipital bone; it inserts on the aponeurosis. The frontalis originates on the aponeurosis and inserts on the soft tissue of the eyebrows.

Action: Contraction of the frontalis causes the forehead to wrinkle horizontally and the eyebrows to elevate. Action of the occipitalis pulls the scalp backward.

Sphincter Action

Sphincters are circular muscles that close openings. Sphincter muscles surround the eyes and mouth (see figure 36.1).

Orbicularis Oris The orbicularis oris is a circular muscle just under the skin of the lips. It originates in various facial muscles, the maxilla, the mandible, and the septum of the nose. It inserts on the lips.

Action: The orbicularis oris closes the lips in various ways: by compressing them over the teeth, or by pouting and pursing them. It is used in kissing.

Orbicularis Oculi An orbicularis oculi surrounds each eye. Each of these sphincters arises from the nasal portion of the frontal bone, the frontal process of the maxilla, and the medial palpebral ligament. It inserts within the tissues of the eyelids (palpebrae).

Action: The orbicularis oculi is involved in blinking and squinting.

Facial Expressions

Facial muscles help you to express such emotions as pleasure, sadness, fear, and anger. The following five muscles play different roles in facial expressions (see figure 36.1).

Zygomaticus The zygomaticus extends diagonally from the zygomatic bone to the corner of the

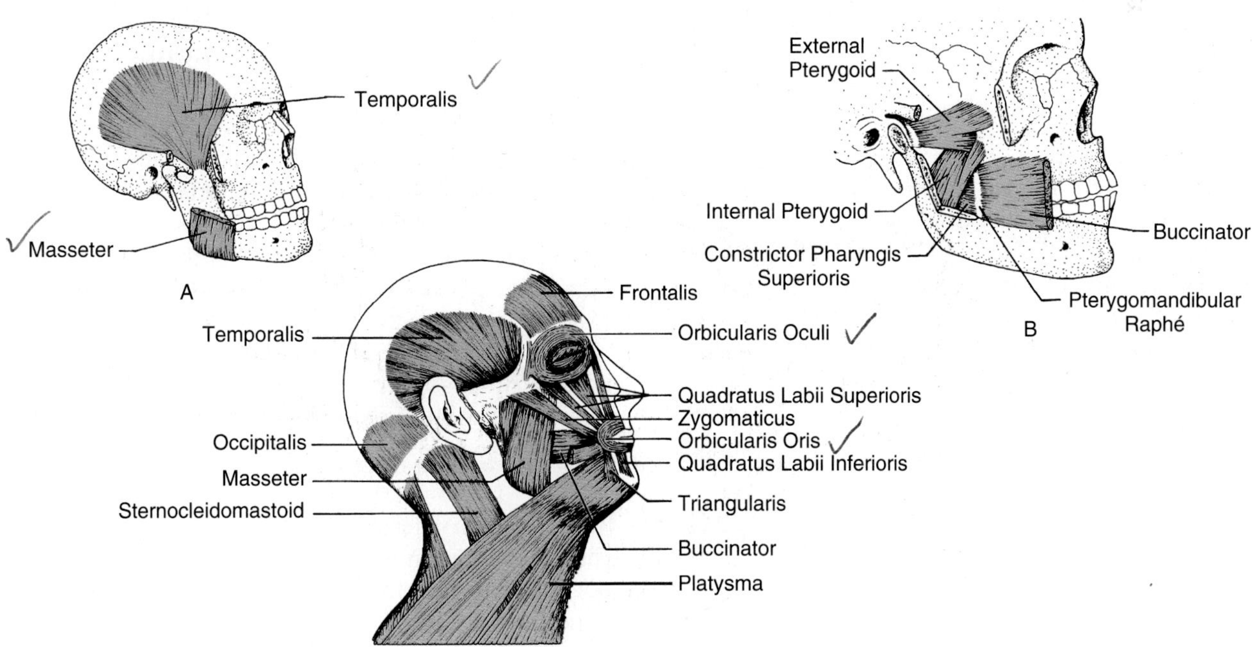

Figure 36.1 Head muscles.

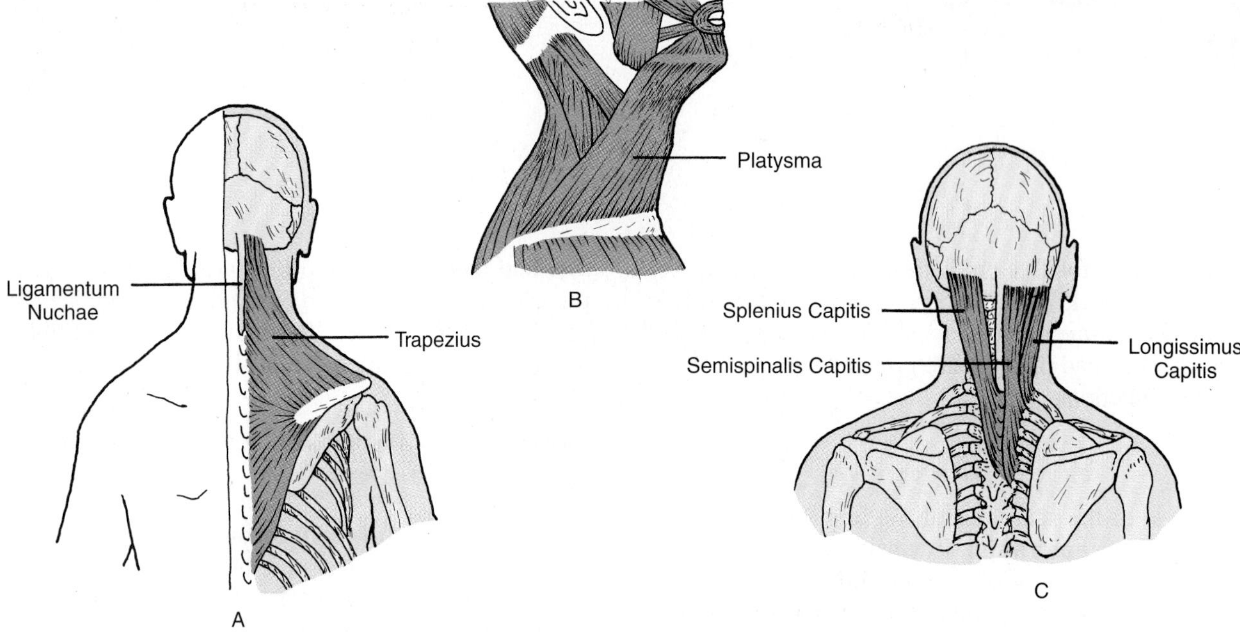

Figure 36.2 Neck muscles, posterior and lateral.

mouth. It originates on the zygomatic bone; it inserts on the edge of the orbicularis oris.

Action: Contraction of the zygomaticus draws the angles of the mouth upward and backward, as in smiling and laughing.

Quadratus Labii Superioris The quadratus labii superioris is a thin muscle, consisting of three heads, between the eye and upper lip. It arises from the upper part of the maxilla and part of the zygomatic bone. It inserts mainly on the superior margin of the orbicularis oris and partially on the alar (Latin: *alaris,* wing) region of the nose.

Action: Contraction of the entire quadratus labii superioris furrows the upper lip, resulting in an expression of disdain or contempt. Contraction of only the infraorbital head produces an expression of sadness.

Quadratus Labii Inferioris The quadratus labii inferioris is a small muscle that extends from the lower lip to the mandible. It originates on the mandible, and it inserts on the lower margin of the orbicularis oris.

Action: The quadratus labii inferioris pulls the lower lip down.

Triangularis The triangularis is lateral to the quadratus labii inferioris. It originates on the mandible and inserts on the orbicularis oris.

Action: Since it is an antagonist of the zygomaticus, the triangularis depresses the corner of the mouth.

Platysma The platysma is a broad, sheetlike muscle that extends from the mandible over the side of the neck. Figure 36.1 only shows a portion of it. Figure 36.2 reveals its entirety. Since it inserts on the mandible and muscles around the mouth, it profoundly affects facial expressions.

Action: The platysma moves the lower lip backward and downward, as in an expression of horror.

Assignment:
Locate all of the preceding muscles on a muscle model or cadaver (the cat is not a suitable specimen for most of these muscles). Also, complete part A of the Laboratory Report.

Masticatory Movements

The chewing of food (mastication) involves five pairs of muscles. Four pairs move the mandible, and one pair helps to hold the food in place. Figures 36.1A and B show all of these muscles.

Masseter Of the four muscles that move the mandible, the masseter is the most powerful. It extends from the angle of the mandible to the zygomatic bone. Figure 36.1A shows only the lower portion of this muscle. It originates on the zygomatic arch and inserts on the mandible.

Action: The masseter raises the mandible.

Temporalis The temporalis is the large, fan-shaped muscle on the side of the skull. It arises on portions of the frontal, parietal, and temporal bones, and it inserts on the coronoid process of the mandible.

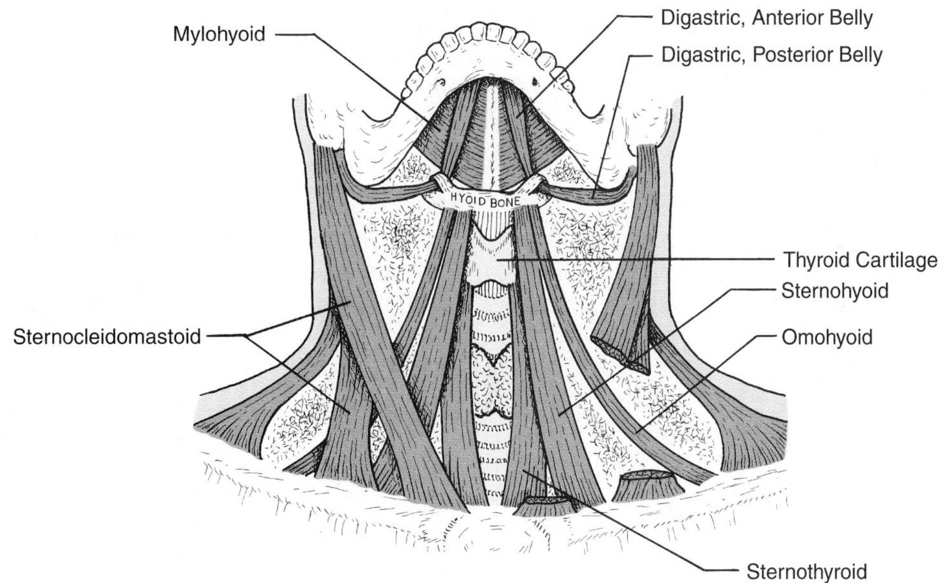

Figure 36.3 **Anterior view of neck muscles with platysmas removed.**

Action: The temporalis acts synergistically with the masseter to raise the mandible. If only its posterior fibers are activated, the temporalis can cause the mandible to retract.

Internal Pterygoid The position of the internal pterygoid has led some to refer to it as the "internal masseter." It lies on the medial surface of the ramus and originates on the maxilla, sphenoid, and palatine bones. It inserts on the medial surface of the mandible between the mylohyoid line and the angle (see figure 36.1B).

Action: The internal pterygoid works synergistically with the masseter and temporalis to raise the mandible.

External Pterygoid The external pterygoid is uppermost in figure 36.1B. It arises on the sphenoid bone and inserts on the neck of the mandibular condyle and the articular disk of the joint.

Action: Acting together, the two external pterygoids can move the mandible forward (protrusion) and downward. Individually, they can move the mandible laterally (to the right and left).

Buccinator The buccinator is the horizontal muscle that partially obscures the teeth in figure 36.1B. It is situated in the cheeks (*buccae*) on each side of the oral cavity. It arises on the maxilla, the mandible, and a ligament, the **pterygomandibular raphe.** This ligament joins the buccinator to the **constrictor pharyngis superioris.** The buccinator inserts on the orbicularis oris.

Action: The buccinator compresses the cheek to hold food between the teeth during chewing.

Assignment:
Locate all the muscles of mastication on a muscle model or cadaver (the cat is not a suitable specimen for most of these muscles). Also, complete part B of the Laboratory Report.

Neck Muscles

Figures 36.2 and 36.3 show most of the major neck muscles. While figure 36.2 is primarily of surface and posterior deep muscles, figure 36.3 reveals the deeper anterior muscles.

Surface Muscles

The principal surface muscles of the neck are the platysma on the anterolateral surfaces and the trapezius on the back of the neck.

Platysma The platysma inserts on the mandible and the muscles around the mouth. Its origin is primarily the fascia that covers the pectoralis and deltoid muscles of the shoulder region.

Action: The platysma primarily draws the lower lip backward and downward. It also assists in opening the jaws.

Trapezius The trapezius is the large, triangular muscle in the upper shoulder region of the back. Only its upper portion functions as a neck muscle. The trapezius arises on the occipital bone, the ligamentum nuchae, and the spinous processes of the seventh cervical and all the thoracic vertebrae. The **ligamentum nuchae** is a ligament that extends from the occipital bone to the seventh cervical vertebra, uniting the spinous processes of all the cervical vertebrae. The trapezius inserts on the spine and acromion of the scapula and the outer third of the clavicle.

Action: The trapezius pulls the scapula toward the median line (adduction). It also raises the scapula, as in shrugging the shoulder. If the shoulders are fixed, the trapezius draws the head backward (hyperextension).

Deeper Neck Muscles, Posterior

Figure 36.2C shows three posterior muscles of the neck, as seen when the trapezius is removed. Another prominent muscle, the levator scapulae, has been omitted here for clarity. It is studied in Exercise 37.

Splenius Capitis The splenius capitis lies just beneath the trapezius and medial to the levator scapulae. Figure 36.2C only shows the left splenius capitis. The splenius capitis originates on the ligamentum nuchae and the spinous process of the seventh cervical and upper three thoracic vertebrae. It inserts on the mastoid process.

Action: Acting singly, the splenius capitis inclines the head and rotates it toward the contracting muscle. If both of the muscles act together, they hyperextend the head.

Semispinalis Capitis The semispinalis capitis is the larger and more medial of the two muscles shown on the right side of the neck in figure 36.2C. Its origin consists of the transverse processes of the upper six thoracic vertebrae and the articular processes of the lower four cervical vertebrae. It inserts on the occipital bone.

Action: The semispinalis capitis works synergistically with the splenius capitis.

Longissimus Capitis The longissimus capitis is lateral and slightly anterior to the semispinalis capitis. This muscle originates on the transverse processes of the first three thoracic vertebrae and the articular processes of the lower four cervical vertebrae. It inserts on the mastoid process.

Action: The longissimus capitis also works synergistically with the splenius capitis.

Deeper Neck Muscles, Anterior

Figure 36.3 shows the anterior muscles that are seen when the two platysma muscles are removed from the neck.

Sternocleidomastoid The sternocleidomastoid is a large neck muscle that derives its name from the skeletal components that anchor it. It originates on the manubrium of the sternum and the sternal (medial) end of the clavicle. It inserts on the mastoid process.

Action: Simultaneous contraction of both sternocleidomastoids flexes the head forward and downward on the chest. Independently, each sternocleidomastoid draws the head down toward the shoulder on the same side as the muscle.

Sternohyoid The sternohyoid is a long muscle that extends from the hyoid bone to the sternum and clavicle. Removal of the lower portion of the sternocleidomastoid exposes the sternohyoid. The sternohyoid arises on a portion of the manubrium and clavicle. It inserts on the lower border of the hyoid bone.

Action: The sternohyoid draws the hyoid bone downward.

Sternothyroid The sternothyroid is somewhat shorter than the sternohyoid. It originates on the posterior surface of the manubrium and inserts on the inferior edge of the thyroid cartilage of the larynx.

Action: The sternothyroid draws the thyroid cartilage of the larynx downward.

Omohyoid The omohyoid is the long, narrow, curving muscle that extends from the hyoid bone into the shoulder region. It arises from the upper surface of the scapula and inserts on the hyoid bone, lateral to the sternohyoid.

Action: The omohyoid draws the hyoid bone downward.

Digastric The digastric is a narrow, V-shaped muscle. Two are visible under the mandible in figure 36.3. Each muscle consists of anterior and posterior bellies, with a fibrous loop attaching the central portion to the hyoid bone.

The posterior belly arises from the medial side of the mastoid process of the temporal bone. The anterior belly arises on the inner surface of the body of the mandible near the symphysis. The point of insertion is the fibrous loop on the hyoid bone.

Action: The digastric raises the hyoid bone and assists in lowering the mandible. Acting independently, the anterior belly can pull the hyoid bone forward; the posterior belly can pull the hyoid bone backward.

Mylohyoid The two mylohyoids extend from the medial surfaces of the mandible to the median line of the head to form the floor of the mouth. They lie just superior to the anterior bellies of the digastric muscles. Each mylohyoid arises along the mylohyoid line of the mandible and inserts on a median fibrous raphe that extends from the symphysis of the mandible to the hyoid bone.

Action: The mylohyoid raises the hyoid bone and tongue.

Assignment:
Complete part C of the Laboratory Report. Then proceed to the dissection of neck muscles in the cat.

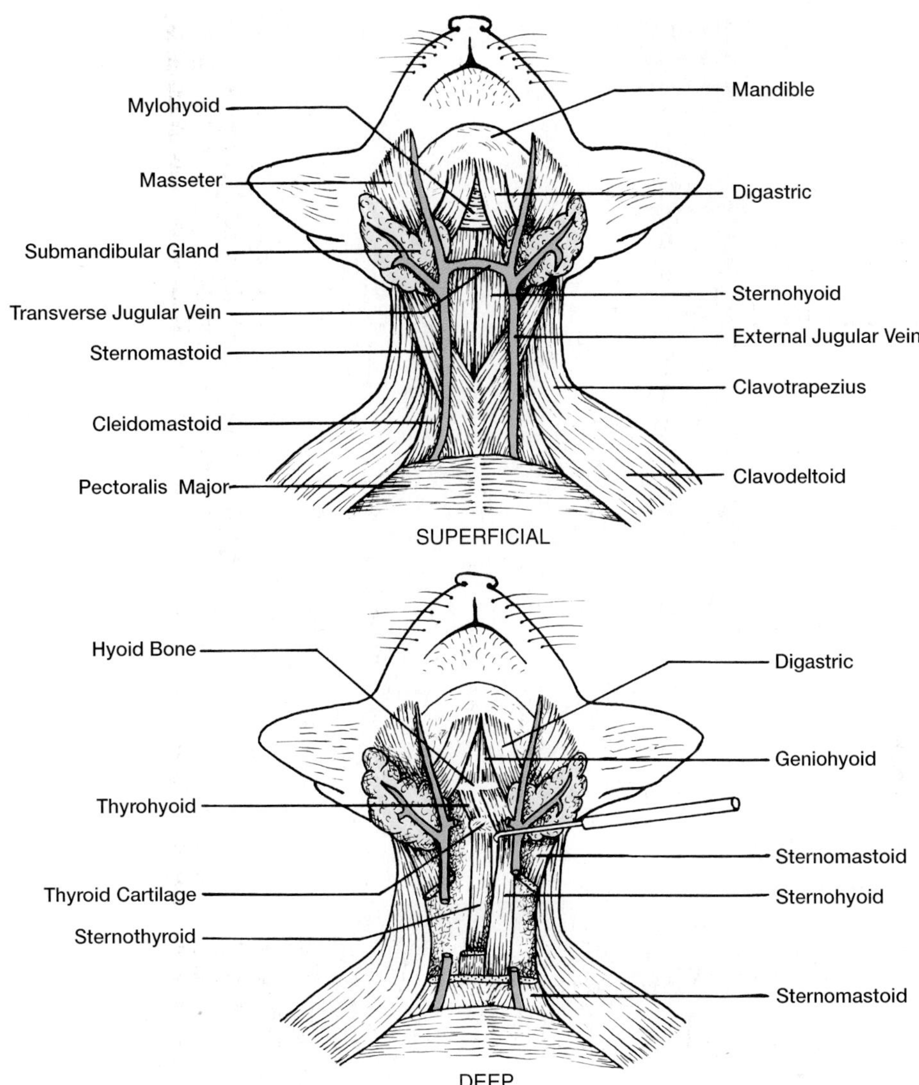

Figure 36.4 Cat neck muscles, superficial and deep.

Cat Dissection

Once the skin has been removed from the cat, place the animal ventral side up on a dissecting tray, and refer to figures 36.4 and 36.5 to identify the neck muscles. Since an embalming incision, which destroys several muscles, has been made along one side of the neck, work on the side of the neck opposite the incision. Three key words in the instructions that follow are:

Dissect: Dissecting a muscle in this laboratory means using a blunt probe to free the muscle from adjacent muscles, exposing its full length. *Do not cut up the muscle.*

Transect: Transection involves cutting the muscle transversely with a scalpel or scissors.

Reflect: Once a muscle has been transected, its cut ends can be lifted out of the way, or reflected, to expose deeper muscles.

Do not mutilate or remove the muscles on your specimen unless directed to do so. Muscle dissection should be methodical and careful. A good way to keep track of muscles is to color them on figure 36.4 with different-colored pencils as you identify them. This may also be helpful for review.

Identify the cat muscles in the following order:

Digastric Like the human digastric, this superficial muscle has two portions: an anterior belly and a posterior belly. The anterior belly attaches to the mandible, and the posterior belly attaches to the mastoid process of the skull.

Mylohyoid The mylohyoid is a thin, flat muscle with fibers that course horizontally below the mandible. It lies deep to the anterior belly of the digastric. Pull the anterior bellies of the digastric

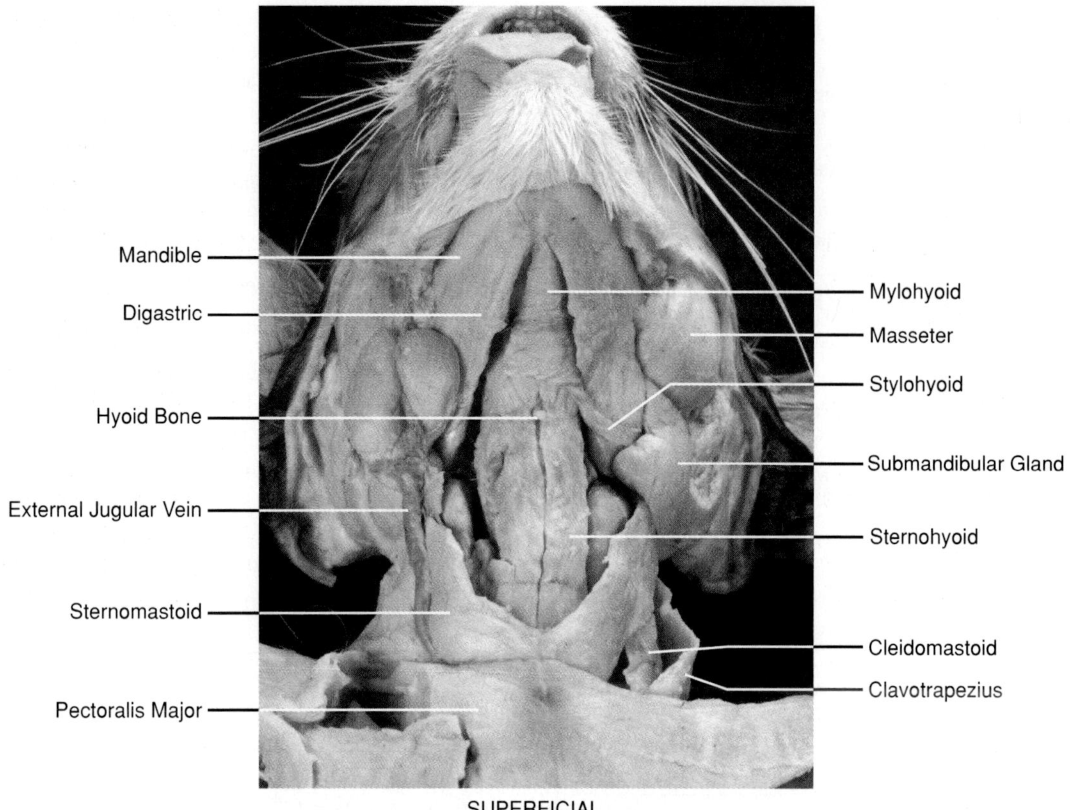

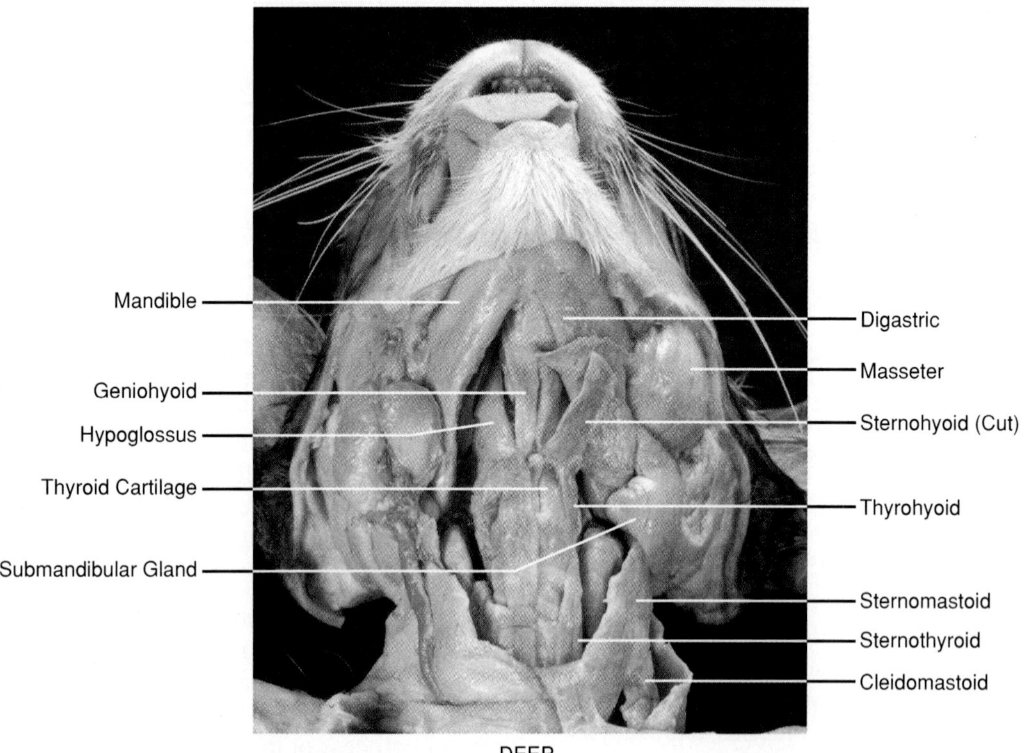

Figure 36.5 Cat neck muscles, superficial and deep.

muscles laterally to see the full extent of the mylohyoid.

Sternomastoid The sternomastoid is a bandlike muscle that extends from the manubrium of the sternum to the mastoid portion of the temporal bone and lies deep to the external jugular vein. Note that the right and left sternomastoids fuse on the midline. Cut through this point of fusion in the caudal direction to the manubrium. *Dissect* the sternomastoid that is on the side opposite the embalming incision.

Cleidomastoid The cleidomastoid is a narrow band of muscle between the sternomastoid and clavotrapezius. *Dissect* it. What human muscle replaces these last two muscles?

Sternohyoid The sternohyoid is a straight, narrow muscle that extends from the manubrium to the hyoid bone, covering the larynx and trachea. *Dissect* it. You will need to cut away the transverse jugular vein at this time.

Sternothyroid The sternothyroid is a narrow band of muscle that is lateral and dorsal to the sternohyoid. It extends from the manubrium to the thyroid cartilage of the larynx. To expose the sternothyroid, lift up the sternohyoid, and push it to one side. *Dissect.*

Thyrohyoid Cranial to the sternothyroid, and extending between the hyoid bone and the lateral portion of the thyroid cartilage, is a very small muscle, the thyrohyoid.

Masseter The masseter, the most powerful muscle of mastication, is a large, rounded muscle in the cheek region. It originates on the zygomatic arch and inserts on the lateral surface of the mandible.

37 Trunk and Shoulder Muscles

This exercise focuses on all of the surface and most of the deeper muscles of the trunk and shoulder. These muscles primarily move the shoulders, upper arms, spine, and head. Some assist in respiration.

Anterior Muscles

Removal of the skin and subcutaneous fat from the body reveals the muscles shown in figure 37.1.

Surface Muscles

Pectoralis Major The pectoralis major is a thick, fan-shaped muscle that occupies the upper quadrant of the chest. It originates on the clavicle, sternum, costal cartilages, and the aponeurosis of the external oblique muscle. It inserts in the groove between the greater and lesser tubercles of the humerus.

Action: The pectoralis major adducts, flexes, and rotates the humerus medially.

Deltoid The deltoid is the principal muscle of the shoulder. It originates on the lateral third of the clavicle, the acromion, and the spine of the scapula. It inserts on the deltoid tuberosity of the humerus.

Action: Activation of the entire deltoid abducts the arm. Activation of only certain parts of the deltoid results in flexion, extension, and rotation (medial and lateral).

Serratus Anterior The serratus anterior covers the upper and lateral surfaces of the rib cage. It originates on the upper eight or nine ribs and inserts on the anterior surface of the scapula near the vertebral border.

Action: The serratus anterior pulls the scapula forward, downward, and inward toward the chest wall.

Deeper Muscles

Pectoralis Minor The pectoralis minor lies beneath the pectoralis major and is completely obscured by the latter. It arises on the third, fourth, and fifth ribs, and it inserts on the coracoid process of the scapula.

Action: The pectoralis minor draws the scapula forward and downward with some rotation.

External Intercostals Between the ribs on both sides are 11 pairs of short muscles, the external intercostals. Each of these short muscles arises from the lower border of a rib and inserts on the upper border of the rib below it. Note that their fibers are directed obliquely forward on the front of the ribs.

Action: The external intercostals pull the ribs closer to each other, raising them, which increases the volume of the thorax and causes inspiration of air.

Internal Intercostals The internal intercostals are antagonists of the external intercostals and lie on the internal surface of the rib cage. These muscles line the complete rib cage. Each muscle arises from a ridge on the inner surface of a rib, as well as the rib's corresponding costal cartilage, and inserts on the upper border of the rib below. The fibers of

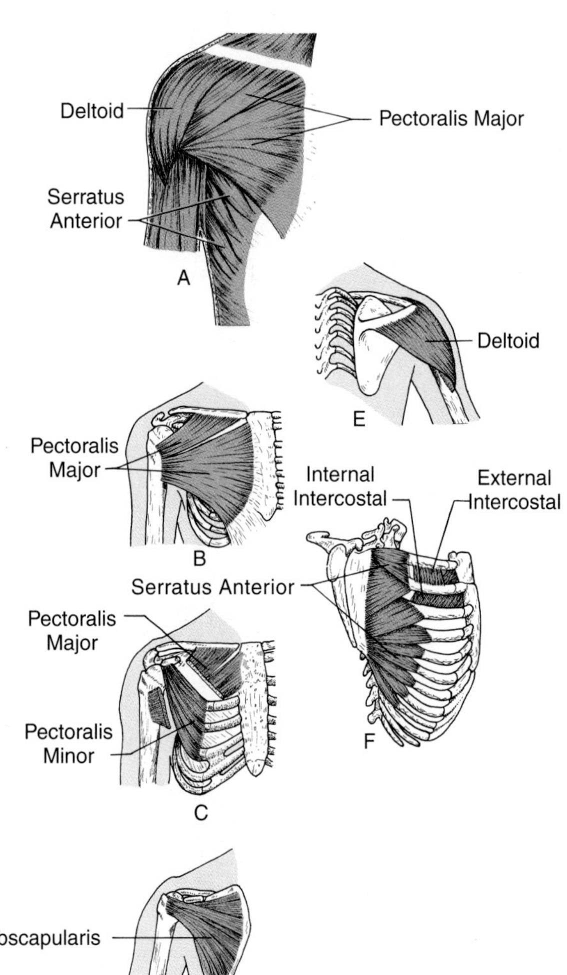

Figure 37.1 Anterior shoulder and trunk muscles.

these muscles are oppositely oriented to the fibers of the external intercostals.

Action: The internal intercostals draw adjacent ribs closer together, which lowers the ribs and decreases the volume of the thoracic cavity, resulting in expiration of air.

Subscapularis The subscapularis is the large, triangular muscle that fills the subscapular fossa of the scapula. Some fibers originate on the vertebral (medial) margin; others are anchored to the lower two-thirds of the axillary margin of the scapula. All fibers pass laterally to insert on the lesser tubercle of the humerus and the anterior portion of the joint capsule.

Action: The subscapularis rotates the arm medially and assists in other directional movements of the arm, depending on the position of the arm.

Assignment:
Complete part A of the Laboratory Report.

Posterior Muscles

Figure 37.2A shows the muscles of the upper back region that are revealed when the skin is removed, while figures 37.2B, C, and D show the deeper muscles of the back and shoulder.

Latissimus Dorsi The latissimus dorsi is the large back muscle that covers the lumbar area. It originates in a broad aponeurosis attached to thoracic and lumbar vertebrae, the spine of the sacrum, the iliac crest, and the lower ribs. It inserts on the intertubercular groove of the humerus.

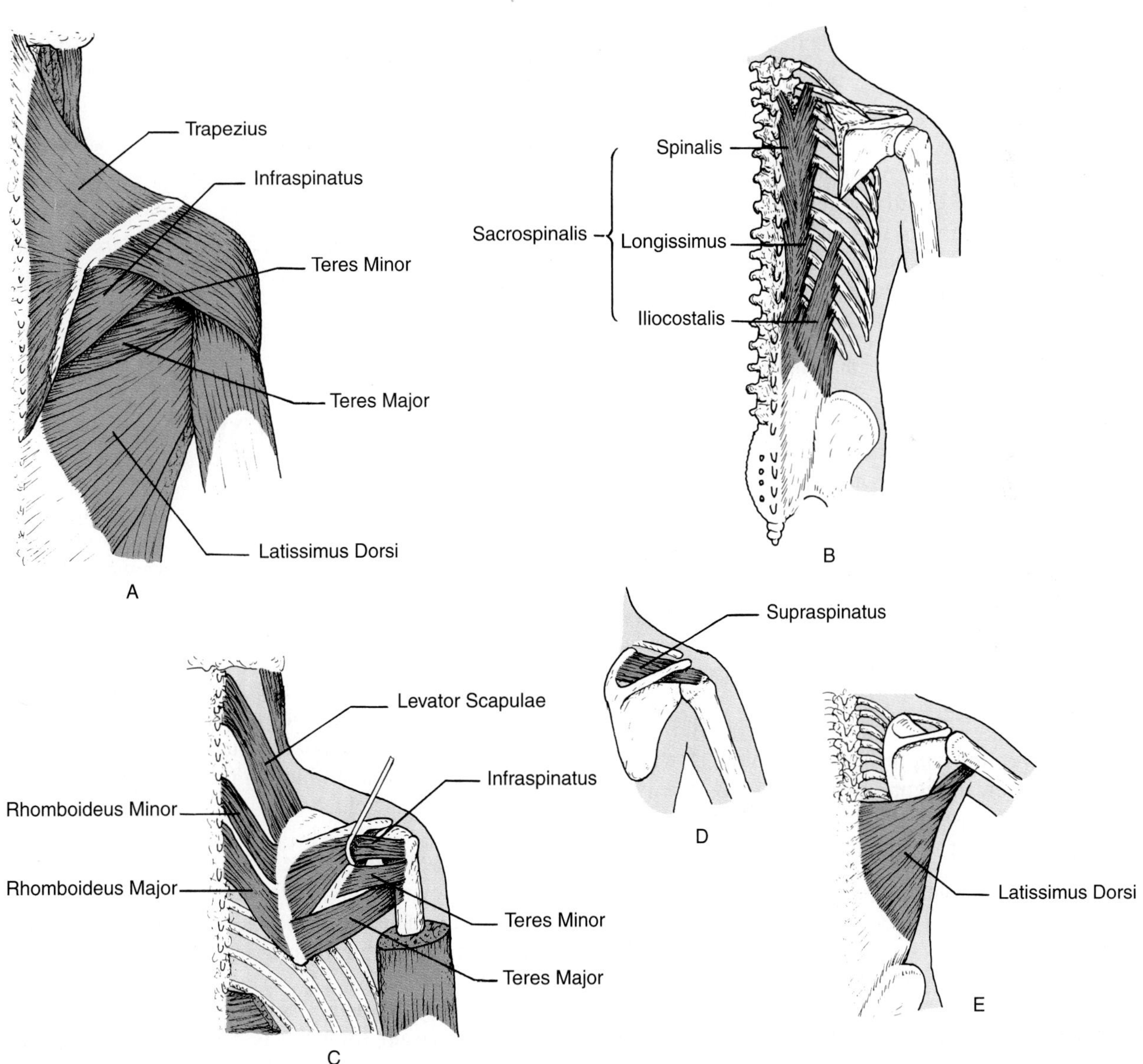

Figure 37.2 Shoulder and back muscles.

Action: The latissimus dorsi extends, adducts, and rotates the arm medially. It also draws the shoulder downward and backward.

Infraspinatus The infraspinatus derives its name from its position: It attaches to the inferior margin of the scapular spine. Although a portion of it can be seen in figure 37.2A, its complete structure is not visible unless the deltoid and trapezius are removed, as in figure 37.2C. It originates on the infraspinous fossa. It inserts on the middle facet of the greater tubercle of the humerus.

Action: The infraspinatus rotates the humerus laterally.

Teres Major Of the three muscles that cover most of the scapula in figure 37.2C, the teres major is the most inferior. Although the latissimus dorsi obscures part of it, some of the teres major is visible in figure 37.2A. It originates near the inferior angle of the scapula and inserts into the crest of the lesser tubercle of the humerus.

Action: The teres major rotates the humerus medially and weakly adducts it.

Teres Minor The teres minor is a small muscle that lies between the infraspinatus and teres major. Only a small portion of it is visible in figure 37.2A. It originates on the lateral margin of the scapula, and it inserts on the lowest facet of the greater tubercle of the humerus.

Action: The teres minor rotates the arm laterally and weakly adducts it.

Levator Scapulae The levator scapulae lies under the trapezius in the neck region. It originates on the transverse processes of the first four cervical vertebrae. It inserts on the upper portion of the vertebral border of the scapula.

Action: The levator scapulae raises the scapula and draws it medially. With the scapula in a fixed position, the levator scapulae can bend the neck laterally.

Supraspinatus The trapezius and deltoid completely cover the supraspinatus, which is why it cannot be seen in figure 37.2A. It arises from the supraspinatus fossa of the scapula and inserts on the greater tubercle of the humerus.

Action: The supraspinatus assists the deltoid in abducting the humerus.

Rhomboideus Major and Minor The rhomboideus major and minor are flat muscles that lie beneath the trapezius. They extend from the vertebral border of the scapula to the spine (see figure 37.2C).

The rhomboideus minor is smaller and lies between the levator scapulae and the rhomboideus major. It arises from the lower part of the ligamentum nuchae and the first thoracic vertebra. It inserts on that part of the scapular vertebral margin where the scapular spine originates.

The rhomboideus major arises from the spinous processes of the second, third, fourth, and fifth thoracic vertebrae. It inserts below the rhomboideus minor on the vertebral border of the scapula.

Action: The rhomboideus major and minor act together to pull the scapula medially and slightly upward. The lower part of the rhomboideus major rotates the scapula to depress the lateral angle, assisting in adduction of the arm.

Sacrospinalis (Erector Spinae) The sacrospinalis is a long muscle that extends over the back from the sacral region to the midshoulder region (see figure 37.2B). It consists of three portions: a lateral **iliocostalis,** an intermediate **longissimus,** and a medial **spinalis.** It arises from the lower and posterior portion of the sacrum, the iliac crest, and the lower two thoracic vertebrae. It inserts on the ribs and transverse processes of the vertebrae.

Action: The sacrospinalis is an extensor, pulling backward on the ribs and vertebrae to maintain erectness.

Assignment:
Complete part B of the Laboratory Report. Then proceed to the dissection of the cat thorax and shoulder muscles.

Cat Dissection

Thorax and Shoulder Muscles (Superficial)

Cat muscles of the thorax fall into three categories: the pectoralis, trapezius, and deltoid groups. Use figures 37.3, 37.4, and 37.5 for reference.

The Pectoralis Group

Pectoantebrachialis The pectoantebrachialis is a narrow (about 1 cm wide), ribbonlike muscle that extends from the upper sternum to the forearm fascia. Since there is no human homolog, *transect* and remove this muscle.

Pectoralis Major The pectoralis major is about 5 cm wide and lies deep to the pectoantebrachialis. With a blunt probe, separate it from the clavodeltoid and the pectoralis minor. *Dissect* it.

Pectoralis Minor Note that the pectoralis minor is larger than the pectoralis major. It arises from the sternum and passes diagonally, cranially, and laterally beneath the pectoralis major to insert on both the scapula and humerus. In

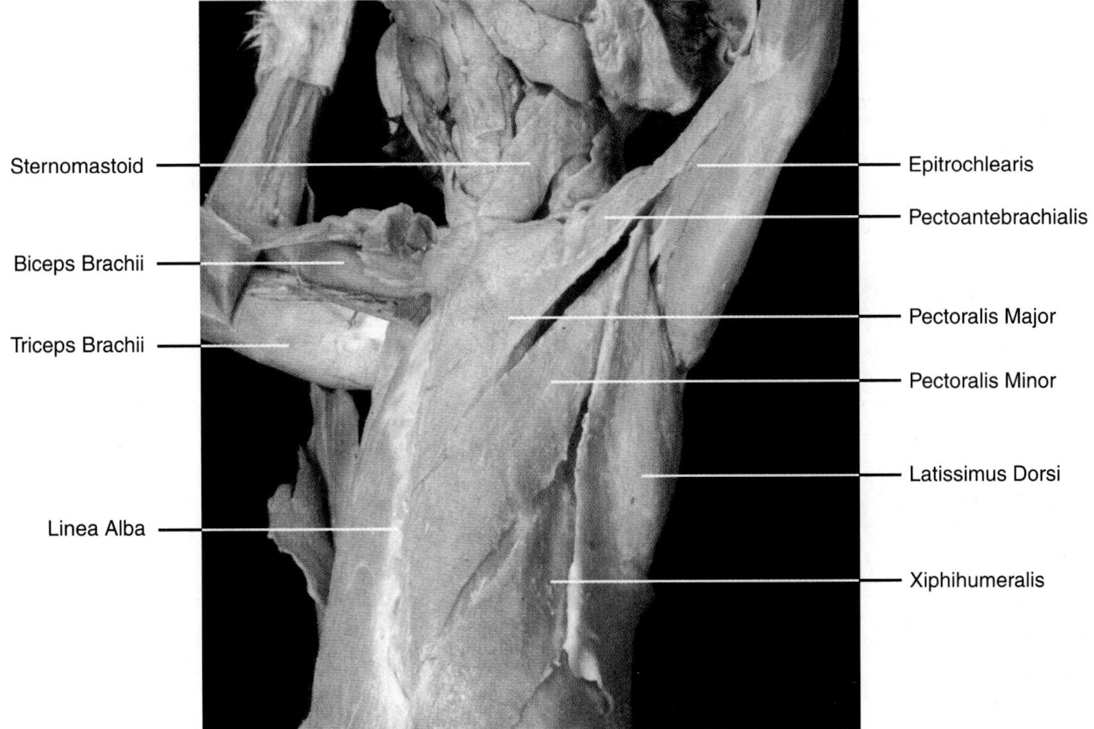

Figure 37.3 Ventrolateral aspect of superficial thoracic muscles of the cat.

humans, this muscle inserts only on the scapula. *Transect* and *reflect.*

Xiphihumeralis The xiphihumeralis is the most caudal muscle of the pectoral group. Its fibers run parallel to the pectoralis minor and eventually pass deep to the pectoralis minor to insert on the humerus. There is no human homolog. Remove this muscle.

The Trapezius Group

Three separate muscles in the cat represent the human trapezius muscle. Use figures 37.4 and 37.5 to identify them on your specimen.

Clavotrapezius The clavotrapezius is the most cranial trapezius muscle on the cat, covering the dorsal surface of the neck. It inserts on the clavicle and unites with the clavodeltoid to form a single muscle, the **brachiocephalicus.** The brachiocephalicus is homologous to the portion of the human trapezius that inserts on the clavicle.

Acromiotrapezius The acromiotrapezius lies caudad to the clavotrapezius. It compares to the part of the human trapezius that inserts on the acromion. *Dissect, transect* by cutting through the tendon of origin on the middorsal line, and *reflect.*

Spinotrapezius The spinotrapezius is caudad to the acromiotrapezius and superficial to the cranial border of the latissimus dorsi. It compares to the part of the human trapezius that inserts on the spine of the scapula. *Dissect, transect,* and *reflect.*

The Deltoid Group

The cat has three deltoid muscles that are homologs of the single human deltoid. Locate each of the following muscles:

Clavodeltoid (Clavobrachialis) The position of the clavodeltoid is best seen in figure 37.5 and on the ventral view in figure 37.4. This muscle is an extension of the clavotrapezius. It originates on the clavicle and inserts, primarily, on the proximal end of the ulna; some of its fibers, however, merge with the pectoantebrachialis to insert on the fascia of the forearm. *Dissect.*

Acromiodeltoid The acromiodeltoid compares to the acromial portion of the human deltoid. *Dissect.* Although freeing up this muscle is difficult, clearly separate at least the cranial and caudal borders from adjacent muscles.

Spinodeltoid The spinodeltoid is a thick muscle that lies caudad to the acromiodeltoid and arises

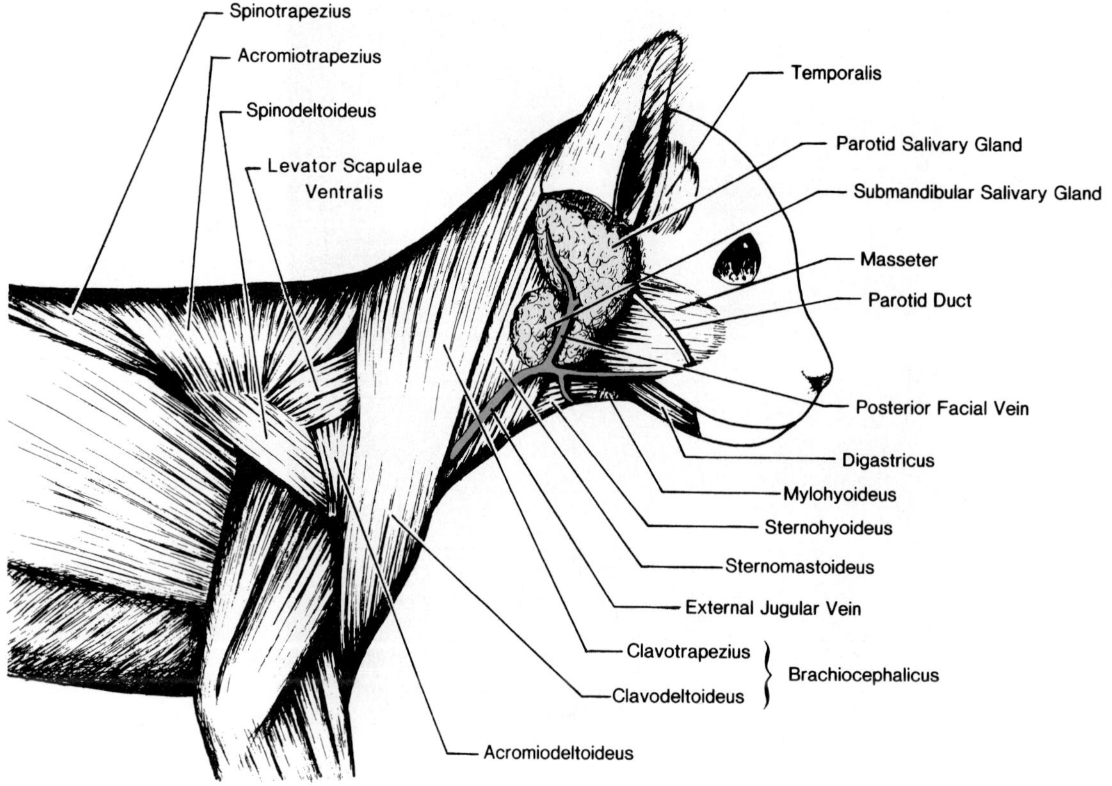

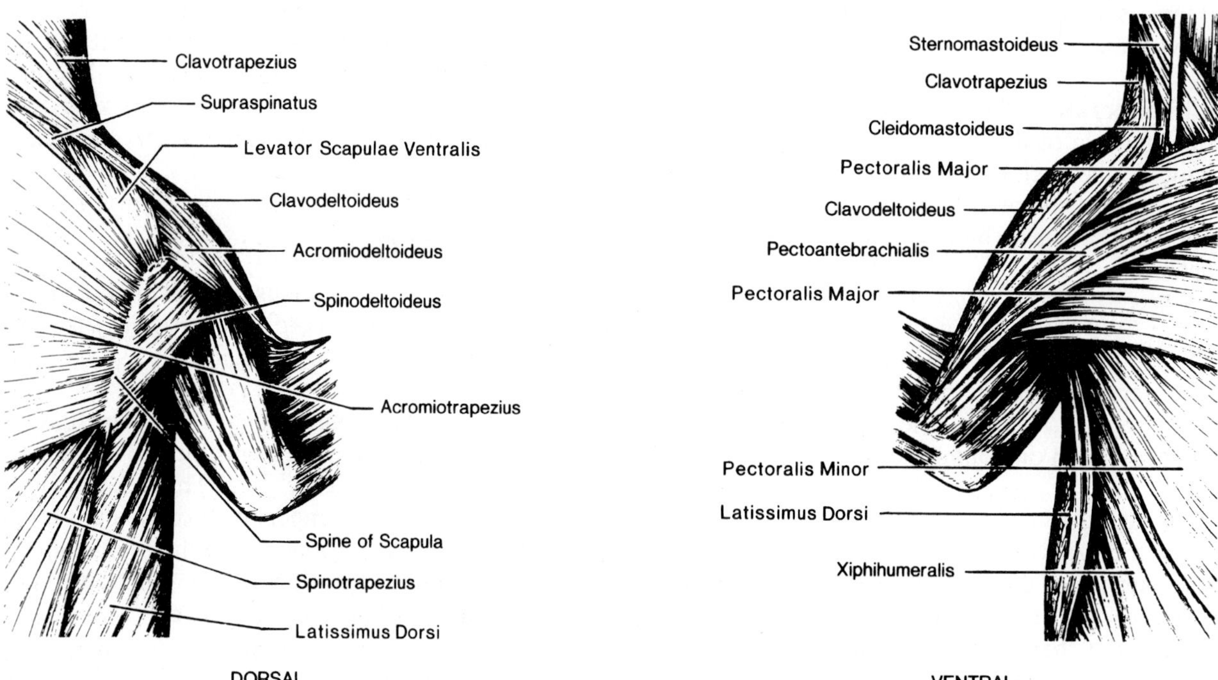

Figure 37.4 Superficial trunk and shoulder muscles of the cat.

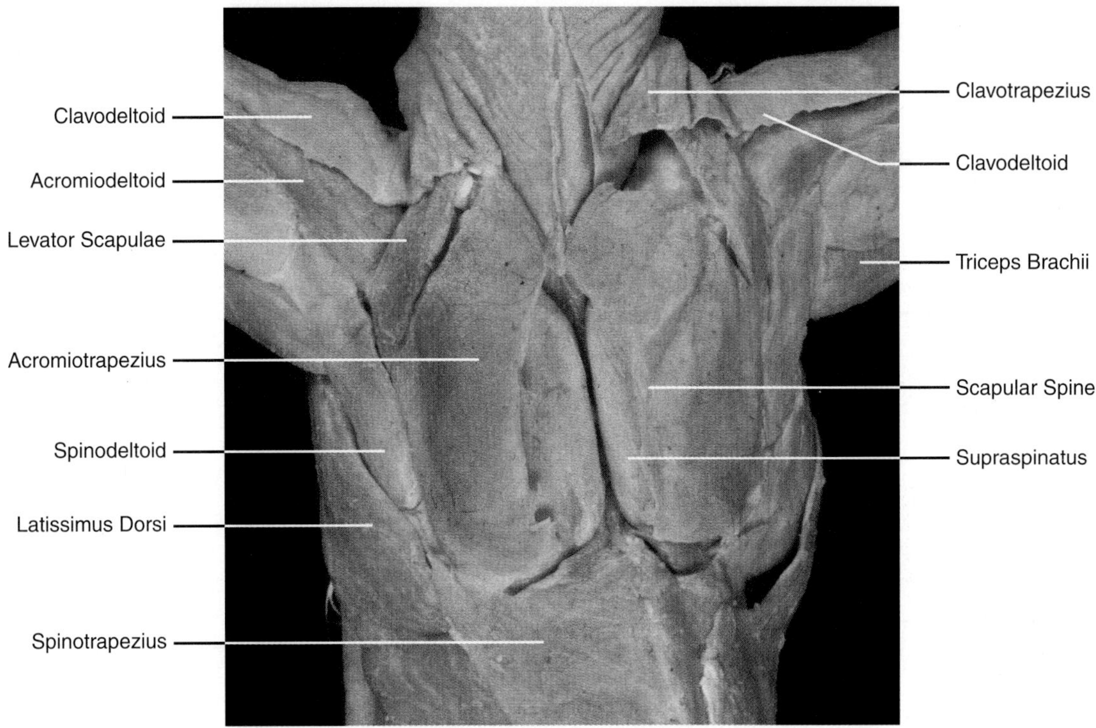

Figure 37.5 Dorsal aspect of superficial thoracic muscles of the cat.

from the spine of the scapula. Use figure 37.5 and the dorsal view in figure 37.4 for reference. *Dissect, transect,* and *reflect.*

Other Superficial Thoracic Muscles

In addition to the previous three muscle groups, two more superficial muscles are important:

Levator Scapulae Ventralis The levator scapulae ventralis is shown on the lateral and dorsal views in figure 37.4. Note that this muscle is caudad to and passes beneath the clavotrapezius. It joins the craniolateral edge of the acromiotrapezius and inserts on the acromial part of the scapular spine. It arises from the transverse process of the atlas and the occipital bone. Since it has no human homolog, identify it and move on to the next muscle.

Latissimus Dorsi The latissimus dorsi is a broad, flat, triangular muscle with an extensive origin on the spines of the last six thoracic vertebrae, the lumbar vertebrae, the sacrum, and the iliac crest. The lumbodorsal fascia, an aponeurosis, contains the fibers of this origin. The muscle inserts on the proximal portion of the humerus. *Dissect, transect,* and *reflect.*

Deep Thoracic Muscles

Use figures 37.6, 37.7, and 37.8 for reference as you examine the following deeper muscles of the thorax:

Serratus Ventralis Pull the scapula away from the body wall, and rotate it dorsad. Clean off the fascia from the body wall. Note the serrate margin of the serratus ventralis; this is due to the origin by digitations from the upper eight or nine ribs.

The cranial portion of the serratus ventralis is homologous to the human levator scapulae, and the caudal portion is homologous to the serratus anterior. *Dissect.*

Rhomboideus The rhomboideus (see figure 37.7) lies deep to the acromiotrapezius and spinotrapezius, which should have been transected previously. Pull the two forelegs together on the ventral side. This abducts the scapulae so that you can see the full extent of the rhomboideus.

The rhomboideus extends from the spines of the upper thoracic vertebrae to the vertebral border of the scapula. The cranial (and larger) part of this muscle is homologous to the rhomboideus minor muscle in humans, and the caudal (and smaller) part is homologous to the rhomboideus major. The rhomboideus major is larger than the rhomboideus minor in humans.

Levator Scapulae The levator scapulae (see figure 37.5) is not a separate muscle but a cranial continuation of the serratus ventralis. It originates from the transverse processes of the caudal five cervical vertebrae, and it inserts on the vertebral border of the scapula, ventral to the insertion of the rhomboideus.

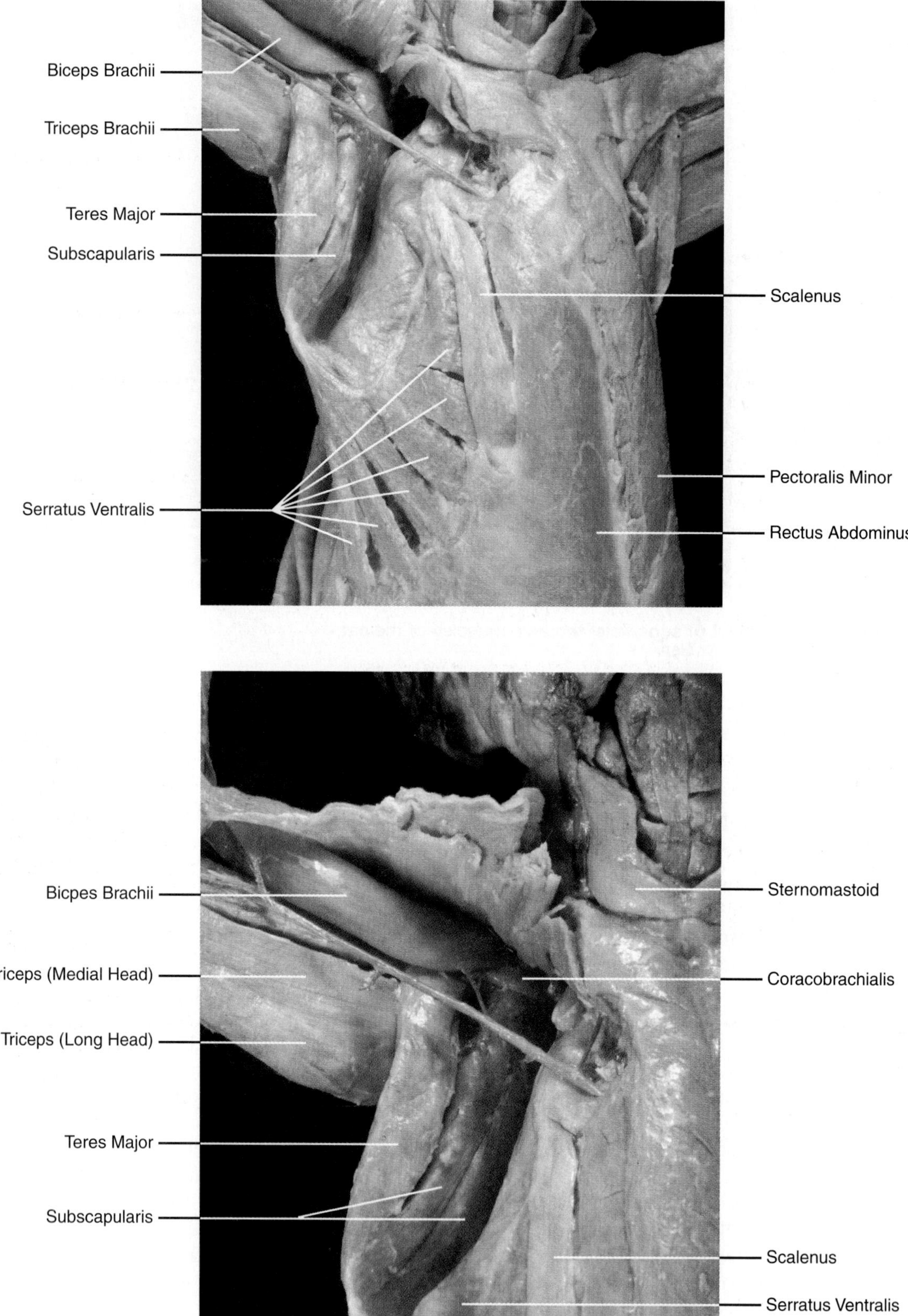

Figure 37.6 Deep muscles of cat thorax and shoulder, ventral aspect.

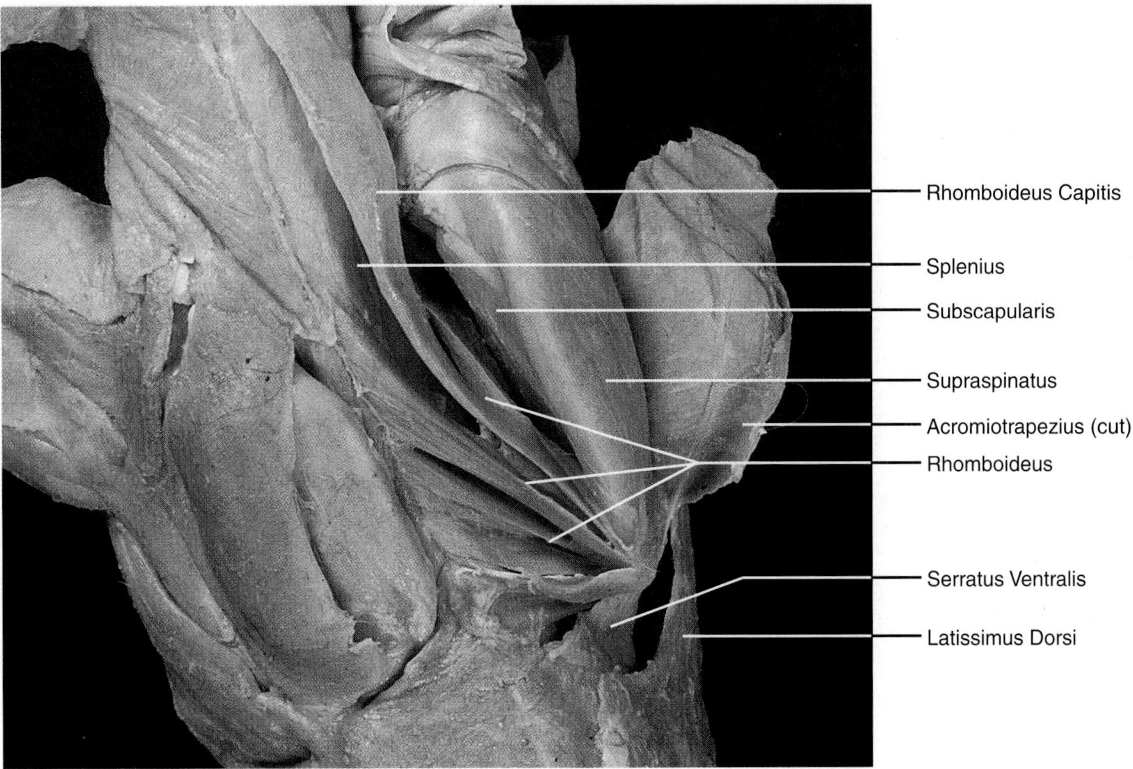

Figure 37.7 Deep muscles of cat shoulder and back, dorsal aspect.

In the human, the insertion is comparable to that of the cat, but the origin is roughly comparable to that of the cat's levator scapulae ventralis.

Supraspinatus With the acromiotrapezius fully reflected, the supraspinatus (see figures 37.7 and 37.8) can be seen filling the supraspinous fossa of the scapula. From its origin in the supraspinous fossa, it passes to the greater tubercle of the humerus.

Infraspinatus *Reflect* the transected spinodeltoid. You can now see the infraspinatus (figure 37.8) extending from its origin in the infraspinous fossa of the scapula to the greater tubercle of the humerus.

Teres Major The teres major (see figures 37.6 and 37.8) is a thick band of muscle that is caudad to the infraspinatus muscle and covers the axillary border of the scapula. It extends forward to insert on the proximal end of the humerus. *Dissect. Reflect* the latissimus dorsi to expose the full extent of the teres major.

Teres Minor The small teres minor is between the infraspinatus and the long head of the biceps brachii. Since it is not labeled on any of the illustrations in this manual, you have to look for it by reflecting the spinodeltoid. The teres minor extends from the axillary border of the scapula to the humerus.

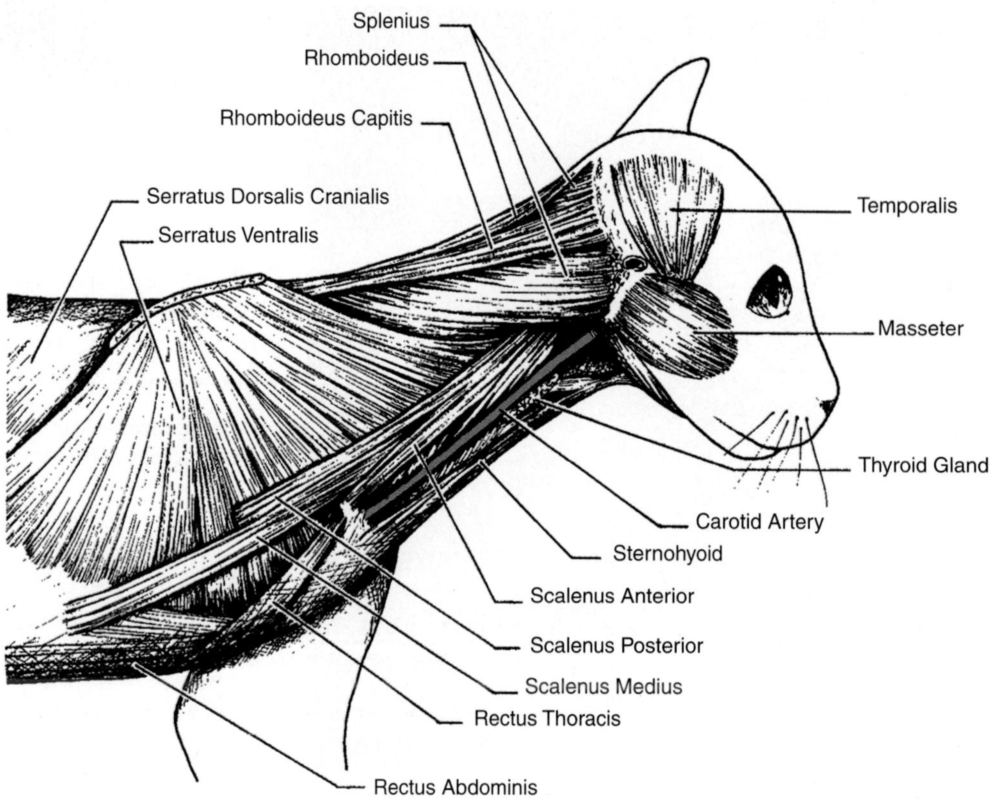

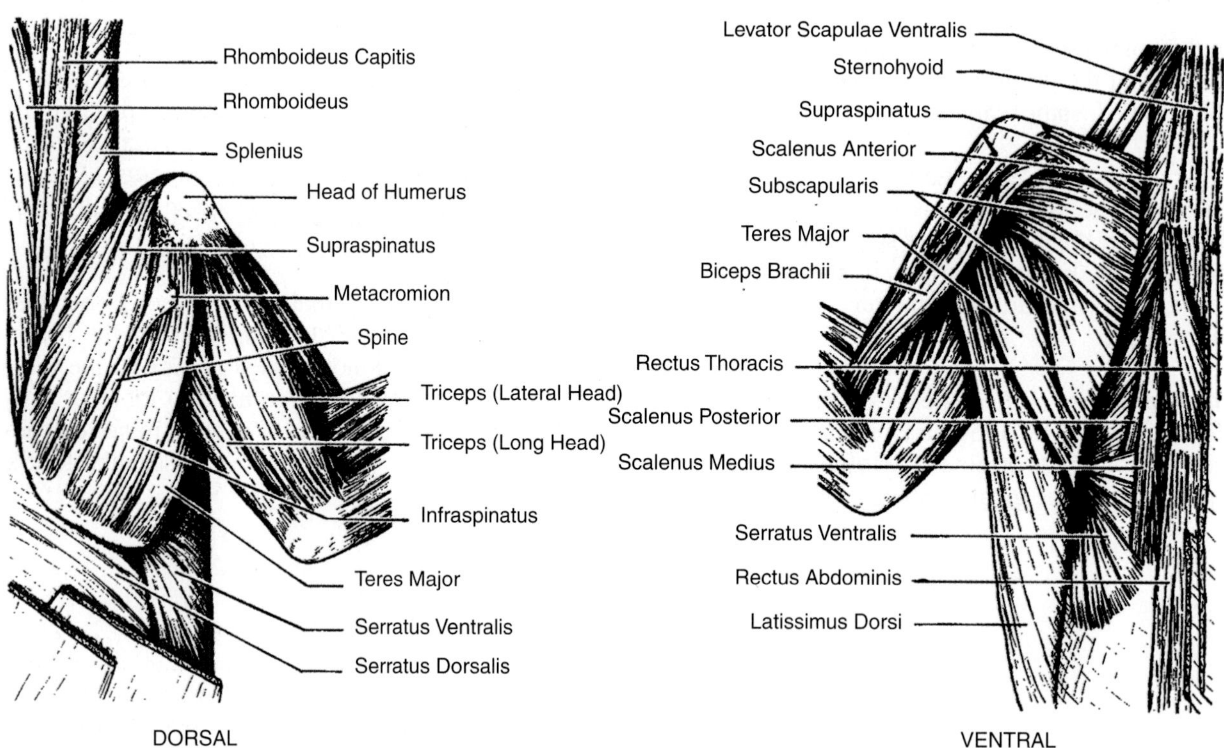

Figure 37.8 Deep trunk, shoulder, and neck muscles of the cat.

Upper Extremity Muscles

38

As discussed in Exercise 37, the muscles of the shoulder and trunk are primarily responsible for moving the upper arm. This exercise focuses on the muscles of the brachium that control forearm movements. One muscle that was not mentioned in Exercise 37 but that does move the upper arm is the **coracobrachialis** (see figure 38.1D).

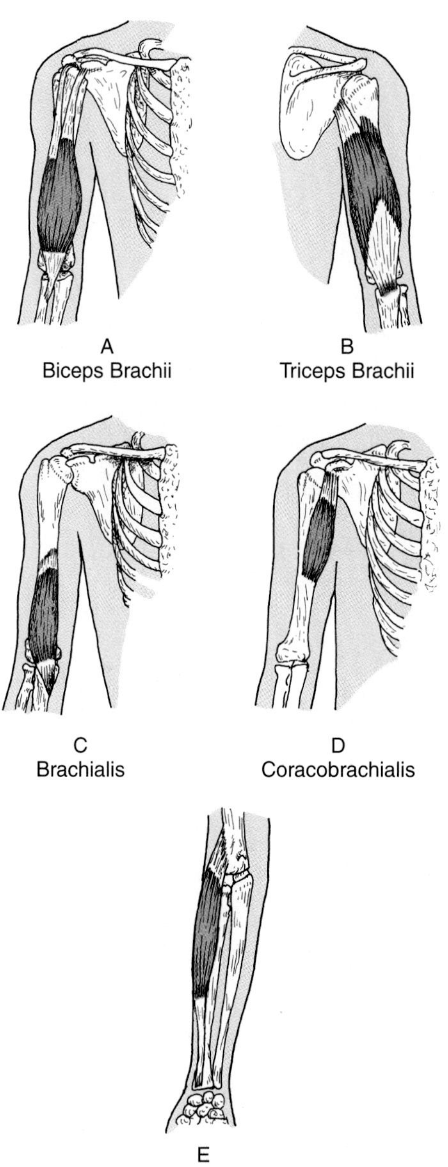

Figure 38.1 Muscles attached to the brachium.

Coracobrachialis The coracobrachialis covers a portion of the upper medial surface of the humerus. It originates on the apex of the coracoid process of the scapula. It inserts in the middle of the medial surface of the humerus.

Action: The coracobrachialis carries the arm forward in flexion and adducts the arm.

Forearm Movements

The principal movers of the forearm are the biceps brachii, brachialis, brachioradialis, and triceps brachii (see figure 38.1). Note that the Latin term, *brachium* (upper arm) is incorporated into all the names of these upper arm muscles.

Biceps Brachii The biceps brachii is the large muscle on the anterior surface of the upper arm that bulges when the forearm flexes. Its origin consists of two tendinous heads: a medial tendon that attaches to the coracoid process and a lateral tendon that fits into the intertubercular groove on the humerus. The lateral tendon attaches to the supraglenoid tubercle of the scapula. At the lower end of the muscle, the two heads unite to form a single tendinous insertion on the radial tuberosity of the radius.

Action: The biceps brachii flexes the forearm and also moves the radius outward to supinate the hand.

Brachialis The brachialis lies immediately under the biceps brachii on the distal anterior portion of the humerus. It originates on the lower half of the humerus. Its insertion attaches to the front surface of the coronoid process of the ulna.

Action: The brachialis flexes the forearm.

Brachioradialis The brachioradialis is the most superficial muscle on the lateral (radial) side of the forearm. It originates above the lateral epicondyle of the humerus and inserts on the lateral surface of the radius, slightly above the styloid process.

Action: The brachioradialis flexes the forearm.

Triceps Brachii The triceps brachii covers the entire back surface of the brachium. It has three heads of origin: a long head from the scapula, a lateral head

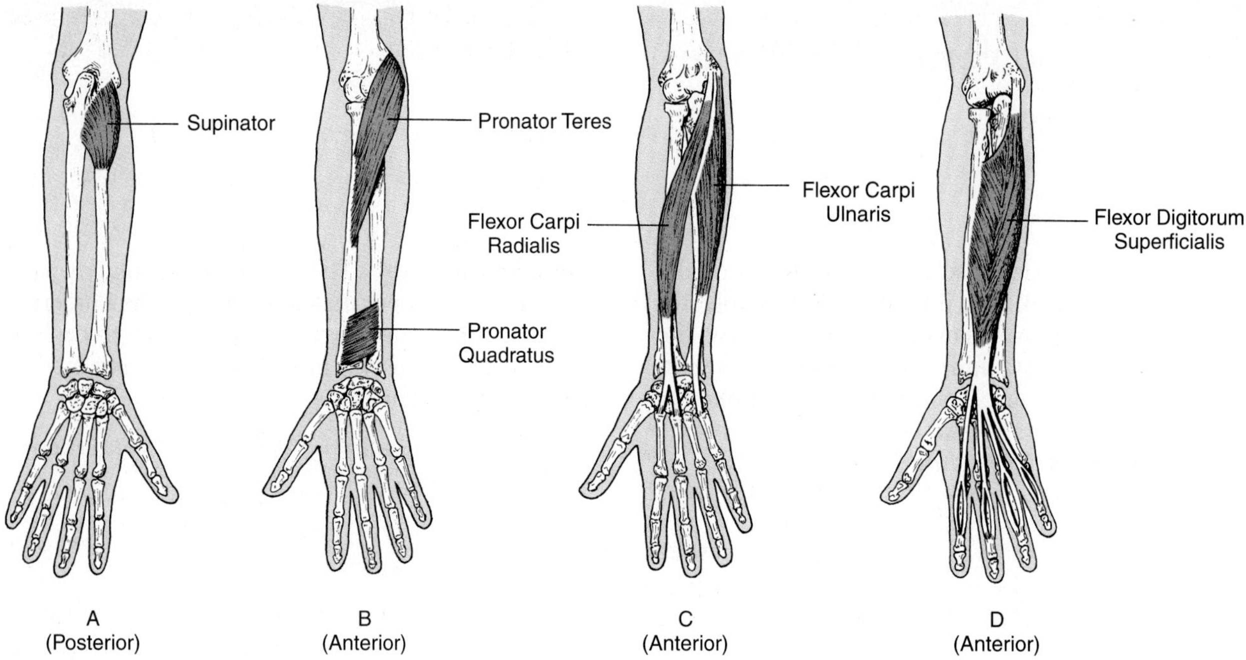

Figure 38.2 Forearm muscles.

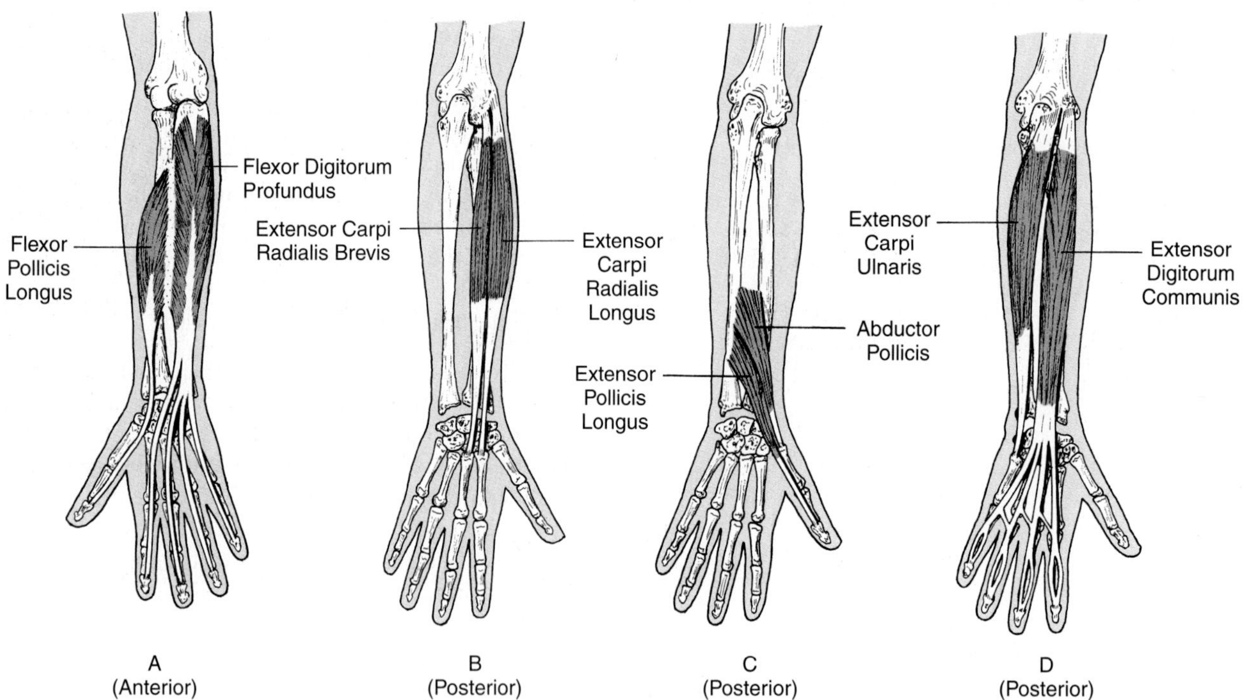

Figure 38.3 Forearm muscles.

from the posterior surface of the humerus, and a medial head from the surface below the radial groove. The tendinous insertion of the muscle attaches to the olecranon process of the ulna.

Action: The triceps brachii extends the forearm and is the antagonist of the flexor muscles.

Assignment:
Complete part A of the Laboratory Report.

Hand Movements

Figures 38.2 and 38.3 show forearm muscles that cause hand movements. All illustrations are of the right arm. Thus, if the thumb points to the left side of the page, you are looking at the anterior aspect; if the thumb points to the right side of the page, you are seeing a posterior view.

Supination and Pronation

Two muscles can cause supination: the biceps brachii and the supinator. The **supinator** is a short muscle near the elbow that arises from the lateral epicondyle of the humerus and the ridge of the ulna. It curves around the upper portion of the radius and inserts on the lateral edge of the radial tuberosity and the oblique line of the radius.

Two pronator muscles shown in figure 38.2B cause pronation. The **pronator teres** arises on the medial epicondyle of the humerus and inserts on the upper lateral surface of the radius. The **pronator quadratus** originates on the distal portion of the ulna and inserts on the distal lateral portion of the radius.

Flexion of the Hand

Figure 38.2C shows two muscles that flex the hand. The **flexor carpi radialis** extends diagonally across the forearm. It arises on the medial epicondyle of the humerus and inserts on the proximal portions of the second and third metacarpals. The **flexor carpi ulnaris** originates on the medial epicondyle of the humerus and the posterior surface (olecranon process) of the ulna. Its insertion consists of a tendon that attaches to the base of the fifth metacarpal.

Action: While both muscles flex the hand, the flexor carpi radialis abducts and the flexor carpi ulnaris adducts the hand.

Flexion of Fingers of the Hand

Three muscles flex the fingers. The **flexor digitorum superficialis** in figure 38.2D flexes all fingers except the thumb. It arises on the humerus, the ulna, and the radius. Its insertion consists of tendons that attach to the middle phalanges of the second, third, fourth, and fifth fingers.

The second flexor of the fingers is the **flexor digitorum profundus** (see figure 38.3A). It lies directly under the flexor digitorum superficialis. It originates on the ulna and the interosseous membrane between the radius and ulna. It inserts with four tendons on the distal phalanges of the second, third, fourth, and fifth fingers. The flexor digitorum profundus flexes the distal portions of the fingers.

The third flexor is the **flexor pollicis longus** (Latin: *pollex,* thumb) (see figure 38.3A). It arises on the radius, the ulna, and the interosseous membrane between these two bones. Its insertion consists of a tendon anchored to the distal phalanx of the thumb. The flexor pollicis longus flexes only the thumb.

Extension of Wrist and Hand

Figures 38.3B and D show the three muscles that extend the hand at the wrist.

Figure 38.3B shows the **extensor carpi radialis longus** and **brevis.** The brevis muscle is medial to the longus. Both muscles originate on the humerus, with the longus taking a more proximal position. The longus inserts on the second metacarpal; the brevis inserts on the middle metacarpal.

The **extensor carpi ulnaris** is the third muscle involved in extending the hand (see figure 38.3D). It arises on the lateral epicondyle of the humerus and part of the ulna. It inserts on the fifth metacarpal.

Extension and Abduction of Fingers

Three muscles extend the fingers; one abducts the thumb. The **extensor digitorum communis** lies alongside the extensor carpi ulnaris (see figure 38.3D). It arises from the lateral epicondyle of the humerus and inserts on the distal phalanges of fingers two through five. It extends all fingers except the thumb.

The **extensor pollicis longus, extensor pollicis brevis,** and **abductor pollicis** move the thumb. The longus muscle (see figure 38.3C) arises on both the ulna and the radius. It inserts on the distal phalanx of the thumb. The brevis muscle, which lies superior to the longus, is not shown in figure 38.3C. It inserts on the proximal phalanx of the thumb and assists the longus in thumb extension.

The origin of the abductor pollicis (see figure 38.3C) is on the interosseous membrane. The abductor pollicis inserts on the lateral portion of the first metacarpal and trapezium. Its only action is to abduct the thumb.

Assignment:
Complete part B of the Laboratory Report for this exercise.

Cat Dissection (Forelimb)

Muscles of the Brachium

The cat has six homologs of the human brachium. Except for the anconeus, all the human muscles are identified on pages 165 and 166.

One muscle of the cat brachium, the *epitrochlearis,* has no homolog in the human and should be removed prior to studying the six homologs. It is a flat, thin, superficial muscle on the medial side of the brachium (see figure 38.4). It

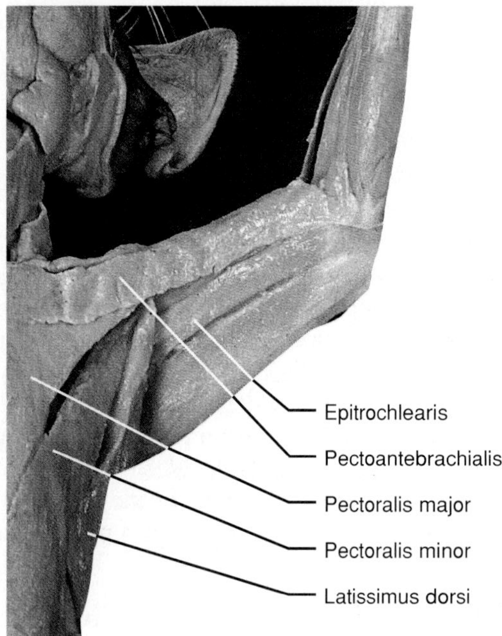

Figure 38.4 Superficial forelimb muscles of the cat.

arises from the latissimus dorsi and is continuous, distally, with the fascia of the forearm. After removing this muscle, identify and free with a probe the six homologs in the following order:

Triceps Brachii The triceps brachii has three heads of origin, which merge to insert by a common tendon on the olecranon process of the ulna. The long head, located on the caudal surface of the arm, is the largest. The lateral head covers much of the lateral surface of the arm. *Dissect.* To reveal the medial head, which is between the long and lateral heads, *transect* and *reflect* the lateral head (see figure 38.5). Only the long head arises from the scapula.

Biceps Brachii In the cat, the biceps brachii lies deep in the pectoral muscles, just medial to their humoral insertions. This muscle is, therefore, on the ventromedial surface of the arm (see figure 38.6). It inserts on the radius. *Dissect.*

Brachialis The brachialis is on the ventrolateral surface of the arm, cranial to the lateral head of the triceps brachii (see figure 38.5).

The insertion of the pectoralis major separates the biceps brachii and brachialis. Since the brachialis arises from the lateral surface of the humerus, you will not be able to run a probe underneath it from its origin to its insertion on the ulna.

Brachioradialis In the cat, the brachioradialis (see figures 38.5, 38.6, and 38.8) is a narrow ribbon that may have been removed with the superficial fascia. It arises on the lateral side of the humerus and passes along the radial side of the forearm to insert on the styloid process of the radius. *Dissect.*

Coracobrachialis The coracobrachialis (see figures 38.6 and 38.7) is on the medial side of the scapula caudad of the proximal (origin) end of the biceps brachii. This muscle is almost impossible to

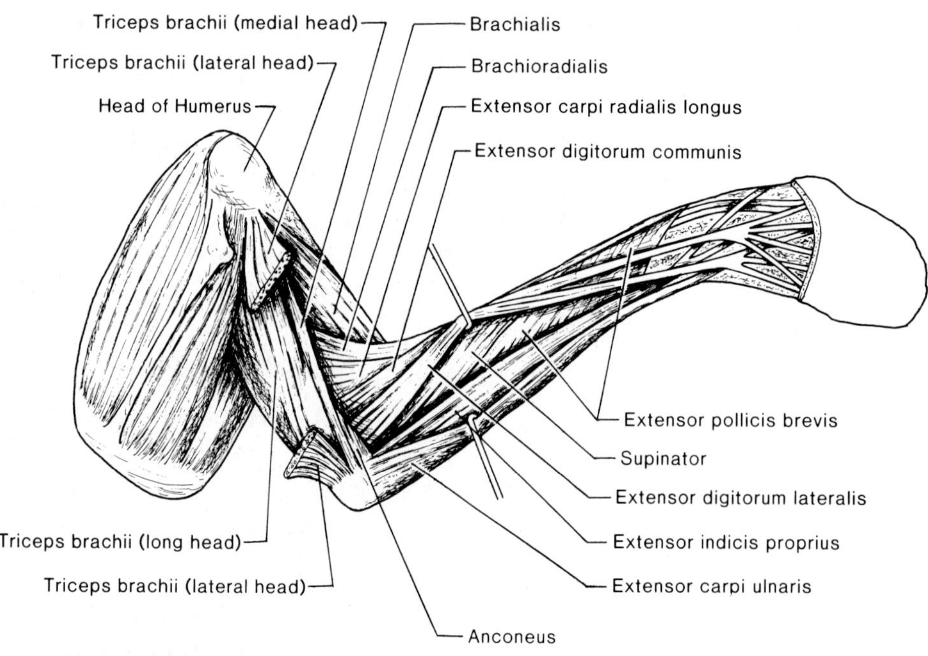

Figure 38.5 Cat forelimb muscles, lateral aspect.

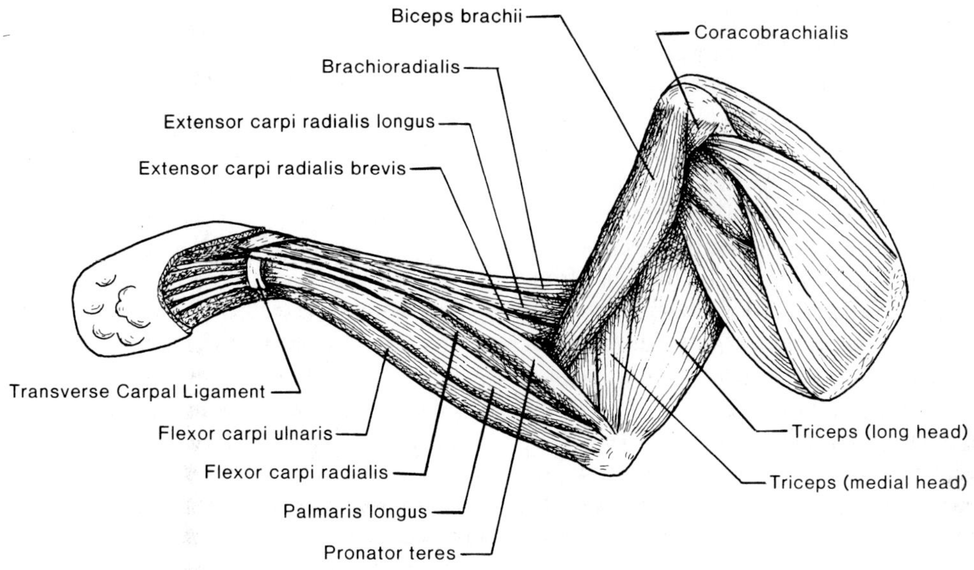

Figure 38.6 Medial aspect of cat forelimb muscles.

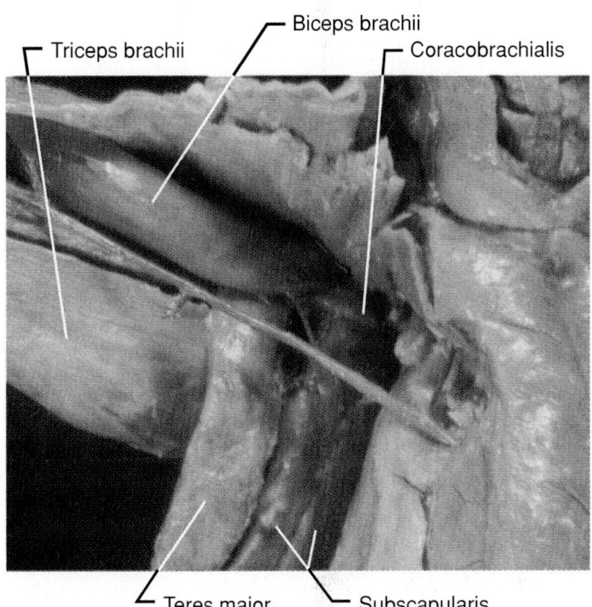

Figure 38.7 Brachium and shoulder muscles of the cat.

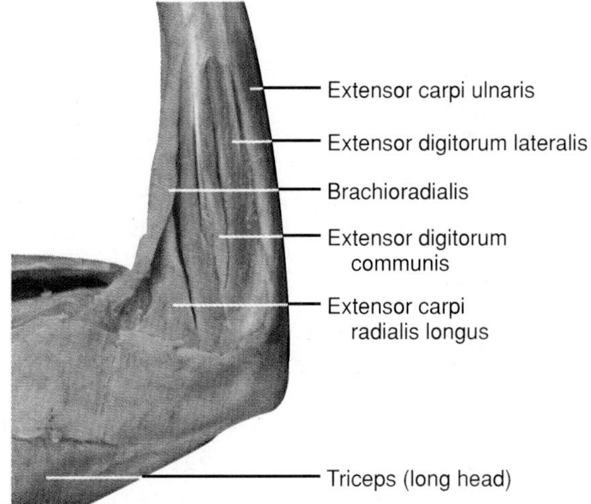

Figure 38.8 Superficial muscles of cat forearm, lateral aspect.

locate if the pectoralis major and minor muscles have not been previously transected and reflected.

Anconeus The anconeus (see figure 38.5) consists of short, superficial fibers that originate near the lateral epicondyle of the humerus and insert near the olecranon. It assists the triceps in both the cat and human in extending the forearm.

Muscles of the Antebrachium

With a probe, free each of the following eight muscles of the cat's forearm:

Pronator Teres The pronator teres (see figure 38.6) extends from the medial epicondyle of the humerus to the middle third of the radius. *Dissect.*

Palmaris Longus The palmaris longus (see figure 38.6) is a large, flat, broad muscle near the center of the medial surface of the forearm. It arises from the medial epicondyle of the humerus and inserts by means of four tendons on the digits. At its distal end, the four insertion tendons pass under a *transverse carpal ligament.* The palmaris longus is not found in about 10% of humans. *Dissect.*

Flexor Carpi Radialis The flexor carpi radialis (see figure 38.6) is a long, spindle-shaped muscle that lies between the palmaris longus and the pronator teres. It arises from the medial epicondyle of the humerus and inserts at the base of the second and third metacarpals.

Flexor Carpi Ulnaris In the flexor carpi ulnaris (see figure 38.4), what appears to be two muscles is actually two heads of origin: One originates from the medial epicondyle of the humerus and the other from the olecranon. At about the middle of the ulna, the two heads join to form a single band of muscle fibers that inserts on the pisiform and hamate bones. *Dissect.*

Extensor Carpi Ulnaris The extensor carpi ulnaris (see figures 38.5 and 38.8) arises from the lateral epicondyle of the humerus and inserts on the base of the fifth metacarpal. *Dissect.*

Extensor Digitorum Communis The extensor digitorum communis (see figure 38.5) arises from the lateral epicondyle of the humerus and extends to the wrist, where it divides into four tendons that insert on the second phalanges of digits two through five. Note that these four tendons lie superficial to the three tendons of the extensor digitorum lateralis. *Dissect.*

Extensor Carpi Radialis Longus The extensor carpi radialis longus (see figure 38.8) lies adjacent to the brachioradialis. It originates on the lateral supracondylar ridge of the humerus and inserts at the base of the second metacarpal. *Dissect.*

Supinator Expose the supinator by reflecting the extensor digitorum lateralis as figure 38.5 shows. The supinator originates from ligaments on the elbow. It inserts on the proximal end of the radius and acts as a lateral rotator of the radius (supination).

Complete exposure of the supinator requires transection and reflection of the overlying muscles. To avoid irreversible mutilation of your cat, do not do this.

Abdominal, Pelvic, and Intercostal Muscles 39

This exercise focuses on the muscles of the abdominal wall and pelvis. Cat dissection is utilized for this muscle study.

Abdominal Muscles

The abdominal wall consists of four pairs of thin muscles. The lower illustration in figure 39.1 shows portions of the right human abdominal wall removed to reveal the nature of the laminations. Identify the following muscles in this illustration:

External Oblique The external oblique is the most superficial layer on the abdominal wall. The **aponeurosis of the external oblique,** which terminates at the linea alba and inguinal ligament, ensheaths the external oblique.

The external oblique originates on the external surfaces of the lower eight ribs. Although it appears to insert on the **linea semilunaris,** its actual insertion is the **linea alba,** where fibers of the left and right aponeuroses interlace on the midline of the abdomen.

The lower border of each aponeurosis forms the **inguinal ligament,** which extends from the anterior spine of the ilium to the pubic tubercle (see figure 39.3).

Internal Oblique The internal oblique lies immediately under the external oblique—that is,

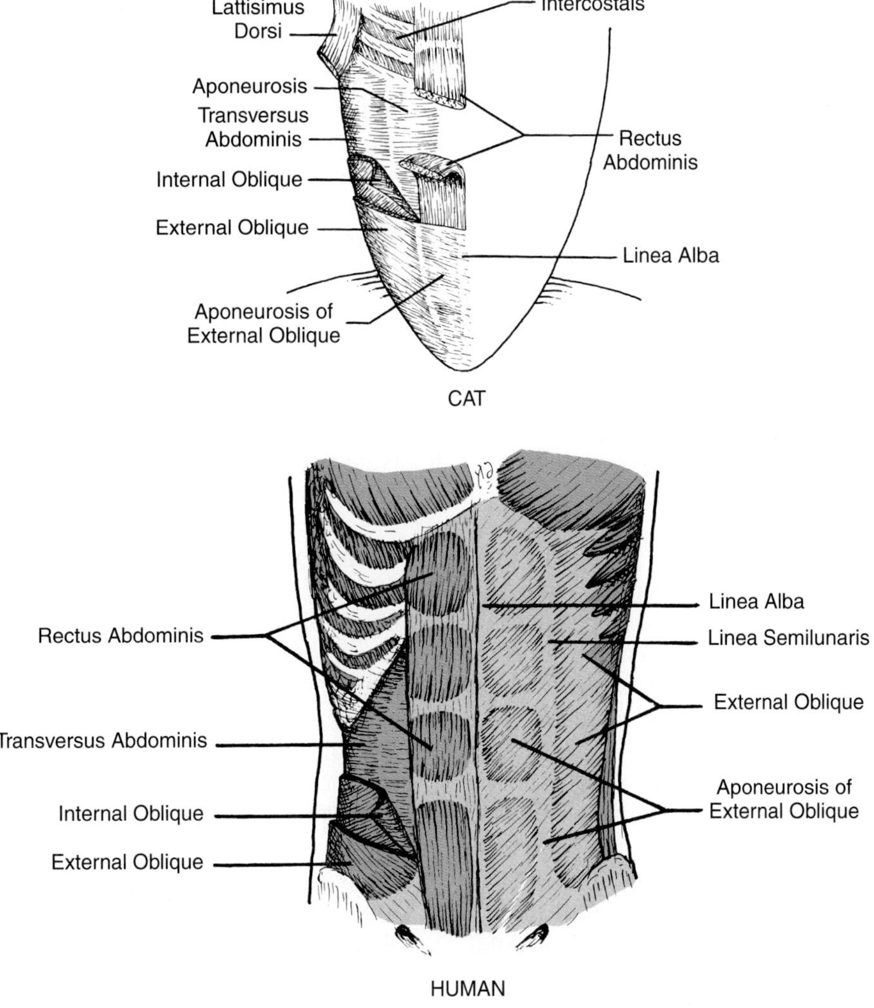

Figure 39.1 Abdominal muscles, cat and human.

between the external oblique and the transversus abdominis. Its origin is on the lateral half of the inguinal ligament, the anterior two-thirds of the iliac crest, and the thoracolumbar fascia. It inserts on the costal cartilages of the lower three ribs, the linea alba, and the crest of the pubis.

Transversus Abdominis The transversus abdominis is the innermost muscle of the abdominal wall. It arises on the inguinal ligament, the iliac crest, the costal cartilages of the lower six ribs, and the thoracolumbar fascia. It inserts on the linea alba and the crest of the pubis.

Rectus Abdominis The right rectus abdominis is the long, narrow, segmented muscle running from the rib cage to the pubic bone. A fibrous sheath that the aponeuroses of the external oblique, internal oblique, and transversus abdominis form encloses the rectus abdominis.

The rectus abdominis originates on the pubic bone. It inserts on the cartilages of the fifth, sixth, and seventh ribs. The linea alba lies between the pair of rectus abdominis muscles. Contraction of the rectus abdominis muscles aids in flexion of the spine in the lumbar region.

Collective Action: The previous four abdominal muscles compress the abdominal organs and assist in maintaining intraabdominal pressure. They also act as antagonists to the diaphragm. When the diaphragm contracts, they relax. When the diaphragm relaxes, they contract to press air out of the lungs. They assist in defecation, urination, vomiting, and parturition (childbirth delivery). These muscles also flex the body at the lumbar region.

Assignment:
Complete part A of the Laboratory Report for this exercise.

Cat Dissection

Abdominal Muscles

As the upper illustration of figure 39.1 and figure 39.2 show, the cat abdominal wall has the same four muscles as that in humans. As you separate each muscle layer, pay particular attention to the direction of the fibers. The different directions of the fibers in each layer greatly enhance body wall strength.

External Oblique The external oblique is the outer muscle of the body wall and arises from the external surface of the lower nine ribs. Note how the cranial portion of this muscle interdigitates with the origin of the serratus ventralis. Insertion is on the iliac crest and by an aponeurosis that fuses with that of the opposite side to form the *linea alba.*

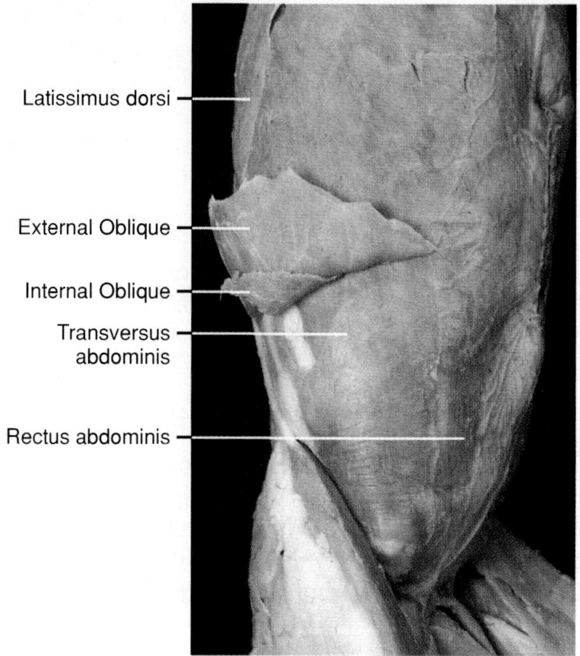

Figure 39.2 **Abdominal muscles of the cat.**

Loosen the cranial border of the muscle with a blunt probe, and slide the probe underneath the muscle in a caudal direction. *Transect* alongside the probe, and *reflect.* (Remember, when you transect a muscle, your cut should be through the midbelly halfway between the origin and insertion.)

Internal Oblique To expose the internal oblique, make a shallow incision on the right side through the external oblique from the ribs to the pelvis. *Reflect* the external oblique. Since this muscle is very thin, take care not to cut too deep. Note that the fibers of the internal oblique course at approximately right angles to those of the external oblique. Note also how the ends of the muscle fibers are continuous with the aponeurosis.

Transversus Abdominis The transversus abdominis is the innermost muscle layer of the ventrolateral abdominal wall and is very difficult to separate from the internal oblique. Try to slide the tip of a blunt probe underneath either a small area of the aponeurosis or under the muscle fibers of the internal oblique. Make a cut to create a small "window," and note the general transverse direction of the fibers of the transversus abdominis.

Rectus Abdominis The rectus abdominis extends from the pubis to the sternum on each side of the linea alba. It extends farther cranially in the cat than in the human. The cranial two-thirds of the aponeurosis of this muscle splits at the lateral border of the rectus abdominis, creating a dorsal leaf that passes dorsal and a ventral leaf that passes ventral to the rectus abdominis. The dorsal and ventral leaves unite at the medial border of the rectus abdominis to help form the linea alba. Because the aponeurosis "wraps" the rectus abdominis in this manner, passing a probe under the rectus abdominis from its origin to the insertion is difficult.

Intercostal Muscles

Although Exercise 37 covered the human intercostal muscles, the same muscles in the cat were not studied at that time. Using figure 39.1, examine the intercostal muscles on your specimen.

These muscles of respiration in the cat are essentially the same as in humans except that the cat has 12 sets, and humans have 11 sets. *Dissect* these muscles as follows:

External Intercostals *Reflect* the external oblique to expose the lateral aspect of the rib cage. The external intercostals occupy the intercostal spaces. Note that the direction of the fibers is forward and downward from the caudal border of one rib to the cranial border of the next rib caudad. Note also that the external intercostals do not reach the sternum and that the internal intercostals can be seen in the interval.

Internal Intercostals The internal intercostals are immediately deep to the external intercostals. Cut a small "window" through an external intercostal between two ribs to expose an internal intercostal. Note that the fibers of the internal muscle course at approximately right angles to the fibers of the external muscle as they pass from the cranial border of the rib to the caudal border of the next rib cranially.

Pelvic Muscles

Figure 39.3 shows the principal muscles of the pelvic region. They are the quadratus lumborum, psoas major, and iliacus. Muscles in this region of the cat are dissected in Exercise 40.

Quadratus Lumborum The quadratus lumborum extends from the iliac crest of the os coxa to the lowest rib, as figure 39.3 shows. Only the left quadratus lumborum in figure 39.3 is completely exposed.

The quadratus lumborum originates on the iliac crest and the iliolumbar ligament that borders the iliac crest. It inserts on the lower border of the last rib and the apices of the transverse processes of the upper four lumbar vertebrae.

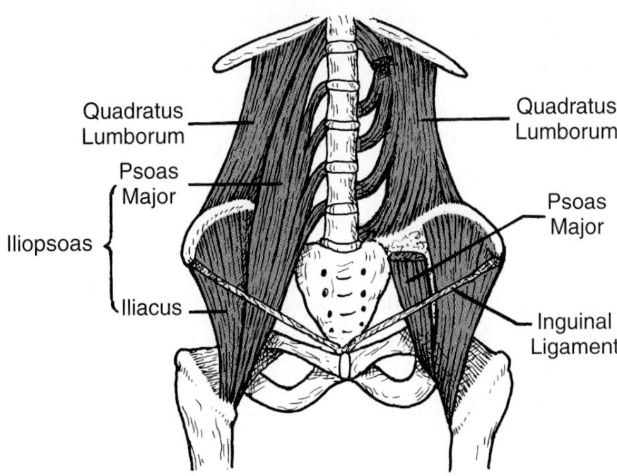

Figure 39.3 Pelvic muscles.

Occasionally, a second portion of the quadratus lumborum is positioned just in front of it. The origin of this part is on the upper borders of the transverse processes of the lower three or four lumbar vertebrae, and it inserts on the lower margin of the last rib. It does not extend down to the iliac crest.

When both portions of the quadratus lumborum are present, the transverse processes of some of the lumbar vertebrae act as both origin and insertion—truly, a unique situation.

Acting together, the right and left quadratus lumborums extend the spine at the lumbar vertebrae. Lateral flexion or abduction results when one acts independently of the other.

Psoas Major The psoas major is the long muscle in figure 39.3 on the right side of the body. A portion of the left psoas major has been cut away in figure 39.3 to reveal the full extent of the quadratus lumborum.

The psoas major arises from the sides of the bodies and transverse processes of the lumbar vertebrae. It inserts with the iliacus on the lesser trochanter of the femur.

The two psoas majors work synergistically with the rectus abdominis muscles to flex the lumbar region of the vertebral column.

Iliacus The iliacus extends from the iliac crest to the proximal end of the femur. Its origin is the whole iliac fossa. Its insertion is the lesser trochanter of the femur. The psoas major and iliacus are jointly referred to as the *iliopsoas* because of their intimate relationship at their insertion. The iliacus works synergistically with the psoas major to flex the femur on the trunk.

Assignment:
Complete part B of the Laboratory Report for this exercise.

40 Lower Extremity Muscles

This exercise examines 27 muscles of the leg. As with the upper extremity muscles in Exercise 38, the lower extremity muscles are grouped according to type of movement.

Thigh Movements

Figure 40.1 shows seven muscles that move the femur. All originate on a part of the pelvis.

Gluteus Maximus The gluteus maximus is covered by a deep fascia, the *fascia lata,* that completely invests the thigh muscles. Emerging downward from the fascia lata is a broad tendon, the **iliotibial tract,** which attaches to the tibia.

The gluteus maximus arises on the ilium, the sacrum, and the coccyx. It inserts on the iliotibial tract and the posterior part of the femur.

Action: The gluteus maximus extends and outwardly rotates the femur.

Gluteus Medius The gluteus medius lies immediately under the gluteus maximus, covering a good portion of the ilium. Its origin is on the ilium, and its insertion is on the lateral part of the greater trochanter.

Action: The gluteus medius abducts and medially rotates the femur.

Gluteus Minimus The gluteus minimus is the smallest of the three gluteal muscles and is immediately under the gluteus medius. It, too, arises on the ilium. It inserts on the anterior border of the greater trochanter.

Action: The gluteus minimus abducts, inwardly rotates, and slightly flexes the femur.

Piriformis The piriformis originates on the anterior surface of the sacrum and inserts on the upper border of the greater trochanter.

Action: The piriformis outwardly rotates, abducts, and extends the femur.

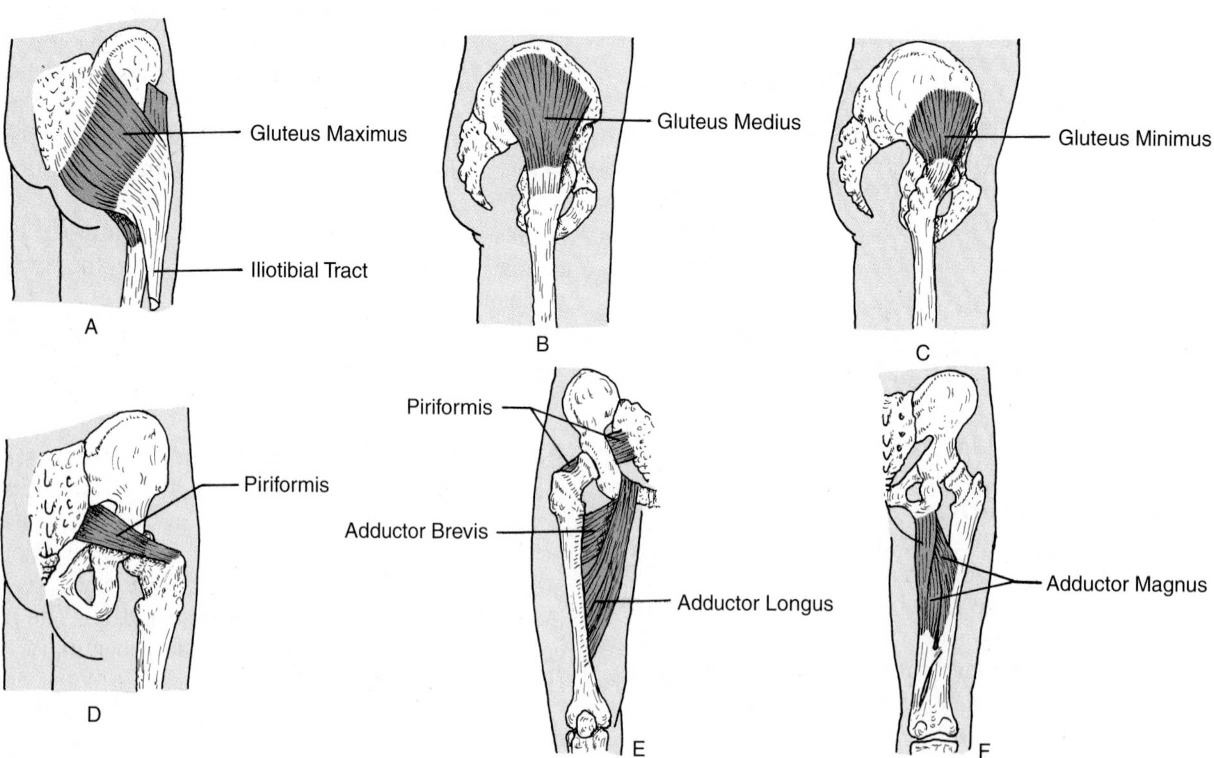

Figure 40.1 Muscles that move the femur.

Adductor Longus and Adductor Brevis The adductor longus originates on the front of the pubis and inserts on the linea aspera of the femur. The adductor brevis arises on the posterior side of the pubis and inserts on the femur above the longus muscle.

Action: In addition to adduction, the adductor longus and adductor brevis flex and rotate the femur medially.

Adductor Magnus The adductor magnus originates on the inferior ischial and pubic rami, and inserts on the linea aspera and the medial condyle of the femur.

Action: The adductor magnus adducts, laterally rotates, and flexes the thigh.

Assignment:

Complete part A of the Laboratory Report for this exercise.

Thigh and Lower Leg Movements

Figure 40.2 shows the muscles that flex and extend the lower part of the leg.

Hamstrings Three muscles are collectively known as the "hamstrings":

Biceps Femoris

The biceps femoris occupies the most lateral position of the three hamstrings. It has two heads: one long and the other short. The long head, which obscures the short one, arises on the ischial tuberosity. The short head originates on the linea aspera of the femur. The muscle inserts on the head of the fibula and the lateral condyle of the tibia.

Semitendinosus

The semitendinosus is medial to the biceps femoris. It arises on the ischial tuberosity and inserts on the upper end of the shaft of the tibia.

Semimembranosus

The semimembranosus is the most medial of the three hamstrings. Its origin consists of a thick, semimembranous tendon attached to the ischial tuberosity. Its insertion is primarily on the posterior medial part of the medial condyle of the tibia.

Collective Action: All of the hamstring muscles flex the calf on the thigh. They also extend and rotate the thigh. The biceps femoris rotates the thigh outward; the semitendinosus and semimembranosus rotate the thigh inward.

Quadriceps Femoris The quadriceps femoris is a large muscle that covers the anterior portion of the thigh. It consists of the following four parts that unite to form a common tendon that passes over the patella to insert on the tibia:

Rectus Femoris

The rectus femoris occupies a superficial central position of the quadriceps. It arises from two tendons: one from the anterior inferior iliac spine and the other from a groove just above the acetabulum. The lower portion of the rectus femoris is a broad aponeurosis that terminates in the tendon of insertion.

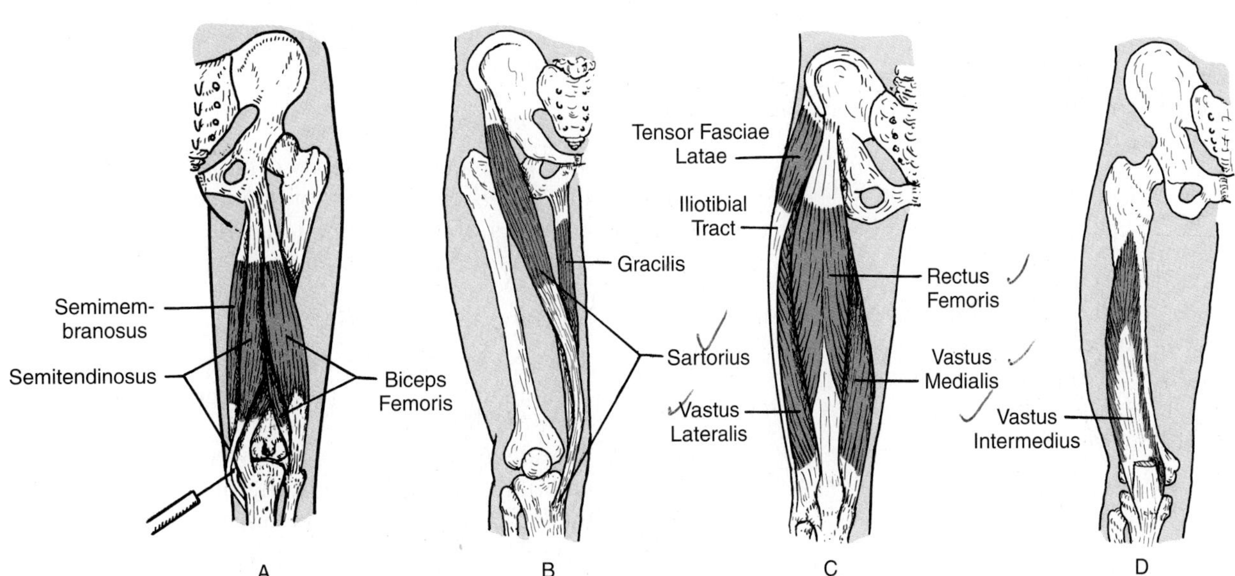

Figure 40.2 Muscles that move the tibia and fibula.

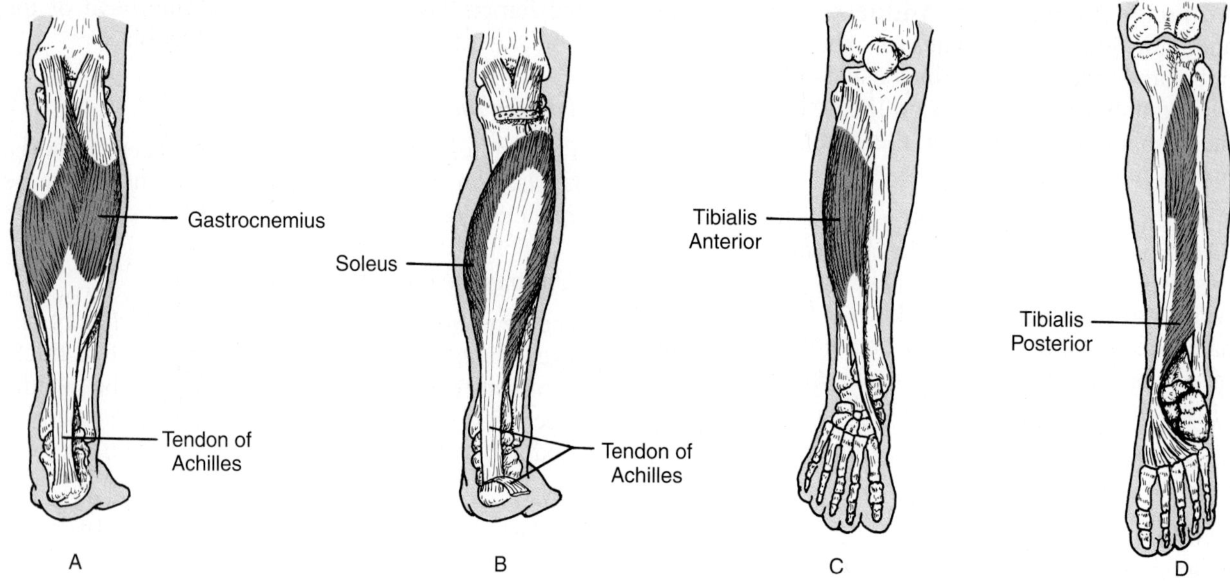

Figure 40.3 Lower leg muscles.

Vastus Lateralis

The vastus lateralis is the lateral portion of the quadriceps and the largest of the four quadriceps muscles. It arises from the lateral lip of the linea aspera.

Vastus Medialis

The vastus medialis is the most medial portion of the quadriceps. It arises from the linea aspera.

Vastus Intermedius

The vastus intermedius is deep to the other three quadriceps muscles. Its origin is on the front and lateral surfaces of the femur.

Collective Action: The entire quadriceps femoris extends (straightens) the leg. The rectus femoris, however, flexes the thigh.

Sartorius The sartorius arises from the anterior superior spine of the ilium and inserts on the medial surface of the tibia.

Action: The sartorius flexes the calf on the thigh, and the thigh on the pelvis; it also rotates the leg laterally.

Tensor Fasciae Latae The tensor fasciae latae arises from the anterior outer lip of the iliac crest, the anterior superior spine, and the deep surface of the fascia lata. It inserts between the two layers of the iliotibial tract at about the junction of the middle and upper third of the thigh.

Action: The tensor fasciae latae flexes the thigh and rotates it slightly medially.

Gracilis The gracilis is on the medial surface of the thigh. It arises from the lower margin of the pubic bone and inserts on the medial surface of the tibia near the insertion of the sartorius.

Action: The gracilis adducts, flexes, and rotates the thigh medially.

Assignment:
Complete part B of the Laboratory Report for this exercise.

Lower Leg and Foot Movements

Figures 40.3 and 40.4 illustrate most of the muscles of the lower leg that cause flexion and foot movements.

Triceps Surae The large, superficial muscle that covers the calf of the leg is the triceps surae. It consists of the gastrocnemius and the soleus, united by a common **tendon of Achilles** that inserts on the calcaneus of the foot.

Gastrocnemius

The gastrocnemius has two heads that arise from the posterior surfaces of the medial and lateral condyles of the femur. It can flex the calf on the thigh as well as cause plantar flexion.

Soleus

The soleus arises on the heads of the fibula and tibia. Plantar flexion is its only action.

Tibialis Anterior The tibialis anterior covers the anterior lateral portion of the tibia. It arises from the lateral condyle and the upper two-thirds of the tibia. Its distal end forms a long tendon that passes over the tarsus and inserts on the inferior surface of the first cuneiform and first metatarsal bones.

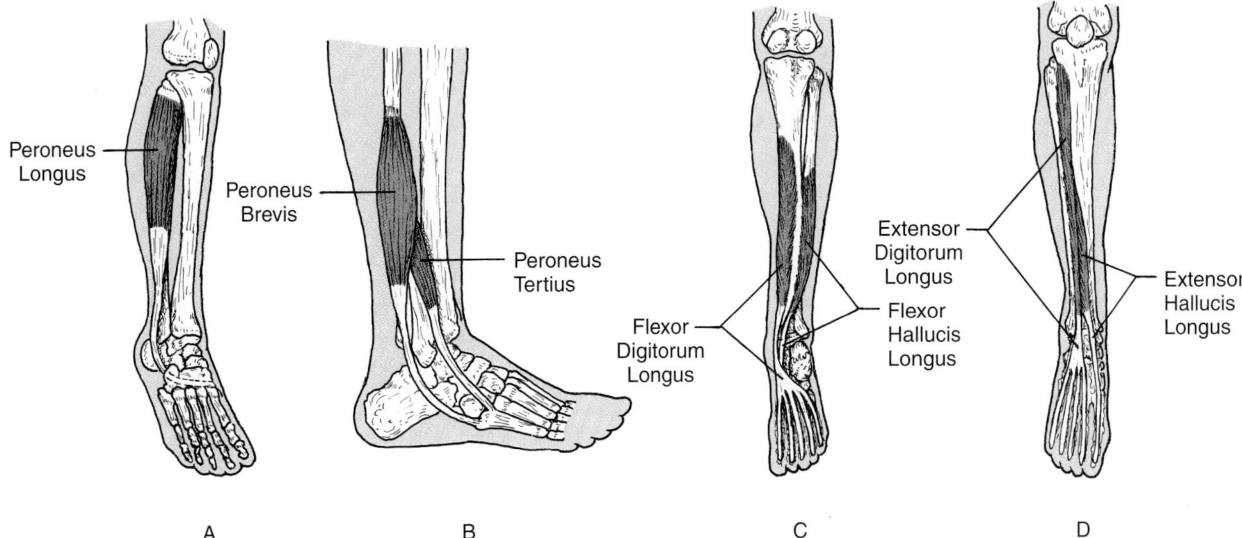

Figure 40.4 Lower leg muscles.

Action: The tibialis anterior dorsiflexes and inverts the foot.

Tibialis Posterior The tibialis posterior lies on the posterior surfaces of the tibia and fibula. It arises from both these bones and from the interosseous membrane. It inserts on the inferior surfaces of the navicular, the cuneiforms, the cuboid, and the second through fourth metatarsals.

Action: The tibialis posterior is involved in plantar flexion and inversion of the foot. It also assists in maintaining the longitudinal and transverse arches of the foot.

Peroneus Muscles Figure 40.4 shows three peroneus (Latin: *peroneus,* fibula) muscles, all of which originate on the fibula. The **peroneus longus** is the longest peroneus muscle and originates on the head and upper two-thirds of the fibula. The **peroneus tertius** is the smallest peroneus muscle, while the **peroneus brevis** is an in-between size. The longus inserts on the first metatarsal and second cuneiform bones. The brevis and tertius insert at different points on the fifth metatarsal bone.

Action: The peroneus longus and brevis cause plantar flexion and eversion of the foot. The peroneus tertius dorsiflexes and everts the foot.

Flexor Muscles Figure 40.4C shows two flexor muscles of the foot. The **flexor hallucis longus** (Latin: *hallux,* big toe) originates on the fibula and intermuscular septa. Its long distal tendon inserts on the distal phalanx of the great toe. It flexes the great toe.

The **flexor digitorum longus** originates on the tibia and the fascia that covers the tibialis posterior. Its distal end divides into four tendons that insert on the bases of the distal phalanges of the second, third, fourth, and fifth toes. It flexes the distal phalanges of the four smaller toes.

Extensor Muscles Figure 40.4D shows two extensors of the foot. The **extensor digitorum longus** originates on the lateral condyle of the tibia, part of the fibula, and part of the interosseous membrane. Its tendon of insertion divides into four parts that insert on the superior surfaces of the second and third phalanges of the four smaller toes. It extends the proximal phalanges of the four smaller toes. It also flexes and inverts the foot.

The **extensor hallucis longus** originates on the fibula and interosseous membrane. Its distal tendon inserts at the base of the distal phalanx of the great toe. This muscle extends the proximal phalanx of the great toe and aids in dorsiflexion of the foot.

Assignment:
Complete part C of the Laboratory Report for this exercise.

Cat Dissection

Hip Muscles

Remove the superficial fat and fascia that encase the muscles of the hip and thigh so that you can determine muscle fiber direction. *Take care not to remove the fascia lata and iliotibial tract.* The latter is a white, tendinous band on the lateral side of the thigh. Although the sartorius muscle is a muscle of the thigh, it is dissected here first to expose the full extent of the tensor fasciae latae.

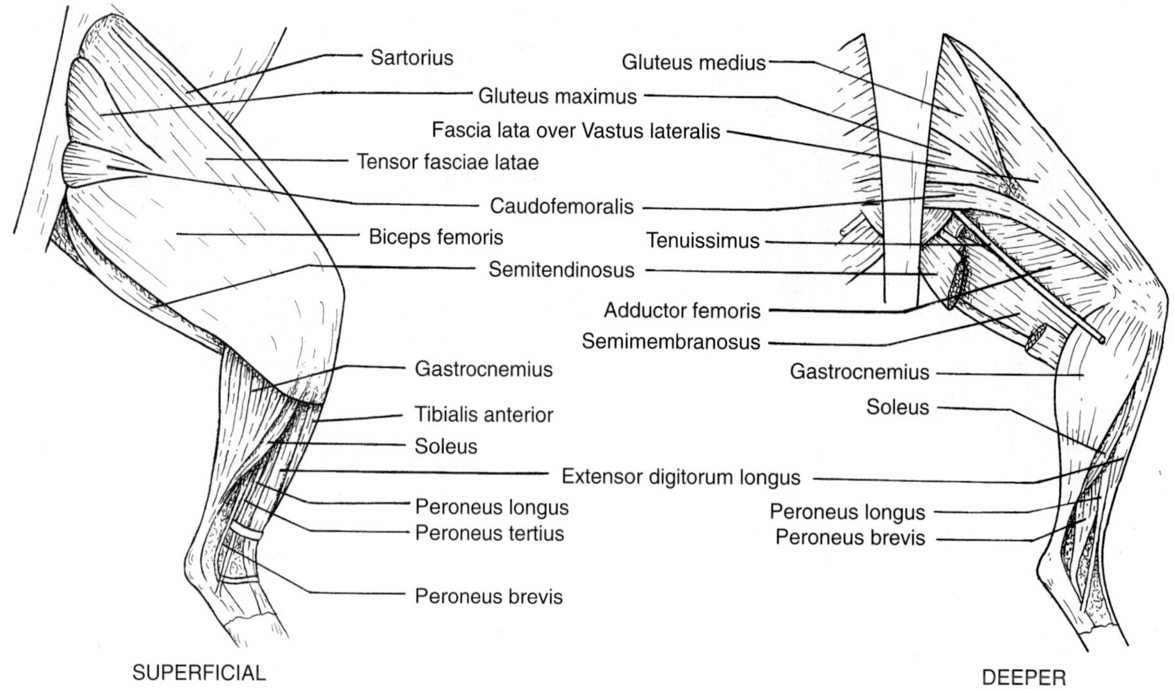

Figure 40.5 Lateral aspect of cat hindlimb.

Sartorius The sartorius is a broad, straplike muscle on the cranial half of the medial side of the thigh (see figures 40.5 and 40.8). *Dissect, transect,* and *reflect.*

Tensor Fasciae Latae The tensor fasciae latae is a fan-shaped muscle on the anterolateral side of the hip region, somewhat ventral to the other muscles in the group (see figures 40.5 and 40.6). It arises from the lateral ilium and inserts on the iliotibial tract portion of the fascia lata. The fascia lata is a tough layer of connective tissue that covers the vastus lateralis muscle.

Slide a blunt probe underneath the fascia lata to separate it from the vastus lateralis, and *transect* the fascia lata so that you can completely *reflect* the tensor fasciae latae. (The tensor muscle usually fuses somewhat with the gluteus maximus. This connection should be cut.)

Gluteus Medius With the tensor fasciae latae fully reflected, you can see the full extent of the gluteus medius (see figure 40.6). This thick muscle arises from the lateral surface of the ilium and inserts on the greater trochanter. In the cat, this muscle is much larger than and cranial to the gluteus maximus.

Gluteus Maximus The gluteus maximus is the small hip muscle just caudad to the gluteus medius (see figure 40.6). It arises from the last sacral and first caudal vertebrae and inserts on the greater trochanter. Therefore, the origin and insertion in the cat are far less extensive than in the human. *Dissect.*

Caudofemoralis The caudofemoralis is a small muscle just caudad to the gluteus maximus (see figure 40.6). It is difficult to see because it "hides"

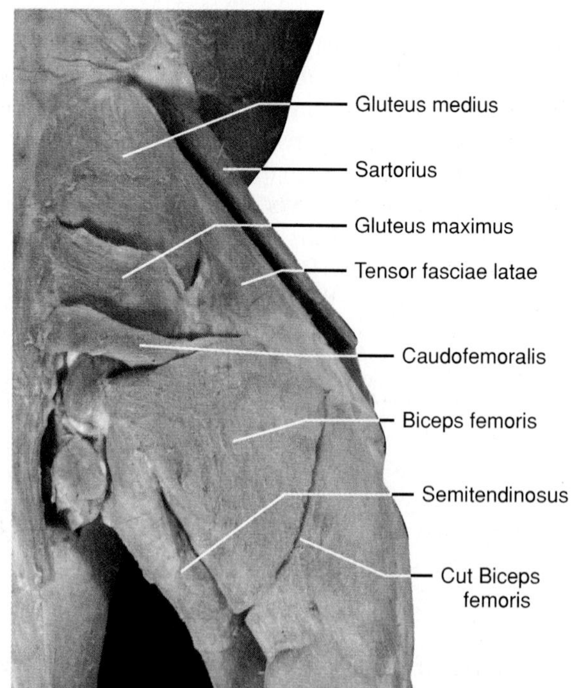

Figure 40.6 Hip muscles of the cat.

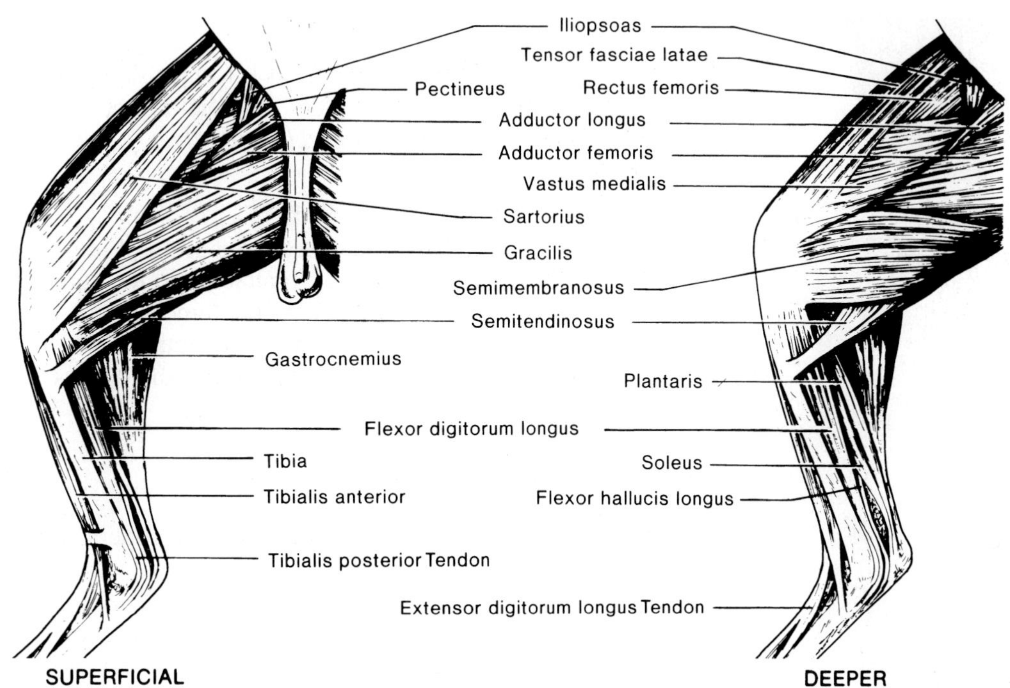

Figure 40.7 Medial aspect of cat hindlimb.

below the cranial border of the biceps femoris. It arises on the caudal vertebrae and inserts by a very thin tendon on the patella.

Humans do not have a caudofemoralis. In the cat, however, the caudofemoralis and the gluteus maximus together are more comparable to the human gluteus maximus than is the cat's gluteus maximus alone. *Dissect.*

Thigh Muscles

Thirteen muscles make up the musculature of the cat's thigh. Identify them in the following order:

Sartorius Previously dissected (see p. 178).

Gracilis The gracilis is a wide, flat muscle covering most of the caudal portion of the medial side of the thigh (see figures 40.7 and 40.9). At the insertion, the fibers end in a very thin flat tendon, part of which becomes continuous with the fascia covering the distal portion of the leg. *Dissect, transect,* and *reflect.*

The Hamstrings The cat has the same group of three muscles called *hamstrings* in humans.

Semimembranosus

The semimembranosus is a large, thick muscle just beneath the gracilis (the gracilis must be reflected to see the full extent of the semimembranosus). Dorsal to the semimembranosus is the adductor femoris, and ventral to it is the semitendinosus.

Dissect by separating this muscle from the adductor femoris and semitendinosus. While the human homolog inserts only on the tibia, this muscle in the cat inserts on both the femur and tibia.

Semitendinosus

The semitendinosus is the large band of muscle on the ventral border of the thigh between the semimembranosus and the biceps femoris. It inserts on the tibia. *Dissect.*

Biceps Femoris

The biceps femoris is a large, broad muscle that covers most of the lateral surface of the thigh. *Dissect, transect,* and *reflect.* When the muscle is transected, be careful not to cut the caudofemoralis and sciatic nerve. The sciatic nerve, shown in figure 40.8, appears as a white cord just beneath the biceps femoris.

Adductor Femoris The adductor femoris lies deep to the gracilis and dorsal to the semimembranosus. It originates on the hipbone and inserts on the femur. The adductor femoris corresponds to the adductor magnus and adductor brevis of the human. *Dissect.*

Adductor Longus The adductor longus lies along the cranial border of the adductor femoris and extends from the hipbone to the femur.

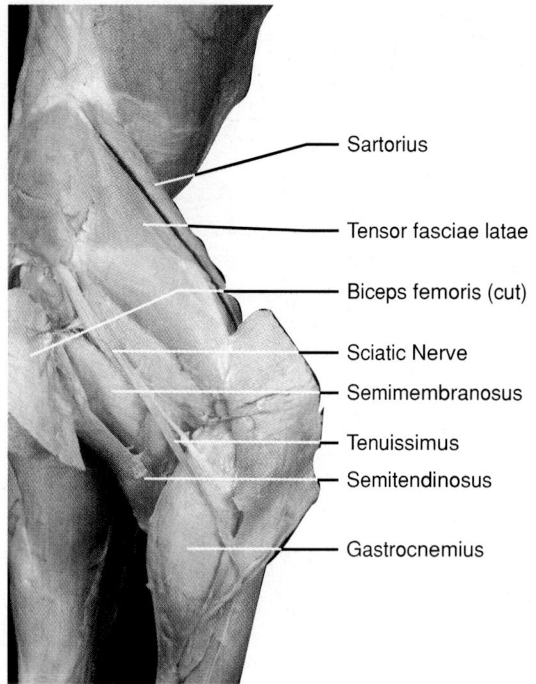

Figure 40.8 Deep muscles of cat thigh, lateral aspect.

Pectineus The pectineus is a very small triangular muscle (see figure 40.7) that lies cranial to the adductor longus and extends from the hipbone to the proximal end of the femur.

Iliopsoas As in humans, the name *iliopsoas* applies to a pair of cat muscles, the *iliacus* and the *psoas major.* Because they have a common insertion on the lesser trochanter, they are grouped together. Only the distal (insertion) end is observed at this time. Note that the muscle fibers of the iliopsoas are oriented almost perpendicular to the fibers of the pectineus and adductor longus.

Quadriceps Femoris As in humans, the cat has a quadriceps femoris that consists of four muscles (see figure 40.9). This great extensor muscle of the knee joint has a common insertion by a large tendon that extends to the patella, attaches around the patella, and passes to the tibial tuberosity to insert.

Vastus Lateralis

The vastus lateralis is the large muscle deep to the tensor fasciae latae that occupies the craniolateral surface of the thigh.

Vastus Medialis

The vastus medialis is the large muscle deep to the sartorius that occupies the craniomedial surface of the thigh.

Rectus Femoris

The rectus femoris is a cigar-shaped muscle that the vastus medialis borders laterally and medially. *Dissect, transect,* and *reflect.*

Vastus Intermedius

The vastus intermedius is a flat muscle deep to the rectus femoris; it attaches to the anterior surface of the femur. The rectus femoris must be reflected to see the full extent of this muscle. Try to separate the vastus intermedius from the vastus lateralis and vastus medialis.

Leg and Foot Muscles

Five muscles of the leg and foot of the cat are identified here:

Gastrocnemius The gastrocnemius (see figures 40.5, 40.7, 40.8, 40.9) on the dorsal side of the leg has two heads of origin: a lateral head from the lateral epicondyle of the femur and a medial head from the medial epicondyle of the femur. The muscle inserts by way of the Achilles tendon onto the calcaneus.

In the cat, a large *plantaris* (see figure 40.7) muscle is between the heads of the gastrocnemius. In humans, the plantaris is a small fusiform muscle of little importance. *Dissect.*

Soleus The soleus (see figures 40.5, 40.7, and 40.9) lies deep to the lateral head of the gastrocnemius. As in humans, the soleus inserts with the gastrocnemius on the calcaneus bone by way of the Achilles tendon.

Tibialis Anterior The tibialis anterior (see figures 40.5 and 40.7) is a tapered band of muscle on the anterolateral (ventrolateral) aspect of the tibia. Follow the tendon of insertion of this muscle as it crosses the ankle obliquely to reach the medial surface of the foot.

Extensor Digitorum Longus The extensor digitorum longus (see figure 40.5) is covered throughout most of its length by the tibialis anterior. Separate these two muscles, and follow the tendon of insertion as it crosses the ankle ventrally and divides into four tendons distributed to the digits. *Dissect.*

Peroneus Muscles Like humans, the cat has three peroneus muscles: peroneus longus, peroneus brevis, and peroneus tertius (see figure 40.5). These muscles are on the lateral side of the leg between the extensor digitorum longus and the soleus.

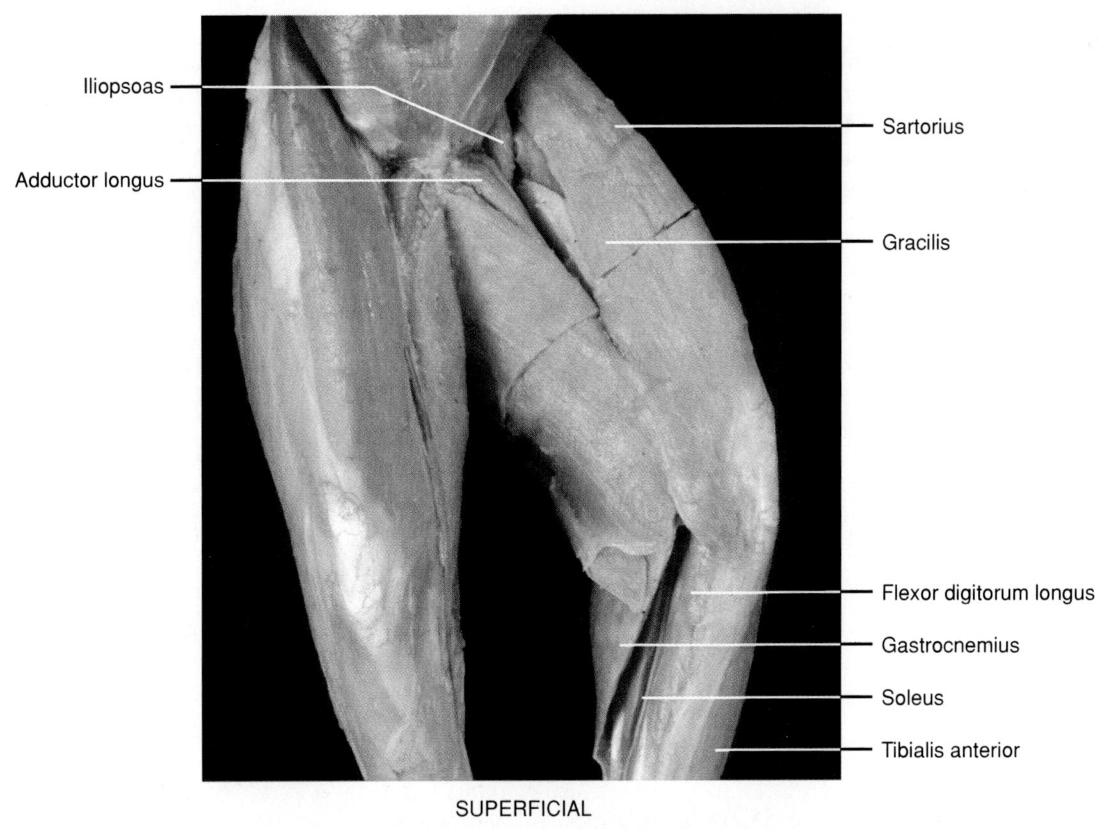

SUPERFICIAL

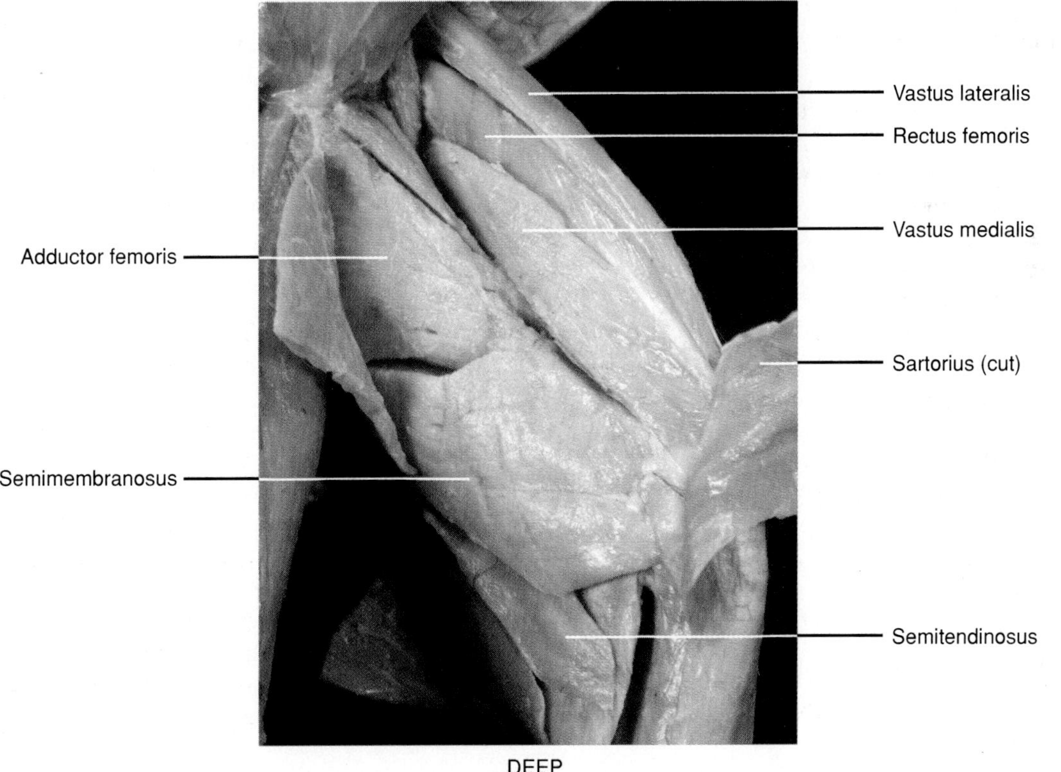

DEEP

Figure 40.9 **Superficial and deep muscles of cat hindlimb, medial aspect.**

The positions of the tendons of insertion in relation to the lateral malleolus in the cat differ from those in humans; also, the insertion points are somewhat different. Therefore, identifying the peroneus muscles simply as a group without trying to distinguish which muscle is which is best.

Assignment:

Complete part D of the Laboratory Report for this exercise. Also complete part E, which reviews the surface muscles presented in Exercises 36 through 40. Try to complete part E without referring back to previous illustrations. This type of self-testing will help you to determine where you need additional study.

PART 8 The Nervous System

The nervous system is an intricate maze of neurons and receptors that coordinates various physiological activities of the body. It includes the brain and spinal cord of the *central nervous system* (CNS) and the various nerves of the *peripheral nervous system* (PNS). The nervous system also divides functionally into the *somatic* (voluntary) and *autonomic* (involuntary) systems.

This part has seven exercises. The first three exercises explore the anatomy and physiology of reflexes. Exercise 41 looks at the anatomy of the spinal cord and its nerves in relationship to the reflex arc. Exercise 42 has two parts: frog experimentation and a study of some of the more common diagnostic human reflexes. Exercise 43 analyzes the patellar reflex, using Flexicomp software by Intelitool Inc.

In Exercise 44, the approximate speed of the nerve impulse in the ulnar nerve of the arm is measured with an electronic stimulator and chart recorder.

Exercises 45 and 46 are in-depth studies of the anatomy of the brain. While you may examine preserved human brains in these laboratory sessions, only sheep brains are dissected. In Exercise 47, brain waves are monitored to produce an electroencephalogram (EEG) of a subject in the laboratory.

41 The Spinal Cord, Spinal Nerves, and Reflex Arcs

Automatic stereotypical responses to various stimuli enable animals to adjust quickly to adverse environmental changes. The nervous system generates these automatic responses, called **reflexes.**

Reflexes can be either somatic or visceral. **Somatic reflexes** involve skeletal muscle responses, while **visceral reflexes** involve the adjustments of smooth muscle, cardiac muscle, and glands.

A **reflex arc** is the neural pathway for a reflex. This pathway includes receptors, spinal nerves, the spinal cord, and effector organs. This exercise examines each of the components involved in both types of reflexes.

The Spinal Cord

The spinal cord is a downward extension of the medulla oblongata of the brain. Figure 41.1 reveals its gross structure, as seen posteriorly, with the posterior portions of the vertebrae and sacrum removed.

The spinal cord starts at the upper border of the atlas and terminates as the **conus medullaris** at the lower border of the first lumbar vertebra.

The fetal spinal cord occupies the entire length of the vertebral canal (spinal cavity), but as the vertebral column elongates during growth, the spinal cord fails to lengthen with it. Thus, the vertebral canal extends down beyond the end of the spinal cord.

Extending down from the conus medullaris is an aggregate of fibers called the **cauda equina** (horse's tail), which fills the lower vertebral canal. The innermost fiber of the cauda equina is located on the median line and is called the **filum terminale interna.** This delicate prolongation of the conus medullaris becomes the **filum terminale externa** after it passes through the **dural sac** (dura mater) that lines the vertebral canal. The dural sac is continuous with the dura mater that surrounds the brain. Figure 41.1 only shows a portion of it.

The Spinal Nerves

Pairs of spinal nerves emerge from the spinal cord through the intervertebral foramina on each side of the spinal cavity. The eight cervical, twelve thoracic, five lumbar, five sacral, and one pair of coccygeal nerves total 31 pairs of spinal nerves.

Anterior and **posterior roots** connect each spinal nerve to the spinal cord (see figure 41.1). Note that the posterior root has an enlargement, the **spinal ganglion** (posterior root ganglion), which contains cell bodies of sensory neurons. Neuronal fibers from these roots pass through the outer **white matter** of the spinal cord into the inner **gray matter** to make neuronal connections (*synapses*).

Cervical Nerves The first cervical nerve (C_1) emerges from the spinal cord in the space between the base of the skull and the atlas. The eighth cervical nerve (C_8) emerges between the seventh cervical and first thoracic vertebrae. Note that the first four cervical nerves unite in the neck region to form a network called the **cervical plexus.** The remaining cervical nerves (C_5, C_6, C_7, and C_8) and the first thoracic nerve unite to form the **brachial plexus** in the shoulder region.

Thoracic (*Intercostal*) Nerves The first thoracic nerve (T_1) emerges through the intervertebral foramen between the first and second thoracic vertebrae. The twelfth thoracic nerve (T_{12}) emerges through the foramen between the twelfth thoracic and first lumbar vertebrae. Each of these nerves is adjacent to the lower margin of the rib above it.

Lumbar Nerves The first lumbar nerve (L_1) emerges from the intervertebral foramen between the first and second lumbar vertebrae. Close to where the nerve emerges from the spinal cavity, it divides to form an upper **iliohypogastric nerve** and a lower **ilioinguinal nerve.**

L_1, L_2, L_3, and most of L_4 unite to form the **lumbar plexus** (see figure 41.1). The largest trunk emanating from this plexus is the **femoral nerve,** which innervates part of the leg. In reality, the union of L_2, L_3, and L_4 forms the femoral.

Sacral and Coccygeal Nerves The remaining nerves of the cauda equina include five pairs of sacral and one pair of coccygeal nerves. The union of roots of L_4, L_5, and the first three sacral nerves (S_1, S_2, and S_3) forms the **sacral plexus.** The largest nerve that emerges from this plexus is the **sciatic**

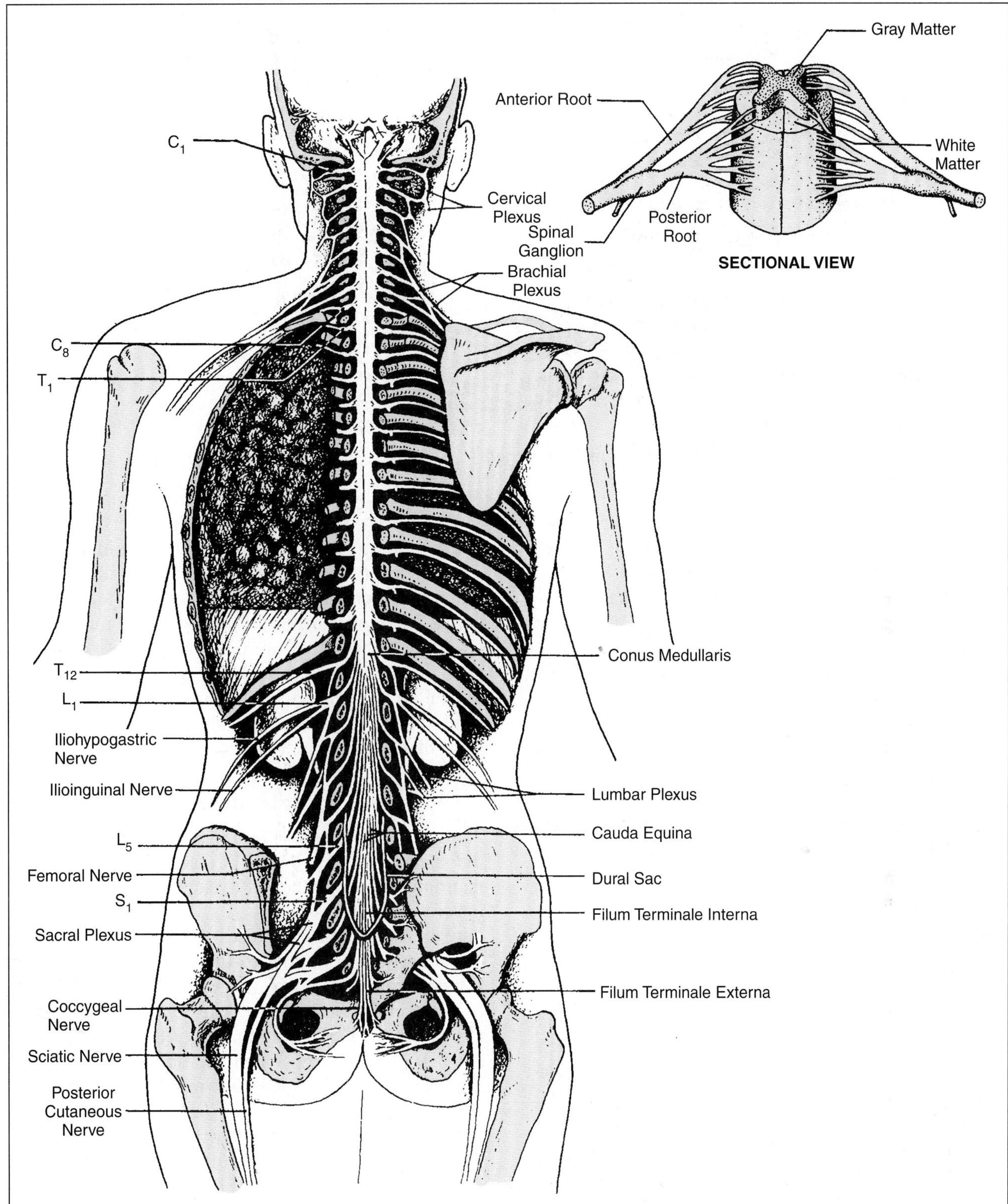

Figure 41.1 The spinal cord and spinal nerves.

nerve, which passes down into the leg. The smaller nerve that parallels the sciatic is the **posterior cutaneous nerve** of the leg. The **coccygeal nerve** contains nerve fibers from the fourth and fifth sacral nerves (S_4 and S_5).

Spinal Cord Sectional Details

Trauma to or severance of the spinal cord may cause permanent damage. Fortunately, the bony spinal cavity, as well as connective tissues, meninges, and cerebrospinal fluid, protect the

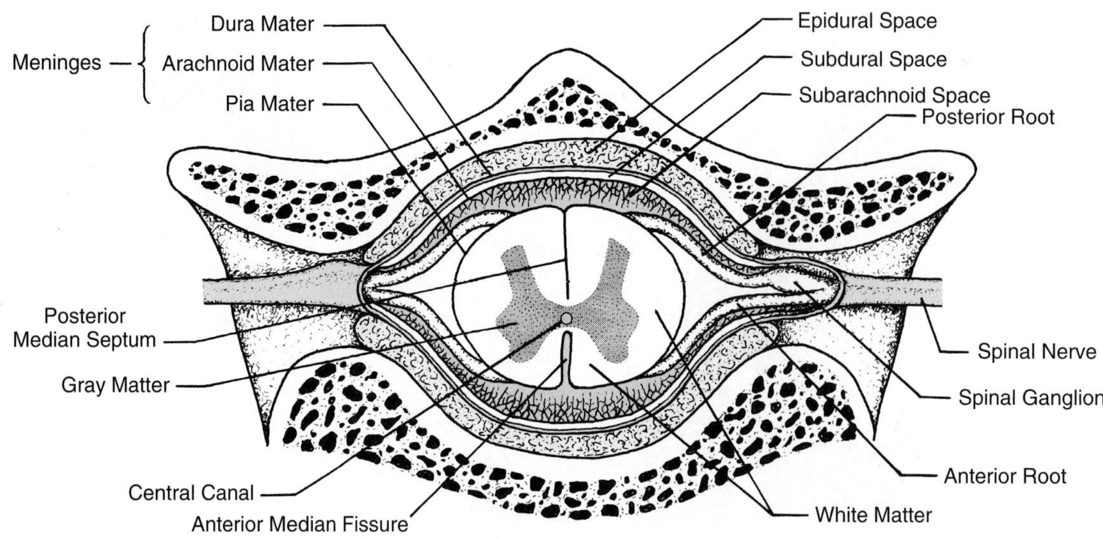

Figure 41.2 Cross section of the spinal cord and meninges.

vulnerable cord. Figure 41.2 shows how these protective layers surround the cord and spinal nerves.

Cord Structure The most noticeable characteristic of the spinal cord is the **gray matter,** which, in cross section, looks somewhat like the outstretched wings of a butterfly (see figure 41.2). The **white matter,** which consists primarily of myelinated nerve fibers, surrounds the gray matter.

A septum and a fissure divide the spinal cord along its median line. A **posterior median septum** of neuroglial tissue extends deep into the white matter and bisects the dorsal portion of the cord. On the ventral surface of the cord is an **anterior median fissure.** In the gray matter on the median line lies a tiny **central canal,** evidence of the spinal cord's tubular nature. This canal is continuous with the ventricles of the brain and extends into the filum terminale. Ciliated ependymal cells line the central canal.

Spinal Nerves Note in figure 41.2 how the spinal nerves emerge from between the vertebrae on each side of the spinal column. Locate the **posterior root** with its **spinal ganglion** and the **anterior root.**

Meninges Three meninges (*meninx,* singular) surround the spinal cord (and brain). The **dura mater** is outermost and also covers the spinal nerves. A cutaway portion of the dura mater on each side in figure 41.2 reveals the inner spinal nerve roots. The dura mater consists of fibrous connective tissue, making it the toughest of the three meninges. Between the dura mater and the bony vertebrae is an **epidural space** that contains loose (areolar) connective tissue and blood vessels.

The innermost meninx is the **pia mater,** a thin, delicate membrane that is actually the outer surface of the spinal cord. Between the pia mater and dura mater is the third meninx, the **arachnoid mater.** The space between the dura mater and arachnoid mater is a potential cavity called the **subdural space.** The inner surface of the arachnoid mater has a delicate, fibrous texture that forms a netlike support around the spinal cord. The space between the arachnoid mater and the pia mater is the **subarachnoid space.** This space is filled with cerebrospinal fluid that cushions and protects the spinal cord.

The Somatic Reflex Arc

As figure 41.3 shows, the principal components of the somatic reflex arc are (1) a **receptor,** such as a neuromuscular spindle or a cutaneous end-organ, that receives the stimulus; (2) a **sensory** (afferent) **neuron,** that carries impulses through a peripheral nerve and posterior root to the spinal cord; (3) an **interneuron** (association neuron) that forms synaptic connections between the sensory neuron and the motor neuron in the gray matter of the spinal cord; (4) a **motor** (efferent) **neuron** that carries nerve impulses from the central nervous system through the anterior root to the effector organ via a peripheral nerve; and (5) an **effector organ** (in this case, a muscle) that responds to the stimulus by contracting.

Polysynaptic reflex arcs have one or more interneurons. *Monosynaptic* reflex arcs, such as the knee-jerk reflex, lack interneurons.

The Visceral Reflex Arc

The **autonomic nervous system** controls muscular and glandular responses of the viscera to internal environmental changes. The reflexes of this system follow a different pathway in the nervous system, called a visceral reflex arc (see figure 41.4).

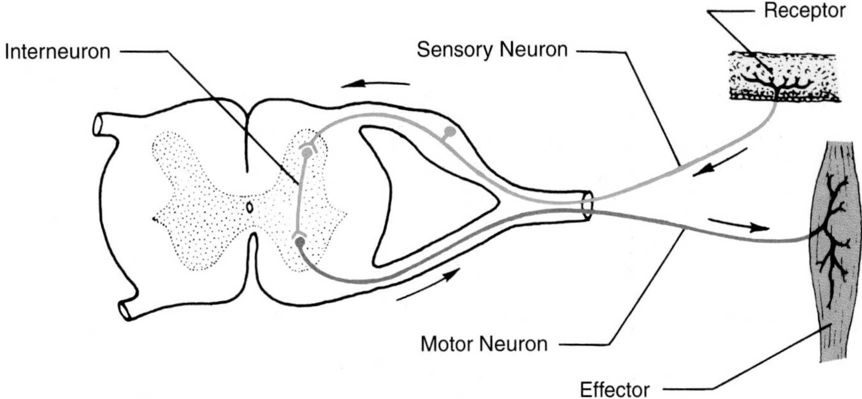

Figure 41.3 The somatic reflex arc.

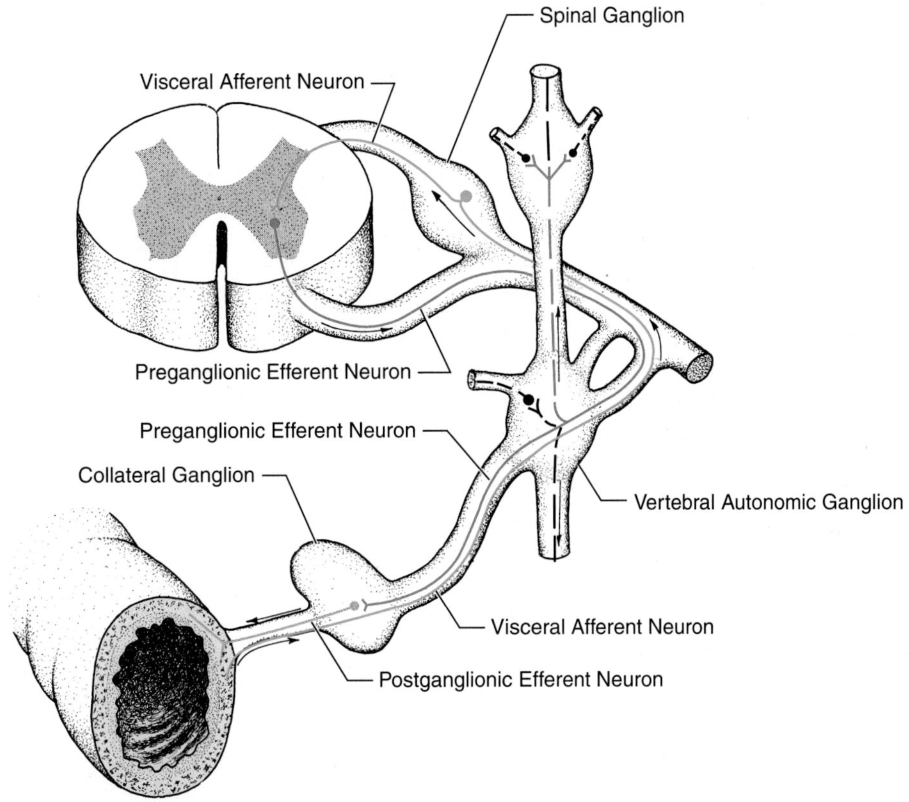

Figure 41.4 The visceral reflex arc.

The principal anatomical difference between somatic and visceral reflex arcs is that visceral reflex arcs have two efferent neurons instead of only one. These two efferent neurons synaptically connect with each other outside of the central nervous system in an autonomic ganglion.

Figure 41.4 shows two types of autonomic ganglia: vertebral and collateral. **Vertebral autonomic ganglia** form a chain along the vertebral column. **Collateral ganglia** are further away from the central nervous system.

A visceral reflex originates with a stimulus acting on a receptor in the viscera. Impulses pass along the dendrite of a **visceral afferent neuron,** whose cell body is in the **spinal ganglion.** The axon of the visceral afferent neuron forms a synapse with the **preganglionic efferent neuron** in the spinal cord. This neuron conveys impulses to the **postganglionic efferent neuron** at synaptic connections in either type of autonomic ganglion (note the short portions of cut-off postganglionic efferent neurons in the vertebral autonomic ganglia). From these ganglia,

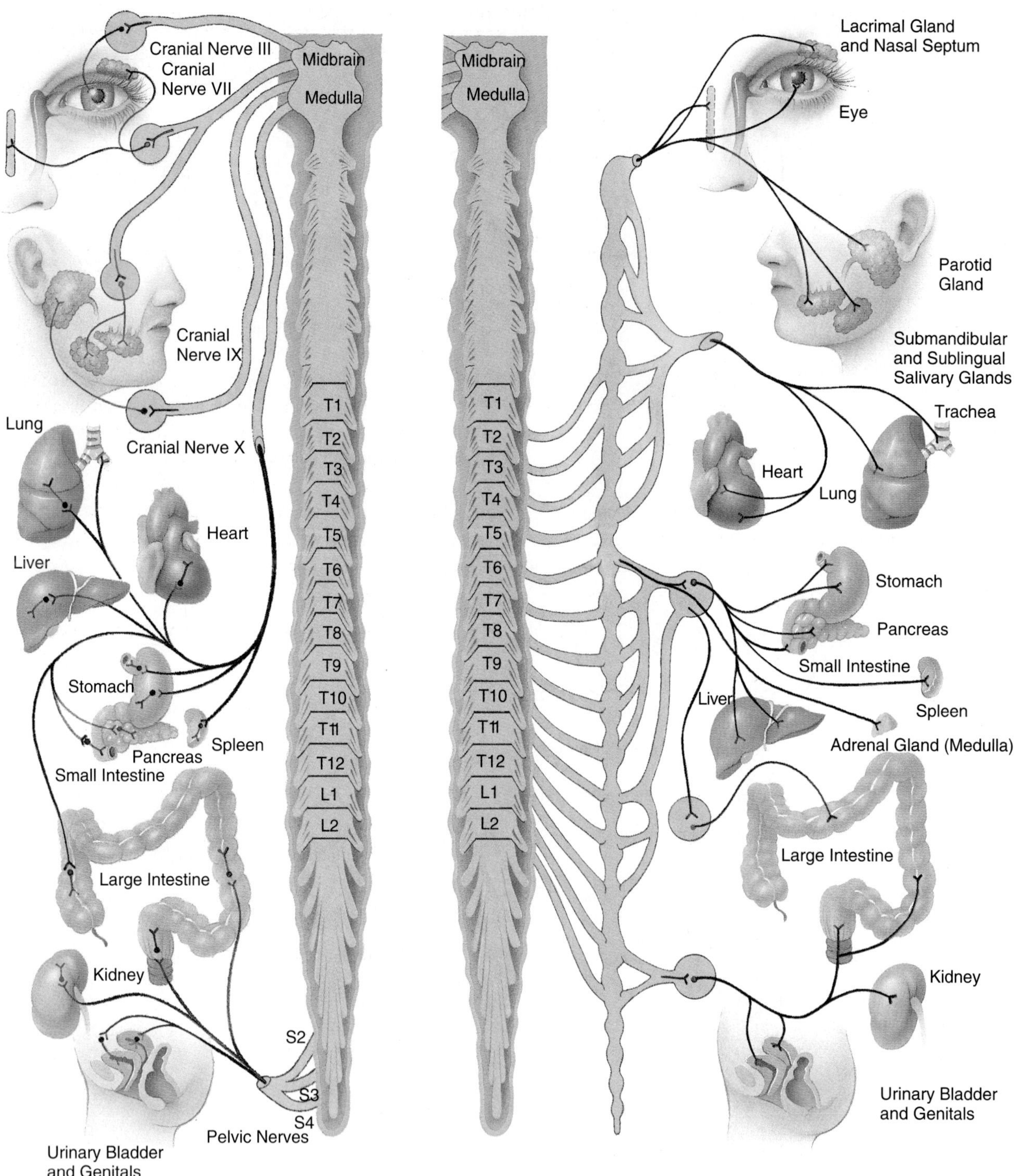

Figure 41.5 Parasympathetic (left) and sympathetic (right) divisions of the autonomic nervous system.

Table 41.1 Effects of Automatic Nervous System Stimulation on Various Organs

Organ	Sympathetic Effect	Parasympathetic Effect
Eyes	Dilation of pupil	Constriction of pupil
Salivary glands	Decreased secretion	Increased secretion
Digestive tract	Decreased motility	Increased motility
Heart	Increased rate and force of contraction	Decreased rate of contraction
Lungs	Dilation of bronchioles	Constriction of bronchioles
Liver	Stimulation of glycogen usage	No effect
Urinary bladder	Maintenance of muscle tone	Contraction

the postganglionic efferent neuron carries the impulses to the innervated organ.

The autonomic nervous system consists of two parts: the sympathetic and parasympathetic divisions. The *sympathetic* (thoracolumbar) *division* includes the spinal nerves of the thoracic and lumbar regions. The *parasympathetic* (craniosacral) *division* incorporates the cranial and sacral nerves. Nerves from both of these divisions innervate most viscera. This double innervation means that an organ can be stimulated by one system and inhibited by the other. Figure 41.5 shows the parasympathetic and sympathetic spinal nerves of the autonomic nervous system.

The parasympathetic system is often called the "housekeeping" system because it maintains normal levels of visceral organ functioning as required to maintain internal homeostasis. The sympathetic system, on the other hand, is called the "fight-or-flight" system. The responses of visceral organs to sympathetic stimulation prepare the body for stressful situations. Table 41.1 lists the opposing effects of the parasympathetic and sympathetic nervous systems on several organs.

Assignment:

Complete the Laboratory Report for this exercise.

42 Somatic Reflexes

In Exercise 41, you learned that somatic reflexes are automatic responses in skeletal muscles to stimuli applied to appropriate receptors. This exercise examines somatic reflexes in more detail—first in the frog, then in humans.

Frog Experiments

Somatic reflexes, whether in amphibians or humans, have five features worthy of study: (1) function, (2) speed of reaction, (3) radiation, (4) inhibition, and (5) synaptic fatigue. A pithed frog is a good laboratory specimen for exploring these various facets of muscular response.

Frog Preparation

"Single pith" a frog to destroy its brain, using the following procedure:

Materials:
small frog
sharp dissecting needle

1. Rinse the frog under cool tap water, and grip it in your left hand, as figure 42.1A shows.
2. Use your left index finger to force the frog's nose downward so that the head makes a sharp angle with the trunk. Now locate the transverse groove between the skull and the vertebral column by pressing down on the skin with your fingernail to mark its location.
3. Push a sharp dissecting needle into the crevice, forcing it forward into the center of the skull through the foramen magnum at the skull's base. Twist the needle from side to side to destroy the brain. This process of "single pithing" produces what is called a *spinal frog.*
4. After 5 or 6 minutes, test the effectiveness of the pithing by touching the cornea of each eye with a dissecting needle. If the lower eyelid on each eye does not rise to cover the eyeball, the pithing is complete.

   This test is valid only after approximately 5 minutes to allow recovery from a temporary state of neural (spinal) shock to the entire spinal cord.
5. Determine whether spinal shock has ended by pinching one of the toes with forceps. If a with-

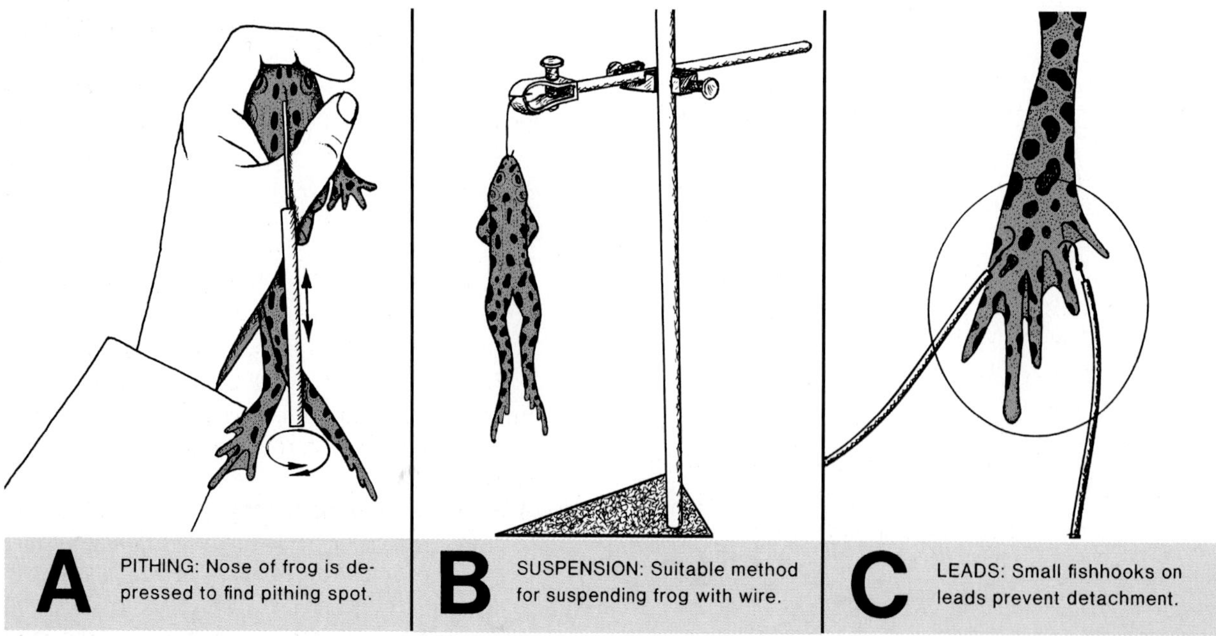

Figure 42.1 Frog pithing and setups.

drawal reflex occurs, spinal shock has ended, and your specimen is ready to use.

Functional Nature of Reflexes

Muscle tone, movement, and coordinated action in a spinal frog would indicate that these physiological activities occur independently of cognition and are the result of reflex activity. *Remember that the frog is now "brain-dead" and no longer consciously aware of its environment.* Test for these physiological phenomena as follows:

Materials:
spinal frog
ring stand and clamp
galvanized iron wire, pliers, forceps
filter paper squares (3 mm)
28% acetic acid
squeeze bottle of 10% $NaHCO_3$
squeeze bottle of tap water
beaker (250 ml size)

1. With the frog in a squatting position, gently draw out one of the hind legs. Does the frog exert any pull? When you release the foot, does it return to its previous position?
2. Squeeze the muscles of the hind legs. Do they feel flaccid or firm? Does muscle tone seem to exist?
3. Can you induce the frog to jump by prodding its posterior?
4. Place the frog in a sink basin filled with room-temperature water. Does the frog attempt to swim? Does it float or sink? Report all answers to questions 1–4 on the Laboratory Report.
5. Suspend the frog with a wire hook through the lower jaw, as figure 42.1B shows.
6. With forceps, dip a small square of filter paper in acetic acid, shake off the excess, and apply the paper to the ventral surface of the frog's thigh for about 10 seconds. Is the frog able to remove the irritant? Do you think that the frog *experiences* the burning sensation the acid causes? Explain.
7. Spray the acid off the leg with 10% $NaHCO_3$. Allow the liquid to drain from the leg into an empty beaker.
8. After 1 or 2 minutes, spray the leg with water.
9. Place another piece of filter paper and acid on another portion of the body (leg or abdomen), and observe the reaction. What happens if you restrain the limb that attempts removal? Does any other adaptive behavior take place?

Assignment:
Complete part A of the Laboratory Report.

Reaction Time of Reflexes

To determine whether or not the reaction time to a stimulus is influenced by the strength of the stimulus, proceed as follows:

Materials:
seven test tubes (20 mm diameter) in test tube rack
three beakers (150 ml size)
graduated cylinder (10 ml size)
squeeze bottle of tap water
squeeze bottle of 1% HCl
squeeze bottle of 1% $NaHCO_3$
Kimwipes

1. Dispense 30 ml of 1% $NaHCO_3$ into a beaker.
2. Label the seven test tubes: 1.0%, 0.5%, 0.4%, 0.3%, 0.2%, 0.1%, and 0.05%.
3. Measure out 10 ml of 1% HCl from the squeeze bottle, and pour it into the first test tube.
4. Into the other six tubes, measure out the correct amounts of 1% HCl and tap water to make 10 ml each of the correct percentages of HCl. (Examples: 5 ml of 1% HCl and 5 ml of water in the 0.5% tube; 4 ml of 1% HCl and 6 ml of water in the 0.4% tube.)
5. Pour the contents from the 0.05% tube into a clean beaker, and immerse the long toe of one of the frog's hind legs in it for 90 seconds. Prevent other parts of the foot from touching the sides of the beaker.

   If the foot is withdrawn, record on the Laboratory Report the number of seconds from immersion to withdrawal. If the foot is not withdrawn, remove the beaker of 0.05% HCl after 90 seconds.
6. Bathe the foot in the beaker of 1% $NaHCO_3$ for 15 seconds, and then spray the foot with water. Use the third beaker to catch any water sprayed on the foot. Dry the foot off with Kimwipes, and wait for 2 or 3 minutes.
7. Make two more tests with the 0.05% HCl, and record the average for the three tests. Be sure to neutralize, wash, and dry between tests.
8. Pour the 0.05% HCl back into its test tube, and empty the tube of 0.1% HCl into the beaker.
9. Place the long toe into this solution, as previously, and again record the withdrawal time in seconds.
10. After neutralization, bathing, drying, and resting the foot, repeat the test twice with this solution. Do not allow any test to exceed 90 seconds.
11. Repeat these procedures for all other tubes of diluted HCl, progressively increasing the acid concentration from low to high.

12. Plot the reaction times against the concentrations on the graph provided on the Laboratory Report.

Assignment:
Complete part B of the Laboratory Report.

Reflex Radiation

Subjecting the foot of a spinal frog to increasing electrical stimulation results in a phenomenon called *reflex radiation.* A demonstration of this effect requires applying a low-level tetanic stimulus first and then gradually increasing it. Proceed as follows:

Materials:
electronic stimulator
two-wire leads with small fishhooks soldered to the leads

Note: Before performing this experiment, you may wish to review Exercise 25 on the electronic stimulator.

1. Connect the jacks of the two-wire leads to the output posts of the stimulator. (These leads have very small fishhooks soldered to one end to allow easy attachment to the frog's skin. Handle them carefully. They are sharp!)
2. Attach the fishhooks to the skin of the frog's left foot, as figure 42.1C shows.
3. Set the stimulator to produce a minimum tetanic stimulus as follows: voltage at 0.1 V, duration at 50 msec, frequency at 50 pps, stimulus switch at Regular, polarity at Normal, and output switch on Biphasic.
4. Press the Mode switch to Repeat, and look for a response. If there is no response, sequentially increase the voltage by doubling (i.e., 0.2, 0.4, 0.8, etc.) until a response occurs.
5. Increase the voltage in steps of 2 or 3 V at a time, and note any changes in the reflex pattern.

Assignment:
Complete part C of the Laboratory Report.

Reflex Inhibition

Just as it is possible to stifle a sneeze or prevent a knee jerk, it should be possible to use electrical stimulation to override the reflex reaction to acid. With the stimulator still hooked up to the frog's left foot from the previous experiment, proceed as follows:

1. Lower the toes of the right foot into a beaker of HCl of a concentration that produced a moderately fast reflex response.
2. As the right foot is lowered into the acid, administer a moderate tetanizing stimulus to the left foot.
3. Adjust the voltage until the foot in the acid is inhibited.

Assignment:
Complete part D of the Laboratory Report.

Synaptic Fatigue

If your specimen is not responding well to stimuli, replace it with a freshly pithed specimen for this experiment.

1. Remove the skin from both thighs (see figure 42.2).
2. Stimulate the sciatic nerve of the *left leg,* and note the muscle contraction in both legs.
3. Continue the stimulation until the muscles in the *right leg* fail to respond.
4. After 30 seconds' rest, stimulate the left leg again to make sure that *no right leg muscles contract.*
5. Now stimulate the sciatic nerve of the *right leg.* Does this cause muscles of the right leg to respond?

Assignment:
Complete part E of the Laboratory Report.

Reflexes in Medical Diagnosis

Reflex testing is a standard, useful clinical procedure that aids in understanding damage to interver-

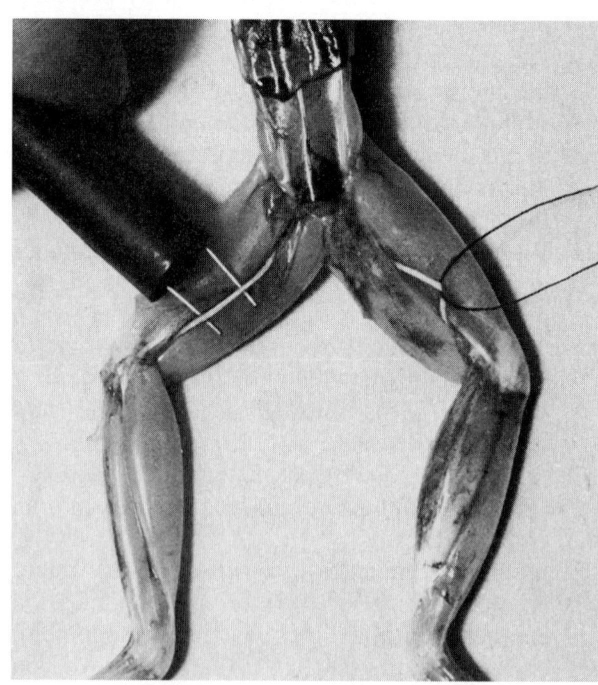

Figure 42.2 Setup for determining onset of synaptic fatigue. Note the string on the nerve to prevent trauma when the nerve is manipulated.

tebral disks, tumors, cerebrovascular damage, and many other conditions. The diagnostic reflexes studied here are the ones that physicians employ most often.

Interpretation of reflex responses is often subjective and requires a highly experienced diagnostician. The purpose in performing the tests here is not to diagnose, but rather to observe and understand why certain tests are performed.

Two Types of Reflexes

Clinically, reflexes are categorized as either deep or superficial. A sharp tap on an appropriate tendon or muscle ellicits a **deep reflex,** also called a *jerk, stretch,* or *myotatic reflex.* The receptors (muscle spindles) for deep reflexes are in the muscle, not the tendon. Tapping the tendon stretches the muscle, which activates the muscle spindle, which triggers the reflex response.

Superficial reflexes are withdrawal reflexes that noxious or tactile stimulation elicit. They are also called *cutaneous reflexes* because the receptors involved are in the skin. To initiate these reflexes, the skin is stroked or scratched.

Reflex Aberrations

Abnormal responses to stimuli may be diminished (hyporeflexia), exaggerated (hyperreflexia), or pathological. **Hyporeflexia** may be due to malnutrition, neuronal lesions, aging, deliberate relaxation, and so on. **Hyperreflexia** is often accompanied by marked muscle tone due to motor cortex loss of inhibitory control. Strychnine poisoning may also cause hyperreflexia. **Pathological reflexes** are reflex responses that occur in one or more muscles other than the muscle where the stimulus originates.

Physicians look for these three reflex aberrations when testing reflexes. They use the following scale to evaluate each response:

++++	Very brisk, hyperactive; often indicative of disease; may be associated with clonus (spasms)
+++	Brisker than average; may or may not be indicative of disease
++	Average; normal
+	Somewhat diminished
0	No response

Reflex Tests Procedure

Perform the seven reflex tests shown in figure 42.3 according to the procedures that follow.

Materials:
reflex hammer

Biceps Reflex The biceps reflex is a deep reflex that flexes the arm. Hold the subject's elbow with your thumb pressed over the tendon of the biceps brachii, as figure 42.3A shows. To produce the desired response, strike a sharp blow to the first phalanx of the thumb with the reflex hammer.

If the reflex is absent or diminished (0 or +), apply reinforcement by asking the subject to clench his or her teeth or squeeze his or her thigh with the other hand. Test both arms, and record the degree of responses on the Laboratory Report.

The biceps reflex functions through spinal nerves C_5 and C_6.

Triceps Reflex The triceps reflex is a deep reflex that extends the arm in normal individuals. To demonstrate this reflex, flex the arm at the elbow, holding the wrist as figure 42.3B shows, with the palm facing the body. Use the pointed end of the reflex hammer to strike the triceps brachii tendon above the elbow.

Use the same reinforcement techniques as with the biceps reflex, if necessary. Test both arms, and record the degree of responses on the Laboratory Report.

The triceps reflex functions through spinal nerves C_7 and C_8.

Brachioradialis Reflex In normal individuals, the brachioradialis reflex flexes and pronates the forearm and flexes the fingers. To demonstrate this reflex, direct the subject to rest the hand on the thigh, as figure 42.3C shows. Use the wide end of the reflex hammer to strike the forearm about 1 inch above the end of the radius (figure 42.3C shows the approximate spot). Use reinforcement techniques, if necessary. The tendon struck here is for the brachioradialis muscle of the arm. Test both arms, and record the degree of responses on the Laboratory Report.

The brachioradialis reflex functions through the same spinal nerves (C_5 and C_6) as the biceps reflex.

Hoffmann's Reflex Hoffmann's reflex is a deep reflex in which the response is not limited to the stretched muscle. A broad response (affecting several fingers, other than the finger stimulated) to the stimulus indicates pyramidal tract (spinal cord) damage.

To induce this reflex, flick the terminal phalanx of the index finger upward, as figure 42.3D shows. Deep reflex hyperactivity and pathological characteristics

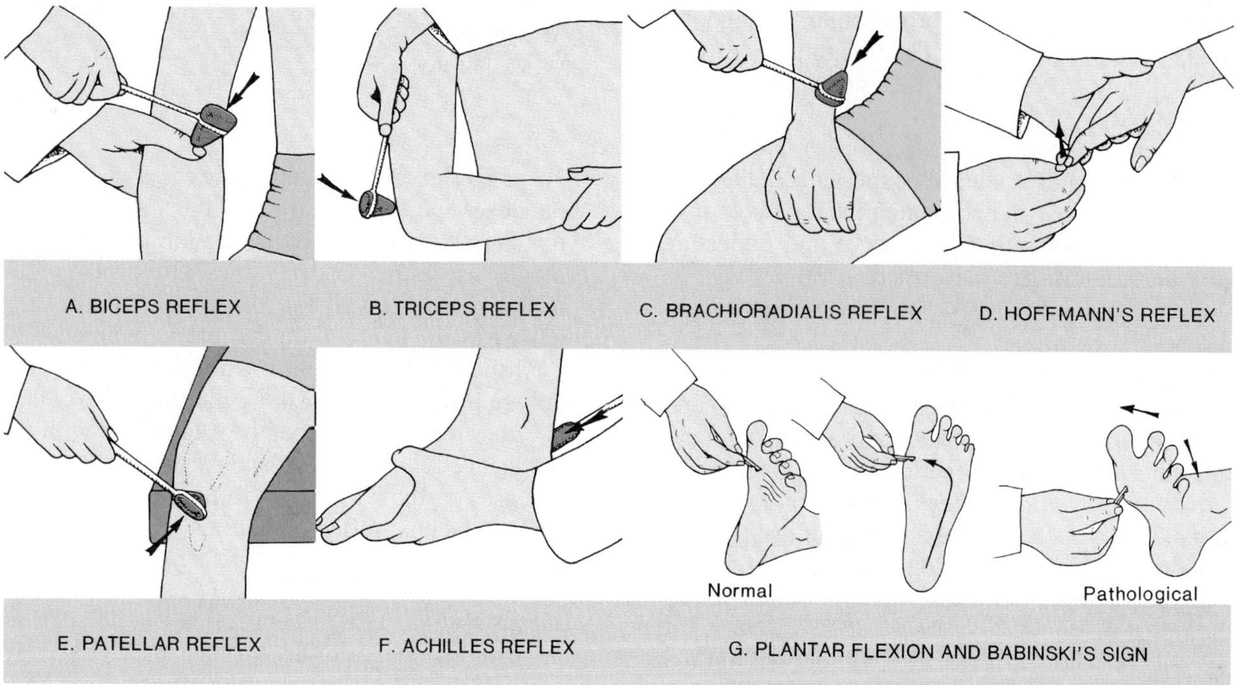

Figure 42.3 Methods for producing seven types of somatic reflexes.

are indicated if the thumb adducts and flexes, and the other fingers exhibit twitchlike flexion.

Test both hands, and record the degree of responses on the Laboratory Report.

Patellar Reflex The patellar reflex, variously referred to as the *knee reflex, knee-jerk reflex,* or *quadriceps reflex,* is monosynaptic. The subject should sit on the edge of a table with the leg suspended and somewhat flexed over the edge. To elicit the typical response, strike the patellar tendon just below the kneecap, as figure 42.3E shows. If the response is negative, utilize the **Jendrassic maneuver,** which involves the subject locking the fingers of both hands in front of the body and pulling each hand against each other isometrically.

Inducing the patellar reflex sometimes produces a phenomenon called **clonus.** Clonus is a succession of jerklike contractions (spasms) that follows the normal response and persists for a period of time. This condition is a manifestation of hyperreflexia and may indicate damage within the central nervous system.

Test both legs, and record the degree of responses on the Laboratory Report. The patellar reflex functions through spinal nerves L_2, L_3, and L_4.

Achilles Reflex The Achilles reflex, also known as the *ankle jerk,* is characterized by plantar flexion when the Achilles tendon is struck a sharp blow. Stretching the Achilles tendon affects the muscle spindles in the triceps surae, causing it to contract.

To perform this test, grip the foot with the left hand, as figure 42.3F shows, forcing the foot upward somewhat. With the subject relaxed, strike the Achilles tendon as shown. Hyporeflexia here is often associated with hypothyroidism. Test both legs, and record the degree of responses on the Laboratory Report.

The Achilles reflex functions through spinal nerves S_1 and S_2.

Plantar Flexion ***(Babinski's Sign)*** Plantar flexion is the only reflex in this series that is superficial. Perform this test using a hard object, such as a key, and following the pattern in the middle illustration of figure 42.3G. Test both feet, and record the degree of responses on the Laboratory Report.

Note in figure 42.3G, that the normal reaction to stroking the sole of an adult's foot is plantar flexion. If dorsiflexion occurs, starting in the great toe and spreading to the other toes (Babinski's sign), fibers in the pyramidal tracts have myelin damage. Dorsiflexion is normal in infants, especially if they are asleep. Babinski's sign disappears in infants once all nerve fibers are myelinized.

The plantar flexion reflex functions through spinal nerves S_1 and S_2.

Assignment:
Complete part F of the Laboratory Report for this exercise.

The Patellar Reflex: A Computerized Evaluation

43

In Exercise 42, you learned that physicians use reflex responses to detect neural pathology. In general, the interpretation of reflex responses is subjective and requires considerable experience. The proper software and hardware, however, can quantify certain reflex responses to confirm suspected damage to the nervous system.

The Intelitool Flexicomp system is used in this exercise to remove some of the subjectivity in observing the patellar reflex. You can also use this system to analyze the biceps and triceps reflexes. The setup, which figure 43.1 shows, accomplishes the following:

- The stimulus response (degree of deflection) can be quantified.
- Time intervals, such as the latent period and time to maximal response, can be quantified.
- Neural facilitation and antagonistic dampening can be demonstrated.
- Responses and response times of the same reflex in opposite limbs can be compared.

In this exercise, the patellar reflexes in both legs are measured to make the previous four observations. Note in the "Materials" list that follows that the *Flexicomp Lab Manual* and *Flexicomp User Manual* are available for cross-referencing of procedures. Proceed as follows:

Materials:
computer, monitor, and printer
displacement transducer with Velcro straps
percussion hammer (computer interfaced)
transducer cable (game port connector)
Flexicomp program diskette
data disk (initialized)
calibration template
Flexicomp Lab Manual
Flexicomp User Manual

Preliminary Preparations

Before starting the experiment, you need to set up the equipment, boot up the software, calibrate the transducer, and attach the Flexicomp transducer to the leg. Proceed as follows:

Start-Up

1. Insert the Flexicomp program disk into drive A and the data disk into drive B.
2. Turn on the computer. If you are using an Apple IIe, IIc, or IIGS, depress the Caps-Lock key.
3. Press any key, and the Flexicomp program will take over. Answer the configuration questions.

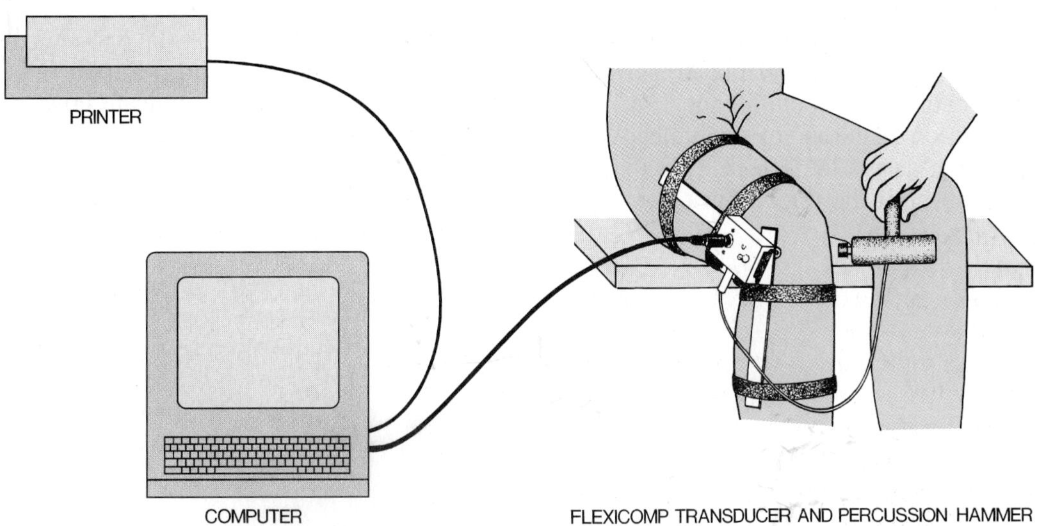

Figure 43.1 Flexicomp setup for monitoring the patellar reflex.

Calibration

After you have described the configuration of the setup, the software will ask you if you wish to use the Flexicomp transducer. Instructions on the screen will tell you how to use the calibration guide to set the angle of the transducer first at 90° and then at 150°. You make adjustments by turning the aluminum knob on the transducer.

Important: Following calibration, loosen the thumbscrew locking tab in the angle of the transducer. Failure to do so may damage the transducer.

If calibration fails for any reason, the software will detect the flaw and ask you to redo the calibration procedure.

Flexicomp Menu

The Flexicomp Menu is as follows:

FLEXICOMP MENU
( ) Experiment Menu
( ) Review/Analyze Menu
( ) Data File/Disk Commands
( ) Preferences
( ) ESCape to Program Menu

Note: "Review/Analyze Menu" and "Data File/Disk Commands" selections only function when experimental data are available.

Experiment Procedure

With the subject seated on the edge of a table, strap the transducer to the right leg in the manner shown in figure 43.1. **The thumbscrew locking tab should be loose.** Note that the transducer box is facing outward and that the hinge of the transducer is aligned with the hinge of the knee.

1. Select "Experiment Menu" from the Main Menu, and select "Real Time Experiment Mode" from the Experiment Menu.
2. Press the button on the end of the mallet. As the trace is going across the screen, turn the aluminum knob until the trace is just below the center of the screen.
3. Tighten the thumbscrew locking tab.
4. Press the C key to clear the computer's memory. You are now ready to begin data acquisition.
5. Strike the subject's patellar ligament hard enough to elicit a good response.
6. When the plot reaches the edge of the screen, note the maximum degree of excursion and the time until the maximum degree of excursion.
7. Strike the knee several times to see if you can establish somewhat uniform results. If your procedure is good, you should get similar results each time.
8. To demonstrate "facilitation," have the subject pull one hand against another with fingers interlocked in front of the chest as the target area is struck. Repeat this procedure several times. Record the frame numbers showing facilitation so that you can compare them with other frames.
9. Press ESC to return to the Main Menu, and select "Review/Analyze Menu."
10. Measure the latent period and any other features your instructor designates by using the following command keys: **F** for forward to next data frame; **B** for backward to previous data frame; **left cursor arrow** key to move the data analysis cursor to the left and **right cursor arrow** key to move the data analysis cursor to the right; **J** to move the data analysis cursor rapidly in direction of last arrow movement; **Z** to set the value in the analysis window to zero and to set the current data marker (vertical line) as the origin or base marker; and **P** for print.

Optional Experiments

For additional experiments on the biceps reflex, triceps reflex, and others, consult the *Flexicomp Lab Manual.*

Assignment:
Since there is no Laboratory Report for this exercise, your instructor will indicate how to write up this experiment.

Action Potential Velocities

44

The velocity of nerve impulse conduction in nerve fibers varies. Small-diameter, unmyelinated fibers in portions of the autonomic nervous system may have velocities as low as 0.5 m per second. Large-diameter, myelinated fibers of the somatic nervous system, on the other hand, may carry nerve impulses at speeds as high as 130 m per second. This high velocity in myelinated fibers is due to the ion flow from one node of Ranvier to another along the fiber. Such ionic movement through the axoplasm and extracellular fluid from node to node is called *saltatory conduction.*

Determining the speed of a nerve impulse along a nerve only requires measuring how long it takes the action potential to traverse a known distance. That distance in meters divided by the time (total time minus latency) yields the velocity in meters per second. An oscilloscope, polygraph, or Duograph can readily make these measurements. Although figure 44.1 shows a Duograph in the setup, a polygraph or oscilloscope works as well. Your instructor may wish to demonstrate the procedure first with a dual-beam oscilloscope.

The Ulnar Nerve

The electrode hookup shown in figure 44.1 monitors action potentials in the ulnar nerve. This nerve has been selected for study because its close proximity to the surface of the arm simplifies its stimulation with an electronic stimulator, such as the Grass SD9. An understanding of the exact location of this nerve is essential to correctly place the various electrodes.

The ulnar nerve arises from the medial cord of the brachial plexus, deriving its fibers from the eighth cervical (C_8) and first thoracic (T_1) nerves. Figure 44.2 illustrates its path through the arm. Note that, in the upper arm, the nerve lies on the medial surface, just under the skin. In the elbow region, it rests in a groove posterior to the medial epicondyle of the humerus. In the forearm, it lies,

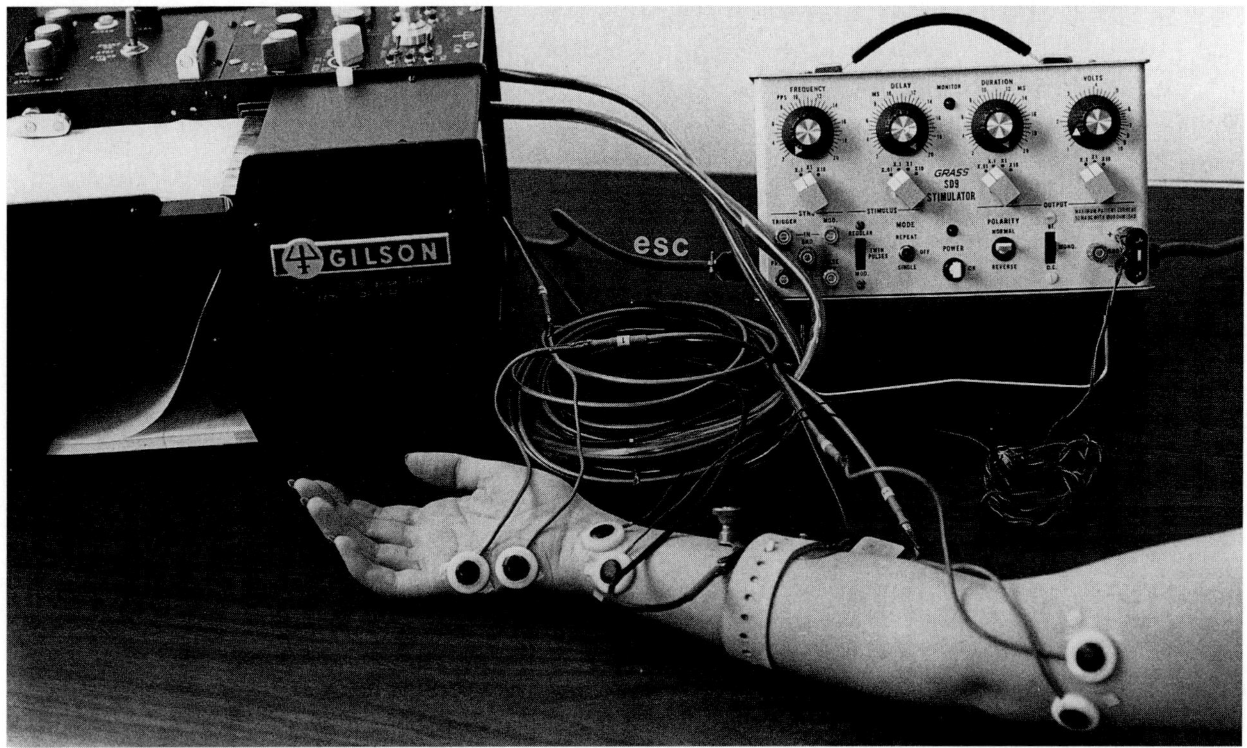

Figure 44.1 Experimental setup.

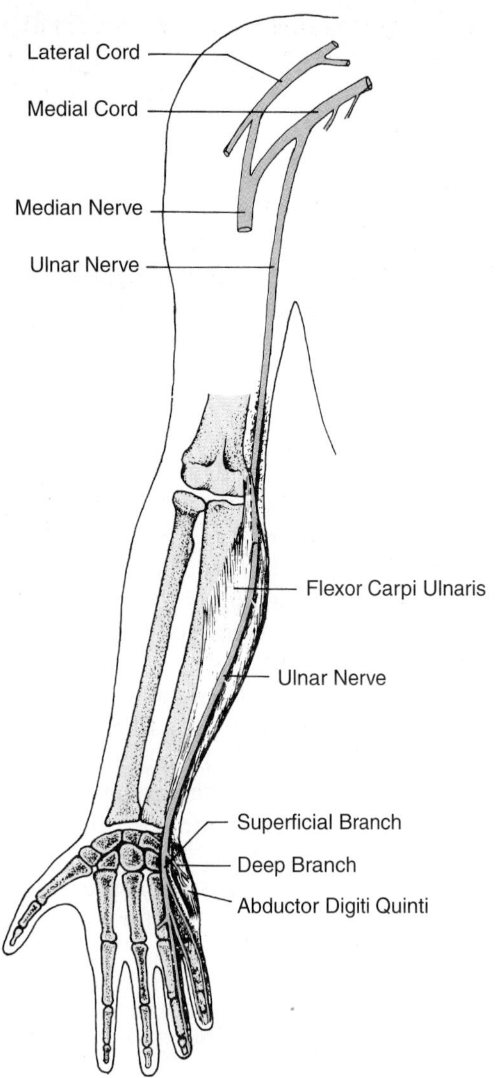

Figure 44.2 The ulnar nerve (right arm).

superficially, on the medial surface and supplies branches to the *flexor carpi ulnaris* and the medial half of the *flexor digitorum profundus.*

In the hand, the ulnar nerve divides to form deep and superficial branches. The superficial branch has branches that extend down both sides of the fifth finger and the medial side of the fourth finger. The deep branch innervates the *abductor digiti quinti* and two other muscles of the small finger. The abductor digiti quinti is the muscle of interest in this experiment.

Electrode Placement

Figure 44.3 is a close-up of the electrode arrangement on the arm. Note that the two stimulator electrodes (anode and cathode) are on the wrist at point A. A pair of recording electrodes that attach at point B on the side of the hand are over the abductor digiti quinti muscle. These electrodes transmit a signal to the Duograph when the muscle contracts.

The two electrodes above the elbow at point C pick up the action potential as it travels from point A through point C. The signal from these electrodes is fed into the other channel on the Duograph.

The ECG electrode strapped on the arm serves as a common ground for the recording electrodes at points B and C. Note that two alligator clamps attach to it.

Artificial electrical stimulation of an action potential in a nerve is called an **evoked potential,** as opposed to potentials produced endogenously during voluntary muscle contraction. When a nerve is stimulated somewhere along its course, evoked potentials move in opposite directions from the point of stimulation.

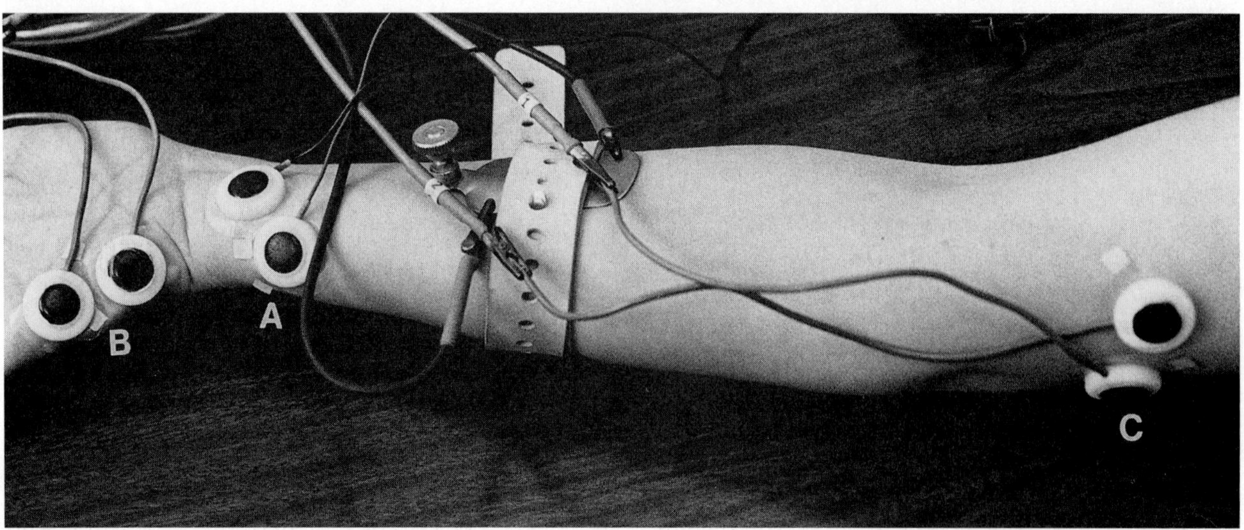

Figure 44.3 Arrangement of electrodes.

Impulses that move along a fiber in the natural direction exhibit **orthodromic conduction.** In this experiment, the evoked potential that moves from point A to point B to stimulate the abductor digiti quinti is orthodromic, since all fibers of the deep branch are motor. The evoked action potential that moves in the opposite direction exhibits **antidromic conduction.** In this experiment, the action potential that moves along motor fibers from point A to point C is antidromic. Any nerve impulses that sensory fibers in the ulnar nerve carry from point A to point B are orthodromic. If the stimulus is strong enough, orthodromic impulses in the sensory fibers trigger a secondary response called a **reflex potential.**

Team Assignments

The class will be divided into four-member teams. If equipment is limited, some teams may need to work on another experiment or project until other teams have finished. Each team will consist of a subject, a director, a stimulator operator, and a Duograph operator. The responsibilities of each position are as follows:

The Subject

The subject must be capable of keeping the experimental arm completely relaxed during the entire procedure. While the low amperage used in this experiment cannot cause injury, the subject may experience slight discomfort occasionally.

Careful selection of the subject is crucial to the success of this experiment. A good subject is not squeamish, timid, coy, silly, or hyperexcitable; nor is a good subject brash, careless, or disinterested.

The Director

The director coordinates team activities. An important role of the director is locating the course of the ulnar nerve and properly placing the electrodes on the subject.

At the conclusion of the experiment, the director disconnects the subject, cleans the skin and electrodes, and returns the electrodes to their proper places. The director and the two equipment operators work together to clean the work area at the end of the period.

The Duograph Operator

The Duograph operator sets up the Duograph, selects the proper chart speeds, makes notations on the chart, adjusts sensitivity, and puts the Duograph away at the end of the experiment.

The Duograph records three events simultaneously: (1) the onset and duration of the stimulus, (2) the evoked muscle potential, and (3) the evoked nerve potential.

The Stimulator Operator

The stimulator operator must understand the functions of the various stimulator controls. Stimuli are manually delivered at predetermined intervals. The stimulator operator makes adjustments that other team members call for.

Experiment Procedure

As soon as all team members have agreed upon their responsibilities, proceed as follows to perform the experiment:

Materials:
Gilson Duograph or polygraph
electronic stimulator
two three-lead patient cables
lead cable for stimulator
event synchronization cable
six skin electrodes (Gilson self-adhering, #E1081K)
ECG plate electrode
electrode paste
Scotchbrite pad (grade #7447)
alcohol swabs
meter stick

Equipment Setup

The Duograph operator and stimulator operator work together to set up the equipment as follows:

1. Plug in the power cords for the Duograph and stimulator.
2. Connect the stimulator to the Duograph with the event synchronization cable (labeled **esc** in figure 44.1).
3. Arrange the equipment as figure 44.1 shows.
4. Attach two leads from the wrist electrodes to the output posts of the stimulator.
5. Plug the two three-lead cables into the Duograph. One of these cables will receive signals from the hand. The other will monitor nerve impulses in the arm. Make the leads of these cables accessible to the director, who will make the electrode hookups.
6. Set the **stimulator controls** as follows: duration at 1 msec, voltage at 1 V, stimulus switch at Regular, mode at Off, power switch On, polarity switch at Normal, and output switch at Mono.

7. Set the **Duograph controls** as follows: power switch On, c.c. (chart control) switch at STBY, stylus heat control knob at the two o'clock position, chart speed at 25 mm per sec, mode control on both channels at EEG, and gain control at 2 mV/cm.

Preparation of Subject

The director prepares the subject in the following manner:

1. Position the subject comfortably so that the arm rests on the table as figure 44.1 shows.
2. If enough table space is available, locate the subject to the right of the stimulator instead of directly in front of it to allow room for the stimulator and Duograph operators.
3. Gently scrub the skin areas with a Scotchbrite pad, and wipe with an alcohol swab.
4. Attach the electrodes to the arm as figure 44.3 shows. Refer to figures 32.3 through 32.6 on page 139–140 for placing the adhesive pads on the self-adhering electrodes.
5. Apply electrode paste to the ECG plate electrode before strapping it in place.
6. Connect the two electrode wires of the hand to the #1 and #2 leads of one three-lead cable.
7. Connect the two electrode wires of the upper arm to the #1 and #2 leads of the other three-lead cable.
8. Clamp the ground wires from each of these cables to the ECG plate electrode in the middle of the arm.
9. With a meter stick, measure the distances between points A and B, and points A and C, and record these measurements on the Laboratory Report.

Stimulation and Recording

Once all electrodes are in place, determine the threshold stimulus that will just barely activate the abductor digiti quinti muscle as follows:

1. Starting with a duration of 1 msec and 1 V, depress the mode switch to Single to see if the muscle reacts.
2. If no response occurs, increase the duration in 1 msec increments until the muscle reacts. You should be able to establish the threshold voltage at some point below 10 msec.
3. If there still is no response at 10 msec duration, try increasing the voltage in 1 V increments at 5 msec duration until the muscle reacts.
4. After you have determined the threshold stimulus, return the voltage to a subthreshold level.
5. With the chart moving at 25 mm per second, depress the mode switch on the stimulator to Single at approximately 1-second intervals as you increase the voltage in uniform increments from subthreshold to the point where you record maximal potentials.
6. Label the tracing with pertinent information.
7. Disconnect the subject, and shut off the instruments.
8. Clean up the work area, and put away all equipment.

Assignment:
Calculate the velocities of the nerve impulse in the ulnar nerve according to the instructions provided on the Laboratory Report. Answer all questions on the report.

Brain Anatomy: External

45

In this exercise, you study the human brain in conjunction with dissecting a sheep brain. The ample size and availability of sheep brains make them preferable to brains of most other animals. Also, the anatomical similarities between human and sheep brains are much greater than their differences. Preserved human brains or models of human brains will also be available for study, but not dissection.

The Meninges

As mentioned in Exercise 41, three protective membranes called the *meninges* surround the spinal cord and brain. They are the dura mater, the arachnoid mater, and the pia mater.

Dura Mater The outermost meninx, the *dura mater,* is a thick, tough membrane of fibrous connective tissue. The lateral view of the opened skull in figure 45.1 shows the dura mater lifted away from the exposed brain. Note that on the median line of the skull, it forms a large **sagittal sinus** that collects blood from the surface of the brain. Observe also that many cerebral veins empty into this sinus.

Extending downward between the two halves of the cerebrum is an extension of the dura mater, the **falx cerebri.** This structure shows up only in the frontal section of figure 45.1. At its anterior inferior extremity, the falx cerebri attaches to the crista galli of the ethmoid bone.

Arachnoid Mater Inferior to the dura mater lies the second meninx, the *arachnoid mater.* Between this delicate, netlike membrane and the surface of the brain is the **subarachnoid space,** which contains cerebrospinal fluid. Note that the arachnoid mater has small projections of tissue, the **arachnoid granulations,** that extend up into the sagittal sinus. These granulations allow cerebrospinal fluid to diffuse from the subarachnoid space into the venous blood of the sagittal sinus.

Pia Mater The third meninx, the *pia mater,* covers the surface of the brain. This membrane is very thin. Note that the surface of the brain has grooves, or **sulci,** that increase the surface of the **gray matter.** As in the spinal cord, the gray matter is darker than the inner **white matter** because it contains neuron cell bodies. The white matter consists of neuron fibers that extend from the gray matter to other parts of the brain and spinal cord.

Assignment:
Complete part A of the Laboratory Report for this exercise.

Sheep Brain Dissection

This sheep brain study consists of two procedures: (1) an in situ study of fresh brain material, and (2) the dissection of a preserved sheep brain. In the in situ dissection, the brain is observed as it appears, undisturbed, in the skull. Getting to the brain requires removing the top of the skull with an autopsy saw. This type of dissection enables you to

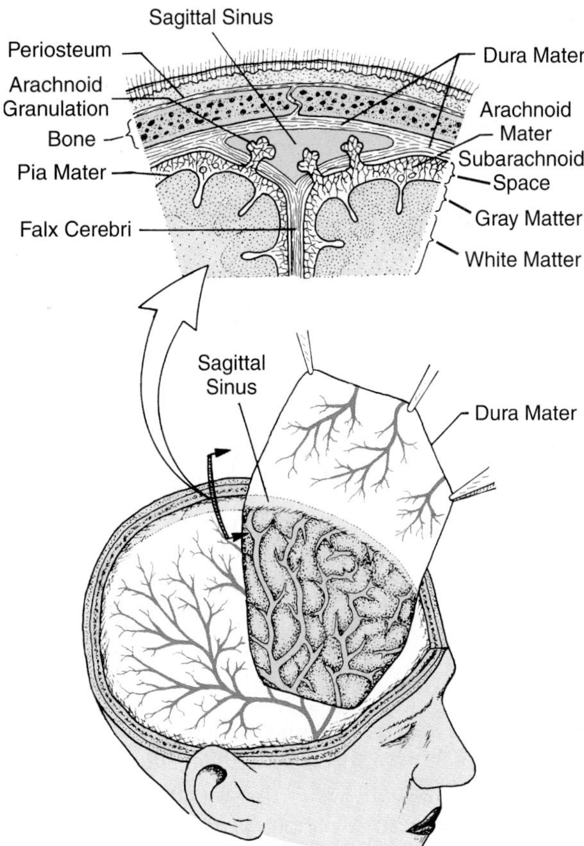

Figure 45.1 The meninges of the brain.

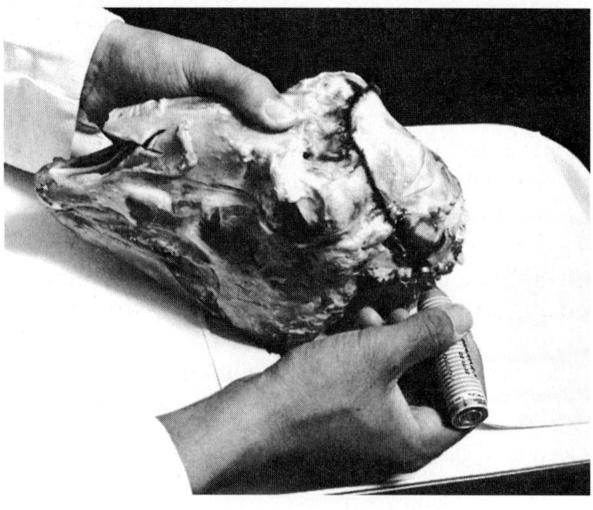

Figure 45.2 Mark the skull with a felt-tip pen to indicate the line of cut.

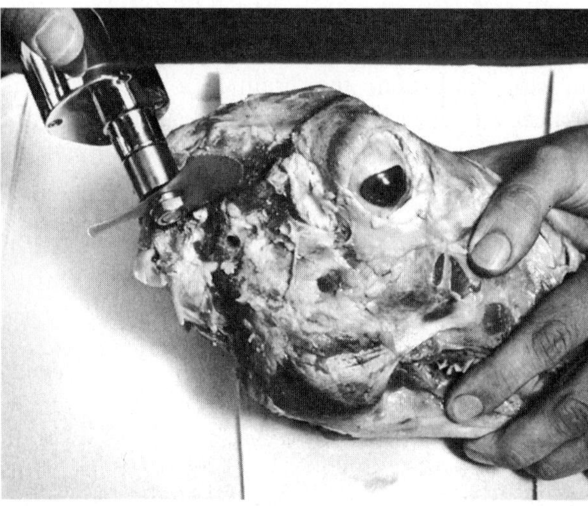

Figure 45.3 While an assistant securely holds the head, make the cut with an autopsy saw.

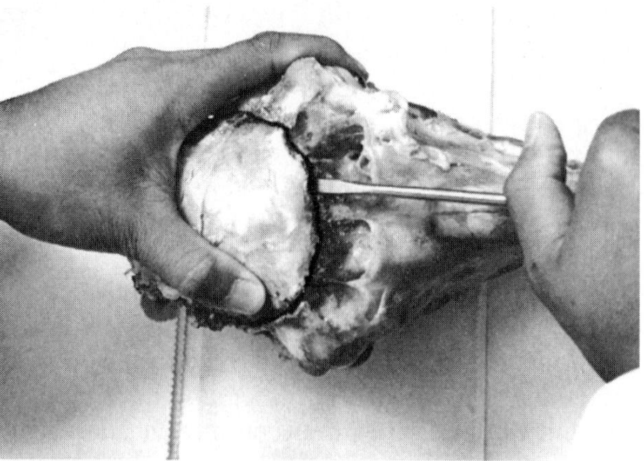

Figure 45.4 Use a large screwdriver to pry the cut section of the skull off the head.

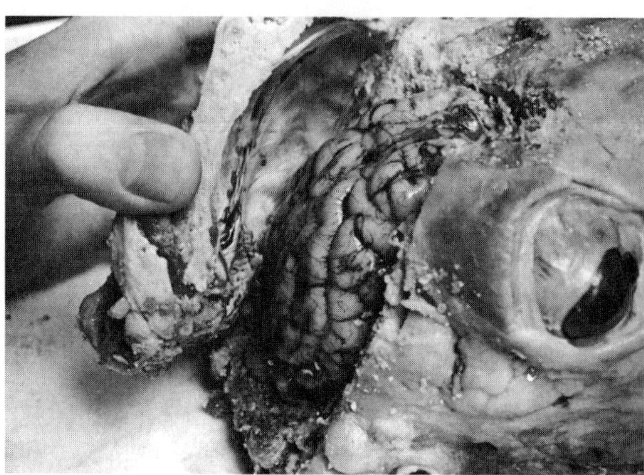

Figure 45.5 Lift off the cut portion of the skull to expose the intact brain.

see many structures not seen in a preserved specimen already removed from the skull.

The value of the second dissection is that preserved brains are hardened from soaking in preservatives and reveal structures that are usually destroyed when fresh brains are removed from skulls. If time restrictions prevent doing both dissections, the second dissection is preferred.

Sheep Head Dissection

Examining a sheep brain within the cranium requires removing the top of the skull with an autopsy saw, as figures 45.2 through 45.5 show. An autopsy saw cuts through bone with a reciprocating action. Since it does not have a rotary action, it is relatively safe to use; however, you must exercise care to avoid accidents. For maximum safety, one person should hold the sheep head while another individual does the cutting. *At no time should the holder's hands be in the cutting path of the saw blade.*

If only one or two saws are available, not everyone will be able to start cutting at the same time. Since it takes 5 to 10 minutes to remove the top of the skull, some students will have to work on other parts of this exercise until a saw is available. Proceed as follows:

Materials:
fresh sheep head
autopsy saw
large screwdriver
dissecting kit
black felt-tip pen

1. With a black felt-tip pen, mark a line on the skull across the forehead between the eyes,

around the sides of the skull, and over the occipital condyles. Refer to figure 45.2.

2. While firmly gripping the saw with both hands, cut around the skull following the marked line. Brace your hands against the table for support. Do not try to make a freehand cut. Use the large cutting edge for straight cuts, the small cutting edge for sharp curves.

   Be sure that your laboratory partner is holding the head securely (see figure 45.3). Cut only deep enough to get through the bone. Try not to damage the brain.

   Rest your hands after cutting an inch or two. To avoid accidents, do not allow your wrists to get too tired. As mentioned earlier, watch out for your partner's hands!
3. After cutting all the way around the skull, place the end of a screwdriver into the saw cut, and pry upward to loosen the bony plate (see figure 45.4). Cut free any remaining portions.
4. Gently lift the bone section off the top of the skull (see figure 45.5). Note how fragments of the **dura mater** prevent you from completely removing the cap. Use a sharp scalpel or scissors to cut the dura mater loose while lifting the skull cap off.
5. Examine the inside surface of the removed bone. Note how collagenous fibers make the dura mater thick and tough.
6. Locate the **sagittal sinus.** Open it with a scalpel or scissors, and look for evidence of **arachnoid granulations.**
7. Note that convolutions called **gyri** (*gyrus,* singular) cover the surface of the brain. Between the gyri are the grooves, or **sulci.**
8. Locate the middle meninx, or **arachnoid mater,** on the surface of the brain. This very thin, netlike membrane appears to be closely attached to the brain. In actuality, a thin **subarachnoid space,** containing **cerebrospinal fluid** is between the arachnoid mater and the surface of the brain. You probably will not be able to detect the subarachnoid space on your specimen.
9. Gently force your second, third, and fourth fingers under the anterior portion of the brain, and lift the brain from the floor of the cranial cavity. Take care not to damage the brain.

   Locate the **olfactory bulbs** (see figures 45.6 and 45.7). Note that they attach to the forepart of the skull. Olfactory nerves pass from these bulbs through the cribriform plate and into the nasal cavity.
10. Free the olfactory bulbs from the skull with your scalpel.
11. Raise the brain further to expose the **optic chiasma** (see figure 45.7). Note how the optic nerves that pass up through the skull are continuous with the optic chiasma. Some of the neurons from the retina of each eye cross over in the optic chiasma.
12. Sever the optic nerves as close to the skull as possible with your scalpel, and raise the brain a little further to expose the two **internal carotid arteries** that supply the brain with blood. Do you remember what foramina in the skull allow these blood vessels to enter the cranium?
13. Sever the internal carotid arteries, and raise the brain a little more to expose the **infundibulum** that connects with the hypophysis. Figure 45.6 illustrates the relationship of the infundibulum to the hypophysis.
14. Carefully dissect the **hypophysis** out of the sella turcica *without* severing the infundibulum.
15. Continue lifting the brain, and identify as many cranial nerves as you can (refer to figures 45.6 and 45.7).
16. Remove the brain completely from the cranium. Section the brain, frontally, to expose the inner material. Note how distinct the **gray matter** is from the **white matter.**
17. Dispose of the remains in the proper waste receptacle, and proceed to the dissection of a preserved sheep brain.

Preserved Sheep Brain Dissection

The preserved sheep brain dissection focuses on the finer details of sheep brain external and internal structure and on comparisons with the human brain. Once you have located a structure on the sheep brain, try to find a comparable structure on the human brain; in some instances, however, direct comparisons are not possible.

Materials:
preserved sheep brains
preserved human brains and/or human brain models
dissecting instruments
dissecting trays

Place a sheep brain on a dissecting tray, and by comparing it to figures 45.6 and 45.7, identify the **cerebrum, cerebellum, pons, midbrain,** and **medulla oblongata.** Study each of these five major divisions to identify the fissures, grooves, ridges, and other structures as follows:

The Cerebral Hemispheres In both sheep and humans, the cerebrum is the predominant portion of

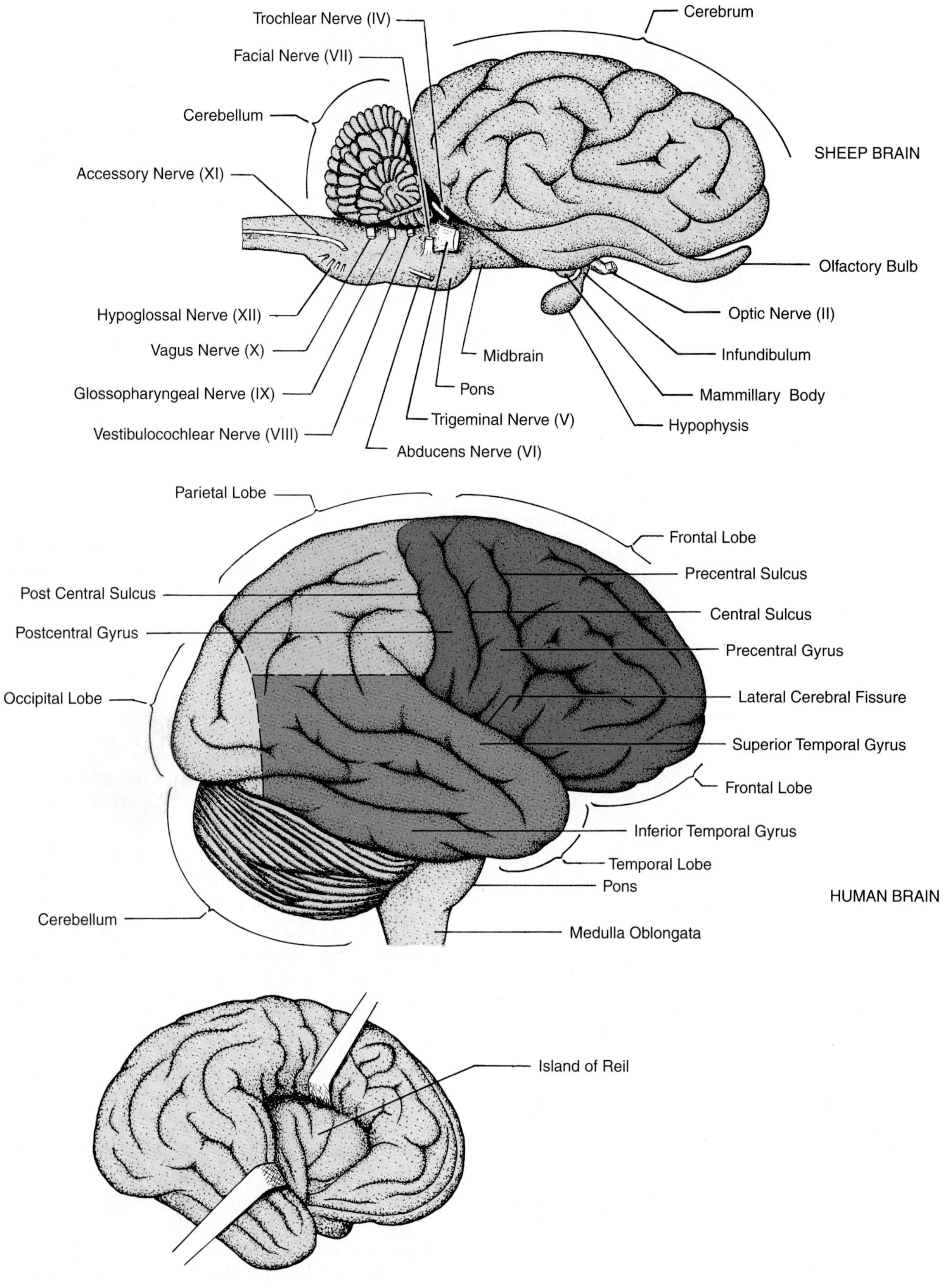

Figure 45.6 Lateral aspects of sheep and human brains.

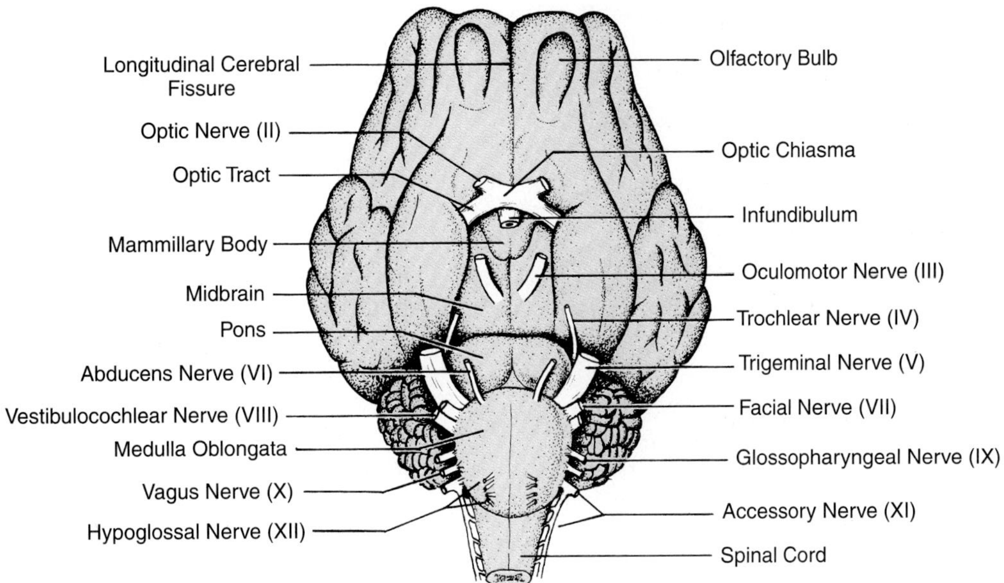

Figure 45.7 **Ventral aspect of sheep brain.**

the brain. Note that when observed from the dorsal or ventral aspect, the cerebrum divides along the **longitudinal cerebral fissure** on the median line to form two *cerebral hemispheres.*

Observe that ridges and furrows of variable depth cover the surface of the cerebrum. The ridges or convolutions are **gyri.** The deeper furrows are **fissures,** and the shallow ones, **sulci.** The infolding of the cerebral surface greatly increases the brain's gray matter.

The principal fissures of the human cerebrum are the **central sulcus lateral cerebral fissure,** and the **parieto-occipital fissure.** The greater portion of the parieto-occipital fissure is seen on the medial surface of the cerebrum (see figure 46.2 in the next exercise). Retraction of the lateral cerebral fissure, as the bottom illustration in figure 45.6 shows, exposes an area called the **island of Reil.**

Each half of the human cerebrum divides into four lobes. The **frontal lobe** is the most anterior portion. Its posterior margin falls on the central sulcus, and its inferior border consists of the lateral cerebral fissure. The **occipital lobe** is the most posterior lobe of the brain. Its anterior margin falls on the parieto-occipital fissure. The **temporal lobe** lies inferior to the lateral cerebral fissure and extends back to the parieto-occipital fissure. The **parietal lobe** is superior to the temporal lobe.

The Cerebellum Note that sulci furrow the surface of the sheep cerebellum. How do these sulci differ from those of the cerebrum? The human cerebellum is constricted in the middle to form right and left hemispheres. Is this also true of the sheep's cerebellum? The cerebellum helps to maintain posture and to coordinate complex muscular movements.

The Brain Stem

The remaining structures to be observed are components of the *brain stem.* This axial portion of the brain includes the midbrain, diencephalon, pons, and medulla oblongata. It excludes the cerebrum and cerebellum.

Midbrain The cerebellum and cerebrum conceal the dorsal portion of the midbrain. Forcing the cerebellum downward with the thumb, as figure 45.8 illustrates, exposes four roundish bodies on the surface of the midbrain that constitute the *corpora quadrigemina.* The two larger bodies are the **superior colliculi;** the two smaller bodies are the **inferior colliculi.** In some animals, the superior colliculi are called "optic lobes" because of their close association with the optic tracts. The superior colliculi contain centers for certain visual reflexes, such as those that allow the eyes to view objects while the head is moving. The inferior colliculi house reflex centers involved in orienting the head in response to sound stimuli.

Diencephalon *(Interbrain)* The diencephalon is difficult to observe because it lies dorsal and anterior to the midbrain and ventral to the cerebrum. The **pineal gland** (see figure 45.8) is a part of the diencephalon, as are the thalamus, hypothalamus,

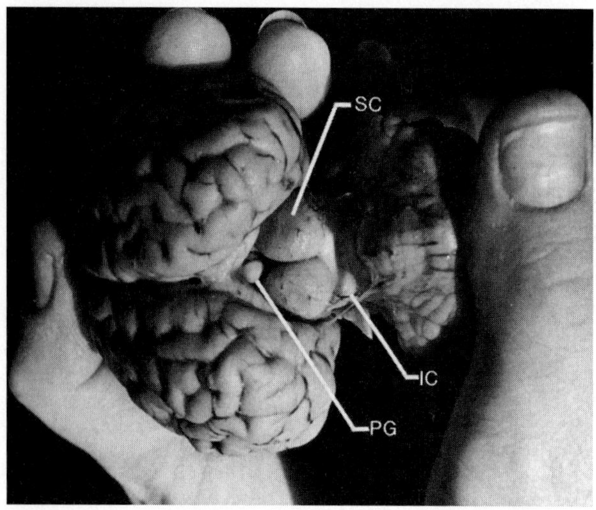

Figure 45.8 Exposing structures of the diencephalon and midbrain: SC—superior colliculus; IC—inferior colliculus; PG—pineal gland.

and mammillary bodies. More is said about these parts later.

The pineal gland is thought to be an evolutionary remnant of the third eye that certain reptiles have. The gland produces a hormone called *melatonin* that in lower animals inhibits the estrus cycle. Melatonin's exact role in humans is not completely understood; it may have something to do with ovarian hormone production. It is probably significant that this structure is, histologically, glandular until puberty, after which it becomes fibrous in nature.

Pons Examine the ventral surface of the sheep brain, and compare it with figures 45.6 and 45.7 to locate the pons. This part of the brain stem contains fibers that connect parts of the cerebellum and medulla with the cerebrum. The pons also contains nuclei of the fifth, sixth, seventh, and eighth cranial nerves.

Medulla Oblongata The medulla oblongata contains centers of gray matter that control the heart, respiration, and vasomotor reactions that regulate blood pressure. The last four cranial nerves emerge from the medulla oblongata.

The Cranial Nerves

Twelve pairs of *cranial nerves* emerge from various parts of the brain and pass through foramina of the skull to innervate parts of the head and trunk. Each pair has a name as well as a position number. Although most of these nerves contain both motor and sensory fibers (*mixed nerves*), a few contain only sensory fibers (*sensory nerves*). The cell bodies of the sensory fibers are in ganglia outside the brain; the cell bodies of the motor neurons, on the other hand, are within nuclei of the brain.

Compare your sheep brain with figures 45.6 and 45.7 to identify each pair of nerves. As you proceed, avoid damaging the brain tissue with dissecting instruments. Keep in mind, also, that once a nerve is broken off, determining where it was is difficult. The pairs of cranial nerves are as follows:

I Olfactory Nerve The olfactory nerve contains sensory fibers for the sense of smell. Since it extends from the mucous membranes in the nose through the ethmoid bone to the **olfactory bulb,** you cannot see it on your specimen. In the olfactory bulb, the olfactory nerve synapses with fibers that extend inward to olfactory areas in the cerebrum. Note that on the human brain an **olfactory tract** extends between the bulb and the brain (see figure 45.9).

II Optic Nerve The optic nerve is a sensory nerve that functions in vision. It contains axon fibers from ganglion cells of the retina of the eye. Part of the fibers from each optic nerve cross over to the other side of the brain as they pass through the **optic chiasma.** From the optic chiasma, the fibers pass through the **optic tracts** to the thalamus and finally to the visual areas of the cerebrum. Identify these structures on the sheep brain (see figure 45.7).

III Oculomotor Nerve The oculomotor nerve emerges from the midbrain and supplies somatic nerve fibers to the *levator palpebrae* (eyelid muscle) and four of the extrinsic ocular muscles (*superior rectus, medial rectus, inferior rectus,* and *inferior oblique*). Cranial nerves IV and VI, respectively, innervate the superior oblique and lateral rectus muscles.

The oculomotor nerve also supplies parasympathetic fibers to the *constrictor pupillaris* of the iris and the *ciliaris muscle* of the ciliary body. These fibers constrict the iris and change the lens shape as needed in accommodation (focusing).

If the oculomotor nerve is not visible on your sheep brain, the hypophysis (pituitary gland) may be obscuring it. The sheep brain in figure 45.7 lacks a hypophysis. To remove the hypophysis without damaging other structures, gently lift it up with forceps and sever the infundibulum with a scalpel or scissors.

IV Trochlear Nerve The trochlear nerve provides muscle sense and motor stimulation of the *superior oblique* muscle of the eye. Like the oculo-

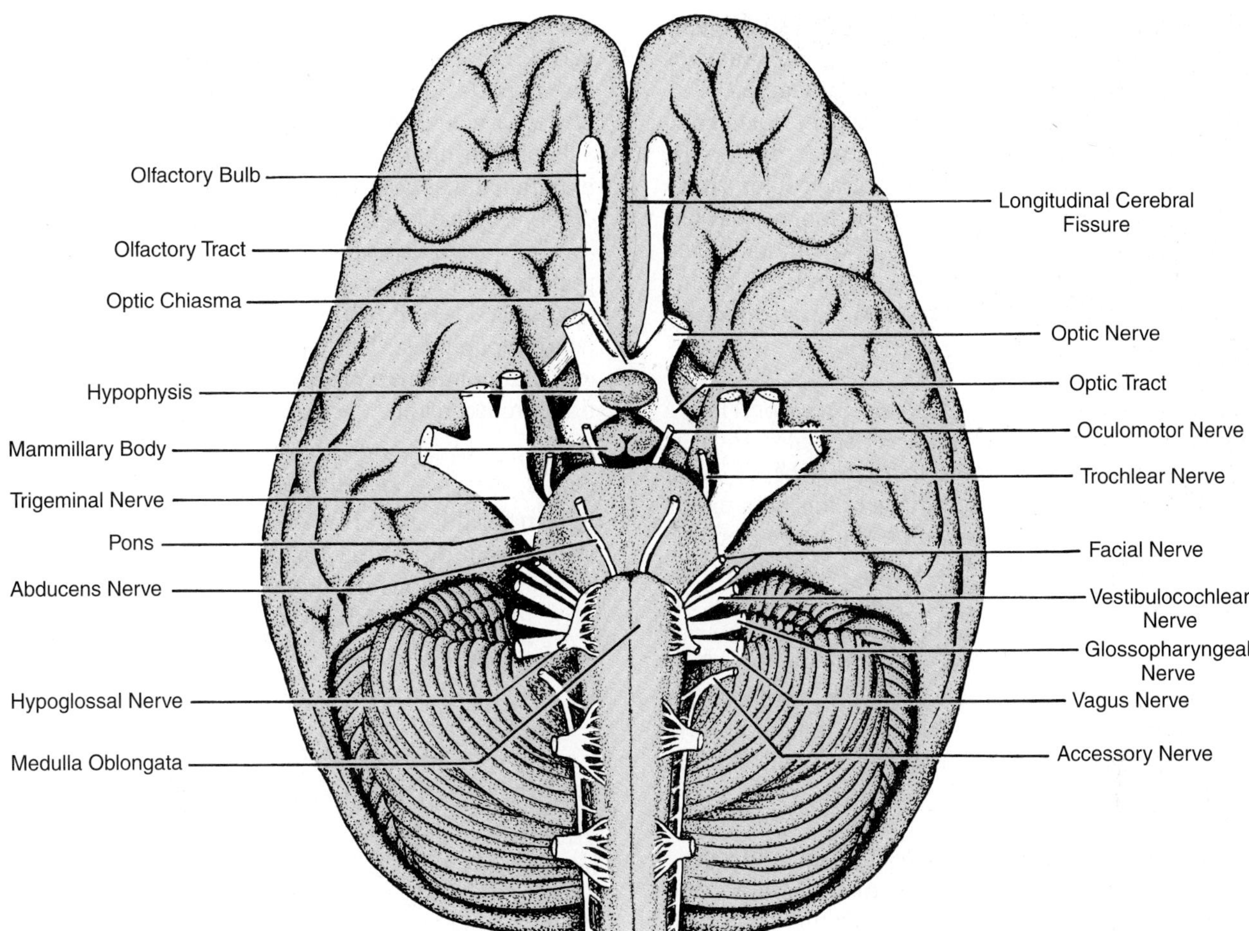

Figure 45.9 Inferior aspect of the human brain.

motor nerve, the trochlear nerve emerges from the midbrain. It is very small in diameter and difficult to locate (see figures 45.6 and 45.7).

V Trigeminal Nerve Just posterior to the trochlear nerve lies the largest cranial nerve, the trigeminal. Although a mixed nerve, its sensory functions are much more extensive than its motor functions. Because of its extensive innervation of the mouth and face, it is described in detail on pages 209 and 210.

VI Abducens Nerve The small abducens nerve innervates the *lateral rectus* muscle of the eye (see figures 45.6 and 45.7). It is a mixed nerve in that it provides muscle sense as well as muscular contraction. On the human brain, it emerges from the lower part of the pons near the medulla oblongata (see figure 45.9).

VII Facial Nerve The facial nerve has motor and sensory divisions. Figure 45.9 shows that this nerve on the human brain has two branches. It innervates facial muscles, salivary glands, and the taste buds of the anterior two-thirds of the tongue.

VIII Vestibulocochlear (*Auditory*) Nerve The vestibulocochlear nerve (see figures 45.6 and 45.7) goes to the inner ear. It has two branches: the *vestibular division,* which innervates the vestibular apparatus to function in equilibrium, and the *cochlear division,* which innervates the cochlea to function in hearing. The vestibulocochlear nerve emerges just posterior to the facial nerve on the human brain (see figure 45.9).

IX Glossopharyngeal Nerve The glossopharyngeal is a mixed nerve that emerges from the medulla oblongata, posterior to the vestibulocochlear nerve (see figures 45.6 and 45.7). It functions in taste, swallowing, and reflexes of the heart. Some of its fibers innervate taste buds on the back of the tongue. Efferent fibers innervate pharynx muscles involved in swallowing and the parotid salivary glands.

X Vagus Nerve The vagus nerve originates on the medulla oblongata and exceeds all of the other cranial nerves in its extensive ramifications (see figures 45.6, 45.7, and 45.9). In addition to innervating parts of the head and neck, it has branches that extend down into the chest and abdomen.

The vagus is a mixed nerve. Sensory fibers go to the heart, external acoustic meatus, pharynx, larynx, and thoracic and abdominal viscera. Motor fibers pass to the pharynx, the base of the tongue, the larynx, and the autonomic ganglia of the thoracic and abdominal viscera. Subsequent discussions of the physiology of the lungs, heart, and digestive organs make reference to the vagus nerve.

XI Accessory Nerve Since the accessory nerve emerges from both the brain and spinal cord, it is sometimes called the *spinal accessory* nerve. Note that on both the sheep and human brains, it parallels the lower part of the medulla oblongata and the spinal cord, with branches going into both structures (see figures 45.6, 45.7, and 45.9). On the human brain, it appears to occupy a more posterior position than the twelfth nerve. Afferent and efferent spinal components innervate the *sternocleidomastoid* and *trapezius* muscles. The cranial portion of the nerve innervates the pharynx, upper larynx, uvula, and palate.

XII Hypoglossal Nerve The hypoglossal nerve emerges from the medulla oblongata to innervate several muscles of the tongue (see figures 45.6 and 45.7). It is a mixed nerve. On the human brain in figure 45.9, it looks like a cut-off tree trunk with roots extending into the medulla oblongata.

The Hypothalamus

That portion of the diencephalon that includes the mammillary bodies, infundibulum, and part of the hypophysis is collectively called the *hypothalamus.* The entire diencephalon, including the hypothalamus, is studied in more detail in Exercise 46, when dissection reveals the internal parts of the brain. The mammillary bodies and hypophysis are discussed here.

Mammillary Bodies The mammillary bodies lie superior to the hypophysis. On the sheep brain, they appear as a single body, while on the human brain, they consist of a pair of rounded eminences just posterior to the infundibulum (see figures 45.6, 45.7, and 45.9). Mammillary bodies receive fibers from the olfactory areas and ascending pathways of the brain. Fibers also exit from these bodies to the thalamus and other brain nuclei.

Hypophysis The hypophysis, also known as the *pituitary gland,* attaches to the base of the brain by a stalk, the **infundibulum** (see figures 45.6, 45.7, and 45.9). The hypophysis has two distinctly different parts: an anterior adenohypophysis and a posterior neurohypophysis. Only the neurohypophysis and infundibulum are considered part of the hypothalamus because they both have the same embryological origin as the brain.

The adenohypophysis originates as an outpouching of the ectodermal stomodeum (mouth cavity) and is quite different histologically from the neurohypophysis. The hypothalamus, hypophysis, and autonomic nervous system have a close functional relationship.

Assignment:
Complete parts B, C, and D of the Laboratory Report for this exercise. You have now finished the sheep brain dissection for this exercise. Study the remaining sections of this exercise, and then complete the Laboratory Report. To continue the sheep brain dissection, proceed to Exercise 46.

Functional Localization of the Cerebrum

Extensive experimental studies on monkeys, apes, and humans have provided considerable knowledge of the functional areas of the cerebrum. Knowing the boundaries (sulci and fissures) of the cerebral lobes (see figure 45.6) will help you to locate the functional areas of the brain in figure 45.10.

Somatomotor Area The somatomotor area occupies the surface of the **precentral gyrus** of the frontal lobe. Electrical stimulation of this portion of the cerebral cortex in a conscious human results in movement of specific muscular groups.

Premotor Area The large premotor area is anterior to the somatomotor area and controls the motor area.

Somatosensory Area The somatosensory area is on the **postcentral gyrus** of the parietal lobe. It precisely localizes those points on the body where sensations of light touch and pressure originate. It also assists in determining organ position. The thalamus, rather than the somatosensory area, localizes other sensations, such as aching pain, harsh touch, warmth, and cold.

Motor Speech Area The motor speech area is in the frontal lobe just above the lateral cerebral fissure, anterior to the somatomotor area. It controls the muscles of the larynx and tongue that produce speech.

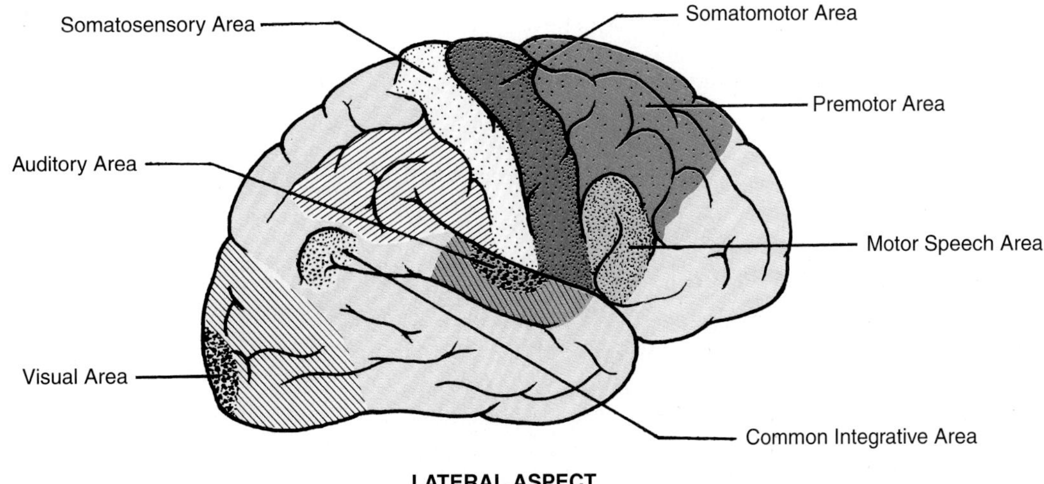

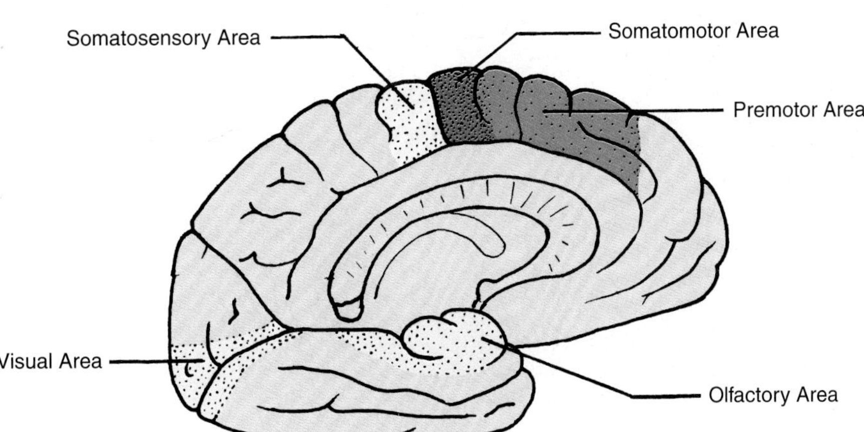

Figure 45.10 Functional areas of the human brain.

Visual Area The visual area is on the occipital lobe. The lateral aspect in figure 45.10 shows only a small portion of the visual area; the greater portion of this area is on the medial surface of the cerebrum. On this surface, it extends and narrows anteriorly to the parieto-occipital fissure. The visual area receives impulses from the retina via the thalamus. Destruction of this region causes blindness, although light and dark are still discernible.

Auditory Area The auditory area receives nerve impulses from the cochlea of the inner ear via the thalamus and is responsible for hearing and speech understanding. It is on the **superior temporal gyrus,** which borders the lateral cerebral fissure.

Olfactory Area The olfactory area on the medial surface of the temporal lobe is responsible for recognition of various odors. Tumors in this area cause individuals to experience nonexistent odors, both pleasant and unpleasant.

Association Areas Adjacent to the somatosensory, visual, and auditory areas are three association areas that lend meaning to what is felt, seen, or heard. These areas are hatched in figure 45.10.

Common Integrative Area The common integrative, or *gnostic,* area integrates information from the three association areas and the olfactory and taste centers. This area is on the **angular gyrus,** which is positioned approximately midway between the three association areas.

Assignment:
Complete part E of the Laboratory Report for this exercise.

The Trigeminal Nerve

Although a detailed study of all cranial nerves is precluded in an elementary anatomy course, the trigeminal nerve is singled out for a thorough study here. This fifth cranial nerve is the largest one that

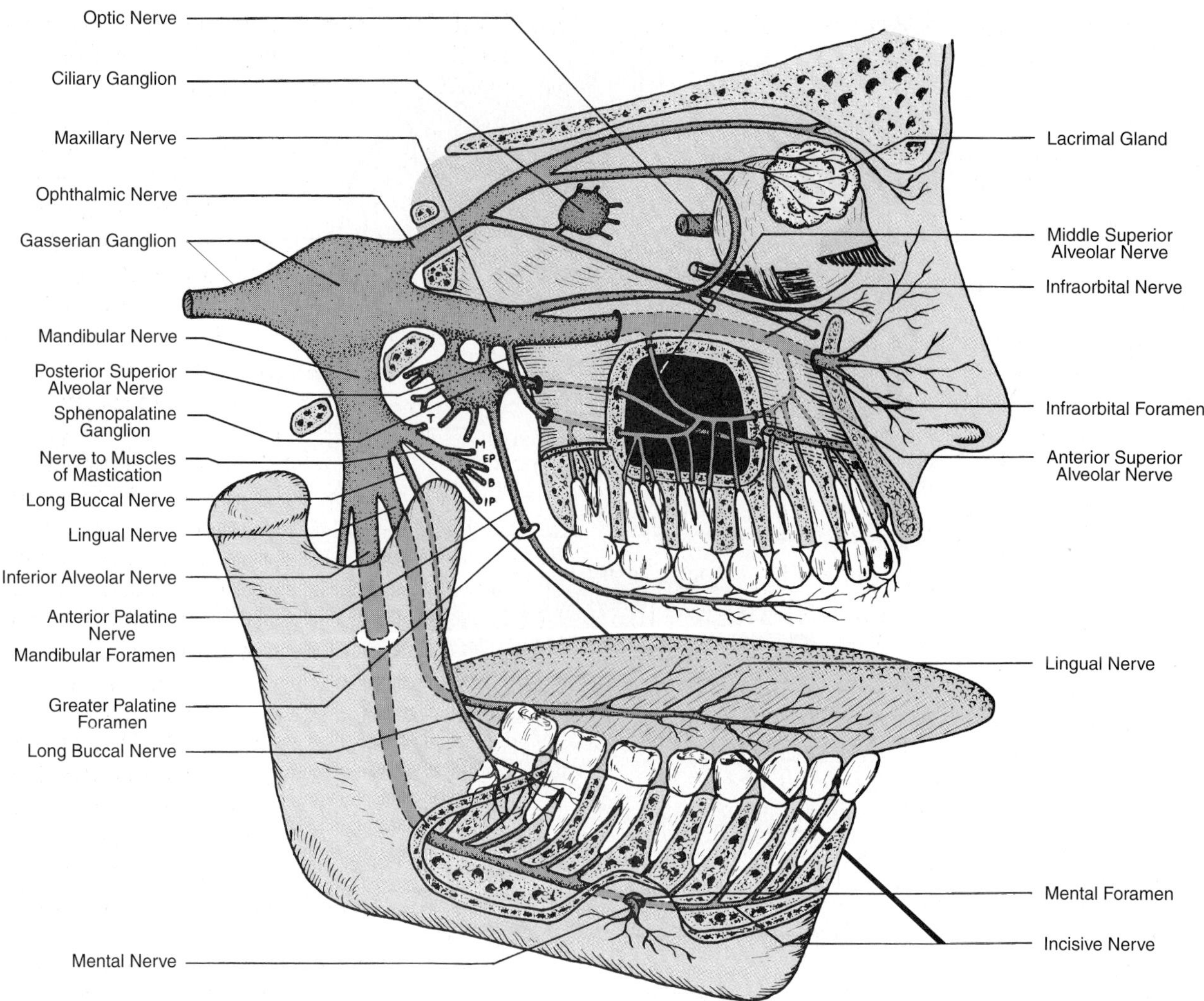

Figure 45.11 The trigeminal nerve.

innervates the head and is of particular medical-dental significance.

Figure 45.11 illustrates the distribution of the trigeminal nerve. Its branches innervate the teeth, tongue, gums, forehead, eyes, nose, and lips. On the left margin of figure 45.11 is the severed end of the trigeminal nerve where it emerges from the brain. Between this cut end and the three main branches is an enlarged portion, the **Gasserian ganglion.** This ganglion contains the nerve cell bodies of the nerve's sensory fibers. The three branches that extend from this ganglion are the *ophthalmic, maxillary,* and *mandibular* nerves. Locate the bony portion of the skull in figure 45.11 through which these three branches pass.

Ophthalmic Nerve The ophthalmic nerve is the upper branch that passes out of the cranium through the *superior orbital fissure.* Its three branches innervate the lacrimal gland, the upper eyelid, and the skin of the nose, forehead, and scalp. One of its branches also supplies the cornea of the eye, the iris, and the ciliary body (a muscle attached to the lens of the eye).

Superior to the lower branch of the ophthalmic is the **ciliary ganglion,** which contains nerve cell bodies that are incorporated into reflexes controlling the ciliary body. Unlike the other two divisions of the trigeminal, the ophthalmic nerve and its branches are not involved in dental anesthesia.

Maxillary Nerve The maxillary nerve, also known as the *second division of the trigeminal nerve,* is a sensory nerve that innervates the nose, upper lip, palate, maxillary sinus, and upper teeth. It branches just inferior to the ophthalmic nerve.

The maxillary nerve has two short branches that extend downward to an oval body, the **sphenopalatine ganglion.** The major portion of the maxillary nerve becomes the **infraorbital nerve,** which passes through the *infraorbital nerve canal* of the maxilla. This canal is shown with dashed lines in figure 45.11. The infraorbital nerve emerges from the maxilla through the **infraorbital foramen** to innervate the tissues of the nose and upper lip.

Three superior alveolar nerves innervate the upper teeth. The **posterior superior alveolar nerve** is a branch of the maxillary nerve that enters the posterior surface of the upper jaw and innervates the molars. The **anterior superior alveolar nerve** is a branch of the infraorbital nerve that innervates the anterior teeth and bicuspids (premolars). The anterior superior alveolar nerve forms a loop with the **middle superior alveolar nerve** (shown in figure 45.11 where a portion of the maxilla has been cut away), which innervates the bicuspids and first molar.

To desensitize all of the upper teeth on one side of the maxilla, a dentist can inject anesthetic near either the maxillary nerve or the sphenopalatine ganglion. Desensitization of the maxillary nerve is a *second division nerve block.* This type of nerve block affects the palate as well as the teeth because it involves fibers of the **anterior palatine nerve.** The latter nerve extends downward from the sphenopalatine ganglion to the soft tissues of the palate through the **greater palatine foramen.**

Mandibular Nerve The mandibular nerve, or *third division of the trigeminal nerve,* is the most inferior branch. It innervates the teeth and gums of the mandible, the muscles of mastication, the anterior part of the tongue, the lower part of the face, and some skin areas on the side of the head.

Below the Gasserian ganglion, the first branch of the mandibular nerve is the **nerve to the muscles of mastication,** which has five branches. In figure 45.11, these five branches have small identifying letters near their cut ends: **T** for temporalis, **M** for masseter, **EP** for external pterygoid, **B** for buccinator, and **IP** for internal pterygoid. The fibers in these nerves control the motor activities of these five muscles.

The largest branch of the mandibular nerve is the **inferior alveolar nerve,** which enters the ramus of the mandible through the **mandibular foramen** and innervates all the teeth on one side of the mandible. At the **mental foramen,** the inferior alveolar nerve becomes the **incisive nerve,** which innervates the anterior teeth. Emerging from the mental foramen is the **mental nerve,** which innervates the soft tissues of the chin and lower lip.

To desensitize the mandibular teeth, a dentist generally injects anesthetic near the mandibular foramen. This type of injection, called a *lower nerve block,* or *third division nerve block,* is common in dentistry because the external and internal alveolar plates of the mandibular alveolar process are too dense to facilitate anesthesia by infiltration of the solid bone. The lower nerve block of one side of the mandible desensitizes all of the teeth on one side except for the anterior teeth. The anterior teeth receive nerve fibers from both the left and right incisive nerves, which prevents complete desensitization.

The **lingual nerve** is a branch of the mandibular nerve that emerges just above the inferior alveolar nerve and innervates the tongue. Like the inferior alveolar nerve, the lingual nerve is sensory in function.

The **long buccal nerve,** another sensory branch of the mandibular nerve, innervates the buccal gum tissues of the molars and bicuspids. This nerve arises from a point that is just superior to the lingual nerve.

Assignment:

Complete parts F and G of the Laboratory Report for this exercise.

46 Brain Anatomy: Internal

The interior of the brain is best seen in longitudinal and frontal sections. In this exercise, you study various sectional views to identify the different cavities (ventricles), as well as deeper portions, such as the thalamus, fornix, and hypothalamus. Proceed as follows:

Materials:
preserved sheep brains
preserved human brains and/or human brain models
dissecting instruments
long, sharp knife for sectioning
dissecting trays

Midsagittal Section

With the preserved sheep brain resting ventral side down on a dissecting tray, gently separate the cerebral hemispheres with a blunt probe (your instructor may demonstrate this procedure to the class). Once the hemispheres are separated, you should be able to run your probe along a white band of tissue that connects the two hemispheres. This is the **corpus callosum.** The corpus callosum is a commissural tract of fibers that connects the right and left cerebral hemispheres, correlating their functions. As figures 46.1 and 46.2 show, this tract is significantly thicker on the human brain than on the sheep brain. Next, with a sharp knife, carefully slice through the corpus callosum and the underlying tissue, cutting the brain into right and left halves on the midline. If only a scalpel is available, cut the tissue as smoothly as possible. Proceed to identify the following structures on the sheep brain and their counterparts on the human brain:

The Diencephalon

The diencephalon (also called the *interbrain*) lies between the corpus callosum of the cerebrum and the midbrain. It contains the fornix, pineal gland, intermediate mass, thalamus, hypothalamus, and third ventricle.

Fornix Locate the fornix on the sheep brain. This body contains fibers that are an integral part of the *rhinencephalon* (olfactory mechanism) of the brain. Note how much larger, proportionately, the fornix is on the sheep brain.

Pineal Gland The pineal gland is just ventral of the posterior aspect of the corpus callosum. If your sheep brain is not sectioned evenly, the pineal gland may be entirely on one half of the brain or the other.

Intermediate Mass The intermediate mass is the only part of the thalamus that midsagittal sections of the sheep and human brains show. The **thala-**

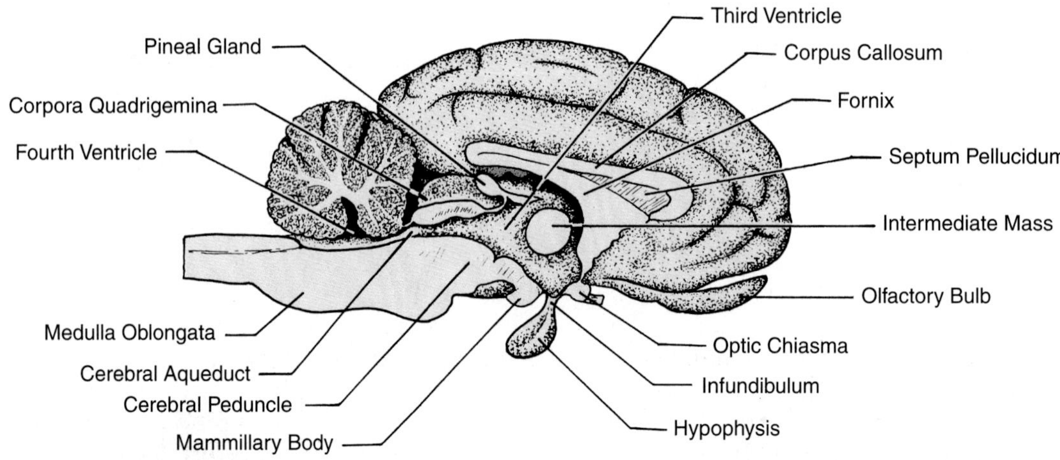

Figure 46.1 Midsagittal section of the sheep brain.

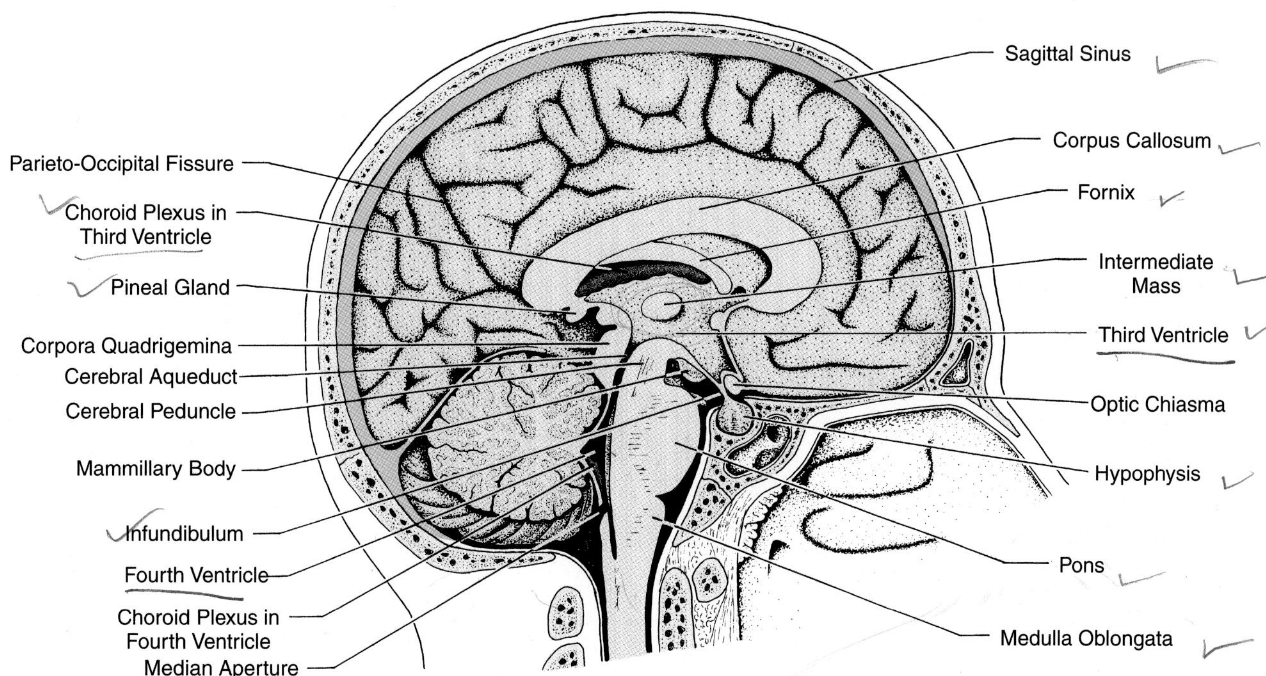

Figure 46.2 Midsagittal section of the human brain.

mus is a large, ovoid structure on the lateral walls of the third ventricle. In this important relay station, sensory pathways of the spinal cord and brain form synapses on their way to the cerebral cortex. The intermediate mass contains fibers that pass through the third ventricle, uniting both sides of the thalamus.

Hypothalamus The hypothalamus extends about 2 cm up into the brain from the ventral surface. Its many nuclei control various physiological body functions (temperature regulation, hypophysis control, coordination of the autonomic nervous system). The only part of the hypothalamus labeled in figures 46.1 and 46.2 is the **mammillary body.** Figure 46.4 shows a better view of the hypothalamus.

Inferior and anterior to the mammillary body in Figures 46.1 and 46.2 lie the **infundibulum** and **hypophysis.** Note how the bony recess of the *sella turcica* encases the hypophysis. The **optic chiasma** is the small, round body just above the hypophysis, anterior to the infundibulum.

The Midbrain

Locate the **corpora quadrigemina, cerebral peduncles,** and **cerebral aqueduct** on the midbrain. The cerebral peduncles of the midbrain are a pair of cylindrical bodies made up largely of ascending and descending fiber tracts that connect the cerebrum with the other three brain divisions. The cerebral aqueduct is a duct that runs longitudinally through the midbrain, connecting the third and fourth ventricles.

The Cerebellum

Examine the cut surface of the cerebellum, and note the gray matter near the outer surface. The cerebellum receives nerve impulses along cerebellar peduncles from motor and visual centers of the brain, the semicircular canals of the inner ears, and muscles of the body. Nerve impulses pass from this region to all the motor centers of the body to maintain posture, equilibrium, and muscle tone. None of the activities of the cerebellum are conscious. Unlike the cerebrum, each hemisphere of the cerebellum controls muscles on its side of the body.

The Pons and Medulla Oblongata

The pons and medulla oblongata make up the greater portion of the brain stem (which also includes the midbrain). The pons is between the midbrain and the medulla oblongata, and it consists primarily of white matter (fiber tracts). Locate the pons and medulla oblongata on both the sheep and human brains.

Within the brain stem is a complex interlacement of nuclei and white fibers (gray and white matter) that constitutes the *reticular formation.* Nuclei of this formation generate a continuous flow of impulses to other parts of the brain to keep the cerebrum in a state of alert consciousness.

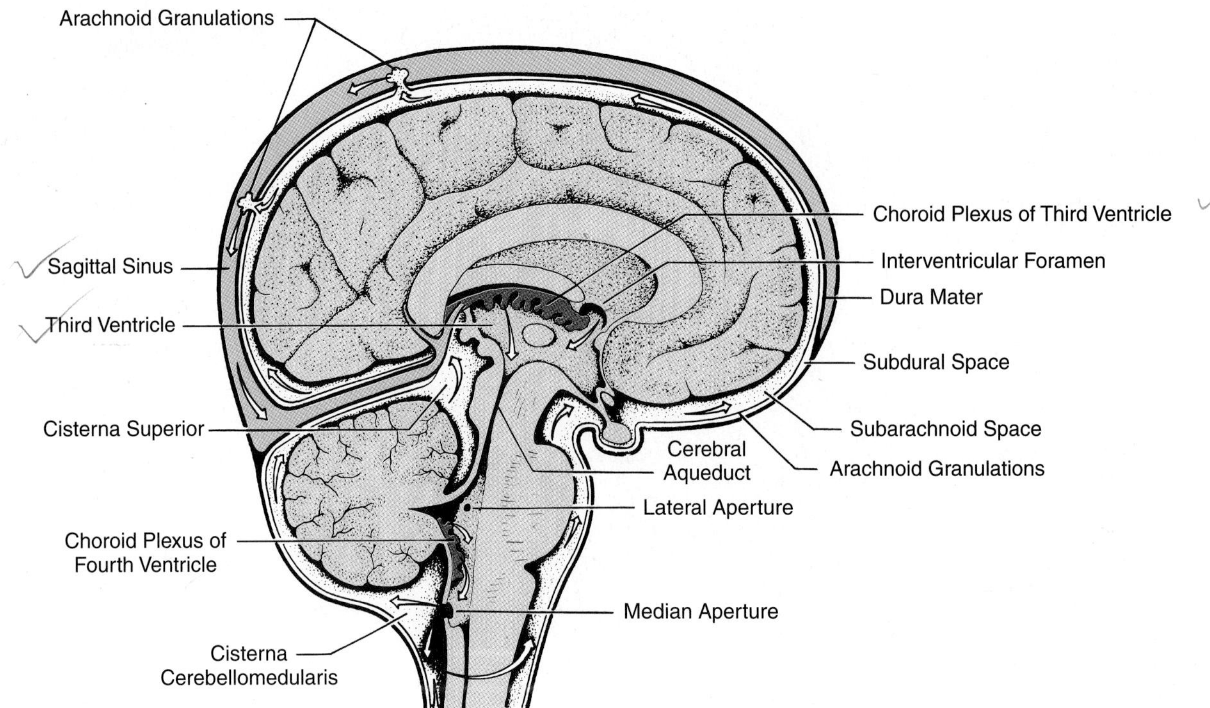

Figure 46.3 Origin and circulation of cerebrospinal fluid.

The Ventricles and Cerebrospinal Fluid

The brain has four cavities called *ventricles* that contain cerebrospinal fluid (CSF). Two **lateral ventricles** are in the lower medial portions of each cerebral hemisphere. Between these two ventricles on the median line is a thin membrane, the **septum pellucidum** (shown in figure 46.1, but not in figure 46.2). Refer to figure 46.4 to better visualize the two lateral ventricles and the septum pellucidum.

The other two ventricles of the brain, the **third** and **fourth ventricles,** are on the median line (see figures 46.1 and 46.2). Locate them on the sheep brain. Note that the **intermediate mass** of the thalamus passes through the third ventricle.

Within the third ventricle is a **choroid plexus** (see figure 46.2) that secretes CSF. Although not shown in any of the human brain illustrations here, the lateral ventricles also have choroid plexuses. Figure 46.2 shows a choroid plexus in the fourth ventricle.

Cerebrospinal Fluid Pathway The choroid plexuses of all the ventricles constantly produce CSF. Figure 46.3 indicates the CSF pathway in the brain.

CSF that originates in the lateral ventricles enters the third ventricle through the **interventricular foramen.** Note in figure 46.3 that this foramen lies anterior to the fornix and choroid plexus of the third ventricle.

From the third ventricle, the CSF passes into the **cerebral aqueduct**, which conveys the CSF to the fourth ventricle. The **choroid plexus of the fourth ventricle** lies on the posterior wall of this ventricle.

From the fourth ventricle, a small amount of CSF enters the vertebral canal of the spinal cord while the majority of the CSF exits into the subarachnoid space that surrounds the cerebellum through three apertures: one aperture on the median line and two lateral apertures.

In figure 46.3, the **median aperture** empties into the **cisterna cerebellomedullaris.** Since the **lateral apertures** are on the lateral walls of the fourth ventricle, figure 46.3 shows only one.

From the cisterna cerebellomedullaris, the CSF moves up into the **cisterna superior** (above the cerebellum) and finally into the subarachnoid space around the cerebral hemispheres. From the subarachnoid space, the CSF escapes into the blood of the **sagittal sinus** through the **arachnoid granulations.** The CSF also passes down the subarachnoid space of the spinal cord on the posterior side and up along the anterior surface.

Frontal Sections

Figures 46.4A and B are frontal sections through two slightly different regions of the sheep brain. Figure 46.4C is a frontal section through a human brain.

Infundibular Section of Sheep Brain

Place a whole sheep brain *dorsal side down* on a dissecting tray. Use a long knife to cut a frontal section that starts at the point of the infundibulum. Cut

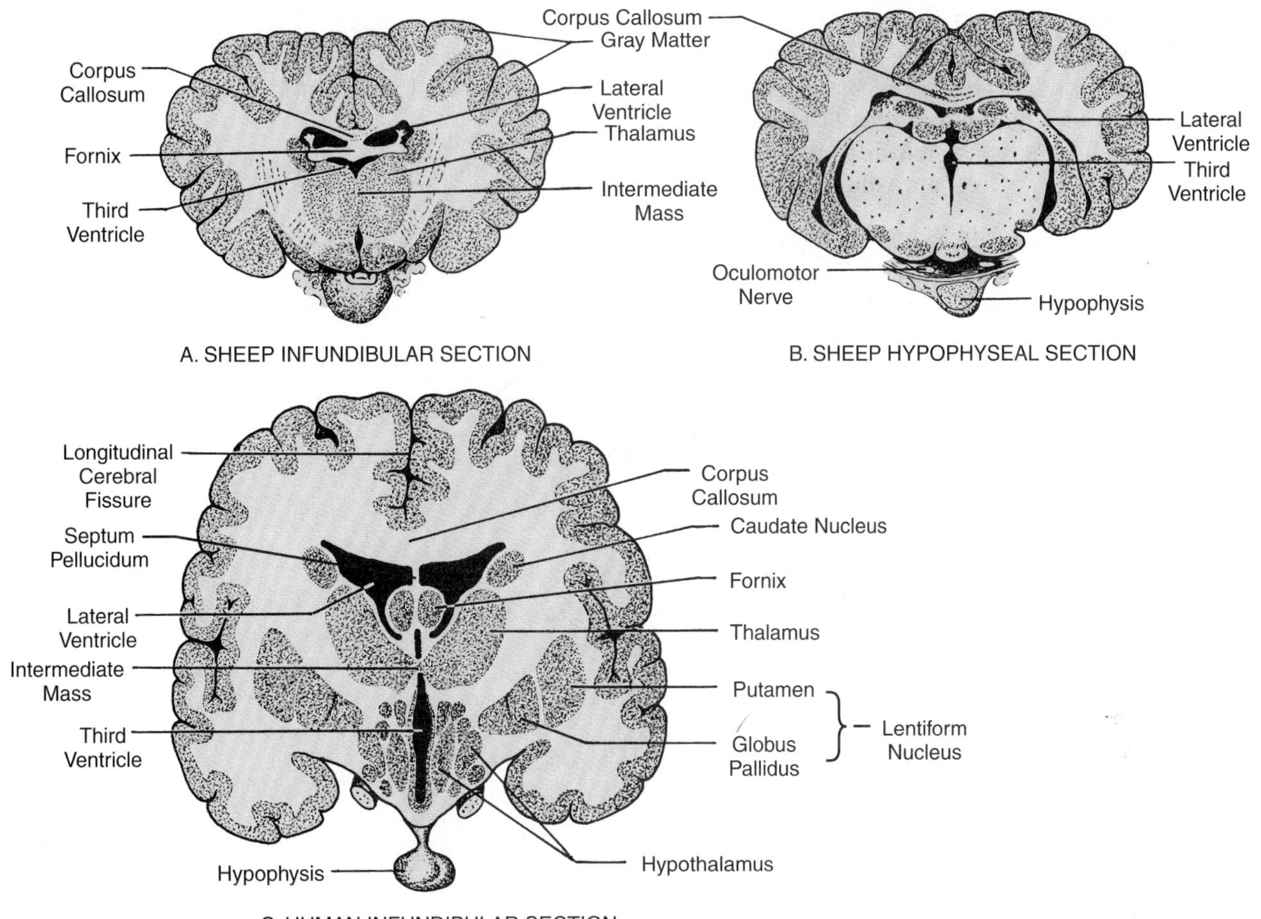

Figure 46.4 Frontal sections of sheep and human brains.

through perpendicularly to the dissecting tray and to the longitudinal axis of the brain. Then, make a second cut parallel to the first cut but 1/4 inch further back. Make the second cut through the center of the hypophysis. Place this slice, *with the first cut upward,* (see figure 46.4A), on a piece of paper toweling for study.

Note the pattern and thickness of the **gray matter** of the cerebral cortex. Force a probe down into a sulcus. What meninx do you break through as the probe moves inward?

Locate the two triangular **lateral ventricles.** Probe one of the ventricles to locate the **choroid plexus.** Identify the **corpus callosum** and **fornix.**

Identify the **third ventricle,** which is on the median line under the fornix. Also, identify the **thalamus** and **intermediate mass.** Does the thalamus consist of white matter or gray matter? What is the function of the white matter between the thalamus and the cerebral cortex?

Hypophyseal Section of Sheep Brain

Prepare a hypophyseal section of the sheep brain by cutting another 1/4-inch slice of the posterior portion of the brain. Place this slice, *anterior face upward* (see figure 46.4B), on paper toweling. On this section, identify the structures labeled in figure 46.4B.

Infundibular Section of Human Brain

Figure 46.4C shows an infundibular section of the human brain. Locate the two triangular **lateral ventricles** and the **third ventricle.** Note how the **longitudinal cerebral fissure** extends downward to the **corpus callosum,** which forms the upper wall of the lateral ventricles. Observe that the **septum pellucidum** separates the two lateral ventricles. Identify the **fornix**—two masses of gray matter just below the septum pellucidum.

Identify the areas on each side of the third ventricle that make up the nuclei of the **hypothalamus.** Approximately eight separate masses of gray matter are seen here. What physiological functions do these nuclei regulate?

Locate the two large areas on the walls of the lateral ventricles that constitute the **thalamus.** Note how the **intermediate mass** that passes through the third ventricle unites these two portions of the thalamus.

Three pairs of ganglia (masses of gray matter) in the cerebral hemispheres are known collectively as the *basal ganglia.* The largest basal ganglia are the **putamen** and **globus pallidus.** Together, they form a triangular mass, the **lentiform nucleus,** and exert a steadying effect on voluntary muscular movements.

The **caudate nuclei** are smaller basal ganglia in the walls of the lateral ventricles superior to the thalamus. These nuclei affect muscular coordination, and surgically produced lesions in these centers can correct certain kinds of palsy.

Assignment:
Complete the Laboratory Report for this exercise.

Electroencephalography

47

Nerve impulse movement from one neuron to another throughout the brain produces background electrical activity that electronic equipment can record. The study of this electrical activity is *electroencephalography.* Needle electrodes inserted directly into the scalp or surface electrodes placed on the scalp monitor electrical activity. The record that either method produces is called an **electroencephalogram,** or **EEG.** Although the electrical activity of the brains of different individuals varies considerably, normal patterns have been established.

EEGs have many medical applications. Physicians use EEGs for diagnosing organic malfunctions, such as epilepsy, brain tumor, brain abscess, cerebral trauma, meningitis, encephalitis, and certain congenital conditions. EEGs are particularly useful in diagnosing cerebral thrombosis and embolism. On the other hand, an individual with considerable brain damage often has a completely normal EEG. Most individuals with nonorganic disorders also have normal EEGs.

Normal Wave Patterns

Most of the time, brain waves are irregular and exhibit no general pattern. At other times, however, distinct patterns are detectable. While some patterns are characteristic of specific abnormalities, such as epilepsy, others are distinctly normal, such as the alpha, beta, theta, and delta waves in figure 47.1.

Alpha Waves A relaxed individual with eyes closed and no particular thought in mind frequently produces *alpha waves.* Alpha waves occur at a rate of 8 to 13 cycles per second (cps).

When the person opens the eyes and attends to a thought or visual object, the alpha waves become fast, irregular, and shallow. This change in wave pattern is called *alpha block.*

Beta Waves *Beta waves* occur at a frequency of 14 to 25 cps.

Theta Waves *Theta waves* have a frequency of 4 to 7 cps. These waves are considered normal in children. In adults, however, theta waves usually indicate tiredness, or disappointment and frustration. Certain brain disorders also exhibit theta waves.

Delta Waves *Delta waves* have frequencies below 3.5 cps. This brain-wave pattern is normal for infants and occurs during sleep in adults.

The brain-wave pattern of infants and children exhibits fast beta rhythms, as well as the slower theta and delta rhythms. As children mature into adolescence, the rhythms develop into a characteristic alpha pattern of adulthood. Physiological conditions, such as low body temperatures, low blood sugar levels, and high levels of carbon dioxide in

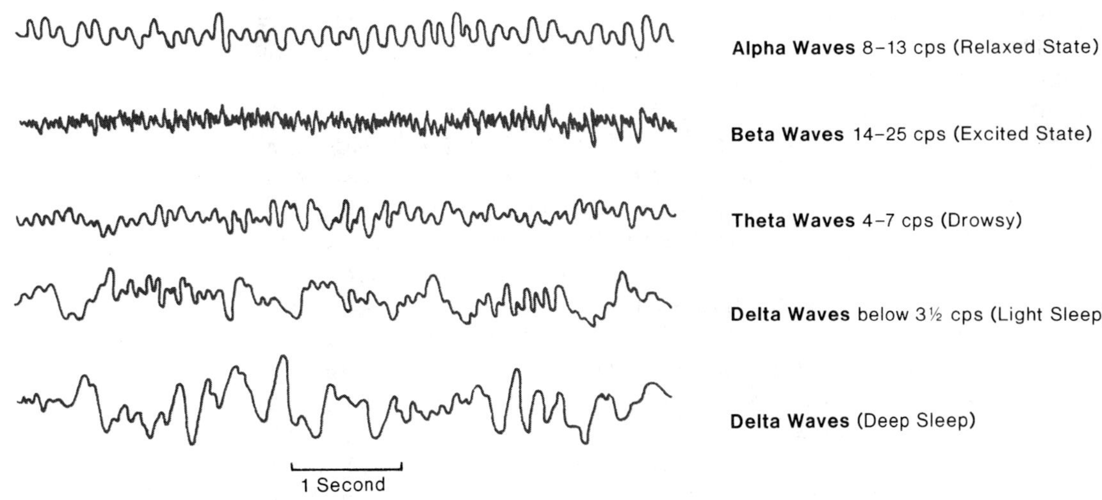

Alpha Waves 8–13 cps (Relaxed State)

Beta Waves 14–25 cps (Excited State)

Theta Waves 4–7 cps (Drowsy)

Delta Waves below 3½ cps (Light Sleep)

Delta Waves (Deep Sleep)

Figure 47.1 Normal electroencephalograms.

the blood, can slow the alpha rhythm; the opposite conditions speed it up.

Pathological Evidence

Figure 47.2 shows five representative EEGs of individuals with brain damage. Note the absence of stability that is typical for the alpha rhythm of figure 47.1.

In evaluating EEGs, neurologists look for slow, fast, and sporadic patterns. Moderately slow or moderately fast EEGs are generally considered *mildly abnormal.* Very slow or very fast EEGs are *definitely abnormal.* The fourth pattern in figure 47.2 is very fast and, in this case, represents an individual who suffered motor seizures and amnesia from barbiturate intoxication.

Considerable extremes in voltage, like that in the first tracing (epilepsy) in figure 47.2, are significant. While epileptic patterns vary with the type of seizure, they characteristically show large spikes that indicate substantial voltage differences.

Electroencephalography Procedure

The clarity of EEG wave patterns is a function of the type of equipment and technique. A good oscilloscope produces excellent results. The procedures outlined here are for the Unigraph.

The electrical potentials the brain emits are of such low magnitude that the Unigraph will be operated at its maximum sensitivity (0.1 mV/cm). Small disk electrodes pressed tightly against the skin will monitor electrical activity.

To overcome skin resistance, the subject's skin will be gently scrubbed with a Scotchbrite pad before the electrodes are applied. This scrubbing, plus the application of an electrode jelly, greatly enhances skin conductivity. An elastic headband will exert constant pressure on the electrodes.

You will work in teams of four. One member of the team will be the subject, another member will prepare the subject, and the remaining two will work with the equipment. If time permits, repeat the experiment with other team members acting as subjects.

Three electrodes will be attached to the subject: two on the forehead and one at the base of the skull in the occiput region. The electrodes will be attached to the three leads of a patient cable, which will be plugged into the Unigraph.

The success of this experiment hinges on environmental conditions and electrode placement. Ideally, the room should be darkened and quiet. A relaxed subject is essential, particularly when trying to record alpha waves. Subjects who are trained in yoga or who are capable of self-hypnosis can produce striking EEG patterns.

While one team member prepares the subject, the other two team members calibrate the Unigraph. Proceed as follows:

Materials:
Unigraph or polygraph
patient cable (three leads)
three EEG disk-type electrodes
three circular Band-Aids
electrode jelly
alcohol swabs
Scotchbrite pad (grad #7447)
elastic wrapping

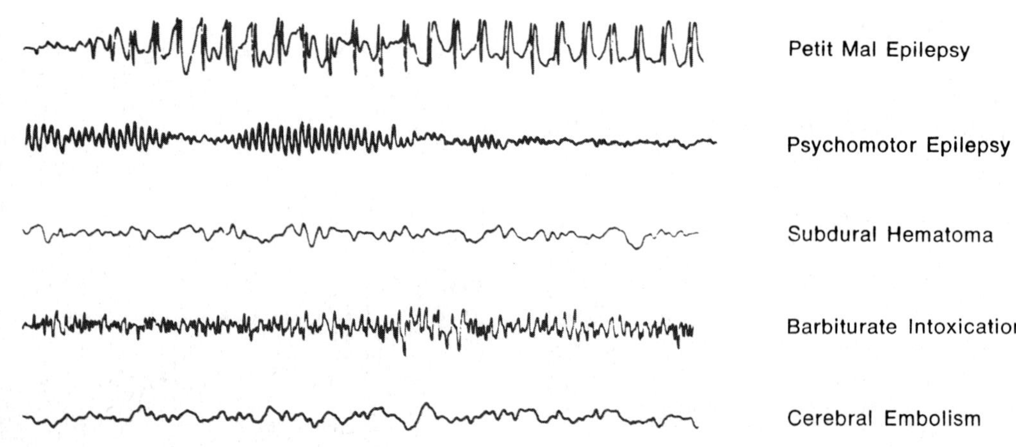

Figure 47.2 Abnormal electroencephalograms.

Calibration

For the EEG tracing to have any significance, you must calibrate the Unigraph so that a 1 cm deflection of the stylus equals 0.1 mV; thus, each millimeter of vertical deflection equals 0.01 mV. To calibrate the Unigraph:

1. Turn the mode selector control to CC/Cal (Capacitator Coupled Calibrate).
2. Turn the power switch to On, and turn the stylus heat control knob to the two o'clock position.
3. Set the speed control lever to the slow position. The free end of the lever should point toward the styluses.
4. Move the chart control switch (c.c.) to Chart On, and observe the width of the tracing on the paper. Adjust the stylus heat control knob to produce the desired width of tracing.
5. Set the gain control knob on 0.1 mV/cm.
6. Push down the 0.1 mV button to determine the amount of deflection. **Hold this button down a full 2 seconds before releasing.** If you release the button too quickly, the tracing will not be perfect. Adjust the sensitivity knob until the deflection is exactly 1 cm in each direction. The unit is now properly calibrated.
7. Set the mode selector control on EEG, and return the c.c. switch to STBY. Place the speed selector lever in the fast position. The Unigraph is now ready for use.

Preparation of Subject

The subject will have two electrodes on the forehead near the hairline and one in the occiput region of the head on the right side. The electrodes attached to the back of the head and the right forehead will pick up the brain waves. The electrode on the left forehead acts as a ground. Proceed as follows:

1. With a Scotchbrite pad, *lightly* scrub the right and left forehead areas four or five times to remove dead surface cells (see figure 47.3). Do not scrub so hard that you draw blood! Examine the surface of the pad for evidence of cleansing (white powder on pad).
2. Disinfect each area with an alcohol swab.
3. For each of the two forehead electrodes, affix the electrode to a round Band-Aid or strip of adhesive tape, add a drop of electrode jelly, and attach to the forehead (see figures 47.4, 47.5, and 47.6). Be sure to position the electrode so that its flat surface is against the skin.
4. Use electrode jelly to position the third electrode on the back of the right side in the occiput region. Do not use a Band-Aid. To hold the electrode in place, put an elastic wrapping over it and the other two electrodes (see figure 47.7).
5. Clamp the leads of the patient cable to the wires of the electrodes as follows:
   Lead 1—to the occiput electrode
   Lead 2—to the right forehead electrode
   Unnumbered lead—to the left forehead electrode
6. Have the subject lie down on a cot or laboratory tabletop. Make sure that all leads are free and not binding under the head or shoulders. Body movements during the test should not disturb the leads. The subject should lie perfectly still.
7. Plug the end of the three-lead cable into the Unigraph. You are now ready to monitor the brain waves.

Monitoring

With the Unigraph set at 0.1 mV/cm in the EEG mode, place the c.c. (chart control) switch at Chart

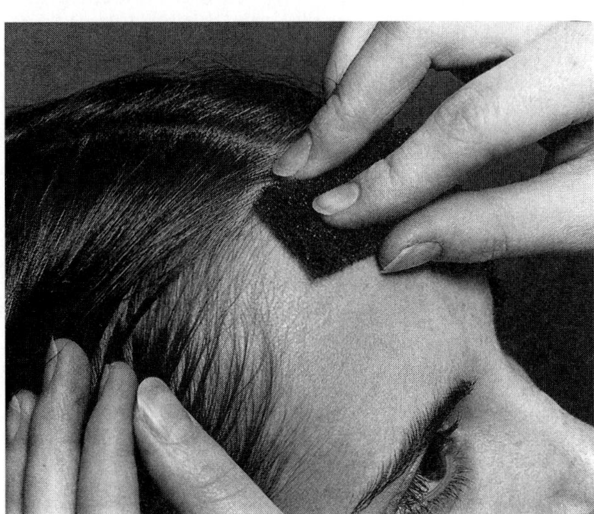

Figure 47.3 Prepare the skin surface by rubbing lightly with a Scotchbrite pad.

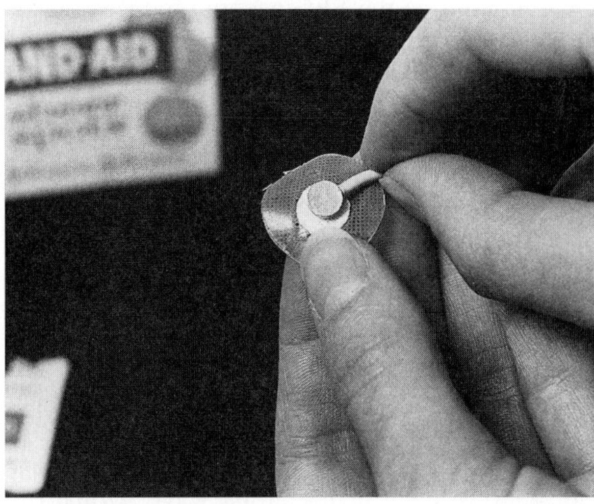

Figure 47.4 Place the electrode—with flat surface up—on an adhesive patch.

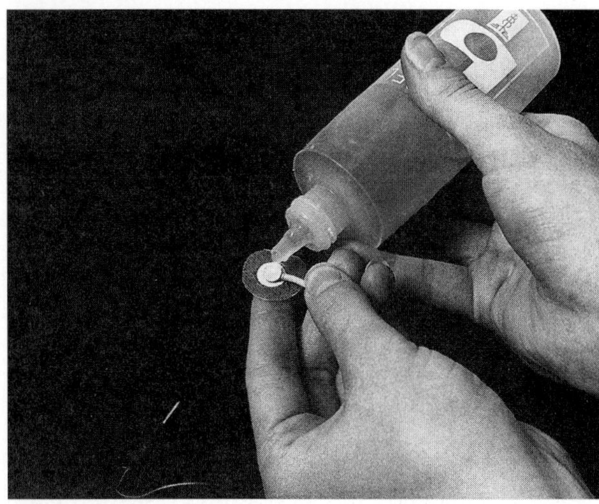

Figure 47.5 Place a small drop of electrode jelly on the electrode.

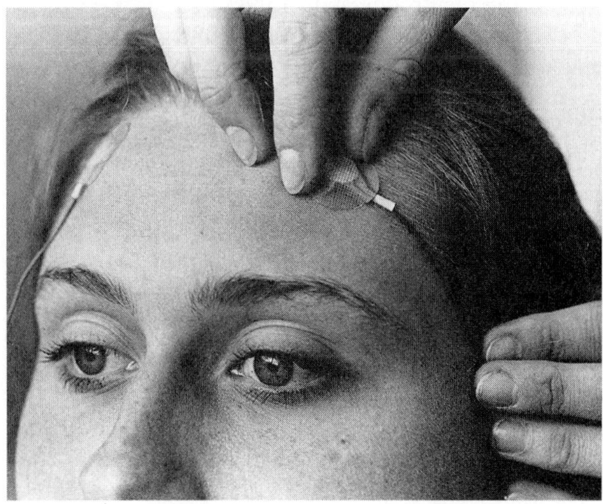

Figure 47.6 Position the forehead electrodes near the hairline.

On, and observe the EEG pattern produced on the chart. The chart should be moving at the fast speed (25 mm per second).

In most cases, the typical wave pattern is irregular. Be sure to keep the subject as quiet as possible. Even eye movements can alter the pattern. After about 2 minutes of recording, attempt to induce alpha waves with the suggestions that follow.

Alpha Rhythm Although alpha waves are the most common waveform in normal adults, they are not always easy to demonstrate in a laboratory situation. Complete relaxation in the subject is important. Extraneous disturbances, such as conversation, slamming doors, furniture moving, and so on, prevent alpha wave formation. With the c.c. switch at Chart On, try to induce

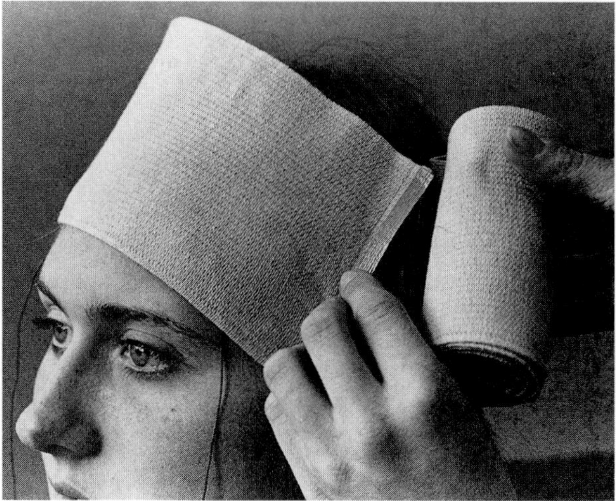

Figure 47.7 Tightly wrap an elastic band over all three electrodes.

relaxation in the subject with the following type of statement:

Close your eyes, and relax all the muscles of your body as much as you can. Try not to concentrate on any one thought. Let your thoughts flit lightly from one thought to another. You have absolutely no problems. Life is beautiful. Everybody is kind, good, and your friend.

Allow the tracing to run intermittently for about 5 minutes; then place the c.c. switch in the Off position. Count the waves per second. Is it an alpha rhythm pattern (8 to 13 cps)? Determine the maximum number of microvolts in a specific wave.

Alpha Block To demonstrate alpha block, place the c.c. switch on Chart On again, and help the subject to relax as before. When the alpha rhythm is present, have the subject open his or her eyes (as you press the event marker button). Then ask the subject some questions that require concentration. Observe the pattern for evidence of alpha block. Continue to present the subject with questions and problems for 4 or 5 minutes. Mark the paper with a pen or pencil where the questioning began.

Hyperventilation Start the chart moving again. Have the subject relax with closed eyes to reestablish the alpha rhythm. Depress the event marker, and ask the subject to take 40 to 50 deep breaths per minute for 2 to 3 minutes. This hyperventilates the lungs, lowering the level of carbon dioxide in the blood. Observe the tracing, and note the changes.

Assignment:

Remove the chart from the Unigraph, and review the results with the subject. Use tape to attach samples of the EEG to the Laboratory Report, and then complete the Laboratory Report.

PART 9 The Sense Organs

This part has four exercises: two about the eye and two about the ear. In Exercise 48, you study the anatomy of the eye through dissection of a cow eye and through visual examination of the eyeball's interior, using an ophthalmoscope. Exercise 49 pertains entirely to eye tests—those that reveal normal physiological phenomena and those that determine deviations from normal vision.

Exercise 50 examines the part of the ear that functions in hearing and describes different kinds of hearing tests for detecting deviations from normal. Exercise 51 focuses on the part of the ear that functions in the maintenance of equilibrium and explores a variety of tests pertaining to equilibrium.

48 Anatomy of the Eye

In preparation for the study of eye function in Exercise 49, the focus here is on learning the anatomy of the eye through dissection, ophthalmoscopy, and microscopy. The exercise begins, however, with basic information on the lacrimal system, internal eye anatomy, and the extrinsic muscles of the eye.

The Lacrimal Apparatus

Each eye lies protected in a bony recess of the skull and is covered externally by the **eyelids** (*palpebrae*). Lining the eyelids and covering the exposed surface of the eyeball is a mucous membrane called the **conjunctiva.** Protective secretions of the lacrimal glands, which lie between the upper eyelid and the eyeball, keep conjunctival surfaces constantly moist.

Figure 48.1 shows the external anatomy of the right eye, with particular emphasis on the lacrimal apparatus. Note that the lacrimal gland consists of two portions: a large **superior lacrimal gland** and a smaller **inferior lacrimal gland.** Secretions from these glands pass through the conjunctiva of the eyelid via six to twelve small ducts.

The flow of fluid from these ducts moves across the eyeball to the *medial canthus* (angle) of the eye, where the fluid enters two small orifices, the **lacrimal puncta** (*punctum,* singular). These openings lead into the **superior** and **inferior lacrimal ducts;** the superior one is in the upper eyelid, and the inferior one is in the lower eyelid. These two ducts empty directly into an enlarged cavity, the **lacrimal sac,** which, in turn, is drained by the **nasolacrimal duct.** The tears finally exit into the nasal cavity through the end of the nasolacrimal duct.

Two other structures—the caruncula and the plica semilunaris—are near the puncta in the medial canthus of the eye. The **caruncula** is a small, red, conical body that consists of a mound of skin containing sebaceous and sweat glands, and a few small hairs. This structure produces a whitish secretion that constantly collects in this region. Lateral to the caruncula is a curved fold of conjunctiva, the **plica semilunaris.** This structure contains some smooth muscle fibers. In the cat and many other animals, the plica semilunaris is more highly developed and is often referred to as the "third eyelid."

Internal Eye Anatomy

Figure 48.2 is a horizontal section of the right eye as seen from above. Note that the wall of the eyeball consists of three layers: an outer **sclera,** a middle **choroid coat,** and an inner **retina.** Only the

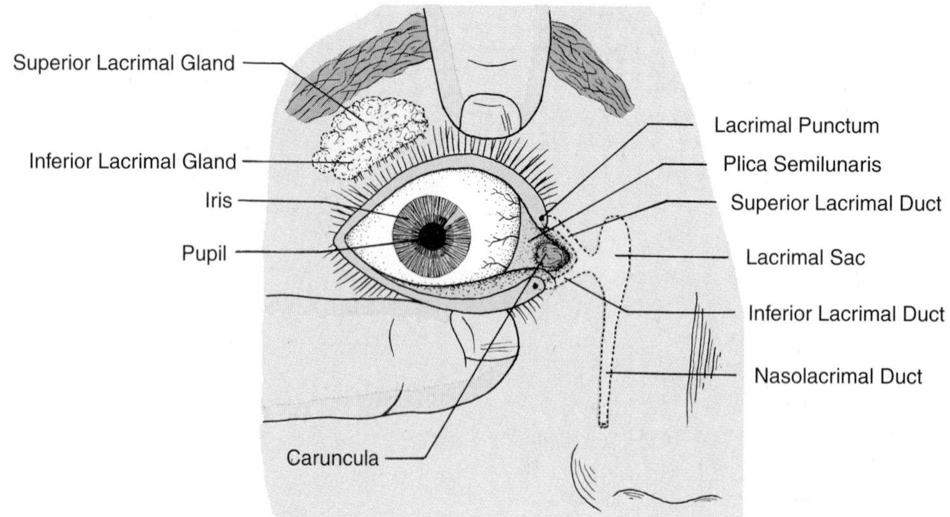

Figure 48.1 The lacrimal apparatus of the eye.

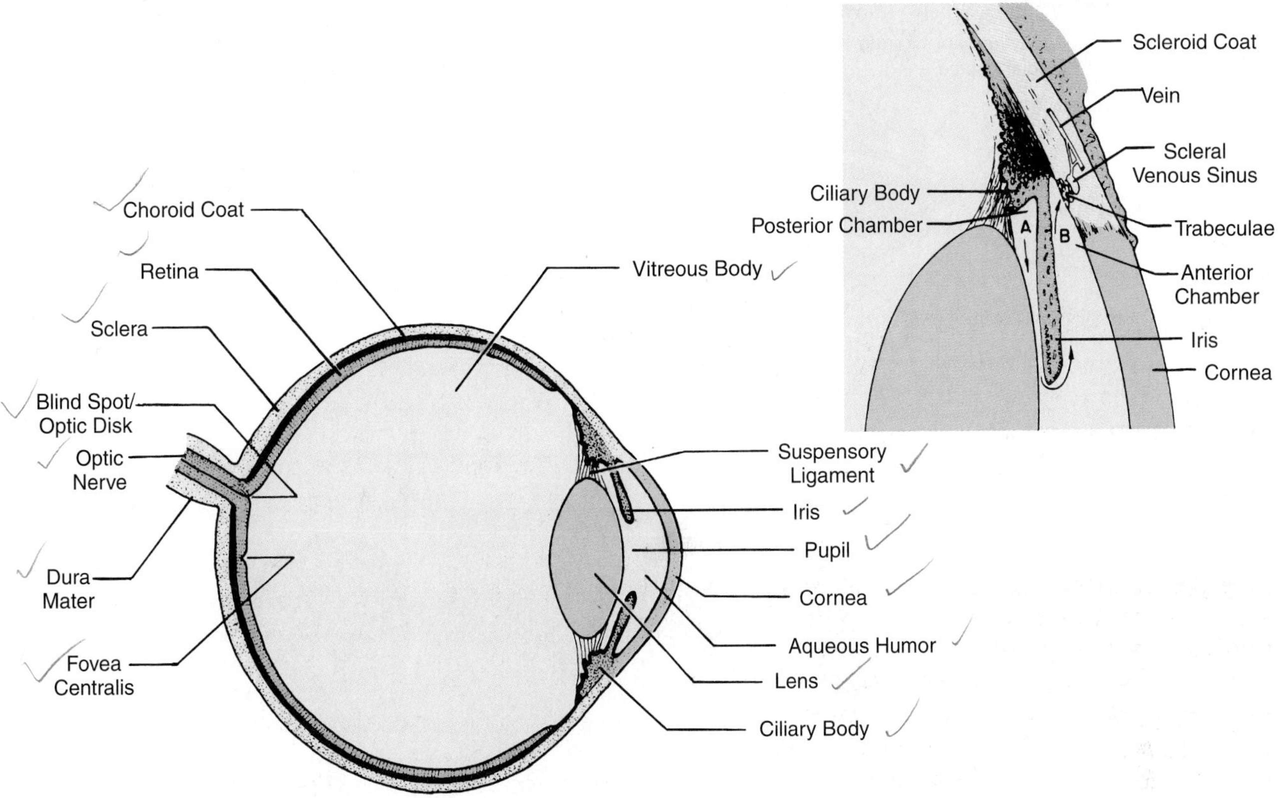

Figure 48.2 Internal anatomy of the eye.

optic nerve and the fovea centralis interrupt the continuity of the retinal surface.

The **optic nerve** contains fibers leading from the rods and cones of the retina to the brain. That point on the retina where the optic nerve enters lacks rods and cones, which makes it insensitive to light; thus, it is called the **blind spot.** It is also known as the **optic disk.**

Note that a sheath, the **dura mater** (an extension of the dura mater that surrounds the brain), wraps around the optic nerve. To the right of the blind spot is a pit, the **fovea centralis,** which is the center of a round yellow spot, the *macula lutea.* The macula, which is only 0.5 mm in diameter, is composed entirely of cones and is the portion of the retina that produces the sharpest vision.

The **lens,** an elliptical, crystalline structure, focuses light entering the eye on the retina. **Suspensory ligaments** hold the lens in place. The **ciliary body,** a smooth, circular muscle that can change the shape of the lens to focus on objects at different distances, attaches to the suspensory ligaments.

In front of the lens is a cavity that contains a watery fluid, the **aqueous humor.** Between the lens and the retina is a larger cavity that contains a more viscous, jelly-like substance, the **vitreous body.** Immediately in front of the lens lies the circular, colored portion of the eye, the **iris.** The circular and radiating muscle fibers of the iris change the size of the **pupil** to regulate the amount of light that enters the eye.

Covering the anterior portion of the eye is the **cornea,** a clear, transparent structure that is an extension of the sclera. It acts as a window to the eye, allowing light to enter.

Aqueous humor is constantly being renewed in the eye. The enlarged inset of the lens-iris area in figure 48.2 illustrates where the aqueous humor is produced (point A) and reabsorbed (point B). At point A (the **posterior chamber**), capillaries of the ciliary body produce aqueous humor. The fluid then passes through the pupil into the **anterior chamber** and is reabsorbed into minute spaces called **trabeculae** at the juncture of the cornea and the iris (point B). From the trabeculae, the fluid passes to the **scleral venous sinus**. This venous sinus drains into the venous system, represented as a short **vein** in figure 48.2.

Assignment:
Complete part A of the Laboratory Report.

The Extrinsic Ocular Muscles

Six *extrinsic ocular muscles* control the eye's ability to move in all directions within its orbit. A

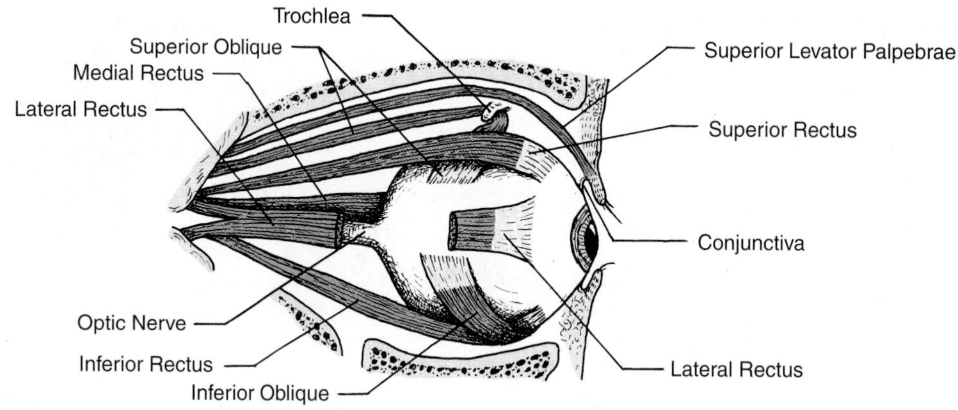

Figure 48.3 The extrinsic muscles of the eye.

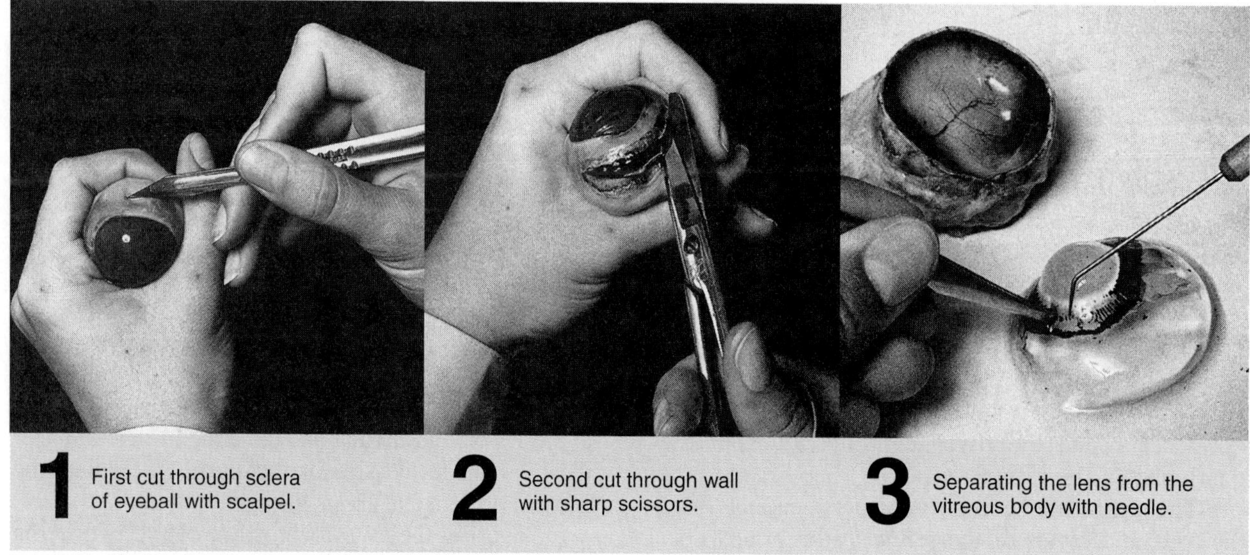

Figure 48.4 Three steps in cow eye dissection.

single muscle, the **superior levator palpebrae,** controls movement of the upper eyelid. Figure 48.3, which is a lateral view of the right eye, reveals the insertions of all seven of these muscles. The muscles originate on various bony formations at the back of the eye orbit.

The muscle on the side of the eyeball is the **lateral rectus** (a portion of this muscle has been removed in figure 48.3). Opposing it on the medial side of the eye is the **medial rectus.** The muscle that inserts on top of the eye is the **superior rectus;** its antagonist on the bottom of the eye is the **inferior rectus.** Just above the stub of the lateral rectus is the insertion of the **superior oblique,** which passes through a cartilaginous loop, the **trochlea.** Opposing the superior oblique is the **inferior oblique** (only a portion of its insertion is seen in figure 48.3).

ssignment:

te parts B and C of the Laboratory Report.

Cow Eye Dissection

Eye tissues are best studied by eye dissection. Figure 48.4 illustrates the steps in dissecting a cow eye.

Materials:
cow eye
dissecting instruments
dissecting tray

1. Examine the external surfaces of the eyeball. Identify the **optic nerve,** and look for remnants of **extrinsic muscles** on the sides of the eyeball.
2. Note the shape of the **pupil.** Is it round or elliptical? Identify the **cornea.**
3. Hold the eyeball as part 1 of figure 48.4 shows. Make an incision into the sclera with a sharp

scalpel about 1/4 inch from the cornea. Note how difficult it is to penetrate.

4. Insert scissors into the incision, and carefully cut all the way around the cornea, *taking care not to squeeze the fluid out of the eye* (see part 2 of figure 48.4).
5. Gently lift the front portion off the eye, and place it on the dissecting tray with the inner surface upward. The lens usually remains attached to the vitreous body, as part 3 of figure 48.4 shows. If the eye is preserved (not fresh), the lens may remain in the anterior portion of the eye.
6. Examine the inner surface of the anterior portion, and identify the thickened, black, circular **ciliary body.** What is the function of the black pigment?
7. Study the **iris** carefully. Can you distinguish **circular** and **radial muscle fibers?**
8. Is any **aqueous humor** between the cornea and iris? Compare its consistency with the **vitreous body.**
9. Separate the lens from the vitreous body by working a dissecting needle around the lens perimeter, as part 3 of figure 48.4 shows.
10. If the cow eye is fresh, attempt the following: (a) Hold the lens up by its edge with a forceps, and look through it at some distant object. What is unusual about the image? (b) Place the lens on printed matter. Are the letters magnified? (c) Squeeze the lens between your thumb and forefinger to compare the consistency of the lens at its center and near its circumference. Do you detect any differences?
11. Locate the **retina** at the back of the eye. It is a thin, off-white, inner coat that separates easily from the pigmented **choroid coat.**

    Now locate the **blind spot** (*optic disk*). This is where the optic nerve forms a juncture with the retina. The retina tightly adheres to this portion of the eye.
12. Note the iridescent nature of a portion of the choroid coat. This reflective surface is called the **tapetum lucidum.** It enables the eyes of some animals to reflect light at night and appears to enhance night vision by reflecting some light back into the retina.

Assignment:
Complete part D of the Laboratory Report.

Ophthalmoscopy

Careful examination of the **fundus,** or interior of the eye, provides information about a patient's general health. The retina of the eye is the only part of the body where relatively large blood vessels can be inspected without surgical intervention.

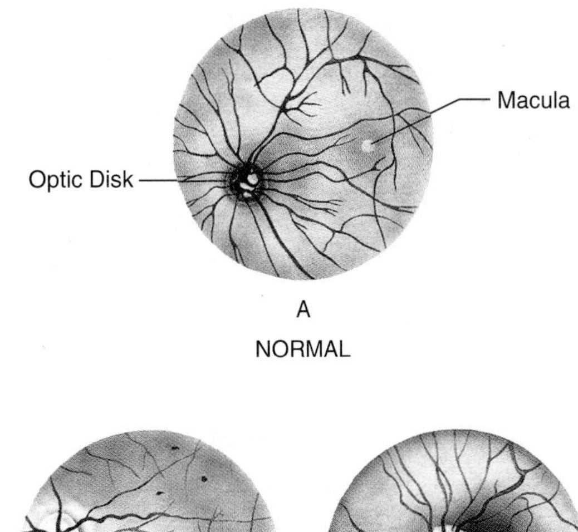

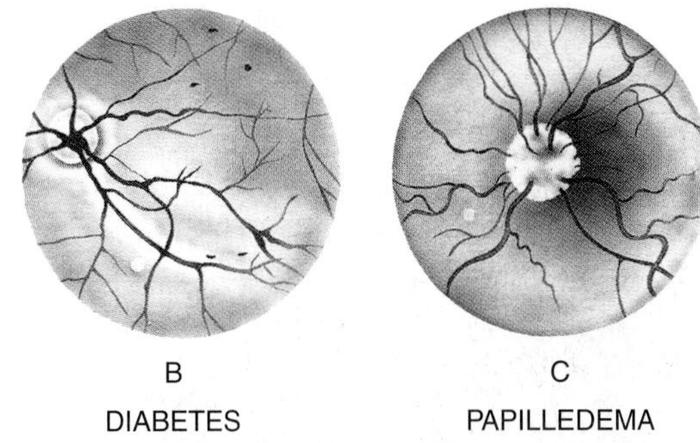

Figure 48.5 Normal and abnormal retinas as seen through an ophthalmoscope. In B (*diabetic retinopathy*), the peripheral vessels are generally irregular and fewer in number, and often exhibit small hemorrhages. In C, *papilledema,* or "choked disk," may be evidence of hypertension, brain tumor, or meningitis. In this case, the veins are considerably enlarged and often hemorrhagic.

Figure 48.5 shows what you might expect to see in a normal eye (A), the eye of a diabetic (B), and the eye of one who exhibits a "choked disk" (C). These are only a few examples of pathology that can be detected in the fundus.

Examining the fundus requires an **ophthalmoscope.** In this exercise, you and a laboratory partner examine the interior of each other's eyes in a darkened room. First, however, you must understand how to operate the ophthalmoscope.

The Ophthalmoscope

Figure 48.6 shows the construction of a Welch Allyn ophthalmoscope. Within its handle are two C-size batteries that power an illuminating bulb in the viewing head. At the top of the handle is a **rheostat control** for regulating the intensity of the light source. You must depress the lock button before this control can be rotated.

The head of the ophthalmoscope has two rotatable notched disks: a large lens selection disk and a smaller aperture selection disk. A viewing aperture is at the upper end of the head.

While looking through the viewing aperture, the examiner can rotate the **lens selection disk** with

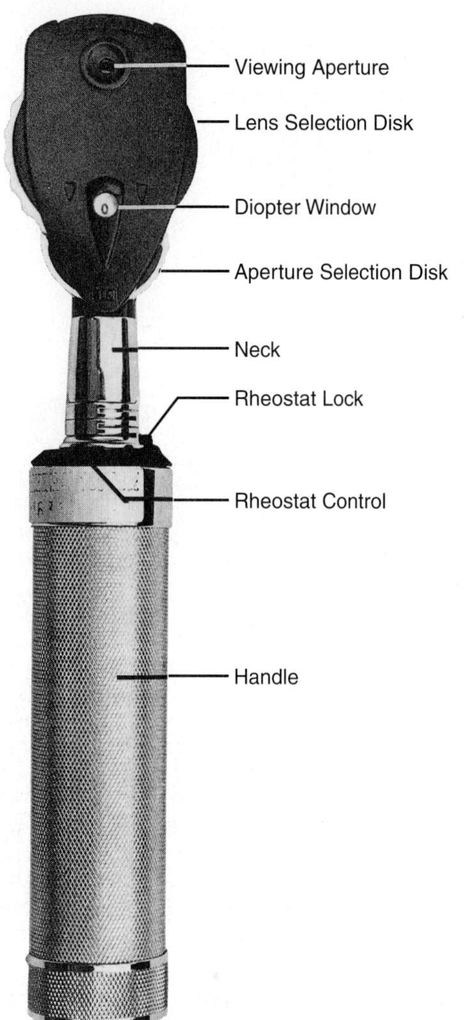

Figure 48.6 The ophthalmoscope.

Figure 48.7 Move the lens selection disk with the index finger to rotate viewing lenses of the ophthalmoscope into position.

the index finger (see figure 48.7) to select one of twenty-three small lenses. Twelve of the lenses on this disk are positive (convex), and eleven are negative (concave). The number in the **diopter window** indicates which lens is in place. The positive lenses have black numbers; the negative lenses have red numbers. A black "0" in the window indicates that no lens is in place.

If the eyes of both subject and examiner are normal (*emmetropic*), no lens is needed. If the eye of the subject or examiner is farsighted (*hyperopic*), the positive lenses are selected. Nearsighted (*myopic*) eyes require the use of negative lenses. The magnitude of the lens number indicates the degree of myopia or hyperopia. Since the eyes of both the examiner and subject affect the lens selection, the selection is entirely empirical.

The **aperture selection disk** enables the examiner to change the character of the light beam projected into the eye. The light can be projected as a grid, straight line, white spot, or green spot. **The green spot is most frequently used** because it is less irritating to the eye of the subject, and because it causes blood vessels to show up more clearly.

Using an ophthalmoscope for the first time is difficult for some, relatively easy for others, and nearly impossible for a few. If the examiner and subject have normal eyes, difficulty is usually minimal. However, when the examiner's eyes are severely myopic, astigmatic, or hypermetropic, difficulties arise. Even examiners with normal vision must deal with the problem of light reflecting from the retina back into the examiner's eye. The best way to minimize this problem is to **direct the light beam toward the edge of the pupil,** rather than through the center.

Avoid subjecting an eye to more than 1 minute of light exposure. After 1 minute of exposure, allow several minutes for the retina to recover.

Ophthalmoscopic Procedure

Before using the ophthalmoscope in a darkened room, you must be completely familiar with its mechanics.

Materials:
ophthalmoscope
metric tape or ruler

Desk Study of Ophthalmoscope

Turn on the light source by depressing the lock button and rotating the rheostat control. Observe how the light intensity varies from dim to very bright. Rotate the aperture selection disk as you hold the

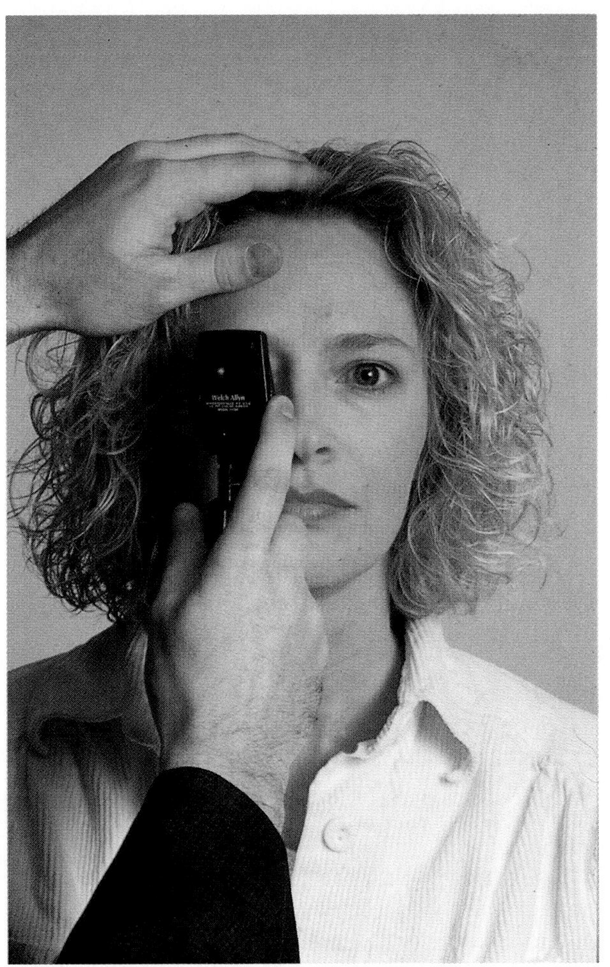

Figure 48.8 When examining the right eye, hold the ophthalmoscope with the right hand.

Figure 48.9 When examining the left eye, hold the ophthalmoscope with the left hand.

ophthalmoscope about 2 inches away from your desktop. How many different light patterns are there? Select the large white spot.

Rotate the lens selection disk until a black "5" appears in the diopter window. While peering through the viewing window, bring the print on this page into sharp focus. Have your lab partner measure the distance in millimeters between the ophthalmoscope and the printed page when the lettering is in sharp focus. Record this distance on the Laboratory Report. Follow the same procedure for 10D, 20D, and 40D lenses, recording all measurements.

Now set the lens selection disk on "0" and the light source on the green spot. The ophthalmoscope is set properly now for the beginning of an eye examination. Read over the following procedure *in its entirety* before entering the darkened room.

The Examination

Figures 48.8 and 48.9 illustrate how to hold the ophthalmoscope when examining different eyes. Note that the examiner uses the right hand when examining the right eye and the left hand when examining the left eye. Proceed as follows:

1. With the "0" in the diopter window and the light turned on, grasp the ophthalmoscope as figure 48.8 shows. Note that the index finger rests on the lens selection disk.
2. Place the viewing aperture in front of your right eye, and steady the ophthalmoscope by resting the top of it against your eyebrow (see figure 48.10). Start viewing the subject's right eye at a distance of about 12 inches. Keep your subject on your right. Instruct him or her to **look straight ahead at a fixed object at eye level.**
3. Direct the beam of light into the pupil, and examine the lens and vitreous body. As you look through the pupil, you will see a red circular area—the illuminated interior of the eye. If the image is not sharp, adjust the focus with the lens selection disk.
4. While keeping the pupil in focus, move to within 2 inches of the subject, as figure 48.11

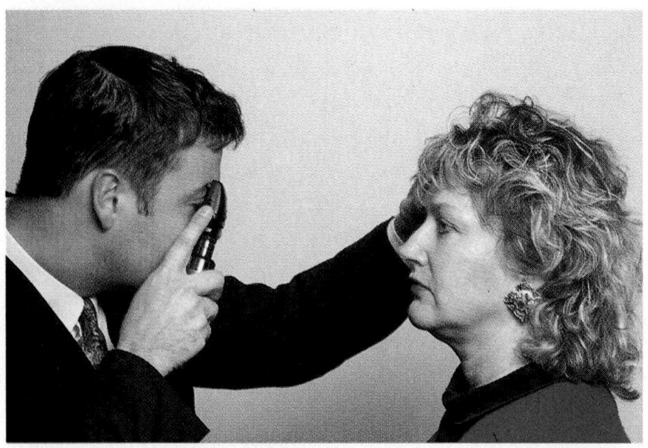

Figure 48.10 When viewing the lens and vitreous body, use this portion, relative to the subject.

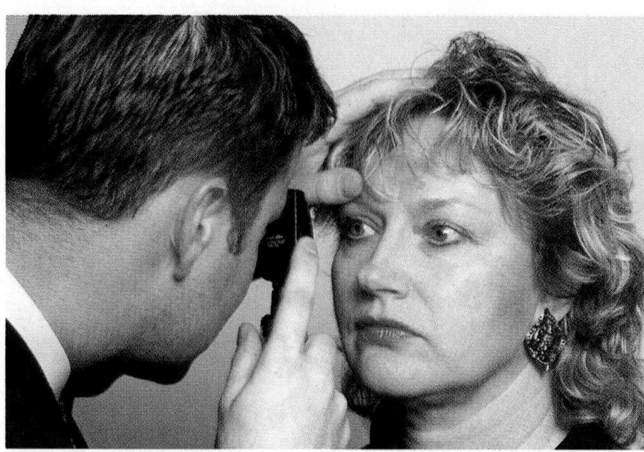

Figure 48.11 When examining the retinal surfaces, move in close.

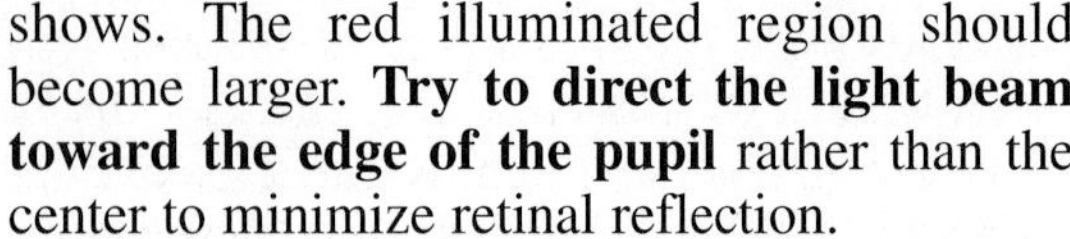

shows. The red illuminated region should become larger. **Try to direct the light beam toward the edge of the pupil** rather than the center to minimize retinal reflection.

5. Search for the **optic disk,** and adjust the focus as necessary to produce a sharp image. Observe how blood vessels radiate from the optic disk. Follow one or more of them to the periphery.

   Locate the **macula** about two disk diameters *laterally* from the optic disk. Note that it lacks blood vessels. The macula is easiest to observe when the subject is asked to look directly into the light, a position that **must be limited to 1 second only.**
6. Examine the entire retinal surface. Look for any irregularities, depressions, or protrusions. A roundish elevation, not unlike a pimple, on the surface is not unusual.

   To examine the periphery, instruct the subject to (1) **look up** for examination of the superior retina, (2) **look down** for examination of the inferior retina, and (3) **look laterally and medially** for those respective areas.
7. **Caution:** Remember to limit the right eye examination to **1 minute!** Then examine the left eye, using the left hand to hold the ophthalmoscope and reversing the entire procedure.
8. Switch roles with your laboratory partner. When you have examined each other's eyes, examine the eyes of other students. Report any unusual conditions to your instructor.

Assignment:
Complete part E of the Laboratory Report.

Histological Studies

Most prepared slides available for laboratory study are made from either rabbit or monkey eyes. Proceed as follows to identify structures depicted in the photomicrographs of figures HA-15 and HA-16 of the Histology Atlas.

Materials:
prepared slides of:
monkey eye, x.s.
rabbit eye, sagittal section

The Cornea With the lowest powered objective on your microscope, scan a cross section of the eye of a rabbit or monkey until you find the cornea (see figures HA-15A,B in the Histology Atlas). Note that the cornea consists of three layers: the epithelium, stroma, and endothelium. For fine detail, as in figure HA-15B, use high-dry magnification.

The **corneal epithelium** is the outer portion and consists of stratified squamous. It is, essentially, the conjunctival portion of the cornea. Note that the outermost cells of this layer are flattened squamous cells and that the basal cells are columnar.

The **corneal endothelium** is the innermost layer of cells, adjacent to the aqueous humor. This thin layer consists of low cuboidal cells.

The **corneal stroma** (*substantia propria*) comprises 90% of the thickness of the cornea and consists of collagenous fibrils, fibroblasts, and cementing substance. The fibrils are arranged in lamellae that run parallel to the corneal surface. A mucopolysaccharide cement holds the components of this stroma together. The chemical structure and fibril arrangement contribute to the transparency of the cornea.

The Lens A homogenous elastic **capsule** to which the **suspensory ligament** attaches encloses the lens. The lens forms from epithelial cells that elongate to a fibrous shape (forming lens fibers) and lose most of their organelles and nuclei. All that remains in a mature lens fiber are a few microtubules and clumps of free ribosomes. These cells are not entirely inert, and they persist throughout life.

The Ciliary Muscle The ciliary muscle is a ring of smooth muscle tissue that is a part of the body wall of the eye. Identify the **ciliary processes** that form ridges on the ciliary muscle and anchor the suspensory ligaments.

Body Wall of the Eye Examine a section through the body wall at the back of the eye. Identify the sclera, choroid coat, and retina. Note that the **sclera** consists of closely packed collagenous fibers, elastic fibers, and fibroblasts.

Melanocytes produce the large amount of pigment you observe in the **choroid coat.** The choroid coat nourishes the retina and sclera. Look for blood vessels.

The **retina** is the inner photosensitive layer of the body wall.

Retinal Layers The retina is composed of five main classes of neurons: the *photoreceptors* (rods and cones), *bipolar cells, ganglion cells, horizontal cells,* and *amacrine cells.* The first three classes form a direct pathway from the retina to the brain. The horizontal and amacrine cells form laterally directed pathways that modify and control the message passed along the direct pathway. Study a section of the retina, and identify the various layers (see figure HA-16 in the Histology Atlas for reference).

At the base of the retina, where it meets the choroid layer, are the **rods** and **cones.** The nuclei for these photoreceptors are in the **outer nuclear layer.** To stimulate these receptors, light must pass through all the other cell layers.

Nuclei of the amacrine and bipolar neurons are in the **inner nuclear layer.** Dendrites of these association neurons make synaptic connections with axons of the rods and cones in the **outer synaptic layer.**

The nuclei of ganglion cells are closest to the exposed surface of the retina—in the **ganglion cell layer.** The dendrites of ganglion cells make synaptic connections with amacrine and bipolar cells in the **inner synaptic layer.** Nonmyelinated axons of the ganglion cells fill up the **nerve fiber layer;** they converge at the optic disk to form the optic nerve. Some neuroglial cell nuclei are in the nerve fiber layer.

The Fovea Centralis If available, study a slide that reveals the structure of the fovea centralis (H10.635). Compare your slide with the photomicrographs in figures HA-16B and C in the Histology Atlas.

49 Visual Tests

This exercise outlines various visual tests for detecting both normal and abnormal conditions. Performing these tests will teach you much about the physiology of vision.

Materials:
penlight (for image suppression)
12-inch ruler or tape measure (for blind spot test)
3-inch-by-5-inch card and pins (for Scheiner's experiment)
Snellen eye charts (for visual acuity test)
laboratory lamp (for reflexes)

Image Suppression

(The Purkinje Tree)

When you examined the fundus of the eye with an ophthalmoscope in Exercise 48, you observed many blood vessels and capillaries. These branching vessels lie very close to receptor cells and cast sharp shadows on most regions of the retina. They do not impair vision, however, because the brain can suppress the disturbing images.

If light rays enter the eyeball at an oblique angle through the eyelid and sclera instead of directly through the pupil, the brain does not suppress the shadows that the blood vessels cast. The branching image seen with this type of illumination is called the *Purkinje tree.*

To perform this experiment on your own eye, proceed as follows: Hold a penlight with your right hand against the eyelid of your right eye at a 45° angle, as figure 49.1 shows. Your eyelid may be open or closed. If it is open, look at a dimly lit wall. If it is closed, do not face a brightly lit window.

Move the penlight slightly (only a few millimeters) from side to side to produce the Purkinje image. Performing this experiment in a dark closet produces the most striking image.

Assignment:
Complete part A of the Laboratory Report.

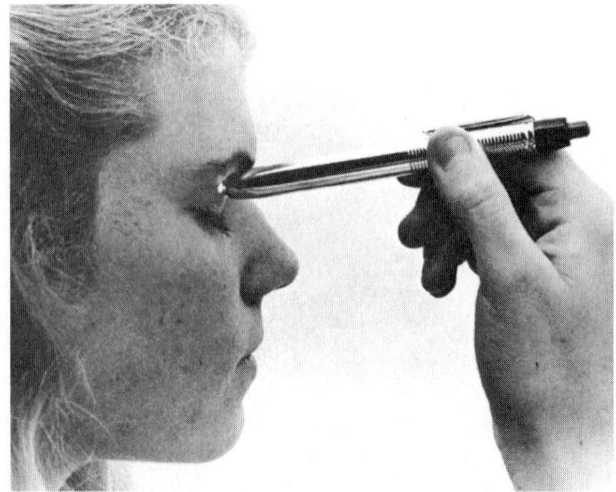

Figure 49.1 Proper position for penlight when observing the Purkinje tree.

The Blind Spot

The fundus examination in Exercise 48 clearly revealed the nature of the optic disk. Where the optic nerve enters the eyeball, rods and cones are absent, which renders that part of the retina insensitive to light. You can use figure 49.2 to detect the presence of this blind spot in each eye.

To test the right eye, close the left eye and stare at the plus sign with the right eye as you move this page from about 18 inches away toward the face. At first, you can see both the plus sign and dot simultaneously. Then, at a certain distance from the eye, the dot disappears as it comes into focus on the optic disk of the retina. Have your laboratory partner measure the distance from your eye to the test chart.

Perform this test on both of your eyes. To test the left eye, look at the dot instead of the plus sign.

Record the measurements in part B of the Laboratory Report.

Figure 49.2 The blind spot test.

Near-Point Determination

(Distance Accommodation)

The lens of the eye can produce a sharp image on the retina partially because of lens elasticity. When focusing on close objects, the lens must be considerably more spherical, or convex, than when focusing on distant objects. For the lens to become more convex, the tension on the lens periphery must relax as the ciliary muscles contract.

In an infant, the eye's ability to accommodate to distance is at its maximum. As an individual ages, the lens gradually becomes less elastic, and the degree of accommodation diminishes.

Generally, accommodation reduces dramatically by age 45 and is nearly nonexistent after age 60. The loss of lens elasticity is probably due to denaturation of protein in the lens, and it causes a form of farsightedness called **presbyopia.** Most individuals over age 45 require bifocal or trifocal glasses.

Lens elasticity can be measured by determining the near point of the eye. The **near point** is the closest distance at which you can see an object in sharp focus. At age 20, the near point is approximately 3.5 inches; at age 30, 4.5 inches; at age 40, 6.5 inches; at age 50, 20.5 inches; and at age 60, 33 inches.

To determine the near point in each eye, use the letter "T" at the beginning of this paragraph. Close one eye and move the page up to the eye until the letter blurs; then move the page away until you see a clear, undistorted image. Have your laboratory partner measure the distance from your eye to the page. The closest distance at which the image is clear is the near point. Test the other eye also, and record your results on the Laboratory Report.

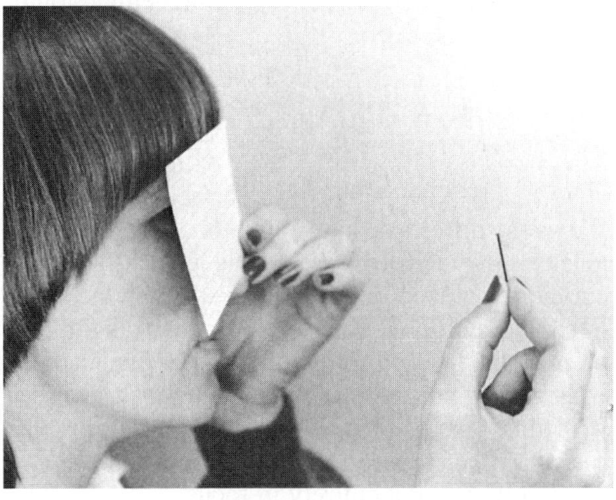

Figure 49.3 Scheiner's experiment. An object is viewed through two pinholes in a card.

Scheiner's Experiment

Another method for determining the near point is with Scheiner's experiment, which involves peering through two small pinholes in a card at an object such as a common pin (see Figure 49.3).

The pinholes are in the center of the card and no more than 2 mm apart, center to center. When you peer through the two holes, the holes should appear to overlap into a single opening. Holes that are too far apart appear as two distinct openings with an opaque bridge between them. Perform this experiment as follows:

1. Make two small pinholes that are no more than 2 mm apart, center to center, in the middle of a card.
2. Peer through the holes at a common pin held up at arm's length. If the two holes merge as a single opening, they are the correct distance apart. **You must view the pin through the space where the two luminous circles overlap.**
3. With the pin still at arm's length, focus on a distant object. The pin should now appear double since the eye is focused on infinity.
4. Now focus on the pin and note the single image.
5. Move the pin slowly toward your eye, keeping it in focus until the image changes from single to double. **Be sure to keep the eye looking through the overlapping area of the two holes.**

   The distance from the eye at which the image changes from single to double is the **near point.** Have your laboratory partner measure this distance.
6. Repeat this experiment several times to establish an average near point. How does this measurement compare with that obtained with the first method of establishing near point? Record the measurements on the Laboratory Report.

Visual Acuity

A normal human eye is able to differentiate, at a distance of 10 m, two points that are only 1 mm apart. Points less than a millimeter apart at this distance are seen as a single spot. The size and proximity of cones in the fovea centralis determine this degree of *visual acuity.*

If pinpoints of light from two different objects strike adjacent cones, you see a single image. This is because the brain cannot differentiate stimuli from adjacent cones. However, if an unexcited cone separates two stimulated cones, the brain recognizes two separate points.

Because the diameter of a cone is approximately 2μm (1/500 mm), the images on the fovea centralis must be at least 2 μm apart. This means that

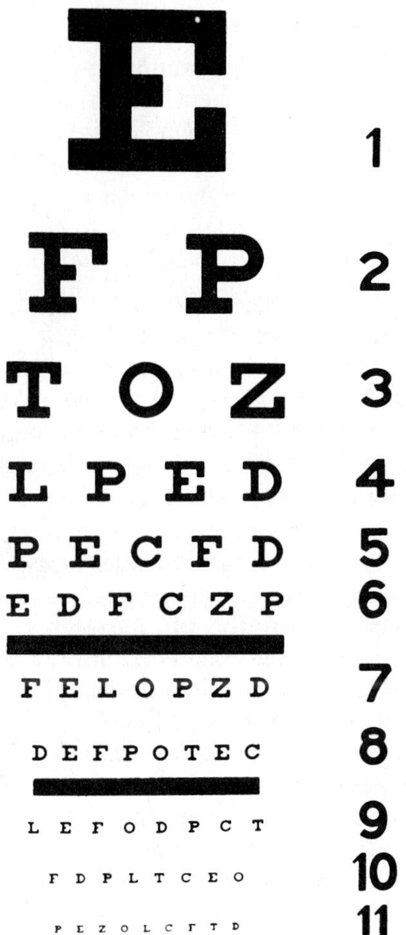

Figure 49.4 The Snellen eye chart.

the brain can differentiate two pinpoints that enter the eye at an angle of less than half of a degree!

The Snellen eye chart (see figure 49.4) has been developed with this mathematics in mind. When you stand at a certain distance from it, usually 20 feet, and are able to read the letters on a line designated to be read at 20 feet, you are said to have 20/20 vision. The ability to read these letters indicates that no aberrations of the lens or cornea interfere with the angle of pinpoints of light reaching the retina of the eye (see figure 49.5A). If you are only able to read the larger letters, such as those that should be read at 200 feet, you are said to have 20/200 vision. This means that, at 20 feet, you can read what a person with completely normal vision could read from a distance of 200 feet.

A Snellen eye chart is on a wall in your laboratory, and a 20-foot mark is also designated on the floor. Work with your laboratory partner, and test each other's eyes. Test one eye at a time, with the other eye covered. Record your results on the Laboratory Report.

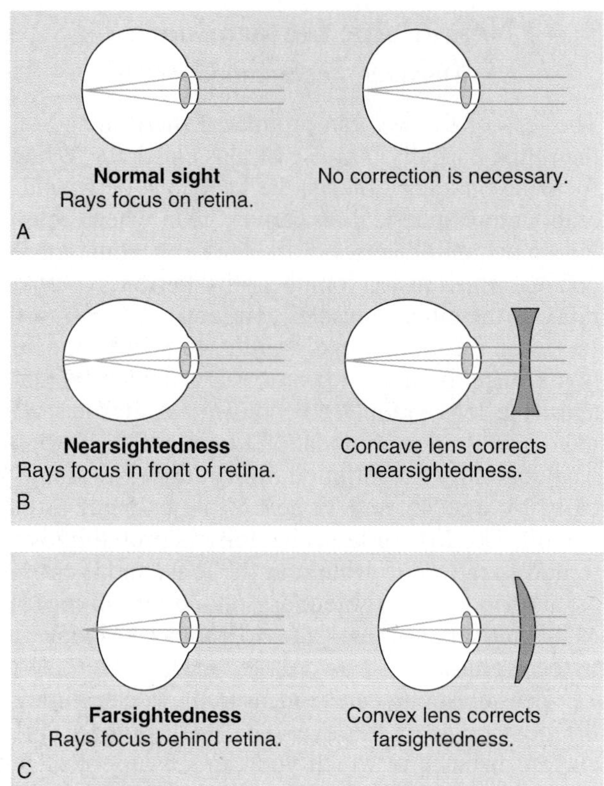

Figure 49.5 (A) Normal sight, (B) nearsightedness, and (C) farsightedness comparison.

Individuals with less than 20/20 vision have myopia or "nearsightedness." In this condition, an elongation of the eyeball results in images being focused in front of the retina (see figure 49.5B). Correction for this condition involves the prescription of concave corrective lenses. Individuals with hyperopia or "farsightedness" can view distant objects but require convex corrective lenses for close vision. Hyperopia results from images being focused behind the retina (see figure 49.5C). Did the visual acuity test demonstrate that either you or your partner is myopic?

Test for Astigmatism

Astigmatism exists if the lens of an eye has an uneven curvature on one of its surfaces. Figure 49.6 shows the difference between normal and astigmatic lenses. Note that the astigmatic lens has a greater curvature on its left upper surface than on the lower left quadrant. This type of lens bends light rays more as they pass through one axis of the lens than when passing through another axis. The image seen with this type of lens is blurred in one axis and sharp in other axes. In other words, the image cast on the retina is not entirely in focus.

To determine the presence of astigmatism, look at the center of the diagram in figure 49.7 with one

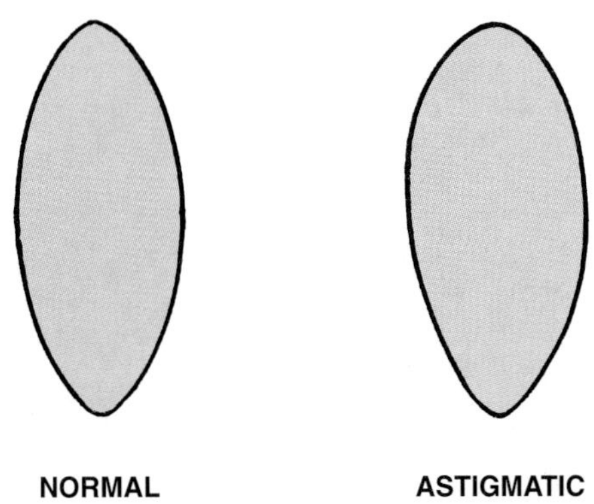

Figure 49.6 Normal and astigmatic lenses.

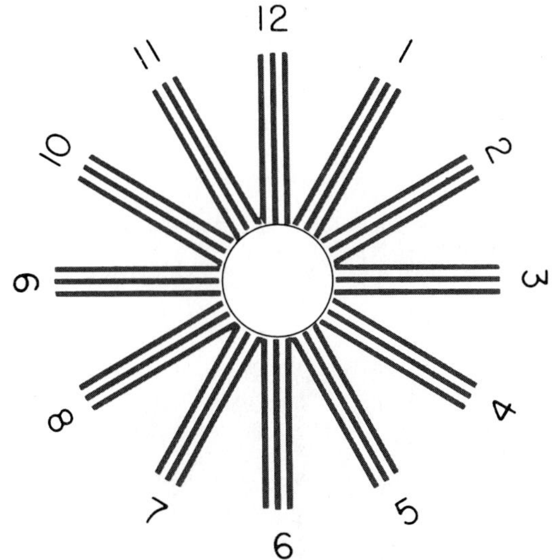

Figure 49.7 Astigmatism test chart.

eye at a time, and note if all radiating lines are in focus and have the same intensity of blackness. If all lines are sharp and equally black, no astigmatism exists. Of course, the presence of other refractive abnormalities can make this test impractical. Record the presence or absence of astigmatism for each eye on the Laboratory Report.

Test for Color Blindness

The retina and the brain work together to perceive color. The receptors of the retina that are sensitive to color are the **cones.** According to the *Young-Helmholtz theory* of color perception, each of the three different types of cones responds maximally to a different color: **red, blue,** and **green.** The degree of stimulation that each type of cone gets from a particular wavelength of light determines what color the brain perceives.

When the retina is exposed to *red* monochromatic light (wavelength of 610 nanometers[nm]), the red cones are stimulated at 75%, the green cones at 13%, and the blue cones not at all. The ratio of stimulation for red is thus 75:13:0 (red:green:blue). The brain interprets this ratio of stimulation from the three types of cones as red.

A monochromatic *blue* light (wavelength of 450 nm) striking the retina does not stimulate the red cones at all (0), stimulates the green cones to a value of 14%, and stimulates the blue cones to a value of 86%. The brain interprets the ratio of 0:14:86 as blue.

For *green,* the ratio is 50:85:15. Orange-yellow produces a ratio of 100:50:0. White light, which has no specific wavelength since it is a mixture of all colors of the spectrum, stimulates the three types of cones equally.

Color blindness is a sex-linked hereditary condition that affects 8% of the male population and 0.5% of females. The most common type is red-green color blindness, in which either the red or green cones are lacking. A lack of red cones results in a condition called **protanopia.** Individuals with this condition see blue-greens and purplish-tinted reds as gray. A lack of green cones is **deuteranopia.**

Although both protanopes and deuteranopes have difficulty differentiating reds and greens, their visual spectrums differ enough that they can be diagnosed with color test charts. Part 4 in figure 49.8 is a test plate for differentiating protanopes and deuteranopes. While a person with normal vision sees the number 96 on this plate, a protanope sees only the number 6, and a deuteranope sees only the number 9.

Rarer forms of color vision deficiencies include blue-color weakness, yellow-color weakness, and total-color blindness. These forms of color vision deficiencies are not as well understood.

Using Ishihara Plates

Obtain the book *Ishihara's Tests for Colour Blindness* from the supply table. Working with your laboratory partner, test each other for color blindness. Hold test plate 1 about 30 inches away from your partner. If sunlight is available, use it. Your partner should say what number he or she sees within the test plate pattern within **3 seconds.** Proceed consecutively through all 14 plates, and record the responses on the chart in part C of the Laboratory Report. After all the plates have been observed and recorded, compare the responses with the correct answers and make a statement at the bottom of the chart (conclusion).

Pupillary Reflexes

The tests and experiments performed so far have focused primarily on the lens and retina. The

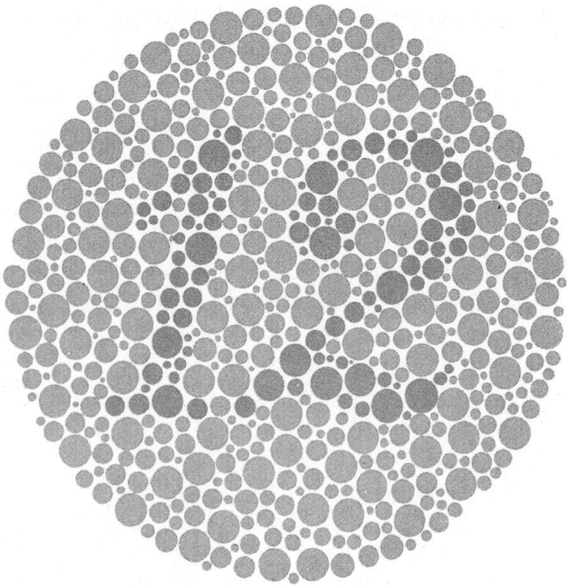

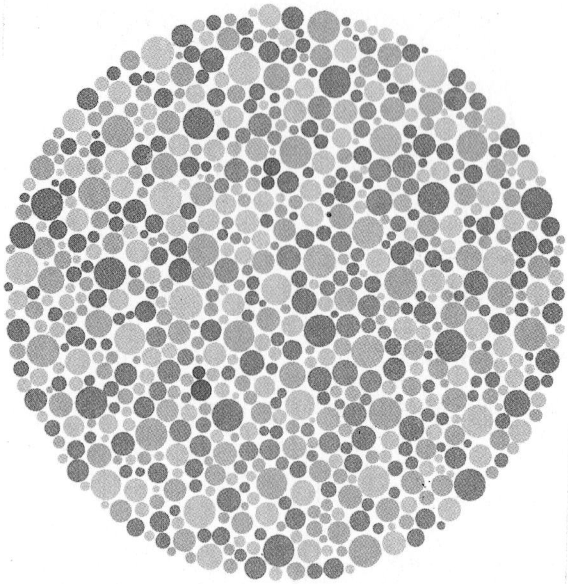

1 This color plate is nonselective. All individuals, normal or color blind, will read this plate as 12.

2 Normal individuals read this plate as 8. Red/green color-blind individuals see a 3 here.

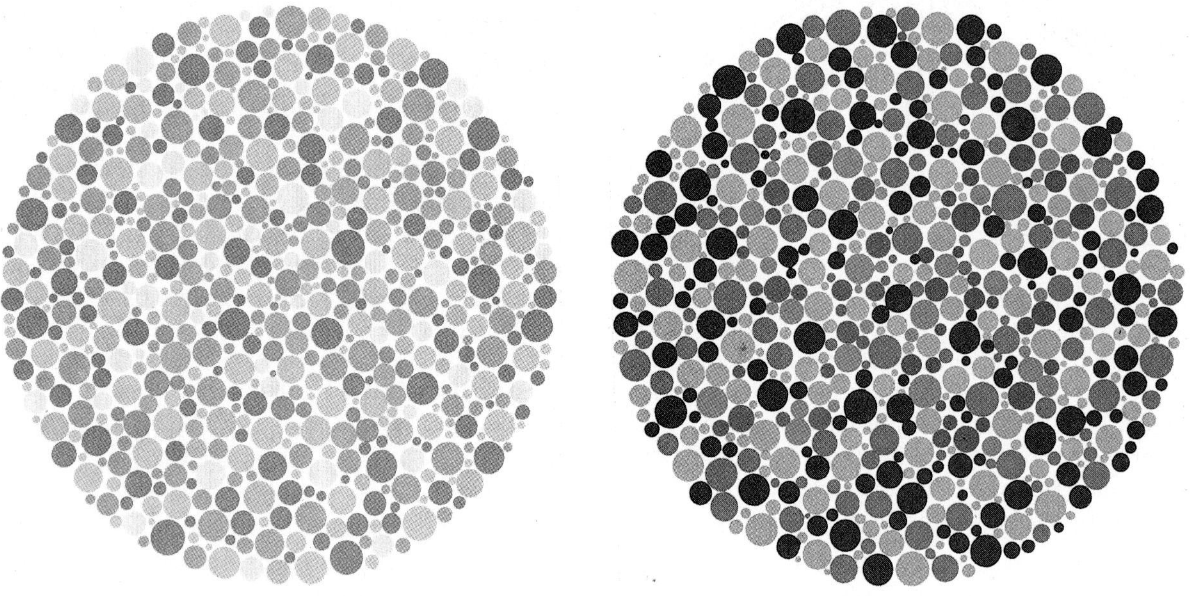

3 A normal person sees a 16 here. Color-deficient individuals read this plate incorrectly or not at all.

4 Normal individuals read a 96 here. Protanopes see only a 6. Deuteranopes see only a 9.

Figure 49.8 Ishihara color test plates.

experiments that follow show how the pupil adjusts to distance and light intensity.

Accommodation to Distance

When the eyes change their focus from a distant object to a close object:

1. Eye muscles react to achieve convergence of the eyes.
2. The lens becomes more convex.
3. The size of the pupil changes.

Working with your laboratory partner, perform the experiment that follows to observe how the pupils respond when the eyes focus on a nearby object after looking at a distant object.

This experiment works best if the subject has **pale blue eyes.** If your laboratory partner does not qualify in this respect, request some other class member to be the subject, and allow several other students to observe the results. Have the subject look at the wall on a side of the room opposite to windows. The subject's eyes should be relaxed and focused on infinity.

While closely watching the subject's pupils, place the printed page of your laboratory manual within 6 inches of the subject's face and ask the individual to focus on the print. The light intensity on the printed page and distant wall should be equal.

Do the subject's pupils remain the same size when looking at the printed page? Repeat the experiment several times, and record your results in part D of the Laboratory Report.

Accommodation to Light Intensity

Sudden exposure of the retina to a bright light causes immediate reflex contraction of the pupil in direct proportion to the degree of light intensity. The pupil contracts to approximately 1.5 mm when the eye is exposed to intense light, and it enlarges to almost 10 mm in complete darkness, making for a 40-fold difference in pupillary area. In this reflex, impulses pass from the retina via the optic nerve through two centers in the brain and then return to the sphincter of the iris through the ciliary ganglion.

Perform this simple experiment to learn more about the pathway of this pupillary reflex. As in the previous experiment, select a subject with pale blue eyes so that the pupil size is more easily observed.

Have the subject hold this laboratory manual vertically, with the spiral binding close to the forehead and extending downward along the bridge of the nose, dividing the face into right and left halves. Now, position an unlighted laboratory lamp about 6 inches from the subject's right eye.

While watching the pupil of the left eye, turn on the lamp for 1 second and then turn it off. Make sure that no light spills over from the right side of the book.

Did the unlighted pupil of the left eye react to the light to which the right eye was exposed? What does this reaction tell you about the pathways of the nerve impulses?

Assignment:
Answer the remaining questions on the Laboratory Report.

50 The Ear: Its Role in Hearing

This exercise focuses on four aspects of the ear: (1) its anatomical components that pertain to hearing, (2) the physiology of hearing, (3) hearing tests, and (4) histological studies. The exercise begins with a brief section on the characteristics of sound. The vestibular apparatus, which functions in equilibrium, is studied in Exercise 51. Complete the sections on ear anatomy and the physiology of hearing before doing the hearing tests and microscopy activities.

Characteristics of Sound

Sound waves move through the air in waveforms with characteristics relative to the vibratory motion that generates them. The ear distinguishes tones that differ in pitch, loudness, and quality. The frequency of vibration determines **pitch. Loudness** pertains to the vibration's intensity. **Quality** relates to the nature of the vibrations as revealed by the wave shape.

Figure 50.1 shows several curves depicting both the shapes of sound waves and the characteristics of the vibrations that produce them. The sine curves A and B in part I differ in frequency, with B producing the tone of higher pitch. Curves A and C in part II have the same frequency but different amplitude, or loudness. Although the amplitude of A is twice that of C, the loudness of A will be four times that of C. This is because the intensity (*I*) of sound is directly proportional to the square of the amplitude (*a*):

$$I = 2\pi^2 V f^2 a^2 d$$

V = velocity of wave propagation
f = frequency
d = density of medium

Waves A and D in part III in figure 50.1 differ in shape, with D having some components of higher frequency that A does not. These curves represent sounds of different quality.

The ear's ability to distinguish differences in pitch, amplitude, and quality depend on (1) the conduction and pressure amplification of sound waves through the ossicles of the middle ear, (2) the stimulation of receptor cells in the cochlea, and (3) the conveyance of action potentials in the cochlear nerve to the auditory centers in the brain for interpretation. Interference with any component of this system results in hearing loss.

Components of the Ear

Anatomically, the human ear has three distinct divisions: the external ear, the middle ear, and the internal ear (see figure 50.2).

The External Ear

The outer or external ear has two parts: the auricle and the external auditory meatus. The **auricle,** or **pinna,** is the outer shell of skin and elastic cartilage that attaches to the side of the head. The **external auditory meatus** is a 1-inch canal that extends from the auricle into the head through the temporal bone. The skin lining this canal contains some small hairs and modified apocrine sweat glands that secrete a waxy substance called *cerumen.* The inner end of the meatus terminates at the **tympanic membrane,** or eardrum. The auricle collects and directs sound waves to the tympanic membrane through the meatus.

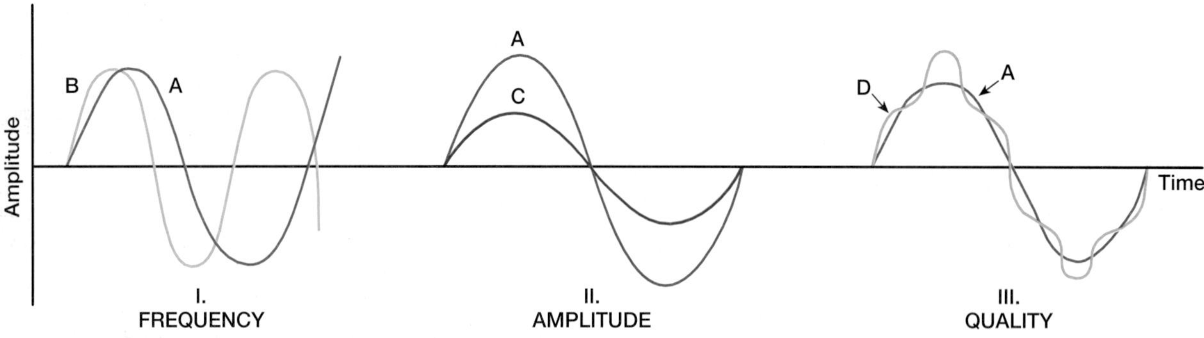

Figure 50.1 Differences in sound waves.

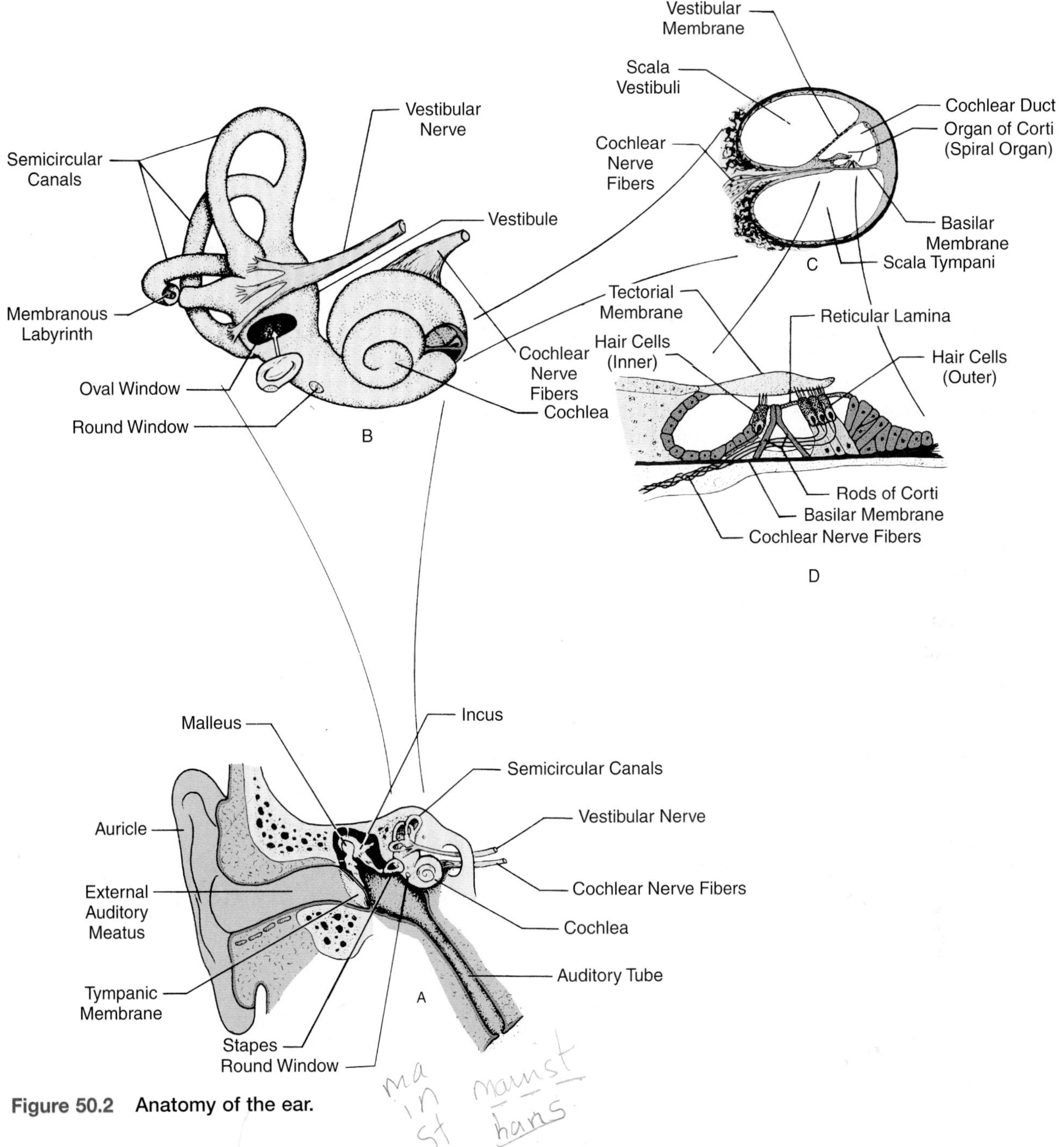

Figure 50.2 Anatomy of the ear.

The Middle Ear

The middle ear consists of a small cavity in the temporal bone between the tympanic membrane and the inner ear. It contains three small bones (*ossicles*) that unite to form a system of levers. The outermost ossicle, which attaches to the tympanic membrane, is the **malleus,** a hammer or club-shaped bone. The middle bone, an anvil-shaped structure, is the **incus.** The innermost bone, the **stapes,** fits into the oval window of the inner ear and is stirrup-shaped.

These ossicles transfer the forces from the eardrum to the cochlear fluids of the inner ear through the **oval window** (*fenestra vestibuli*). Although most sound waves reach the inner ear via the ossicles (*ossicular conduction*), some sounds, specifically very loud ones, reach the cochlea through the bones of the skull (*bone conduction*).

The **auditory** (eustachian) **tube** leads downward from the middle ear to the nasopharynx and allows air pressure in the middle ear to equalize with the outside atmosphere. A valve at the nasopharynx end of the tube keeps the tube closed. Acts of yawning or swallowing cause the valve to open temporarily for pressure equalization. You may have noticed the resulting "pop" in your ears while taking off or descending in an airplane.

The Internal Ear

The internal ear consists of two labyrinths: the osseous and membranous labyrinths. The **osseous labyrinth** is the hollowed-out portion of the temporal bone that contains an inner tubular structure of membranous tissue, the **membranous labyrinth.** Figure 50.2B shows some portions of the membranous labyrinth in cutaway portions of the osseous labyrinth. Figure 51.1 in the next exercise shows the entire membranous labyrinth.

The **endolymph** is a fluid within the membranous labyrinth. Between the membranous and osseous labyrinths is a different fluid, the **perilymph.** Both fluids are conduction media for the forces involved in hearing and maintaining equilibrium.

The osseous labyrinth consists of three semicircular canals, the vestibule, and the cochlea. The oval window into which the stapes fits is on the side of the **vestibule.** The three **semicircular canals** branch off the vestibule to one side, and the **cochlea,** which is shaped like a snail's shell, emerges from the other side.

On the osseous labyrinth are two nerves, which are branches of the eighth cranial (vestibulocochlear) nerve. The **vestibular nerve** is the upper nerve branch that passes from the sensory areas of the semicircular ducts, saccule, and utricle. The other branch is the **cochlear nerve,** which emerges from the cochlea.

The cochlea is a coiled, bony tube with three chambers extending along its full length. Figure 50.2C is a cross section of the cochlear tube. The upper chamber, or **scala vestibuli,** is so named because it is continuous with the vestibule. The lower larger chamber is the **scala tympani.** The **round window** (*fenestra cochlea*) is a membrane-covered opening on the osseous wall of the scala tympani. The **cochlear duct** lies between these two chambers and appears triangular in cross section. This duct is bounded on its upper surface by the **vestibular membrane** and on its lower surface by the **basilar membrane.** On the upper surface of the basilar membrane lies the **organ of Corti** (*spiral organ*), which contains the receptor cells of hearing. All of these chambers contain fluid: perilymph in the scala vestibuli and scala tympani, and endolymph in the cochlear duct.

Figure 50.2D shows the detailed structure of the organ of Corti. Note how the stereocilia of the **hair cells** are embedded in a gelatinlike flap, the **tectorial membrane.** Leading from each hair cell is a nerve fiber that passes through the basilar membrane and becomes part of the cochlear nerve. The **recticular lamina** holds the upper margins of the hair cells in place. The **rods of Corti** are reinforcing structures between the reticular lamina and the basilar membrane.

The Physiology of Hearing

Hearing occurs when hair cells in the organ of Corti initiate action potentials that move along the cochlear nerve fibers of the eighth cranial nerve to the auditory centers of the brain. Activation of these sensory hair cells depends on four factors: (1) forces within the cochlear fluids, (2) basilar membrane structure, (3) secondary energy transfer, and (4) endocochlear potential. A discussion of these factors follows, as does information on how the structures of the cochlea function in pitch discrimination and volume determination.

Role of Cochlear Fluids Figure 50.3 shows the cochlea uncoiled to reveal the relationship of various chambers to each other. The perilymph here is yellow, and the endolymph in the cochlear duct is green.

Because perilymph is incompressible, each time the stapes pushes in the oval window, the round window accommodates for the pressure change by bulging out. The sound wave energy generated at the oval window moves through the cochlea and is absorbed by the round window. The sound wave energy can travel through the cochlea in two ways (see arrows in figure 50.3):

1. Sound wave energy can move through the perilymph of the scala vestibuli to a small opening at the end of the cochlea called the helicotrema. From the helicotrema, sound wave energy moves through the perilymph of the scala tympani to the round window.
2. Sound wave energy moving through the scala vestibuli can also reach the scala tympani by passing through the flexible cochlear duct.

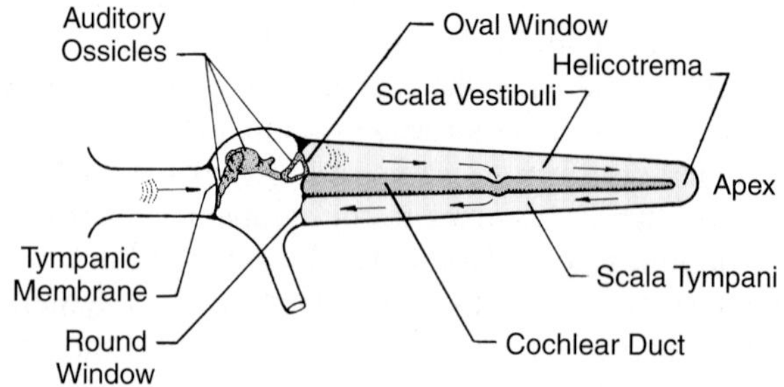

Figure 50.3 Pathway of sound wave transmission in the ear.

Basilar Membrane Structure The first structure of the organ of Corti that reacts to vibrations of different sound frequencies is the basilar membrane. Within this membrane, approximately 20,000 fibers project from the bony center of the cochlea toward the outer wall. The fibers closest to the stapes are short (0.04 mm) and stiff; they vibrate in response to high-frequency sounds. The fibers at the end of the cochlea near the helicotrema are longer (0.5 mm) and more limber, and vibrate in response to low-frequency sounds. These fibers are elastic, reedlike structures that are free at their distal ends, which allows them to vibrate to specific frequencies.

Secondary Energy Transfer Once a specific sound frequency stimulates fibers at a particular point on the basilar membrane, the rods of Corti transfer the energy from the basilar membrane to the reticular lamina, which supports the upper portion of the hair cells (see figure 50.2D). This secondary energy transfer causes all these components to move as a unit, bending and stressing the stereocilia of the hair cells. The back-and-forth bending of the hairs, in turn, causes alternate changes in the electrical potential across the hair cell membrane. This alternating potential, known as the *receptor potential of the hair cell,* stimulates the nerve endings at the base of each hair cell and initiates an action potential in the cochlear nerve fibers.

Endocochlear Potential The differences in the chemical compositions of perilymph and endolymph greatly enhance the sensitivity of the hair cells to depolarization. Perilymph has a high sodium-to-potassium ratio; endolymph, conversely, has a high potassium-to-sodium ratio. These ionic differences result in an *endocochlear potential* of 80 mV. The tops of the hair cells project through the reticular lamina into the endolymph of the scala media, while the bottoms of these cells lie bathed in perilymph. The potential difference of 80 mV between the top and bottom of each hair cell greatly sensitizes the cell to slight hair movements.

Frequency Discrimination The variability in length and stiffness of the fibers within the basilar membrane enables the ear to differentiate sound waves as low as 30 cycles per second (cps) near the apex of the cochlea and as high as 20,000 cps near the base of the cochlea, and a full range of frequencies between these two extremes. The basilar membrane's method of pitch localization is called the *place principle.*

Loudness Determination As noted in figure 50.1, the loudness of a sound is expressed in the amplitude of the sine wave. Increased loudness of sounds in the cochlear fluids increases the amplitude of vibration of the basilar membrane. This stimulates more and more hair cells, causing *spatial summation of impulses;* that is, more nerve fibers are carrying more impulses, and the brain interprets this as increased loudness.

Assignment:
Complete parts A and B on the Laboratory Report.

Hearing Tests

Although deafness may have many different origins, there are essentially two principal kinds of deafness: nerve and conduction deafness. **Nerve deafness** results from damage to the cochlea or cochlear nerve. **Conduction deafness** results from damage to the eardrum or ossicles. Conduction deafness can usually be remedied with surgery or hearing aids. Hearing aids occasionally are helpful with nerve deafness, but most often, the condition cannot be corrected.

The three kinds of hearing tests outlined here are the ones most frequently used, and each has a specific function. Perform those tests for which equipment is available.

Watch-Tick Method (Screening Test)

The watch-tick method is one of the simplest (and oldest) for determining hearing acuity. The only equipment required is a spring-wound pocket watch with an audible ticking sound. A patient's inability to hear a ticking watch at prescribed distances from the ear can signal a hearing problem. The watch-tick method is for screening purposes only. Further tests with tuning forks or an audiometer are necessary if a problem is detected.

Materials:
spring-wound pocket watch
cotton earplugs
meterstick or measuring tape

1. Seat the subject comfortably in a chair, and plug one of the subject's ears with cotton.
2. Instruct the subject to signal with the index finger when sound is first heard as the watch approaches the ear or when sound disappears as the watch moves away from the ear.
3. With the subject looking straight ahead, position the watch out of hearing range, usually about 3 feet. Keep the face of the watch parallel to the side of the head.

4. Move the watch toward the ear very slowly—at **a speed of about 0.5 inch per second.** The subject should signal when he or she hears the **first tick.**
5. Measure the distance between the watch and the ear, record the information on the Laboratory Report, and repeat the test two or three times. On repeat tests, change the angle of approach to the ear to prevent subject anticipation.
6. Reverse the procedure by starting close to the ear and moving away from it. In this test, the subject signals the **last tick** heard.
7. Transfer the cotton plug to the other ear, and follow the same procedure.

Tuning Fork Methods

Nerve and conduction deafness can be readily distinguished with two tuning fork methods: the *Rinne* and *Weber* tests. In the Rinne test, the base of the tuning fork is applied to the mastoid process behind the ear; in the Weber test, the fork is placed on the forehead. In both tests, the tuning fork is activated by striking the heel of the hand. Perform these tests alone or with the aid of your laboratory partner.

Materials:
tuning fork
cotton earplugs

The Rinne Test Perform the Rinne test, which utilizes the mastoid process to differentiate the two types of deafness, as follows:

1. Plug one ear with cotton.
2. Activate a tuning fork by striking the heel of the hand, as part 1 in figure 50.4 shows.
3. Place the tuning fork 3 to 6 inches away from the unplugged ear with the tine facing the ear, as part 2 in figure 50.4 shows. If hearing loss is minimal, you will hear the vibrating sound immediately. If you do not hear anything, go to step 4.

   The sound intensity will diminish until the sound finally disappears. As soon as the sound disappears, place the stem of the tuning fork against the mastoid process, as part 3 in figure 50.4 shows. If the sound vibrations reappear while in contact with the mastoid process, **conduction deafness is present.**
4. If hearing loss is considerable and the tuning fork vibrations cannot be easily heard with the tuning fork near the ear, place the stem of the vibrating fork against the mastoid process, and note if the sound can be heard through bone conduction. If the sound is loud through the bone, **conduction deafness is present.** If no sound is heard while in contact with the skull, **nerve deafness is present.**
5. Repeat the test on the other ear.
6. Record the results on the Laboratory Report.

The Weber Test In the Weber test, the stem of a vibrating tuning fork is placed on the forehead. In a person with normal hearing, the sound of the fork is heard with equal intensity in both ears. An individual with **conduction deafness** hears the sound **louder in the deaf ear than in the normal ear.**

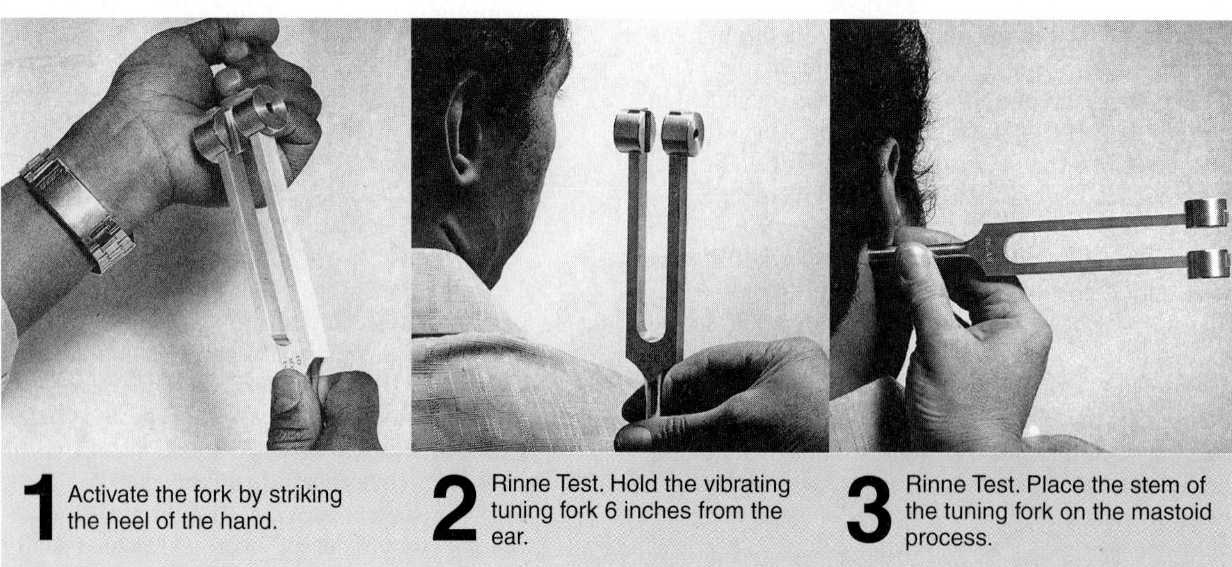

1 Activate the fork by striking the heel of the hand.

2 Rinne Test. Hold the vibrating tuning fork 6 inches from the ear.

3 Rinne Test. Place the stem of the tuning fork on the mastoid process.

Figure 50.4 Tuning fork manipulations in hearing tests.

The deaf ear, which sound waves through the air normally cannot activate, is more acutely attuned to sound waves being conducted to the cochlea through bone. A person with one normal ear and **one ear with nerve deafness** hears the sound **more intensely in the normal ear than in the deaf ear.**

Place a vibrating tuning fork on your forehead, and compare the sounds heard in each ear. Record the results on the Laboratory Report.

Audiometry

The audiometer determines hearing losses in the audible range of normal speech. It measures the ear's ability to hear sounds in the *frequency range of 125 to 8000 cps.* Although the hearing range of the human ear may be as broad as 30 to 20,000 cps, the 125 to 8000 cps range is important in hearing the spoken word.

The audiometer also measures the *level of intensity* of sounds in the audible range that you can hear. The sensation of loudness the ear experiences is not related in a simple way to the intensity of sound that strikes the tympanic membrane. While the range of sensitivity between a faint whisper and the loudest noise is 1 trillion times, the ear interprets this great difference as approximately a 10,000-fold change. The action of the tympanic membrane, ossicles, and cochlear duct compresses the scale of intensity, giving the ear a much broader sensitivity range.

Because of this extreme range in sound intensities, the intensity is expressed in terms of the logarithms of the actual intensities. The basic unit is called a **bel** (after Alexander Graham Bell). Sounds that differ by one bel in intensity differ by ten times. A **decibel** (dB) is a tenth of a bel.

Figure 50.5 The audiometer.

Hearing is measured in decibels mainly because this is the smallest unit that the human ear can differentiate.

The audiometer is, essentially, an electronic oscillator with earphones. As figure 50.5 shows, one control regulates the frequency, and another control regulates the intensity, or loudness. Two separate switches, or tone controls, near the bottom release the tone into the earphones. The left switch is for the left earphone, and the right one is for the right earphone.

The hearing threshold level control is calibrated so that the zero intensity level of sound at each frequency is the minimum volume that a normal person can hear. The numbers on this control represent decibels. If this control must be adjusted to 30 at 125 cps frequency, it means that the patient has a hearing loss of 30 dB at this frequency.

In this experiment, work with your laboratory partner to plot an audiogram on the Laboratory Report for each ear. Proceed as follows:

> ***Materials:***
> audiometer
> red and blue pencils

1. Place the headset securely on the subject's head so that the ear cushions make good contact over the ears and are not obstructed by clothing, earrings, and so on. **Important:** Be certain to place the right earphone over the right ear.
2. To enable the subject to become familiar with the tone, set the frequency control on 1000 cps and the hearing threshold level control at 50 dB. Now, depress the right tone control so that the subject can hear the tone in the right ear.
3. Rotate the frequency control through all frequencies at the 50 dB level to let the subject hear all frequencies prior to testing. **Important:** Depress the tone control *only* when the dial is set on each frequency.
4. Return the frequency control to 1000 cps and, while depressing the tone control, rotate the hearing threshold level control slowly until the subject is just barely able to recognize the tone.

   If, for instance, the subject can hear the tone at 30 dB, but not at 25 dB, the subject's threshold is established at 30 dB for that frequency setting. Have the subject signal with a hand when he or she hears the tone.
5. Record this decibel measurement on the audiogram chart in part E of the Laboratory Report by marking a red "O" where the measurement intersects the vertical line for 1000 cps. (In the

previous example, the red "O" would be where the 30 dB line intersects the 1000 cps line.)

6. Now rotate the frequency control to 2000 cps, following the same procedures as before. After testing at this frequency, test the right ear at the remaining higher frequencies up to 8000 cps.
7. Finish the test on the right ear by testing at frequencies of 250 and 500 cps.
8. Test the left ear, following steps 2 through 7. Remember to use the left tone control switch for testing the left ear.
   Record each threshold level with a blue "X" on the appropriate frequency line on the audiogram chart in the Laboratory Report.
9. On the chart, connect the "O's" with red lines and the "X's" with blue lines to produce a hearing graph for the ear.

Histological Study

Examine microscope slides as instructed to study the parts of the cochlea and crista ampullaris shown in figure HA-17 of the Histology Atlas. Both of these slides are made from guinea pig tissue. In particular, try to identify all the organ of Corti structures that figure HA-17C shows. Note in figure HA-17B that the collapsed wall of the ampulla obscures the cupula.

The Ear: Its Role in Equilibrium

51

The semicircular canals, saccule, and utricle maintain equilibrium. These three components combine into a unit called the *vestibular apparatus.*

In addition to the vestibular apparatus, three other factors are important in maintaining equilibrium: (1) visual recognition of the horizontal position, (2) proprioceptor sensations in joint capsules, and (3) cutaneous sensations through certain exteroceptors. The brain continuously and subconsciously integrates sensations from these four sources to provide the correct responses necessary to maintain equilibrium.

This exercise evaluates the relative significance of visual, proprioceptive, and vestibular sensations in postural control. Some clinical tests for evaluating static and dynamic equilibrium mechanisms are also explored.

Equilibrium is of two types: static and dynamic. **Static equilibrium** pertains to the effect of gravity on receptors. **Dynamic equilibrium** compensates for angular movements of the body in different directions. The physiology of each type follows.

Static Equilibrium

The sensory receptors of the vestibular apparatus that respond to gravitational forces are in the saccule and utricle. Note in figure 51.1 that the **saccule** connects directly to the cochlear duct and that the **utricle** lies at the base of the semicircular canals. These two structures are part of the membranous labyrinth and contain **endolymph.**

On the inner walls of the saccule and utricle lie two sensory structures, the **maculae.** Each macula consists of **hair cells,** a **gelatinous matrix,** and calcium carbonate crystals called **otoliths** embedded in the gelatinous matrix. Fibers of the vestibular nerve carry impulses from the hair cells to the brain.

When the head tilts slightly in any direction, gravitational forces pull the otoliths. The action of the gelatinous matrix against the otoliths bends the cilia of the hair cells. Bending of the cilia in one direction markedly increases impulse traffic in the vestibular nerve fibers; bending in another direction decreases the impulse traffic to a point of no

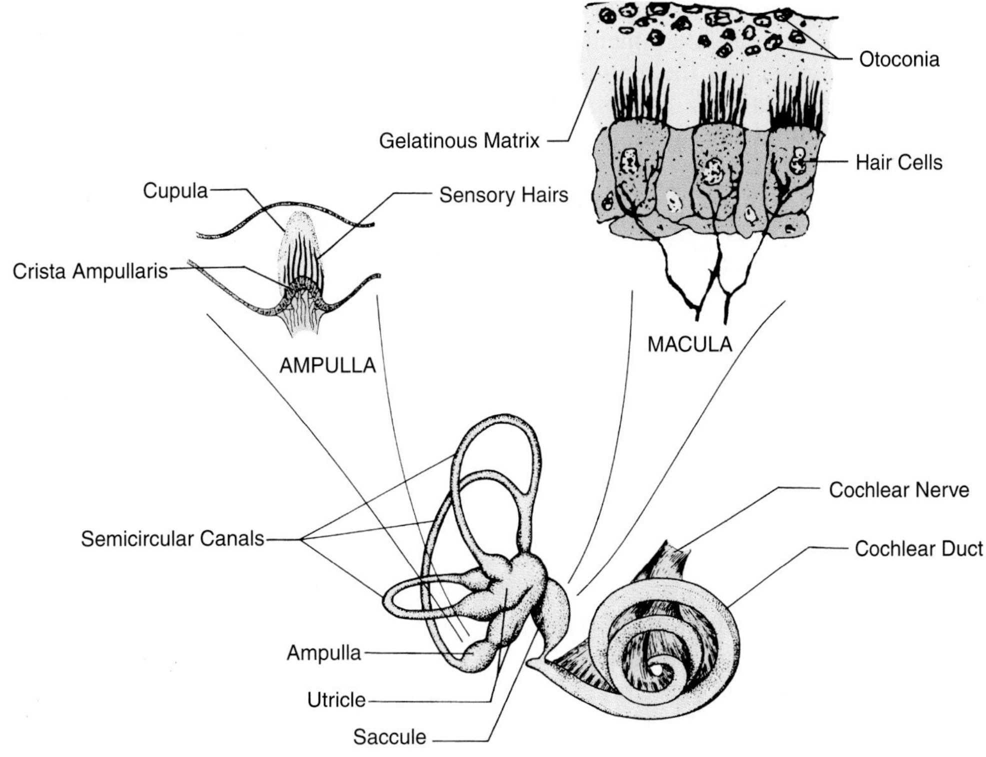

Figure 51.1 The membranous labyrinth.

reaction at all. This system is so sensitive to malequilibrium that shifting the head as little as half a degree in any direction from vertical is detectable.

Simply tilting the head, however, does not bring on a sense of malequilibrium. Fortunately, proprioceptors in the joints of the neck transmit inhibitory signals to the brain stem that neutralize the effects of the vestibular receptors. It is only when the whole body becomes disoriented that the hair-cell initiated impulses are not inhibited.

Balancing Test

One of the simplest tests for determining the integrity of this static equilibrium mechanism is to have the subject stand perfectly still with eyes closed. If the static equilibrium system is damaged in any way, the subject will waver and tend to fall. Individuals with long-term damage, however, are often able to stand fairly well, due to well-developed proprioceptive mechanisms. Work with your laboratory partner to test each other's sense of balance.

Dynamic Equilibrium

The receptor hair cells for dynamic equilibrium are in the **ampullae** of the **semicircular canals.** These hair cells are arranged in a crest, the **crista ampullaris,** within each ampulla. The hair tufts, in turn, are embedded in a gelatinous mass to form a structure called the **cupula.** See figure 51.1.

The three semicircular canals are arranged at right angles to each other in three different planes. When the head moves in any direction, the semicircular canal in the plane of directional movement moves relative to the endolymph within it; in other words, *the fluid is stationary as the canal moves.*

This movement causes the cupula to move in one direction or the other due to the force of the endolymph. The hair cells in this structure act much like the hair cells of the macula, in that bending in one direction increases action potentials, and bending in the other direction is inhibitory. Nerve impulses along the vestibular nerve reflexly excite the appropriate muscles to maintain equilibrium.

Assignment:
Complete part A of the Laboratory Report.

Nystagmus

The semicircular canals maintain equilibrium *in the initial stages* of angular or rotational movement. As soon as you begin to change direction, the semicircular canals trigger reflexes to the proper muscles in anticipation of malequilibrium. A reflex movement of the eyes, called *nystagmus,* occurs during this movement to produce fixed momentary images on the retina.

Rapid and slow jerky eye movements as the body rotates characterize nystagmus. Reflexes transmitted through the vestibular nuclei, the cerebellum, and the medial longitudinal fasciculus to the ocular nuclei cause it. *The function of nystagmus is to permit fixed images rather than blurred ones during head movements.*

If the head slowly rotates to the right, the eyes will, at first, move slowly to the left. This reflex eye movement is the **slow component** of nystagmus. As soon as the eyes have moved as far to the left as they can, they quickly shift to the right, producing the **fast component** of nystagmus. The vestibular apparatus initiates the slow component; the fast component involves the brain stem.

Group Demonstration of Nystagmus

Work in teams of five students, with one team member as a subject. The subject is seated in a swivel chair that the other four team members rotate. Having several individuals anchor and rotate the chair provides the subject with maximum safety. Prior to the test, **securely tighten the chair's back to prevent backward tilting.**

Materials:
swivel-type armchair

1. Have the subject sit cross-legged in the chair, gripping the chair arms.
2. The other four team members form a circle around the chair, with each member placing one foot against the base of the chair to prevent chair base movement.
3. Rotate the subject in a **clockwise** direction, and observe the eye movement as the subject gazes straight ahead during the rotation. Make 10 revolutions at a rate of 1 revolution per second.
4. Allow the subject to remain seated for 1 to 2 minutes after the test to regain stability.
5. Repeat the procedure with another subject, but rotate the individual in a **counterclockwise** direction.
6. Follow the same procedure with the remaining three team members, but have the subjects close their eyes during rotation. Immediately upon stopping the chair, have the subjects open their eyes for observation.

Assignment:
Record all observations in part B of the Laboratory Report.

Ice Water Test

Caution: Do not perform this test without special permission. It may induce nausea and vomiting! A simple test to determine whether or not the vestibular apparatus of one ear is functioning properly is to perfuse the external ear canal with a small amount of ice water. If the vestibular apparatus is intact and functional, the subject will experience a discomforting sensation of rotation, and *nystagmus will immediately be initiated.* If the subject experiences no discomfort or nystagmus, the vestibular apparatus or the vestibular nerve is damaged.

The physiological explanation of this reaction is that the cold water increases the density of the endolymph in a portion of the semicircular canals, making some endolymph heavier than other endolymph. This creates convection currents in the semicircular canals that cause a sense of malequilibrium.

This ice water test is often used to test for vestibular nerve damage resulting from overdosage of certain antibiotics, such as streptomycin.

Proprioceptive Influences

To observe the relative roles of proprioceptors, vision, and vestibular reflexes in equilibrium, perform the experiments that follow. Retain the same five-member teams as before.

Materials:
swivel-type armchair
pencil
blindfold

At-Rest Reactions

Direct the subject to sit in the swivel chair and to do the following to establish norms for comparison:

1. **With eyes closed,** the subject places the heel of the right foot on the toes of the left foot.
2. **With eyes closed,** the subject brings the index finger of the right hand to the tip of the nose from an extended arm position. How do proprioceptors in the appendages function to accomplish these two feats? Record your conclusions in part C of the Laboratory Report.
3. **With eyes open,** the subject raises his or her hand from the right knee vertically and forward to point his or her finger at the eraser of a pencil held about 2 feet directly in front of the subject. The subject repeats this pointing six times, once per second on command of the person holding the pencil. After each pointing, the subject rests the hand on the right knee.
4. **With eyes closed,** the subject repeats step 3. How does the subject approach the eraser with eyes closed? Record the results on the Laboratory Report.

Effects of Rotation

The role of the eyes in equilibrium can be determined in tests similar to the preceding. Proceed as follows:

1. Direct the subject to sit cross-legged in the swivel chair with eyes open. Rotate the chair at 1 revolution per second for 10 full turns. Halt the chair in exactly the same starting position.
2. Ask the subject to point to the pencil eraser held in the same position as in the previous test. As before, the subject should point to the eraser one time per second six times. After each pointing, the subject rests the hand on the right knee. Record your observations on the Laboratory Report.
3. Blindfold the subject, orient the chair to the same starting position, and by feel, show the subject the location of the pencil eraser.
4. Rotate the subject for 10 full turns, stopping the chair at the same starting point.
5. As soon as the chair has come to rest, ask the subject to point to the spot where he or she thinks the eraser is, **but do not let the subject touch the eraser.**
6. Repeat the pointing six times. Remember, the pencil eraser must be held in the same spot as previously and must not be touched.
7. Repeat the entire procedure with another subject, but reverse the direction of rotation.

Assignment:
Complete the remainder of the Laboratory Report.

HISTOLOGY ATLAS

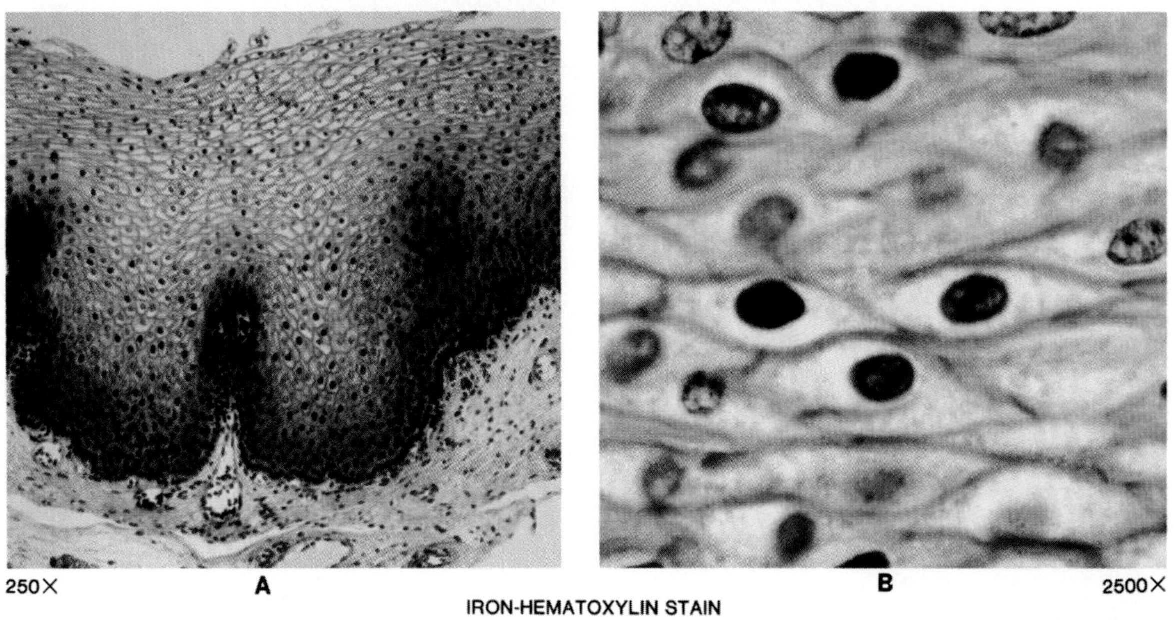

250× A B 2500×

IRON-HEMATOXYLIN STAIN

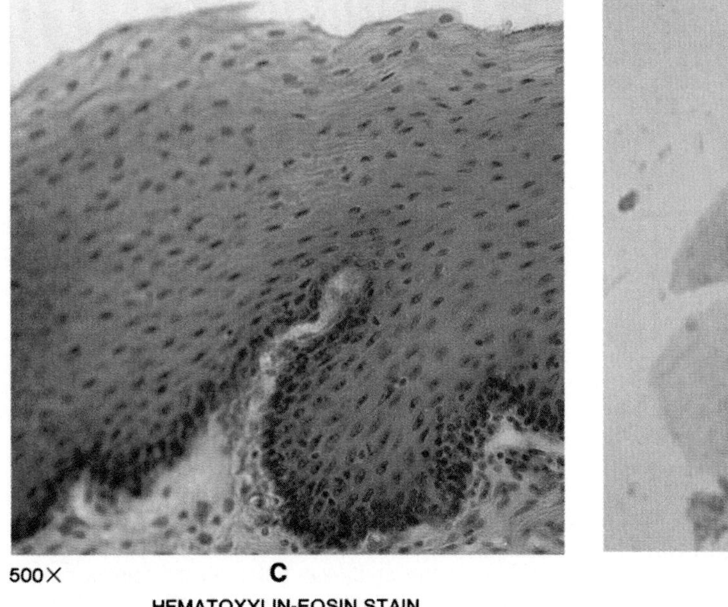

500× C

HEMATOXYLIN-EOSIN STAIN

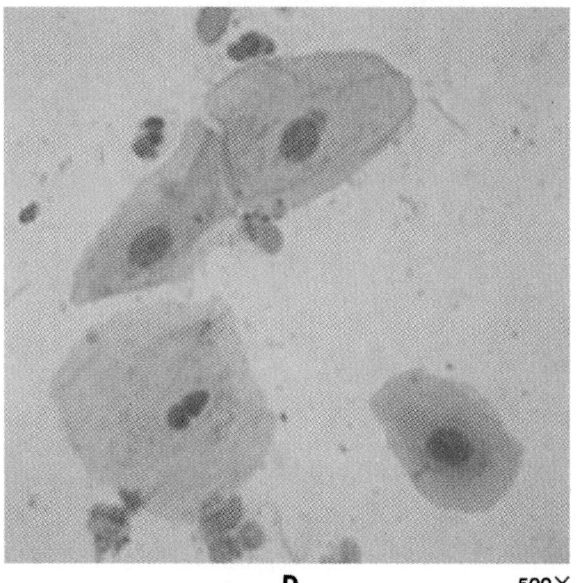

D 500×

EXFOLIATED CELLS

Stratified squamous epithelium is characterized by a transition in cell shape from cuboidal or low columnar in the deepest layer to flattened squamous cells on the surface. Between the squamous and deepest layers lie several layers of irregular and polyhedral cells. Like all epithelial cells, stratified squamous epithelial tissue lacks extracellular matrix and vascularization.

Micrograph B is an enlarged section of the deeper flattened cells in micrograph A. The exfoliated cells in micrograph D are scrapings from the inner cheek of the mouth.

The tissues in micrographs A and C are of the lining of the esophagus.

Figure HA-1 Stratified squamous epithelium.

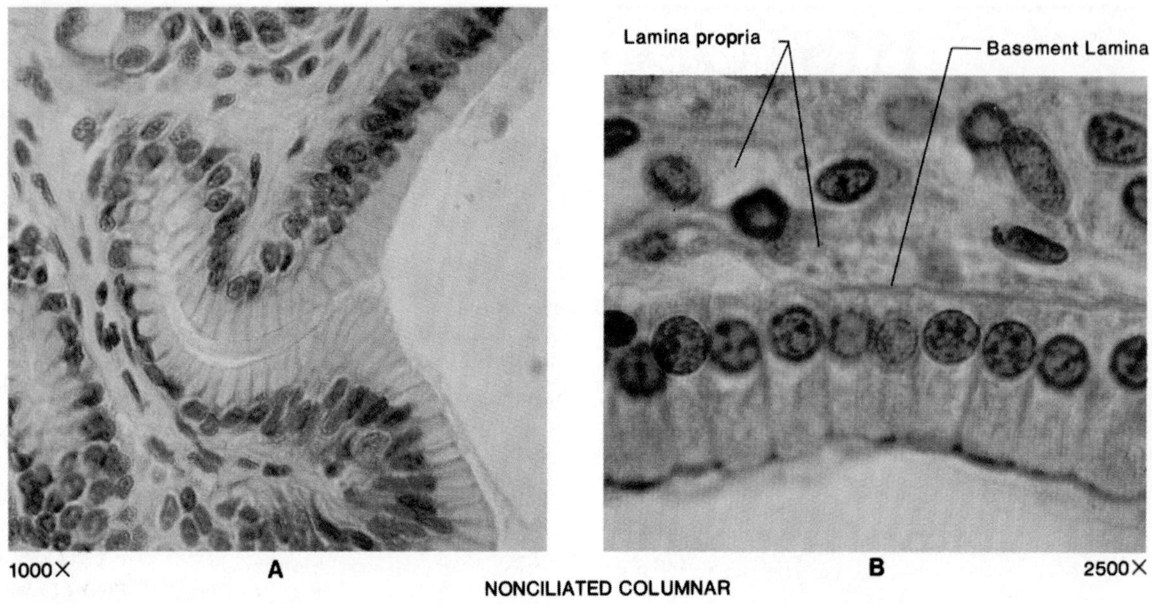

NONCILIATED COLUMNAR

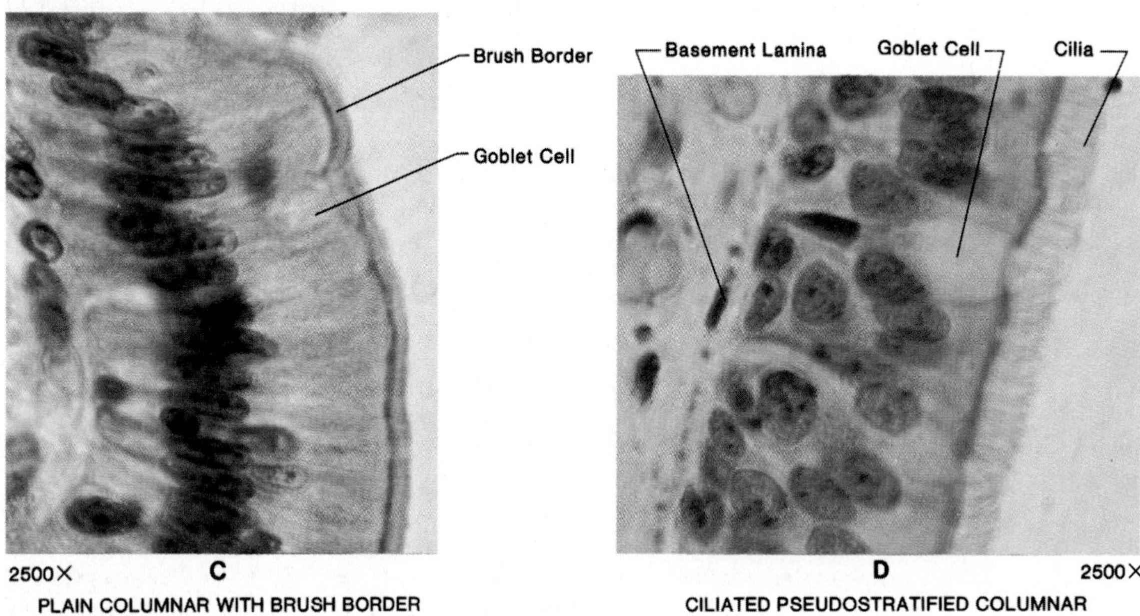

PLAIN COLUMNAR WITH BRUSH BORDER

CILIATED PSEUDOSTRATIFIED COLUMNAR

Columnar epithelia of the digestive and respiratory tracts have many mucus-producing goblet cells. Goblet cells are labeled in micrographs C and D, but are also in micrograph A.

Micrographs C and D show the distinct differences between cilia and a brush border. While cilia form from centrioles, brush borders are modified microvilli. Another modification of microvilli are cilialike structures called stereocilia. These nonmotile organelles are on columnar cells that line the vas deferens (see figure HA-34).

Note the distinct line of demarcation that constitutes the basement lamina in micrographs B and D. This thin layer between the epithelial cells and the lamina propria consists of a colloidal complex of protein, polysaccharide, and reticular fibers.

The lamina propria, to which all epithelial tissues connect, consists of connective tissue, vascular and lymphatic channels, lymphocytes, plasma cells, eosinophils, and mast cells.

Figure HA-2 Columnar epithelium.

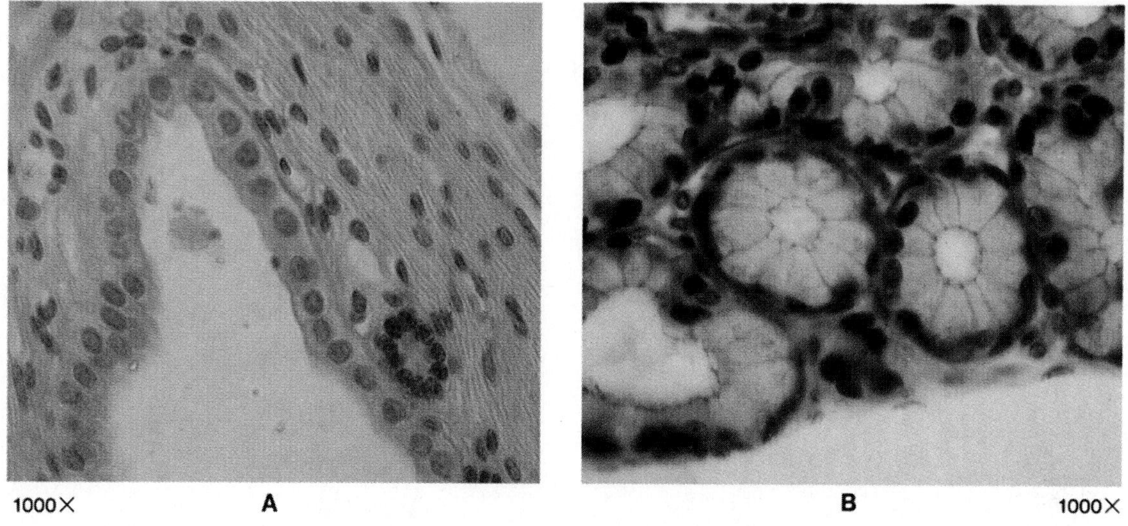

CUBOIDAL EPITHELIUM

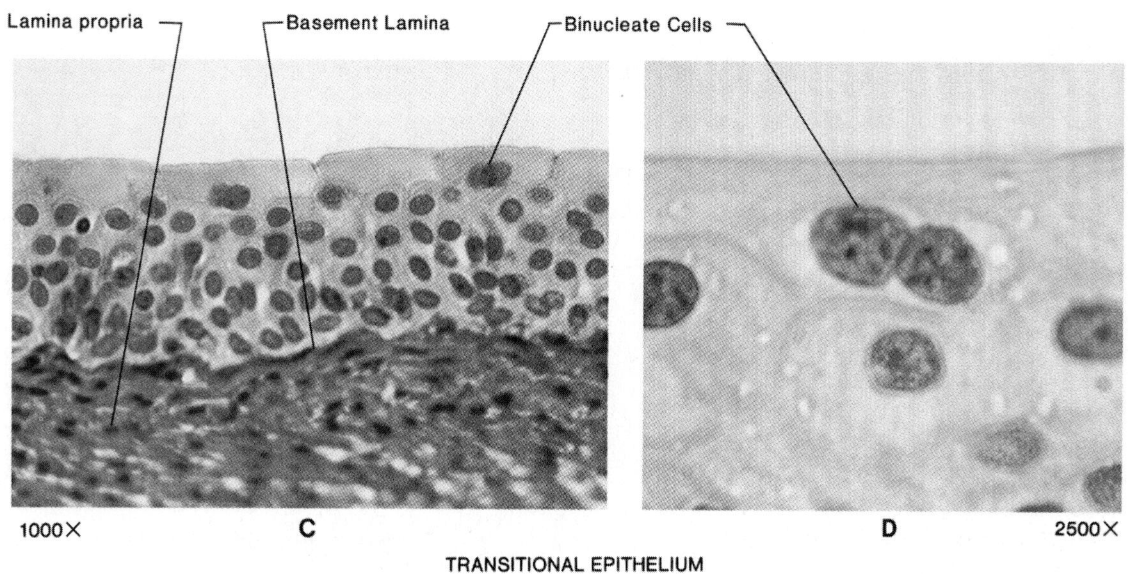

TRANSITIONAL EPITHELIUM

Cuboidal cells often appear squarish, as in micrograph A, but may take on a pyramidal structure surrounding the lumen of a duct or a small gland, as in micrograph B. Cuboidal epithelia may serve both absorptive and secretory functions, as in the case of tubules in the kidney.

Transitional epithelium is a stratified epithelium whose surface cells do not fall into squamous, cuboidal, or columnar categories. Note that the surface cells are dome-shaped and often binucleate, while the basal cells are more like stratified columnar cells. Between the surface cells and basal cells are layers of loosely configured pear-shaped cells. This type of tissue is in the wall of the urinary bladder, the urethra, and certain places in the kidneys, where organ distention demands elasticity of tissue.

Figure HA-3 Cuboidal and transitional epithelium.

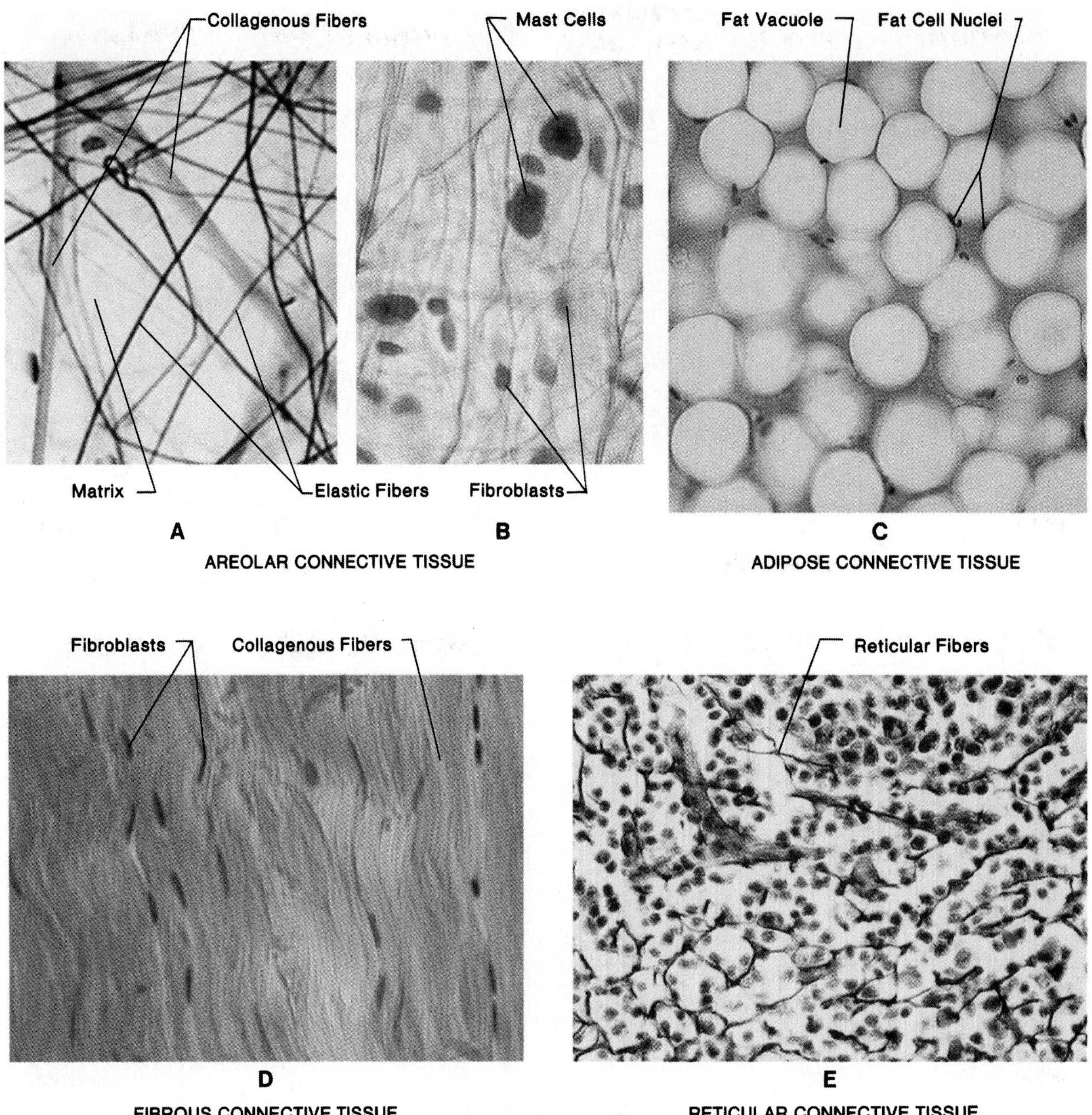

Micrographs A and B show two different preparations of areolar connective tissue. Although both exhibit elastic and collagenous fibers, micrograph A reveals them in better detail, due to a slightly different staining technique. The mast cells in micrograph B are believed to produce heparin and histamine, since both substances have been identified within these cells. Other cells often seen in areolar tissue are macrophages and plasma cells.

Micrograph C shows adipose connective tissue. Note the large fat vacuoles that fill the bulk of the cytoplasm and push the nuclei near the cell membranes.

Note in micrograph D the density of the fibers in fibrous connective tissue, compared to the other three tissues. The closely packed collagenous fibers in this tissue provide the tremendous tensile strength that ligaments and tendons require.

Loose connective tissue is beneath epithelia, around and within muscles and nerves, and part of serous membranes. Reticular tissue (micrograph E) is in the liver, lymphatic structures, and basement laminas.

Figure HA-4 Connective tissues (1000x).

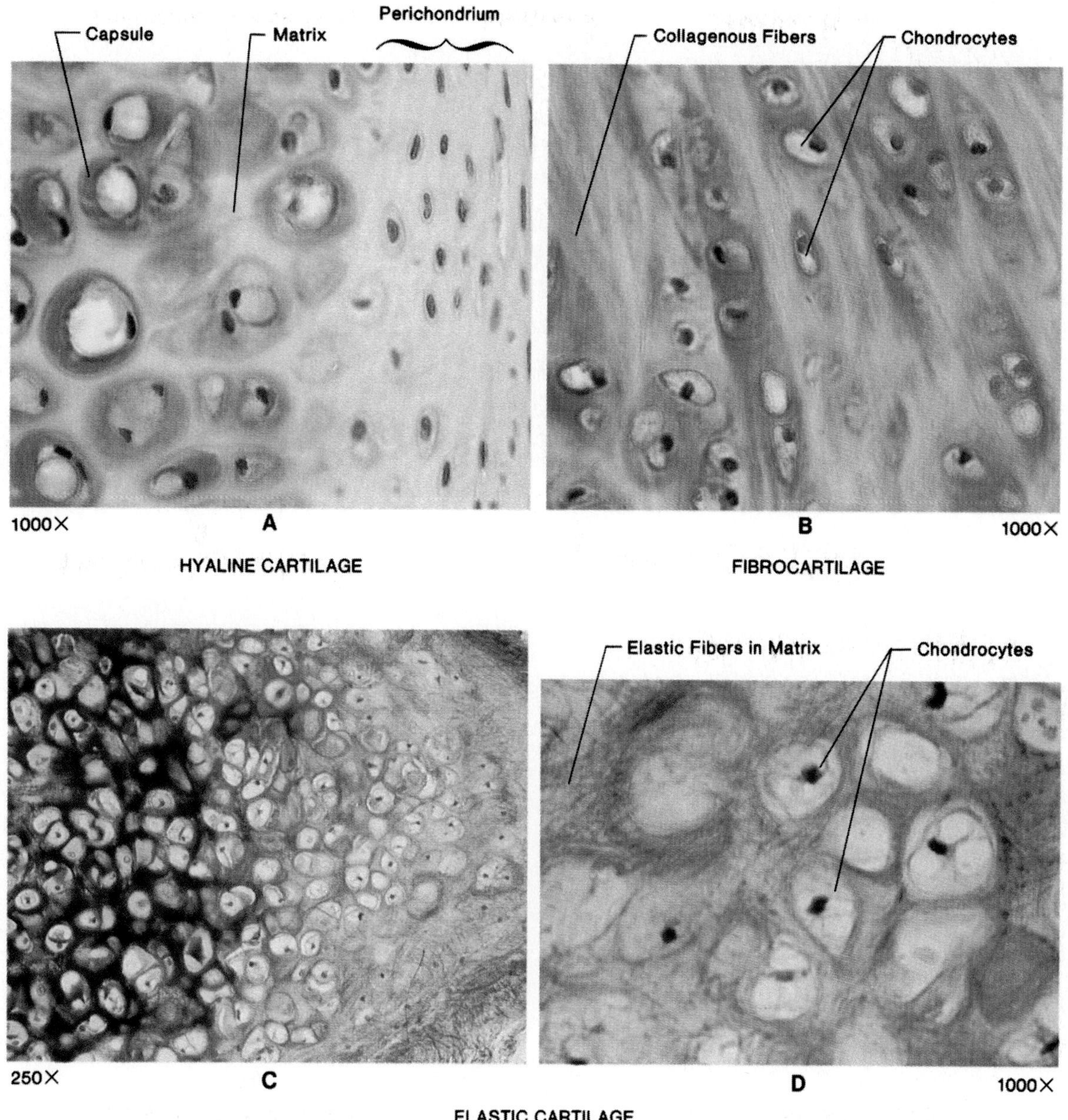

All types of cartilage are similar in that their chondrocytes are contained in smooth-walled spaces called lacunae. Lacunae walls, called capsules, are quite dense and stain darker than surrounding matrix. Note that chondrocytes appear to have large vacuoles. This apparent vacuolation is an artifact of tissue preparation, due to the leaching out of fat and glycogen by solvents used in the staining process. Note that micrograph A shows a dense layer of connective tissue, the perichondrium.

Except for the free surfaces of articular cartilage, all other hyaline cartilage structures are invested with a perichondrium.

The firmness of the matrix in hyaline cartilage is due to condroitin sulfate and collagen. Fibrocartilage (micrograph B) is much stronger than hyaline cartilage because of the preponderance of collagenous fibers in its matrix. Elastic fibers as well as collagenous fibers make elastic cartilage (micrographs C, D) more flexible.

Figure HA-5 Types of cartilage.

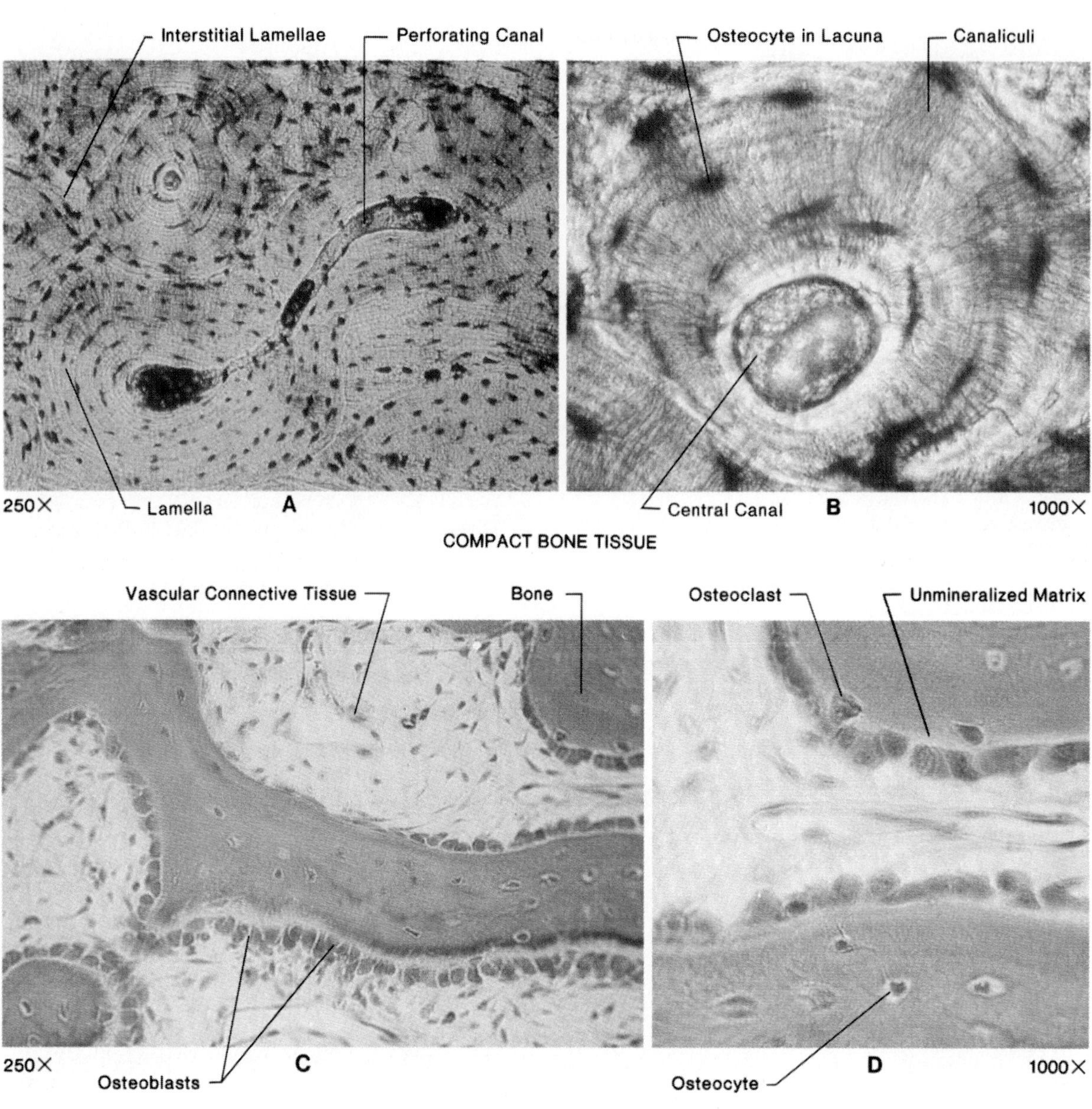

Bone forms (1) from cartilage and (2) from osteogenic mesenchymal connective tissue. Compact bone (micrographs A, B) of the long bones forms from cartilage. Bones of the skull, on the other hand, develop as shown in micrographs C and D and are often called membrane bones.

Note the loci of the different types of bone cells in the micrographs. Micrographs C and D reveal that osteoblasts in membranous bone formation are on the leading edge of the forming bone and secrete matrix. Surrounding mesenchymal cells then add reticular fibers to the matrix. Finally, osteoblastic activity is responsible for mineralization.

Osteoclasts are larger, multinucleated cells that shape bone structure by bone resorption. A clear area (Howship's lacuna) surrounds osteoclasts and is evidence of mineral resorption. The osteoclast in micrograph D is in an early stage of development; thus, Howship's lacuna is not very large. Once bone completely surrounds a bone cell, the cell is called an osteocyte. Nourishment in mature compact bone reaches osteocytes through canaliculi.

Figure HA-6 Bone histology.

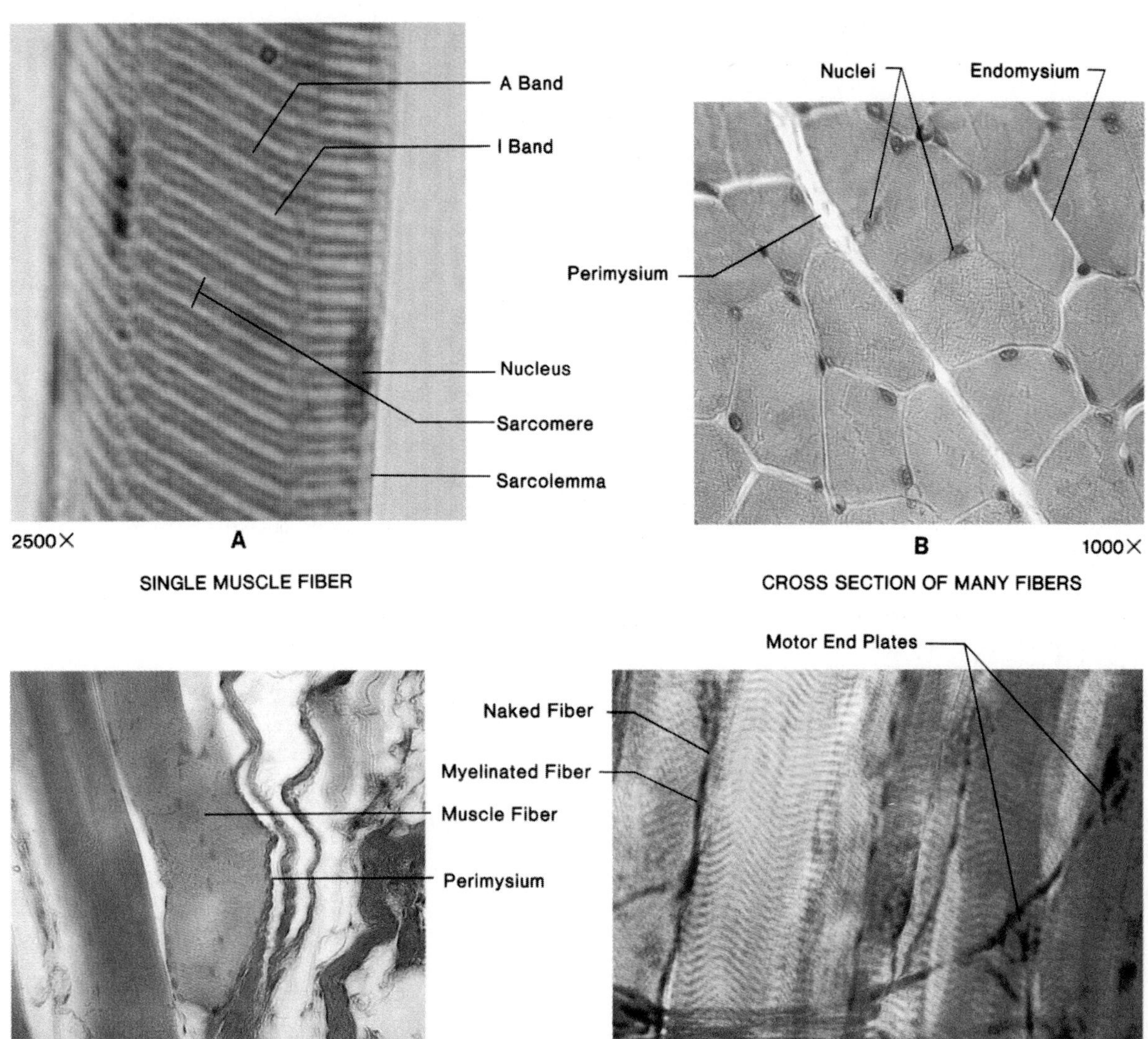

Note in micrograph A that one sarcomere consists of one A band and two I bands. These bands are visual manifestations of the intricate arrangement of actin and myosin myofilaments. Interaction between these filaments causes muscle contraction.

A significant characteristic of skeletal muscle cells is that they are multinucleated, or syncytial. Note that the nuclei are elongated and situated near the sarcolemma of the cell. The peripheral location of the nuclei shows up well in micrograph B.

Micrograph B shows that endomysium separates muscle fibers and that muscle fibers are grouped into bundles called fasciculi. Surrounding each fascicle is a layer of fibrous connective tissue called the perimysium.

Micrograph C shows how the fibrous connective tissue of the endomysium, perimysium, and epimysium is continous with the tendons that attach muscles to bone. The connective tissue of tendons, in turn, is continuous with the periosteum of bone.

Note in micrograph D how the myelinated motor nerve fibers lose their myelin sheaths and become "naked" where they join the motor end plates.

Figure HA-7 Skeletal muscle microstructure.

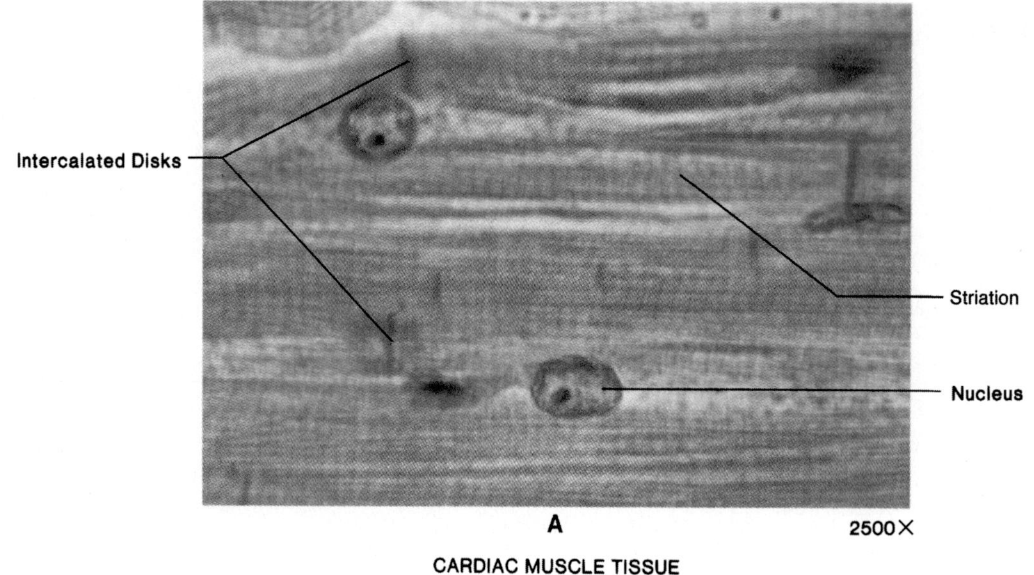

CARDIAC MUSCLE TISSUE

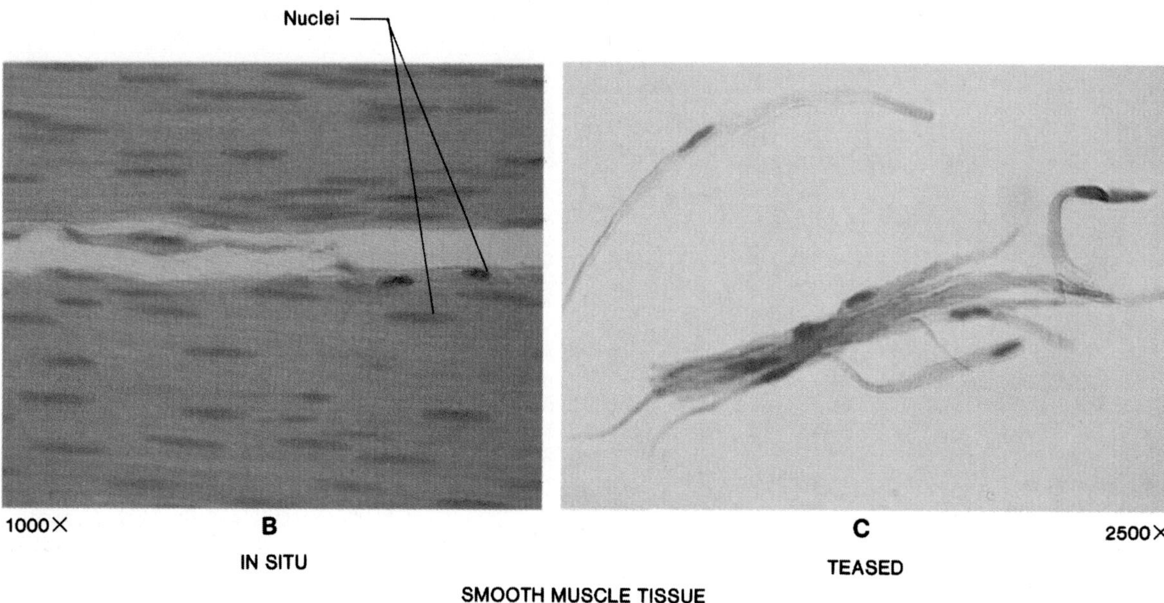

SMOOTH MUSCLE TISSUE

The distinct contrast of the striations in the cardiac muscle preparation in micrograph A is evidence that these fibers are in a state of complete relaxation; fibers prepared from tissue in contraction generally do not reveal striations as distinctly. Although this tissue may appear to be syncytial, it is not. Cardiac fibers have only one nucleus per fiber, with intercalated disks forming the limits for each unit.

The smooth muscle tissue in micrograph B is of a portion of the intestinal wall. Due to the closely packed nature of the cells in such structures, seeing individual cells is very difficult. Micrograph C shows how cells of this type of tissue appear when separated with a probe prior to staining. Each cell has a single nucleus and lacks the cross-striations of skeletal and cardiac muscle tissue.

Figure HA-8 **Cardiac and smooth muscle tissue.**

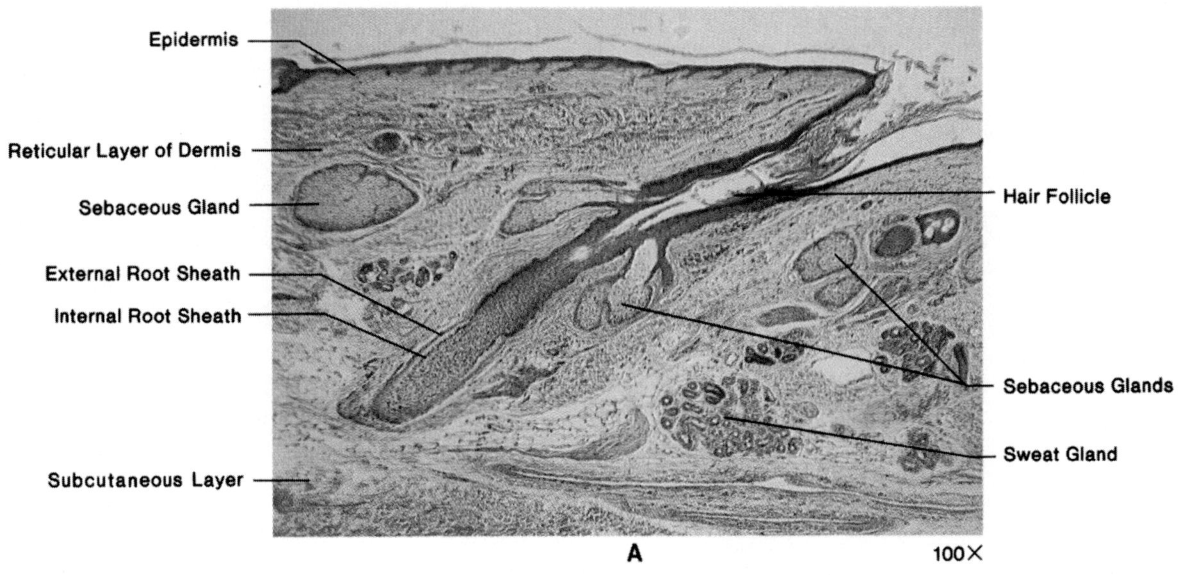

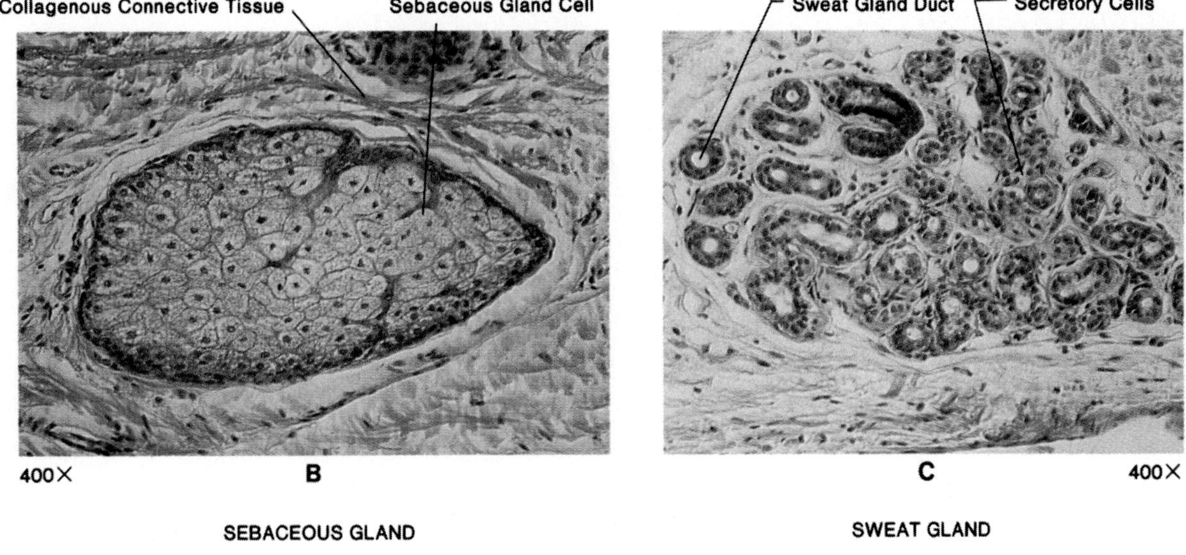

SEBACEOUS GLAND

SWEAT GLAND

These micrographs show the relationship of the sebaceous and sweat glands to other skin structures. The low magnification of micrograph A reveals that the epidermis is very thin, compared to the dermis. Except for the palms and soles of the feet, the epidermis is usually only 0.1 mm thick. The dermis or corium, which underlies the epidermis, varies from 0.3 to 4.0 mm thick in different parts of the body. Note that both the sebaceous and sweat glands are in the reticular layer of the dermis.

Sebaceous glands (micrograph B) consist of rounded masses of cells, cuboidal at the periphery and polygonal in the center. The secretion of sebaceous glands forms from the breakdown of the central cells into an oily complex that is forced out into the space between the follicle and hair shaft.

Sweat glands (micrograph C) are tubuloalveolar structures lined with cuboidal or columnar epithelium. Although not shown in micrograph A, sweat glands open directly onto the skin surface.

Figure HA-9 Scalp histology.

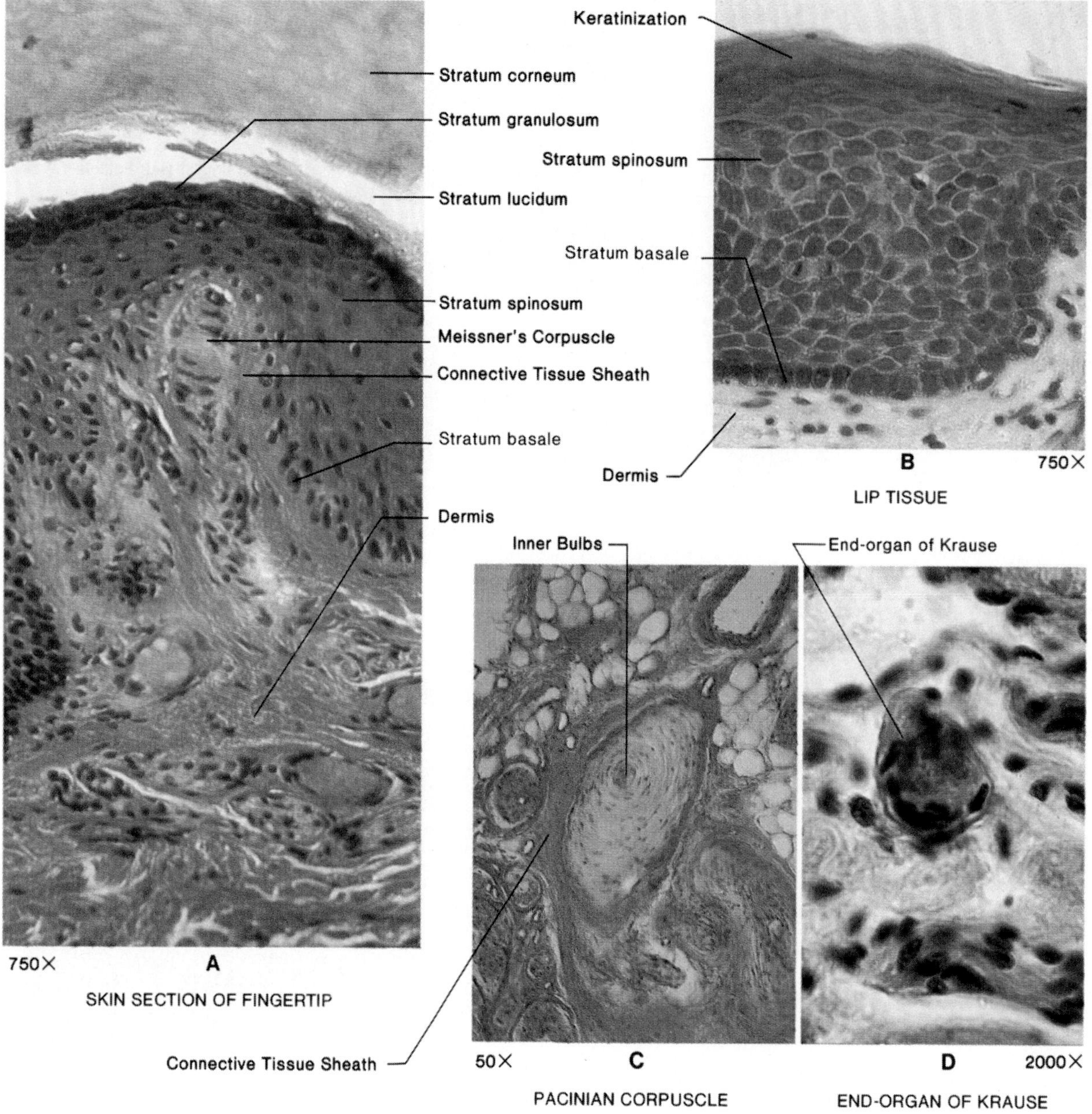

A comparison of the skin of the scalp (figure HA-9), a fingertip (micrograph A), and the lip (micrograph B) reveals that the integument differs in construction in different regions of the body. These differences are due to the presence or absence of hair and to the degree of keratinization. The fingertips, palms, and soles of the feet have epidermal layers with a thick stratum corneum and no hair follicles. Skin on the lips has some keratinization, but to a much lesser degree than the fingertips. Three receptors of the skin are illustrated here: Meissner's, Pacinian, and Krause corpuscles. Meissner's corpuscle (micrograph A) is in the papillary layer of the dermis. Note that a sheath of connective tissue encases it. Meissner's corpuscles are sensors of discriminative touch.

Pacinian corpuscles (micrograph C) consist of laminated collagenous material and several inner bulbs. Pressure on the lamina triggers impulses in nerve endings in the bulbs. Pacinian corpuscles are so large they can be seen without magnification.

End-organs of Krause are small corpuscles that are sensitive to touch. They are found in mucous membranes (e.g. mouth, lips).

Figure HA-10 Skin structure.

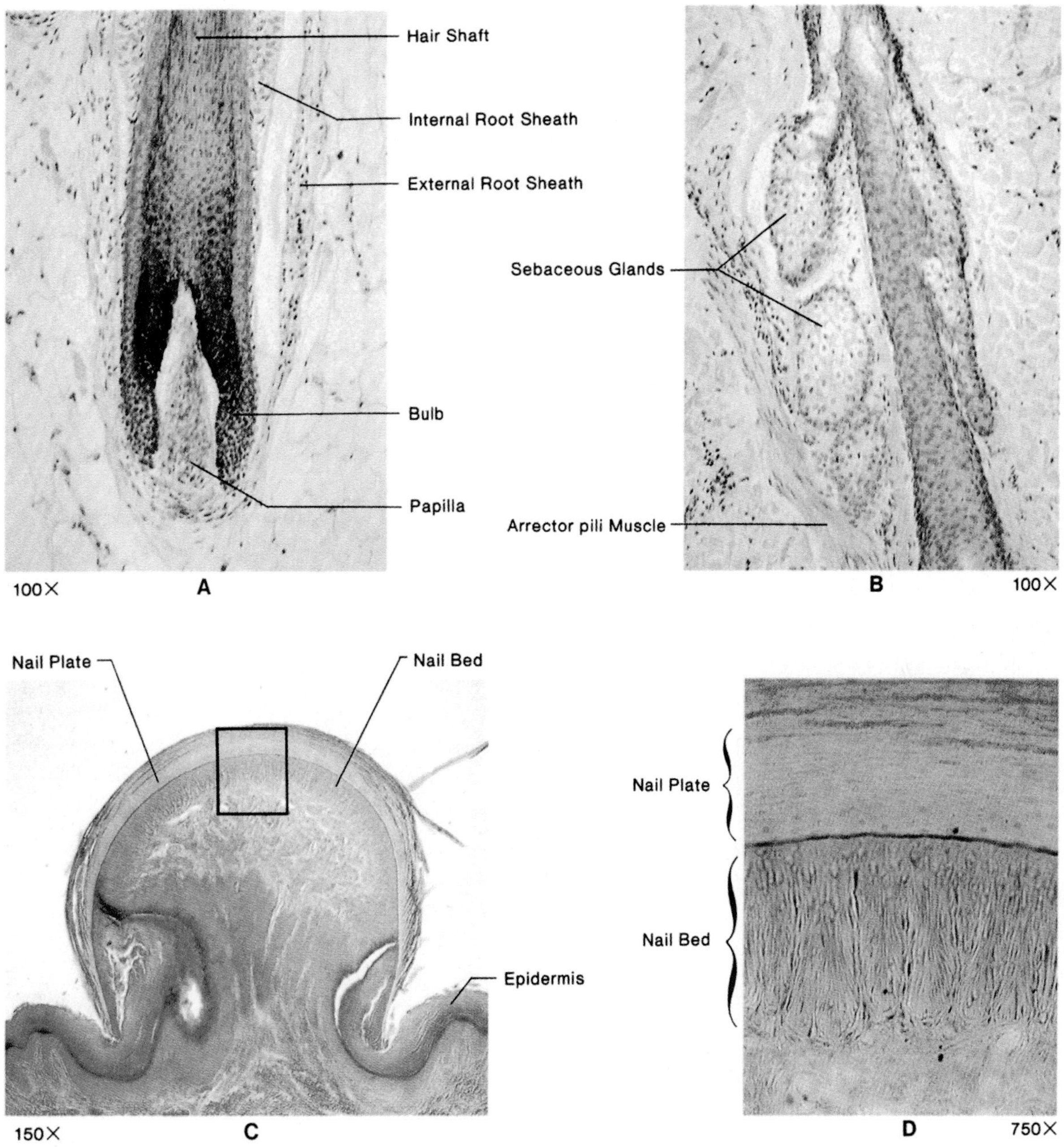

Hair consists of horny threads derived from the epidermis. A tube, the follicle envelops the hair shaft. Three components of the hair follicle (micrograph A) are the internal root sheath, an external root sheath, and a connective tissue sheath. The two inner sheaths are derived from the epidermis; the outer connective tissue sheath forms from the dermis.

Micrograph B shows the relationship of the arrector pili muscle to the sebaceous glands. These small muscles, which consist of smooth muscle tissue, help force sebum out into the hair follicle.

Nails are a modification of the epidermis. The fingernail in micrograph C is from an infant. Microscopically, the body of the nail consists of several layers of flattened, clear cells that take the place of the stratum corneum of skin. These cells are harder than stratum corneum layers and have shrunken nuclei.

The nail bed under the nail plate is the epidermis, but lacks the stratum granulosum and stratum lucidum; only the deeper epidermal layers are present. Micrograph D reveals its structure.

Figure HA-11 Hair and nail structure of the integument.

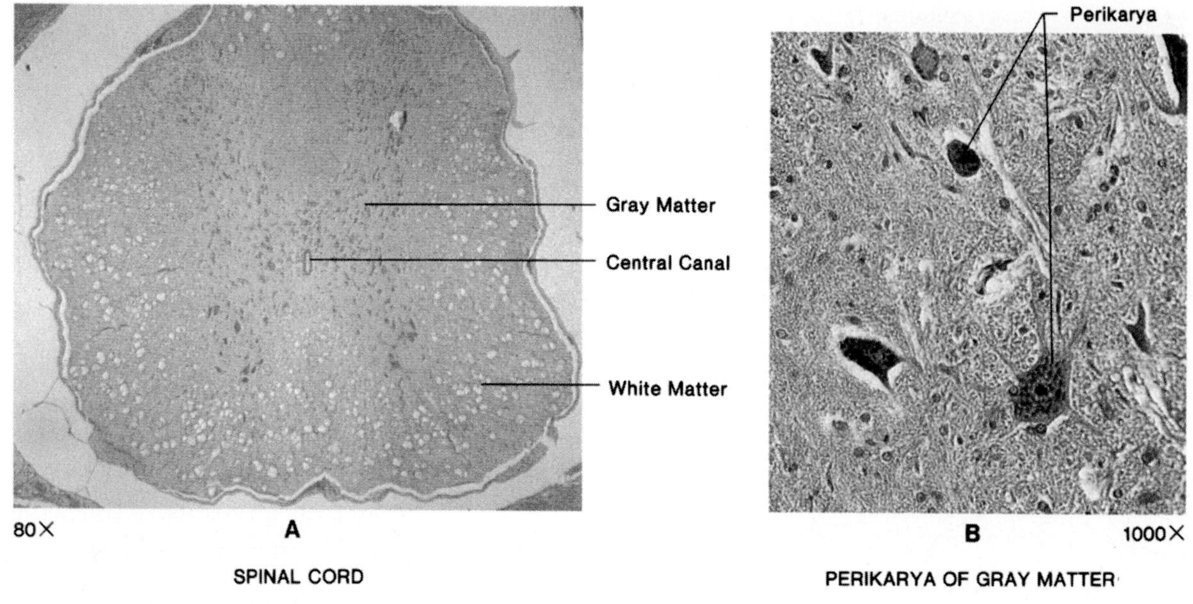

SPINAL CORD

PERIKARYA OF GRAY MATTER

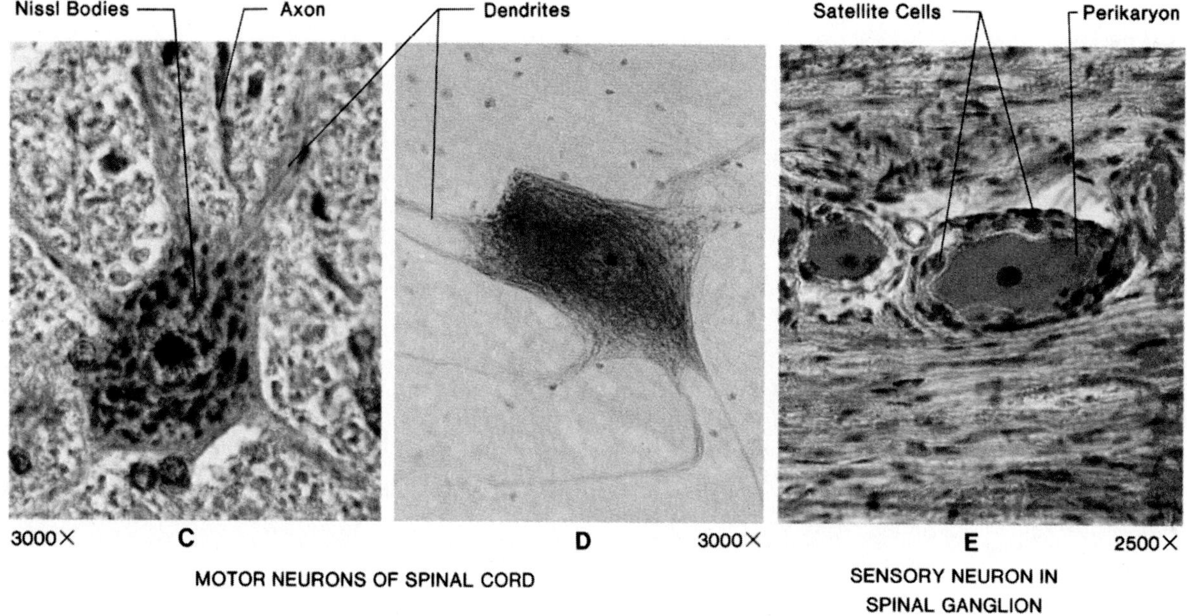

MOTOR NEURONS OF SPINAL CORD

SENSORY NEURON IN SPINAL GANGLION

The perikarya seen in micrographs B, C, and D are all of the gray matter of the spinal cord. While the one in micrograph C is from a cord section, the one in micrograph D is of a smear of spinal cord tissue. Note in micrograph C that the neuron has only one axon and several dendrites. This neuron also exhibits a large number of Nissl bodies. The netlike appearance of the cytoplasm in micrograph D is due to the large number of neurofibrils present.

Micrograph E is a section through a sensory (spinal) ganglion that shows a sensory neuron. The satellite cells that surround the perikaryon are often referred to as capsule cells.

Figure HA-12 Neurons of the spinal cord and spinal ganglia.

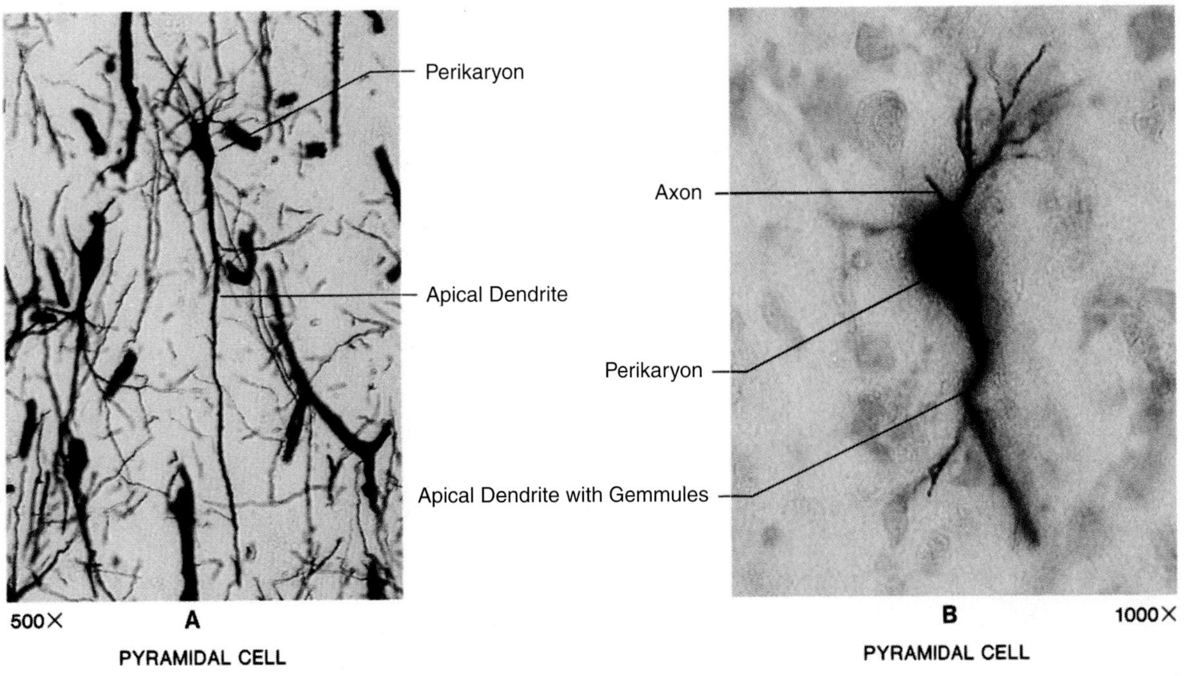

PYRAMIDAL CELL

PYRAMIDAL CELL

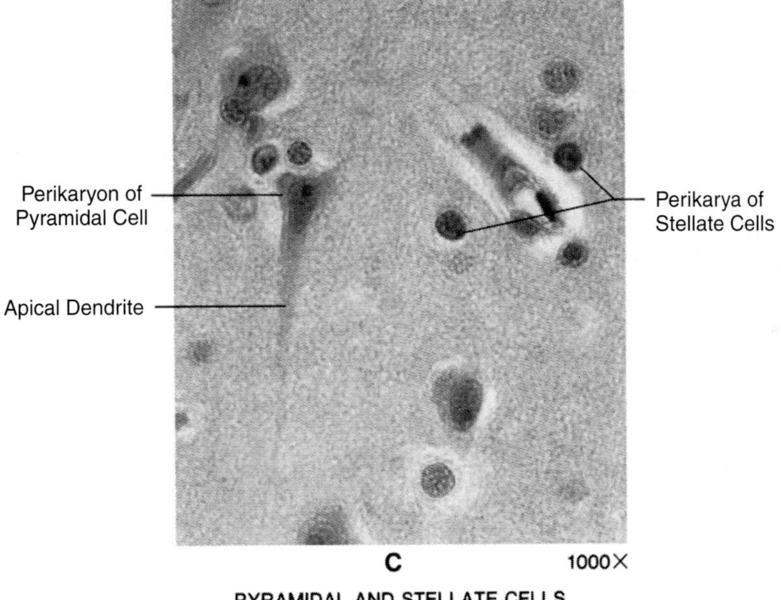

PYRAMIDAL AND STELLATE CELLS

The principal nerve cells of the cerebrum are pyramidal cells. Note in each of the photomicrographs that a long apical dendrite extends from the perikaryon. The apical dendrite is always oriented toward the surface of the cerebrum. Micrograph B shows evidence of "gemmules" on the surface of the dendrite. Gemmules are small processes that greatly increase the surface area of dendrites, allowing large neurons to receive as many as 100,000 separate axon terminals or synapses.

Note the difference in size between the axon and dendrites in micrograph B. The axon is much smaller in diameter and has a smoother surface due to the absence of gemmules.

The stellate cells in micrograph C are association neurons that provide connections between pyramidal cells of the cerebrum. These interneurons are also present in the cerebellum, where they link Purkinje cells.

Figure HA-13 Neurons of the cerebrum.

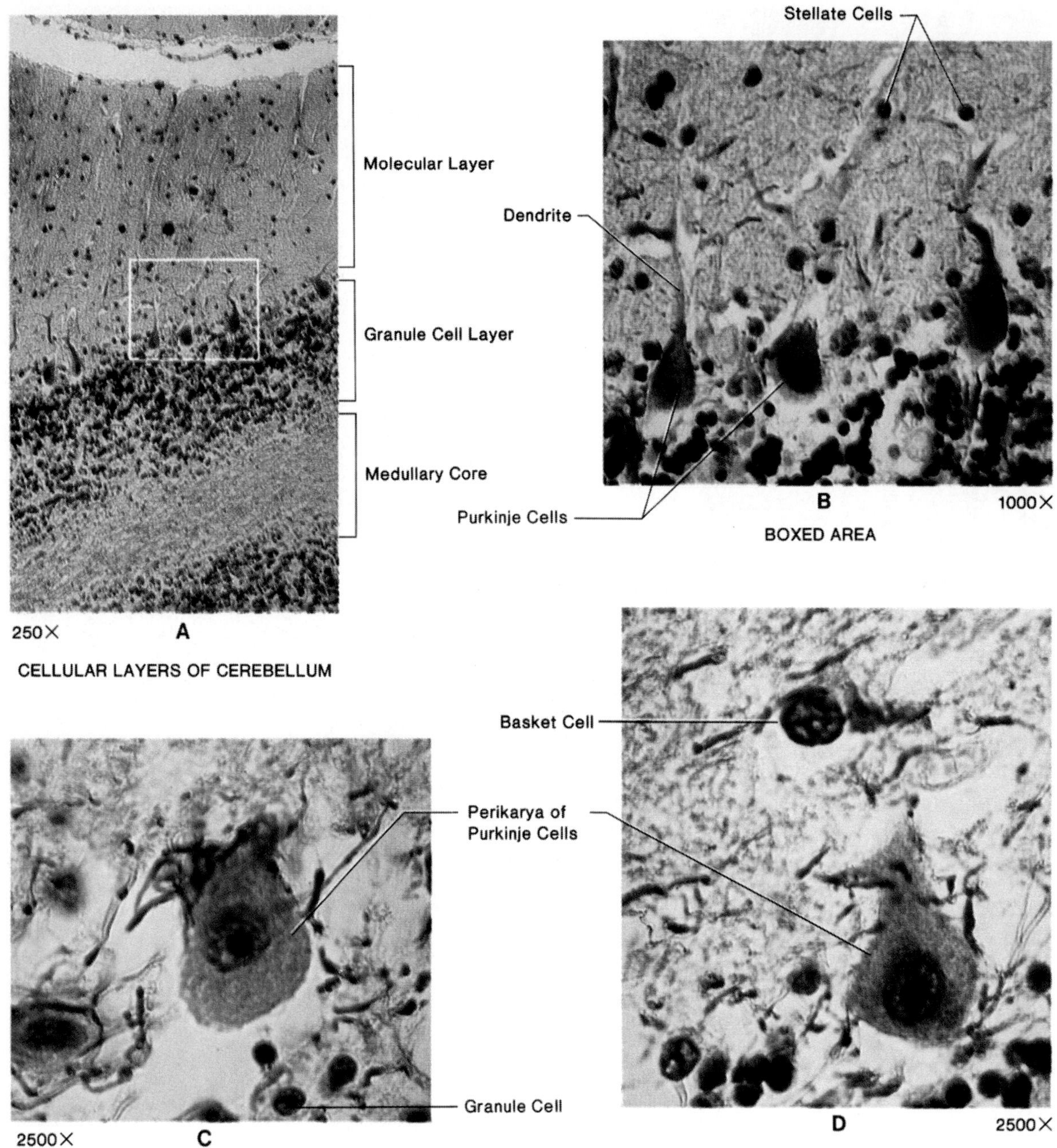

CELLULAR LAYERS OF CEREBELLUM

PURKINJE, BASKET, AND GRANULE CELLS

Micrograph A is a section through the cerebellum that depicts an outermost molecular layer, a middle granule cell layer, and an inner medullary core.

The most distinctive neurons of the cerebellum are the Purkinje cells. Note in micrograph A that these neurons form a layer deep within the molecular layer near the edge of the granule cell layer. Each flask-shaped cell has a thick dendrite directed toward the cerebellar cortex. A short distance from the perikaryon, the dendrite divides into two thick branches, which, in turn, divide many times to arborize into the molecular layer. The axon of each Purkinje cell extends from the perikaryon through the granular and medullary layers, and finally, through collaterals, reenters the molecular layer to contact other Purkinje cells.

Stellate, basket, and granule cells are all multipolar association neurons. Note the proximity of the perikarya of the basket cells to the Purkinje cells.

Figure HA-14 Neurons of the cerebellum.

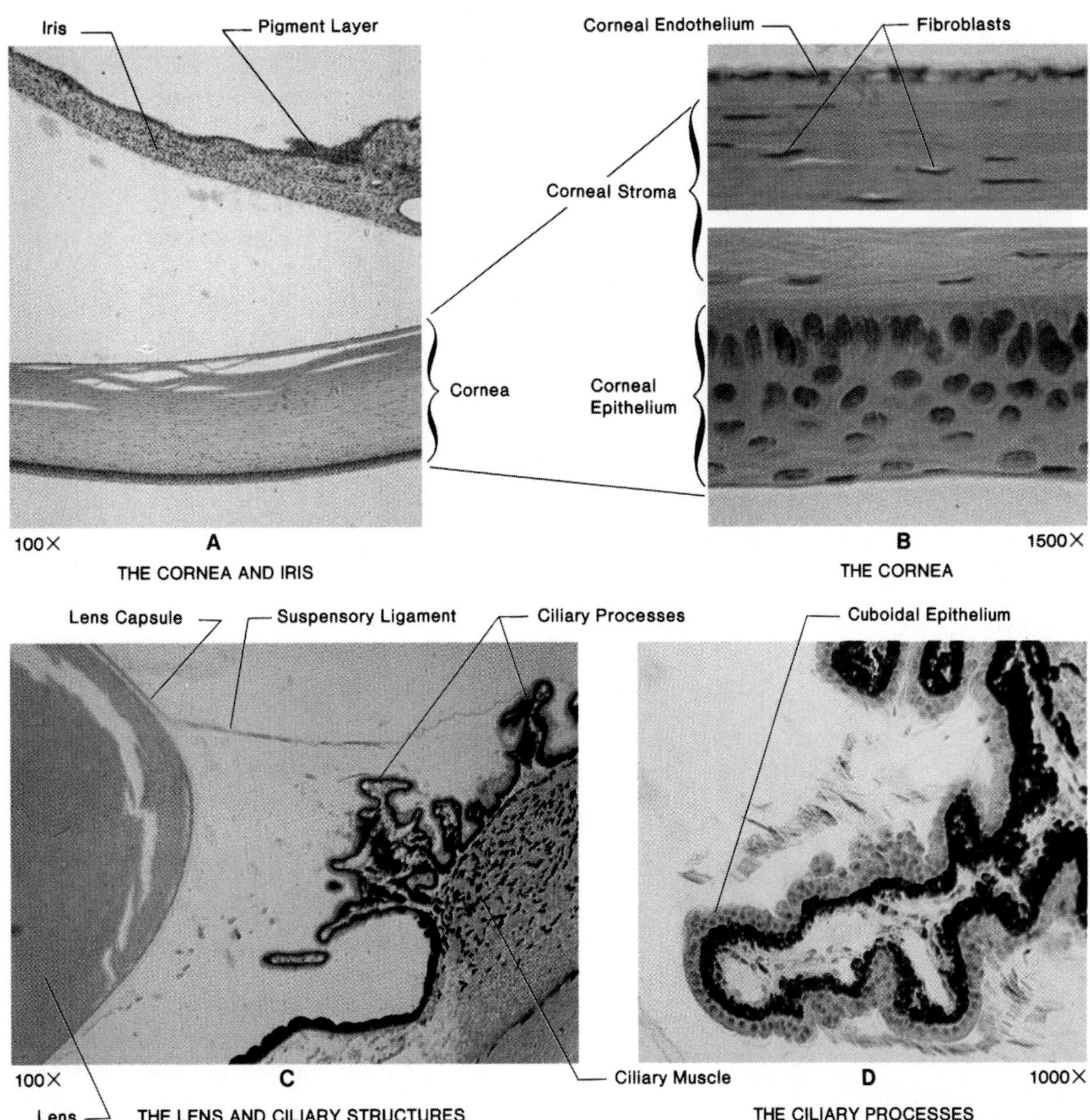

Micrograph A shows the relationship of the iris to the cornea. If a portion of the lens were shown in this photomicrograph, it would be above the iris, since the iris lies between the lens and the cornea.

Micrograph B is a compacted enlargement of the cornea, in which many of the middle layers have been removed so that all significant layers can be seen. Note that the outer corneal epithelium consists of stratified squamous epithelium and that the inner corneal endothelium consists of a single layer of low cuboidal cells. Between these two layers is the corneal stroma (*substantia propria*), which makes up 90% of the thickness of the cornea. This transparent stroma is made up of collagenous fibrils and fibroblasts, all held together with a mucopolysaccharide cement.

Note in micrograph C that the lens is encased in an elastic capsule to which the suspensory ligament attaches. The lens forms from epithelial cells that elongate and lose their nuclei. Note also that cuboidal epithelial cells cover the ciliary processes in micrograph D. These cells produce the aqueous humor that fills the space between the lens and the cornea.

Figure HA-15 The cornea, lens, iris, and ciliary structures.

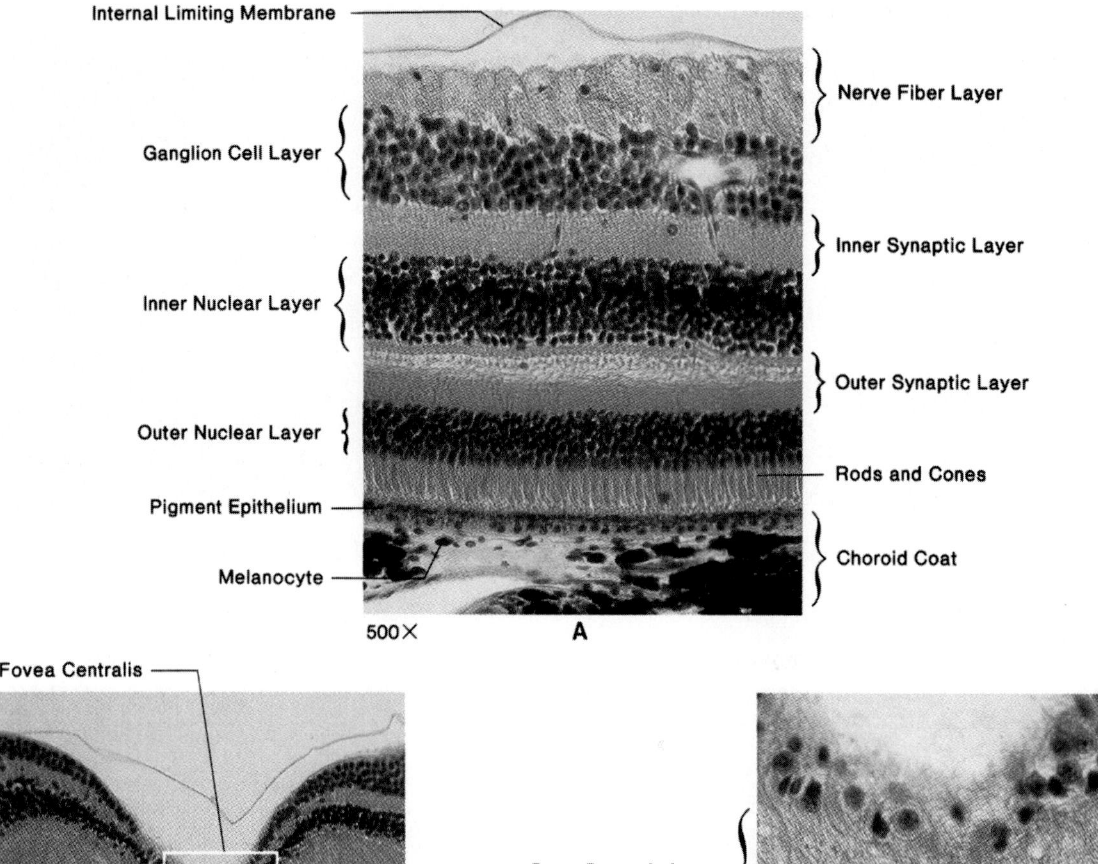

Micrograph A shows the various neuronal layers of the retina. Note that the photoreceptors (rods and cones) are in the deepest layer next to the choroid coat. Light striking the retina must pass through all of the upper layers to reach the receptors.

The nuclei in the outer nuclear layer are of the rods and cones. The outer synaptic layer is where axons of the rods and cones synapse with dendrites of bipolar cells and processes of horizontal cells. The inner nuclear layer contains nuclei of bipolar neurons and association neurons (horizontal and amacrine cells). The inner synaptic layer is where synaptic connections are made between the cells of the inner nuclear layer and the ganglion cells. The nerve fiber layer consists of nonmyelinated axons of the ganglion cells. These fibers converge on the optic disk to form the optic nerve.

Micrographs B and C reveal the nature of the fovea centralis. Note that the receptors in the fovea consist only of cones and that the inner synaptic and nerve fiber layers are lacking.

Figure HA-16 The retina of the eye.

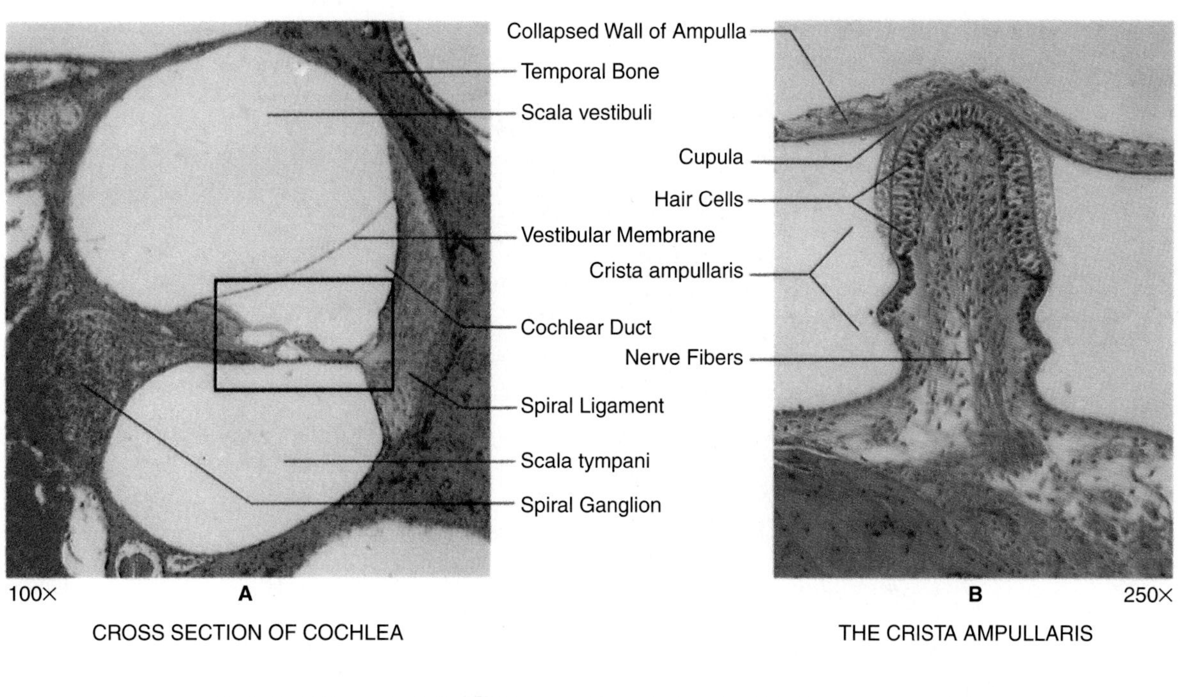

CROSS SECTION OF COCHLEA

THE CRISTA AMPULLARIS

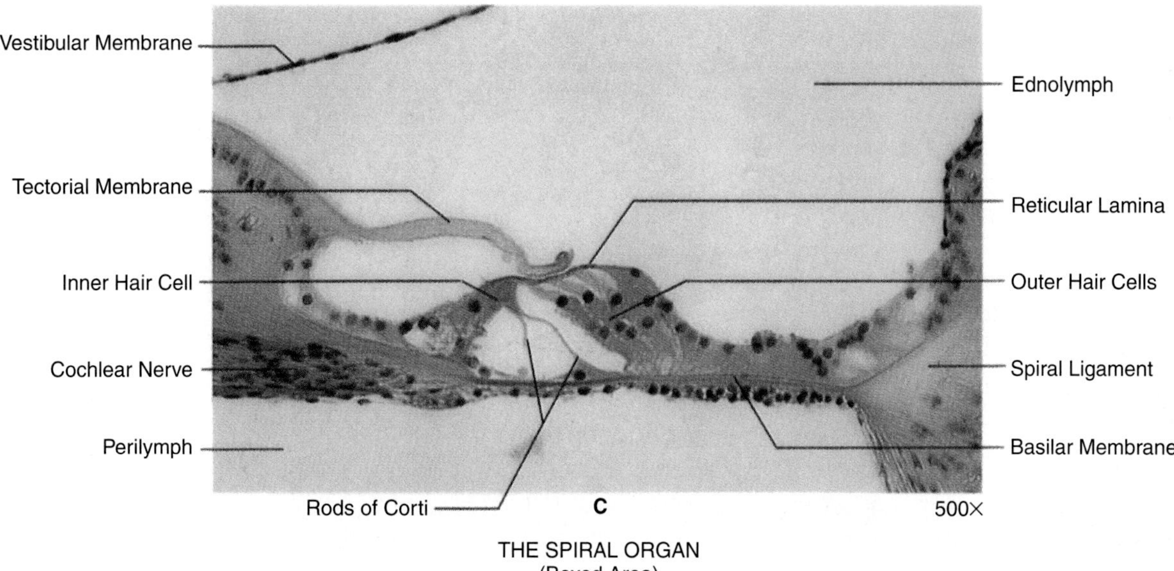

THE SPIRAL ORGAN
(Boxed Area)

The cochlea of the ear lies within the temporal bone. Along its outer wall is a spiral ligament (micrograph A) that forms from periosteum of the bone. Note that fibers of the spiral ligament are continuous with the basilar membrane. Observe that the cochlea has three chambers: scala vestibuli, scala tympani, and cochlear duct. The cochlear duct is bounded on one side by the vestibular membrane and on the other side by the basilar membrane. Endolymph is present in the cochlear duct; perilymph fills the scala vestibuli and scala tympani.

The spiral organ, or organ of Corti, shown in micrograph C is an enlargement of the boxed area in micrograph A. Note how the gelatinous flap of the tectorial membrane lies in contact with hairs that emerge from the hair cells. Discrimination of sound frequencies results from nerve impulses sent to the brain via the cochlear nerve, in conjunction with vibrations in the endolymph and basilar membrane.

In micrograph B, the cupula was destroyed during slide preparation. Other structures in the crista ampullaris are normal.

Figure HA-17 The cochlea and crista ampullaris.

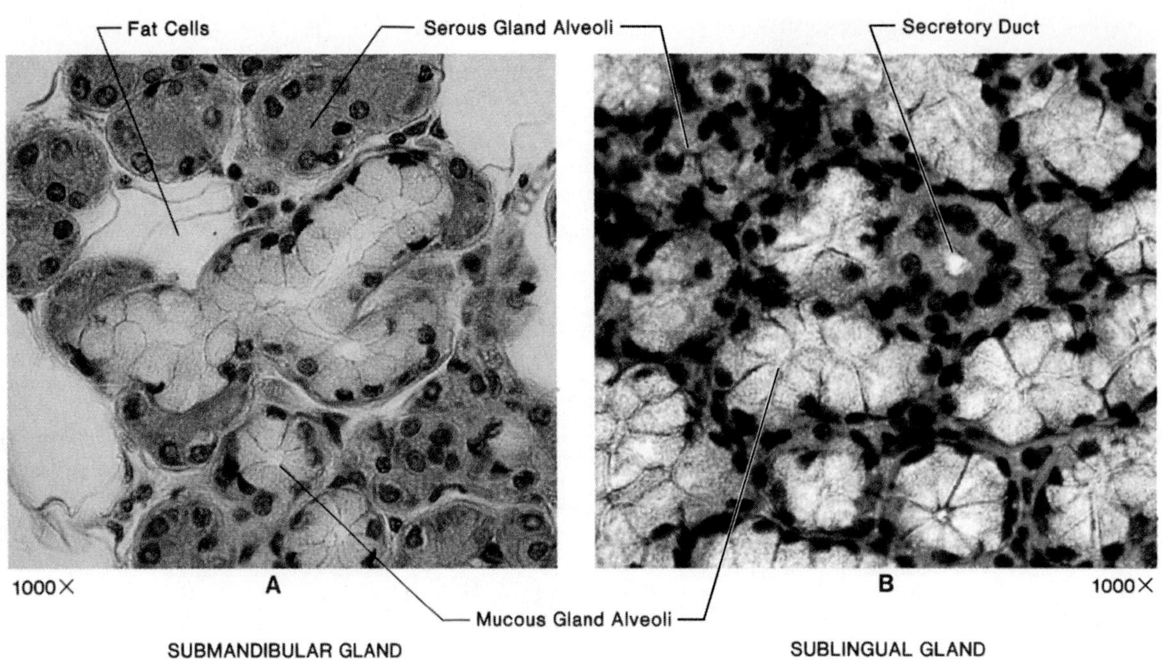

SUBMANDIBULAR GLAND

SUBLINGUAL GLAND

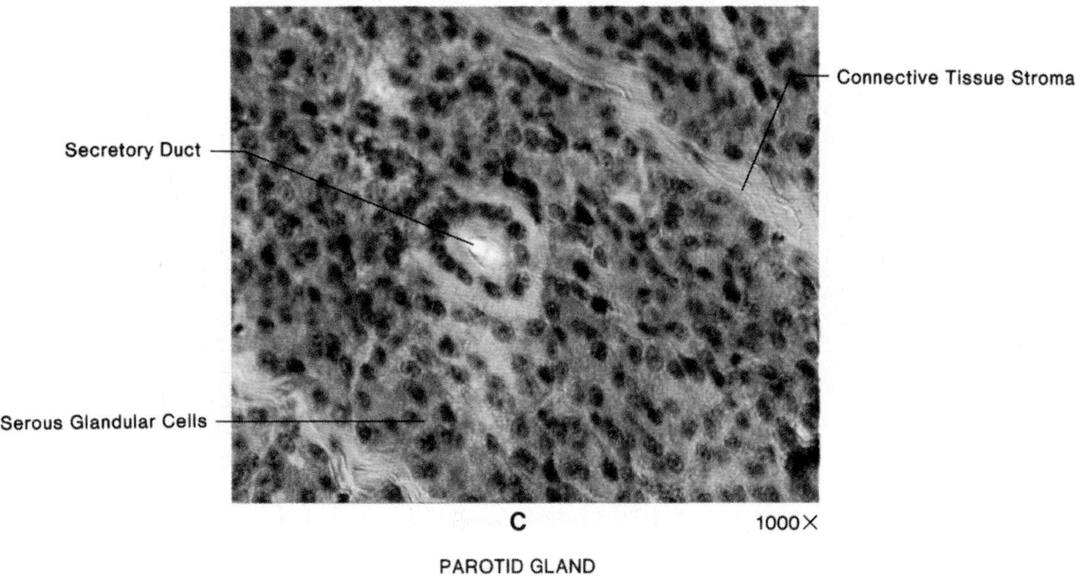

PAROTID GLAND

Microscopic observation easily determines whether a salivary gland produces mucus or the less viscous serous fluid. Note in micrographs A and B that mucus-producing cells are much paler and larger than cells that produce serous fluid. The presence of both types of secretory cells in the submandibular and sublingual glands indicates that these glands secrete a mixture of mucus and serous fluid into the oral cavity. Although the submandibular gland is primarily serous, the sublingual produces mostly mucus. The absence of mucous acini in the parotid glands (micrograph C) indicates that these glands only produce serous fluid.

Figure HA-18 Histology of major salivary glands.

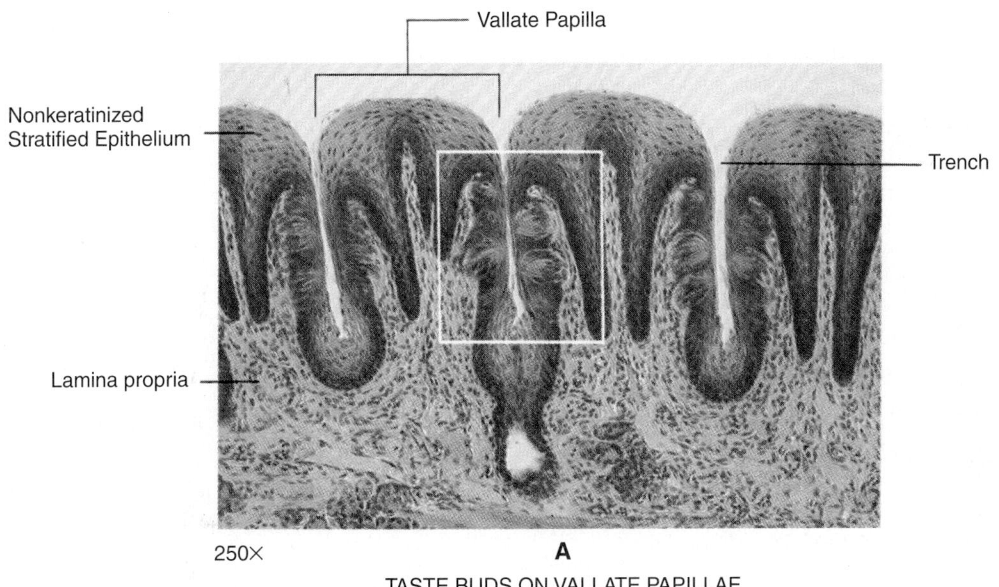

A
TASTE BUDS ON VALLATE PAPILLAE

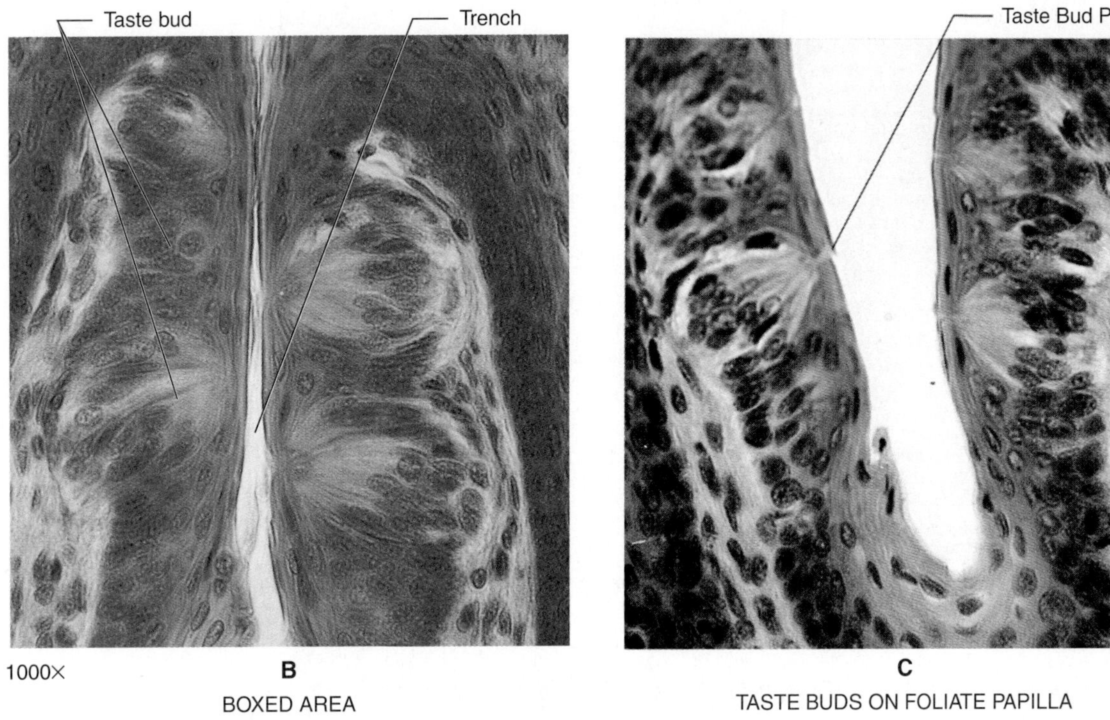

B
BOXED AREA

C
TASTE BUDS ON FOLIATE PAPILLA

The tongue may have as many as 20–30 vallate papillae. A deep trench surrounds each papilla (micrograph A). Since the walls of each papilla have taste buds covering the entire surface of the trench, hundreds of taste receptors may be present on these structures.

Note that each taste bud has a pore (micrograph C) that opens into the trench. At the base of each trench are duct openings from serous glands (glands of von Ebner) that keep the trench moist and assist in dissolving substances that stimulate the taste buds. To see nerve fibers leading into the sensory cells of a taste bud requires staining the tissue with a special silver dye. No such dye was used on these slide preparations.

Foliate papillae (micrograph C) are on the sides of the tongue. The superior clarity of the taste buds in this photomicrograph is due to superb tissue preparation.

Figure HA-19 Taste buds.

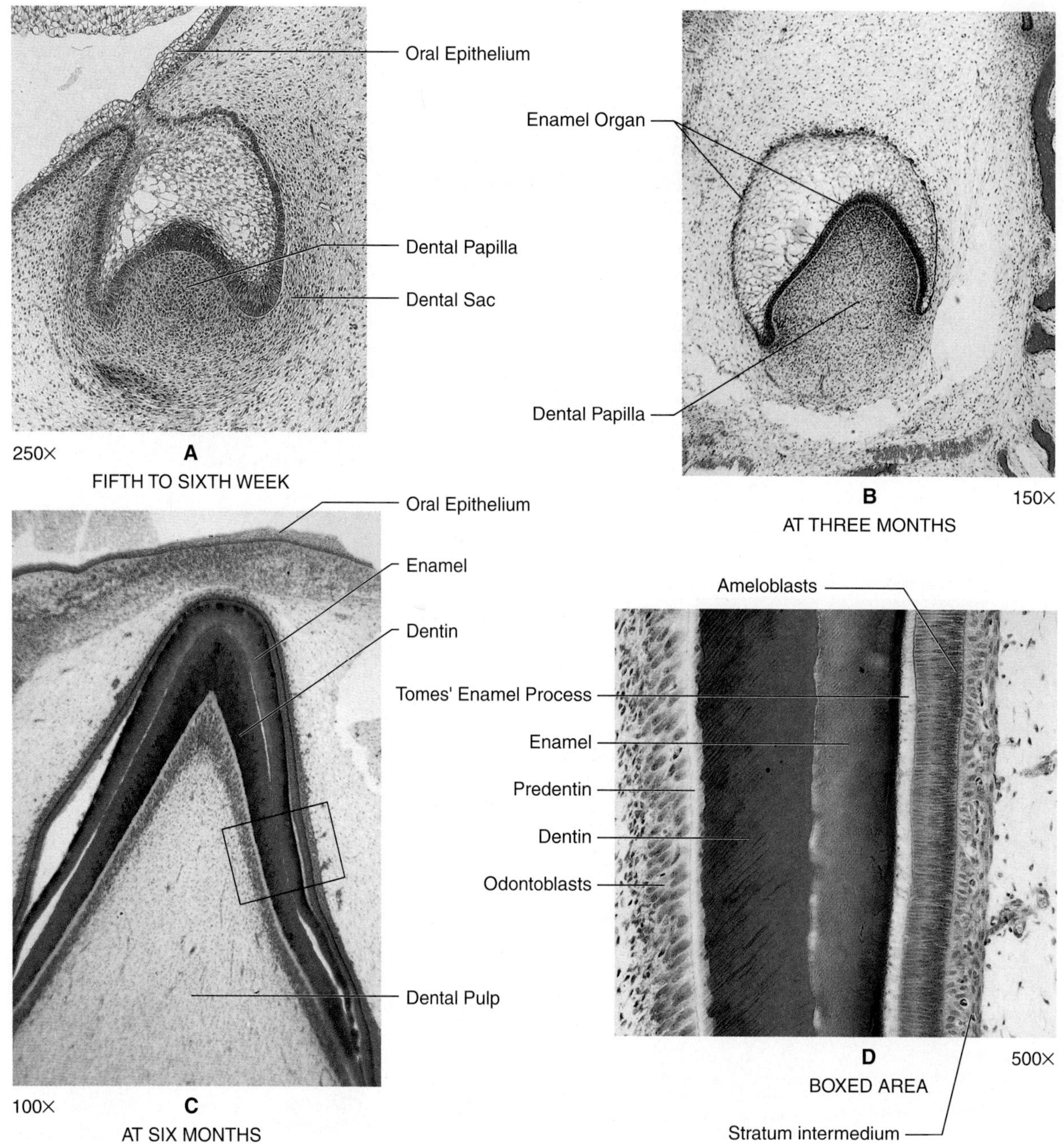

In each dental arch of the embryo, 10 tooth buds form by proliferation of cells in the oral epithelium. At first, the buds are solid and rounded, but as the cells multiply, they push inward (invaginate), as micrograph A shows. The oral epithelium (ectoderm) forms a two-layered structure called the enamel organ (micrograph B). Mesenchymal cells (mesoderm) in the distal portion of the bud push in to form the dental papilla. Micrograph C reveals that by the fifth or sixth month the enamel organ produces the enamel of the tooth, and the cells of the dental papilla differentiate to form dentin.

Dentin is laid down just before the appearance of enamel. A layer of elongated cells called odontoblasts form from mesenchymal cells of the dental papilla and produce dentin. Note in micrograph D that uncalcified dentin, or predentin, forms first. Predentin quickly converts to mature dentin, which consists of collagenous fibers and a calcified component called apatite.

Cells called ameloblasts produce enamel. Note in micrograph D that these cells produce a clear layer called Tomes' enamel process, which converts to mature enamel by calcification with mineral salts. As new enamel forms, the ameloblasts move outward from the dentin and odontoblasts. When the tooth erupts, the ameloblasts are sloughed off.

Figure HA-20 **Embryological stages in tooth development.**

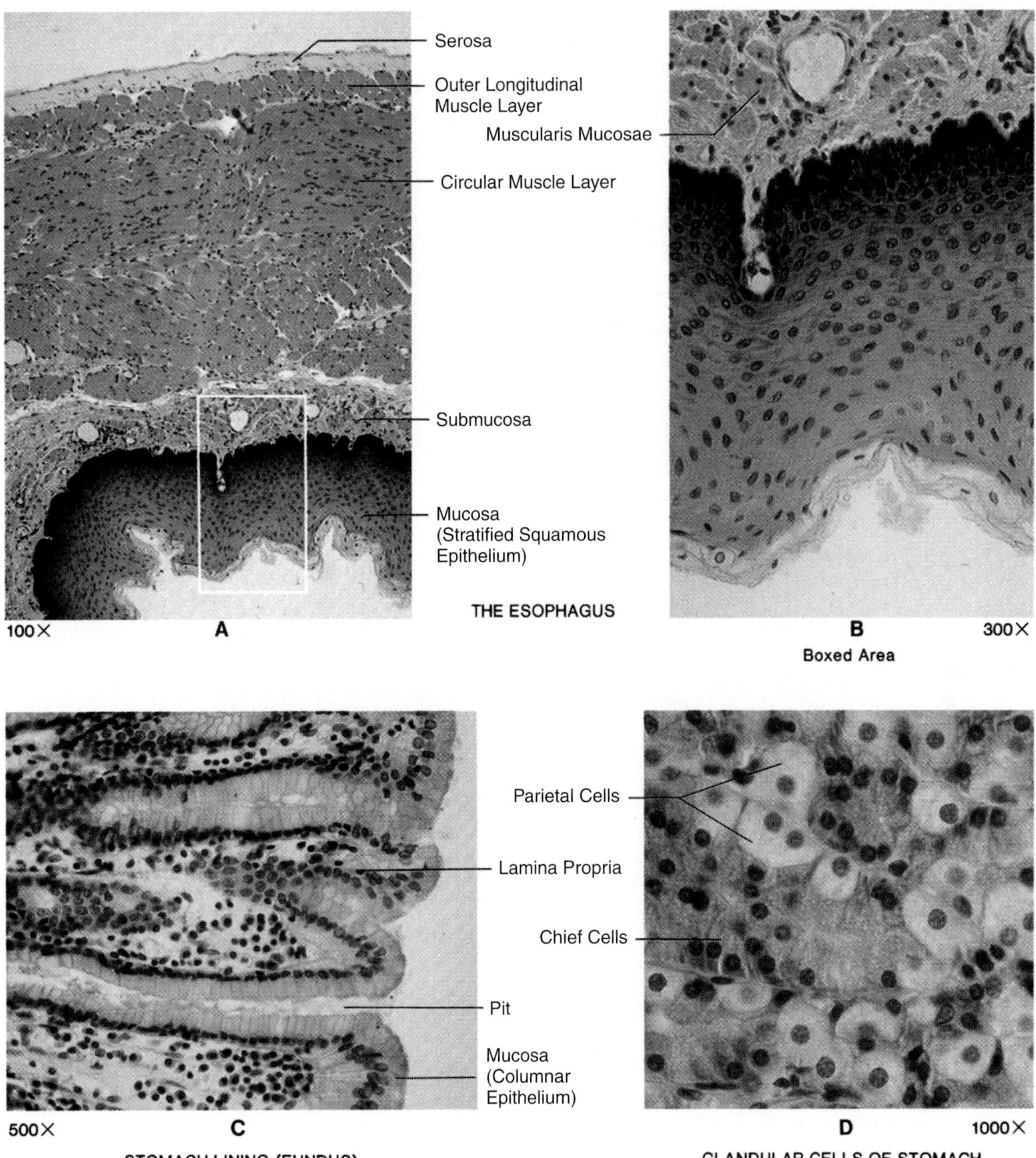

Throughout its length, the digestive tract consists of four coats: mucosa, submucosa, muscularis, and serosa (micrograph A). These layers differ in structure in different parts of the tract.

Note that the mucosa of the esophagus consists of stratified squamous epithelium, similar to that in the oral cavity. The submucosa is a layer of areolar tissue that contains mucous glands, nerves, blood vessels, and some muscle tissue. The muscularis layer consists of longitudinal and circular fibers.

The serosa consists of areolar and adipose tissue with a covering of mesothelium.

The mucosa of the stomach (micrograph C) differs from that of the esophagus in consisting of simple columnar epithelial cells. The surface of the mucosa contains numerous pits. Deep in the pits lie glands that contain parietal and chief cells (micrograph D). The chief cells are serous cells that produce the precursor of pepsin; the parietal cells produce the antecedent of hydrochloric acid and the intrinsic factor gastrin.

Figure HA-21 The esophagus and stomach.

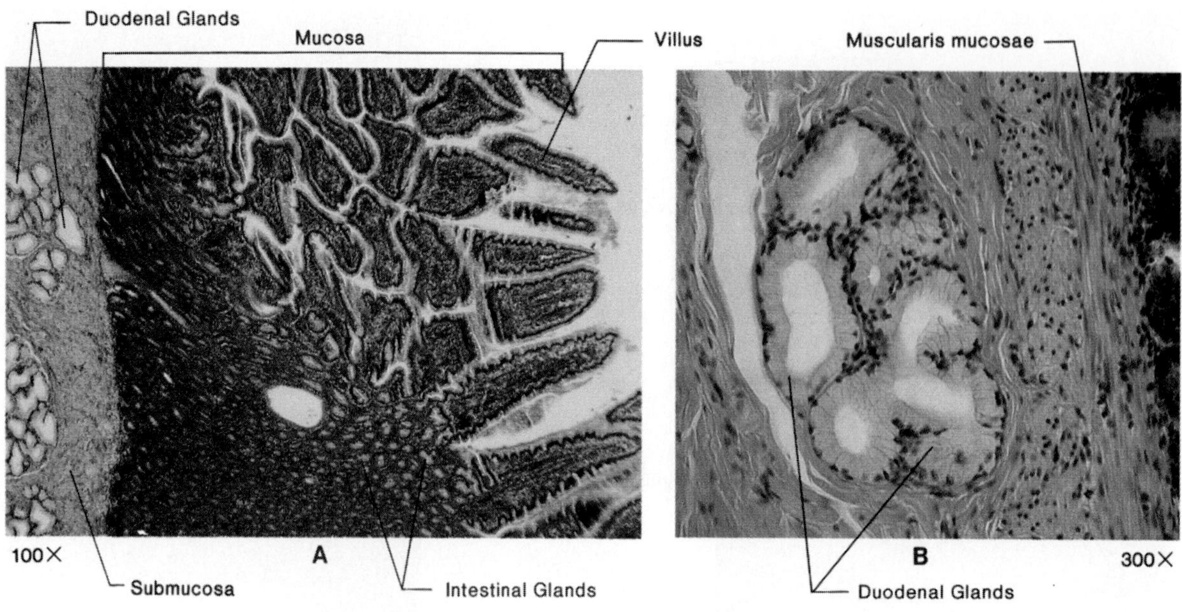

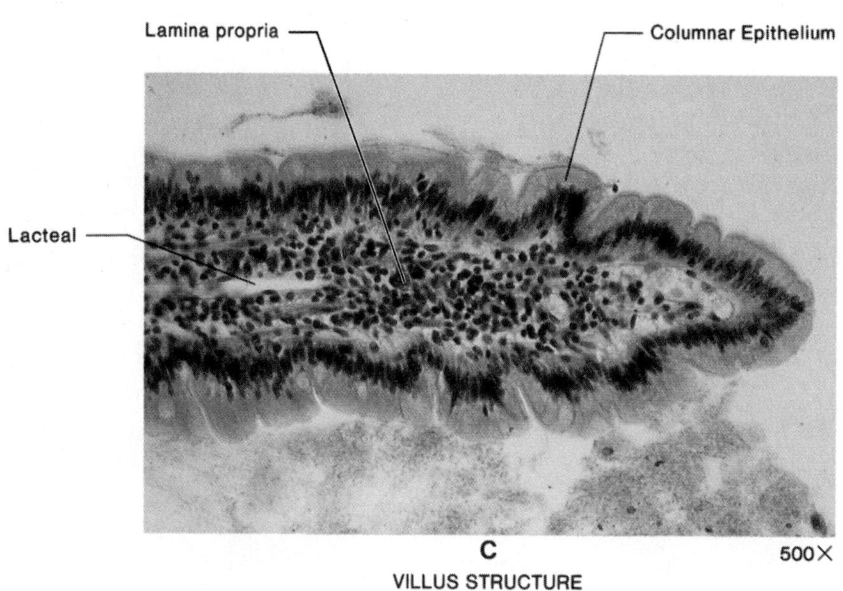

VILLUS STRUCTURE

The mucosa of the duodenum consists of the epithelium and the lamina propria. Within the lamina propria are numerous intestinal glands (micrograph A). Millions of small, fingerlike projections called villi cover the entire mucosal surface.

Where the lamina propria meets the muscularis mucosae, the mucosa ends. Note the large number of mucus-secreting duodenal glands (micrograph B) in the submucosa.

Micrograph C shows the structure of a single villus. The entire epithelium consists of simple columnar cells interspersed with goblet cells. The core of the villus is the lamina propria, which contains loose connective tissue, smooth muscle fibers, blood vessels, and a lymphatic vessel, the lacteal. The muscularis layer (not shown in micrograph C) contains an inner circular layer of smooth muscle tissue and an outer longitudinal layer of smooth fibers. The serosa covers the outer surface of the duodenum.

Figure HA-22 The duodenum.

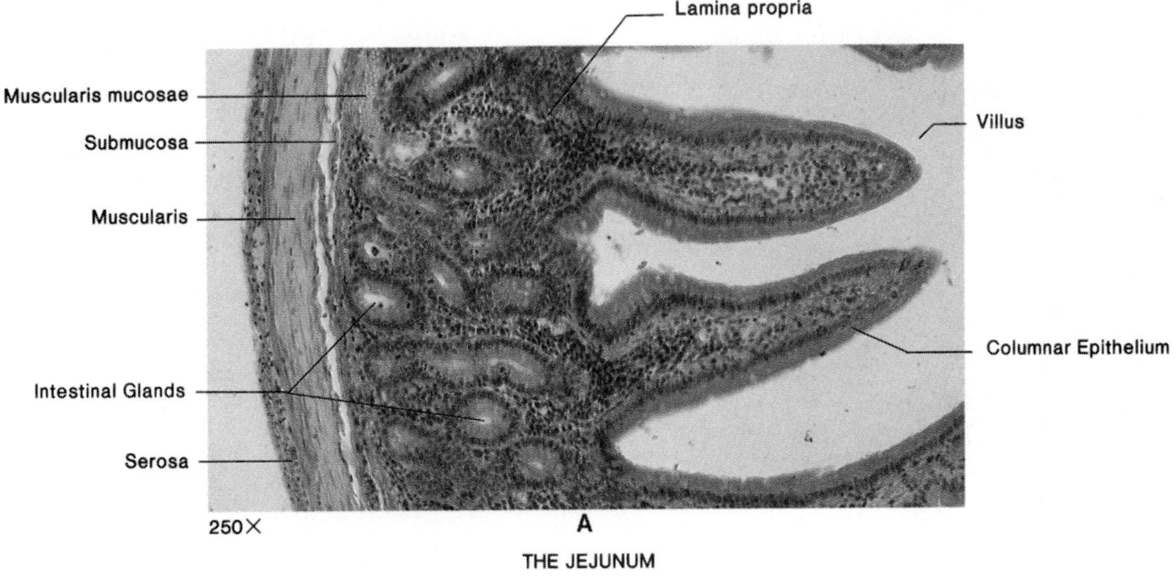

THE JEJUNUM

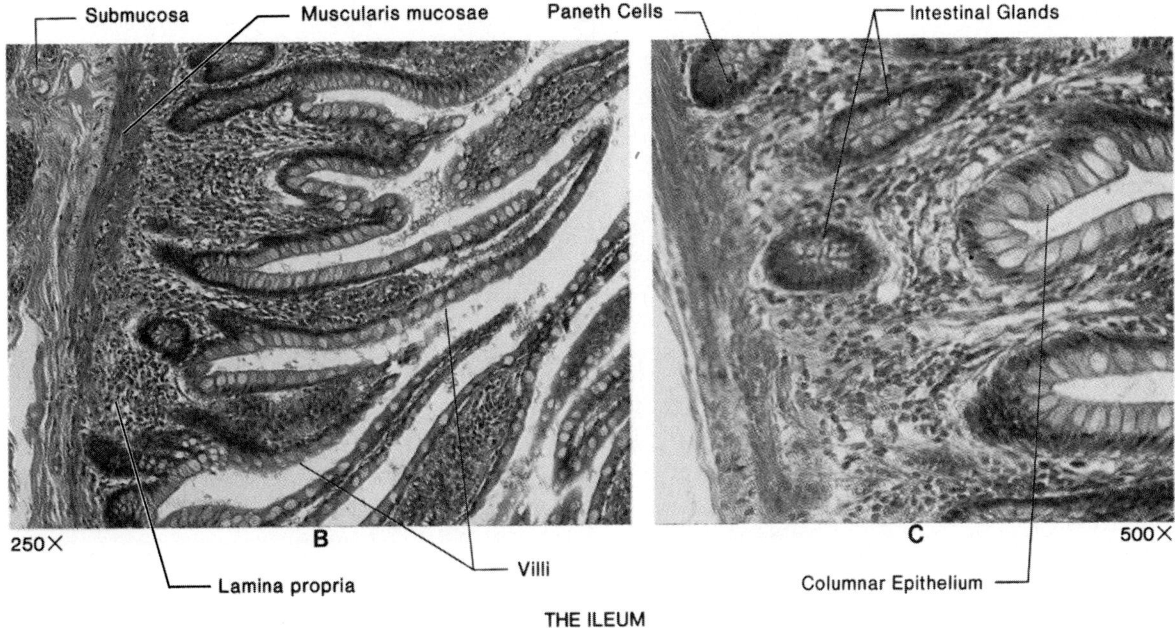

THE ILEUM

Note in micrograph A that the submucosa of the jejenum is very thin and lacks the digestive glands present in the duodenum. The mucosa, however, contains the same intestinal glands as the duodenum.

Observe that the intestinal glands in micrograph C have two kinds of cells: Paneth cells and simple columnar. Although the Paneth cells produce no digestive enzymes, they may have something to do with the production of lysozyme, an antibacterial enzyme. The columnar cells, which resemble the columnar cells of the epithelium, provide new cells for the surfaces of the villi as old cells are shed into the lumen of the intestine. The surfaces of the villi are renewed every few days in the human intestine.

The villi of the ileum (micrograph B) tend to be shorter and more club-shaped than those of the jejunum. Villi of the jejunum are often forked instead of leaflike as in micrograph A.

The epithelial cells of the mucosa are of two types: columnar absorbing cells and goblet cells. The absorbing cells are of the tall columnar variety with nuclei at the bases of the cells. These cells exhibit distinct brush borders (see figure HA-2). The goblet cells produce protective mucus.

Figure HA-23 The jejunum and ileum.

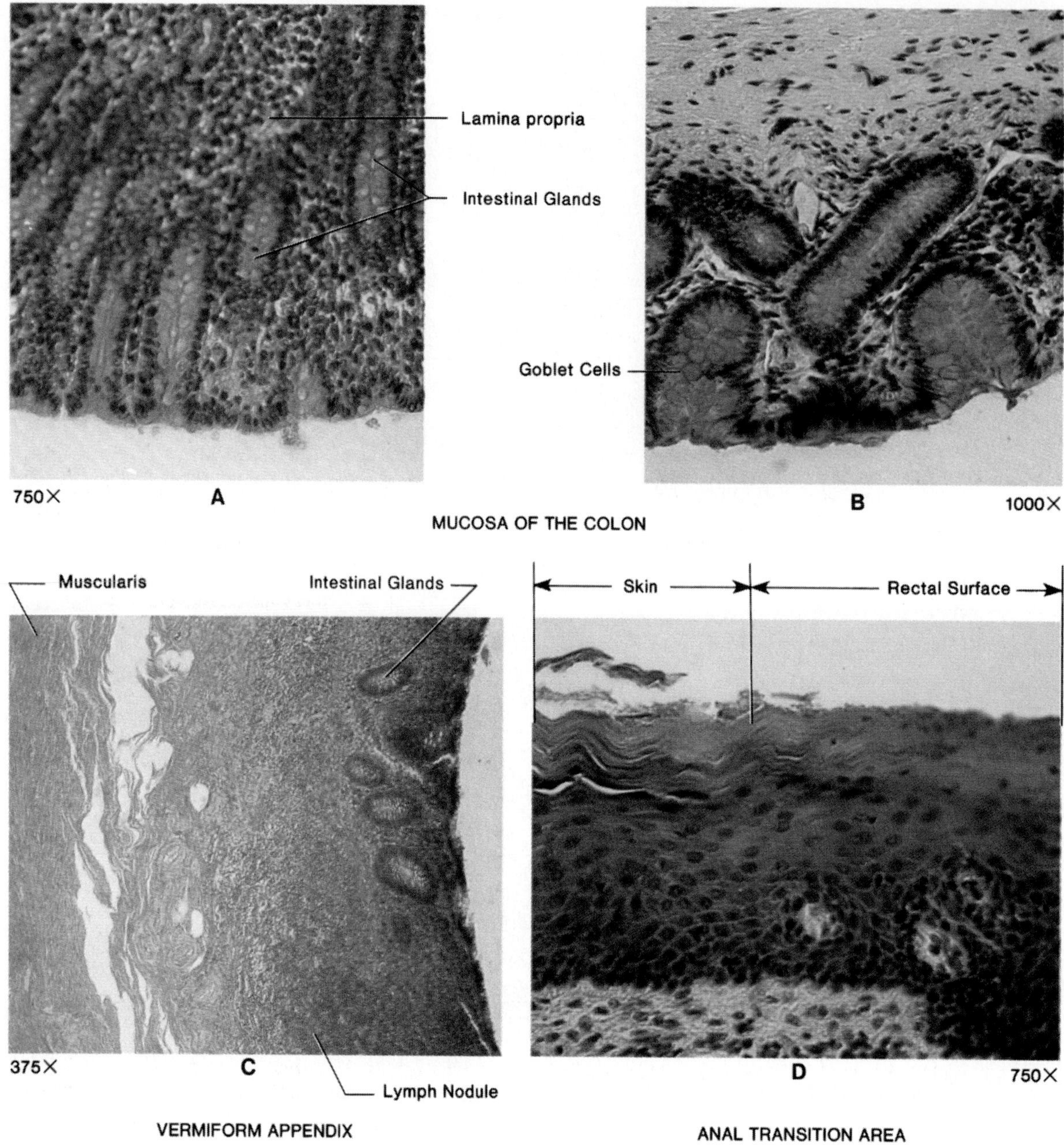

As micrographs A and B show, the absence of villi differentiates the mucosa of the large intestine from that of the small intestine. Note that the epithelium consists of simple columnar absorbing cells and numerous goblet cells. The red areas in micrograph B indicate mucus. Although the lamina propria contains many intestinal glands, no Paneth cells are present. Lymph nodules (not shown) are quite abundant.

The wall of the appendix (micrograph C) has a histological structure quite similar to that of the colon, except that the glands are less numerous and the lymph nodules are more prevalent. Micrograph C reveals a large lymph nodule.

The upper part of the rectum has a histological structure similar to that of the colon; its lower region has many longitudinal folds, or rectal columns. In the anal region, the epithelium becomes stratified squamous of a nonkeratinized (noncornified) variety. Note in micrograph D how this epithelium changes to the characteristic keratinized epithelium of the skin on the outside of the body.

Figure HA-24 The lower digestive tract.

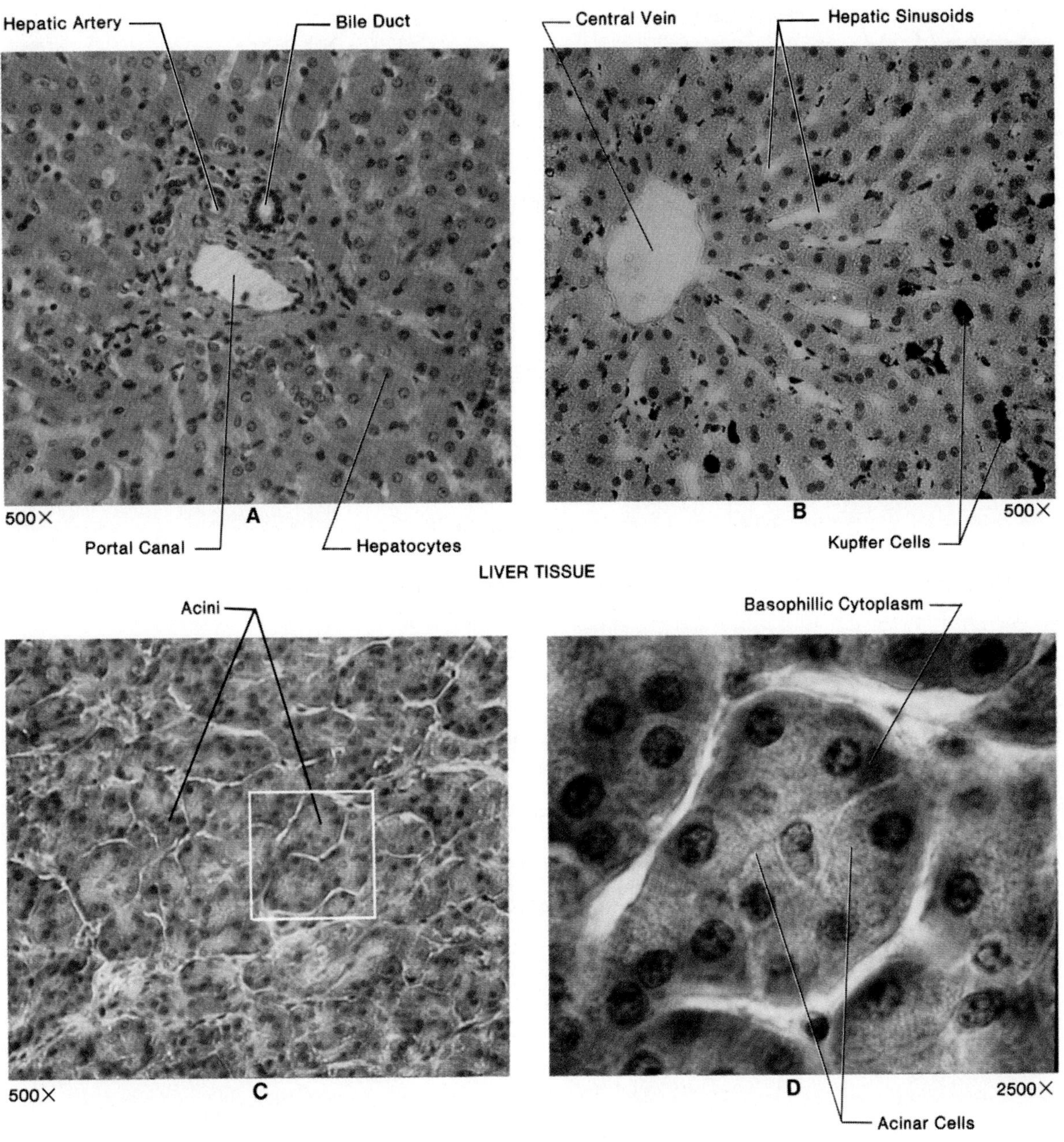

The liver consists of many lobules that contain two kinds of cells: hepatocytes and Kupffer cells. The hepatocytes are arranged in plates that radiate outward around a central vein (micrograph B) and have vascular channels, called sinusoids, between them. The hepatocytes synthesize bile components and several blood components, such as albumin and blood-clotting proteins. The secretions of these cells drain into the blood in the central vein of the lobule by way of the sinusoids. The central veins of each liver lobule, in turn, empty into a portal canal (micrograph A) between the lobules. The Kupffer cells are stationary phagocytic cells of the reticuloendothelial system that filter out of the blood any bacteria or other foreign material passing through the liver. The Kupffer cells in micrograph B have been blackened by the uptake of carbon particles injected into the animal prior to sacrifice for tissue preparation.

Micrographs C and D illustrate the exocrine cells of the pancreas. Note how the pyramid-shaped cells are arranged in alveoli (acini). The basophilic portion of the cells is specialized for the production of zymogen, an inactive enzyme precursor.

Figure HA-25 Liver and pancreas histology.

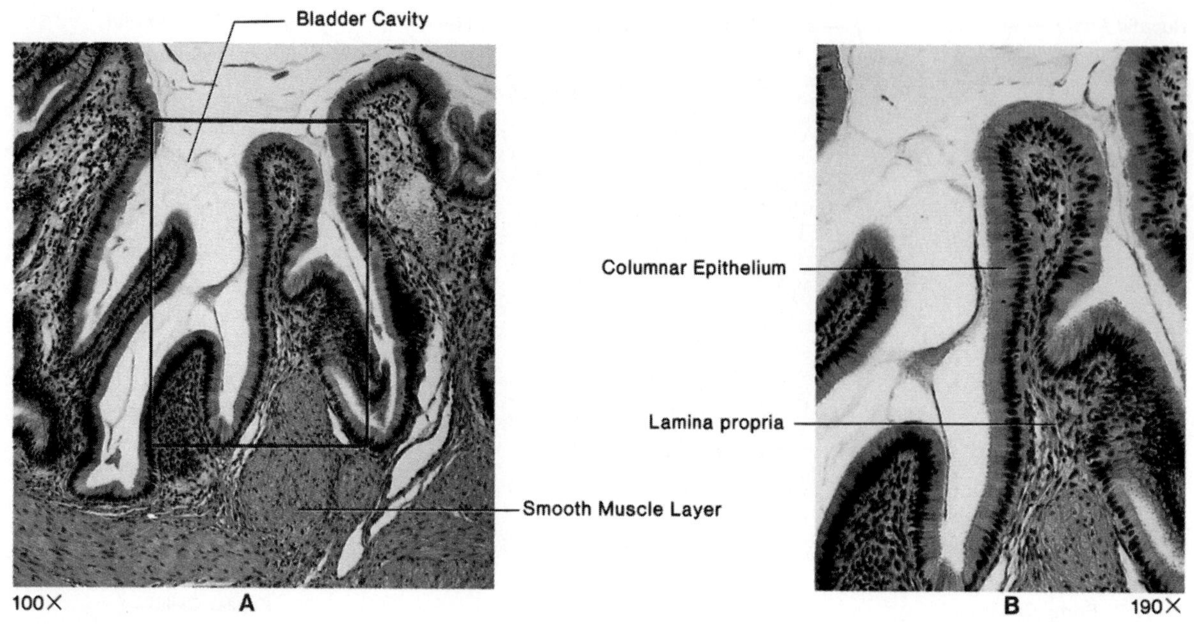

GALLBLADDER

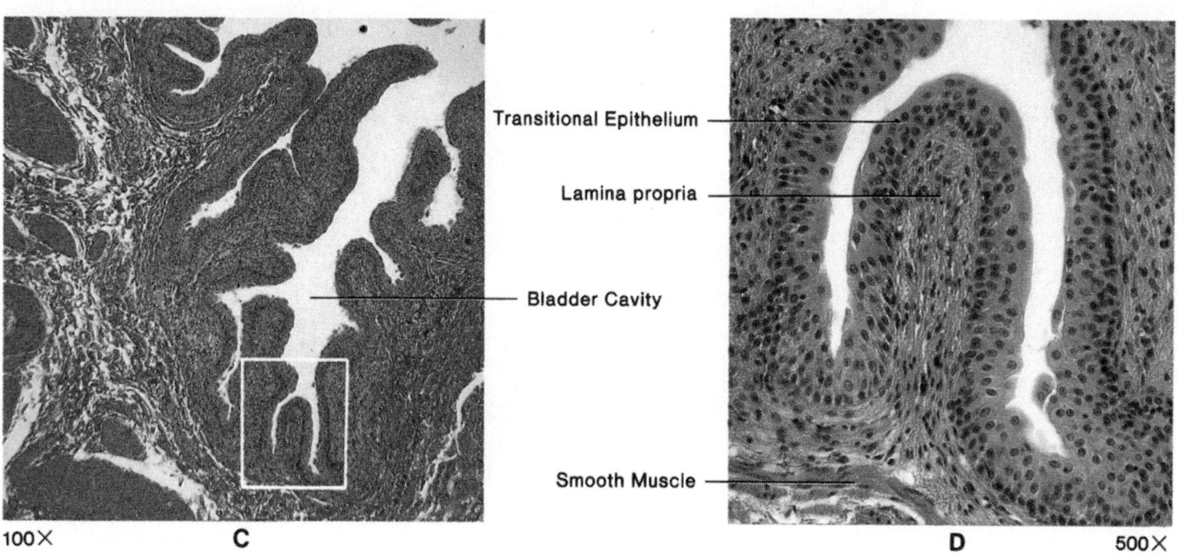

URINARY BLADDER

Although the gallbladder and urinary bladder are organs of two different systems, they have similarities, such as the considerable infolding of the mucosa of both bladders. As both bladders fill with bile or urine, the extensive infolding allows for expansion to contain the fluid.

Close examination of the inner epithelial linings of the two bladders, however, reveals distinct differences. The epithelium of the gallbladder consists of a single layer of columnar cells, and the urinary bladder mucosa is much thicker, consisting of transitional epithelium. These differences are expected: columnar cells in most of the digestive tract and transitional epithelium in the urinary system.

Figure HA-26 The gallbladder and urinary bladder.

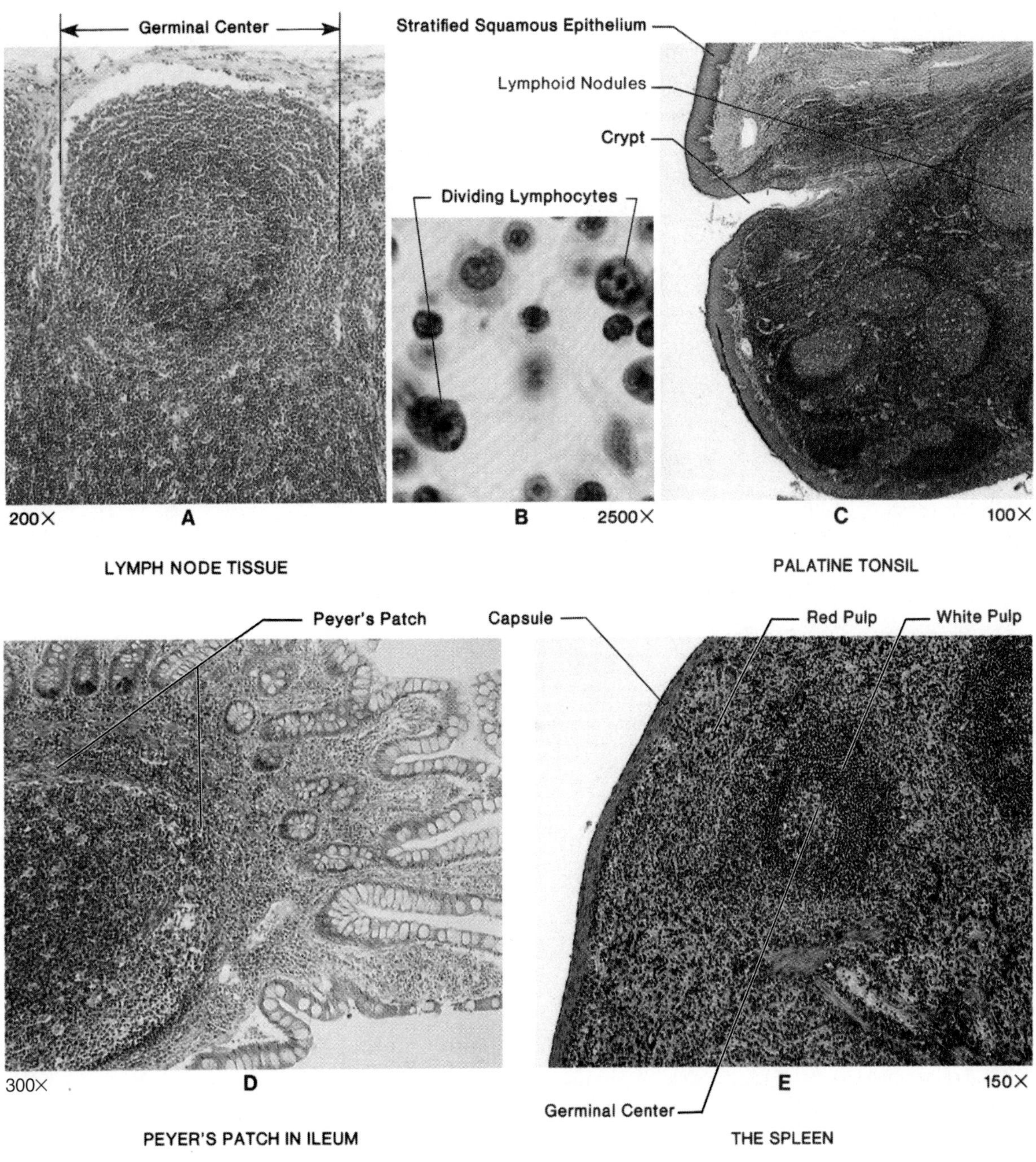

Lymphoid tissues, in strategic places along the most likely invasion routes into the body, extract foreign substances from body fluids. The lymph nodes, tonsils, Peyer's patches, and spleen are all a part of this system.

Micrograph A shows a single germinal center in a lymph node. If the cells in one of these centers is observed with oil immersion optics (micrograph B), multiplying cells are visible.

Micrograph C reveals that tonsils contain masses of lymphoid tissue and are covered by stratified squamous epithelium. Deep crevices, called crypts, increase the exposure surface of these organs. Peyer's patches (micrograph D) are modified lymph nodes in the mucosa and submucosa of the intestine.

The lymphoidal portion of the spleen (micrograph E) consists of the "white" pulp, which is white only in unstained tissue. Within the white pulp are diffuse germinal centers. The spleen also stores blood and destroys worn-out erythrocytes.

Figure HA-27 Lymphoid tissues in lymph gland, palatine tonsil, ileum, and spleen.

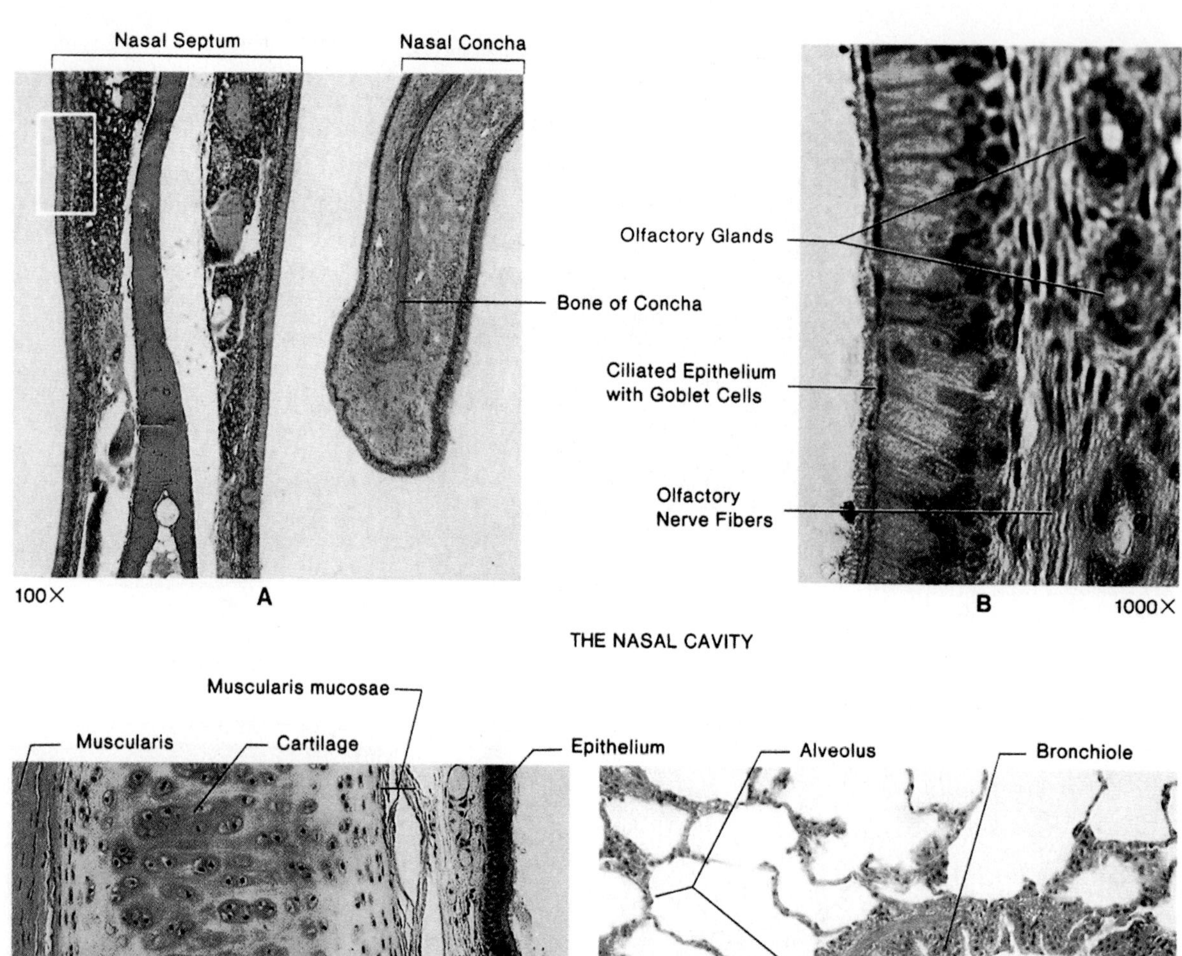

THE NASAL CAVITY

Muscularis mucosae
Muscularis
Cartilage
Epithelium
Alveolus
Bronchiole
250×
C
Blood Vessel
D
500×

TRACHEAL WALL

LUNG TISSUE

Micrograph A is a section through the nasal cavity revealing the structure of the nasal septum and a portion of a nasal concha. Note in micrograph B (enlargement of the boxed area in micrograph A) that the nasal epithelium consists of ciliated columnar cells. Interspersed between these columnar cells are olfactory receptors that are not readily visible here due to the absence of silver dye staining. The olfactory (Bowman's) glands in the lamina propria produce a secretion of mucus and serous fluid that helps to keep the epithelium moist and dissolve molecules that stimulate the olfactory receptors.

Micrograph C shows the structure of the tracheal wall. Hyaline cartilaginous rings provide tracheal rigidity. The epithelium consists of pseudostratified ciliated epithelium (see figure HA-2D). The section of lung tissue in micrograph D depicts a cluster of alveoli, a bronchiole, and a small artery. Several alveoli open into a larger atrium, which empties ultimately into a bronchiole (not shown). Observe that the inner wall of a bronchiole consists of cuboidal or low columnar epithelium. Cilia are present in the proximal portion of bronchioles and absent distally.

Figure HA-28 Histology of respiratory structures.

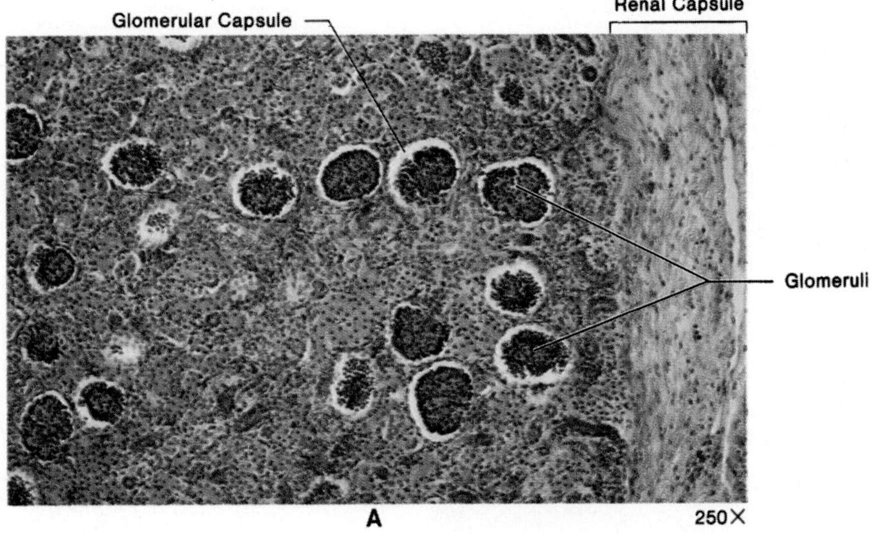

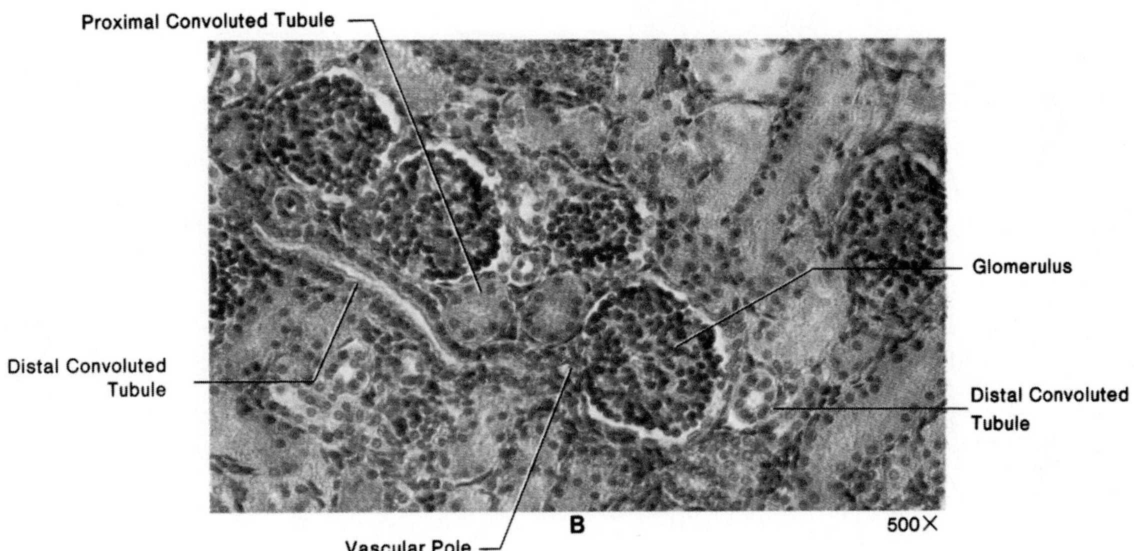

Micrograph A shows the structure of the outermost portion of the cortex of the kidney. A thick, fibrous capsule encloses the entire organ.

Blood enters each glomerulus through its vascular pole (shown in micrograph B). High vascular pressure in the glomeruli results in the production of a glomerular filtrate consisting of water, glucose, amino acids, and other substances. The glomerular capsule collects the filtrate, which then passes down the length of the nephron collecting tubule, being altered in composition as it approaches the calyx of the kidney. Note that the proximal convoluted tubules can be differentiated from the distal convoluted tubules by the size of the cuboidal cells that comprise their walls: large in the proximal tubule and small in the distal tubule. Eighty percent of water absorption occurs in the proximal convoluted tubules.

Figure HA-29 Histology of the cortex of the kidney.

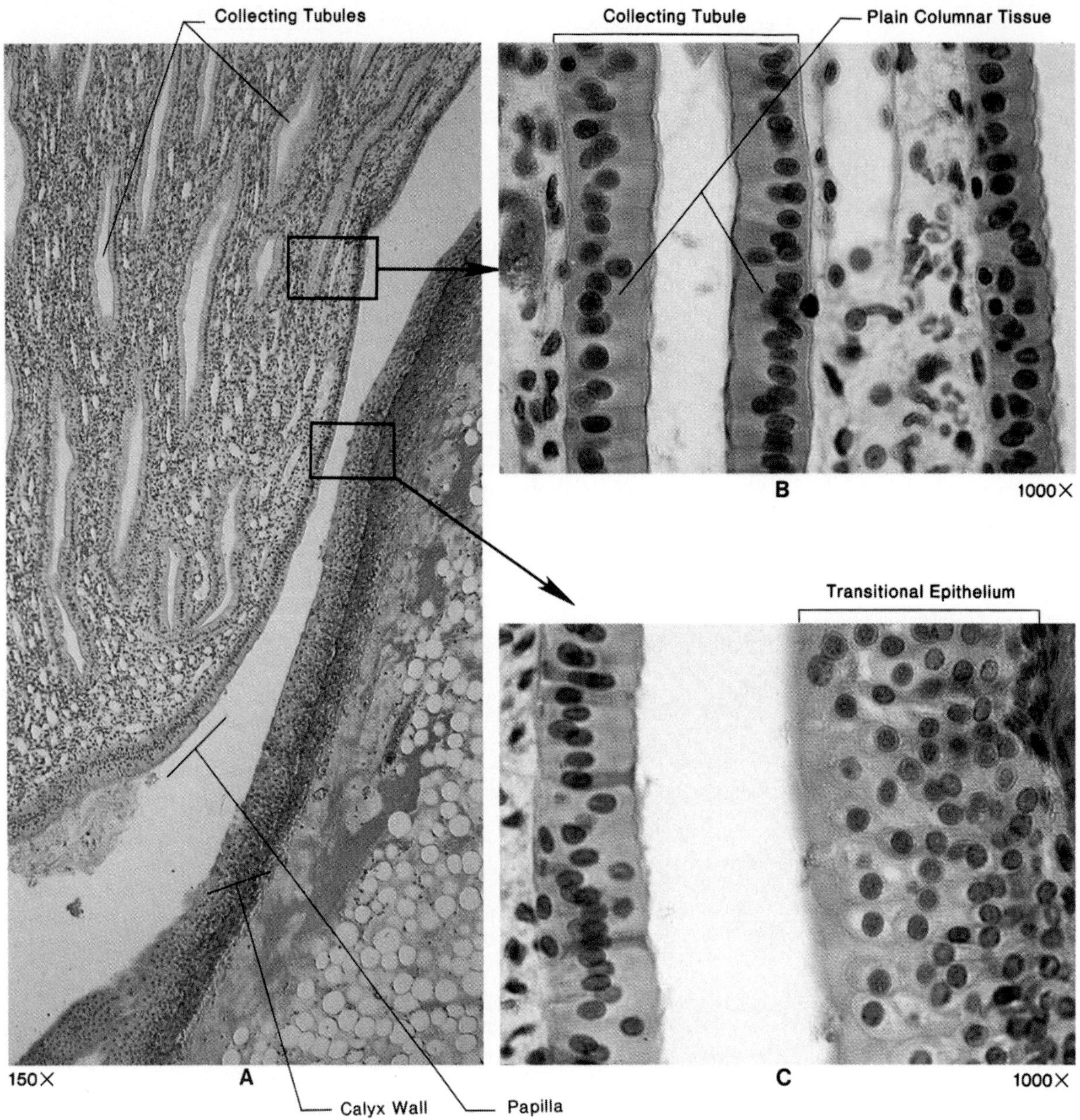

Micrograph A shows a portion of the tip of a pyramid and the wall of the calyx. Micrographs B and C show the histological nature of these two structures. The collecting tubules, which collect urine from several nephrons, are lined with distinct low columnar epithelium. Examination of these cells with oil immersion optics reveals the existence of a distinct brush border. Both absorption and secretion take place through these columnar cells.

Note in micrograph C that transitional epithelium lines the wall of the calyx. This same type of tissue lines the ureters, and the urinary bladder (See figures HA-26 C, D).

Figure HA-30 Histology of the papilla and calyx of the kidney.

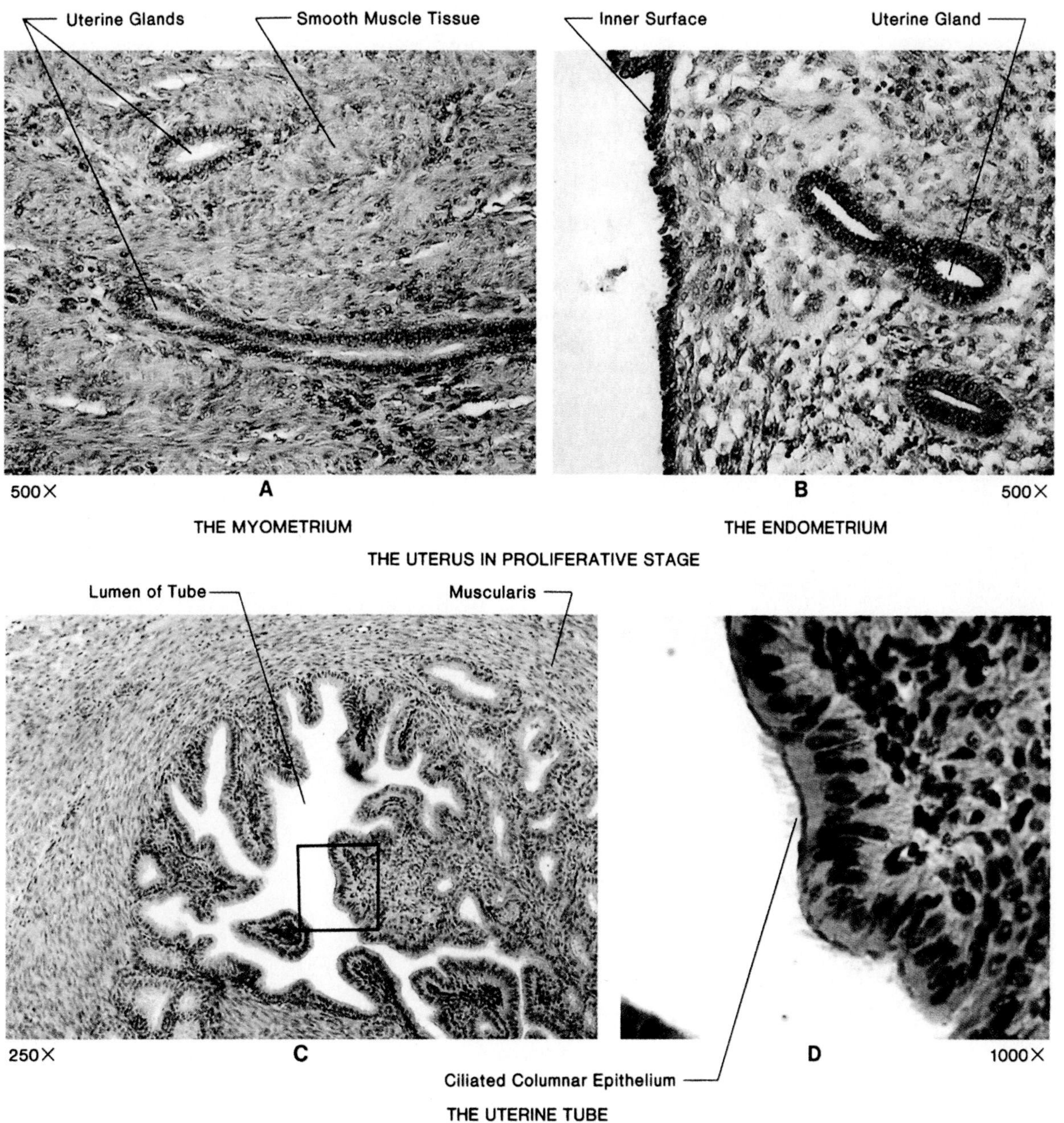

The wall of the uterus has three layers: (1) the endometrium, which corresponds to the mucosa and submucosa; (2) the myometrium, or muscularis; and (3) the perimetrium, a typical serous membrane. The myometrium (micrograph A), which forms 75% of the uterine wall, consists of three layers of smooth muscle fibers: an inner longitudinal layer, a middle circular layer, and an outer oblique layer. Some glands extend into it from the endometrium. The endometrial surface cells are columnar and partially ciliated During the proliferative stage (immediately after menstruation), the endometrium undergoes regeneration of the columnar cells and increased growth of the mucosal glands. The tissue also becomes more vascular. The proliferative stage terminates on the thirteenth or fourteenth day of the menstrual cycle.

The uterine tube (micrograph C) consists of an inner mucosa, a middle muscularis, and an outer serosa. The mucosa epithelium has a mixture of ciliated and nonciliated columnar cells (see micrograph D). The muscularis has inner circular and outer longitudinal layers of smooth muscle fibers. The muscularis mucosae seen in the digestive tract is not present here.

Figure HA-31 The uterus and uterine tube.

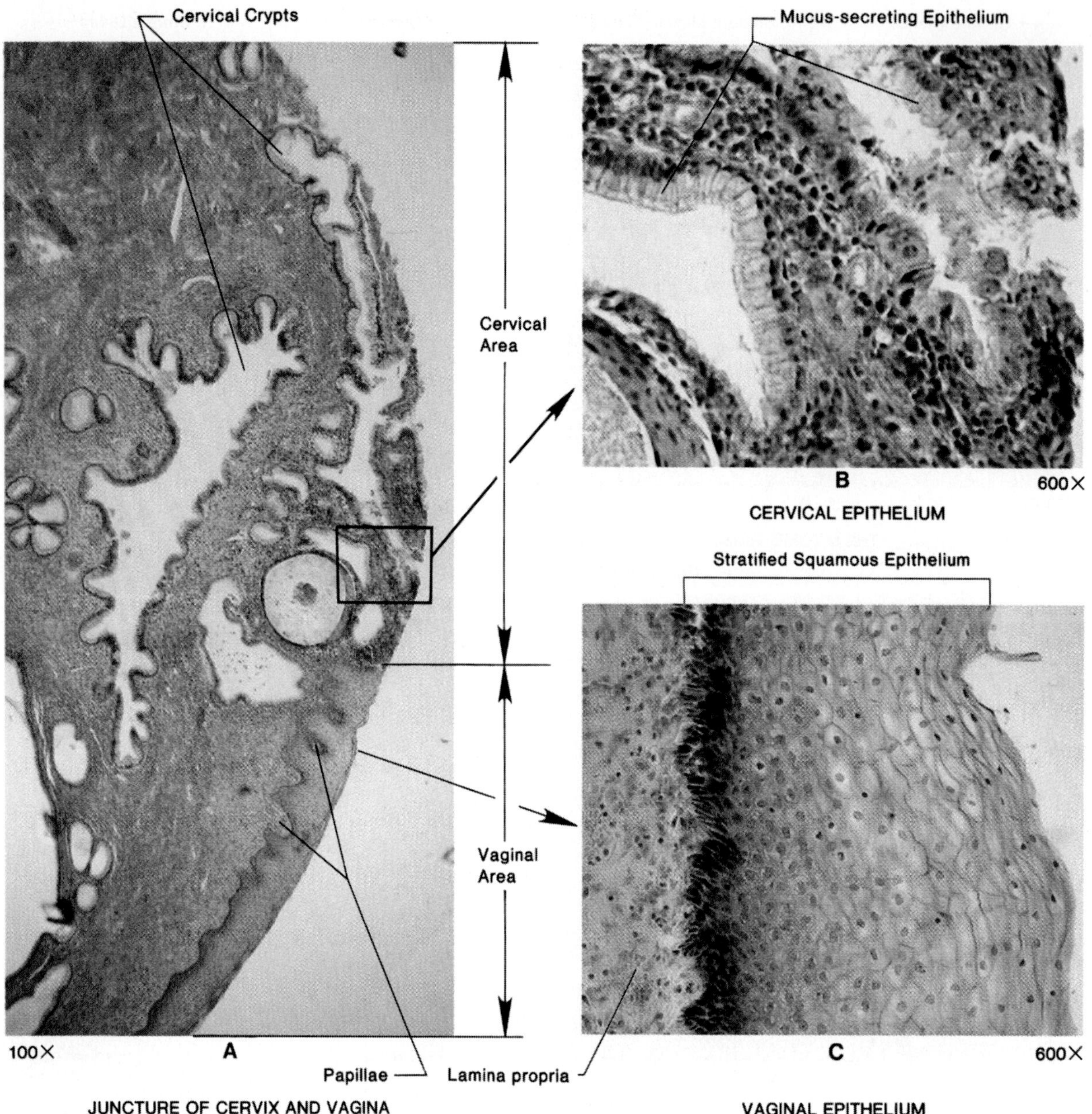

Micrograph A shows the contrasting histological differences between the epithelia of the vagina and cervix. Note in micrograph C that the vaginal epithelium consists of nonkeratinized stratified squamous epithelium. Examination of the lamina propria of the vaginal mucosa with high-dry or oil immersion optics reveals many lymphocytes. Note in micrograph A that the epithelium has many papillae.

The most distinguishing characteristics of the cervical portion of the uterus are the approximately 100 mucus-secreting cervical crypts. As indicated in micrograph B, the epithelium of these crypts consists, primarily, of plain columnar cells; some ciliated cells are also present, however. Approximately 20-60 mg of mucus are produced daily. During ovulation, mucus production increases to over 700 mg daily. Mucus plays an important role in fertility since it is the first secretion the sperm meet upon entering the female tract. Occasionally, the exit of a crypt becomes occluded, causing a cyst to form. The large, round structure near the box in illustration A is cyst.

Figure HA-32 The cervix and vagina.

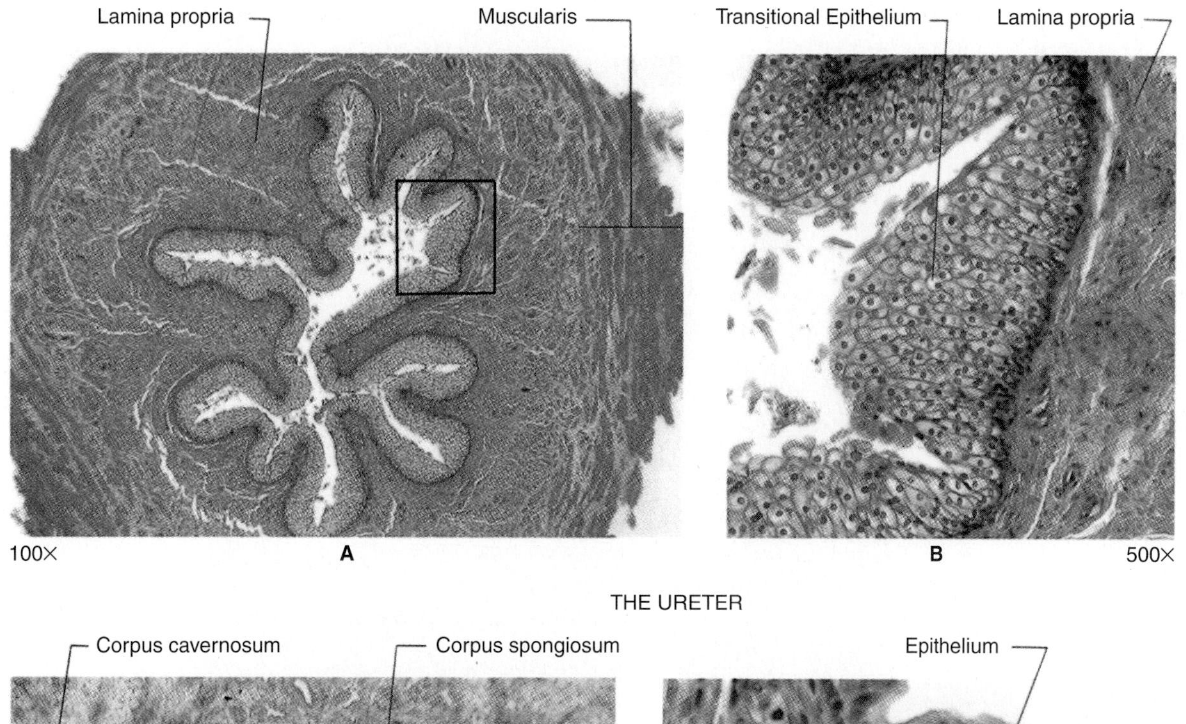

THE URETER

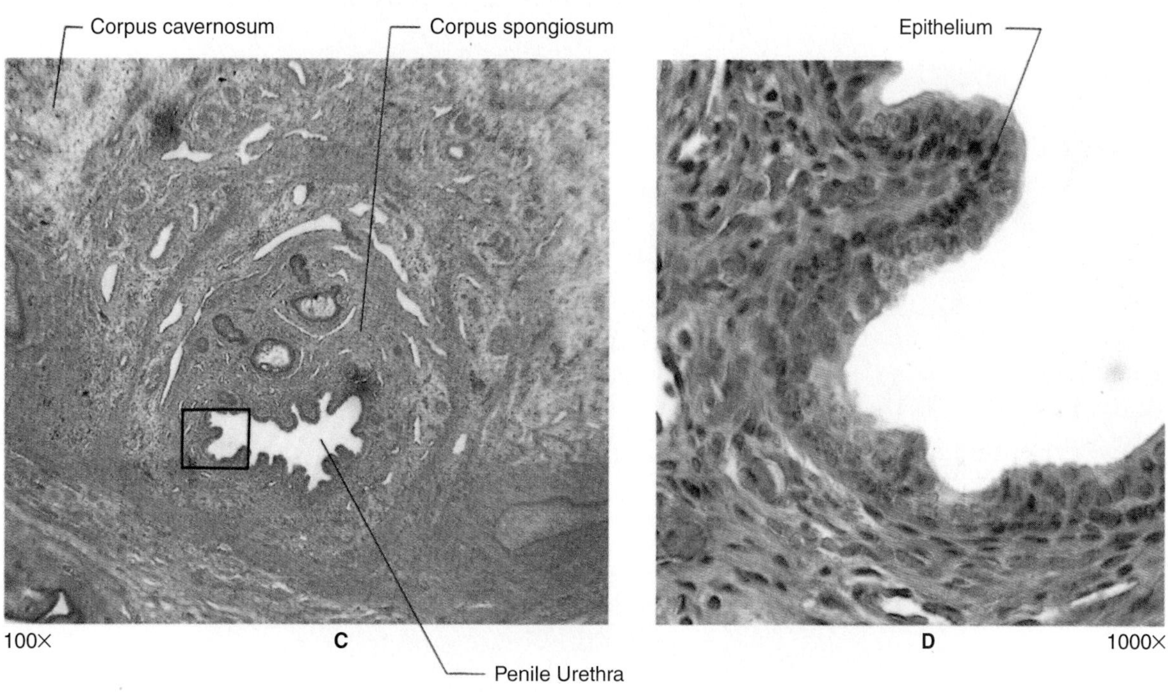

URETHRAL PORTION OF THE PENIS

Each ureteral wall consists of a mucosa, muscularis, and adventitia. As micrograph B shows, the mucosal epithelium is made up of transitional cells, backed up by the lamina propria. The muscularis is composed of two layers: circular fibers in the outer portion and longitudinal fibers in the inner layer. Since the ureters are retroperitoneal, there are no true serosas; instead, the adventitia consists of loose connective tissue.

Micrograph C is of only the ventral portion of the penis, revealing a portion of one of the corpora cavernosa, the corpus spongiosum, and the penile urethra. The type of epithelium in the urethra depends on where the section is cut. In the prostatic region and near the bladder, the epithelium is transitional. At the distal end, the urethra is stratified squamous. The remainder is pseudostratified columnar. The epithelium of the female urethra consists of transitional cells.

Figure HA-33 The ureter and penile urethra.

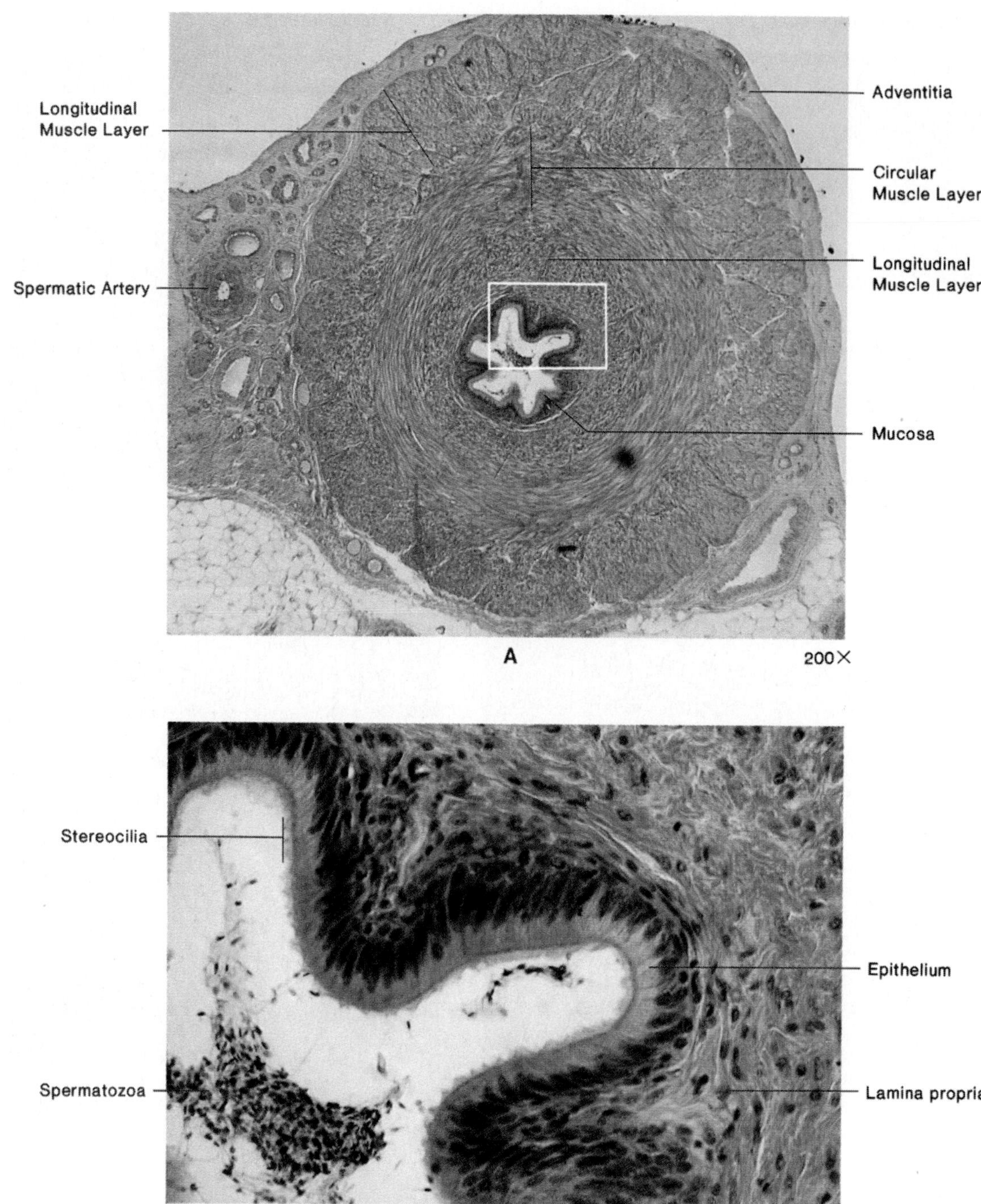

The vas deferens (spermatic duct) is a thick-walled, muscular tube that transports spermatozoa from the epididymis to the ejaculatory duct, where fluid from the seminal vesicles joins the spermatozoa. The wall has three distinct areas: an outer adventitia, a middle muscularis (three-layered), and an inner mucosa. Note that the muscularis is very thick, with two layers of longitudinal fibers and a middle layer of circular fibers. These muscles produce peristaltic waves that help propel the spermatozoa during ejaculation.

The mucosa, consisting of the lamina propria and epithelium, is unique in that the columnar cells that make up the epithelium have stereocilia. These structures are nonmotile, enlarged microvilli that cover the exposed surfaces of the cells.

Figure HA-34 The vas deferens.

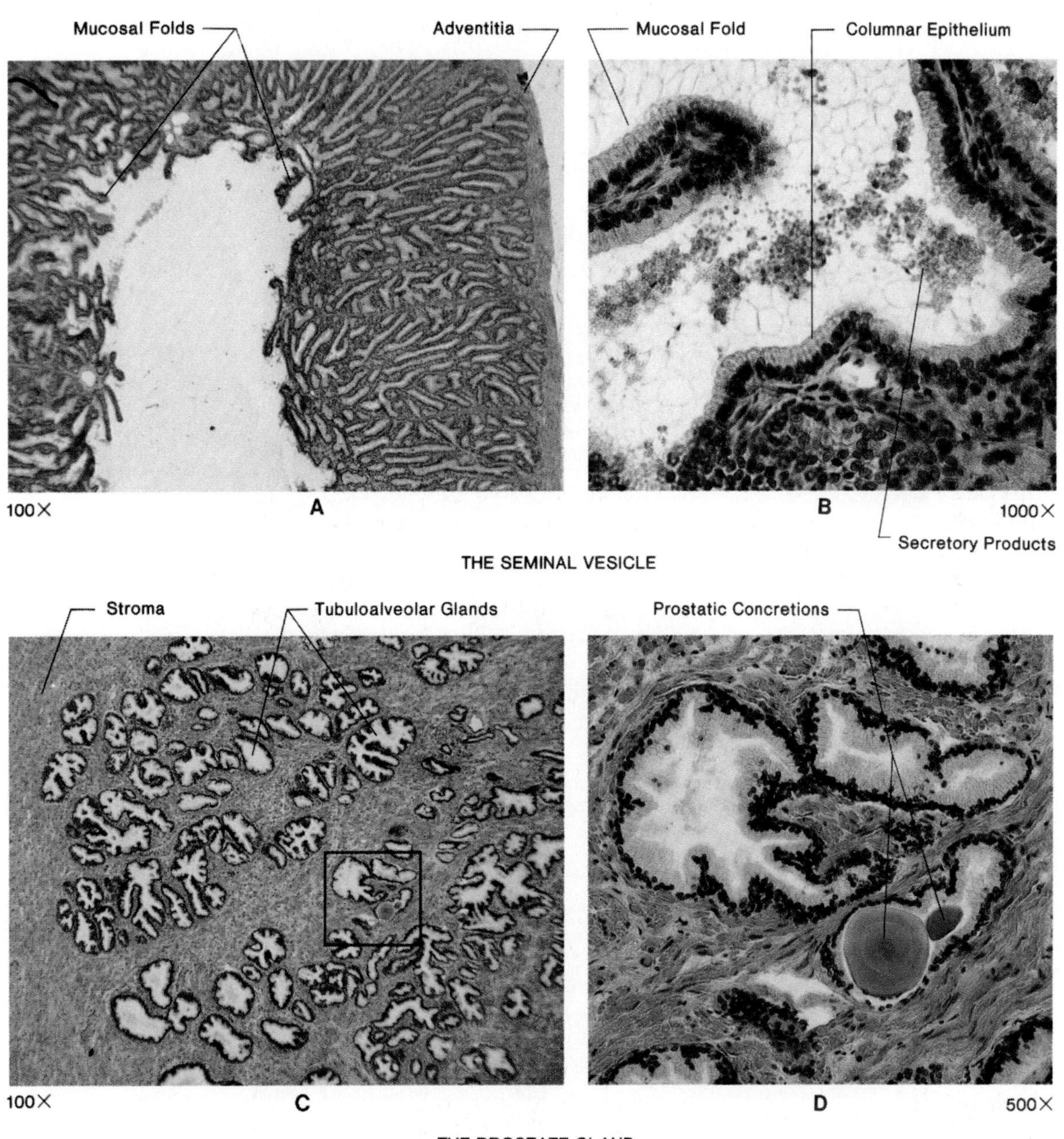

The mucosa of the paired seminal vesicles is unique in its extensive ramification of folds, as micrograph A shows. Note in micrograph B that the epithelium consists of nonciliated columnar cells. The seminal vesicles contribute a thick, yellowish fluid that is alkaline and rich in fructose, a source of energy for the spermatozoa. It also contains coagulating enzymes that cause the seminal fluid to congeal in the female tract after it is deposited. The prostate gland consists of 30-50 tubuloalveolar glands. These glands exhibit considerable infolding to accommodate distention during storage of prostatic fluid. The secretory cells of the epithelium consist of columnar cells superimposed on a few flattened cells. The stroma (50% of the gland) is made up of equal parts of smooth muscle fibers and fibrous tissue. The prostatic concretions shown in micrograph D are lamellar bodies that often become calcified in older men.

Figure HA-35 Seminal vesicle and prostate gland.

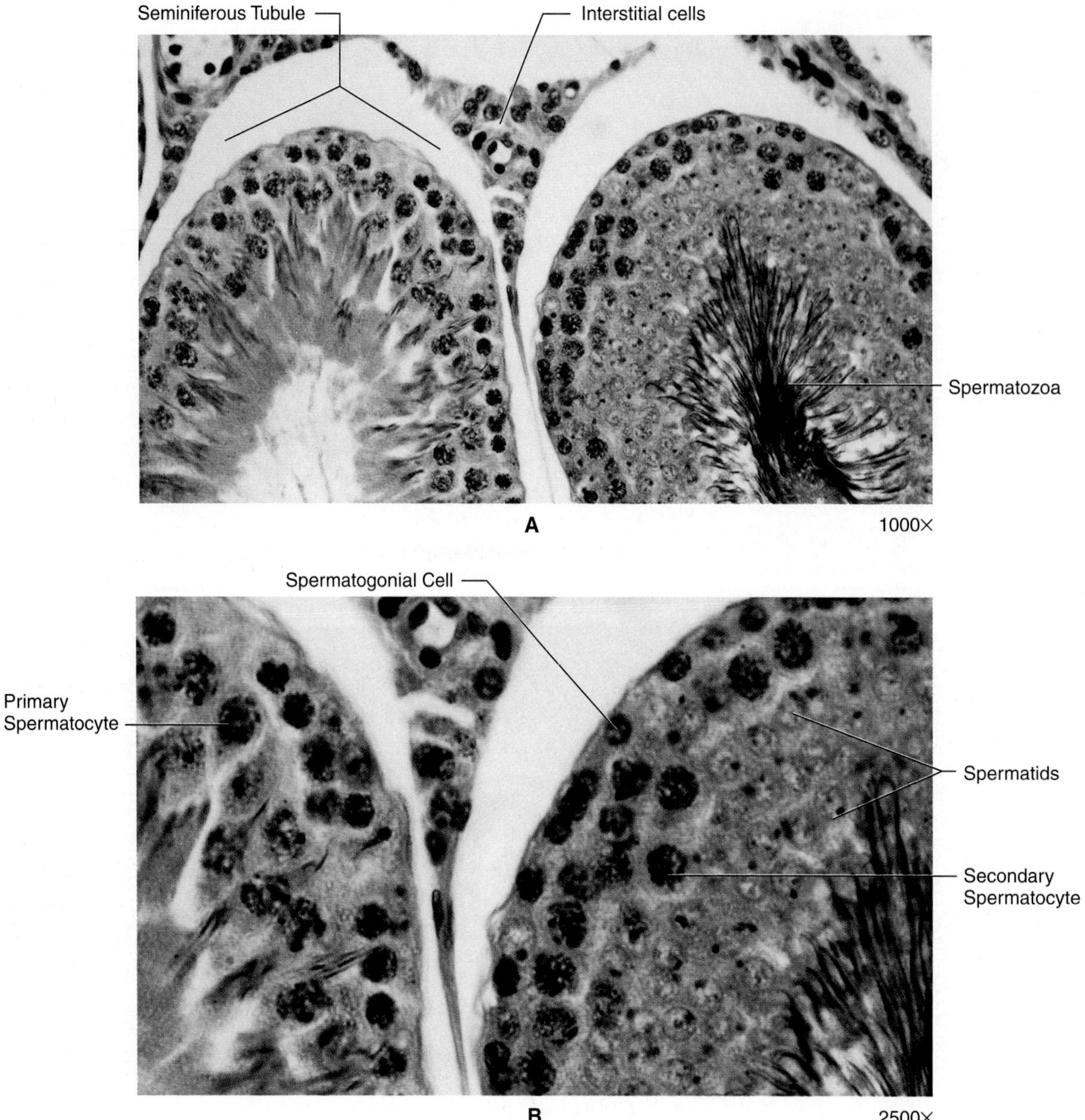

The production of spermatozoa, known as spermato genesis, occurs in the seminiferous tubules of the testes. Micrograph A shows the structure of two tubules. Between the tubules are the interstitial cells that secrete testosterone into the circulatory system.

All spermatozoa originate from the primordial germ cells, or spermatogonia at the periphery of the seminiferous tubule. These cells have a diploid complement of 46 chromosomes. The first step in this process is the synapsis (union) of the homologous chromosomes to form tetrads in the primary spermatocyte. Note the proximity of the primary spermatocyte to the spermatogonial cells in micrograph B.

The next step is for the primary spermatocyte to divide meiotically to produce two secondary spermatocytes with a haploid number (23) of chromosomes. Note in micrograph B that the secondary spermatocytes are closer to the lumen of the seminiferous tubule. Finally, the two secondary spermatocytes divide meiotically again to produce a total of four spermatids that contain 23 chromosomes each. The spermatids continue to develop into mature spermatozoa.

Figure HA-36 Spermatogenesis.

PART 10 Hematology

Hematology is the scientific study of blood and the focus of Exercises 52–58. Hematological information accumulated from a series of blood tests can be valuable in many ways. Prior to surgery, this information can reveal potential problems that might occur in the operating room. In diagnostic studies, certain blood tests can indicate the presence of infection, the nature of an infection, and the presence or absence of anemia.

Many kinds of blood tests are performed in laboratories today, but only seven of the simpler, more common ones are performed here. These tests yield considerable information about a patient's condition.

Perform each blood test on your own blood, recording all the information in Tables I and II of the Laboratory Report for Exercises 52–58. Table II gives normal values for each test. These tests do not require venipuncture; instead, you use blood from a punctured finger for each test.

Precautions

Two hazards must be avoided in doing blood tests: (1) self-infection from environmental contaminants, and (2) transmission of infections (e.g., hepatitis, HIV, etc.) from person to person. You can perform all the tests in this manual safely if you rigidly observe the following safeguards:

- Puncture only your finger, and handle only your blood.
- Sponge your tabletop at the beginning of the period with an appropriate disinfectant.
- Wash your hands with soap and water before and after blood tests.
- Use disposable lancets only one time. Deposit used lancets into a sharps container for biohazardous materials. *Never* toss used lancets into the wastebasket.
- Before perforating the finger for a blood sample, disinfect the skin with alcohol.
- Immerse all contaminated glassware in a basin that contains bleach solution, and deposit all other contaminated items (cotton balls, alcohol swabs, toothpicks) in a receptacle for biohazardous materials.
- Scrub the tabletop at the end of the period, and wash your hands again with soap and water. A final rinse of the hands with disinfectant provides additional protection.

52 A White Blood Cell Study: The Differential WBC Count

Blood is an opaque, rather viscous fluid that is bright red when oxygenated and dark red when depleted of oxygen. Its specific gravity is normally around 1.06, and its pH is normally between 7.35 and 7.45, so it is slightly alkaline. When centrifuged, blood separates into a dark red portion made up of **formed elements** and a clear, straw-colored portion called **plasma.** The formed element content of blood is around 47% in men and 42% in women.

The formed elements of blood consist of **erythrocytes** (red blood cells), **leukocytes,** (white blood cells [WBC]), and **blood platelets.** As figure 52.1 shows, erythrocytes are flattened, nonnucleated cells with concave centers. Erythrocyte cytoplasm contains an abundance of the protein hemoglobin, which transports oxygen gas. Erythrocytes are the most numerous elements in blood with 4.5 to 5.5 million cells per cubic millimeter.

Leukocytes are of two types: granulocytes and agranulocytes (see figure 52.2). *Granulocytes* have conspicuous granules in their cytoplasm; **neutrophils, eosinophils,** and **basophils** fall into this category. *Agranulocytes* lack pronounced granules in the cytoplasm; **monocytes** and **lymphocytes** are of this type. Blood platelets are noncellular elements in the blood that assist in the blood-clotting process.

This exercise focuses on leukocytes. You can study a prepared stained microscope slide or a slide made from your own blood. If you are going to prepare a slide from your own blood, *be sure to review the sanitary precautions on page 283.*) Scanning an entire slide and counting the various types of leukocytes allows you to calculate the percentage of each type of leukocyte. The erythrocytes and blood platelets are ignored.

A white blood cell count that determines the relative percentages of each type of leukocyte is called a **differential white blood cell count.** The purpose of this exercise is learning to distinguish the various types of white blood cells, not becoming proficient in differential WBC counts.

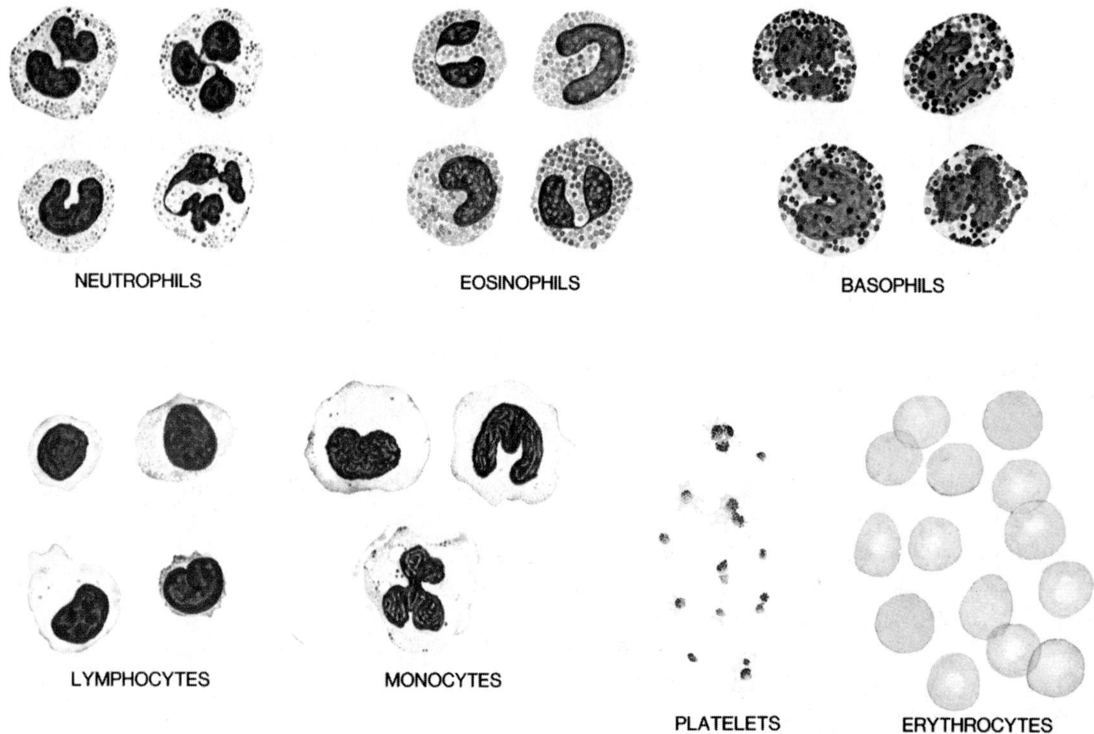

Figure 52.1 Formed elements of blood.

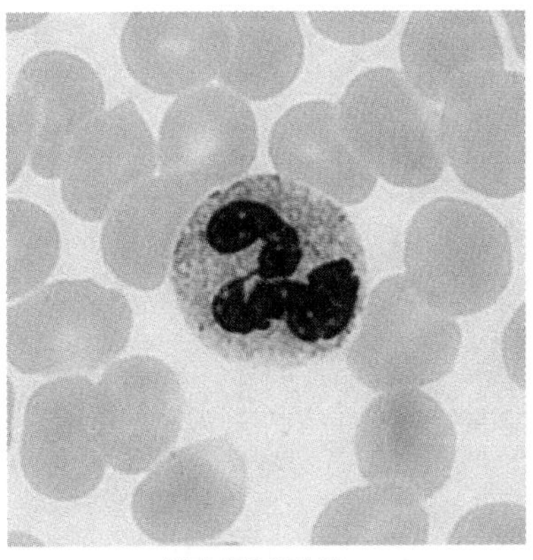

NEUTROPHIL

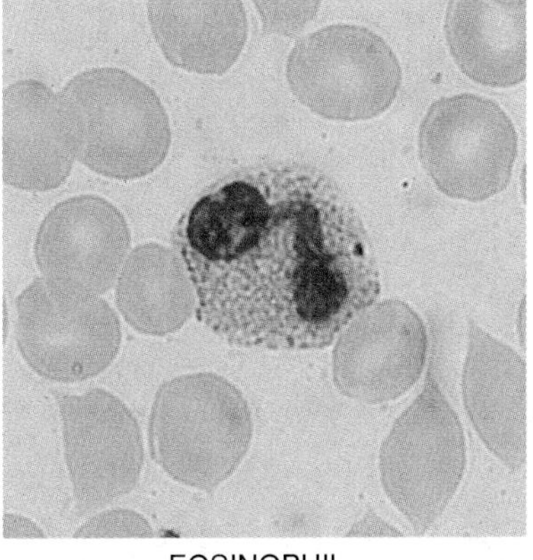

EOSINOPHIL

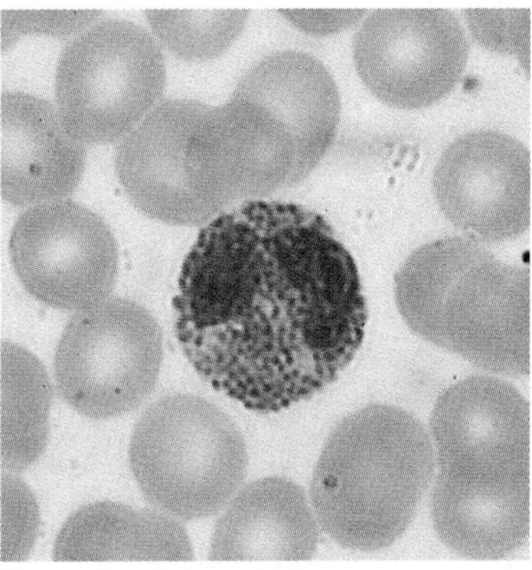

BASOPHIL

GRANULOCYTES

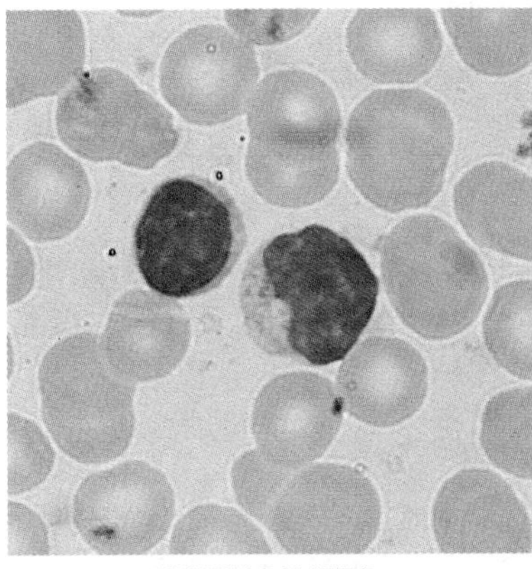

LYMPHOCYTES

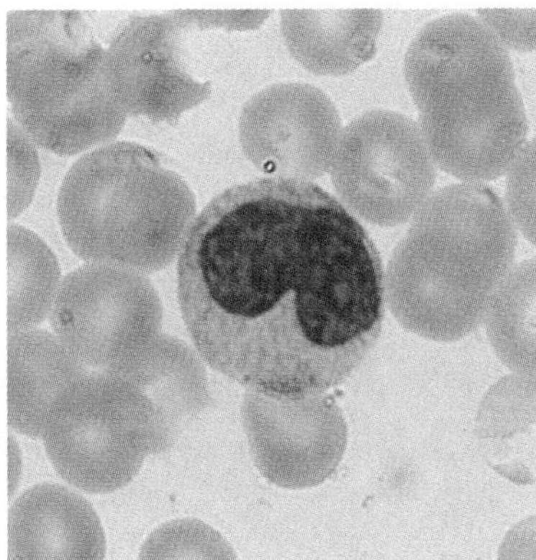

MONOCYTE

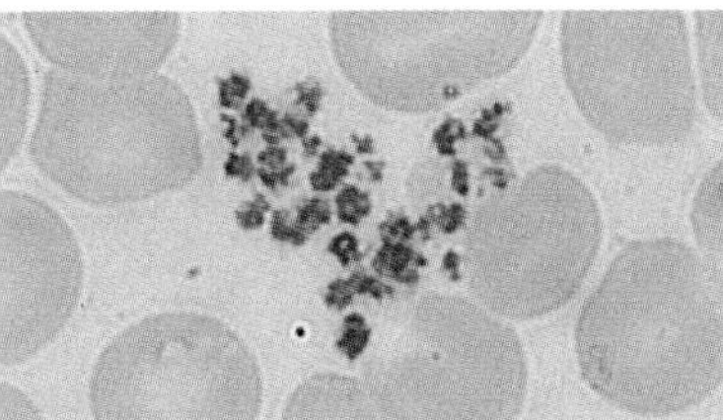

BLOOD PLATELETS

AGRANULOCYTES

Figure 52.2 Neutrophils differ from the other two granulocytes in having smaller and paler granules in the cytoplasm. Nuclei of these leukocytes are characteristically lobulated, with thin bridges between the lobules (see figure 52.1). During infections, when the bone marrow is producing many neutrophils, the nuclei are horseshoe-shaped; these juvenile cells are often called "stab cells."

Eosinophils generally have bilobed nuclei; however, juveniles have horseshoe-shaped nuclei (see figure 52.1). Note that the eosinophilic granules in the cytoplasm are large and pronounced.

Dark basophilic granules in the cytoplasm distinguish the few basophils from the other granulocytes. The nuclei of these cells are large and varied in shape. These cells probably secrete heparin into the blood.

Large and small lymphocytes are shown above and in figure 52.1. The small ones, which are more abundant, have large, dense nuclei surrounded by a thin layer of basophilic cytoplasm. The large lymphocytes have indented nuclei and more cytoplasm than small lymphocytes.

Monocytes are simple to identify because they are the largest cells in the blood. The nuclei of monocytes can be round, oval, lobed, or indented.

Blood platelets, which form from megakaryocytes in the bone marrow, help form blood clots.

The normal percentage ranges for the various leukocytes are as follows: neutrophils 50–70%; lymphocytes, 20–30%; monocytes, 2–6%; eosinophils, 1–5%; and basophils 0.5–1%. Because of the extremely low percentages of eosinophils and basophils, you must examine at least a hundred leukocytes to increase the possibility of encountering one or two eosinophils or basophils on a slide.

Deviations from the previous percentages may indicate serious pathological conditions. High neutrophil counts, or *neutrophilia,* often signal localized infections, such as appendicitis, or abscesses in some part of the body. *Neutropenia,* characterized by a marked decrease in the number of neutrophils, occurs in typhoid fever, undulant fever, and influenza. *Eosinophilia* (high eosinophil count) may indicate allergic conditions or invasions by parasitic roundworms, such as *Trichinella spiralis,* the pork worm. Eosinophil counts may rise to as much as 50% of the total leukocytes in cases of

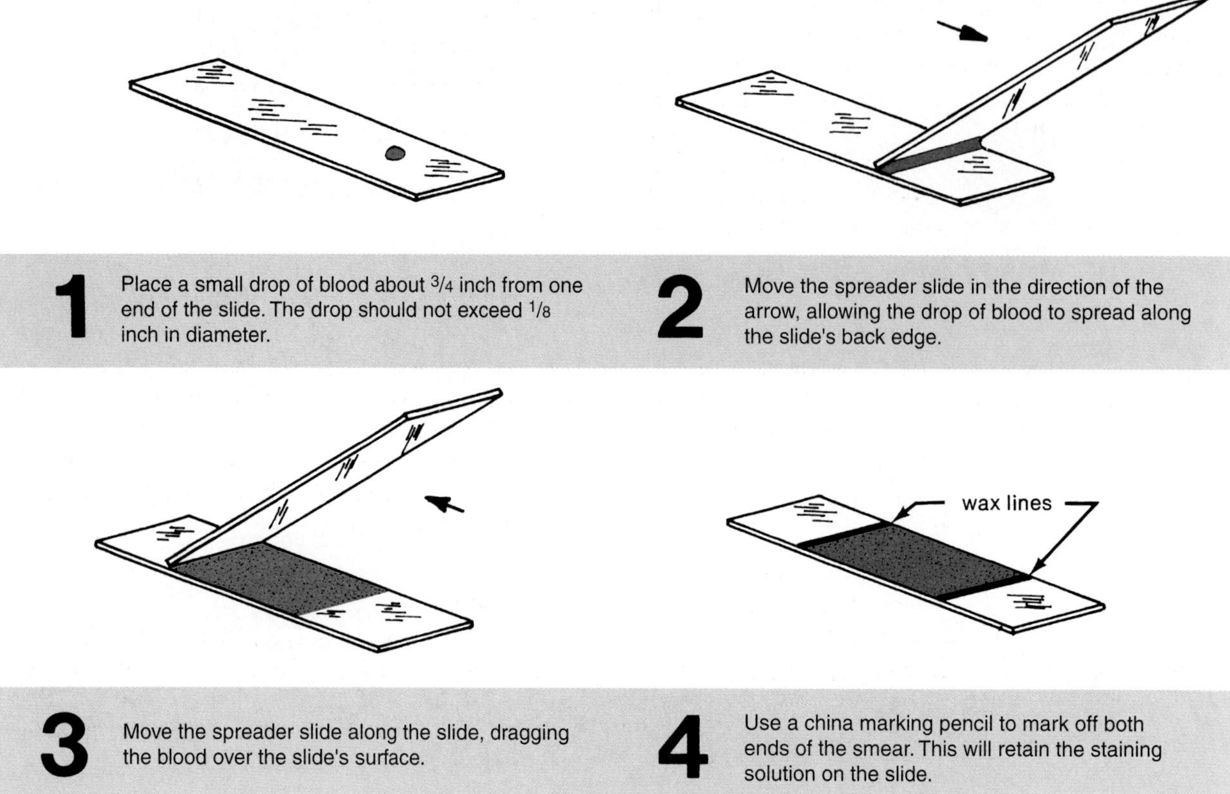

Figure 52.3 **Smear preparation technique for differential white blood cell count.**

trichinosis. High lymphocyte counts, or *lymphocytosis,* are present in whooping cough and some viral infections.

If you are using a prepared slide, proceed to "Performing the Cell Count."

Materials:
prepared blood slide stained with Wright's or Giemsa's stains

or for staining a blood smear:
two or three clean microscopic slides (with polished edges)
sterile, disposable lancets
alcohol swabs
Wright's stain
distilled water in dropping bottle
wax pencil
bibulous paper
immersion oil
paper toweling

Slide Preparation

Figure 52.3 illustrates the procedure for making a stained slide of a blood smear. The most difficult step is spreading the blood thick at one end and thin at the other end. A properly done smear has a gradient of cellular density that allows you to choose an ideal area for study. The angle at which you hold the spreading slide determines the thickness of the smear. You may need to make more than one smear to get one that is suitable.

1. Clean three or four slides with soap and water. Handle them with care to avoid soiling their flat surfaces. Only two slides may be needed for this procedure. The extra slides are for repeating the spreading process, if necessary.
2. Scrub your middle finger with an alcohol swab, and perforate it with a lancet.
3. Put a drop of blood on the slide about 3/4 inch from one end, and spread the blood with another slide as figure 52.3 shows. *Drag the blood over the slide; do not push it.* Do not pull the slide over the smear a second time. If you do not get an even smear the first time, repeat the process on a fresh, clean slide. To get a smear of the proper thickness, hold the spreading slide at an angle somewhat greater than 45°.
4. Draw a wax pencil line on each side of the smear to confine the stain that is to be applied to the slide.

5. Cover the blood smear with Wright's stain, counting the drops as you add them. Stain for **4 minutes,** and then add the same number of drops of distilled water to the stain and let stand for another **10 minutes.** Blow gently on the mixture every few minutes to keep the solutions mixed.
6. Gently wash off the slide under running water for **30 seconds,** and shake off the excess. Blot dry with paper toweling.

Performing the Cell Count

Whether you are using a prepared slide or one that you have just stained, the WBC counting procedure is essentially the same. Ideally, you should use an immersion lens for this procedure; you can use high-dry optics, but it is not as reliable for best results. Proceed as follows:

1. Scan the slide with the low-power objective to find the best area of cell distribution. In a good area, the cells are not jammed together or scattered too far. After selecting an ideal area, use the high-dry objective to focus on the cells.
2. If you are using the high-dry objective for the differential WBC count, move on to step 3. If you are using the oil immersion objective, place a drop of immersion oil on the selected location of the slide, and lower the oil immersion objective into the oil. If you are using a slide you have just prepared, placing oil directly on the stained smear will not cause a problem.
3. Systematically scan the slide, following the pathway that figure 52.4 indicates. As you encounter each leukocyte, use figures 52.1 and 52.2 to identify it.

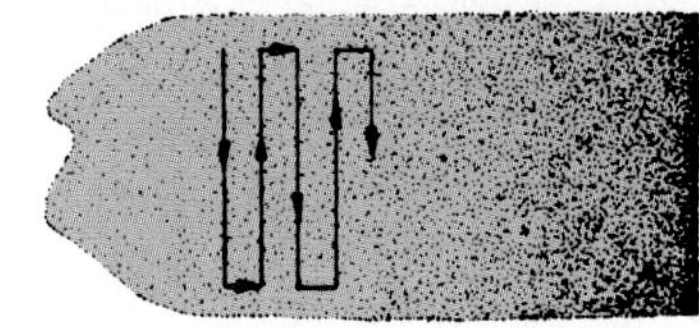

Figure 52.4 Examination path for a differential count.

Assignment:
Tabulate your count on part A of the Laboratory Report for Exercises 52–58 according to the instructions there. Remove your Lab Report sheet from the back of the manual for this identification and tabulation procedure.

53 Total Blood Cell Counts

Diagnostic blood analysis routinely determine the number of leukocytes (white blood cells [WBC]) and erythrocytes (red blood cells) in a cubic millimeter of blood. While the differential WBC count performed in Exercise 52 is useful, it alone cannot give a complete picture of the patient's health. The total number of white and red blood cells per unit of volume (a cubic millimeter) also must be known.

In this exercise, you make accurate blood cell counts, using a special cell-counting slide called a **hemacytometer.** While this device works well, the procedure is tedious. Modern medical laboratories use automated electronic counting devices that are much more rapid.

Total WBC Count

The number of white blood cells in a cubic millimeter of normal blood ranges from 5000 to 9000 cells. The exact number of cells varies with the time of day, exercise, and other factors. If a patient has a high WBC count, such as 17,000 leukocytes per cubic millimeter, and an abnormally high neutrophil count, as determined by a differential count, an infection likely is present somewhere in the body.

Determining the number of leukocytes in a cubic millimeter of blood requires diluting the blood and counting the WBCs on a hemacytometer, using the low-power objective of a microscope. Figures 53.1 and 53.2 illustrate the various steps in diluting the blood and charging a hemacytometer for this count.

Note in figure 53.1 that blood is drawn into a special pipette called a Unopette and then diluted in the Unopette reservoir with a weak acid solution. The sample is then allowed to flow under the cover glass of the hemacytometer, as figure 53.2 shows. When examined under the low-power objective, the diluted blood will reveal only white blood cells, since the acid solution destroys all red blood cells. Figure 53.2 also shows the four "W" areas on the hemacytometer that are examined for a WBC total count. Mathematical calculations determine the number of leukocytes per cubic millimeter.

Materials:
Unopette apparatus for WBC count
hemacytometer and cover glass
hand counter
sterile, disposable lancets
alcohol swabs
clean cloth or Kimwipes
microscope

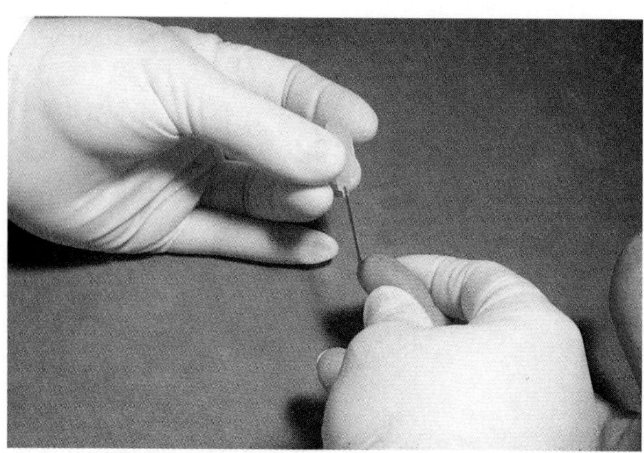

Figure 53.1 The unopette reservoir system for blood counts. The system consists of a reservoir containing a premeasured amount of diluent and a capillary tube pipette with a shield used for puncturing the reservoir top. Draw blood to fill the Pipette.

Collecting and Diluting the Blood Sample

1. Hold the reservoir containing the acid solution firmly on a tabletop. Use the protective shield on the capillary pipette to puncture the diaphragm in the neck of the reservoir; then remove the shield.
2. Scrub a finger with an alcohol swab, and produce a free flow of blood at the tip of the finger using a lancet as your instructor describes. Remove the Unopette shield, and hold the capillary tube nearly horizontal as you touch the tip of the tube to the blood. Capillary action draws the blood into and to the end of the tube. Fill the capillary tube completely and without air bubbles because the volume of the diluting acid solution in the reservoir is appropriate for a filled capillary tube. If your tube contains air bubbles or does not fill, ask your instructor if you need to start the procedure again.
3. Squeeze the reservoir slightly to expel some air (do not expel any liquid), and while maintaining your pressure on the reservoir, seat the capillary tube pipette in the reservoir neck (the capillary tube will be inside the reservoir). As you release pressure on the reservoir, your blood sample is drawn into the acid solution.
4. Slightly squeeze and release the reservoir several times to force the acid solution up into the capillary tube and to rinse out any remaining blood. Do not squeeze so hard as to expel the acid solution out of the open end of the pipette.
5. Cover the open end of the pipette with your index finger, and gently invert several times to thoroughly mix the blood in the reservoir. Place the reservoir on the tabletop, and remove and invert the pipette, reseating the pipette in the neck of the reservoir (the capillary tube is now pointing upward).

Charging the Hemacytometer

1. Wash the hemacytometer and cover glass with soap and water, rinse well, and dry with a clean cloth or Kimwipes.
2. Place the hemacytometer with cover glass in place on the microscope stage, and use the 4× objective lens to find the grid that figure 53.2 shows.
3. Using the Unopette as a dropper, gently squeeze the reservoir, and release and discard three or four drops of the solution.
4. While holding the pipette as figure 53.2 shows, squeeze out one drop of diluted blood on the V-shaped groove next to the edge of the cover glass. If a chamber is filled properly, diluted blood will fill only the space between the cover glass and counting chamber. No fluid should run down into the moat.
5. Charge the other side if the first side is overfilled.

Performing the Count

1. Bring the grid lines into focus under the **low-power** (10×) objective. With the coarse adjustment knob, reduce the lighting to make both the cells and lines visible.
2. Locate one of the "W" (white) sections that figure 53.2 shows. Since the diluting fluid contains acid, all erythrocytes have been destroyed (hemolyzed); only the leukocytes will show up as very small dots.
3. If the cells seem to be evenly distributed, count all the cells in the four "W" areas, using a hand

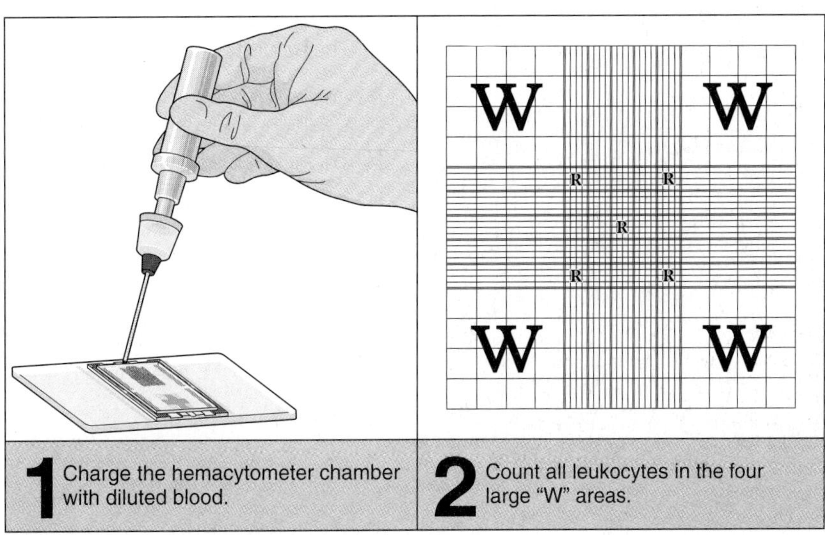

Figure 53.2 Draw blood into the Unopette to the 0.5 mark.

counter. To avoid overcounting of cells at the boundaries, **count cells that touch the lines on the left and top sides only.** Ignore cells that touch the boundary lines on the right and bottom sides. This applies to the boundaries of each "W" area. If the cells are not evenly distributed, charge the other half of the counting chamber after further mixing of the solution. If the other chamber was previously charged unsuccessfully, wash off the hemacytometer and cover glass, and recharge it.

4. Dispose of the Unopette in a receptacle for biohazardous materials.

Calculations

To determine the number of leukocytes per cubic millimeter, multiply the total number of cells counted in the four "W" areas by 50. The factor of 50 is the product of the volume correction factor and dilution factor:

$$2.5 \times 20 = 50$$

The volume correction factor of 2.5 is determined in this way: Each "W" area is exactly 1 mm^2 by 0.1 mm deep. Therefore, the volume of each "W" is 0.1 mm^3. Since four "W" sections are counted, the total amount of diluted blood examined is 0.4 mm^3. And since you are concerned with the number of cells in 1 mm^3 instead of 0.4 mm^3, you must multiply your count by 2.5, derived by dividing 1.0 by 0.4.

Record your results on the Laboratory Report.

Red Blood Cell Count

Normal red blood cell (RBC) counts for adult males average around 5,400,000 (±600,000) cells per cubic millimeter. The normal average for women is 4,600,000 (±500,000) per cubic millimeter. The difference between sexes does not exist prior to puberty. At high altitudes, higher values are normal for all individuals.

Low RBC counts result in anemia. Individuals can be anemic and still have normal RBC counts, however, if the cells are small or lack sufficient hemoglobin. Thus, **anemia** is, simply, a condition in which the blood's oxygen-carrying capacity is reduced.

Polycythemia, a condition characterized by above-normal RBC counts, may be due to living at high altitudes (*physiological polycythemia*) or red marrow malignancy (*polycythemia vera*). In physiological polycythemia, counts may run as high as 8,000,000 per cubic millimeter. In polycythemia vera, counts of 10 to 11 million per cubic millimeter are not uncommon.

Although the general procedures for the RBC count are essentially the same as for the WBC count, the Unopette, diluting solution, and calculations are different. The RBC diluting fluid may be one of several isotonic solutions, such as physiological saline or Hayem's solution.

Materials:
Unopette apparatus for RBC count
hemacytometer and cover glass
hand counter
sterile, disposable lancets
alcohol swabs
clean cloth or Kimwipes
microscope

1. Follow the same procedure for collecting and diluting the blood sample, and for charging the hemacytometer, as for the WBC count.
2. Observe the grid on the hemacytometer as before, but use the high-power lens (40×), and count only the RBCs in the five "R"-marked areas (see figure 53.2).
3. Multiply the number of cells counted by 10,000 (dilution factor is 200; volume correction factor is 50).
4. Dispose of the Unopette in a receptacle for biohazardous materials.

Assignment:
Record your results in part B and table II (section H) of the Laboratory Report for Exercises 52–58.

Hemoglobin Percentage Measurement

54

Since hemoglobin is the essential oxygen-carrying ingredient of erythrocytes, its quantity in the blood determines whether or not a person is anemic. As mentioned in Exercise 53, a person can have a normal RBC (red blood cell) count and still be anemic if the erythrocytes are smaller than normal. This *microcytic anemia* results from having an insufficient amount of hemoglobin. In other cases, an individual can have a normal RBC count but a low hemoglobin percentage in the erythrocytes (*hypochromic cells*). This also can result in anemia. Thus, the amount of hemoglobin in a unit volume of blood determines whether or not an individual is anemic.

Various methods determine blood hemoglobin content. One of the oldest— and least accurate— methods is the Tallquist scale. In this technique, blotting paper saturated with a drop of blood is compared to a color chart to determine the hemoglobin percentage. Very few, if any, physicians use this method today. A rapid, accurate method of measuring hemoglobin involves inserting a cuvette of blood into a photocolorimeter calibrated for blood samples. Photocolorimeters, however, are expensive. In addition, this method requires a considerable quantity of blood.

A relatively inexpensive instrument for measuring blood hemoglobin is the *hemoglobinometer.* Figures 54.1 through 54.4 illustrate the procedure for using a hemoglobinometer. Your instructor will assign which procedure you will use to determine the hemoglobin content of your blood—either the Tallquist method or the hemoglobinometer procedure.

Blood hemoglobin is expressed in terms of grams of hemoglobin per 100 ml of blood, or as percentage of hemoglobin concentration. For adult males, the normal range is 14–18 g/100 ml. For adult females, the normal range is 12–16 g/100 ml. For newborns, the normal range is 21.5 ± 3 g/100 ml. For four-year-old children, the normal range is 13 ± 1.5 g/100 ml. The hemoglobinometer may have a conversion scale for determining hemoglobin percentages from grams of hemoglobin per 100 ml. Proceed as follows to determine your own hemoglobin percentage:

Materials:
for Tallquist method
Tallquist hemoglobin scale
sterile, disposable lancets
alcohol swabs

for using a hemoglobinometer
hemoglobinometer
hemolysis applicators
sterile, disposable lancets
alcohol swabs
Kimwipes

Tallquist Method

1. Scrub a finger with an alcohol swab, and use a lancet to obtain blood from the finger as your instructor describes.
2. Place a drop of blood on the absorbent paper provided with the Tallquist scale. **Note:** Do not press your finger against the paper; just bring the paper into contact with the blood, and let the paper slowly absorb the blood.
3. Set the paper aside to air dry.
4. As soon as the blood has dried, compare its color, under natural light, to the color standards of the Tallquist scale, and record your percentage of hemoglobin saturation.

Using a Hemoglobinometer

1. Disassemble the blood chamber assembly of the hemoglobinometer by pulling the two pieces of glass from the metal clip. Note that one piece of glass has an H-shaped moat cut into it. This piece will receive the blood. The other piece of glass has two flat surfaces and serves as a cover plate.
2. Clean both pieces of glass with alcohol and Kimwipes. Handle them by their edges to keep them clean.
3. Reassemble the glass plates in the clip so that the grooves on the moat plate face the cover plate. Insert the moat plate only halfway to provide an exposed surface to receive the drop of blood (see figures 54.1 and 54.2).
4. Scrub a finger with an alcohol swab, and use a lancet to obtain blood from the finger as your instructor describes.
5. Place a drop of blood on the exposed surface of the moat plate, as figure 54.1 shows.

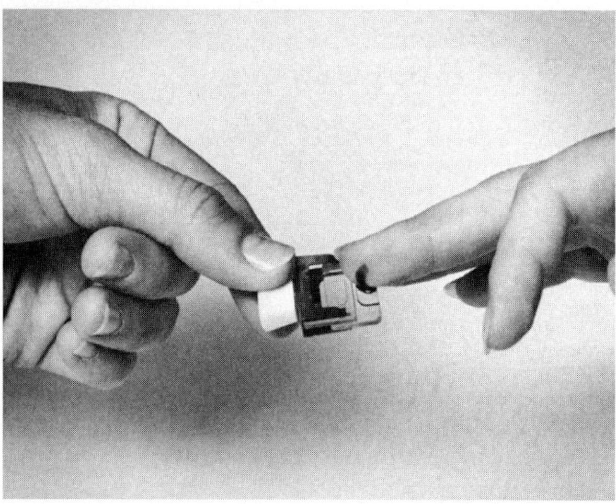

Figure 54.1 Add a fresh drop of blood to the moat plate of the blood chamber assembly. The blood should flow freely without excessive finger pressure.

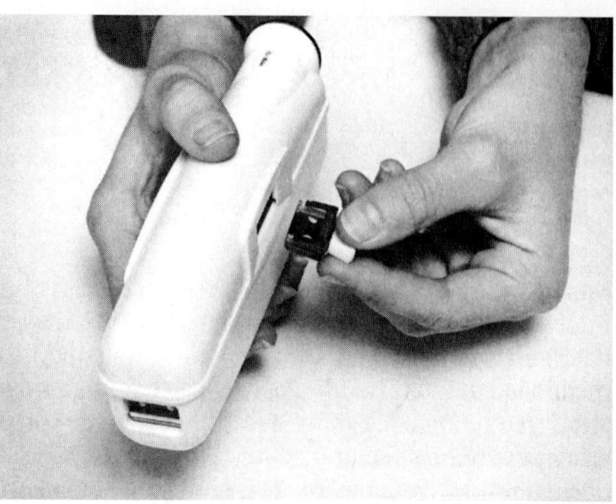

Figure 54.3 Insert the charged blood chamber into the slot of the hemoglobinometer. Before insertion, test the unit to make certain the batteries are active.

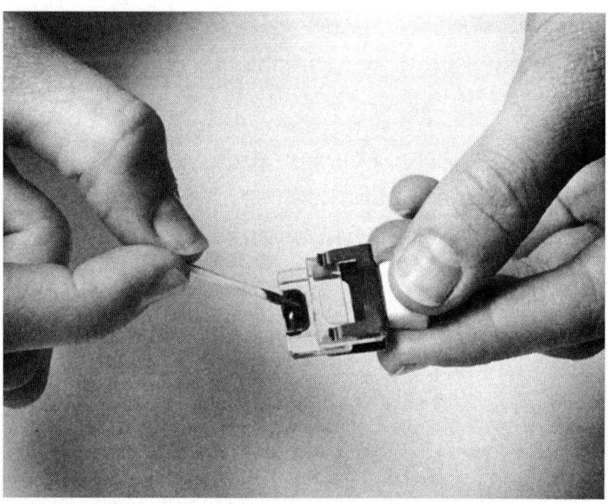

Figure 54.2 Hemolyze the blood sample on the moat plate with a wooden hemolysis applicator. Complete hemolysis requires 35 to 45 seconds.

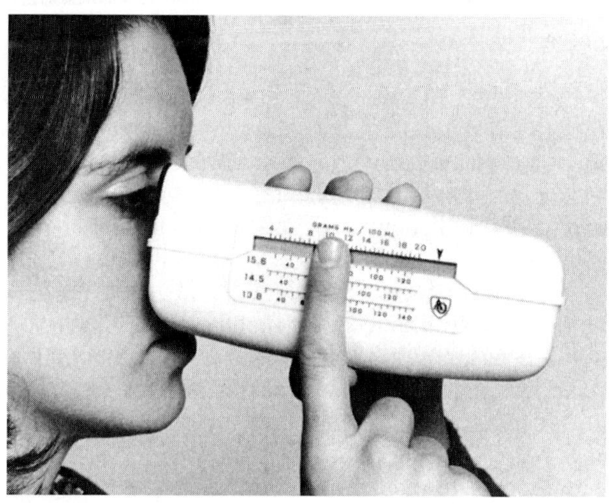

Figure 54.4 Analyze the blood sample by moving the slide button with your right index finger. When the two colors match in density, read the grams of hemoglobin per 100 ml on the scale.

6. Hemolyze the blood on the moat plate by mixing it with the pointed end of a hemolysis applicator, as figure 54.2 shows. It takes 30 to 45 seconds for all the red blood cells to rupture. Hemolysis is complete when the blood loses its cloudy appearance and becomes a transparent red liquid.
7. Push the moat plate in flush with the cover plate, and insert the sample into the side of the hemoglobinometer, as figure 54.3 shows.
8. Place the eyepiece to your eye with the left hand. Rest your left thumb on the light switch button on the bottom of the hemoglobinometer (see figure 54.4).
9. While pressing the light button with the left thumb, move the slide button on the side of the hemoglobinometer back and forth with the right index finger until the two halves of the split field match. The index mark on the slide knob indicates the grams of hemoglobin per 100 ml of blood. Read the percent hemoglobin on the scale.

Assignment:

Record your results in table II in section H of the Laboratory Report for Exercises 52–58.

Hematocrit (Packed Red Cell Volume) 55

Exercises 53 and 54 employed two different tests for determining the presence or absence of anemia: (1) the RBC (red blood cell) count and (2) hemoglobin percentage measurement with a hemoglobinometer. As noted earlier, anemia may be due to a dilution of the number of erythrocytes or to a hemoglobin deficiency, and the most significant factor is the amount of hemoglobin in a given volume of blood. A third method for diagnosing anemia is to determine the volume percentage of blood that red blood cells packed by centrifugation occupy. This is called the **hematocrit** or **packed cell volume (PCV).**

The hematocrit is determined by placing a blood sample in a heparinized capillary tube within a centrifuge specifically designed to accommodate the capillary tube. Following centrifugation, the volume of the packed cells is compared to the volume of the blood plasma.

This exercise uses a micromethod for determining the hematocrit. As figure 55.1 shows, capillary tubing that has been heparinized to prevent blood coagulation is used to collect the blood sample. The blood is centrifuged at high speed for 4 minutes. (Macromethods for determining the hematocrit require a greater volume of blood and usually centrifuge the sample for 30 minutes at a much slower speed.) After centrifugation, the percentage of total volume the packed red blood cells occupy is determined by one of the following three methods: (1) Depending on the type of centrifuge, the percentage of packed cell volume may be read directly from the centrifuge itself. (**Note:** For this reading to be accurate, the capillary tube must be completely filled with no air bubbles.) (2) If available, a special tube reader (see figure 55.6) can be used. (**Note:** Again, the capillary tube must be completely filled.) (3) A ruler can be used to measure the length of the portion of the tube filled with packed cells. Then this number is divided by the length of the tube filled with the entire blood sample (packed cells plus plasma), as figure 55.7 describes. Multiplying the result by 100 gives the percentage of total volume that the packed red blood cells occupy.

Interestingly, the number that represents the hematocrit is usually three times the grams of hemoglobin per 100 ml. In men, the normal hematocrit range is 40 to 54%, with 47% as average. In women, the normal range is 37 to 47%, with 42% as average. The principal advantages of the hematocrit over total RBC counts and hemoglobinometer measurements are simplicity, speed, and accuracy. Although this micromethod is not as precise as macromethods, its accuracy is ± 2% for most blood samples. Proceed as follows to determine the hematocrit of your blood:

Materials:
sterile, disposable lancets
alcohol swabs
hematocrit centrifuge
tube reader (optional)
heparinized capillary tubes
clay for sealing the hematocrit tubes (e.g., Seal-Ease or Critoseal)

1. Scrub a finger with an alcohol swab, and use a lancet to produce a free flow of blood from the finger. Wipe away the first drop of blood.
2. Place the unmarked end of a capillary tube into the blood flow, and allow the blood to be drawn up to the red line (see figure 55.1). Hold the open end of the tube downward from the blood source to fill the tube more rapidly.

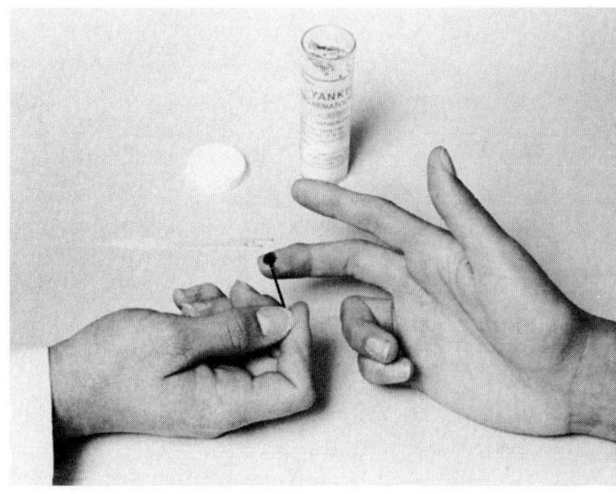

Figure 55.1 Draw blood into a heparinized capillary tube for hematocrit determination.

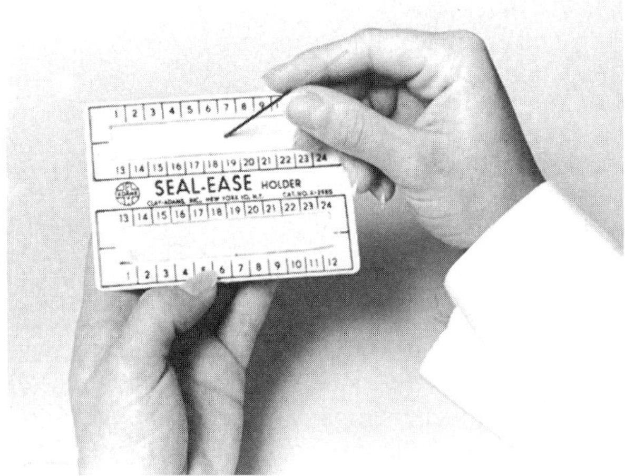

Figure 55.2 When the capillary tube is filled with blood, seal the end of the tube with clay.

Figure 55.3 Place the capillary tube in the centrifuge, with the sealed end toward the perimeter.

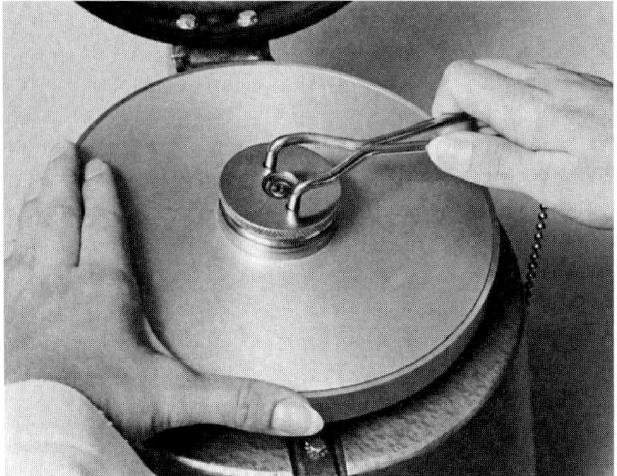

Figure 55.4 Tighten the inner safety lid with a lock wrench.

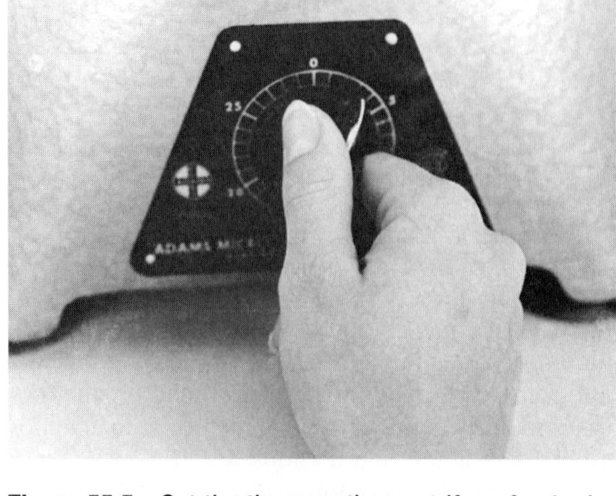

Figure 55.5 Set the timer on the centrifuge for 4 minutes by turning the dial clockwise.

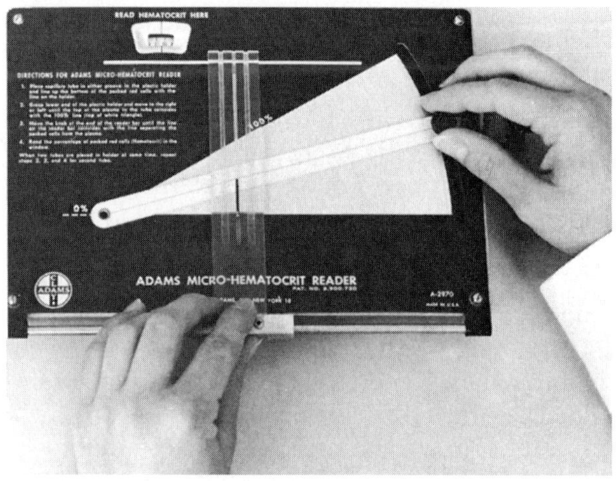

Figure 55.6 Place the capillary tube in the tube reader to determine the hematocrit.

3. Seal the blood end of the capillary tube with clay (see figure 55.2).
4. Place the capillary tube in the centrifuge, with the sealed end toward the outside (see figure 55.3). Load the centrifuge with an even number of tubes to properly balance the load.
5. Secure the inside cover carefully, and be sure it is locked in place (see figure 55.4).
6. Lower the outside cover, and securely latch it.
7. Turn on the centrifuge, setting the timer for 4 minutes (see figure 55.5).
8. Remove the sample from the centrifuge and determine the hematocrit by one of the three previously described methods (see figures 55.6 and 55.7).

Assignment:

Record your results in table II in section H of the Laboratory Report for Exercises 52–58.

Capillary Tube After Centrifugation

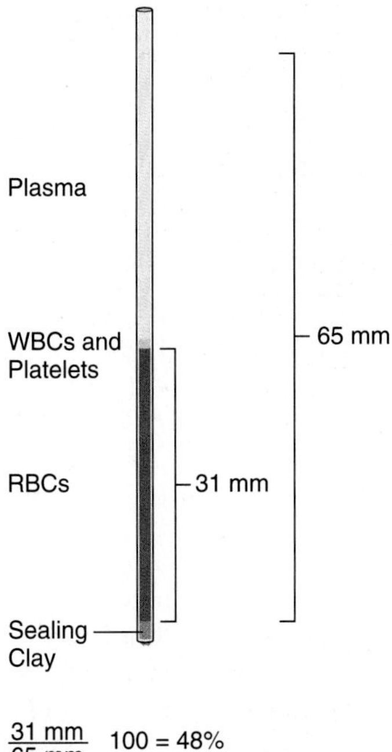

$\frac{31 \text{ mm}}{65 \text{ mm}}$ 100 = 48%

Figure 55.7 Calculate the hematocrit by dividing the length of the packed red blood cells by the length of the entire specimen.

56 Coagulation Time

Blood coagulation is a complex phenomenon involving over 30 substances. Most of these substances inhibit coagulation and are called *anticoagulants;* the remainder promote coagulation and are called *procoagulants.* Whether or not the blood coagulates depends on which group dominates in a given situation. Normally, the anticoagulants dominate, but when a vessel ruptures, the procoagulants in the affected area assume control, causing a clot (*fibrin*) to form in a relatively short time (normally, 2 to 6 minutes).

The method outlined here for determining the rate of blood coagulation is easy to perform and very reliable. Figures 56.1 through 56.3 illustrate the general procedure. Proceed as follows:

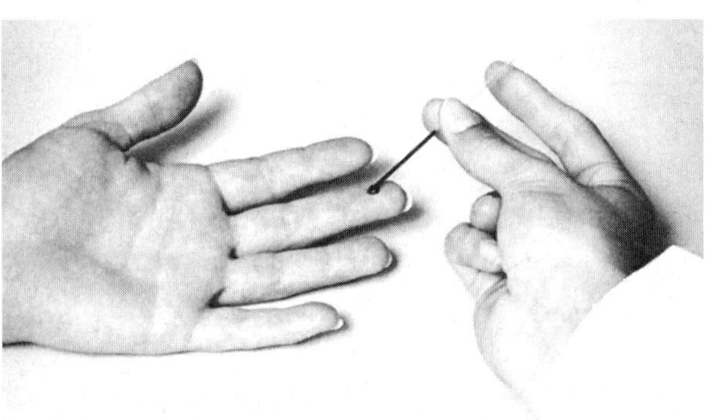

Figure 56.1 Draw blood into a nonheparinized capillary tube.

Materials:
sterile, disposable lancets
alcohol swabs
nonheparinized capillary tubes (0.5 mm diameter)
three-cornered file

1. Scrub a finger with an alcohol swab, and use a lancet to puncture the finger and produce a free flow of blood. **Record the time.**
2. Place one end of the capillary tube into the blood flow (see figure 56.1). Hold the open end of the tube downward from the blood source to fill the tube more rapidly.
3. At **1-minute intervals,** break off small portions of the tube by scratching the glass first with a file (see figure 56.2). **Important:** Separate the broken ends slowly and gently while looking for coagulation. The blood has coagulated when fibrin threads span the gap between the broken ends, as figure 56.3 shows.
4. **Record the time** from when the blood first appeared on the finger to the formation of fibrin.

Assignment:
Record your results in table II in section H of the Laboratory Report for Exercises 52–58.

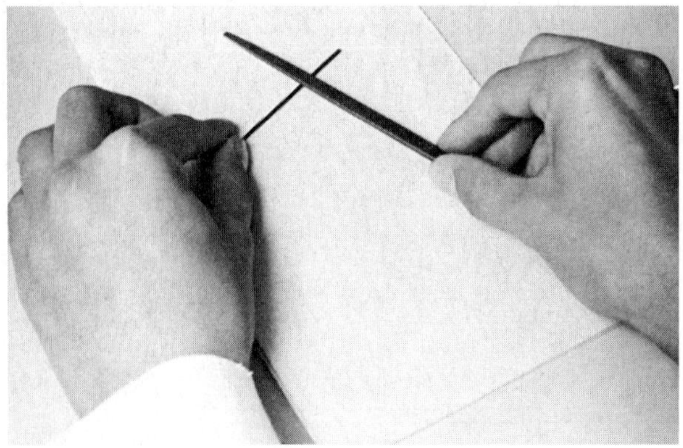

Figure 56.2 At 1-minute intervals, file and then gently break off small sections of the tube for the test.

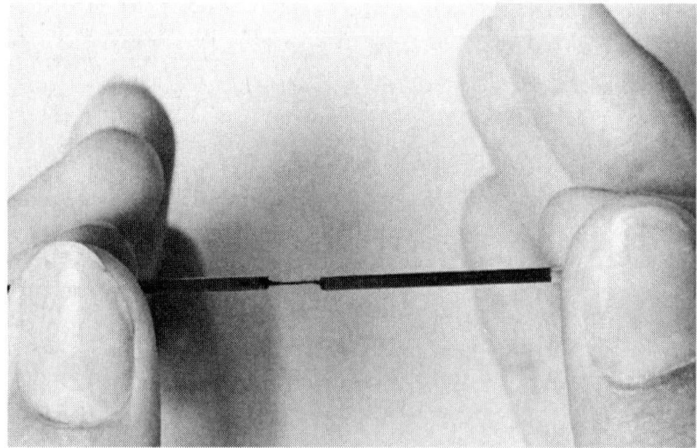

Figure 56.3 When the blood has coagulated, strands of fibrin extend between the broken ends of the tube.

Sedimentation Rate

57

If blood that has been mixed with sodium citrate, an anticoagulant, stands vertically in a tube for a time, the red blood cells fall to the bottom of the tube, leaving clear plasma at the top. The distance that the cells fall in 1 hour is the **sedimentation rate** (more commonly, *sed rate*). You will use the Landau micromethod, which uses only one drop of blood, to determine sed rate. The sed rate in normal adults is 0 to 6 mm; in children, the normal rate is 0 to 8 mm. During menstruation and pregnancy, the sed rate invariably exceeds 8 mm in women. More significantly, the sed rate in all acute general infections and malignancies is accelerated; pulmonary tuberculosis, rheumatic carditis, rheumatoid arthritis, and Hodgkin's disease all have high sedimentation rates.

The factor that most influences sedimentation rate is the size of sedimenting particles in the blood. *Rouleaux formation,* which results from erythrocytes adhering to form stacks of cells like poker-chip piles, is the principal cause of high sedimentation rates. Such vertical stacking of the cells increases the particle mass without greatly increasing surface area resistance to settling. When inflammation exists in the body, the negative surface charge of erythrocytes is reduced, causing the cells to lack essential repulsive forces. The result is rouleaux formation.

The negative charge of erythrocytes is a function of sialic acid groups in the plasma membrane. These acid groups, in turn, are attenuated by the presence of such macromolecules as fibrinogen and gamma globulin in plasma. Currently, high levels of fibrinogen are believed to promote rouleaux formation and thereby cause high sedimentation rates.

The sedimentation rate is an important diagnostic tool. Its significance to the physician matches that of the pulse, blood pressure, temperature, and urine analysis.

In performing the test that follows, take care to produce an ample puncture. Most first-time attempts to perform this test fail due to insufficient blood flow from the punctured finger.

Materials:
sterile, disposable lancets
alcohol swabs
pipette cleaning solutions (four-bottle kit)
Landau sed-rate pipette
Landau rack
mechanical suction device
5% sodium citrate
clean microscope slide (optional)
plastic millimeter ruler

1. Draw sodium citrate solution up to the first line that completely encircles the pipette. Use the mechanical suction device figure 57.1 shows.
2. Scrub a finger with an alcohol swab, and use a lancet to puncture the finger and produce a free flow of blood. Wipe off the first drop with cotton. Then draw the blood into the pipette until the blood reaches the second line (see figure 57.2). Take care not to get any air bubbles into the blood. **Note:** If air bubbles do occur in the blood, carefully expel the mixture onto a clean microscope slide, and draw it up again.
3. Mix the blood and citrate by drawing the two fluids up into the bulb and expelling them into the

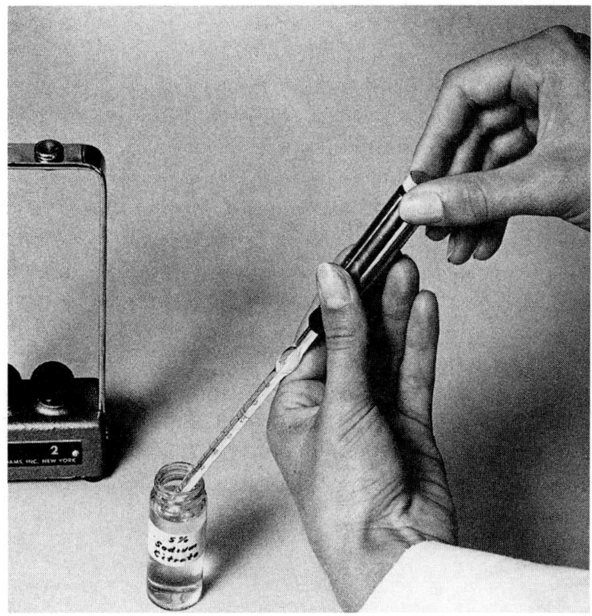

Figure 57.1 Draw anticoagulant sodium citrate into the pipette to the first line.

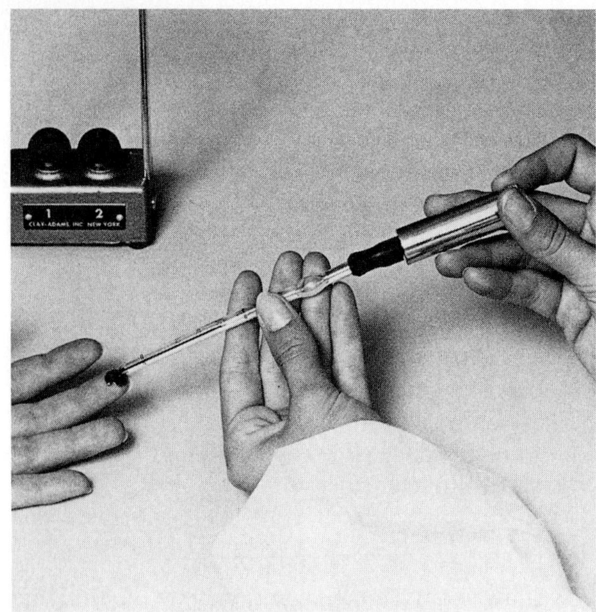

Figure 57.2 Draw blood into the pipette, and mix it with the sodium citrate.

Figure 57.3 After the blood and anticoagulant are completely mixed, place the pipette in the rack.

lumen of the tube. **Do this six times.** If any bubbles occur in the tube lumen at this time, use the microscope slide technique described in step 2.

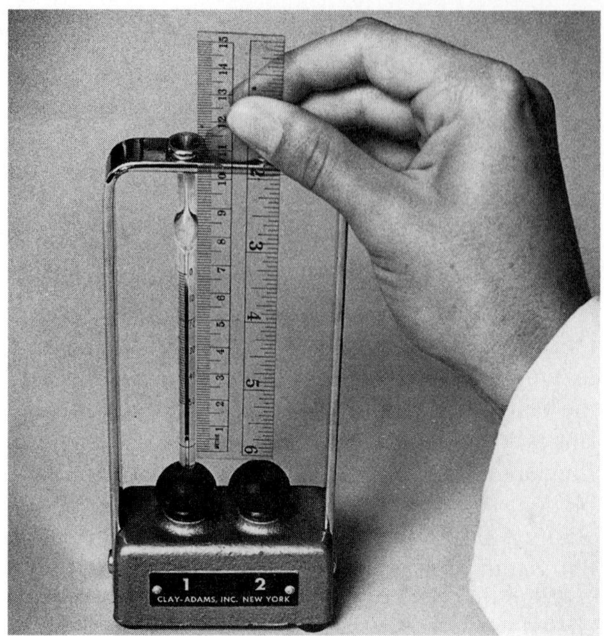

Figure 57.4 After 1 hour, measure the fall of red blood cells.

4. Adjust the top level of the blood as close to zero as possible. (Getting it to stop exactly on zero is almost impossible.)
5. Place the lower end of the pipette on the index finger of the left hand, and then remove the suction device by pulling it off the other end. If the lower end is not completely sealed, this step will pull the blood out of the pipette.
6. Place the lower end of the pipette on the base of the pipette rack and the other end at the top of the rack (see figure 57.3). Be sure that the tube is exactly vertical. Record the time at which you put the tube in the rack.
7. After 1 hour, measure the distance from the top of the plasma to the top of the red blood cells with a plastic millimeter ruler (see figure 57.4).
8. Use the mechanical suction device to rinse the pipette with bleach, distilled water, alcohol, and acetone.

Assignment:

Record your results in table II in section H of the Laboratory Report for Exercises 52–58.

Blood Typing

58

The surfaces of red blood cells possess various antigenic molecules designated by such letters as A, B, C, D, E, c, d, e, M, and N. Although the exact chemical composition of all these substances is undetermined, some appear to be carbohydrate residues (oligosaccharides).

The presence or absence of these various substances determines an individual's blood type. Since the chemical makeup of cells is genetic, an individual's blood type is the same in old age as it is at birth. It never changes.

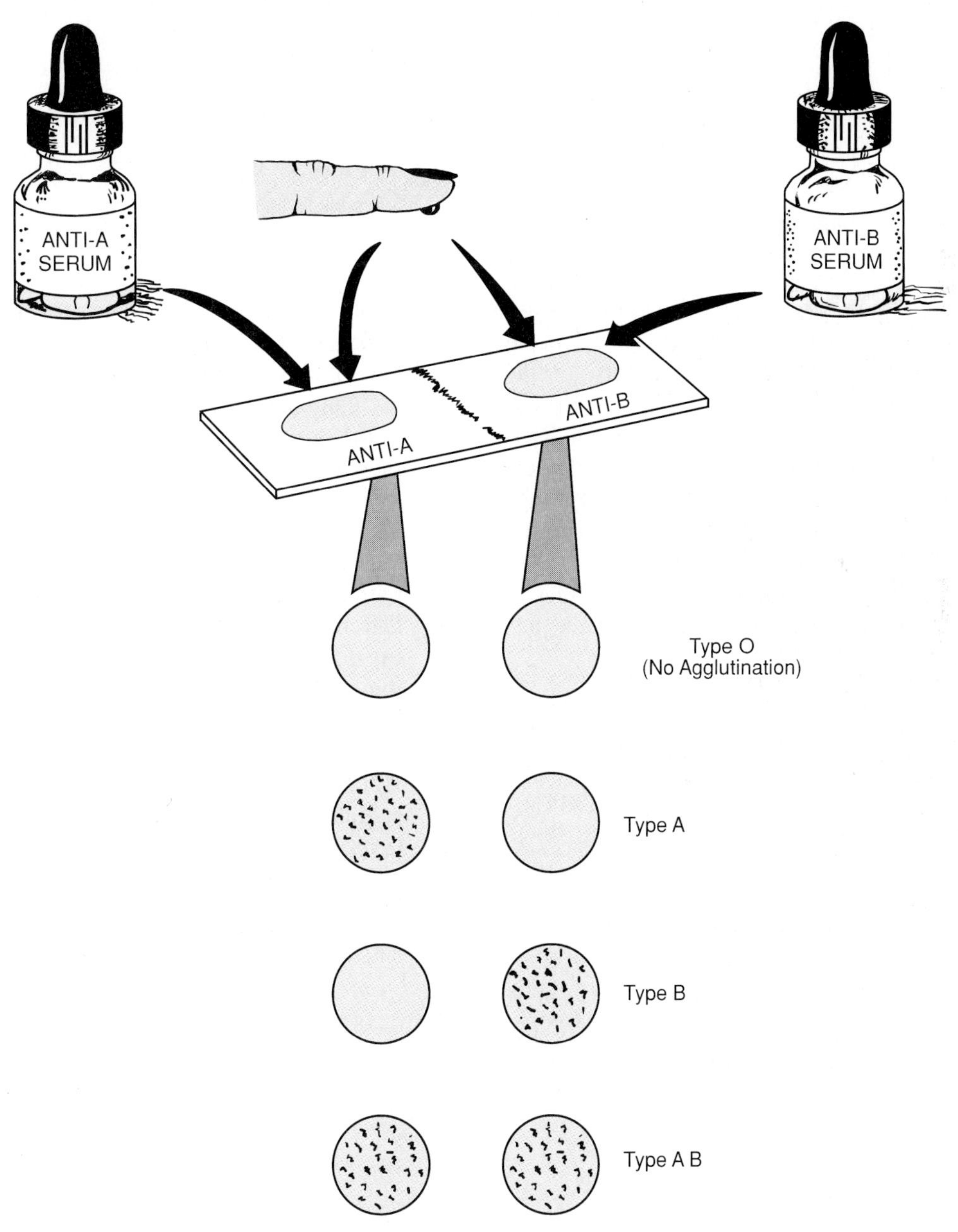

Figure 58.1 Blood typing ABO groups.

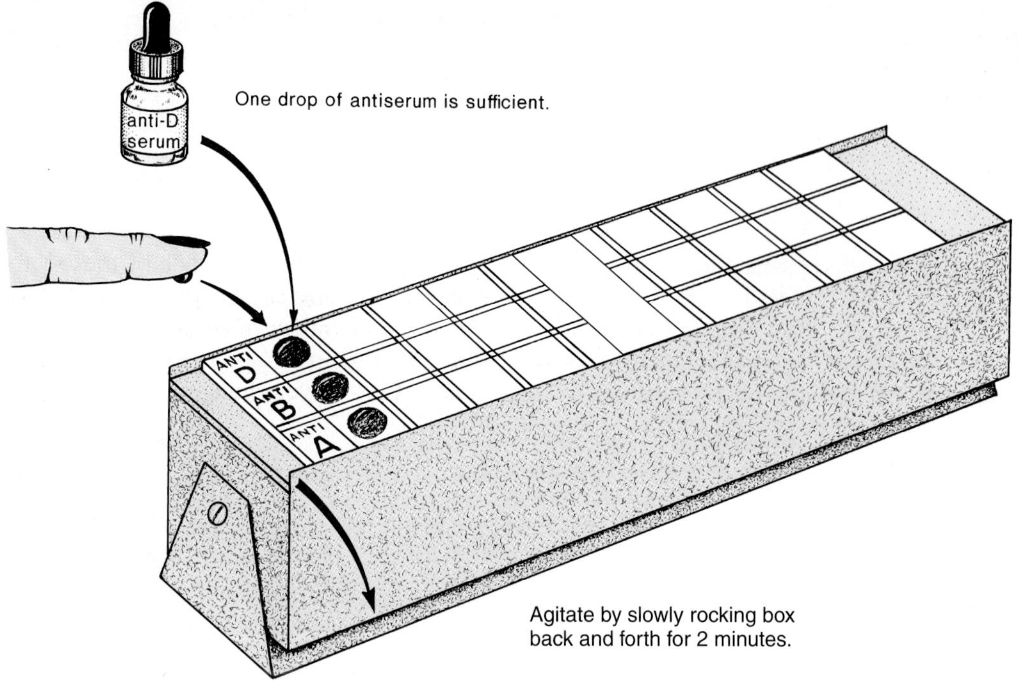

Figure 58.2 Blood typing with a warming box.

The only antigenic factors of concern in this exercise are A, B, and D (Rh), since they are most commonly involved in transfusion reactions.

To determine an individual's blood type, drops of blood-typing sera containing A, B, and D (Rh) antibodies are added to drops of blood to detect **agglutination** (clumping) of the cells. ABO typing (see figure 58.1) may be performed at room temperature. Rh typing for the D factor, on the other hand, requires a higher temperature (around 50°C) and is done on an Rh typing box that warms the slide (see figure 58.2).

Materials:
sterile, disposable lancets
alcohol swabs
wax pencil and two microscope slides
typing sera (anti-A, anti-B, and anti-Rh)
toothpicks
Rh typing box

1. Divide slide 1 down the middle with a wax pencil. Write Anti-A on the left side and Anti-B on the right side. Write Anti-Rh on the bottom of slide 2.
2. Add a drop of anti-A serum to the left side of slide 1 and a drop of anti-B serum to the right side of slide 1. Place a drop of anti-Rh serum in the center of slide 2.
3. Scrub the middle finger with an alcohol swab, pierce the finger with a lancet, and wipe away the first drop of blood.
4. Then place a drop of blood next to each of the three drops of sera.
5. Quickly mix each blood-serum sample, using a different toothpick for each sample. Dispose of the toothpicks, alcohol swabs, and lancets in the appropriate containers.
6. Place the Rh slide on the Rh typing box, and gently rock the box back and forth (see figure 58.2).
7. After 2 minutes, observe the slides for evidence of agglutination. Positive results for the Rh test are difficult to observe because the clumps that form are very small. If you are unsure about the results of the Rh test, look at the slide under the microscope.
8. Record your blood type on the Laboratory Report.

Assignment:
Now complete section C of the Laboratory Report for Exercises 52–58. Then, as a review, complete sections D through I.

PART 11 The Circulatory System

This part consists of twelve exercises. The first eight exercises relate directly to the anatomy and physiology of the heart. The remainder concern general circulation. The extensive use of electronic equipment in many of the experiments may necessitate a review of certain exercises in Part 5. A cat whose arteries and veins have been injected with colored latex is dissected in Exercise 68. The role of the thymus gland in cellular and humoral immunity is studied along with the lymphatic system in Exercise 70.

59 Anatomy of the Heart

This study of the anatomy of the human heart uses a sheep heart for dissection and comparison. Before beginning the dissection, however, study figures 59.1 and 59.2.

Internal Anatomy

Figure 59.1 shows the internal anatomy of the human heart. Red vessels carry oxygenated blood, and blue ones carry deoxygenated blood.

Chambers of the Heart

The heart has four chambers: two small upper atria and two larger ventricles. The atria receive all the blood that enters the heart, and the ventricles pump it out. Note that the **right atrium** in the frontal section is on the left side of the illustration, and the **left atrium** is on the right. Observe also that although the **right ventricle** is somewhat larger than the left ventricle, the **left ventricle** has a thicker wall. Separating the two ventricles is a partition, the **interventricular septum.** An **interatrial septum** separates the two atria but is not visible in either figure 59.1 or 59.2.

Wall of the Heart

The wall of the heart has three layers: the myocardium, the endocardium, and the epicardium. The **myocardium** is composed of cardiac muscle tissue. Note in the enlarged heart wall section in figure 59.1 that the myocardium makes up the bulk of the wall thickness.

The **endocardium** lines the heart's inner surface. This thin serous membrane is continuous with the endothelial lining of the arteries and veins. An infection of this membrane is called *endocarditis.*

Attached to the outer surface of the heart is another serous membrane, the **epicardium,** or **visceral pericardium.** This membrane produces serous fluid that enables the heart to move freely within the pericardial sac. The *pericardial sac,* or **parietal pericardium,** has two layers: an inner serous layer and an outer fibrous layer. Between the heart and the parietal pericardium is the **pericardial cavity.** Its dimension is exaggerated in figure 59.2.

Vessels of the Heart

The major vessels of the heart are the two venae cavae, the pulmonary trunk, the four pulmonary veins, and the aorta. The **superior vena cava** is the upper blue vessel on the right atrium; it conveys deoxygenated blood to the right atrium from the head and arms. The **inferior vena cava** is the lower blue vessel that empties deoxygenated blood from the trunk and legs into the right atrium. From the right atrium, blood passes to the right ventricle, from which it leaves the heart via the **pulmonary trunk.** This large vessel branches into the **right** and **left pulmonary arteries,** which carry blood to the lungs.

The four **pulmonary veins** drain blood from the lungs and carry it to the left atrium. From the left atrium, the blood passes into the left ventricle, where it exits the heart through the *aorta.* Although the initial emergence of the aorta is obscured in figure 59.1, you can see it as a large, curved, red vessel, the **aortic arch,** that has three smaller arteries emerging from its superior surface. The aorta carries oxygenated blood to the entire systemic circulatory system.

The **ligamentum arteriosum** appears to be a short vessel between the pulmonary trunk and the aortic arch. During prenatal life, this ligament is a functional blood vessel, the *ductus arteriosus,* that allows blood to pass from the pulmonary trunk to the aorta, bypassing the lungs.

Valves of the Heart

The heart has two atrioventricular and two semilunar valves. Between the right atrium and the right ventricle is the **tricuspid valve.** Between the left atrium and the left ventricle is the **bicuspid,** or *mitral,* **valve.** The superior sectional view of figure 59.1 shows the differences between these two valves. Note that the tricuspid valve has three flaps, or cusps, and that the bicuspid valve has only two cusps. The edges of the cusps have fine cords, the **chordae tendineae,** which anchor to **papillary muscles** on the wall of the heart. These cords and muscles prevent the cusps from being forced up into the atria during systole (ventricular contraction).

The other two valves are the **pulmonary semilunar** and **aortic semilunar valves** at the bases of

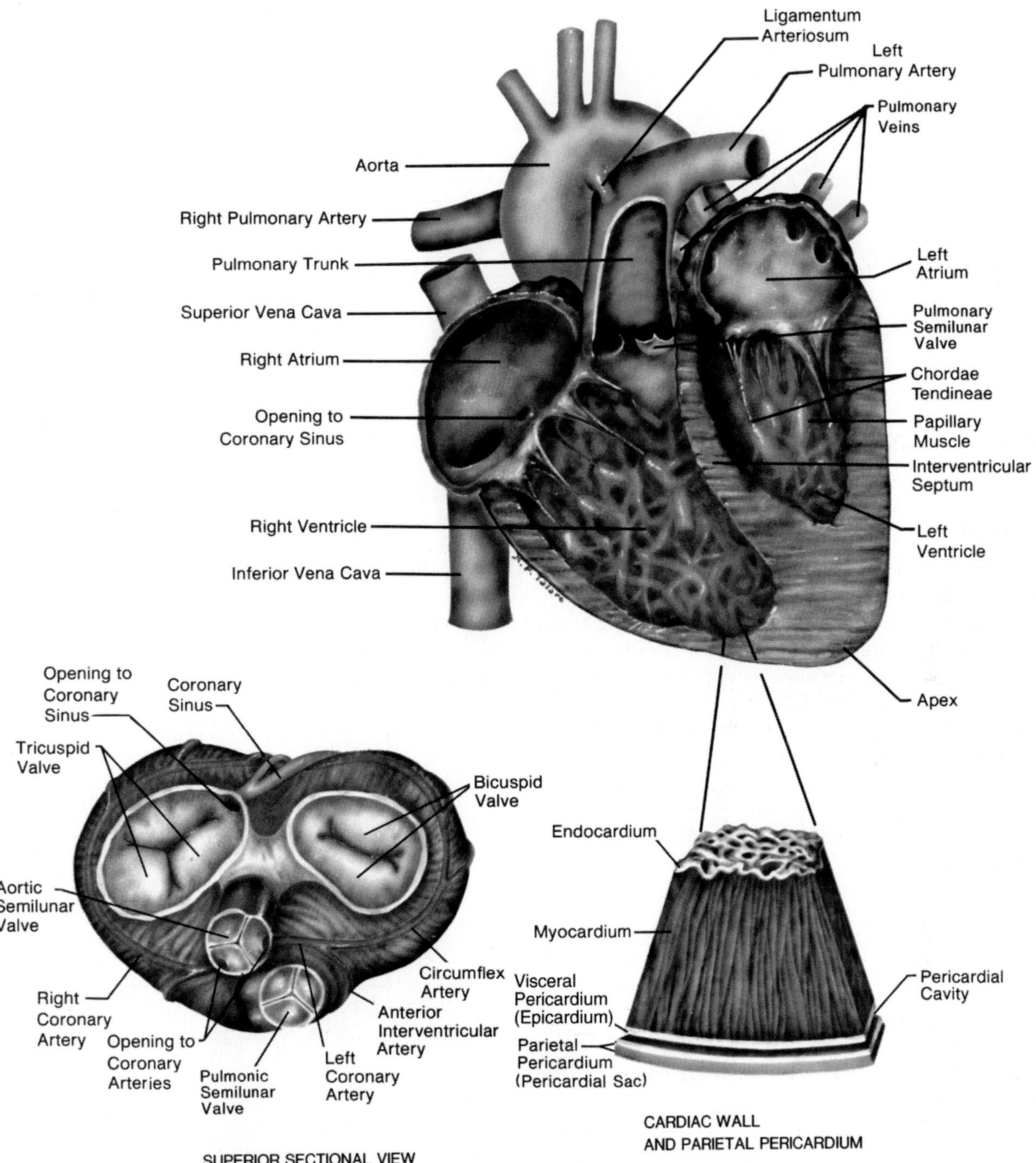

Figure 59.1 Internal anatomy of the heart.

the pulmonary trunk and the aorta. They prevent blood in those vessels from flowing back into the heart during diastole (relaxation phase). Note in the superior sectional view in figure 59.1 that each of these valves has three small cusps.

Coronary Circulation

The vessels that supply the heart muscle with blood comprise the *coronary circulatory system.* Oxygenated blood in the aorta passes through two **openings to the coronary arteries** just superior to the aortic semilunar valve. One opening leads into the **left coronary artery,** and the other leads into the **right coronary artery.**

The left coronary artery has two branches: a **circumflex artery** that passes around the heart in the left atrioventricular sulcus, and an **anterior interventricular artery** that passes down the interventricular sulcus on the heart's anterior surface. The

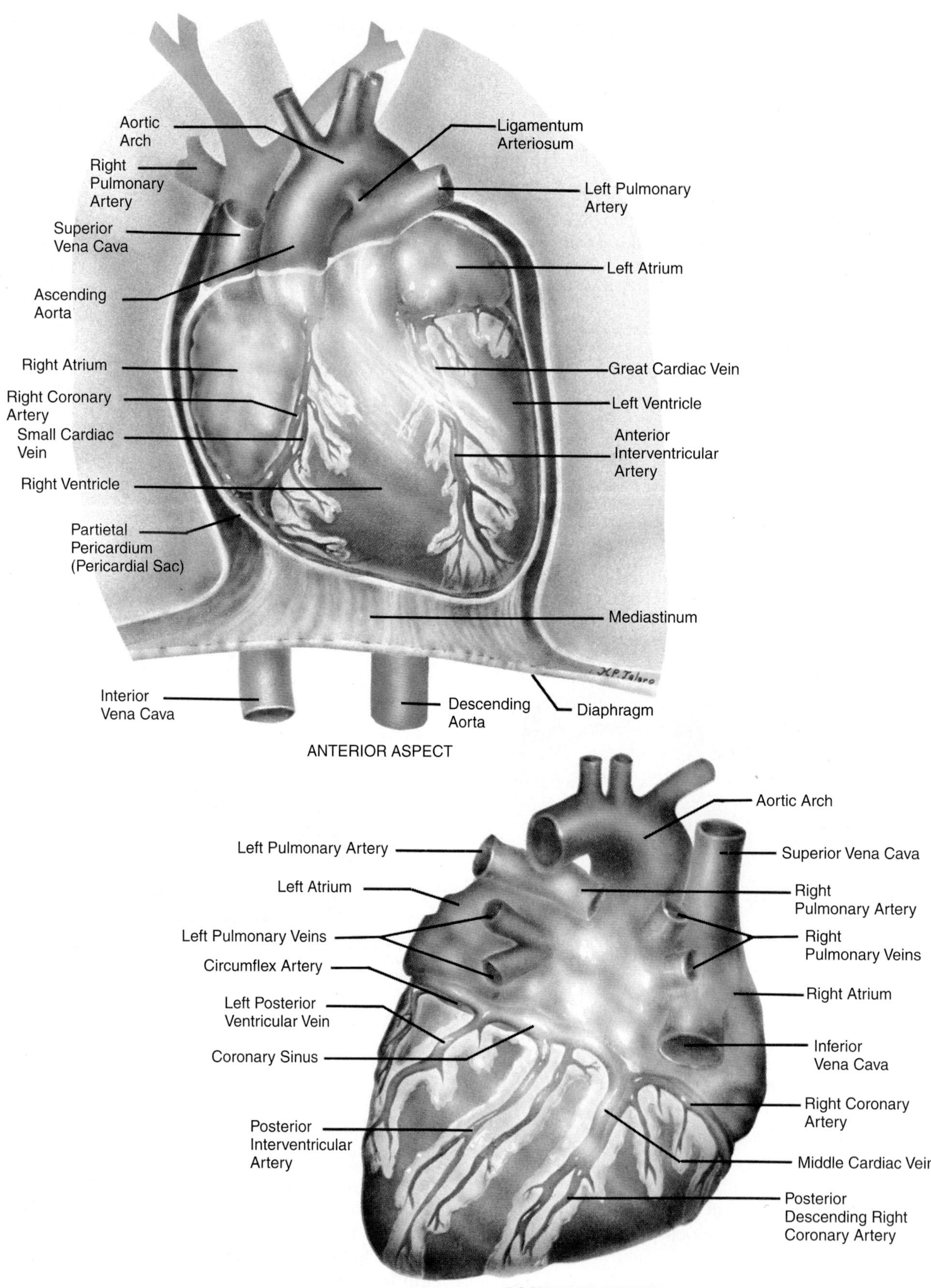

Figure 59.2 External anatomy of the heart.

right coronary artery lies in the right atrioventricular sulcus and has branches that supply the posterior and anterior surfaces of the ventricular muscle.

Once the coronary blood has been relieved of its oxygen and nutrients in the heart muscle, the various veins that parallel the arteries empty the blood into a large vein, the **coronary sinus.** Blood in the coronary sinus empties into the right atrium through a portal, the **opening to the coronary sinus.**

External Anatomy

The external anatomy of the heart is relatively simple once you understand the internal anatomy. Figure 59.2 shows two external views of the heart.

Anterior Aspect

The heart tilts slightly to the right within the **mediastinum** so that the *apex,* or tip of the heart, is somewhat on the left side. The obscured appearance of the coronary vessels in the anterior view in figure 59.2 is due to the **parietal pericardium** lying intact over the organ. The clarity of the structures in the posterior view in figure 59.2 is due to the removal of the parietal pericardium.

Locating the *interventricular sulcus* helps to differentiate the right ventricle from the left ventricle. This sulcus is the slight depression over the interventricular septum that contains the **anterior interventricular artery** and the **great cardiac vein.** The yellow material in the sulcus consists of fat deposits.

Three portions of the aorta are labeled on the anterior aspect in figure 59.2: a short **ascending aorta** that emerges from the left ventricle, a curving **aortic arch,** and a portion of the **descending aorta** that passes down through the mediastinum behind the heart.

Locate the **right** and **left pulmonary arteries** that emerge from behind the aortic arch. The aorta obscures most of the pulmonary trunk that gives rise to these arteries. Also identify the **ligamentum arteriosum.** The coronary vessels in the atrioventricular sulcus are the **right coronary artery** and the **small cardiac vein.**

Posterior Aspect

The posterior view of the heart in figure 59.2 reveals more clearly the position of the four pulmonary veins and additional coronary vessels. The two **right pulmonary veins** are near the superior vena cava. The two **left pulmonary veins** are to the left of the right pulmonary veins. The single blue vessel lying below the aorta is where the pulmonary trunk divides into the **right** and **left pulmonary arteries.**

The posterior view in figure 59.2 shows the following coronary arteries: the **right coronary artery,** which lies in the right atrioventricular sulcus; the **posterior descending right coronary artery,** which is a downward extension of the right coronary artery; the **posterior interventricular artery;** and the **circumflex artery.**

The major coronary veins in the posterior view in figure 59.2 are the **middle cardiac vein,** which parallels the posterior descending right coronary artery; the **left posterior ventricular vein;** and the **posterior interventricular vein,** which lies in the interventricular sulcus with the similarly named artery. All of these coronary veins empty into the large **coronary sinus** that lies in the atrioventricular sulcus.

Assignment:
Complete parts A and B of the Laboratory Report for this exercise.

Sheep Heart Dissection

While dissecting the sheep heart, try to identify as many of the structures that figures 59.1 and 59.2 show as possible.

> ***Materials:***
> sheep heart, fresh or preserved
> dissecting instruments and tray

1. Rinse the heart with cold water to remove excess preservative or blood. Allow water to flow through the large vessels to remove blood clots.
2. Look for evidence of the **parietal pericardium.** This fibroserous membrane is usually absent from laboratory specimens, but remnants of it may still be attached to the large blood vessels of the heart.
3. Try to isolate the **visceral pericardium** (**epicardium**) from the outer surface of the heart. Since it consists of only one layer of squamous cells, it is very thin. With a sharp scalpel, try to peel a small portion of the visceral pericardium away from the myocardium.
4. Identify the **right** and **left ventricles** by squeezing the walls of the heart, as part 1 in figure 59.3 shows. The wall of the right ventricle is thinner.
5. Identify the **anterior interventricular sulcus,** which lies between the two ventricles. Fat tissue usually covers it. Carefully trim away the fat in this sulcus to expose the **anterior interventricular artery** and the **great cardiac vein.**

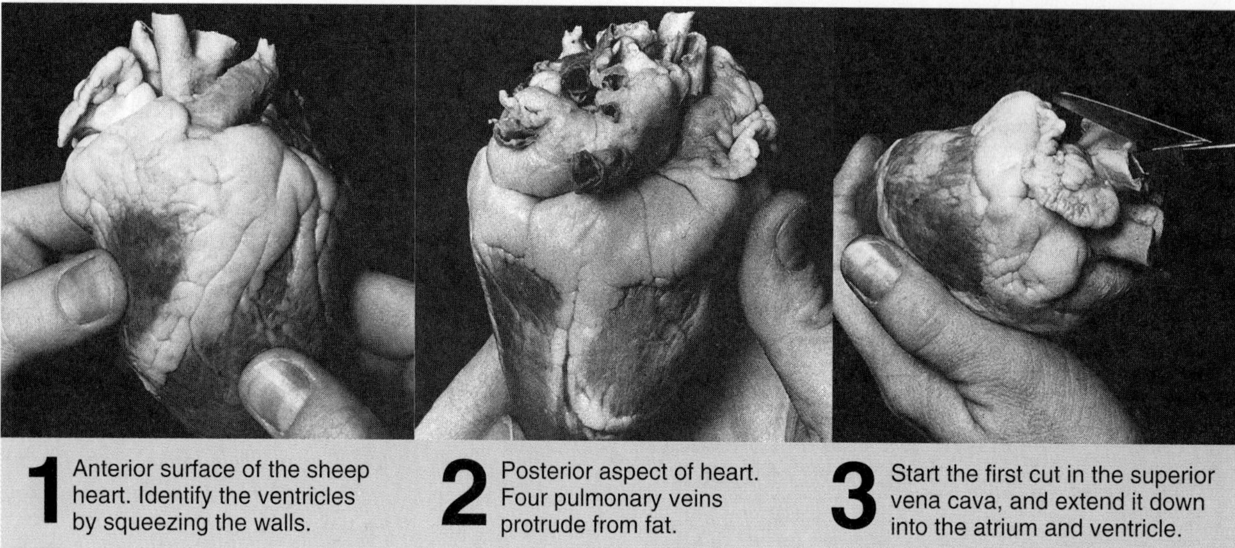

1 Anterior surface of the sheep heart. Identify the ventricles by squeezing the walls.

2 Posterior aspect of heart. Four pulmonary veins protrude from fat.

3 Start the first cut in the superior vena cava, and extend it down into the atrium and ventricle.

Figure 59.3 Sheep heart dissection.

6. Locate the **right** and **left atria.** Part 1 of figure 59.3 shows the right atrium.
7. Identify the **aorta,** which is the large vessel just to the right of the right atrium (see part 1, figure 59.3). Carefully peel away some of the fat around the aorta to expose the **ligamentum arteriosum.**
8. Locate the **pulmonary trunk,** which, on the anterior side of the heart, is the large vessel between the aorta and the left atrium. If the vessel is long enough, trace it to where it divides into the **right** and **left pulmonary arteries.**
9. Examine the posterior surface of the heart (see part 2, figure 59.3). Only the right atrium and two ventricles can be seen on this side. Fat obscures the left atrium. Look for the four thin-walled **pulmonary veins** embedded in the fat. Probe these vessels to see that they lead into the atrium.
10. Locate the **superior vena cava,** which attaches to the upper part of the right atrium. Insert one blade of your dissecting scissors into this vessel (see part 3, figure 59.3), and cut through it into the atrium to expose the **tricuspid valve** between the right atrium and right ventricle. Do not cut into the ventricle at this time.
11. Pour water through the tricuspid valve to fill the right ventricle. Gently squeeze the ventricle's walls to note the closing action of the valve's cusps.
12. Drain the water from the heart, and use scissors to continue the cut from the right atrium through the tricuspid valve down to the apex of the heart.
13. Open the heart, and flush it again with cold water. Examine the interior (see part 1, figure 59.4).
14. Examine the interior wall of the right atrium. This inner surface has ridges that give it a comblike appearance; thus, it is called the **pectinate muscle** (*pecten,* comb).
15. Look for the **opening to the coronary sinus** (see arrow in part 1, figure 59.4). Blood of the coronary circulation returns to the venous circulation through this opening.
16. Insert a probe under the cusps of the tricuspid valve. Are three flaps evident?
17. Locate the **papillary muscles** and **chordae tendineae** in the right ventricle.
18. Look for the **moderator band,** a reinforcement cord that extends from the ventricular septum to the ventricular wall. It prevents excessive stress in the myocardium of the right ventricle.
19. With scissors, cut the right ventricular wall up along its lower margin, parallel to the anterior interventricular sulcus, to the pulmonary trunk. Continue the cut through the exit of the right ventricle into the **pulmonary trunk.** Spread the cut surfaces of this new incision to expose the **pulmonary semilunar valve.** Wash the area with cold water to dispel blood clots.
20. Insert one blade of your scissors into the left atrium, as part 2 in figure 59.4 shows. Cut through the atrium to the left ventricle. Also cut from the left ventricle into the aorta, slitting this vessel longitudinally.
21. Examine the **bicuspid** (mitral) **valve** between the left atrium and left ventricle. With a probe, reveal the two cusps.
22. Examine the pouches of the **aortic semilunar valve.** Compare the number of pouches of this valve with that of the pulmonary semilunar valve.

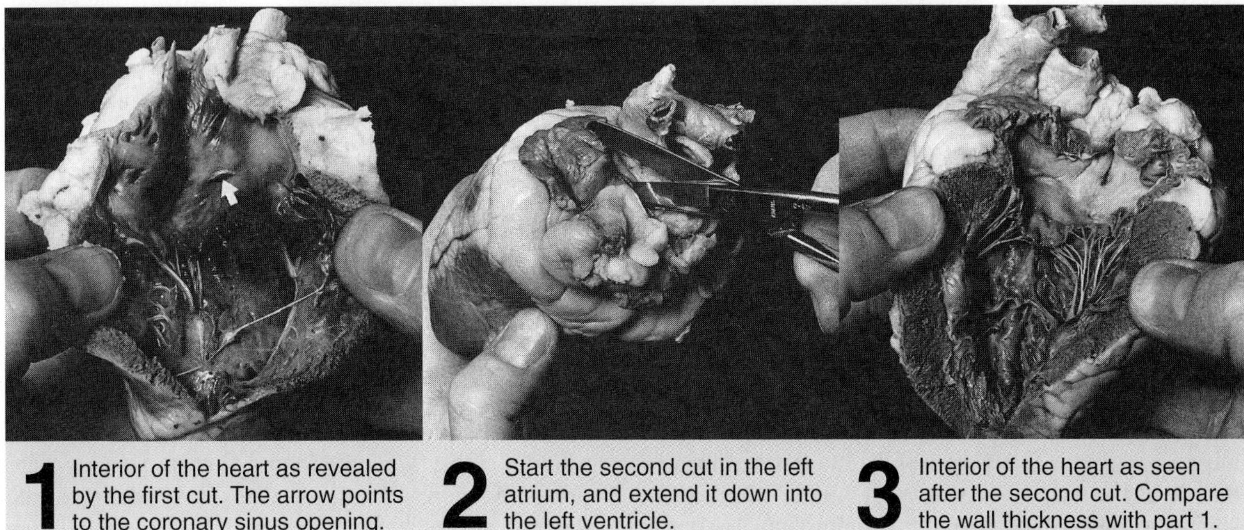

1 Interior of the heart as revealed by the first cut. The arrow points to the coronary sinus opening.

2 Start the second cut in the left atrium, and extend it down into the left ventricle.

3 Interior of the heart as seen after the second cut. Compare the wall thickness with part 1.

Figure 59.4 Sheep heart dissection.

23. Look for the two **openings to the coronary arteries** in the walls of the aorta just above the aortic semilunar valve. Force a blunt probe into each hole. Note how the openings lead into the two coronary arteries.

Assignment:

Complete part C of the Laboratory Report for this exercise.

60 Cardiovascular Sounds

Myocardial contraction and valvular movement create three distinct cardiovascular sounds. The **first sound** (lub) occurs simultaneously with the contraction of the ventricular myocardium and the closure of the mitral and tricuspid valves.

The **second sound** (dup) follows the first one after a brief pause. It differs from the first sound in that it has a snapping quality of higher pitch and shorter duration. The virtually simultaneous closure of the aortic and pulmonary semilunar valves at the end of ventricular systole causes the second sound.

The **third sound** is low pitched and is loudest near the apex of the heart. It is more difficult to detect but can be heard most easily when the subject is lying down. The vibration of the ventricular walls and the closing of the atrioventricular (A-V) valves during atrial systole cause the third sound.

Abnormal heart sounds, collectively referred to as **murmurs,** are usually due to damaged valves. Mitral valve damage due to rheumatic fever is the most frequent cause of heart murmur; the aortic valve is next most vulnerable to damage. Damaged valves may result in *stenosis,* the narrowing of a blood passageway, or *regurgitation* of blood flow.

Examination Procedures

Auscultation is listening to sounds generated within the body. Valvular sounds are best monitored at four specific spots in the thorax, as part 1 of figure 60.1 shows. Note that these auscultatory areas do not coincide with the anatomic locations of the various valves; this is because valvular sounds project to different spots on the rib cage.

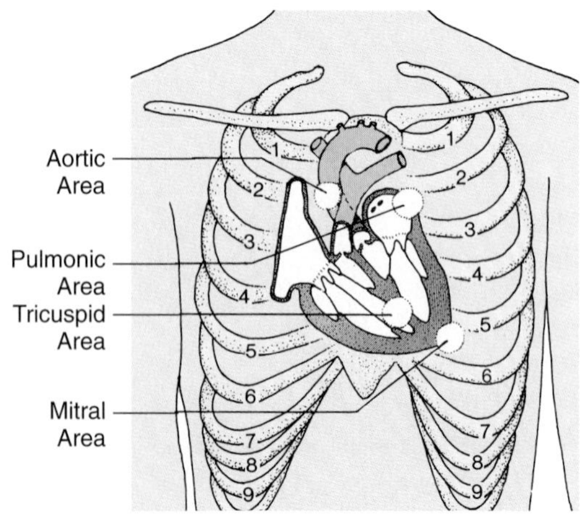

Figure 60.1 Auscultatory areas and two electronic setups.

Heart sounds can be monitored with a stethoscope, with electronic recording, and with an audio monitor. Only the procedures for the first two are presented here. Your instructor may demonstrate the use of an audio monitor, if one is available. *Since noise is distracting in these experiments, everyone must keep their voices low.*

Audio Monitor Demonstration (by Instructor)

With a student in a recumbent position as a subject, the instructor uses an audio monitor to demonstrate heart sounds to the class.

In demonstrations of the splitting of second sounds, hand signals should replace talking. The subject should inspire when the instructor's hand is raised and expire when the hand is lowered.

Stethoscope Auscultation

Work with your laboratory partner to auscultate each other's heart with a stethoscope, using the following procedure:

Materials:
stethoscope
alcohol swabs

 **Laboratory 5:** Electrocardiogram

1. Clean and disinfect the earpieces of a stethoscope with alcohol swabs. Fit them to your ears, directing them inward and upward.
2. Try to maximize the **first sound** by placing the stethoscope on the tricuspid and mitral auscultatory areas (see of figure 60.1) of your laboratory partner, who is sitting erect. Note that the mitral area coincides with the apex of the heart (fifth rib) and that the tricuspid area is 2 to 3 inches medial to this spot.
3. Now move the stethoscope to the aortic and pulmonic areas to hear the **second sound** more clearly.

Note that the aortic area is on the right border of the sternum, between the second and third ribs, and that the pulmonic area is 2 to 3 inches to the left of the aortic area.

4. Try to detect the splitting of the second sound while your laboratory partner is inhaling. Use hand signals instead of talking to communicate. Have your laboratory partner inspire when you raise your hand and expire when you lower your hand.

   This splitting characteristic of the second sound during inspiration is normal. During inspiration, more venous blood is forced into the right side of the heart, delaying the closure of the pulmonic valve during systole. Thus, you hear aortic and pulmonic closure sounds separately.
5. Now listen to the heart sounds with your laboratory partner lying down, face upward.
6. Can you detect the **third sound?**
7. Is there any evidence of **murmurs?**
8. Ask your laboratory partner to run up and down stairs or jog for a short distance. Compare the heart sounds before and after exercise, recording your observations on the Laboratory Report.
9. Before ending the examination, check all the auscultatory areas in part 1 of figure 60.1 to make sure that you have identified all valvular sounds.

Electronic Recording (Unigraph Setup) of Heart Sounds

Produce a phonocardiogram of heart sounds of a subject lying down, face upward.

Materials:
Unigraph
Trans/Med 6605 adapter
Statham P23AA pressure transducer

1. Attach a Trans/Med 6605 adapter to the input end of the Unigraph.
2. Attach a Statham pressure transducer and microphone jack to the 6605 adapter. Refer to part 3 of figure 60.1.
3. Turn on the Unigraph, and set the chart control (c.c.) switch at STBY, the heat control at the two o'clock position, the speed control at slow, the gain control at 1 mV/cm, and the mode control at TRANS.
4. Place the microphone against one of the auscultatory areas, and start the chart moving by placing the c.c. switch at Chart On. Adjust the temperature control knob to get a good recording. Center the tracing with the centering control knob, and adjust sensitivity to produce at least 5 mm stylus displacement.
5. Move the speed control lever to the high speed position.
6. Make recordings with the microphone over each auscultatory area. Be sure to mark the chart to identify the areas being monitored.
7. Try to record the splitting of the second sound, using the same techniques that you used with a stethoscope.
8. Attach phonocardiograms to the Laboratory Report.
9. Deactivate all controls on the Unigraph at the end of the experiment.

Assignment:
Complete the Laboratory Report for this exercise.

61 Heart Rate Control

Internal and external factors govern the heartbeat rate. Specialized centers of modified muscle tissue within the heart establish a particular rhythm that extrinsic nerves supplying the heart can accelerate or slow down. The **cardioregulator center** of the medulla oblongata directs these extrinsic nerves. Chemical and physical factors, acting on parts of the heart, certain blood vessels, and the cardioregulator center, reflexly alter the heart rate.

Intrinsic Tissues

Figure 61.1 shows the locations of the intrinsic and extrinsic tissues responsible for heart rate regulation. The intrinsic conducting tissues of the heart are the sinoatrial (S-A) and atrioventricular (A-V) nodes. The **S-A node** is a small mass of tissue in the wall of the right atrium near the superior vena cava. This node initiates contractions in the right atrium and for this reason is often called the *pacemaker.*

The **A-V node** in the lower part of the right atrium acts as a relay station. It transmits impulses of the S-A node to the ventricles.

Fibers of the **atrioventricular bundle** (bundle of His) in the ventricular septum carry the impulses from the A-V node to a network of interlacing fibers, called the **Purkinje fibers** (conduction myofibers) in the walls of the ventricles.

Thus, impulses originate in the S-A node and spread to the ventricles via the A-V node, the atrioventricular bundle, and the Purkinje fibers, causing the ventricles to contract after the two atria have contracted.

Extrinsic Nerve Tissue

The extrinsic nerves consist of parasympathetic fibers that comprise parts of the **vagus nerves** and **accelerator nerves** of the sympathetic system. Fibers of the right vagus nerve innervate the S-A node, and those of the left vagus pass to the A-V node. The vagus is an inhibitory nerve in that it slows the heart rate and weakens systole. Extremely strong stimulation may bring the heart to a temporary standstill. Accelerator nerve fibers arise from cervical and thoracic sympathetic ganglia. The accelerator nerves of the right side innervate the atrial region of the S-A node, while those of the left side terminate in the atrial region of the A-V node. Stimulation of these sympathetic fibers increases the heartbeat by shortening the systolic

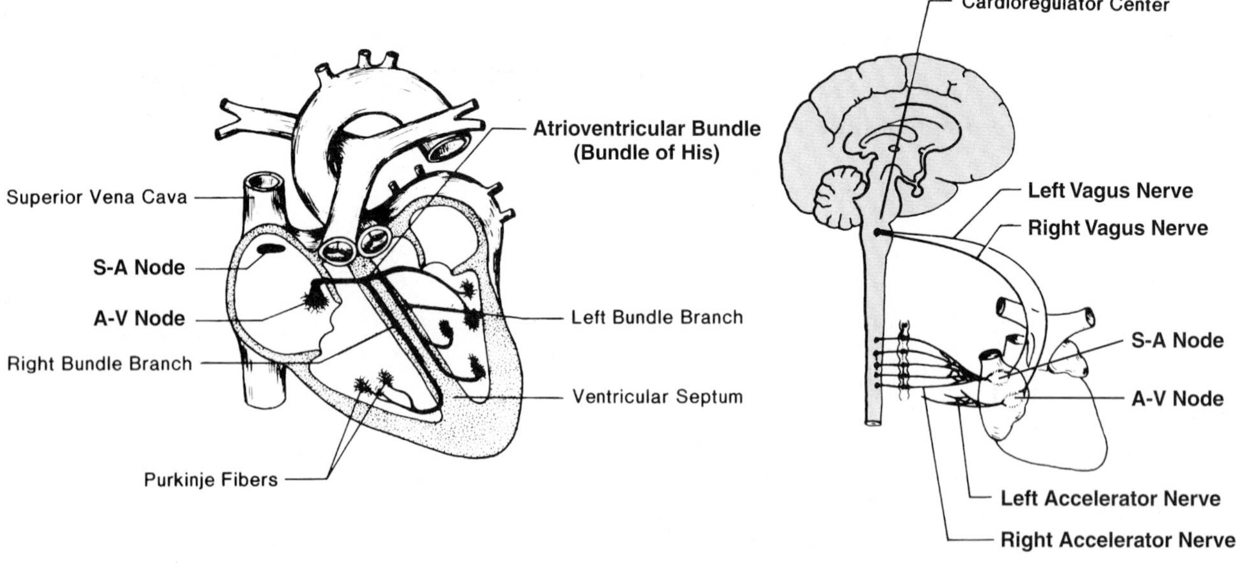

Figure 61.1 Regulatory tissues of the heart.

period and increasing the force of atrial and ventricular contractions.

Neurotransmitter Control

Two neurotransmitter also affect the heart rate. Norepinephrine production at the ends of the accelerator nerves increases the rate of heart muscle contraction. Acetylcholine production at the ends of vagus nerve fibers decreases the rate of contraction.

Heart Rate Experiment

In this experiment, you administer epinephrine (a chemical that functions in the same way as norepinephrine) and acetylcholine to a frog via the liver to note the effects on the heart rate. Figure 61.2 illustrates the equipment setup.

This experiment has three steps: (1) animal dissection, (2) equipment setup, and (3) monitoring. If time is limited, work in teams of four or more, with each team member assigned to one of the three operational steps.

Except for the absence of an electronic stimulator, this experiment requires essentially the same equipment setup as Exercise 29, where you studied frog skeletal muscle contraction.

Materials:

for dissection:
small frog, one per team
dissection pan
decapitation scissors
scalpel, small scissors, forceps
squeeze bottle of frog Ringer's solution
spool of cotton thread
small fishhook
dissecting pins

for Unigraph setup:
Unigraph, T/M Biocom #1030 transducer
ring stand and two double clamps
isometric clamp (Harvard Apparatus, Inc.)

for Physiograph setup:
Physiograph IIIs with transducer coupler
transducer stand
myograph
myograph tension adjuster

reagents:
epinephrine, 1:10,000
acetylcholine, 1:10,000
hypodermic syringes

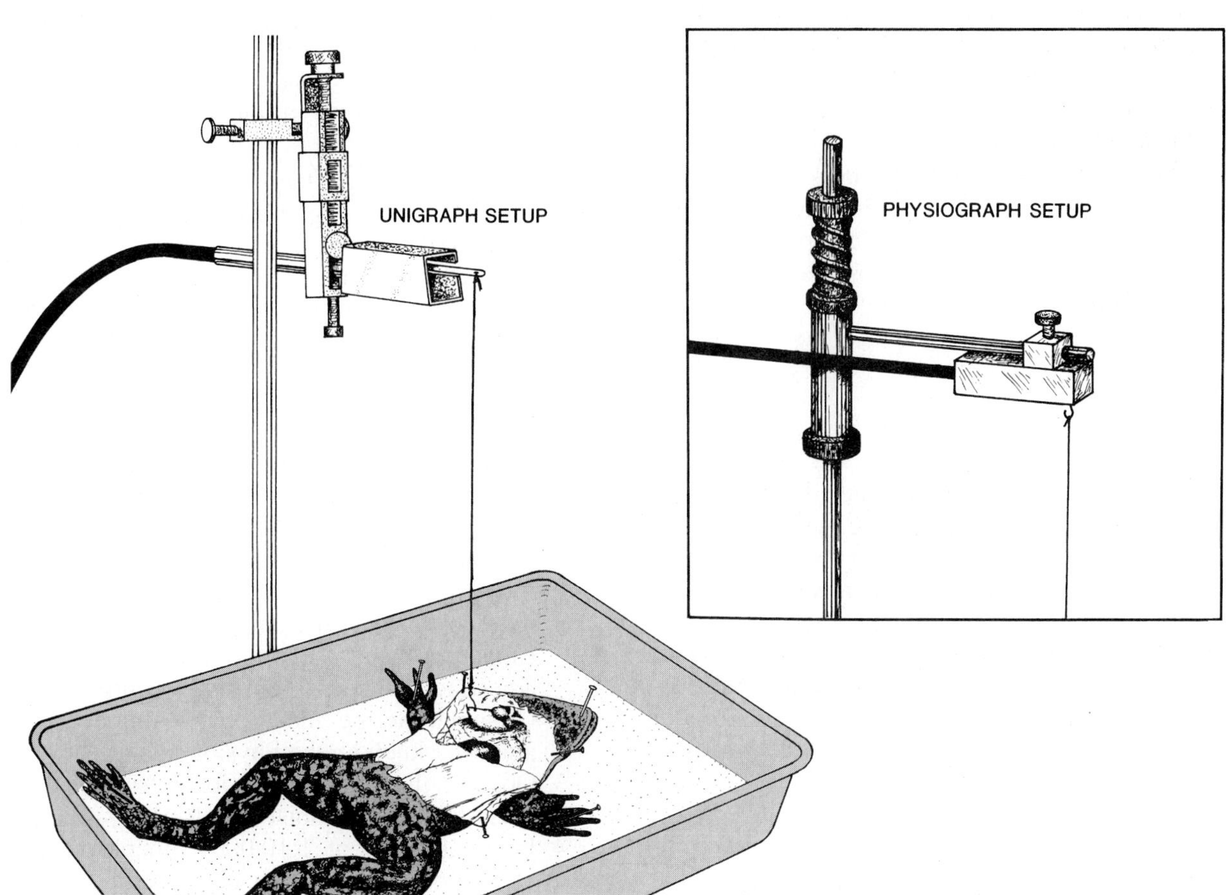

Figure 61.2 Frog heart monitoring.

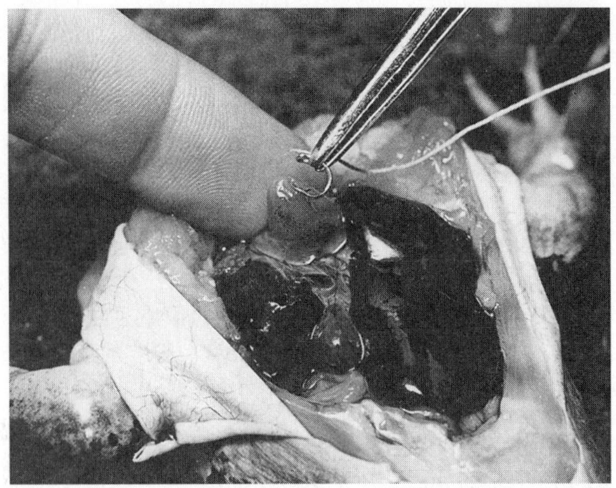

Figure 61.3 Support the heart with the index finger of the left hand when forcing the hook into the ventricle.

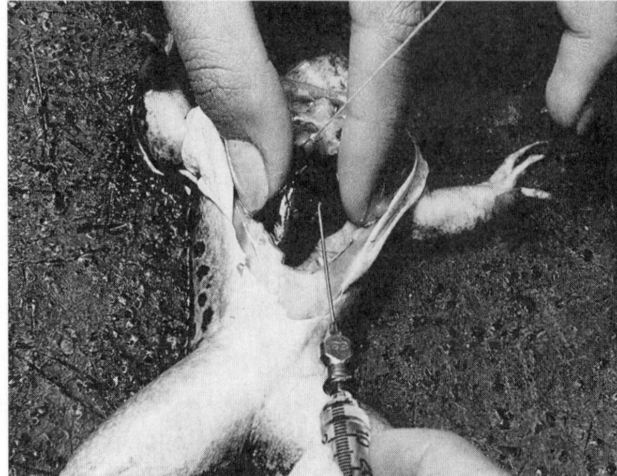

Figure 61.4 While injecting drugs into the liver, firm up the liver with the fingers of the left hand.

 **Virtual Physiology Lab #4:** Effect of Drugs on the Frog's Heart

Dissection

1. Decapitate and spinal pith a frog in the manner described in Exercise 29 (page 124).
2. With sharp scissors, remove the skin from the thorax region, exposing the underlying muscle layer.
3. Remove the muscle layer with scissors, exposing the underlying organs without injuring them. Also cut away the sternum to expose the beating heart within the thin pericardium.
4. Lift the pericardium with forceps, and slit it open with scissors to expose the heart. The heart is now ready to accept the fishhook.
5. Moisten the heart with Ringer's solution. The heart must be kept moist.

Equipment Setup

1. Note in figure 61.2 that in the Gilson Unigraph setup, a ring stand holds the isometric clamp and transducer. In the Narco Physiograph setup, a transducer stand and myograph tension adjustor support the myograph.

   Hook up the various components as figure 61.2 shows, and plug the cord for the transducer (myograph) into the appropriate socket of the recorder. Turn on the recorder's power switch.
2. **Unigraph Settings:** Mode control at TRANS, gain control at 1 mV/cm, sensitivity knob completely counterclockwise, speed control lever at slow position (2.5 mm/sec), and c.c. switch at STBY.
3. **Physiograph Settings:** Amplifier control at 1 mV/cm with inner knob turned completely clockwise, chart speed at 0.25 cm/sec.
4. Tie an 18-inch length of thread to a small fishhook or bent pin. Connect the free end of the thread to the transducer (myograph). For the Gilson setup, use only the single stationary leaf on the transducer.
5. As figure 61.3 shows, support the back of the heart with an index finger as you use forceps to force the fishhook through the apex of the frog's heart.
6. Adjust the thread tension with the isometric clamp or the myograph tension adjustor. The thread tension should be just enough to take up the slack, without pulling the heart straight up.
7. Continue to moisten the heart with Ringer's solution.

Monitoring

1. Start the chart moving to observe the trace, and center the tracing to a desired position. (Use the centering control or the Trans-Bal control on the Unigraph for centering.)
2. Increase the gain control settings to 2, 5, or more mV/cm to get 1 cm stylus deflection with each heartbeat. For in-between sensitivity adjustments, use the sensitivity control on the Unigraph or the inner control knob on the amplifier knob on the Physiograph.
3. Observe the contractions for 3 to 5 minutes, keeping the heart moist with Ringer's solution.
4. Inject 0.05 ml of 1:10,000 epinephrine *very slowly* into the liver with a hypodermic syringe (see figure 61.4). *Mark the chart* to indicate this injection. Observe the trace for 3 to 5 min-

utes. Does the number of beats per minute change? Does the contraction strength change?

5. With a different syringe, inject 0.05 ml of 1:10,000 acetylcholine into the liver, marking the chart to indicate the time of injection. How long does it take for the heart to respond to this injection? Observe the trace for 3 to 5 minutes.
6. Now inject another 0.05 ml of epinephrine into the liver, and mark the chart again. Observe the trace for 3 minutes. Does the heart seem to recover completely?
7. Inject 0.25 ml of epinephrine into the liver, and mark the chart again. Observe the trace for 3 to 5 minutes. How does the heart respond to this massive dose?
8. Finally, inject 0.25 ml of acetylcholine, mark the chart, and observe the trace for another 3 to 5 minutes.

Assignment:
Complete the Laboratory Report for this exercise.

62 Electrocardiogram Monitoring: Using Chart Recorders

In this exercise, you learn how to make an electrocardiogram utilizing a chart recorder such as the Gilson Unigraph or the Narco Physiograph. Instructions are provided for both types of equipment. If the Cardiocomp (Exercise 63) is to be used instead, you should read the applications provided in this exercise, since this background information is lacking in the Cardiocomp exercise.

The ECG Waveform

Myocardial contractions of the heart originate with depolarization of the sinoatrial (S-A) node of the right atrium. As the myocardium of the right atrium is depolarized and contracts in response to depolarization of the S-A node, the atrioventricular (A-V) node is activated and sends a depolarization wave via the atrioventricular bundle and Purkinje fibers to the ventricular myocardium. Monitoring of the electrical potential changes from this depolarization and repolarization produces a record called an **electrocardiogram,** or **ECG.**

Producing an ECG requires attaching a minimum of two and as many as ten electrodes to different portions of the body to record differences in electrical potential. If one electrode is placed slightly above the heart and to the right, and another is placed slightly below the heart and to the left, the waveform is recorded at its maximum potential.

Figure 62.1 shows a typical ECG as recorded on a Unigraph, and figure 62.2 illustrates the characteristics of an individual normal ECG waveform. The initial depolarization of the S-A node, which causes atrial contraction, manifests itself as the **P wave.** Repolarization of the atria immediately follows this depolarization. Surface electrodes do not detect atrial repolarization because it occurs while the ventricles are depolarizing.

Depolarization of the ventricles via the atrioventricular bundle and Purkinje fibers produces the **QRS complex.** This depolarization causes ventricular systole.

As soon as the ventricular muscle is depolarized, repolarization produces the **T wave.** Some ECG waveforms show a **U wave** after the T wave. The small U wave is attributed to the gradual repolarization of the papillary muscles.

The final result of this depolarization wave is that the myocardium of the ventricles contracts as a unit, forcing blood out through the pulmonary trunk and aorta. Repolarization of the affected tissues quickly follows in preparation for the next cycle of contraction (cardiac cycle).

The ECG has immeasurable value to cardiologists. They can detect almost all serious heart muscle abnormalities by analyzing the contours of the different waves. Interpretation of ECG waveforms to determine the nature of heart damage is a highly developed technical skill involving vector analysis and is considerably beyond the scope of this course; however, figure 62.3 shows three examples of abnormal ECGs.

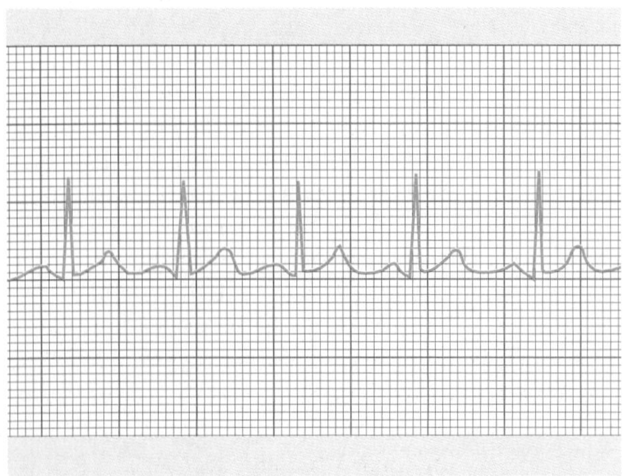

Figure 62.1 An electrocardiogram (Unigraph tracing).

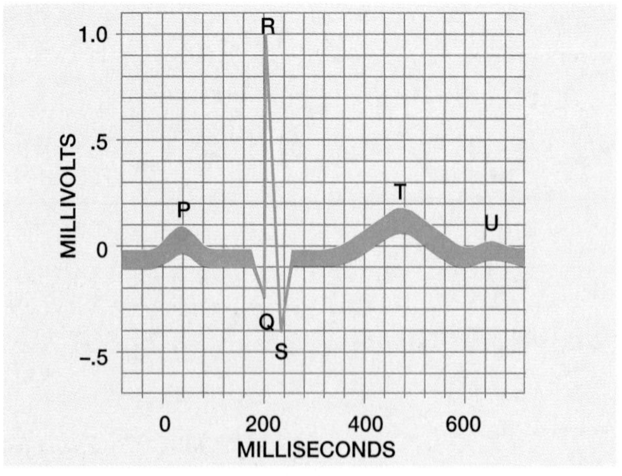

Figure 62.2 The ECG waveform.

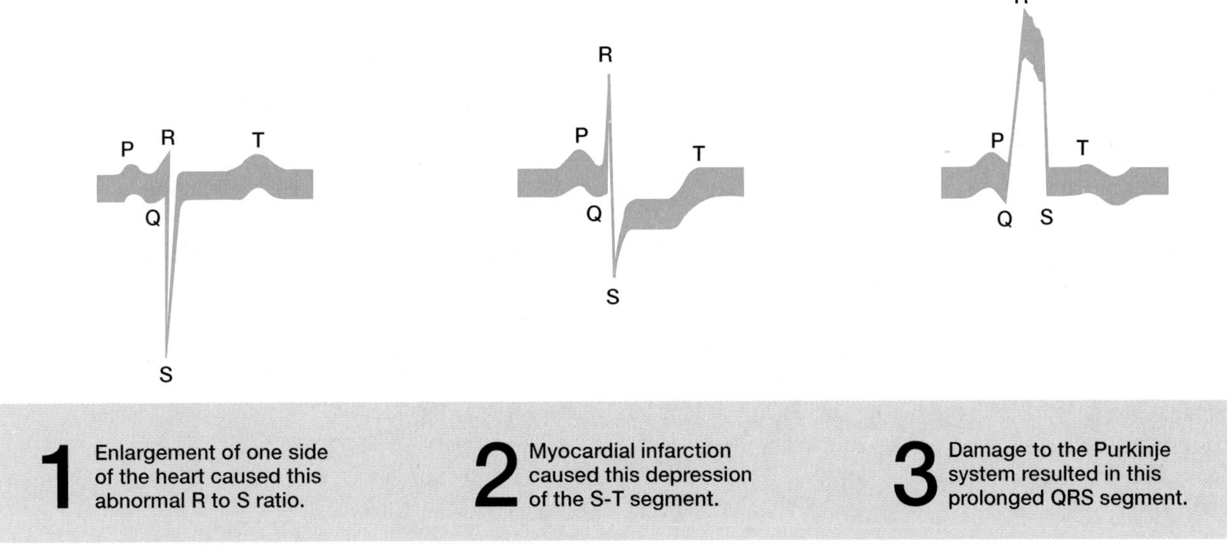

Figure 62.3 Abnormal electrocardiograms.

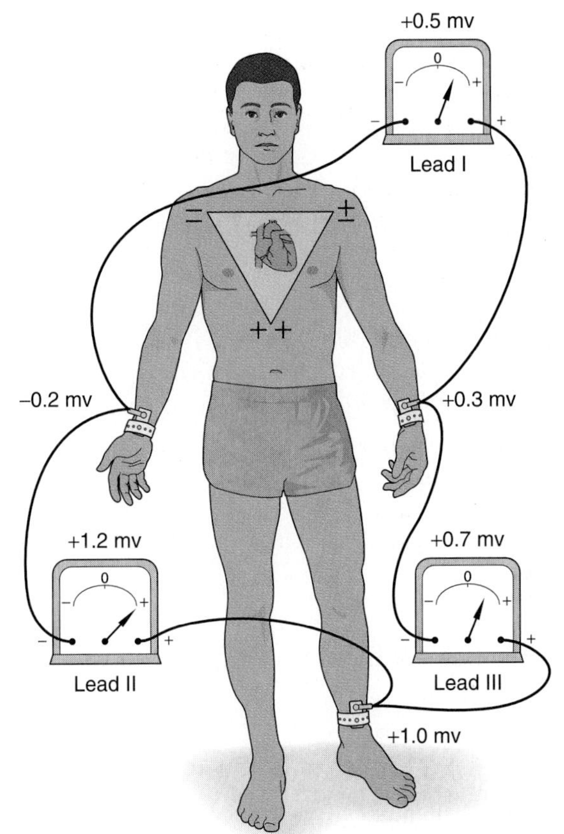

Figure 62.4 In the conventional three-lead hookup, the two arms and left leg form apexes of a triangle (Einthoven's) surrounding the heart. Note that the sum of the voltages in leads I and III equals the voltage in lead II.

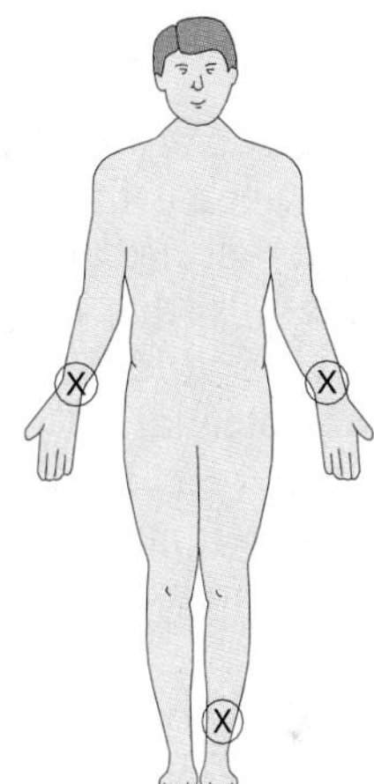

Figure 62.5 Three-lead hookup for Unigraph setup.

Electrodes and Leads

As indicated earlier, producing an ECG requires as few as two and as many as ten electrodes. Einthoven's triangle (figure 62.4) helps to clarify the significance of the different electrode positions and the meaning of leads I, II, and III. (Einthoven was the first person to produce an electrocardiogram, around the end of the nineteenth century.)

The Unigraph requires three **electrodes**—for the two wrists and the left ankle, as figure 62.5 shows. The Physiograph setup requires five electrodes—for the four appendages and for the chest over the heart, as figure 62.6 shows. The right leg acts as a ground in the five-electrode setup.

Three leads, designated leads I, II, and III, are derived from these electrode hookups. Einthoven defined a **lead** as *the electrical potential in the fluids around the heart between two electrodes during depolarization.* On the Physiograph and Cardiocomp setups, you can make lead selections.

Note in figure 62.4 that the apices of the upper part of the triangle represent the points at which the

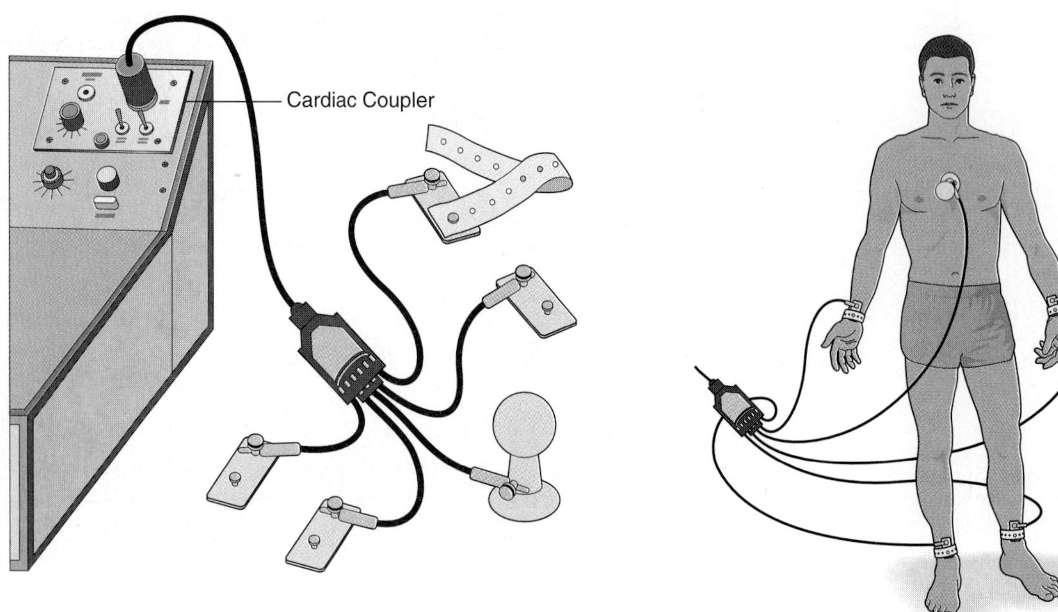

Figure 62.6 Physiograph setup for ECG monitoring.

two arms connect electrically with the fluids around the heart. The lower apex is the point at which the left leg (+) connects electrically with pericardial fluids. **Lead I** records the electrical potential between the right (−) and left (+) arm electrodes. **Lead II** records the potential between the right arm (−) and the left leg (+). **Lead III** records the potential between the left arm (−) and the left leg (+). These potentials are recorded graphically on the ECG in millivolts.

When leads I, II, and III are placed together on one diagram, as in figure 62.4, the three leads form a triangle around the heart, known as **Einthoven's triangle.** The significance of this triangle is that if the potentials of two of the leads are recorded, the third one can be determined mathematically. Or, stated another way: The sum of the voltages in leads I and III equals the voltage of lead II. This principle is known as **Einthoven's law:**

Lead II = Lead I + Lead III

Experiment Procedure

You will work in teams of three or four, with one individual being the subject and the others making the electrode hookups and operating the controls. Calibrate the recorder prior to recording the ECG; follow the calibration instructions for the type of recorder being used.

Failure in this experiment usually results from (1) poor skin contact, (2) damaged electrode cables, or (3) incorrect hookup. Proceed as follows:

Materials:

for Unigraph setup:
Unigraph
three-lead patient cable
ECG plate electrodes (three) and straps

for Physiograph setup:
Physiograph with cardiac coupler
five-lead patient cable
four ECG plate electrodes and straps
one suction-type chest electrode

for both setups:
Scotchbrite pad
70% alcohol
electrode paste

 **Lab #5:** Electrocardiogram

Unigraph Calibration

While one member of your team attaches the electrodes to the subject, calibrate the Unigraph so that 1 mV produces 1 cm of stylus deflection. Follow these steps:

1. With the main power switch in the Off position, set the controls as follows: c.c. (chart control) switch at STBY, speed control lever at the slow position, gain control at 1 mV/cm, sensitivity knob completely counterclockwise, mode control at CC-Cal, and the hi filter and mean switches on Norm.

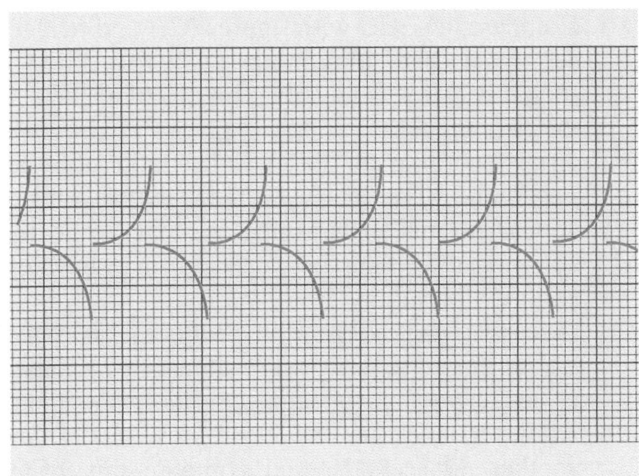

Figure 62.7 Calibration tracing for Unigraph.

2. Turn on the power switch, and rotate the stylus heat control to the two o'clock position.
3. Place the c.c. switch at Chart On. Observe the trace and center the stylus, using the centering control.
4. Increase the sensitivity by rotating the sensitivity control clockwise a half turn.
5. Depress the 1 mV button for about 2 seconds, and then release it. The upward deflection should measure 1 cm; the downward deflection should equal the upward deflection (see figure 62.7). If the travel is not 1 cm, adjust the sensitivity control and depress the 1 mV button until you achieve exactly 1 cm.
6. Place the c.c. switch on STBY, and label the chart with the date and the subject's name.
7. Place the mode control on ECG, and move the speed control lever to the high speed (25 mm/sec) position. The Unigraph is now ready to accept the patient cable.

Physiograph Calibration

While one member of your team attaches the electrodes to the subject, calibrate the channel sensitivity of the Physiograph so that 1 mV produces 2 cm of pen deflection. Follow these steps:

1. Place the time constant switch of the cardiac coupler in the 3.2 position.
2. Place the gain switch on the coupler to the ×2 position.
3. Place the lead selector control in the calibrate (Cal) position.
4. Set the outer knob of the channel amplifier sensitivity control on the 10 mV/cm position, and rotate the inner sensitivity knob completely clockwise until it clicks.
5. Start the chart moving at 0.25 cm/sec, and lower the pens onto the paper. Adjust the pen to the center of its arc with the position control.

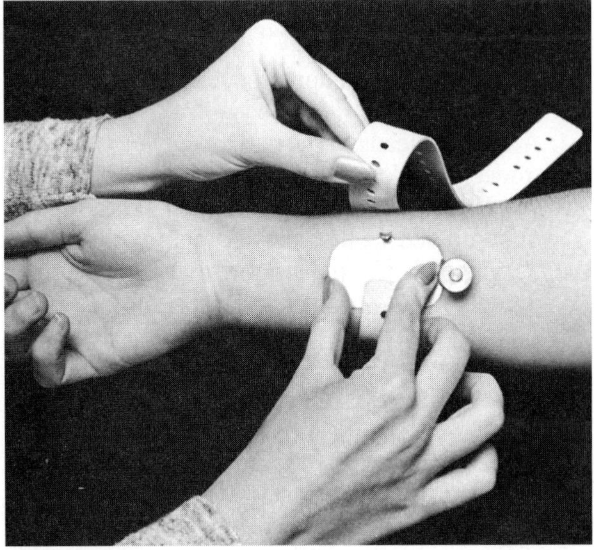

Figure 62.8 Apply electrode paste to the electrode and then strap it in place.

6. Place the channel amplifier record button into the On position.
7. Place the CAL toggle switch on the cardiac coupler into the 1 mV position. Observe that the channel recording pen deflects *upward* 2 cm.
8. Hold the CAL switch in the 1 mV position until the recording pen returns to the center line; then release the switch. Observe that, upon release, the channel recording pen now deflects *downward* 2 cm.
9. If the upward and downward deflections are not exactly 2 cm, make adjustments with the amplifier sensitivity controls.
10. Stop recording. The channel is now calibrated.

Preparation of the Subject

While one person calibrates the recorder, another team member attaches the electrodes to the subject. Although not essential, having the subject lie on a comfortable cot is preferable.

As figure 62.5 shows, electrodes attach to both wrists and the left ankle for the Unigraph setup. Electrodes attach to both wrists, both ankles, and the heart region for the Physiograph setup. Proceed as follows to attach the electrodes to the subject:

1. To achieve good electrode contact on the wrists and ankles, rub the contact areas with a Scotchbrite pad, disinfect the skin with 70% alcohol, and add a little electrode paste to the contact surface of each electrode.
2. Attach each of the plate electrodes with a rubber strap, as figure 62.8 shows.

3. With the five-electrode setup, use an ample amount of electrode paste on the contact surface, and affix the suction-type electrode over the heart region.
4. Attach the leads of the patient cable to the electrodes as follows:

   Unigraph Setup: Attach the black, unnumbered lead to the left ankle, one of the numbered leads to the right wrist, and the other numbered lead to the left wrist.

   Physiograph Setup: Attach each lead according to its label designation: LL—left leg, RL—right leg, LA—left arm, RA—right arm, and C—cardiac.
5. Verify that the leads securely connect to the electrode binding posts by firmly tightening the knobs on the binding posts. If the patient cable has alligator clamps instead of straight plugs, verify that the clamps cannot slip off easily.
6. If the subject is sitting in a chair instead of lying down, *make sure that the chair is not metallic.*
7. Connect the free end of the patient cable to the appropriate socket on the Unigraph or Physiograph cardiac coupler.
8. The subject is now ready for monitoring.

Monitoring

Tell the subject to relax and be still. Turn on the recorder, and observe the trace. Then refer to the instructions that follow for the type of recorder you are using.

Unigraph Adjust the stylus heat control to achieve optimum trace.

1. Compare the ECG with figures 62.1 and 62.2 to see if the QRS spike is up (positive) or down (negative). If it is negative, reverse the leads to the two wrists.
2. After recording for about 30 seconds, stop the paper, and study the ECG. Does the waveform appear normal? Determine the pulse rate by counting the spikes between two margin lines and multiplying by 20.
3. Disconnect the leads to the electrodes, and have the subject exercise for 2 to 5 minutes by running up and down stairs or jogging outside the building.
4. After reconnecting the electrodes, record for another 30 seconds, and compare this ECG with the previous one. Save the record for attachment to the Laboratory Report.

Physiograph Place the lead selector control at the Lead I position. With the paper moving, lower the pens onto the recording paper. Verify that the timer is activated.

1. Place the record button in the On position, and record one sheet of ECG activity.
2. If the pen goes off scale, press the trace reset switch on the cardiac coupler, and the pen will return to the center line.
3. Place the record button in the Off position, and turn the lead selector control to the Lead II position.
4. Place the record button back in the On position, and record another sheet of ECG activity.
5. Repeat step 4 for the Lead III position on the lead selector control.
6. Stop the recording, and compare your results with the sample tracing in figure 62.9.

Assignment:
Complete the Laboratory Report for this exercise.

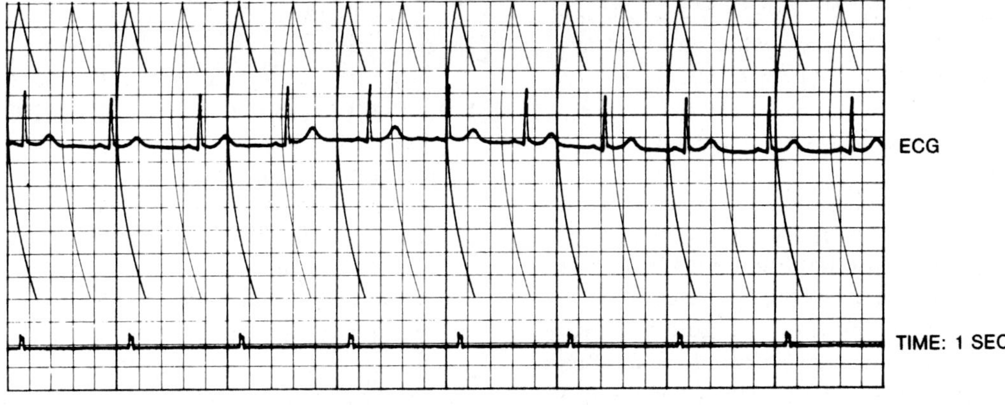

LEAD I SAMPLE RECORD

ECG

TIME: 1 SEC

LEAD II SAMPLE RECORD

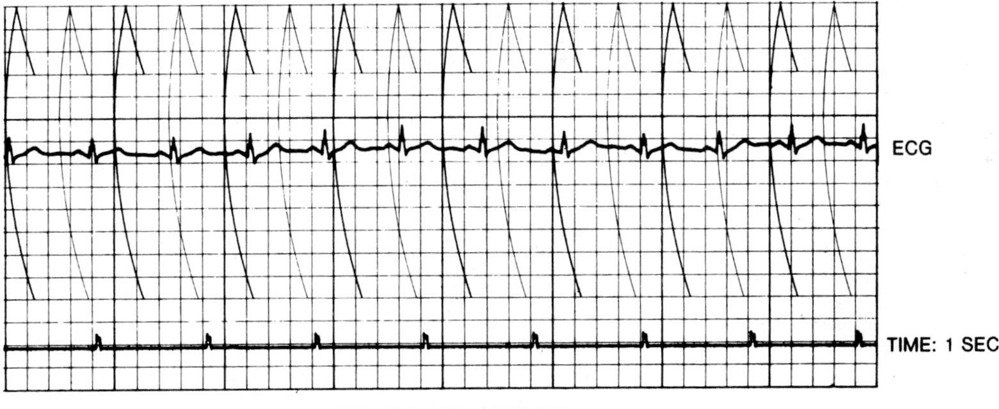

LEAD III SAMPLE RECORD

Figure 62.9 Physiograph electrocardiogram sample recordings at different lead settings.

63 Electrocardiogram Monitoring: Using Computerized Hardware

In this exercise, you use the Intelitool Cardiocomp System with a computer to produce electrocardiograms (ECGs) of class members. Figure 63.1 illustrates the setup. Before proceeding, read pages 314–316 in Exercise 62 so that you understand the characteristics of the ECG waveform, Einthoven's law, and leads I, II, and III.

Augmented Unipolar Leads

As explained in Exercise 62, the three bipolar leads (I, II, and III) are based on Einthoven's triangle. Figure 63.2 illustrates the relationship of these leads to the triangle. Arrows in the diagram represent the potential gradient direction for a positive deflection on the screen.

In addition to these three bipolar leads are three augmented unipolar leads. In an **augmented unipolar lead,** *the electrical potential is between one of the positively charged limbs and the average of the other two limbs that are negative.*

Figure 63.3 shows how these three leads are superimposed on the triangle with bipolar leads I, II, and III. Note that the arrowhead of each lead points to the positive (+) electrode of an appendage and that the line is perpendicular to one side of Einthoven's triangle, bisecting that side. The three augmented leads are as follows:

- **aVR:** The aVR lead bisects the side of the triangle that goes from the left arm (LA) to the left leg (LL). It points toward the electrode of the right arm (RA).
- **aVL:** The aVL lead bisects the side of the triangle that goes from the right arm (RA) to the left leg (LL). It points toward the positive electrode of the left arm (LA).
- **aVF:** The aVF lead bisects the side of the triangle that extends from RA to LA. It points toward the positive electrode of the left leg (LL).

Figure 63.4 clarifies the directional nature of the six leads. Note that six leads allow monitoring of the direction of the electrical potential through the heart radially, in 30° segments. By scanning the six leads, cardiologists can determine the direction of the potential *by simply noting in which lead the R wave has the strongest deflection.* If the strongest deflection is in lead II, as is usually the case, cardiologists know that the potential is in the direction from the right shoulder to the left foot, or about 60°. Using six leads instead of only three bipolar leads

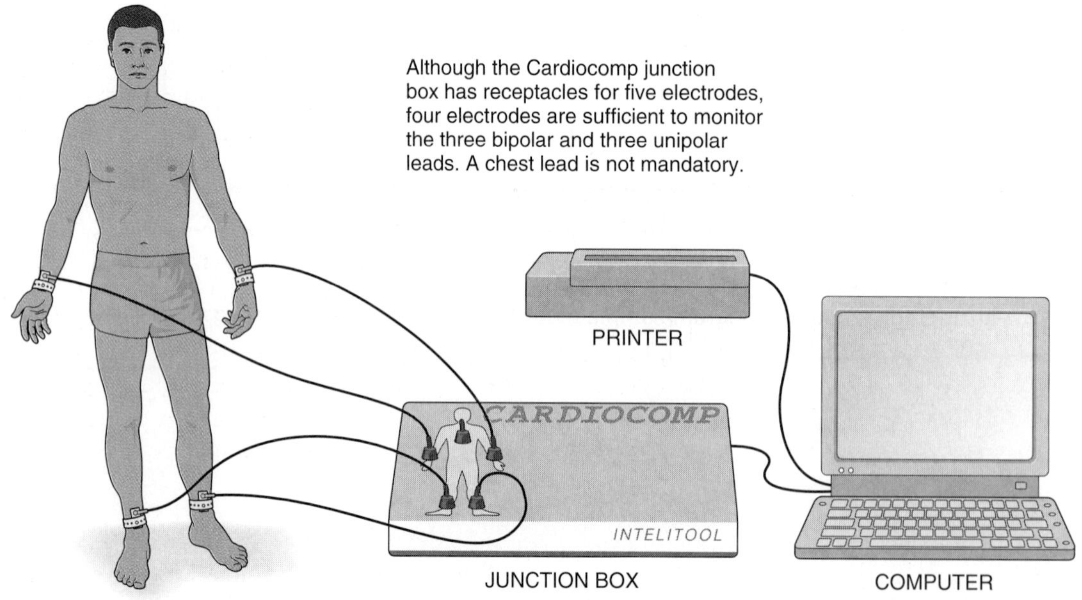

Figure 63.1 Cardiocomp setup for ECG monitoring.

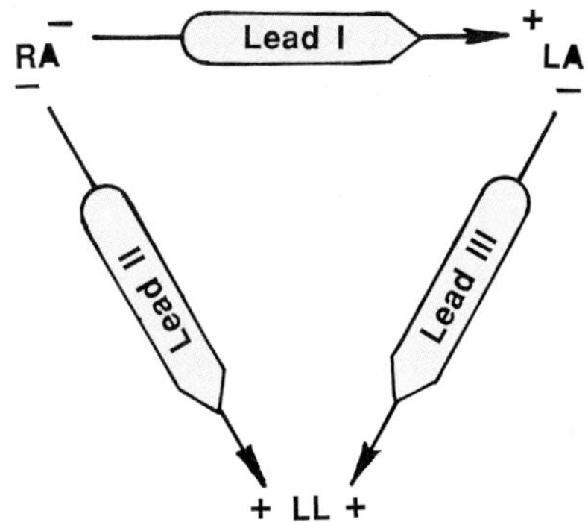

Figure 63.2 Bipolar lead orientation according to Einthoven's triangle. (LA = left arm; LL = left leg; RA = right arm.)

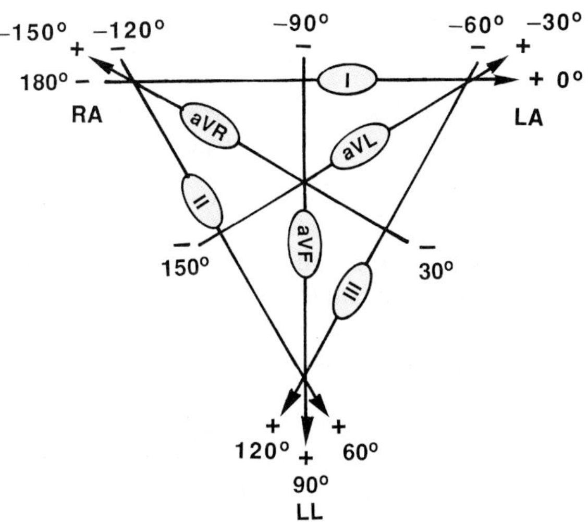

Figure 63.3 The relationship of bipolar to unipolar augmented leads. (LA = left arm; LL = left leg; RA = right arm.)

also lessens the chances of missing some unusual cardiac event.

Chest Leads

Note in figure 63.1 that the experiment setup uses only four electrodes, even though the Cardiocomp has one receptacle for a chest electrode. In clinical practice, most cardiologists rely on six standardized chest leads, in addition to the ones studied here. While chest leads are invaluable to cardiologists, they reveal nothing about the waveform or arrhythmias that cannot be determined with the bipolar and unipolar augmented leads. Thus, for simplicity, this experiment does not require a chest electrode.

Experiment Procedure

In groups of two or more students, use the Cardiocomp to do a series of ECG measurements on

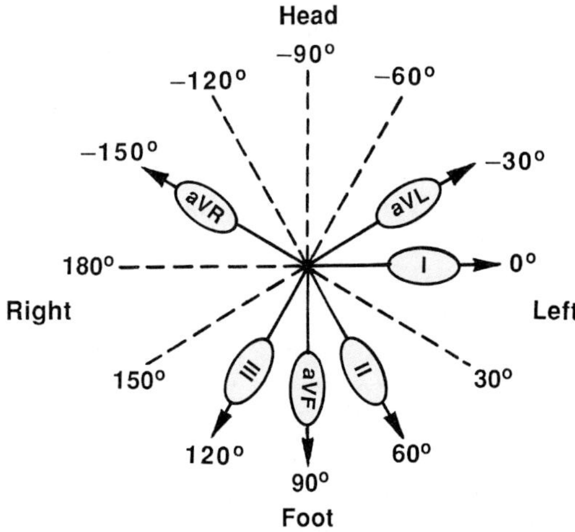

Figure 63.4 Directional orientation of the six most commonly used leads in ECG monitoring.

class members. While one student is being monitored, another student is wired for testing. Note that two Cardiocomp manuals are available for reference

Materials:
computer and printer
Cardiocomp junction box
Cardiocomp program diskette
blank initialized diskette
four electrode wires and four flat-plate electrodes
Scotchbrite pad, alcohol swabs, electrode gel
Cardiocomp User Manual
Cardiocomp Lab Manual

Electrode Attachment

1. To achieve good electrode contact on the wrists and ankles, rub the contact areas with a Scotchbrite pad, disinfect the skin with an alcohol swab, and add a little electrode gel to the flat contact surface of each electrode.
2. Attach each of the plate electrodes to the limbs with the rubber straps, and snap the wires to the electrodes. Be sure to match the labels to the limbs.
3. Plug the electrode wires into the proper receptacles of the Cardiocomp junction box.
4. Ask the subject to lie down and to avoid movement. **Arrange the wires parallel to the body, separate from each other and away from any power cables, to avoid "noise."**

Six-Lead Data Acquisition

When the subject is properly hooked up to the electrodes and in a relaxed, horizontal position, monitor the ECG as follows:

1. Insert the Cardiocomp program diskette into drive A and the data diskette into drive B, and turn on the computer. If you are using an Apple IIe, IIc, or IIGS, be sure to push down the Caps-Lock key.
2. Touch any key, and the Cardiocomp program will take over. After you have answered the system configuration questions, the Main (cardiocomp) Menu appears on the screen.

( ) Experiment Menu
( ) Review / Analyze Menu
( ) Data File / Disc Commands
( ) Preferences
( ) ESCape to Program Menu

Note: The last two lines of the Main Menu may not appear on the computer screen.

3. Select "Experiment Menu" from the Main Menu. The Experiment Menu then appears on the screen.
4. Select "Quick Exam Mode" from the Experiment Menu. In this mode you must provide the following answers: (1) six frames, (2) save each exam, (3) print a hard copy, and, (4) no interval analysis.
5. Select the data analysis features to be run for your class as described by your instructor. This will only have to be done once for the entire class.
6. Press the S key and enter the subject's initials.
7. When the subject is ready, press any key to begin the exam. When the exam is done, start to hook up the next subject while the printer is printing out the exam.
8. Run the exam for the next subject.

Review-Screen Analysis

Select "Review / Analyze Menu" from the Main Menu, and select "Review / Time Analysis" from the Review / Analyze Menu.

Look carefully at your ECG strip. Scan the six leads, and **determine which lead has the highest R wave.** Use figure 63.4 to determine the angle of that particular lead. The angle of the lead in which the R wave deflection is the greatest is close to the angle of the R wave. If the magnitude of the R wave deflection is about the same in two leads, then the R wave axis is between the angles that those two leads describe.

Interval Analysis

Using the Review-Analyze Menu (Full Screen), determine the duration of the following:

- PR interval ________ (0.12–0.2 seconds expected)
- QT interval ________ (less than 0.38 seconds expected)
- QRS complex ________ (less than 0.10 seconds expected)

Measure the interval between R waves of several heartbeats. Add them, and divide by the number of intervals to determine the ventricular rate. Are the intervals the same length?

Additional Experiments

You may wish to monitor ECG waveforms of an individual in different body positions, such as sitting up straight, hunched over the back of a chair, or standing up. In any case, take care to avoid muscle "noise." If you wish to do a cardiovectorcardiogram, consult Lab 2 of the *Cardiocomp Lab Manual.*

Assignment:
Since there is no Laboratory Report for this exercise, your instructor will indicate what form of report is required.

Pulse Monitoring

64

When the ventricles of the heart undergo systole, blood surges into the arterial system, manifesting itself as the **pulse** in the extremities. Since the rate and strength of the pulse indicate cardiovascular function, physicians always monitor it in routine medical examinations by placing fingers over one of the patient's arteries in the wrist or neck region.

In this exercise, a photoelectric plethysmograph on the index finger records the pulse. The recording that this transducer produces is called a **plethysmogram** (see figure 64.1). As figures 64.2 and 64.3 show, you can perform this experiment with either a Unigraph or a Physiograph.

The construction and operation of the plethysmograph are based on a light beam that is projected into the soft tissues of the finger (see p. 97 in Exercise 21). A surge of blood passing through the finger alters the beam. The scattered light activates a photoresistor to produce a signal that can be recorded on an oscilloscope or chart paper.

In addition to determining the pulse rate, the plethysmogram can detect indirect and relative blood pressure differences, but not precise blood

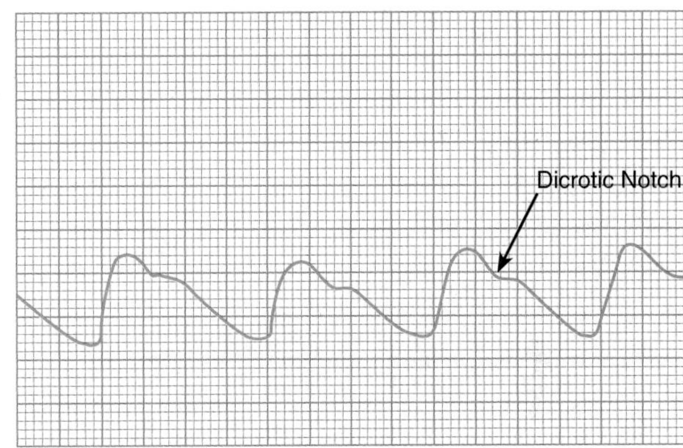

Figure 64.1 A plethysmogram (Unigraph tracing).

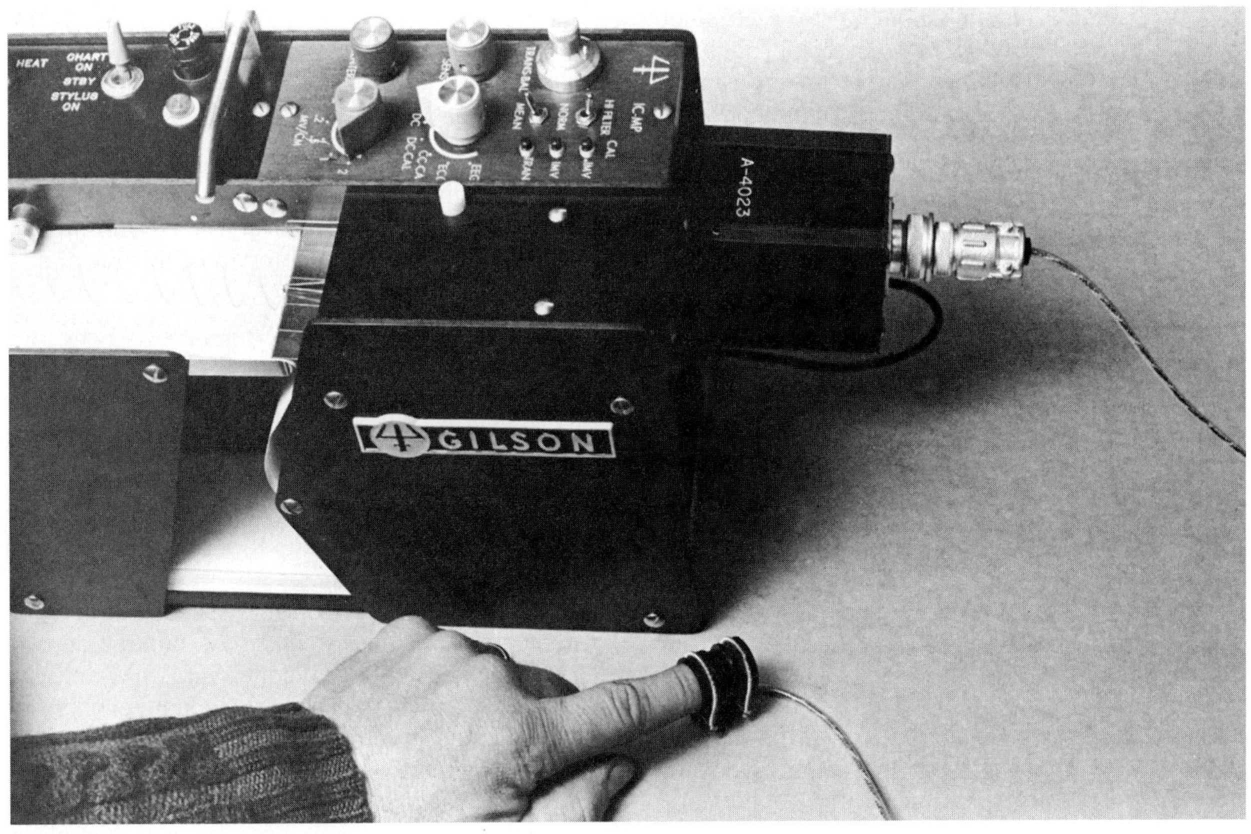

Figure 64.2 Pulse monitoring with Unigraph.

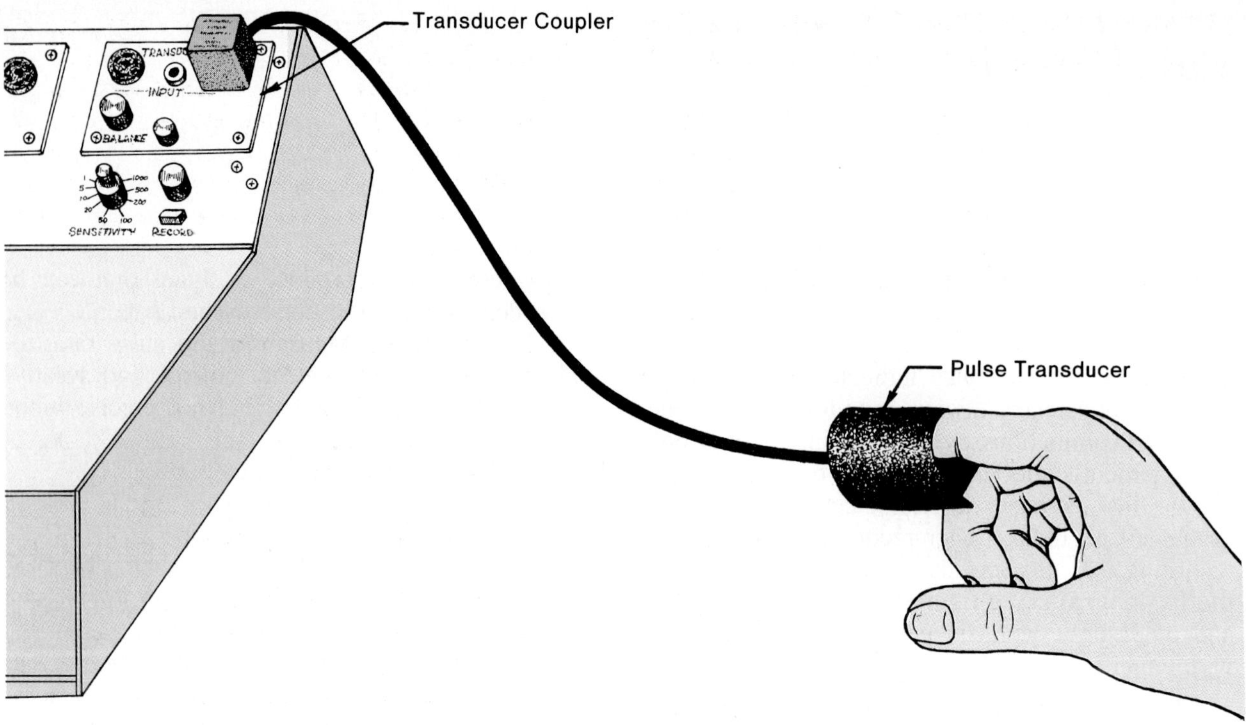

Figure 64.3 Pulse monitoring with Physiograph.

pressures. Calibrating recorders to note the volume of blood surging through the finger at any given moment, which represents an exact blood pressure, is very difficult. This limitation, however, does not prevent the plethysmogram from being a useful tool. As you shall see in this experiment, using a plethysmogram can teach you much about the mechanics of blood pressure.

The objectives of this experiment are to (1) determine the exact pulse rate, (2) explore factors that affect pulse rate, (3) identify the dicrotic notch and its significance, and (4) determine the range of pulse rates among class members. Proceed as follows:

Materials:
for Unigraph setup:
Unigraph
A4023 adapter (Gilson)
Photoresistor pulse pickup (Gilson T4020)

for Physiograph setup:
Physiograph with transducer coupler
pulse transducer (Narco 705-0050)

Preliminaries

Prepare the subject, and make the preliminary adjustments on the recorder according to the following suggestions:

Subject Preparation

1. Position the subject on a laboratory tabletop, face upward and with both arms parallel to the body.
2. Attach the pulse transducer to the index finger of one hand. Be sure that the light source faces the pad of the fingertip. This hand should be near the edge of the table so that the subject can lower it toward the floor or extend it upward toward the ceiling.
3. Make certain that the subject is comfortable and relaxed. The laboratory must be free of distracting stimuli.

Unigraph Setup

1. Attach the A4023 adapter to the receiving end of the Unigraph. Verify that the locknut is tightened securely and that the phone jack is also connected.
2. Secure the pulse pickup to the adapter. Plug in the power cord.
3. Set the controls as follows: speed control lever at slow position, stylus heat control at two o'clock position, gain control at 2 mV/cm, sensitivity control completely counterclockwise, and mode selector at DC.
4. Turn on the power switch, and set the c.c. (chart control) switch at Stylus On. Wait about 15 seconds for the stylus to warm up, and then place the c.c. switch at Chart On.

5. Observe the trace, and adjust the centering control to position the trace near the center of the paper.
6. Increase the sensitivity by turning the sensitivity control clockwise until pen deflection is approximately 2 cm. If you cannot achieve this amount of deflection, turn the sensitivity control down again, and increase the gain to 1 mV/cm. Now increase the sensitivity with the sensitivity control again to get the desired deflection.
7. Proceed to the heading "Investigative Procedure."

Physiograph Setup

1. Plug the end of the transducer cable into the proper socket on the transducer coupler.
2. Turn on the power switch, and set the paper speed at 0.5 cm/sec.
3. Start the chart moving, and lower the pens onto the paper. Verify that the timer is activated.
4. Use the position control to position the pen so that it is recording 1 cm below the centerline.
5. Place the record button in the On position.
6. Adjust the amplifier sensitivity control to produce pulse waves that are approximately 2 cm high.

Investigate Procedure

Now that the subject's pulse is being recorded, proceed as follows with your observations:

1. With the subject lying still and the hand with the pulse transducer lying flat on the tabletop, record the pulse for **2 minutes.**
2. Stop the recording, and calculate the pulse rate. (On the Unigraph chart, the space between two margin marks represents 30 seconds; thus, you count the spikes and the fractions thereof between two marks and multiply by two.)
3. Refer to figure SR-2A, in Appendix D for a sample Physiograph record.
4. Resume recording, and have the subject slowly raise his or her arm to a full vertical position above the body. Ask the subject to hold that position for **15 seconds** and then to slowly return the arm to its original position on the table. Allow the pulse to stabilize. Refer to figure SR-2B in Appendix D for a sample Physiograph record.
5. Now have the subject slowly lower his or her arm over the edge of the table until the arm hangs down vertically. Ask the subject to hold this position for **15 seconds** and then to slowly return the arm to its former resting position. Allow the pulse to stabilize. Refer to figure SR-2C in Appendix D for a sample Physiograph record.
6. Ask the subject to take a deep breath and to hold it for **15 seconds,** before exhaling and resuming normal breathing. Refer to figure SR-3A in Appendix D for a sample Physiograph record.
7. Locate the subject's brachial artery below the biceps muscle. Occlude (block) this vessel for **15 to 20 seconds,** and then release the pressure. Allow the pulse to stabilize. Refer to figure SR-3B in Appendix D for a sample Physiograph record.
8. Increase the speed to 2.5 cm/sec (fast speed on Unigraph), and record for **1 minute.** Do you see a pronounced **dicrotic notch** on the curve's descending slope (see figure 64.1)? This interruption of the curve is due to sudden closure of the heart's aortic valve during diastole.
9. Return the speed to 0.25 cm/sec (slow speed on Unigraph), and mildly frighten or startle the subject with an unexpected, sudden hand clap or other loud noise. The stimulus must be totally unanticipated. Refer to figure SR-3C in Appendix D for a sample Physiograph record.
10. Terminate the recording after sufficient sample recordings have been made for all team members.

Assignment:
Complete the Laboratory Report for this exercise.

65 Blood Pressure Monitoring

When the ventricles of the heart contract, they propel blood. The force that this blood exerts within vessels throughout the body is blood pressure. The heart pumps blood approximately 70 times a minute while a subject is at rest. Blood pressure over a short time interval fluctuates up and down.

Systolic pressure is the force on the walls of blood vessels during maximum contraction. When the heart relaxes and fills up with blood in preparation for another contraction, the pressure falls to its lowest value, called **diastolic pressure.**

You can monitor blood pressure with several different methods. Probably the most sensitive and precise way is to insert a hollow needle into a vessel, which allows the force of the blood pressure to act on a pressure transducer. The signal from such a device can be fed into an amplifier and recorder for monitoring. This method, however, is impractical for this laboratory.

This exercise outlines procedures for two methods of blood pressure monitoring: (1) the conventional stethoscope–pressure cuff (sphygmomanometer) method that medical personnel use routinely, and (2) an electronic recording method utilizing a Physiograph or Unigraph to record pressure changes via a pressure cuff. Your instructor may also demonstrate a third method that uses an audio monitor.

A Demonstration with Audio Monitor

When you monitor blood pressure with a sphygmomanometer and stethoscope, you listen for the sounds of blood rushing through a partially occluded brachial artery of the arm. The sounds, called **Korotkoff** (Korotkow) **sounds,** can readily be demonstrated with an audio monitor, utilizing a setup similar to the one figure 65.1 shows.

Materials:
microphone
audio monitor
pressure cuff (sphygmomanometer)

The instructor wraps a pressure cuff around the upper arm of a subject, inserts a microphone under the cuff over the brachial artery, and demonstrates the Korotkoff sounds for determining systolic and diastolic pressures. To initiate the process, the instructor pumps air into the cuff until all blood flow to the brachial artery is shut off. While watching the pressure gauge, the instructor *slowly* releases air from the cuff until he or she hears the first sound of blood rushing through the occluded artery. The pressure at which the sound is *first heard* is called the **systolic pressure.** The sound slowly diminishes. The pressure at which the sound disappears is the **diastolic pressure.**

By correlating the sounds heard through the audio monitor with the pressure gauge, a large number of students can, in a short period of time, quickly grasp the concept.

Stethoscope-Sphygmomanometer Method

You and a lab partner will take each other's blood pressures. To determine the broad range of normal blood pressures, individual systolic pressures will be recorded on the chalkboard and a median systolic pressure for the class determined. You will also study the effects of exercise on blood pressure.

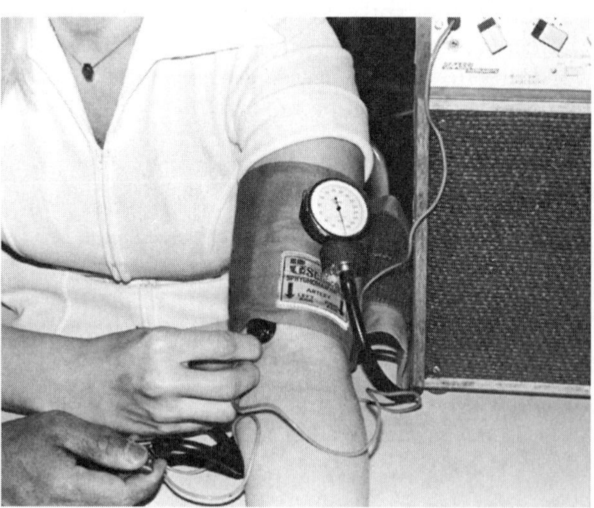

Figure 65.1 Auscultation of Korotkoff sounds with an audio monitor is possible by slipping a microphone under the cuff over the brachial artery.

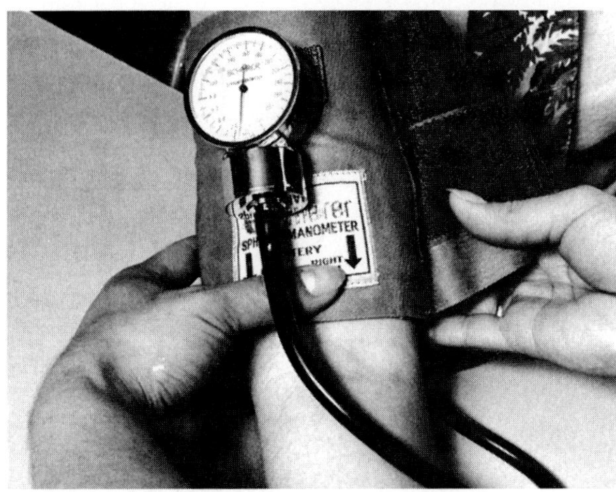

Figure 65.2 Wrap the sphygomomanometer cuff around the upper arm, keeping the lower margin of the cuff above the line through the epicondyles of the humerus.

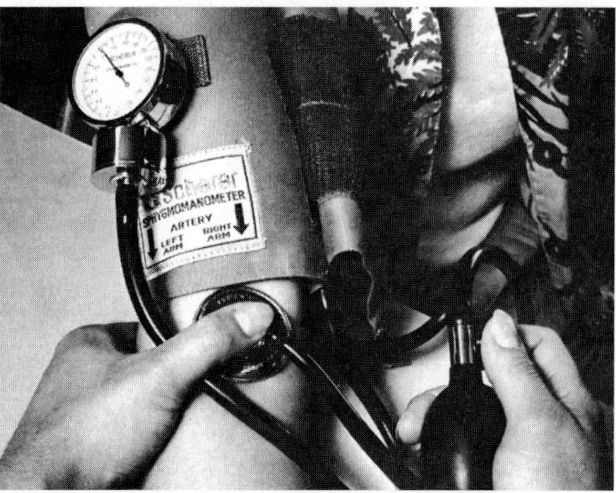

Figure 65.3 The inflated cuff shuts off the flow of blood in the brachial artery. Listen for Korotkoff sounds as you gradually lower the pressure.

Materials:
stethoscope
sphygomomanometer

1. Wrap the cuff around the right upper arm of your lab partner (see figure 65.2). The method of attachment depends on the type of equipment. Your instructor will show you the best way to use the specific type available in your laboratory. Make certain that your partner's forearm rests comfortably on the table.
2. Close the metering valve on the neck of the rubber bulb. **Important:** *Do not twist it so tight that you will not be able to open it!*
3. Pump air into the sleeve by squeezing the bulb in your right hand. Watch the pressure gauge. Allow the pressure to rise to about 180 mm Hg while you continue to hold the metering valve closed.
4. Position the bell of the stethoscope just below the cuff at a point that is *midway between the epicondyles of the humerus* (see figure 65.3). This is the lowest extremity of the brachial artery. About 2.5 cm distal to this point, the artery bifurcates to become the radial and ulnar arteries.
5. Slowly release the valve to gradually decrease the pressure. Listen carefully as you watch the pressure fall. When you just begin to hear the Korotkoff sounds, note the pressure on the gauge. **This is the systolic pressure.**
6. Continue listening as the pressure falls. Just when you are unable to hear the Korotkoff sounds anymore, note the pressure on the gauge. **This is the diastolic pressure.**
7. Record the systolic pressure on the chalkboard and the systolic/diastolic pressures in part A of the Laboratory Report.
8. Take the blood pressure two or three times, switching arms each time, to see if you get consistent results.
9. Now have your lab partner do some exercise, such as running up and down stairs a few times. Measure the blood pressure again, and record your results on the Laboratory Report.

Using a Chart Recorder

Either a Unigraph or a Physiograph can make a chart recording of blood pressure. Figure 65.4 shows the Physiograph setup. A Unigraph setup requires a Statham pressure transducer similar to the one figure 74.3 shows. Procedures are provided here for using either type of recorder.

In this portion of the exercise, you (1) record systolic and diastolic pressures, (2) determine mean arterial and mean pulse pressures, and (3) observe the effects of postural changes and physical exertion on arterial pressure.

Materials:
for Unigraph setup:
Unigraph
adult pressure cuff
Statham T-P231D pressure transducer
transducer stand and clamp for transducer

for Physiograph setup:
Physiograph with ESG (electro-sphygmomanometer) coupler
adult pressure cuff (Narco PN 712-0016)

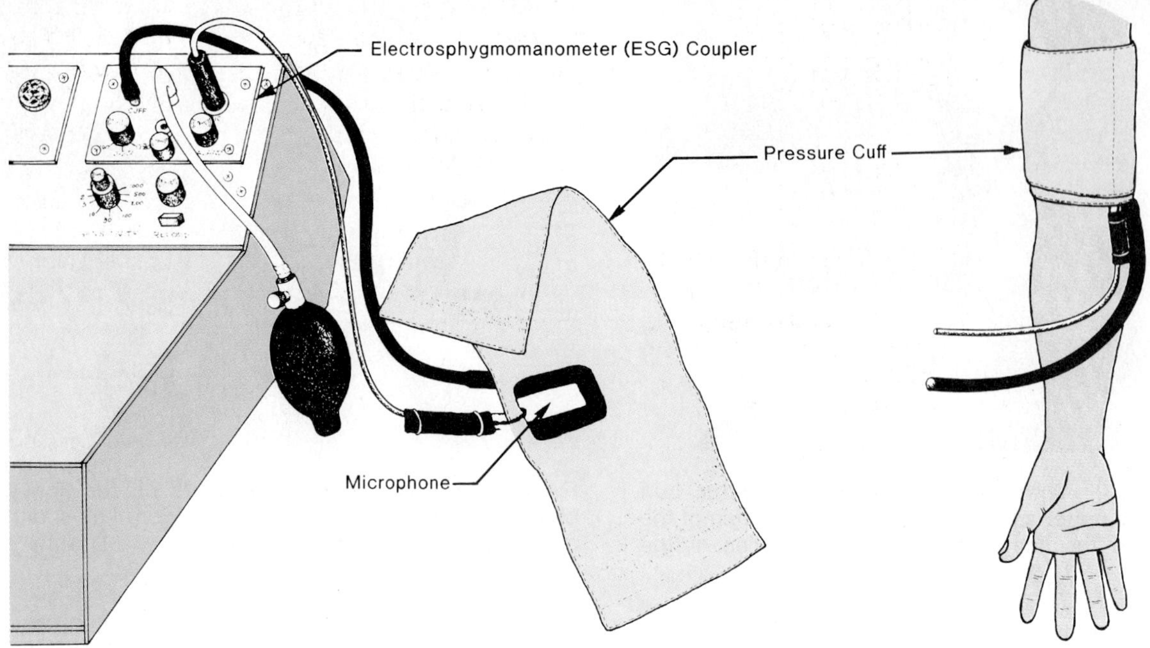

Figure 65.4 Blood pressure monitoring setup with a Physiograph.

Preliminary Preparations

Separate instructions are provided here for the Unigraph and Physiograph. The Unigraph setup requires a separate pressure transducer. In the Physiograph setup, a pressure transducer is built into the ESG coupler.

Unigraph Setup Refer to figure 74.3 on page 000 to see how the Statham pressure transducer mounts to a transducer stand with a transducer clamp (the transducer is in the background, and a hose attaches to it). If no support is available for this unit, place it on its side on the table, in a spot where it cannot roll off onto the floor.

1. Insert the end of the pressure transducer cord into the receptacle of the Unigraph. No adapter is necessary.
2. Place the pressure cuff on the subject's upper right arm, securing it tightly in place.
3. Attach the open end of the tube from the cuff to the fitting on the pressure transducer.
4. Test the cuff by pumping some air into it and noting if the pressure holds when the metering valve on the bulb is closed. Release the pressure for the comfort of the subject while you make other adjustments.
5. Plug in the Unigraph, and turn on the power. Set the speed control lever at the slow position and the stylus heat control knob in the two o'clock position.
6. Set the gain control at 1 mV/cm, and turn the sensitivity control completely counterclockwise to its lowest value. Set the mode control on TRANS.
7. **Calibrate the Statham transducer** by depressing the TRANS button and simultaneously adjusting the sensitivity knob to get 2 cm deflection. The calibrated line represents 100 mm Hg.
8. Proceed to the "Investigative Procedure" section of this experiment.

Physiograph Setup Figure 65.4 shows the Physiograph setup. Note that the pressure bulb tube affixes to the "Bulb" fitting of the ESG coupler, that the pressure cuff attaches to the "Cuff" fitting on the coupler, and that the microphone cable inserts into the "Mic" receptacle of the ESG coupler. You must balance and calibrate the Physiograph before running any tests.

1. **Balance the ESG coupler** as follows:
   a. Start the chart moving at 0.25 cm/sec, and lower the pens onto the paper.
   b. Keep the amplifier record button in the Off position.
   c. Use the position control knob to orient the recording pen to the centerline.
   d. Start recording by pressing the record button to the On position (down).
   e. Using the balance control knob on the ESG coupler, return the recording pen to the centerline established in step *c*.
   f. Check the balance by placing the amplifier record button in the Off position. If the

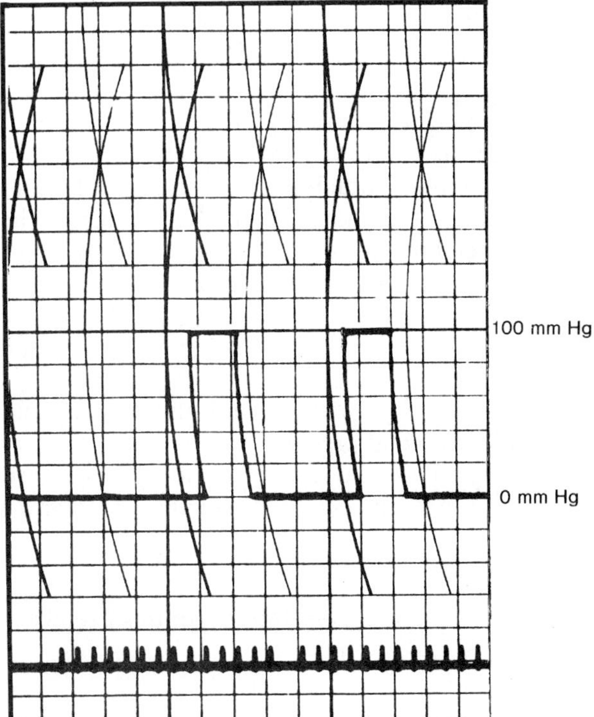

Figure 65.5 Calibrate the zero ("0") pressure line so that it is 2.5 cm below the centerline. Each block on the recording paper represents 20 mm Hg; thus, the centerline represents 100 mm Hg.

system is balanced, the pen will remain on the centerline.

g. If the pen does not remain on the centerline, repeat steps *e* and *f* until the system is balanced.

2. **Calibrate the coupler** to produce 2.5 cm of pen deflection, which is equivalent to 100 mm Hg. Figure 65.5 is a sample recording of the procedure. Proceed as follows:
   a. Start recording by pressing the record button to the On position.
   b. With the position control knob, set the recording pen on a baseline that is exactly 2.5 cm below the centerline—that is, 2.5 cm above the zero pressure line.
   c. While depressing the "100 mm Hg CAL" button on the coupler, rotate the inner knob of the amplifier sensitivity control until the pen moves back to the centerline—that is, 2.5 cm above the zero pressure line.
   d. Release the CAL button, and see if the pen returns exactly to the zero pressure line.
   e. Repeat steps *c* and *d* several times to ensure exact calibration.

Note: The ESG coupler is now calibrated so that each block on the recording paper represents 20 mm Hg pressure and so that each centimeter of pen deflection is equivalent to 40 mm Hg.

3. *From this point on, do not touch* the balance control knob, the position control knob, or the inner knob of the amplifier sensitivity control. Adjustments to these controls will produce inaccurate results.
4. Place the pressure cuff on the subject's right arm so that the microphone is over the brachial artery. Secure the cuff tightly around the upper arm.
5. **Calibrate the cuff microphone sensitivity** as follows:
   a. Start the chart moving at 0.5 cm/sec, and lower the pens to the paper.
   b. Place the record button in the On position.
   c. Close the valve on the hand bulb by rotating it clockwise.
   d. With the subject's forearm resting on the table in a relaxed position, pump up the cuff until the recording pen indicates approximately 160 mm Hg (4 cm of pen deflection).
   e. Release the bulb valve slightly so that the recorded pressure begins to fall at a rate of approximately 10 mm Hg/sec. Note the Korotkoff sounds on the recording.
   f. Taking care not to touch the inner knob of the amplifier sensitivity control, adjust the outer knob of this control so that you get an amplitude of 1.0 to 1.5 cm on the recorded sound. Usually, a setting of 10 or 20 on the outer knob produces good sound recordings. With some individuals, the setting may be as low as 5 or 2; with others, it may be as high as 100.
6. You are now ready to perform your investigation.

Investigative Procedure

You will make four recordings of the subject's blood pressure: (1) while the subject sits in a chair, (2) after the subject has been standing for 4 or 5 minutes, (3) while the subject is prone, and (4) after the subject has been exercising for 5 minutes. Use the following procedure:

With Subject in Chair

1. Start the paper advance at 0.5 cm/sec, and lower the pens onto the chart. Activate the timer.
2. Press the amplifier record button to the On position.
3. Tighten the pressure bulb valve (turn fully clockwise), and inflate the cuff until you are recording a pressure of approximately 40 mm Hg above the expected systolic pressure. Usually, 160 to 180 mm Hg is sufficient.
4. Release the hand bulb valve slightly so that the recorded pressure begins to fall at a rate of approximately 10 mm Hg/sec.

5. Watch for the appearance of the **first** and **last Korotkoff sounds** on the recording. These pressures are your subject's "normal" systolic and diastolic pressures.
6. Terminate the recording. Refer to figure SR-4A in Appendix D for a sample Physiograph record.

With Subject Standing

Have the subject stand for 4 or 5 minutes. Then repeat steps 1 through 6 of the previous procedure while the individual is still standing. Refer to figure SR-4B in Appendix D for a sample Physiograph record.

With Subject Reclining

Have the subject lie on a laboratory table or cot for 4 or 5 minutes. Then repeat steps 1 through 6 of the earlier procedure while the individual continues to recline. Refer to figure SR-4C in Appendix D for a sample Physiograph record.

With Subject After Exercising

Remove the pressure cuff, and have the subject perform 5 minutes of strenuous exercise, such as running, jumping, or climbing stairs.

After the exercise, seat the subject in a chair and *immediately* attach the pressure cuff. Record the systolic and diastolic pressures, using the same procedures as before. Remember, you must inflate the cuff to a higher pressure (240 mm Hg) to occlude the brachial artery. Refer to figure SR-5A in Appendix D for a sample Physiograph record.

Make a series of recordings 1 minute apart until the subject's blood pressure returns to what it was in the original (seated) test. Refer to figures SR-5B and C in Appendix D for sample Physiograph records.

Assignment:
Complete the Laboratory Report for this exercise.

A Polygraph Study of the Cardiac Cycle

66

Over a lifetime, the heart is responsible for an amazing automatic sequence of events. At intervals of less than a second, the heart is automatically triggered, pressures generate, valves open and close, blood flows, and the heart refills. This sequence repeats 100,000 times a day to pump 2000 gallons of blood and continues, faithfully and consistently, usually into old age.

In Exercises 60, 62, and 64, you performed individual studies of heart sounds, the electrocardiogram (ECG), and the pulse. In this exercise, you examine these three cardiac manifestations simultaneously on a polygraph to see how the ECG relates to the other two phenomena.

The **cardiac cycle** divides into two major events: systole and diastole. These events further subdivide into five phases.

Five Phases of the Cardiac Cycle

An arbitrary starting point of the cardiac cycle is at the very end of diastole, when the ventricles are nearly filled with blood, the atrioventricular (A-V) valves are still open, and the aortic and pulmonary valves are still closed. Figure 66.1 shows the relationship of the three polygraph recordings to the cardiac cycle. Note the headings identifying the five phases of the cardiac cycle at the top of the diagram.

Atrial Systole

As mentioned in Exercise 62, a depolarization wave that starts in the S-A node and moves over both atria initiates atrial systole. Its electrical potential creates the **P wave** on the ECG.

Note in figure 66.1 that *no heart sounds are heard during this phase,* even though both atria contract simultaneously. Heart sounds result from the closure of the heart valves, and no valves close during atrial systole.

During atrial systole, the ventricles relax, and the intraventricular pressure is less than the pressure in the atria, aorta, or pulmonary trunk. As a result, the semilunar valves are closed, and the A-V valves are open.

Isometric Ventricular Systole

When the depolarization wave reaches the A-V node, it moves down the atrioventricular bundle into the conduction myofibers to initiate depolarization of the ventricular myocardium. This depolarization manifests itself as the **QRS complex,** which effectively masks the weaker repolarization wave that occurs in the atria.

Immediately after depolarization, the ventricles contract, compressing the blood in these chambers. This closes the A-V valves without significantly reducing ventricular size. This phase is *isometric ventricular contraction.* The closure of A-V valves produces a vibration that is conducted through the chest wall and is heard as the **first heart sound.** This event marks the beginning of ventricular systole.

Ventricular Ejection

Ventricular ejection is the third phase in the cardiac cycle. It starts after the conducting system and ventricles have repolarized. The process of ventricular repolarization produces the **T wave.**

The ventricles eject blood through the aorta and pulmonary trunk as ventricular pressures exceed diastolic pressures in the aorta and pulmonary trunk. In figure 66.1, the pulse pressure is at its lowest level just as the aortic and pulmonary valves open. The up-slope of the pulse pressure corresponds to the increasing systolic pressure in the left ventricle as it ejects blood from the heart.

During ventricular ejection, the A-V valves remain closed, while the aortic and pulmonary valves are open. **No heart sounds** are heard during this interval.

Early Diastole (Isometric Ventricular Relaxation)

In early diastole, the fourth phase in the cardiac cycle, the conducting system and myocardium are nearly repolarized, and the ECG curve is at the baseline. As the ventricles start to relax, isometrically at first, the intraventricular pressure falls below that of the aorta and pulmonary trunk. This creates a backward flow of blood in the arteries, abruptly closing the aortic and pulmonary valves. The closure of these two valves produces vibrations heard as the **second sound.** On the pulse curve, the sudden closure of the aortic valve manifests itself as the **dicrotic notch.**

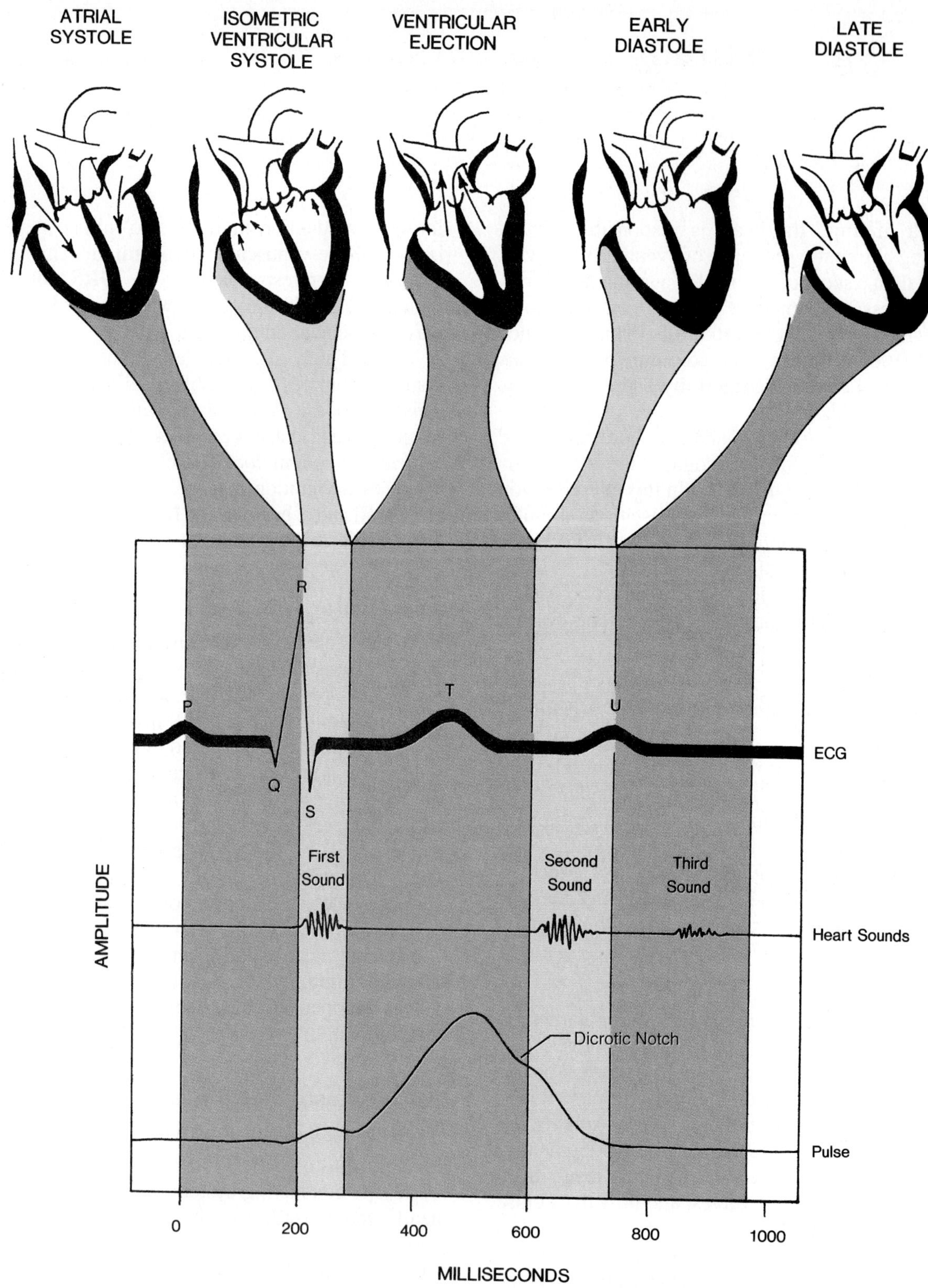

Figure 66.1 Correlation of ECG, heart sounds, and pulse to the cardiac cycle.

Late Diastole

During late diastole, repolarization is virtually complete except, perhaps, for the endocardium and papillary muscles. Late diastole corresponds to the **U wave** of the ECG, which is usually not seen. Isometric ventricular relaxation extends from the dicrotic notch to the bottom of the pulse curve. It ends when the intraventricular pressure falls below the pressure within the atria, thus causing the A-V valves to open and the ventricles to refill. Initially, ventricular filling is rapid, and turbulence may produce a faint **third sound.**

Experiment Demonstration

Since this is the first use of a multichannel recorder, your instructor may decide to demonstrate the procedure of this three-channel experiment. If time and equipment are available, you can then repeat the experiment, working in three- and four-member teams. Alternatively, your instructor may furnish you with a recording of the results so that you can complete the Laboratory Report.

Experiment Procedure

In setting up this experiment, refer back to Exercise 60 for the phonocardiogram procedure, Exercise 62 for the ECG procedure, and Exercise 64 for the plethysmogram procedure. Note that materials are listed separately for Gilson and Narco equipment.

Materials:

for Gilson setup:
Gilson polygraph
Trans/Med 6605 adapter
Gilson A4023 adapter
Statham P23AA pressure transducer
microphone
Gilson T4020 finger pulse pickup
ECG patient cable
ECG electrodes (three)

for Narco setup:
Physiograph with the following three couplers:
- cardiac coupler
- transducer coupler
- ESG coupler

pulse transducer (Narco 705–0050)
five-lead patient cable
four ECG plate electrodes and straps
suction-type chest electrode
adult pressure cuff (Narco PN 712–0016)

for both setups:
Scotchbrite pad
alcohol swabs
electrode paste

After attaching the three ECG electrodes to the subject and the finger pulse pickup to the index finger of one hand, direct the subject to hold the microphone at a designated spot on his or her chest. Produce sufficient tracing so that each team member has a sample for the Laboratory Report.

Assignment:
Complete the Laboratory Report for this exercise.

67 Peripheral Circulation Control (Frog)

An intricate network of capillaries unites the arterial and venous divisions of the circulatory system. **Capillaries** are short, thin-walled vessels approximately 9 μm in diameter. They are so numerous that most body cells are no more than two or three cells away from a capillary.

Blood slows as it passes through capillaries because of capillaries' large combined diameter. This slow blood flow allows sufficient time for blood and tissue cells to exchange materials. The pathway of exchange is as follows:

capillary blood ↔ tissue fluid ↔ tissue cells

The size of the preceding arteriole and the action of the precapillary sphincter muscle determine whether or not blood enters a capillary (see figure 67.1). The arteriole can vasodilate and vasoconstrict to affect blood flow through it. The **precapillary sphincter** regulates the entrance of blood into the capillary and thus can also restrict or permit blood flow. Arterioles and sphincters are under the neural control of the vasomotor center of the medulla oblongata through the autonomic nervous system. Although many factors influence the vasomotor center, the blood's carbon dioxide concentration is very important.

Locally, carbon dioxide (CO_2), histamine, and epinephrine may influence blood flow volume. High CO_2 concentration in certain tissues results in increased blood flow in these tissues and decreased blood flow in other tissues. Increased epinephrine levels in the blood cause vasodilation in some areas (muscles, heart, lungs) and vasoconstriction in other areas. Histamine, a compound that is present in large quantities during allergic reactions, causes extensive vasodilation in peripheral circulation. Such reactions may cause a precipitous drop in blood pressure, and death.

In this exercise, you study the capillaries in the webbing between a frog's toes under a microscope to learn more about the factors that influence capillary circulation.

Materials:
small frog
frog board
4-inch-by-12-inch cloth straps
continued on next page

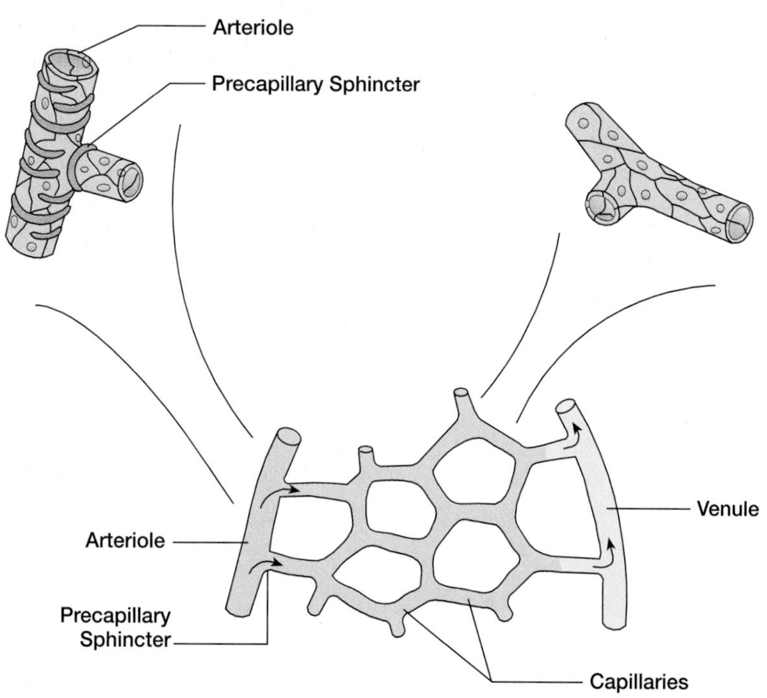

Figure 67.1 Capillary circulation.

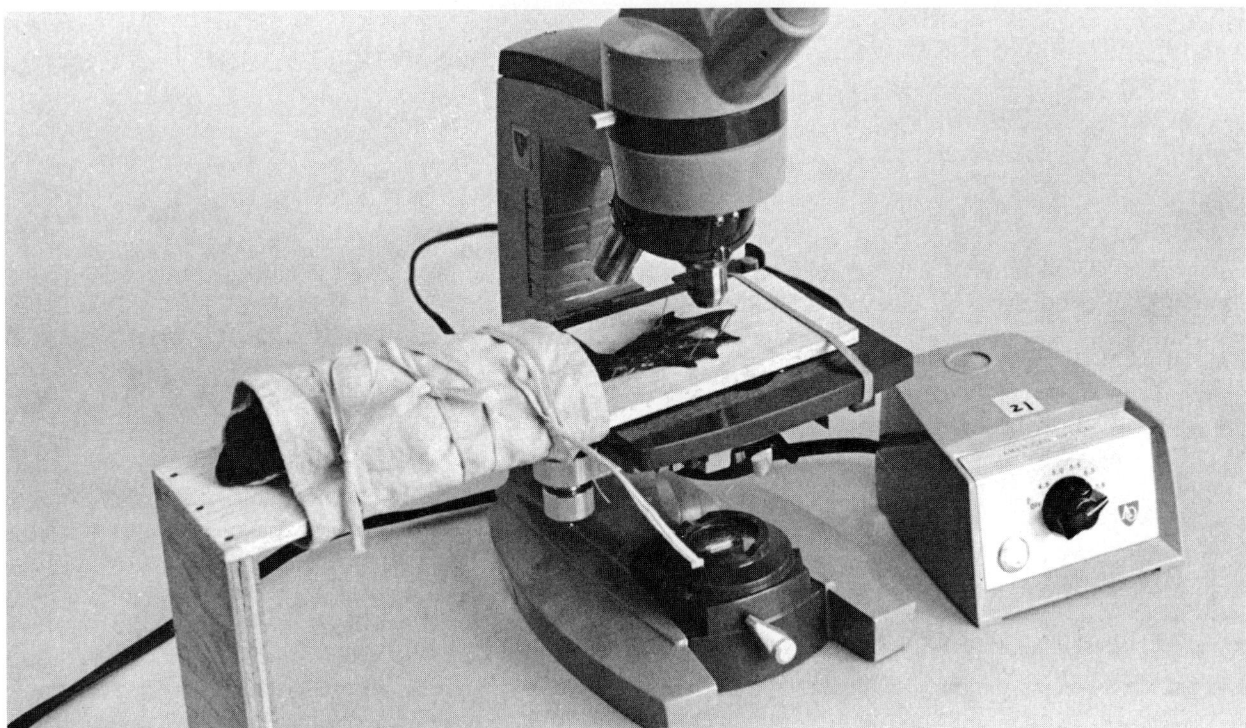

Figure 67.2 Blood flow setup.

string, pins, rubber bands
dropping bottles of epinephrine (1:1000)
dropping bottles of histamine (1:10,000)

1. Strap a frog to a frog board, as figure 67.2 shows. Note that the webbing of one foot is positioned over the hole in the board.
2. Pin the webbing of the foot over the viewing hole. Keep the webbing moistened with water.
3. If necessary, secure the board to the microscope stage with a large rubber band.
4. Position the foot over the light source, and focus on it with the 10× objective.
5. Observe the blood flowing through the vessels. Locate a pulsating **arteriole,** with its rapid blood flow, and a **venule** with its slower, steady blood flow.
6. Identify a **capillary,** with its smaller diameter. Note that blood cells move through it slowly and in single file.
7. Study a capillary closely to locate a **precapillary sphincter** at the juncture of the capillary and arteriole. You probably will not be able to see the sphincter at this low magnification, but you can observe the irregular flow of blood cells into the capillary, which results partly from the sphincter muscle's action.
8. Blot the frog's foot with paper toweling, and apply several drops of 1:10,000 **histamine** solution. Observe the change in blood flow, and record your observations on the Laboratory Report.
9. Wash the foot with water, blot it dry, and then add several drops of 1:1000 **epinephrine** solution. Observe any change in rate of blood flow, and record your results on the Laboratory Report.
10. Wash the foot again, remove the frog from the board, and return the frog to the stock table. Clean the microscope stage, if necessary.

Assignment:
Complete the Laboratory Report for this exercise.

68 The Arteries and Veins

This exercise examines the principal arteries and veins of the circulatory system. You will dissect a cat whose arteries and veins have been injected with colored latex. Before performing the cat dissection, however, study all of the human vessels as follows.

The Human Circulatory Plan

The figure in part A of the Laboratory Report is an incomplete flow diagram of blood through the heart and major body regions. Complete this diagram by filling in the proper blood vessels and providing correct labels *after* you study this section. Students have most difficulty understanding blood flow to the intestines and liver. Starting at the heart, blood enters the right atrium (left side of the Laboratory Report illustration) from the **superior** and **inferior venae cavae,** which collect blood from all parts of the body. This blood is dark colored because it is low in oxygen and high in carbon dioxide.

From the right atrium, blood passes to the right ventricle. When the heart contracts, blood leaves the right ventricle through the **pulmonary trunk** and pulmonary arteries (one vessel on the Laboratory Report illustration) and goes to the lungs, where it picks up oxygen and gives off carbon dioxide. Blood leaves the lungs by way of the **pulmonary veins** (one vessel on the Laboratory Report illustration) and returns to the left atrium of the heart. Blood in the pulmonary veins is brightly colored due to its high oxygen content. The pulmonary arteries, veins, and lung capillaries constitute the *pulmonary system.*

From the left atrium, blood passes to the left ventricle. When the heart contracts, blood leaves the left ventricle through the **aortic arch.** This blood, rich in oxygen, passes to all parts of the body. The aortic arch has branches (one vessel on the Laboratory Report illustration) that go to the head and arms.

Passing downward on the right side of the Laboratory Report illustration, the aortic arch becomes the **descending aorta,** which branches to the liver, digestive organs, kidneys, pelvis, and legs. The branch that enters the liver is the **hepatic artery.** The **superior mesenteric artery** supplies blood to the intestines, and the kidneys receive blood through the **renal arteries.** Several arteries supply blood to the pelvis and legs, but the Laboratory Report illustration only shows one, for simplicity.

Blood leaving most of the organs of the trunk (including the pelvis and legs) empties directly into a large collecting vein, the inferior vena cava. The renal veins (one vessel on the Laboratory Report illustration) pass from the kidneys, and the hepatic vein leads from the liver. Both empty directly into the inferior vena cava.

Note that blood from the intestines does not go directly into the vena cava; instead, it passes to the liver by way of the **hepatic portal vein.** This route of blood from the intestines to the liver is an example of *portal circulation.*

Assignment:
Complete part A of the Laboratory Report by drawing in all of the necessary blood vessels. Color the oxygenated vessels red and the deoxygenated ones blue, and provide arrows to show direction of blood flow. Label all of the vessels.

The Arterial Division

Figure 68.1 shows the principal arteries of the body. Blood leaves the heart through the **ascending aorta,** which curves to the left to form the **aortic arch.** From the upper surface of the aortic arch emerge the brachiocephalic, left common carotid arteries, and left subclavian.

The **left subclavian** artery passes from the aortic arch into the left shoulder behind the clavicle. In the armpit (*axilla*), the subclavian becomes the **axillary** artery. The axillary, in turn, becomes the **brachial** artery in the upper arm. The brachial artery divides to form the **radial** and **ulnar** arteries that follow the radial and ulnar bones of the forearm.

The **brachiocephalic** (*innominate*) artery comes off the aortic arch on the right side of the body. It gives rise to two vessels: the right common carotid and the right subclavian. The branch that extends upward to the head is the **right common carotid** artery. It furnishes blood to the right side of the head. The **right subclavian** artery is an outward extension of the brachiocephalic from the base of the right common carotid. It passes behind the right clavicle into the shoulder.

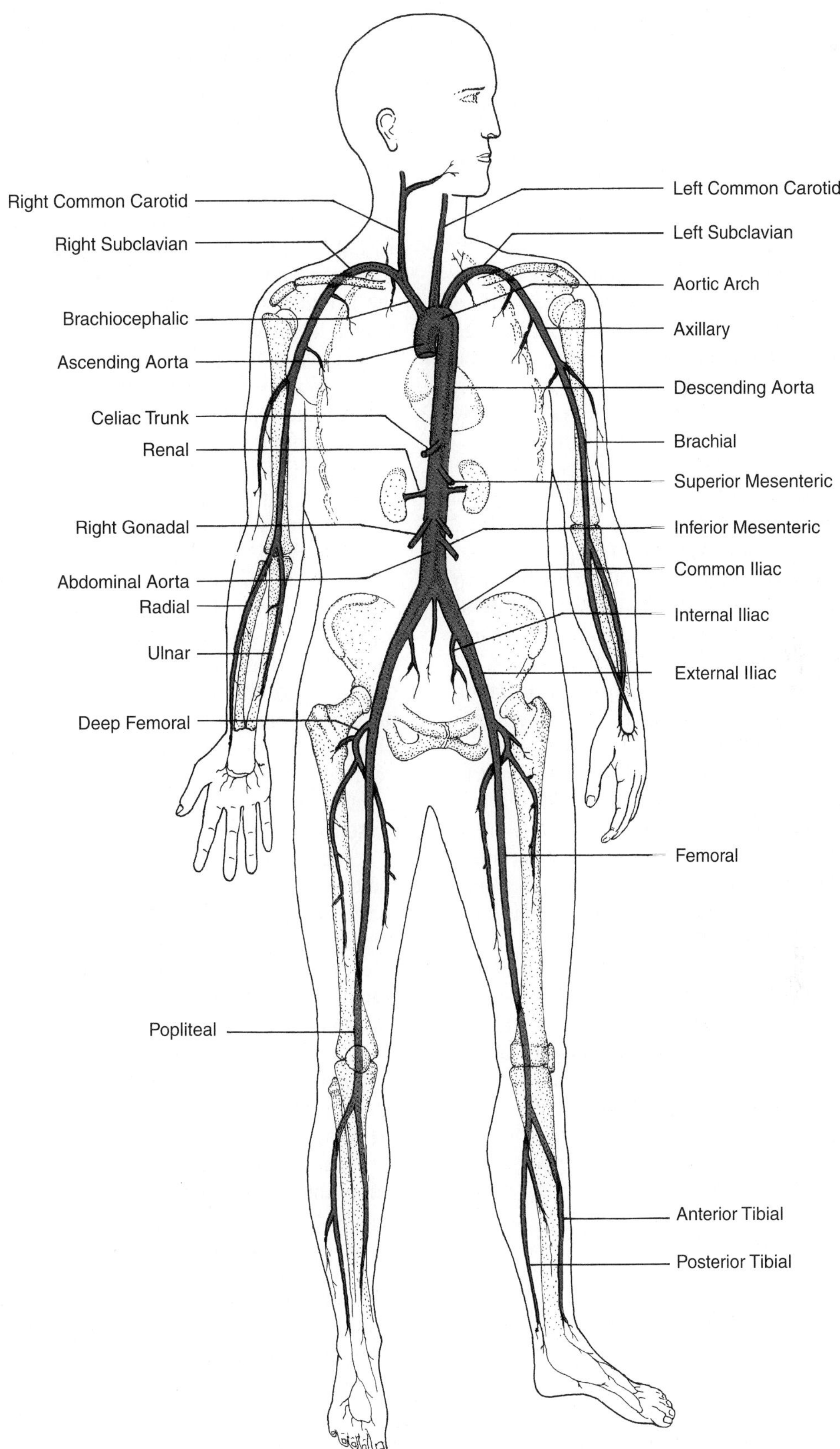

Figure 68.1 Major arteries of the body.

The third branch of the aortic arch is the **left common carotid.** It supplies the left side of the head with oxygenated blood.

As it passes down through the thorax, the aortic arch becomes the **descending** (*thoracic*) **aorta.** Below the diaphragm, it is called the **abdominal aorta.** Although the aorta has many branches leading to various organs, figure 68.1 only shows the larger ones.

The first branch of the abdominal aorta emerging just below the diaphragm is the **celiac trunk,** a very short (about 1.25 cm) artery. It has three branches—the *left gastric,* the *hepatic,* and the *splenic*—that supply blood to the stomach, liver, and spleen, respectively. Figure 68.1 does not show these branches.

Just below the celiac trunk is the **superior mesenteric** artery, which supplies blood to most of the small intestine and part of the large intestine. Inferior to the superior mesenteric are a pair of **renal** arteries that supply the kidneys, and emerging from the anterior surface of the aorta inferior to the renal arteries are two **gonadal** (*spermatic* or *ovarian*) arteries that supply blood to the testes or ovaries.

Inferior to the gonadal arteries is the **inferior mesenteric,** which supplies part of the large intestine and rectum with blood.

In the lumbar region, the abdominal aorta divides into the right and left **common iliac** arteries. Each common iliac passes downward a short distance and then divides into a smaller inner branch, the **internal iliac** (*hypogastric*) artery, and a larger branch, the **external iliac** artery, that continues down into the leg.

The external iliac becomes the **femoral** artery in the upper three-fourths of the thigh. Near the origin of the femoral artery, the **deep femoral** branches off and passes backward and downward along the medial surface of the femur. In the knee region, the femoral becomes the **popliteal** artery. Just below the knee, the popliteal divides to form the **posterior tibial** and **anterior tibial** arteries.

Assignment:
Complete part B of the Laboratory Report for this exercise.

The Venous Division

Figure 68.2 illustrates most of the larger veins of the body. Starting at the heart in figure 68.2 is the **superior vena cava,** which empties into the upper part of the right atrium, and the **inferior vena cava,** which leads into the lower part of the right atrium.

Four veins are in the neck region: two medial **internal jugular** veins and two lateral **external jugulars.** The internal jugulars empty into the brachiocephalic veins, and the smaller externals empty into the **subclavian** veins that pass behind the clavicles. The **brachiocephalic** (*innominate*) veins empty directly into the superior vena cava.

Three veins—the basilic, brachial, and cephalic—collect blood in the upper arm. The **basilic** is on the medial side of the arm, and the **brachial** lies along the posterior surface of the humerus. The basilic and brachial veins unite to form the **axillary** vein in the armpit region. The **cephalic** vein courses along the lateral aspect of the arm and enters the axillary vein at its proximal end near the subclavian vein.

Between the basilic and cephalic veins in the elbow region is the **median cubital** vein. An **accessory cephalic** vein lies on the lateral portion of the forearm and empties into the cephalic in the elbow region.

Superficial and deep sets of veins return blood in the legs to the heart. The superficial veins are just beneath the skin. The deep veins accompany the arteries. Both sets have valves, which are more numerous in the deep veins.

You can vividly demonstrate the presence of valves in veins on the back of your hand. If you use the middle finger of your right hand to apply pressure to a prominent vein on the back of your left hand near the knuckles, and if you then strip the blood in the vein away from pressure point with your right index finger, blood does not flow back toward the point of pressure. This indicates the presence of valves that prevent backward flow.

The **posterior tibial** vein is a deep vein behind the tibia. It collects blood from the calf and foot. In the knee region, the posterior tibial becomes the **popliteal** vein, and above the knee, this vessel becomes the **femoral** vein, which, in turn, empties into the **external iliac** vein. The external and **internal iliacs** empty into the **common iliac** vein.

The **great saphenous** is a large superficial vein of the leg. It originates from the **dorsal venous arch** on the superior surface of the foot and enters the femoral vein at the top of the thigh.

The hepatic, renal, and right gonadal veins empty into the inferior vena cava. The **hepatic** vein carries blood from the liver to the inferior vena cava. The two **renals** that drain the kidneys are inferior to the hepatic. The left renal vein has a downward extending branch, the **left gonadal** (*spermatic* or *ovarian*), that drains blood from the left testis. The **right gonadal** empties directly into the inferior vena cava.

Human Hepatic Portal Circulation

Figure 68.3 shows the venous system that comprises the human *hepatic portal circulation.* All the veins shown here that lead from the stomach, spleen,

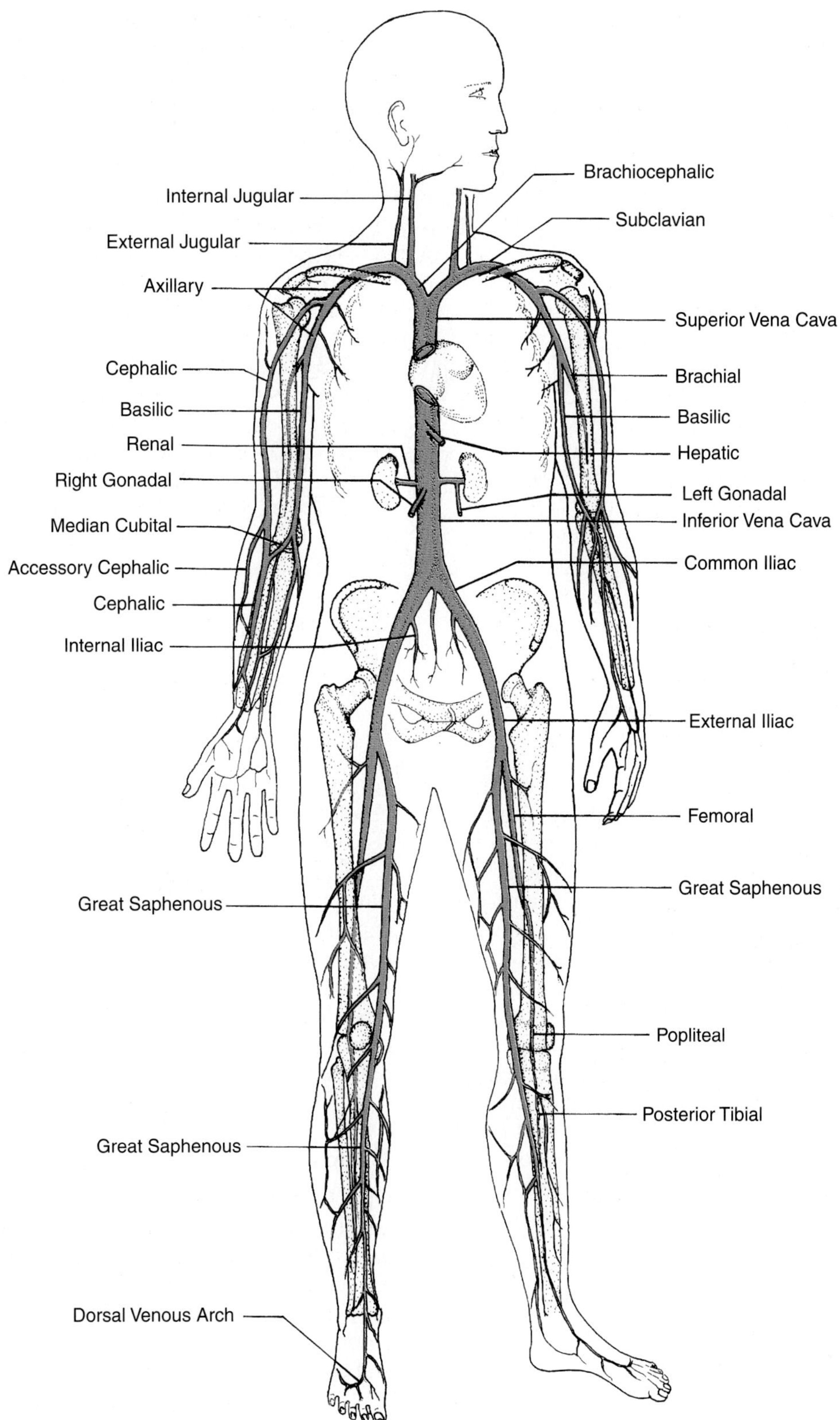

Figure 68.2 Major veins of the body.

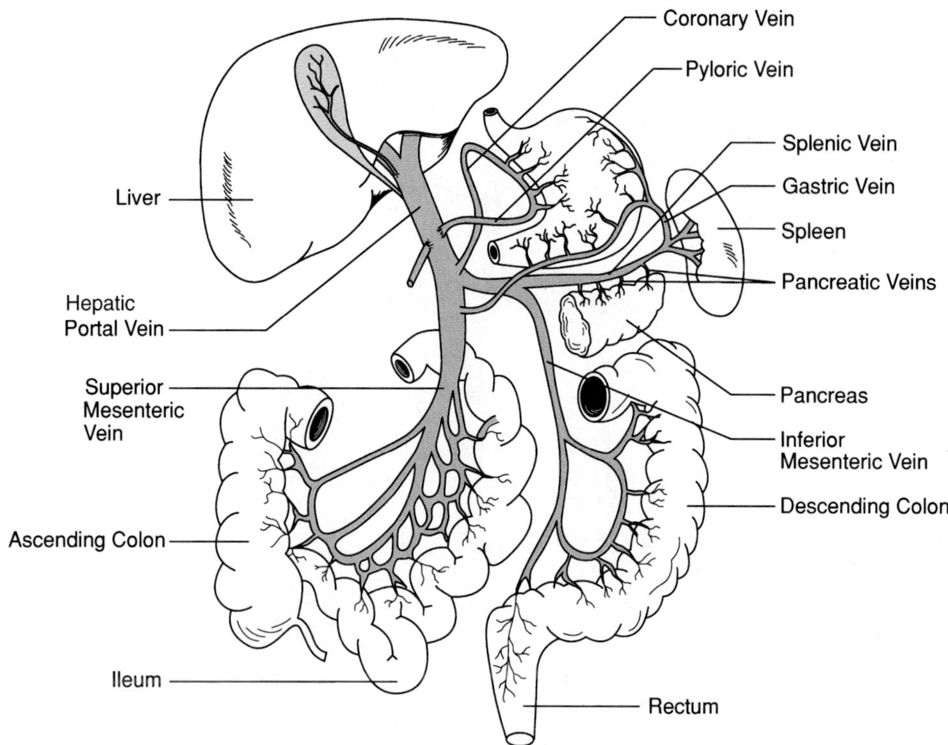

Figure 68.3 Portal circulation, human.

pancreas, and intestine drain into the **hepatic portal** vein, which, in turn, empties into the liver. Thus, all blood from the digestive organs passes through the liver before entering the general circulation. The liver, with its multiplicity of metabolic functions, can then balance blood composition before allowing blood to be transported throughout the body.

The large vessel that collects blood from the ascending colon and the ileum is the **superior mesenteric** vein. It empties directly into the hepatic portal vein. Blood from the descending colon and rectum empties into the **inferior mesenteric,** which empties into the **splenic** (*lienal*) vein. The splenic vein parallels the length of the pancreas, receiving blood from the pancreas via several short **pancreatic** veins. Near the spleen, the splenic vein also receives blood from the stomach through the short **gastric** vein.

A venous loop empties blood from the medial surface of the stomach into the portal vein. The upper portion of the loop is the **coronary** vein; the lower portion is the **pyloric** vein.

Assignment:
Complete part C of the Laboratory Report for this exercise.

Cat Dissection

To differentiate arteries and veins, the cat has been injected with red and blue latex: red for arteries and blue for veins. When tracing blood vessels, free each vessel from adjacent tissue with a sharp dissecting needle. Avoid accidentally severing the blood vessels. Once cut, they become difficult to follow.

During this dissection, you will need to refer to illustrations in other portions of this manual that pertain to other organ systems. When you have completed this circulatory study, you will be familiar with the respiratory, digestive, excretory, and reproductive organs of the cat. Avoid damaging the organs of the various systems, because they are required for later study.

Exposing the Organs

Open the cat's ventral surface by using a sharp scalpel or scissors to start a longitudinal incision about 1 cm to the right or left of midline in the thoracic wall. This cut should pass through the rib cartilages. Extend the incision anteriorly to the apex of the thorax and posteriorly to the pubic region. In addition, make two lateral cuts posterior to the diaphragm.

Spread the walls of the thorax to expose its viscera. Cut away part of the chest wall on each side—using heavy-duty scissors to break through the ribs—to provide better exposure. Separating the diaphragm from the body wall (do not remove the diaphragm) also enhances organ exposure. In addition, trim away portions of the abdominal wall on each side of the original longitudinal incision to expose the abdominal organs.

Organ Identification

Compare your dissection with figure 71.4 to identify the **lungs, thymus gland, thyroid gland, diaphragm, trachea,** and **larynx.** In addition, squeeze the surface of the **heart** between your thumb and forefinger, noting the thickness and texture of the **pericardium** that envelops it. Now refer to figure 81.8 to identify the **liver, stomach, spleen, pancreas,** and the various portions of the **small** and **large intestines.**

Veins of the Thorax

Since the veins of the thoracic region lie over most of the arteries, studying the veins first is best. Refer to figures 68.4 and 68.5 to identify the veins in the sequence that follows.

Veins Entering the Heart

The **superior vena cava, inferior vena cava,** and three groups of **pulmonary** veins empty blood into the heart. Figure 68.4 only shows the superior vena cava. Probe around the heart to find all these vessels. Each group of pulmonary veins, which empty into the left atrium, is composed of two or three veins.

Veins Emptying into the Superior Vena Cava

Locate the large **azygous** vein, which empties into the dorsal surface of the superior vena cava (s.v.c.), close to where the s.v.c. enters the right atrium of the heart. Since the azygous is obscured by the heart and lungs, and not shown in figure 68.4, you need to force these organs to the cat's left to locate it. Note that the azygous lies along the right side of the vertebral column, collecting blood from the **intercostal, esophageal,** and **bronchial** veins. Trace the azygous vein to these branches. All of these veins are present in humans.

Identify the **internal mammary** (*sternal*) vein that enters the s.v.c. opposite the third rib.

Examine the dorsal surface of the s.v.c. to identify where the **right vertebral** and **right costocervical** veins enter. These two veins unite to form a short vessel before entering the s.v.c. Figure 68.4 shows a stub of this vessel, labeled VE. Note that figure 68.4 shows the **left costocervical** vein entering the left brachiocephalic. The figure also shows a stub of the **left vertebral** joining the left costocervical.

Identify the right and left **brachiocephalic veins** that empty into the s.v.c. These veins receive blood from the head and arms.

Drainage into the Brachiocephalic Veins

The right brachiocephalic vein receives blood from two vessels, and the left brachiocephalic accepts blood from four veins. The **external jugular** and **subclavian** veins empty into both brachiocephalics. The left brachiocephalic also receives blood from the **left vertebral** and **left costocervical.** Identify all of these vessels on your specimen.

Drainage into the External Jugulars

The principal veins that empty into the external jugulars are the **transverse scapular** and **internal jugulars.** Locate them first in figure 68.4 and then on your specimen. Note that the internal jugulars parallel the common carotid arteries.

Refer to figure 68.5 to locate the **transverse jugular** vein that connects the two external jugular veins in the chin region; then identify it on your specimen. Also locate the **posterior** and **anterior facial** veins in figure 68.5.

The **thoracic duct,** a part of the lymphatic system, empties into the left external jugular vein. Its point of entrance is very close to where the left external jugular enters the left subclavian. The thoracic duct is difficult to find. Push the left lung to the right, and look at the back of the left thoracic cavity near the vertebral column. Valves give the thoracic duct a beaded appearance.

Drainage into the Subclavian Vein

The principal veins that empty into the subclavian are the **subscapular** and **axillary.** Locate them on figure 68.4 and on your specimen. Note that the axillary receives blood from the **long thoracic** and **brachial** veins. Locate these veins also.

Arteries of the Thorax

Once you have identified the veins of the thorax, examine the thoracic arteries by carefully moving the veins to one side. Do not cut away the veins, however. Refer to figures 68.4 and 68.6 to identify the arteries in the sequence that follows.

Arteries Emerging from the Heart

Locate the **aortic arch** and **pulmonary trunk** in figures 68.4 and 68.6, and on your specimen. Trace the pulmonary trunk to where it divides into the right and left pulmonary arteries.

Dissect the connective tissue at the base of the aorta to locate the **coronary** arteries. Note their pathways on the surface of the heart.

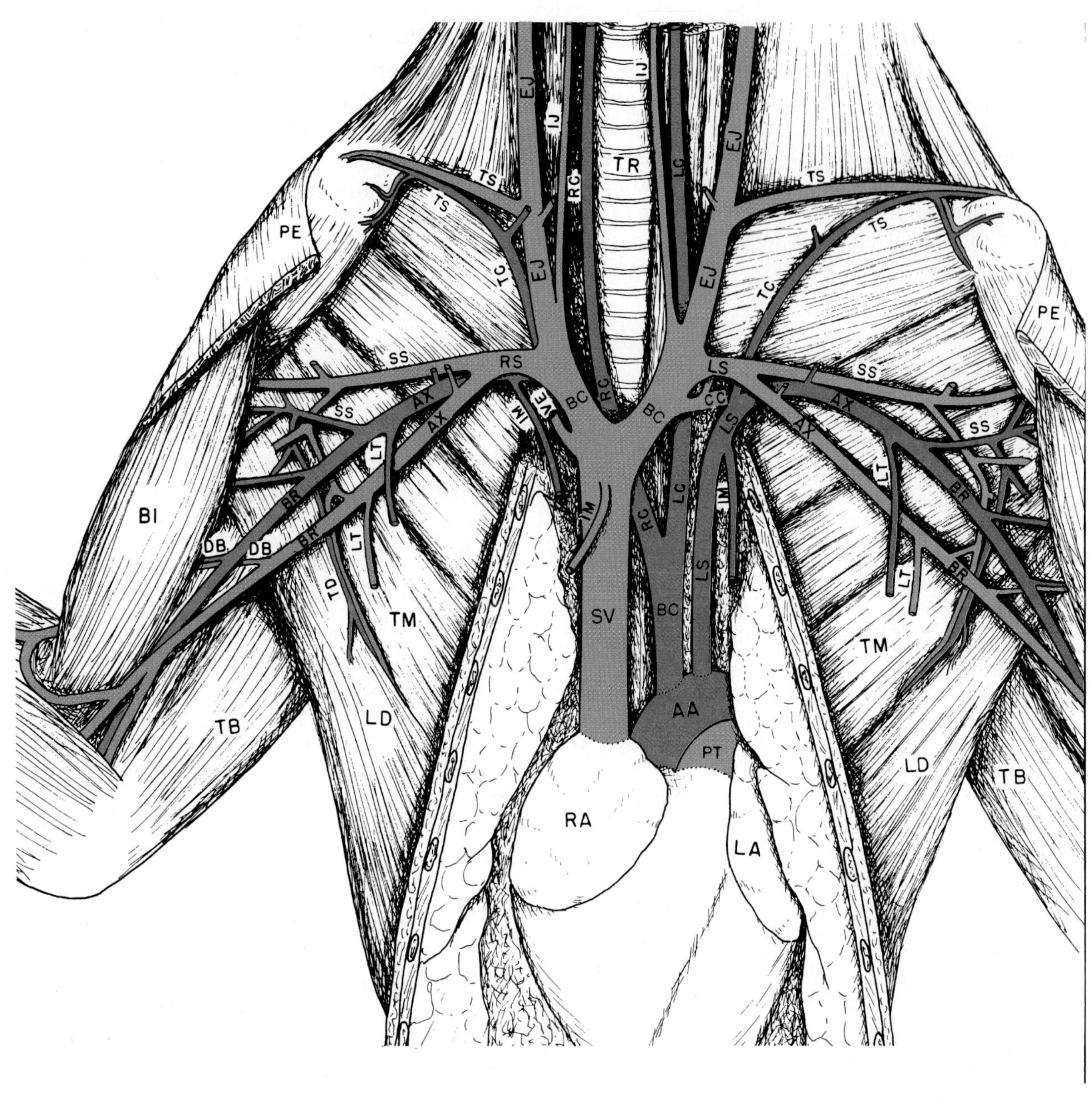

AA—Aortic Arch
AX—Axillary
BC—Brachiocephalic
BI—Biceps Muscle
BR—Brachial
CC—Costocervical
DB—Deep Brachial
EJ—External Jugular
IJ—Internal Jugular
IM—Internal Mammary
LA—Left Atrium
LC—Left Common Carotid
LD—Latissimus dorsi Muscle
LS—Left Subclavian
LT—Long Thoracic
PE—Pectoralis Muscles
PT—Pulmonary Trunk
RA—Right Atrium
RC—Right Common Carotid
RS—Right Subclavian
SS—Subscapular
SV—Superior Vena Cava
TB—Triceps brachii Muscle
TC—Thyrocervical
TD—Thoracodorsal
TM—Teres major Muscle
TR—Trachea
TS—Transverse Scapular
VE—Vertebral

Figure 68.4 Veins and arteries of the thorax.

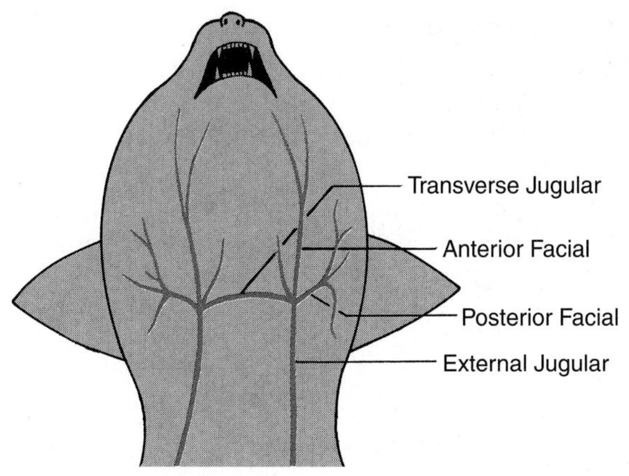

Figure 68.5 Branches of the external jugular vein.

Branches of the Aortic Arch

Observe that the cat has only two branches emerging from the upper surface of the aortic arch: the **brachiocephalic** (*innominate*) and the **left subclavian.** How does this compare with humans? Note also that the **right common carotid, right subclavian,** and **left common carotid** arteries branch off the brachiocephalic.

Branches of the Subclavian Arteries

In figures 68.4 and 68.6, and on your specimen, locate the vertebral, costocervical, internal mammary, and thyrocervical arteries that branch off the subclavian arteries. The **vertebral** carries blood to the brain, passing through the transverse foramina of the cervical vertebrae. The **costocervical** supplies blood to deep muscles of the back and neck. The **internal mammary** arteries supply blood to the ventral body wall, and the **thyrocervical** arteries supply blood to the neck and shoulder.

Note that the thyrocervical becomes the **transverse scapular** and that the subclavian becomes the **axillary.** The axillary, in turn, eventually becomes the **brachial** artery.

Branches of the Axillary Artery

In figures 68.4 and 68.6, and on your specimen, identify the **ventral thoracic, long thoracic,** and **subscapular** arteries that branch off the axillary artery. These branches supply blood to the latissimus dorsi and pectoral muscles. Note also that the

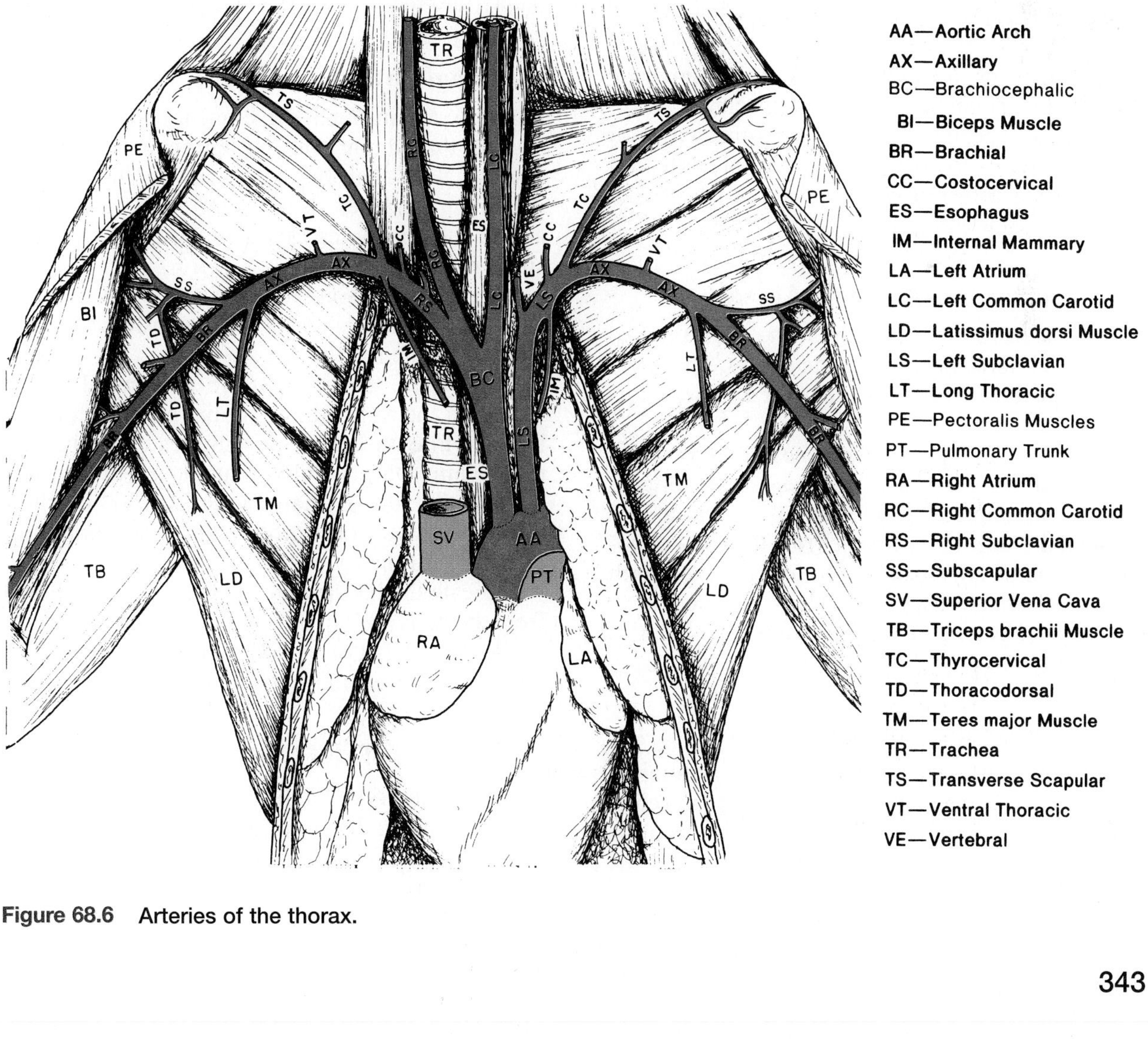

AA—Aortic Arch
AX—Axillary
BC—Brachiocephalic
BI—Biceps Muscle
BR—Brachial
CC—Costocervical
ES—Esophagus
IM—Internal Mammary
LA—Left Atrium
LC—Left Common Carotid
LD—Latissimus dorsi Muscle
LS—Left Subclavian
LT—Long Thoracic
PE—Pectoralis Muscles
PT—Pulmonary Trunk
RA—Right Atrium
RC—Right Common Carotid
RS—Right Subclavian
SS—Subscapular
SV—Superior Vena Cava
TB—Triceps brachii Muscle
TC—Thyrocervical
TD—Thoracodorsal
TM—Teres major Muscle
TR—Trachea
TS—Transverse Scapular
VT—Ventral Thoracic
VE—Vertebral

Figure 68.6 Arteries of the thorax.

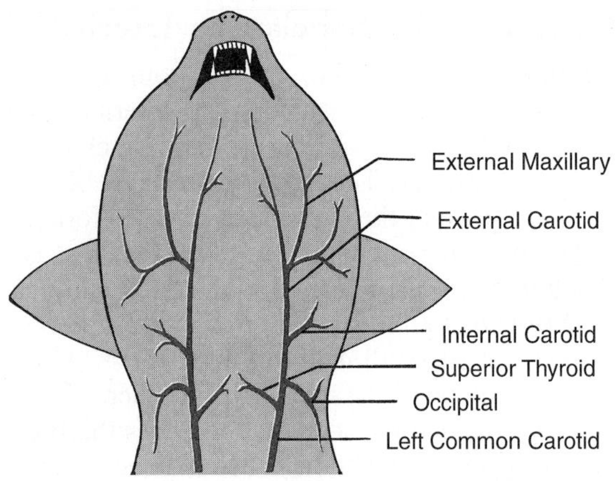

Figure 68.7 Branches of the carotid artery.

subscapular becomes the **thoracodorsal** and that the axillary becomes the **brachial.** The brachial becomes the **radial** artery at the elbow.

Branches of Each Common Carotid Artery

Each common carotid has an inferior and a superior thyroid artery. The small **inferior thyroid** originates near the base of the common carotid. Probe around this area to locate it. Figure 68.7 shows the **superior thyroid.** Locate this artery on your specimen also.

If your specimen is properly injected, you should be able to locate the **occipital, external carotid,** and **internal carotid** that branch off the common carotid (see figure 68.7).

Branches of the Thoracic Aorta

Pull the viscera of the thorax to the cat's right to expose the aorta in the thoracic cavity. The aorta lies dorsal to the parietal pleura. Peel away this membrane to expose the aorta.

Locate the 10 pairs of **intercostal** arteries that emerge from the thoracic portion of the aorta to supply the intercostal muscles. Also locate the **bronchial** arteries that emerge from the aorta opposite the fourth intercostal space or from the fourth intercostal artery. They accompany the bronchi to the lungs. Next, identify the **esophageal** arteries that supply blood to the esophagus.

Branches of the Abdominal Aorta

Remove the peritoneum from the dorsal abdominal wall just below the diaphragm to expose the **celiac trunk.** Note in figure 68.8 that the celiac trunk is the first branch of the abdominal portion of the aorta. Follow it out to its three branches: **hepatic, left gastric,** and **splenic** arteries.

In figure 68.8 and on your specimen, identify the **superior mesenteric, renal, spermatic** or **ovarian (gonadal), inferior mesenteric,** and **iliolumbar** arteries, in that order, noting the organs that they supply. Note that a pair of **adrenolumbar** arteries emerge caudad to the superior mesenteric. These arteries supply the adrenal glands, diaphragm, and muscles of the body wall.

Seven pairs of small **lumbar** arteries assist the iliolumbar arteries in supplying blood to the abdominal body. Can you locate them?

Observe that the cat has no common iliac arteries; instead, two large branches, the **external iliac** arteries, pass into the hind legs. Each external iliac eventually becomes the **femoral** in the thigh region.

Caudad to the external iliacs is a pair of **internal iliac** arteries branching off the aorta. The aorta terminates as the **caudal** artery.

Veins of the Abdomen

You have now identified the arteries and veins of the upper part of the cat. Move on to the veins that empty into the inferior vena cava.

Drainage into the Inferior Vena Cava

Figure 68.9 illustrates the branches of the venous system that drain into the **inferior vena cava** (i.v.c.). Note that the i.v.c. passes through the diaphragm and lies to the right of the abdominal aorta. The tributaries of this vein usually parallel similarly named arteries.

Scrape away some of the tissue on the liver's anterior surface to locate the **hepatic** veins that empty into the i.v.c.

Study the **adrenolumbar** and **renal** veins. Note that the right adrenolumbar empties into the i.v.c., while the left one empties into the left renal vein. Observe also that the **right** and **left spermatic** (or ovarian) veins differ in their drainage, with only the right one emptying directly into the i.v.c.

Locate the **iliolumbar** veins that convey blood from the body wall in the lumbar region to the i.v.c.

Identify the **right** and **left common iliac** veins that convey blood from the legs to the inferior vena cava.

Tributaries of the Common Iliac Veins

Note in figure 68.9 that the **caudal** vein empties into the left common iliac instead of being a direct extension of the i.v.c. This is not always the case, however.

Trace the common iliac vein in one leg down to where it branches to form the **external iliac** and the

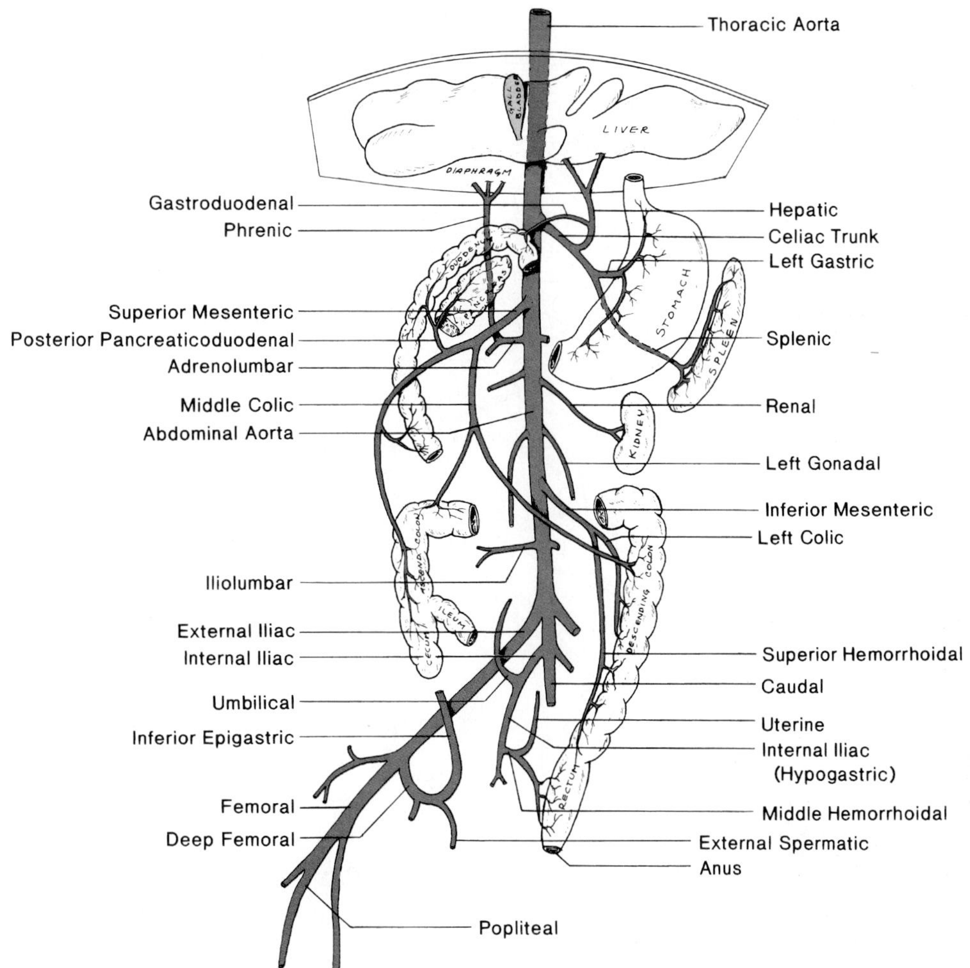

Figure 68.8 Arteries of the abdomen and leg.

internal iliac. Note that, as in humans, the external iliac becomes the **femoral** in the thigh region. Follow the femoral further into the leg to locate the **greater saphenous** and **popliteal** veins.

The Hepatic Portal System of the Cat

Figures 68.9 and 68.10 show the veins of the hepatic portal drainage system. Observe that, as in humans, blood from the stomach, spleen, pancreas, small intestine, and large intestine empties into the **hepatic portal** vein. Locate the following tributaries that feed into this large vessel: **gastrosplenic, superior mesenteric, inferior mesenteric,** and **pancreaticoduodenal** veins. Pay particular attention to the organs that supply these tributaries with venous blood.

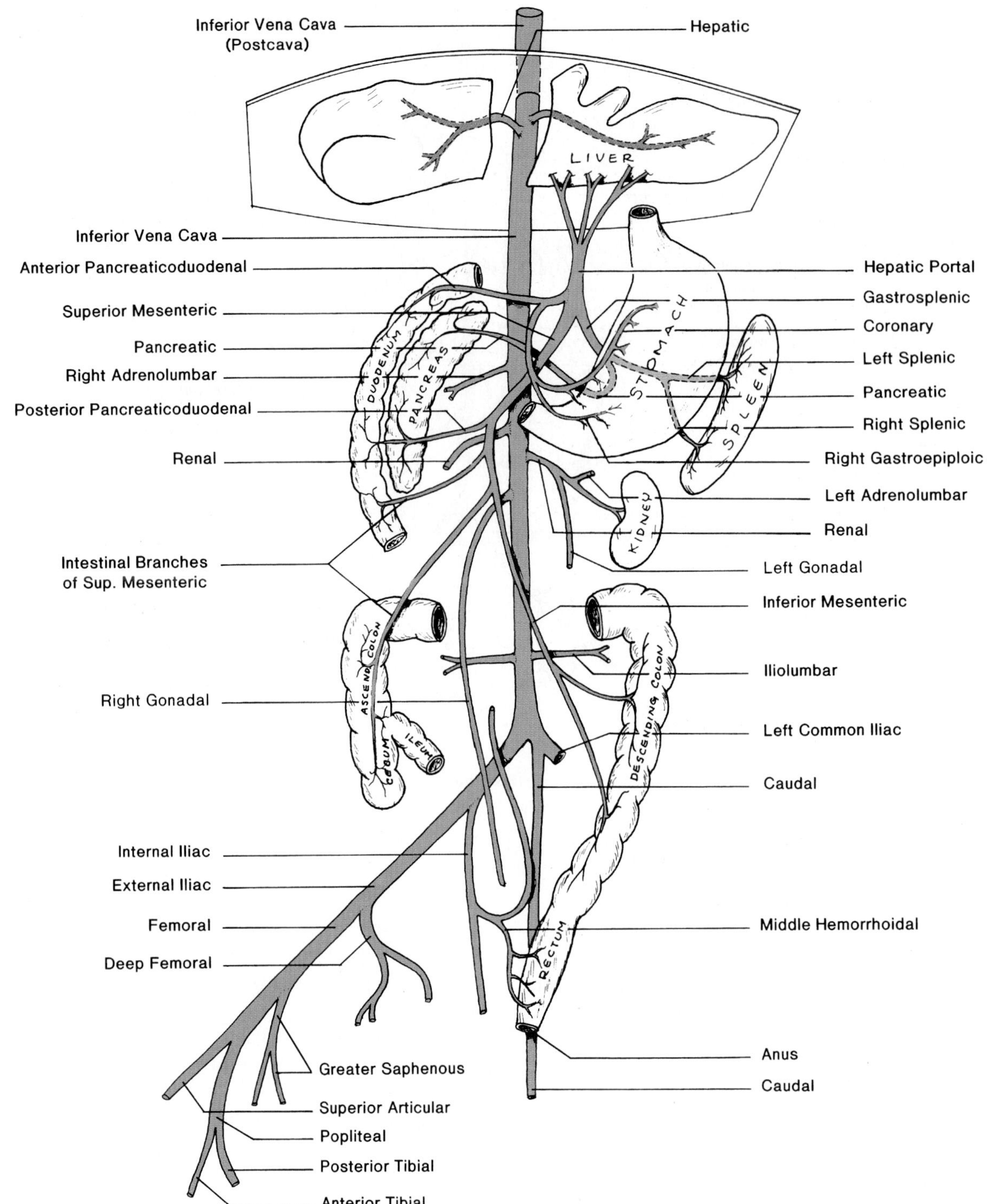

Figure 68.9 Veins of abdomen and leg.

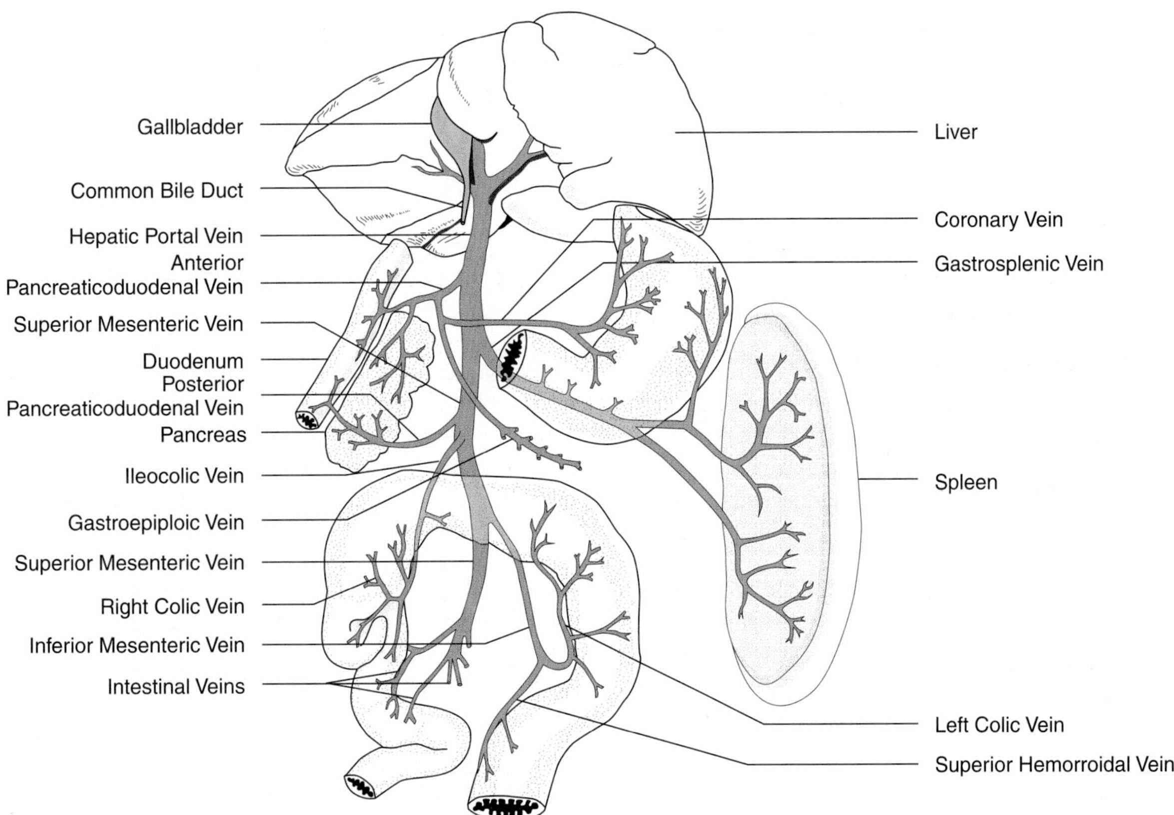

Figure 68.10 **Hepatic portal circulation, cat.**

69 Fetal Circulation

During human embryological development, the circulatory system is necessarily somewhat different from that after birth. The lungs are nonfunctional before birth, which dictates that less blood should go to the lungs. Also, because the mother provides the food and oxygen supply for the fetus by way of the placenta, blood vessels to and from the placenta through the umbilical cord must be present and functional.

In this exercise, you learn (1) how the blood flow pattern in a fetus differs from that in an adult, (2) how the flow pattern changes after birth, and (3) what forces control the changes after birth.

Figure 69.1 shows the pathway of blood through the fetal heart. Fetal blood supply passes from the placenta in the uterus through the **umbilical cord** via the **umbilical vein.** The blood in the umbilical vein is rich in nutrients and oxygen. After circulating throughout the body of the fetus, the blood, laden with carbon dioxide and metabolic wastes, returns to the placenta via a pair of **umbilical arteries** wrapped around the umbilical vein in the umbilical cord.

The umbilical vein divides as it approaches the **liver:** One branch enters the liver, and the other branch becomes the **ductus venosus** that continues upward to join a branch from the liver and then enters the **inferior vena cava.** Note that a **sphincter muscle** constricts the entrance to the ductus venosus. This sphincter does not restrict blood flow during growth and development; it comes into play only after parturition (birth).

Oxygen-rich blood of the ductus venosus mixes with venous blood in the inferior vena cava before it enters the right atrium of the fetal heart. In the fetal right atrium, blood can move in two directions. Some passes to the right ventricle, as it would in an adult, and the remainder passes through a valve, the **foramen ovale,** into the left atrium. Blood is shunted to the left side of the heart through the foramen ovale because the fetal lungs are nonfunctional and thus require less blood.

During ventricular systole, blood in the right ventricle exits the heart through the **pulmonary trunk,** and blood in the left ventricle exits through the **aortic arch.** Blood flowing from the right ventricle through the pulmonary trunk flows in three directions: Some blood goes to the right lung through the **right pulmonary artery,** some goes to the left lung through the **left pulmonary artery,** and the remainder goes to the aortic arch through a very short vessel, the **ductus arteriosus.** Thus, blood in the left atrium is a mixture of deoxygenated blood from the lungs (via four pulmonary veins) and partially oxygenated blood from the right atrium. This mixture passes from the left atrium to the left ventricle and out to all parts of the body through the aortic arch.

Blood flowing through the aortic arch passes down the **abdominal aorta** to the pelvic region, where it divides into two **common iliac arteries.** Each common iliac artery, in turn, branches into **external iliac** and **internal iliac arteries.** The internal iliac arteries pass by the **urinary bladder** to the umbilical cord, where they become the umbilical arteries that return blood to the placenta.

Once the umbilical cord is cut, separating the fetus from the placenta at birth, changes in the heart, veins, and arteries provide a new route for the blood. One of the first changes is that the sphincter muscle closes the umbilical vein at the ductus venosus to minimize blood loss. Blood also coagulates in this vessel. The attending physician also prevents further blood loss by ligature. Within 5 days after birth, the umbilical vein converts to the *ligamentum teres hepatis,* which extends from the umbilicus to the liver in the adult. The ductus venosus becomes a fibrous band, the *ligamentum venosum* of the liver.

As soon as the newborn infant begins to breathe, the pulmonary arteries carry more blood to the lungs. The ductus arteriosus then collapses and begins to gradually transform to connective tissue of the *ligamentum arteriosum.*

The increased blood volume and blood pressure in the left atrium, due to a greater blood flow from the lungs, close the flap on the foramen ovale. Eventually, connective tissue permanently seals this valve.

Assignment:
Complete the Laboratory Report for this exercise.

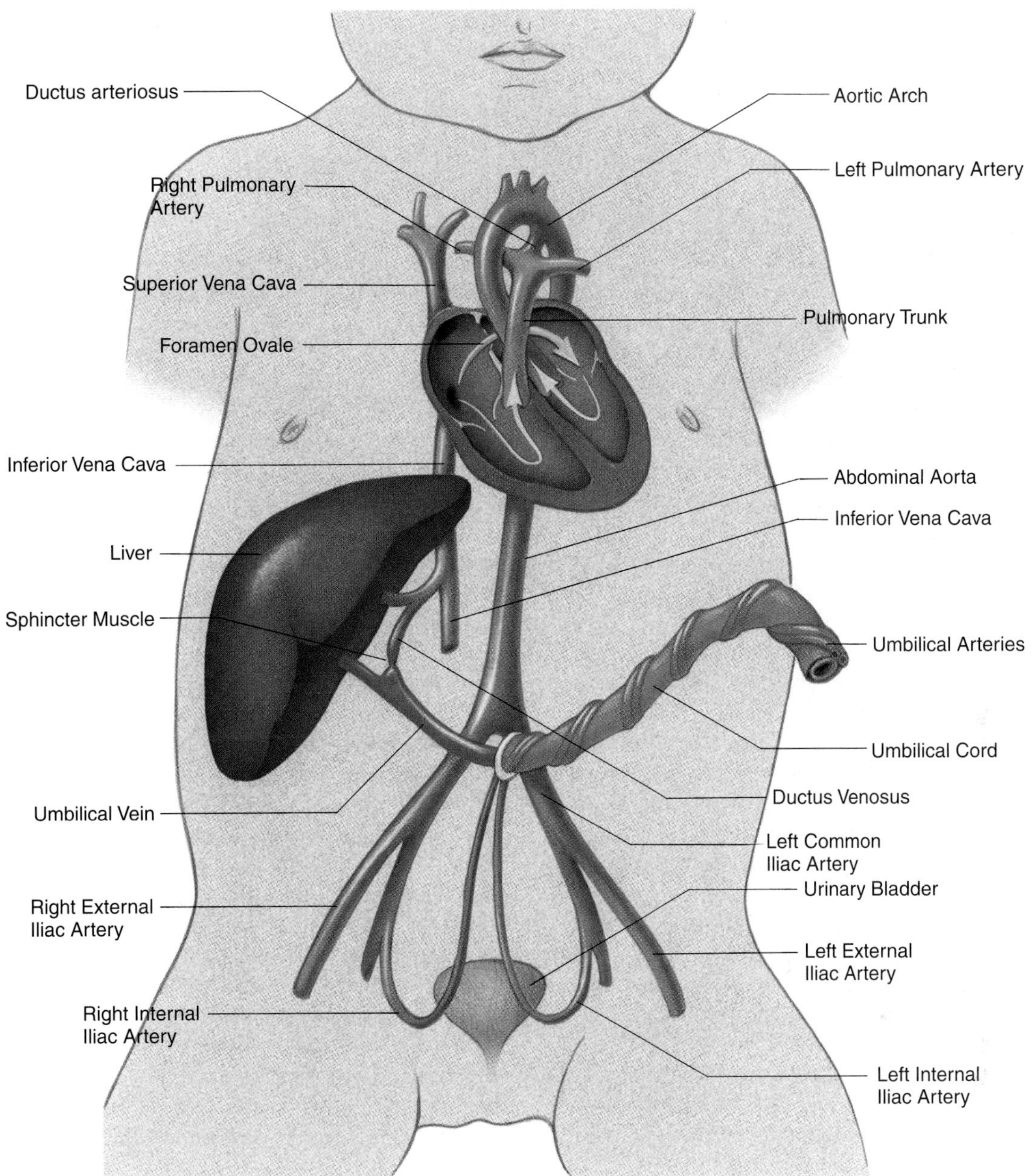

Figure 69.1 Fetal circulation.

70 The Lymphatic System and the Immune Response

The lymphatic system consists of a drainage system of lymphatic vessels that return tissue fluid from interstitial cell spaces to the blood. On the journey back to the blood, macrophages in the lymph nodes clean the lymph (the fluid in lymphatic vessels) of bacteria and other foreign material. Lymph nodes are also packed with lymphocytes, which play a leading role in cellular and humoral immunity. This exercise explores the genesis and physiology of lymphocytes, as well as the overall anatomy of the lymphatic system.

Lymph Pathway

Lymph capillaries are the smallest vessels of the lymphatic system. These microscopic vessels have closed ends and are distributed among the cells of the various body tissues. Lymph capillaries unite to form **lymphatic vessels** (*lymphatics*) (see figure 70.1). Lymphatic vessels contain many valves to restrict the backward flow of lymph.

As lymph moves toward the center of the body through lymphatic vessels, it eventually passes through one or more small, oval **lymph nodes.** These nodes cluster in the neck, groin, axillae, and abdominal cavity, and in smaller groups in other parts of the body. They may be as small as a pinhead or as large as an almond.

The **thoracic duct** is a large vessel that collects lymph from the legs, abdomen, left half of the thorax, left side of the head, and left arm. Its lower extremity consists of a saclike enlargement, the **cisterna chyli.** Lymph from the intestines passes through lymphatic vessels in the mesentery to the cisterna chyli. Lymph from this region contains a great deal of fat and is usually called *chyle.* The thoracic duct empties into the left subclavian vein near the left internal jugular vein. The **right lymphatic duct** is a short vessel that drains lymph from the right arm and right side of the head into the right subclavian vein near the right internal jugular.

Lymph Node Structure

The enlarged lymph node in figure 70.1 has a slight depression, called the **hilum,** at its upper end. Blood vessels and an **efferent lymphatic vessel** emerge through this depression. Although each node has only one hilum and one efferent lymphatic vessel, two or more **afferent lymphatic vessels** usually carry lymph into the node (figure 70.1 shows two).

Surrounding the entire lymph node is a **capsule** of fibrous connective tissue, and just inside the capsule is a **cortex** that consists of many **germinal centers** (*nodules*) and of sinuses reinforced with reticular tissue. Germinal centers are the basic structural units of a lymph node. They arise from small nests of lymphocytes or lymphoblasts, and they reach about 1 mm in diameter. The central portion of a lymph node is the **medulla.**

Role of Lymphocytes in Immunity

The integrated action of an army of different cell types, including monocytes, macrophages, eosinophils, basophils, and lymphocytes, comprises the immune response to foreign invaders. Although each of these cells plays a different role, the different cell types interact, to the extent of regulating each other's activities. The "commander" and predominant "foot soldier" of this defensive force is the lymphocyte.

Although all lymphocytes are morphologically similar, individual cells differ considerably. They are present in blood, lymph, and lymph nodes, as well as in bone marrow, the thymus gland, the spleen, and lymphoid masses associated with the digestive, respiratory, and urinary passages.

Lymphocytes exist as either B cells or T cells. **B cells** (*B* for bone marrow) govern *humoral* or *antibody-mediated immunity.* **T cells** (*T* for thymus) are responsible for *cell-mediated immunity.*

Each B cell recognizes only a single foreign antigen, be it on a bacterium, virus, or some other invader. Activated B cells produce *circulating antibodies* that bind to antigens or antigen-bearing targets and mark them for destruction by other components of the immune system.

T cells are of three principal types: *helper, suppressor,* and *cytotoxic* or *"killer" cells.* A T cell that recognizes and binds to an antigen on the surface of another cell becomes activated. An activated T cell can multiply, and if it is a cytotoxic cell, it can kill the bound cell. For example, cancer cells that display antigens not found on healthy cells can potentially activate T cells, which destroy the cancer cells.

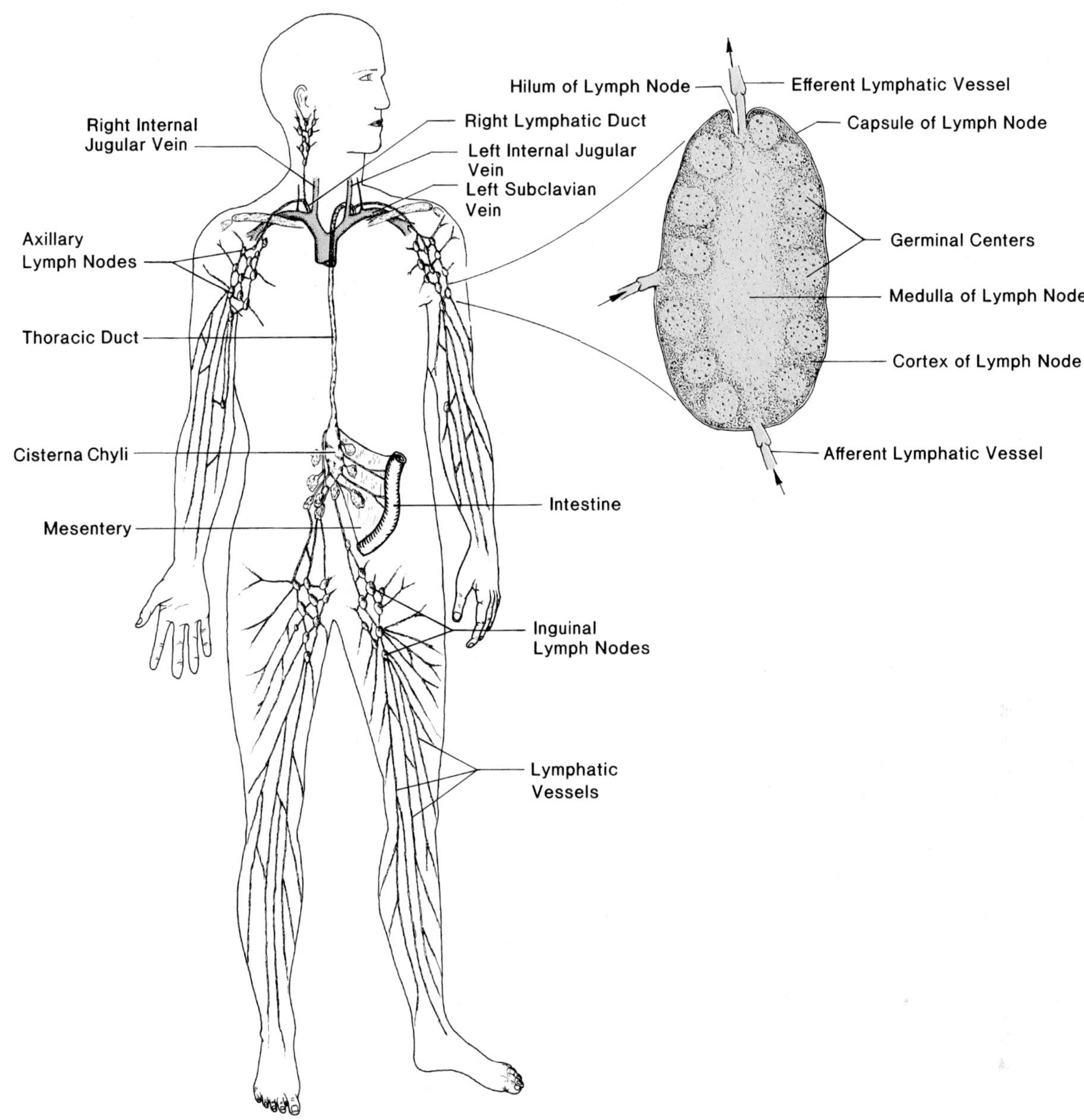

Figure 70.1 The lymphatic system.

Cells of the immune system regulate each other by secreting potent chemicals called **cytokines.** Lymphocytes produce cytokines called *lymphokines.* Monocytes and macrophages produce cytokines called *monokines.*

Lymphocytes, as well as *all* other blood cells, develop from stem cells in the bone marrow before and shortly after birth. Lymphocyte stem cells released into the blood follow one of two pathways for "processing," as figure 70.2 shows.

Stem cells that settle in the thymus for modification become T cells. Other stem cells are processed in an unknown area and become B cells. They may form spontaneously, at random. They were originally named B cells because, in birds, they are processed in the *bursa of Fabricius,* a lymphoidal structure on the hindgut. At present, researchers have not found an equivalent structure in humans, and the *B* is now said to refer to bone marrow.

Mature T cells leave the thymus via the blood and colonize in specific loci of the lymph nodes and spleen. T cells may survive for years or possibly even a lifetime. When antigens activate them, T cells transform to **lymphoblasts** to perform their specific functions.

B cells also colonize in specific regions of the lymph nodes and spleen. When antigens activate them, B cells transform into **plasma cells** that produce the antibodies specific for the foreign protein and into **memory cells** that survive and maintain immunity to a specific disease for many years.

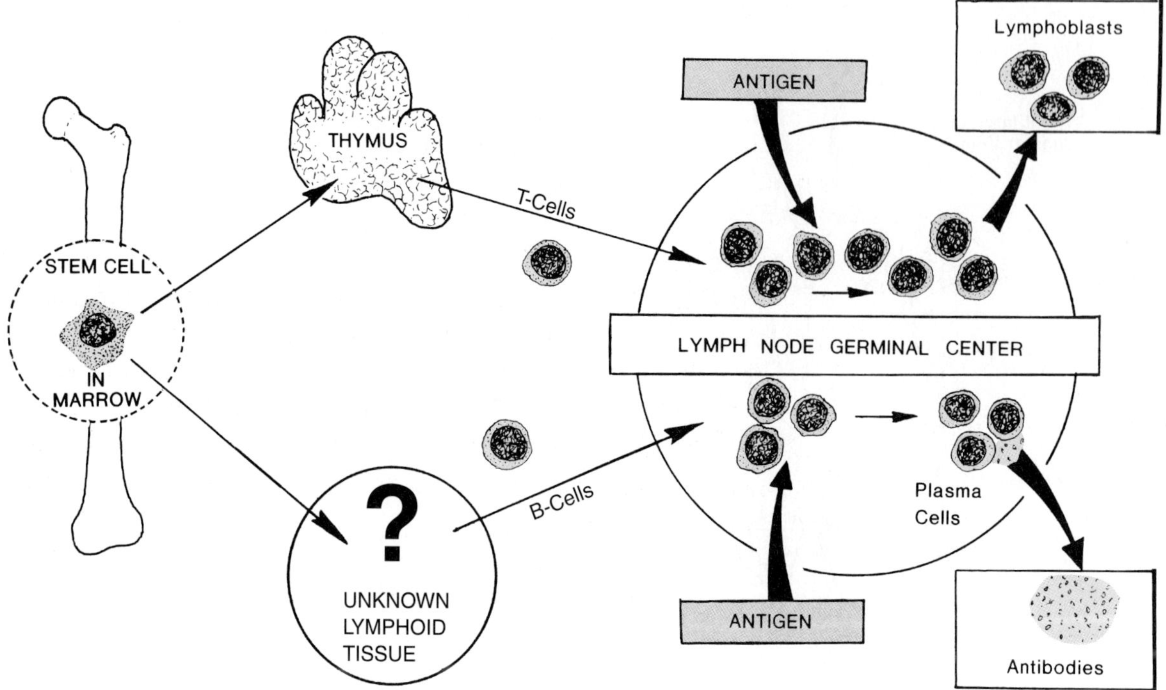

Figure 70.2 The genesis of lymphoblasts and plasma cells.

Laboratory Assignment

Materials:
prepared slide of a lymph node
microscope

1. Examine a prepared slide of a lymph node, and compare it with the photomicrographs in figure HA–27 of the Histology Atlas. Locate a **germinal center.**
2. Examine the **lymphocytes** under high-dry or oil immersion to identify dividing cells.
3. Make drawings, if required.

Assignment:
Complete the Laboratory Report for this exercise.

PART 12 The Respiratory System

This part is comprised of seven exercises that cover the anatomy of the respiratory system, hyperventilation, the diving reflex, the Valsalva maneuver, and spirometry. Four of the experiments involve using chart recorders. One experiment utilizes computerized hardware to monitor spirometry experiments. If you perform all the recording experiments in Part 12, the protocol for performing the multichannel experiments in Part 13 should be relatively easy.

71 The Respiratory Organs

This exercise explores both gross and microscopic anatomy of the respiratory system. Although the cat is the primary dissection specimen here, sheep and frog materials are also available for certain observations. Before performing any laboratory dissections, however, study the first two sections of this exercise, and become familiar with the labeled structures in figures 71.1, 71.2, and 71.3.

The Upper Respiratory Tract

The upper respiratory tract (see figure 71.1) functions in breathing, eating, and speech. A study of the respiratory organs in this region must include those organs concerned with all of these activities.

The *palate* separates the upper **nasal cavity,** which is the air passageway, from the lower **oral cavity.** The anterior portion of the palate, or **hard palate,** is reinforced with bone. The posterior portion, or **soft palate,** terminates in a fingerlike projection called the **uvula.**

The oral cavity consists of two parts: the **oral cavity proper,** which contains the tongue, and the **oral vestibule** between the lips and teeth.

During breathing, air enters the nasal cavity through the nostrils, or **nares** (*naris,* singular). The lateral walls of this cavity have three pairs of fleshy lobes: the superior, middle, and inferior **nasal conchae.** These lobes warm the air as it enters the body.

Above and behind the soft palate is the **pharynx.** That part of the pharynx posterior to the tongue, where swallowing is initiated, is called the **oropharynx.** Above it, and posterior to the nasal cavity, is the **nasopharynx.** Two openings to the **auditory** (*Eustachian*) **tubes** are on the walls of the nasopharynx.

After air passes through the nasal cavity, nasopharynx, and oropharynx, it passes through the **larynx** and **trachea** to the lungs. The cartilages of the larynx vary considerably in size and histology. The **epiglottis,** which is the uppermost cartilage, is composed of elastic cartilage. It prevents food from entering the respiratory passages during swallowing. Inferior to the epiglottis is the **thyroid cartilage** that forms the sidewalls of the larynx and protrudes externally as the "Adam's apple" in the throat region. Just below the thyroid cartilage is the **cricoid cartilage,** shaped like a signet ring. The thyroid and cricoid cartilages are composed of hyaline cartilage.

Both sides of the larynx have a **vocal fold.** These vocal folds are, essentially, the true vocal cords of the larynx and are actuated by muscles through the **arytenoid cartilages** (see figure 71.3) to produce sounds. The slit like opening between the vocal folds is called the *glottis.* Do not confuse the vocal folds with the **ventricular folds** ("false vocal cords") that lie superior to the vocal folds. Posterior to the larynx and trachea lies the **esophagus,** the tube that carries food from the pharynx to the stomach.

In several areas of the oral cavity and nasopharynx are islands of lymphoidal tissue called *tonsils.* The **pharyngeal tonsils,** or *adenoids,* are on the roof of the nasopharynx; the **lingual tonsils** are on the posterior inferior portion of the tongue; and the **palatine tonsils** are on each side of the tongue on the lateral walls of the oropharynx. These masses of tissue are as important as the lymph nodes in protecting against infection. The palatine tonsils are often removed surgically.

Assignment:

Study models of median sections of the head and larynx in the laboratory. Be able to identify all structures.

The Lungs, Trachea, and Larynx

Figure 71.2 shows the respiratory passages from the larynx to the lungs. Note that the larynx consists of three cartilages: an upper **epiglottis;** a large, middle **thyroid cartilage;** and a smaller **cricoid cartilage.** Figure 71.3 shows these cartilages in greater detail.

The **Trachea** extends below the larynx to a point in the center of the thorax where it divides to form two short **primary bronchi.** Note that rings of hyaline cartilage reinforce both the trachea and bronchi. Each primary bronchus divides further into **secondary** and **tertiary bronchi.** These bronchi continue to branch, giving rise to many smaller tubes called **bronchioles.** Each bronchiole terminates in a cluster of tiny sacs, the **alveoli,** where gas exchange with the blood takes place. Each lung has thousands of alveoli.

The pleural membranes facilitate free movement of the lungs in the thoracic cavity. A **pulmonary (visceral) pleura** covers each lung, and a

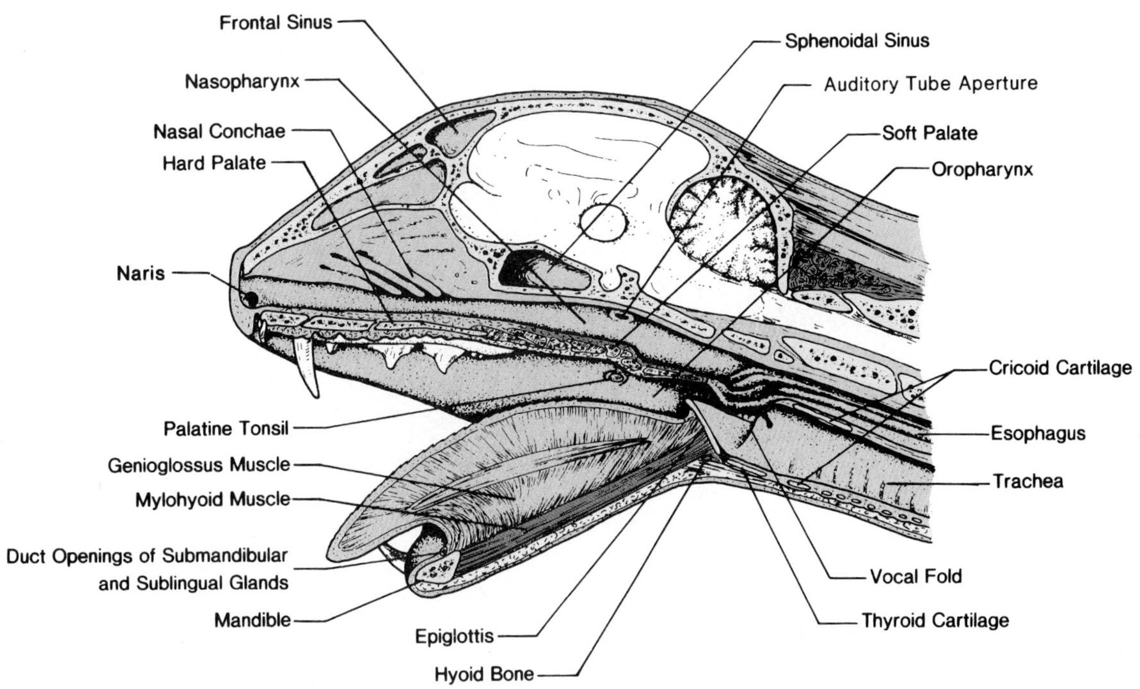

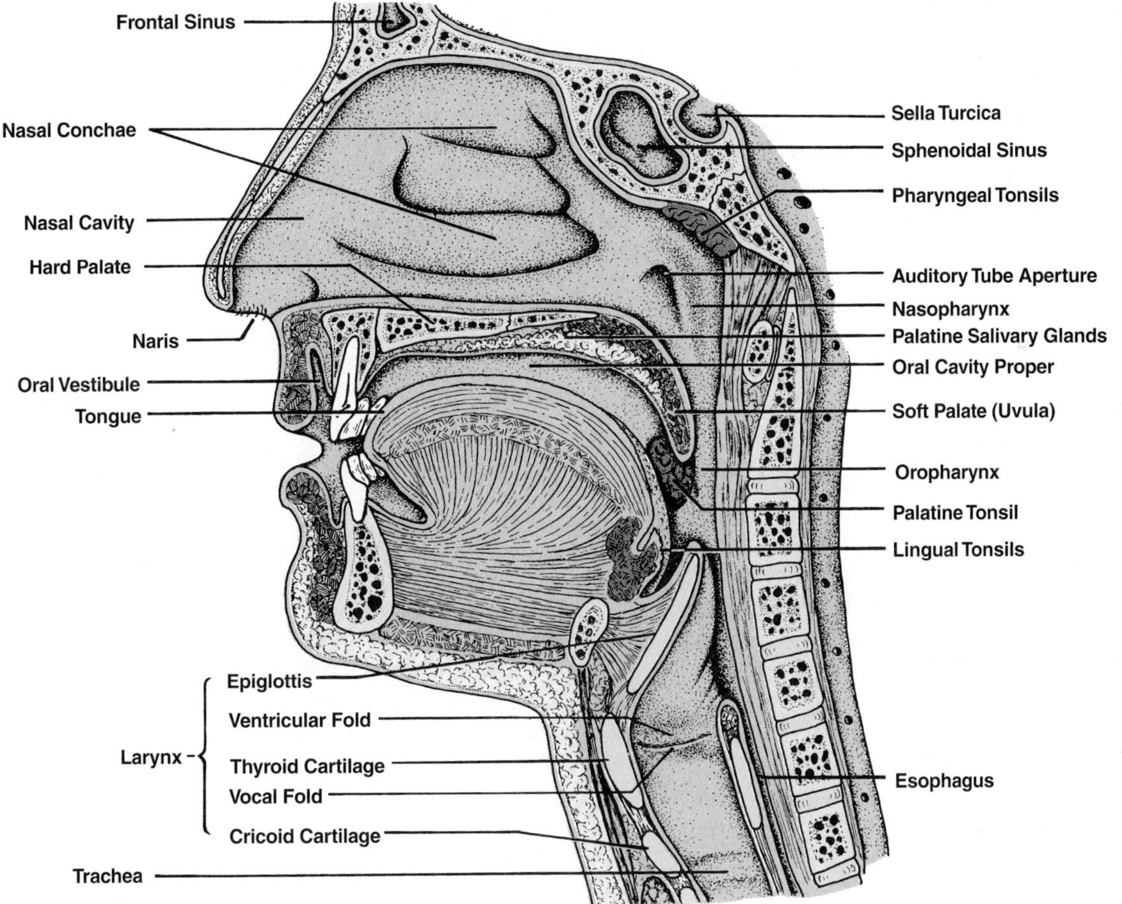

Figure 71.1 Upper respiratory passages, cat and human.

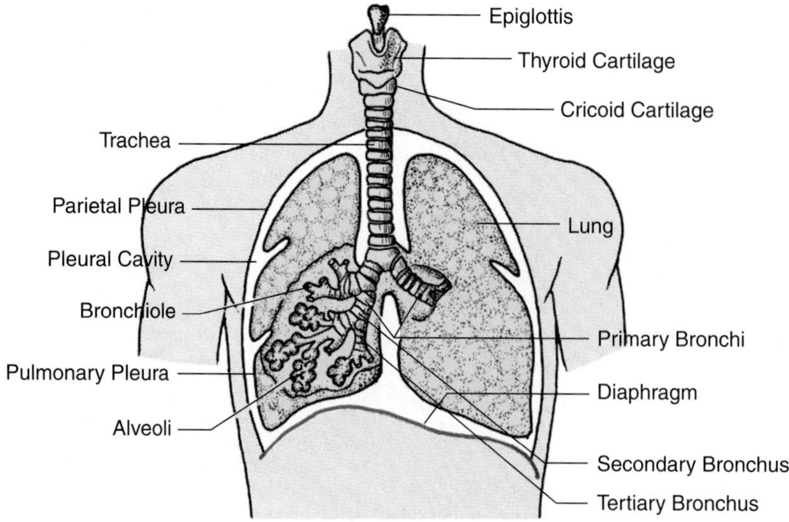

Figure 71.2 The human respiratory tract.

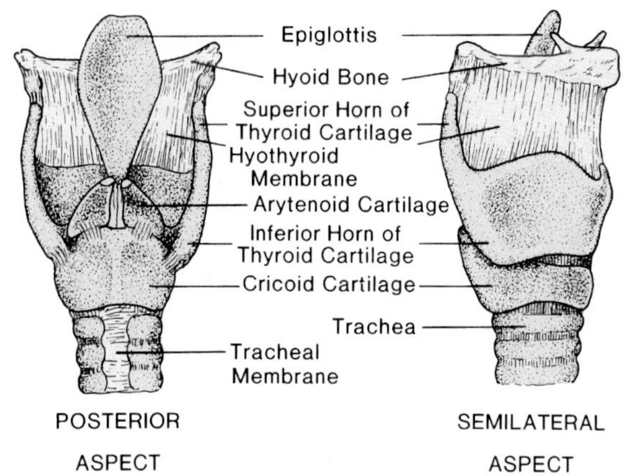

Figure 71.3 The human larynx.

parietal pleura attaches to the thoracic wall. Between these two pleurae is a potential cavity, the **pleural** (*intrapleural*) **cavity.** Normally, the lungs firmly press against the body wall, with little or no space between the two pleurae. The bottom of the lungs rests against the muscular **diaphragm,** the principal muscle of respiration. The parietal pleura attaches to the diaphragm also.

Cat Dissection

Examination of the upper and lower respiratory passages of the cat requires some dissection, depending on what systems have been previously studied on your specimen. Use figures 71.1, 71.4, and 71.5 for reference.

Upper Respiratory Tract

To identify all the cavities and structures of the cat's head shown in figure 71.1, you need a mechanical bone saw or an electric autopsy saw to cut the head down the median line. Your instructor may designate certain students to make sagittal sections that all class members can study. Most specimens are left intact to facilitate the study of other systems.

If an autopsy saw is used, one student must hold the head of the specimen while another student does the cutting. Your instructor will demonstrate the procedure. Although an electric autopsy saw is relatively safe because of its reciprocating action, you must observe certain precautions.

The principal rules of safety are: (1) always cut *away* from the assistant's hands; (2) while cutting, brace your hands against the table for steadiness: don't try to do it freehand; and (3) rest your hands frequently.

Use the large cutting edge for straight cuts and the small cutting edge for sharp curves. Cut only deep enough to get through the bone; use the scalpel for soft tissues. Once you have cut through the head, wash away loose debris. Now identify the following structures, using figure 71.1 for reference:

1. First, identify the **nares** and **nasal cavity.** As in humans, the nasal cavity is superior to the palate.
2. Locate the **nasal conchae,** which are shaped somewhat differently in cats than in humans.
3. Identify the **nasopharynx.** and the small **auditory tube aperture** located there. Insert a probe into the aperture.
4. Press against the palate with a blunt probe to note where the **hard palate** ends and the **soft palate** begins.
5. Locate the **oropharynx** at the back of the mouth near the base of the tongue.
6. Observe that the cat has a very small **palatine tonsil** on each side of the oropharynx. Does it lie in a recess, as does the human palatine tonsil?

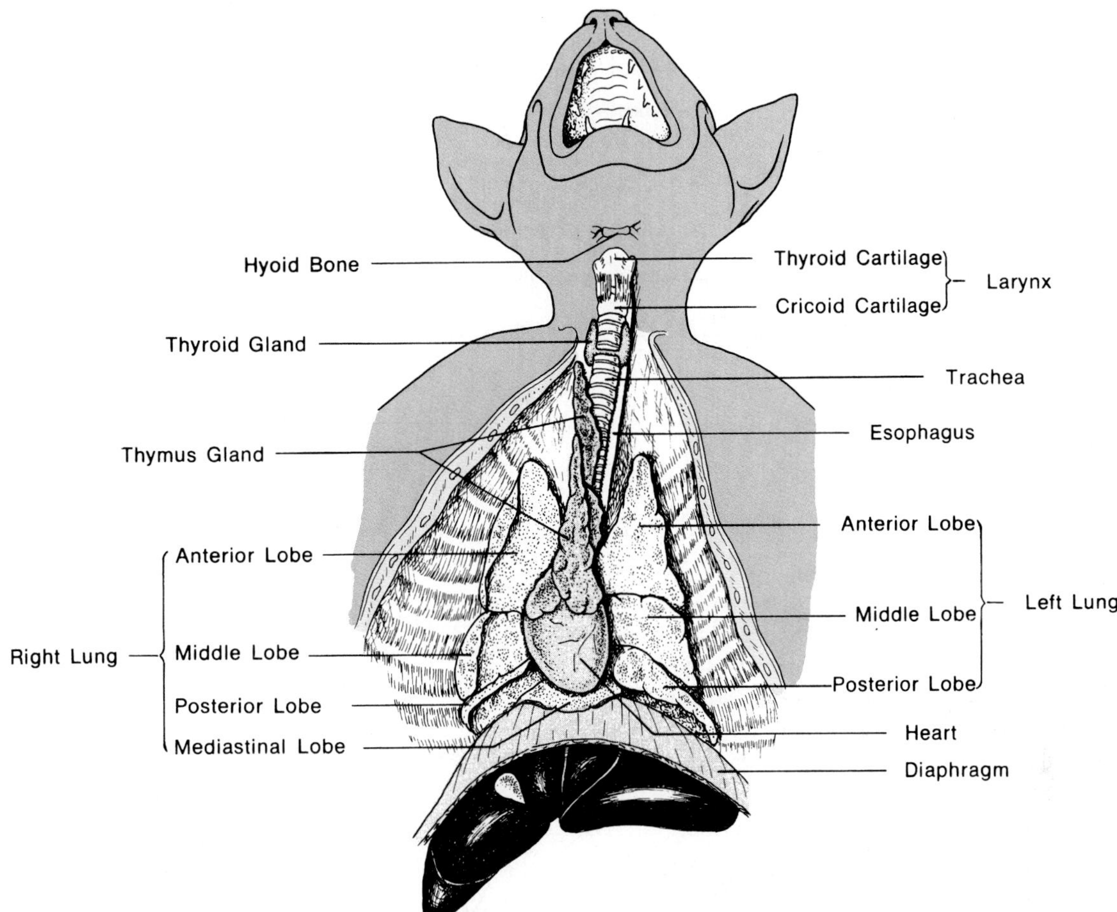

Figure 71.4 Thoracic organs of the cat.

Lower Respiratory Passages

After identifying the aforementioned structures on the sagittal section, open the thoracic cavity if it is not already open. (If the circulatory system has been studied, it will already be open.) Figure 71.4 shows the exposed organs you should see.

To open the chest cavity, make a longitudinal incision about 1 cm to the right or left of the midline. Such a cut is primarily through muscle and rib cartilage, rather than bone. Cut partially through with a scalpel, and then complete the incision with scissors. Avoid damaging the internal organs. Extend the cut up the throat to the mandible, exposing the larynx and trachea.

Continue the incision down the abdominal wall past the liver. Make two cuts laterally from the midline in the region of the liver. Spread the thoracic walls, and sever the diaphragm from the wall with scissors. To keep the thoracic walls open, make a shallow longitudinal cut with a scalpel along the inside surface of each side. Make the cut sufficiently deep to weaken the ribs. When you fold back the thoracic walls, they should break at the cuts.

Finally, use a bone scissors to cut through the angles of the jaw. This allows you to open the mouth wide enough to view the posterior portion of the oropharynx and the epiglottis.

Now identify the organs of the lower respiratory passages, using figures 71.4 and 71.5 for reference:

1. Examine the larynx to identify the large, flap-like **epiglottis,** the **thyroid cartilage,** and the **cricoid cartilage** (see figure 71.5).
2. Locate the paired **vocal cords** on the larynx.
3. Trace the **trachea** down into the **lungs.** In humans, the right lung has three lobes, and the left lung has only two. How does this compare with the cat?
4. Cut out a short section of the trachea, and examine a cross section through the **cartilaginous rings.** Are the rings continuous all the way around the trachea?
5. Can you see where the trachea branches to form the **bronchi?** Make a frontal section through one lung, and look for further branching of the bronchus.

Assignment:
Complete part A of the Laboratory Report.

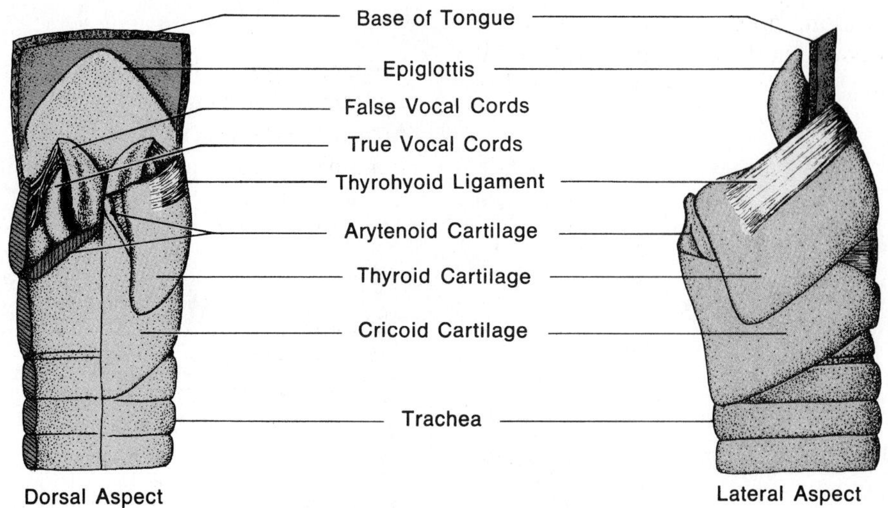

Figure 71.5 Larynx of the cat.

Sheep Pluck Dissection

A "sheep pluck" consists of the trachea, lungs, larynx, and heart of a sheep as removed during routine slaughter. Since it is fresh material rather than formalin-preserved, it is more lifelike than the organs of an embalmed cat. If sheep plucks are available, proceed as follows to dissect one:

> *Materials:*
> sheep plucks, less the liver
> dissecting tray and instruments
> drinking straws or sterile serological pipettes

1. Lay out a fresh sheep pluck on a tray. Identify the major organs, such as the **lungs, larynx, trachea, diaphragm** remnants, and **heart.**
2. Examine the larynx more closely. Can you identify the **epiglottis, thyroid,** and **cricoid cartilages?** Look into the larynx. Can you see the **vocal cords**?
3. Cut a cross section through the upper portion of the trachea, and examine a sectioned cartilaginous ring. Is it similar to the cat in structure?
4. Force your index finger down into the trachea, noting the smooth, slimy nature of the inner lining. What kind of tissue lines the trachea and makes it so smooth and slippery?
5. Note that each lung divides into lobes. How many lobes make up the right lung? The left lung? Compare with the human lung (see figure 71.2).
6. Rub your fingers over the surface of the lung. What membrane on the surface of the lung makes it so smooth?
7. Free the connective tissue around the **pulmonary trunk,** and expose its branches leading into the lung. Also locate the **pulmonary veins** that empty into the left atrium. Can you find the **ligamentum arteriosum** that is between the pulmonary trunk and aorta?
8. With a sharp scalpel, free the trachea from the lung tissue, and trace it down to where it divides into the **bronchi.**
9. With scissors, cut off the trachea level with the top of the heart.
10. Now, cut the trachea down its posterior surface (opposite the heart) on the median line with scissors. Extend this cut down to where the trachea branches into the two primary bronchi. (Observe that a separate bronchus leads into the upper right lobe. This bronchus branches off some distance above where the primary bronchi divide from the trachea.)
11. Continue opening up the respiratory tree until you are deep into the center of the lung. Note the extensive branching.
12. Insert a plastic drinking straw or a serological pipette into one of the bronchioles, and blow into the lung. **If you use a pipette, put the mouthpiece of the pipette into the bronchiole, and blow on the small end.** Note the lung's great expansive capability.
13. Cut off a lobe of the lung, and examine the cut surface. Note the sponginess of the tissue.
14. Record all your observations in part B of the Laboratory Report.

Important: Wash your hands with soap and water at the end of this dissection and at the end of the period.

Frog Lung Observation

Microscopic examination of a frog lung shortly after the frog has been pithed and dissected shows you the nature of living lung tissue in an animal. A frog lung is less complex than a human lung, but it has essentially the same kinds of tissue. Proceed as follows:

Materials:
frog, recently double pithed
dissecting instruments and dissecting trays
medicine droppers (optional)
dissecting microscope
frog Ringer's solution in wash bottle

1. Pin the frog, ventral side up, in a wax-bottomed dissecting pan.
2. With scissors, make an incision through the skin along the midline of the abdomen.
3. Make transverse cuts in both directions at each end of the incision, and lay back the flaps of skin.
4. Carefully make an incision through the right or left side of the abdomen over the lung, parallel to the midline. Take care not to cut too deeply. The inflated lung should now be visible.
5. With a probe, gently lift the lung out through the incision. *Do not perforate the lung!* From this point on, keep the lung moist with Ringer's solution.
6. If the lungs of your frog are not inflated, insert a medicine dropper into the slitlike glottis on the floor of the oral cavity, and pump air into them by squeezing the bulb. The lungs may be deflated from excessive squeezing during pithing.
7. Observe the basically saclike shape and general appearance of the frog lung.
8. Place the frog under a dissecting microscope, and examine the lung surface.
9. Locate the network of ridges on the inner walls. The thin-walled regions between the ridges represent the **alveoli.**
10. Look carefully to detect blood cells moving slowly across the alveolar surface. Careful examination of the lungs may reveal parasitic worms that are quite common in frogs.
11. When you have completed this study, dispose of the frog as your instructor directs, and clean the pan and instruments. Wash your hands with soap and warm water.

Histological Study

Prepared slides of the nasal cavity, trachea, and lung tissue are available for study. The tissues may be from monkey or human organs, or from other mammals, such as the mouse or rabbit.

Use figure HA–28 in the Histology Atlas for reference, and follow these suggestions in your study:

Materials:
prepared slides of nasal septum, trachea, and lung tissue

Nasal Septum Scan the nasal septum slide first with low power to locate the **nasal septum** and **nasal concha** (see figure HA–28A in the Histology Atlas). Identify the bony tissue and nasal epithelium.

Study the ciliated epithelium and the tissues beneath it with high-dry magnification. Look for the **olfactory glands** (see figure HA–28B in the Histology Atlas) in the lamina propria. These glands secrete a substance that keeps the epithelial surface moist and dissolves the materials that stimulate the olfactory receptors.

Observe that a large number of unmyelinated **olfactory fibers** lie between the glands and the ciliated epithelium. These fibers carry nerve impulses from the olfactory receptors dispersed among the cells of the epithelium.

Trachea On the slide of the trachea, identify the hyaline cartilage, ciliated epithelium, lamina propria, and muscularis mucosae. Look for goblet cells on the epithelial layer. Refer to figures HA–28B and C in the Histology Atlas.

Lung Tissue Examine the lung tissue slide with the high-dry objective. Look for the thin-walled **alveoli,** a **bronchiole,** and blood vessels (see figure HA–28D in the Histology Atlas).

Assignment:
Complete the Laboratory Report for this exercise.

72 Hyperventilation and Rebreathing

During breathing, the respiratory muscles increase and decrease the volume of the thoracic cavity to ventilate the alveoli sufficiently to satisfy the oxygen requirements of all body cells. The breathing control center is the respiratory center in the medulla oblongata.

Carbon dioxide (CO_2), hydrogen ion (H^+), and oxygen (O_2) concentrations in the blood and cerebrospinal fluid are the chemical stimuli that act directly, or indirectly, on the respiratory center to regulate the respiratory muscles. Of the three stimuli, H^+ concentration influences the respiratory rate the most, while O_2 concentration affects it the least. Carbon dioxide facilitates H^+ concentration by forming carbonic acid in the blood and cerebrospinal fluid.

The pH of blood and tissue fluid during normal breathing is around 7.4. Forced deep breathing **(hyperventilation)** can elevate the pH of these fluids to 7.5 or 7.6 as CO_2 is blown off. The reduced H^+ concentration depresses the respiratory center, lessening the desire for increased alveolar ventilation.

Hyperventilation from unconscious deep breathing or sighing can cause a drop in blood pressure, extreme discomfort, dizziness, and even unconsciousness. All of the symptoms are due to the washing out of CO_2 from the blood, resulting in *alkalosis.* **Rebreathing** into a paper bag for several minutes quickly restores CO_2 to the blood.

This exercise uses nonrecording procedures as well as Physiograph or Duograph recordings to examine the symptoms of hyperventilation and how to compensate for hyperventilation. Both experiments have unique characteristics.

Nonrecording Experiment

In this experiment, you observe the characteristics of hyperventilation and compensate for it with proper rebreathing techniques.

Materials:
paper bag

1. Breathe very deeply at a rate of about 15 inspirations per minute for **1 to 2 minutes.** Do not hurry the rate; concentrate on breathing deeply. Be aware of the following symptoms:
   - It becomes increasingly difficult to breathe deeply.
   - Dizziness develops.

   Two events account for the dizziness: (1) Dilation of the splanchnic vessels causes blood pressure to drop, which lessens blood to the brain; and (2) the reduced CO_2 content of the blood causes vasoconstriction of the cerebral blood vessels, further reducing the blood supply.
2. Now tightly hold a paper bag over your nose and mouth and breathe deeply into it for about **3 minutes.** Note how much easier it is to breathe into the bag than into free air. Why?
3. Remove the paper bag, and allow your breathing rate to return to normal. Breathe normally for **3 to 4 minutes.** At the end of a normal inspiration, *without deep inhaling,* pinch your nose shut with the fingers of one hand, and hold your breath as long as you can. Do this three times, timing each duration carefully. Record these times, and calculate the average on the Laboratory Report.
4. Now breathe deeply, as in step 1, for **2 minutes,** and then hold your breath as long as you can. Record your results on the Laboratory Report.
5. Exercise by running in place for **2 minutes,** and then hold your breath as long as you can. Record your results on the Laboratory Report.

Recording Experiment

A bellows transducer (see figures 72.1 and 72.2) is required to produce a record of ventilation during hyperventilation and rebreathing. Separate instructions are provided for the Physiograph and Duograph. The Duograph is preferable to the Unigraph here for Gilson equipment users because, if your laboratory period is long enough, you can use the Duograph to perform the two-channel recording experiments in Exercises 73 and 74.

Materials:
for Physiograph setup:
Physiograph with transducer coupler
bellows pneumograph transducer (PN 705-0190)
cable for bellows pneumograph transducer
continued on next page

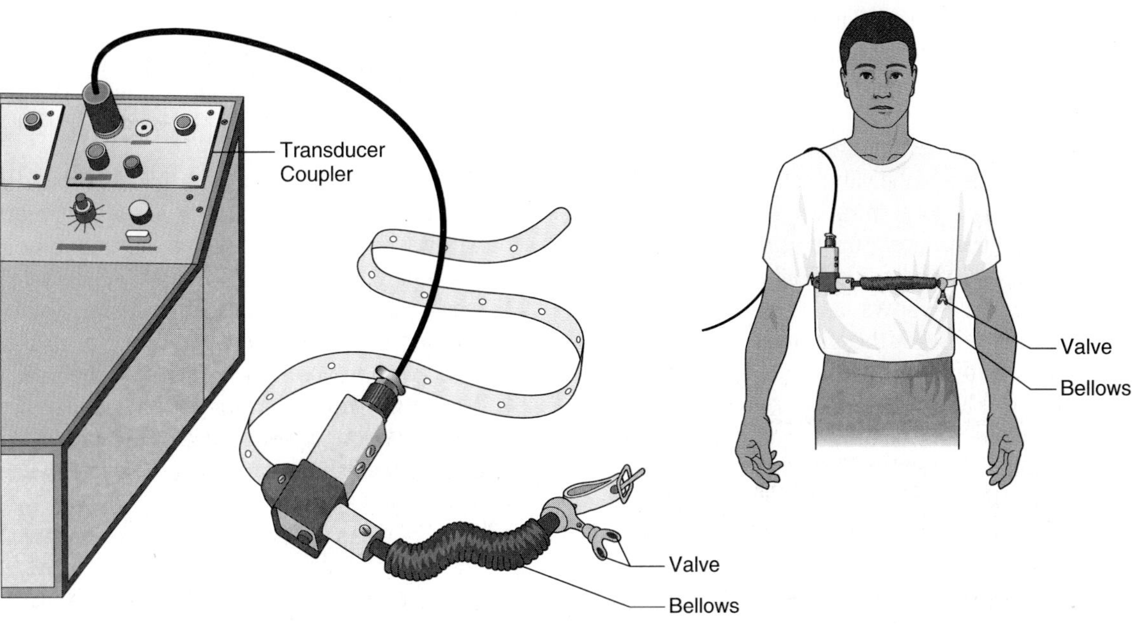

Figure 72.1 Physiograph setup for monitoring respiration.

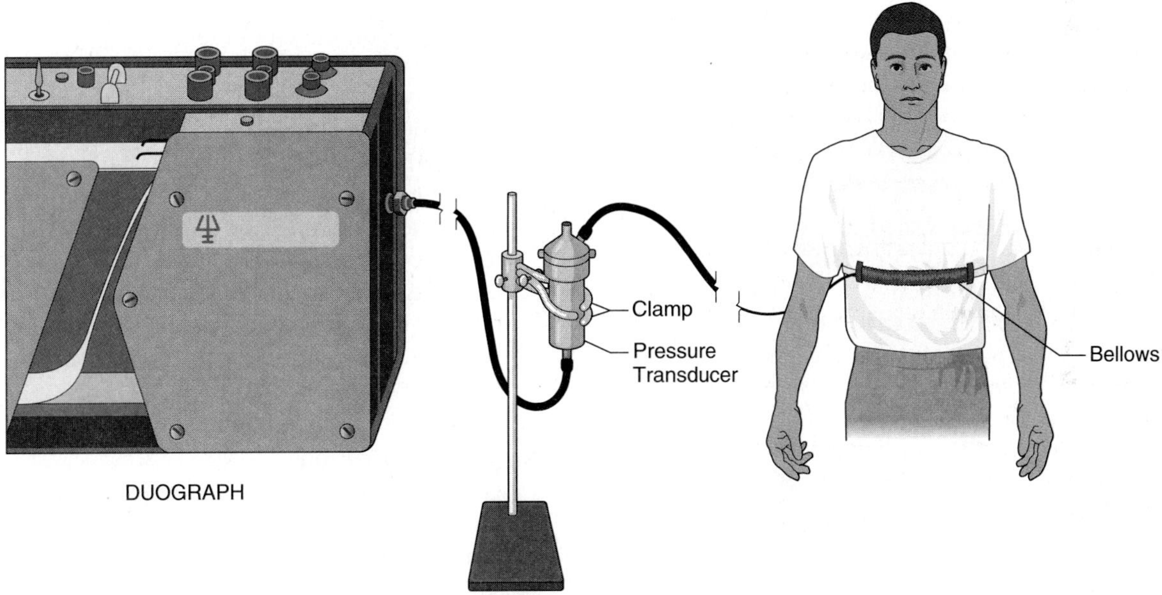

Figure 72.2 Duograph setup for monitoring respiration.

for Duograph setup:
Duograph
Statham T–P231D pressure transducer
Gilson T–4030 chest bellows
stand and clamp for transducer

for both setups:
noseclip
paper bag

Physiograph Setup

Set up one channel, using the transducer coupler and bellows, as described in the following procedure (see figure 72.1):

1. Connect the bellows pneumograph transducer to the transducer coupler with a transducer cable. Be sure to match up the yellow dots on the connectors.

2. Balance the recording channel according to the instructions in Appendix C.
3. Attach the bellows pneumograph to the subject as follows:
   a. With the channel amplifier record button in the Off position, and the valve at the end of the rubber bellows open (turned counterclockwise), place the transducer on the subject's chest, and fasten it with the attached leather strap.
   b. Locate the bellows as high as possible, where it will get the greatest chest movement during breathing. Position the transducer cable over the shoulder to help support the weight of the transducer.
4. Start the paper advance at 0.25 cm/sec, and lower the pens to the recording paper. Activate the timer.
5. Position the recording pen with the position control to 2 cm below the center line.
6. Close the valve on the bellows by rotating it clockwise while you pinch the bellows slightly.
7. Place the channel amplifier record button in the On (down) position.
8. Starting at the 1000 position, rotate the outer channel amplifier sensitivity control knob to higher sensitivities (lower numbers) until normal shallow ventilation by the subject gives pen deflections of approximately 2 cm. Proceed to perform the investigations that follow.

Normal Respiration

1. With the subject seated and relaxed, start the paper moving at 0.25 cm/sec, lower the pens, and activate the timer.
2. Place the amplifier record button in the On position.
3. Record one page of normal respiratory activity.
4. Increase the paper speed to 5 cm/sec, and record several pages of normal respiratory activity.
5. Compare your tracings with the sample Physiograph records in figures SR–6A and B in Appendix D.

Rebreathing

1. Place a noseclip on the subject.
2. Set the paper speed at 0.25 cm/sec, and continue recording.
3. Record about a half page of normal ventilation. Then have the subject place a paper bag tightly over his or her mouth.
4. Allow the subject to breathe into and out of the bag until the inspiratory rate and depth visibly change.
5. Have the subject remove the bag. Allow the subject's respiratory rate to return to normal.
6. Compare your tracings with the sample Physiograph records in figures SR–6C and D in Appendix D.

Hyperventilation

1. While you record at 0.25 cm/sec, have the subject hyperventilate by inhaling and exhaling as deeply as possible (about 15 inspirations a minute) for 1.5 to 2 minutes.
2. **When the subject begins to feel discomfort,** tell the subject to stop the deep breathing. Allow the subject's respiratory movements to normalize.
3. During normalization, look for changes in the rate and amplitude of respiratory movements as compared with the prehyperventilation recordings.
4. Terminate the recording.
5. Compare your tracings with the sample Physiograph records in figure SR–7 in Appendix D.

Duograph Setup

Use one channel of the Duograph to hook up a subject according to the following instructions (see figure 72.2):

1. Mount the Statham pressure transducer on the transducer stand with a clamp, and plug the jack from the transducer into the receptacle of channel 1 on the Duograph.
2. Attach the chest bellows to the subject at the highest point possible, where respiratory movements are greatest.
3. Attach the tubing from the bellows to the Statham transducer.
4. Activate the Duograph at slow speed, noting the trace.
5. Adjust the excursion for normal breathing at about 1 cm. Proceed as follows to record normal, rebreathing, and hyperventilation activities.

Normal Respiration

1. With the subject seated and relaxed, record normal breathing at the slow speed (2.5 mm/sec) for 1.5 to 2 minutes.
2. Increase the speed to fast (25 mm/sec), and record for another 1.5 to 2 minutes.

Rebreathing

1. Place a noseclip on the subject.
2. Return the paper speed to slow, and continue recording.
3. For about 30 seconds, record normal ventilation. Then have the subject place a paper bag tightly over his or her mouth.
4. Allow the subject to breathe into and out of the bag until the respiratory rate and depth visibly change.
5. Have the subject remove the bag. Allow the subject's respiratory rate to return to normal.

Hyperventilation

1. Continue recording at slow speed, and have the subject hyperventilate as described previously.
2. **When the subject begins to feel discomfort,** tell the subject to stop the deep breathing. Continue the recording as the respiratory movements normalize.
3. During normalization, look for changes in the rate and amplitude of respiratory movements as compared with the prehyperventilation recordings.
4. Terminate the recording.

Assignment:
Complete the Laboratory Report for this exercise.

73 The Diving Reflex

Many aquatic, air-breathing animals have evolutionary adaptations that enable them to remain submerged under water for long periods. One of these adaptations—the "diving reflex" (or dive reflex)—consists of prompt cardiovascular and respiratory adjustments triggered by partial entry of water into the air passages. These adjustments include a change in the heart rate, the shunting of blood from less essential tissues, and a cessation of breathing. While these changes are less pronounced in humans, they do account for the survival of some individuals, usually young children, who have fallen through the ice and remained underwater for unusually long periods.

In this exercise, you try to observe the diving reflex in humans by monitoring cardiovascular changes with a pulse transducer while a subject's face is submerged in cold water. Needless to say, the subject for this experiment must have no aversion to facial submersion.

The equipment setup requires two channels on a recorder. One channel records the respiratory rate from a chest bellows (see Exercise 72 on hyperventilation). A second channel detects changes in the heart rate from a pulse transducer. Figures 73.1 and 73.2 illustrate the respective setups for a Physiograph and a Duograph.

The submersion maneuver may take place at your laboratory station if large pans are available or in a sink at a perimeter counter. Set up the recording equipment at the appropriate submersion site.

Before submersion, recordings establish the subject's pulse rates during (1) normal breathing and (2) while holding his or her breath. After norms are established, the subject submerges his or her face in water, first at 70°F and then at 60°F (cooled with ice), to note changes in the heart rate. The effects of subject hyperventilation are also recorded. These experiments determine if any cardiovascular changes are detectable during submersion, and to what extent, if any, temperature affects heart rate.

Equipment Setup

If you are performing this experiment along with Exercise 72, the only additional equipment you need is a pulse transducer. Separate instructions are provided for the Physiograph and the Duograph.

Physiograph Setup

Refer to figure 73.1 and the instructions that follow for the Physiograph setup.

Materials:
Physiograph with two transducer couplers
bellows pneumograph transducer (Narco PN 705–0190)
cable for bellows pneumograph transducer
pulse transducer (Narco 705–0050)
large pan of water (or sink basin)
ice, large bath towel, thermometer

1. Attach the finger pulse transducer and chest bellows to the subject. As in Exercise 72, locate the bellows high enough to get maximum excursion during breathing.
2. Connect all cables in such a way that the subject's submersion movement does not tug on the cables.
3. Fill the pan or sink with tap water, adjusting the temperature to 70°F.
4. Turn on the power switch, and set the paper speed at 0.25 cm/sec.
5. Start the chart paper moving, and lower the pens to the paper. Verify that the timer is activated.
6. Use the position controls on both channels to position the pens to record about 1 cm below the centerline.
7. Place the record button in the On position.
8. Adjust the amplifier sensitivity controls so that the pulse wave amplitude is 2 cm and the tidal ventilation (normal breathing) amplitude is 1 cm. Proceed to the section entitled "Experiment Procedure."

Duograph Setup

Refer to figure 73.2 and the instructions that follow for the Duograph setup.

Materials:
Duograph
Statham T–P231D pressure transducer

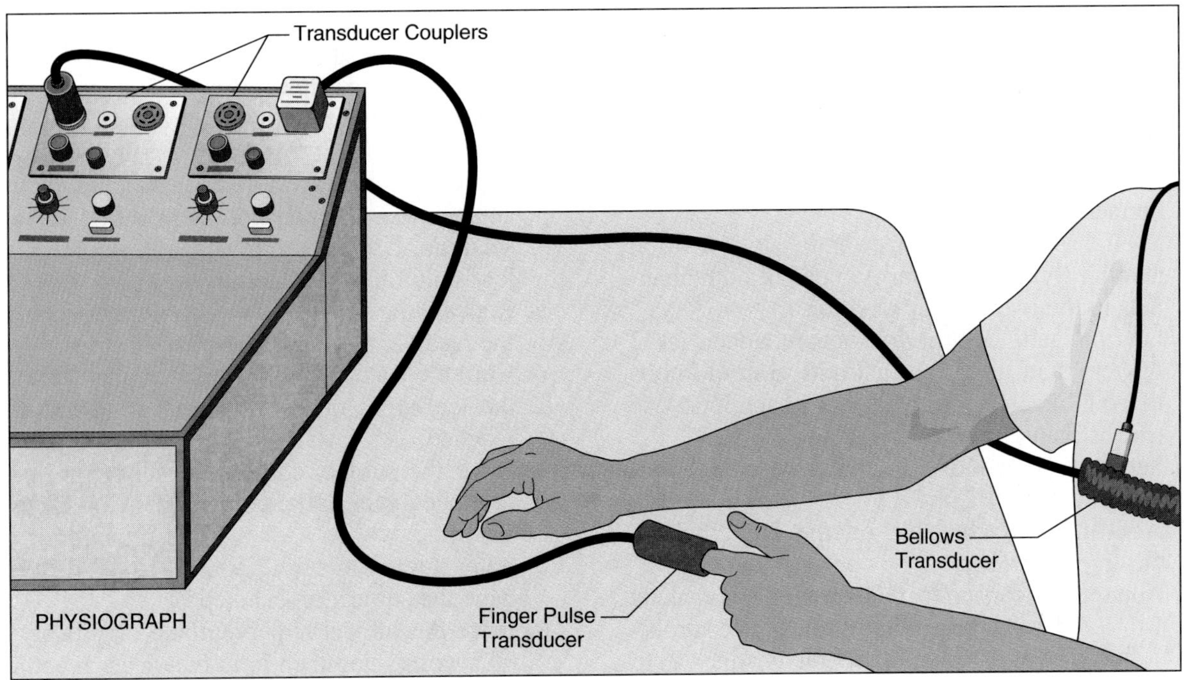

Figure 73.1 Physiograph setup for monitoring the diving reflex.

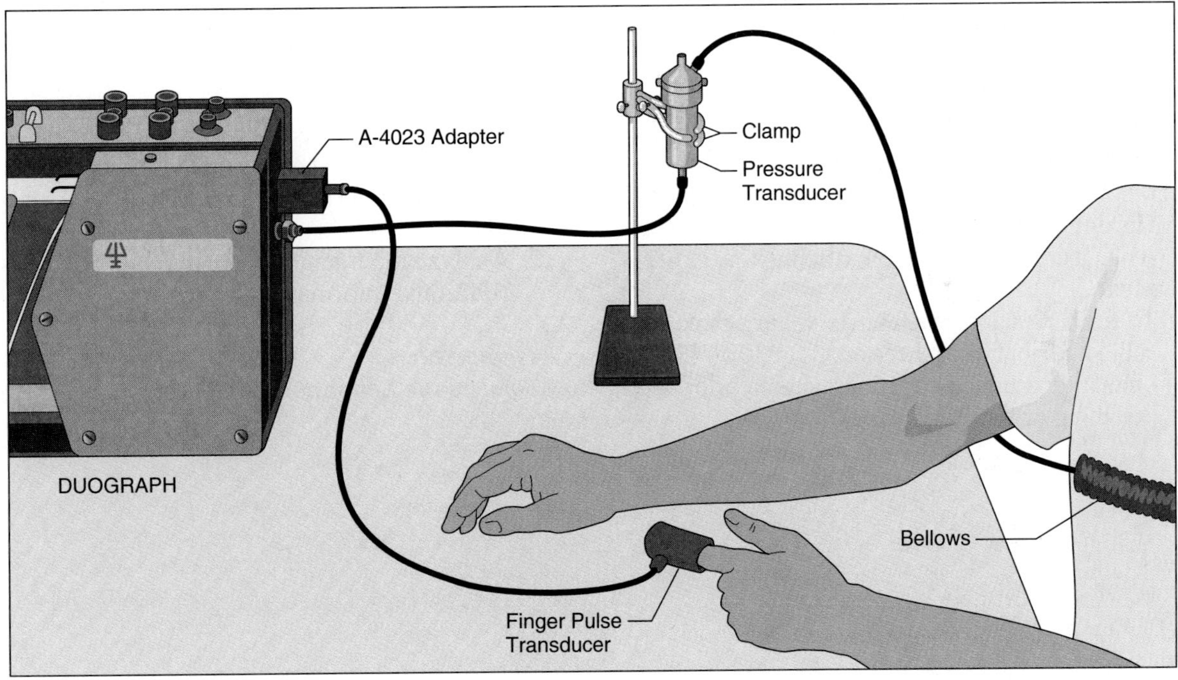

Figure 73.2 Duograph setup for monitoring the diving reflex.

stand and clamp for Statham T–P231D transducer
Gilson T–4030 chest bellows
Gilson T–4020 finger pulse pickup
Gilson A–4023 adapter for pulse pickup
large pan of water (or sink basin)
ice, large bath towel, thermometer

1. Attach the tubing from the chest bellows to the Statham transducer, and plug the transducer into the receptacle for channel 1 of the Duograph.
2. Plug the A–4023 adapter into the receptacle for channel 2 of the Duograph. Verify that the lock-nut is tightened securely and that the pulse pickup is plugged into the A–4023 adapter.

3. Attach the finger pulse pickup and chest bellows to the subject. Keep the chain of the bellows as high into the armpits as possible.
4. Connect all cables in such a way that the subject's submersion movement does not tug on the cables.
5. Fill the pan or sink with tap water, adjusting the temperature to 70°F.
6. Set the speed control lever in the slow position and the stylus heat control knobs of both channels in the two o'clock position.
7. Set the gain control for both channels at 2 mV/cm, and turn the sensitivity control knobs of both channels completely counterclockwise (lowest value).
8. Set the mode control for channel 1 on TRANS and for channel 2 on DC.
9. Activate both channels on the Duograph to observe the traces.
10. Adjust the excursion for tidal breathing on channel 1 to 1 cm, using the sensitivity and gain controls.
11. Adjust the pulse amplitude on channel 2 to about 2 cm.

Experiment Procedure

Now that the equipment is properly adjusted and activated, proceed as follows:

1. Establish a reference by recording for at least **1 minute** before facial submersion. During this time, the subject breathes regularly (tidal inspirations) without speaking or hyperventilating. **Do not readjust the gain or sensitivity controls from this point on during the experiment.**
2. To further establish reference values, direct the subject to hold his or her breath for **30 to 45 seconds** after a tidal inspiration, without prior deep breathing or hyperventilating.
3. Allow the subject to recover for **2 to 3 minutes** by normal breathing.
4. After recovery, have the subject completely submerge his or her face into the 70°F water. Mark the chart at this point. On the Duograph, depress the event button at the exact moment of submersion. **Submersion should last 15 to 30 seconds.** Use a large towel to protect the subject's clothing from getting wet.
5. Repeat the submersion procedure to confirm the results.
6. Chill the water to 60°F with ice cubes. Remove the ice cubes once the correct temperature is reached.
7. Have the subject completely submerge his or her face again in the colder water for **15 to 20 seconds.**
8. After approximately 2 to 3 minutes' recovery, have the subject repeat step 7.
9. Have the subject hyperventilate vigorously for **30 seconds** and then hold his or her breath for **as long as possible** (without submersion). Mark the chart for this activity.
10. Once regular breathing has resumed, have the subject hyperventilate again for **30 seconds.** Then have the subject submerge his or her face for **as long as possible.** Be sure to mark the chart.
11. Terminate the recording, disconnect the subject, put away the apparatus, clean the work area, and return all materials to their proper places.
12. Analyze the tracing as a team, and label it with additional information as needed.

Assignment:
Complete the Laboratory Report for this exercise.

The Valsalva Maneuver

74

Antonio Valsalva, an Italian anatomist, discovered in 1723 that expiring forcibly against a closed mouth and nostrils filled the middle ear with air. This action became known as the *Valsalva maneuver.* The term also applies to attempting to expire air from the lungs against a closed glottis. In this maneuver, the abdominal muscles and internal intercostal muscles greatly increase intraabdominal and intrathoracic pressures. This momentary increase in intrathoracic pressure affects both the pulse and the blood pressure.

Figure 74.1 illustrates what happens to the blood pressure and pulse during the Valsalva maneuver. Blood pressure increases at the start of the straining and then falls a short time later. After the maneuver, blood pressure increases considerably, and the pulse rate slows somewhat.

Intrathoracic pressure on the thoracic aorta causes the initial rise in blood pressure. Restricted blood flow into the right side of the heart through the compressed venae cavae reduces venous return. This decreases cardiac output and causes the subsequent drop in blood pressure. The final surge of increased cardiac output is a compensation for the reduced prior output.

The Valsalva maneuver is more than just a stunt to observe in the laboratory. Its real merit is in the diagnosis of two pathological conditions: autonomic insufficiency and hyperaldosteronism. Individuals with *autonomic insufficiency* exhibit no pulse changes during the Valsalva maneuver. The cause of this condition is not well understood. Individuals with *hyperaldosteronism* (in which the adrenal cortex produces excess aldosterone) fail to show pulse rate changes and a blood pressure rise after the Valsalva maneuver. Physicians often use the Valsalva maneuver to diagnose adrenal cortex tumors.

You will work in teams of four or five to monitor the blood pressure and pulse of one team member during the Valsalva maneuver. *Caution:* Although the Valsalva maneuver is a safe, normal physiological phenomenon for healthy individuals, team members with a history of heart disease should not be used as subjects.

Equipment Setup

Exercise 73 used a chest bellows and a pulse transducer. In this exercise, an arm cuff replaces the chest bellows. Make the following hookups and adjustments:

Physiograph Setup

Refer to figure 74.2 and the instructions that follow for the Physiograph setup. Note that an ESG coupler replaces one of the transducer couplers used in Exercise 73.

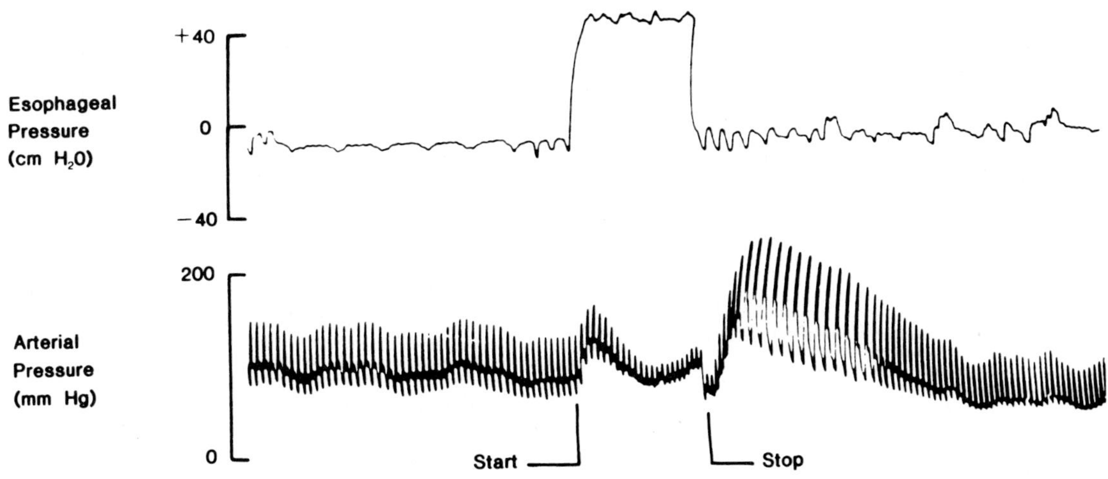

Figure 74.1
Blood pressure and pulse response during the Valsalva maneuver.

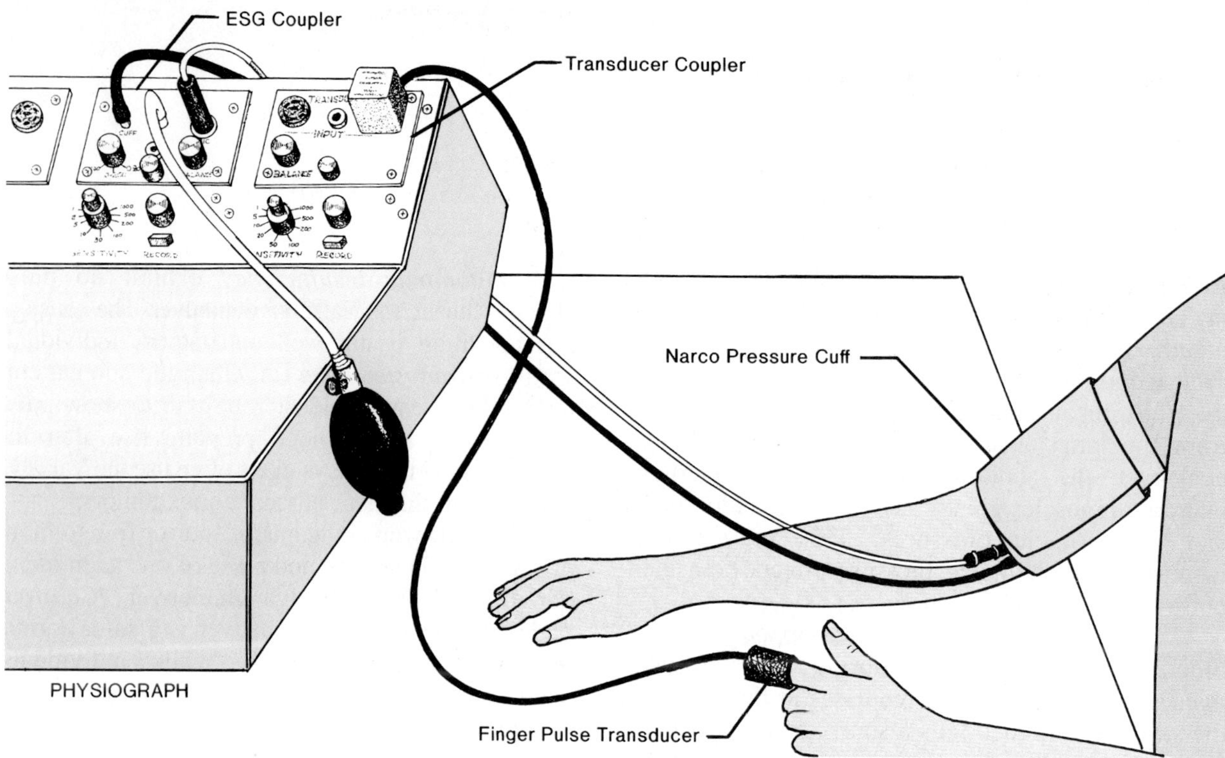

Figure 74.2 Physiograph setup for monitoring the effects of the Valsalva maneuver.

Materials:
Physiograph with transducer and ESG (electrosphygmomanometer) couplers
adult pressure cuff (Narco PN 712–0016)
pulse transducer (Narco 705–0050)

1. Connect the pressure cuff to the ESG coupler (see figure 74.2).
2. Turn on the power switch, and set the paper speed at 0.25 cm/sec.
3. Balance and calibrate the ESG coupler according to the instructions on page 362.
4. From this point on, do not touch the balance and position control knobs.
5. Place the cuff on the subject's upper right arm, positioning the microphone over the brachial artery.
6. Calibrate the cuff microphone according to the instructions on page 362.
7. Insert the subject's left index finger into the pulse transducer, and plug the pulse transducer into the transducer channel.
8. Start the chart paper moving again, lower the pens to the paper, and verify that the timer is activated.
9. Adjust the amplifier sensitivity controls on the transducer channel so that the pulse wave amplitude is 1.5 to 2 cm. Proceed to the section entitled "Experiment Procedure."

Duograph Setup

Refer to figure 74.3 and the instructions that follow for the Duograph setup. Note that the same Statham pressure transducer used in Exercise 73 with a chest bellows is used here with an arm cuff.

Materials:
Duograph
Statham T–P231D pressure transducer
adult arm cuff
stand and clamp for Statham T–P231D transducer
Gilson T–4020 finger pulse pickup
Gilson A–4023 adapter for pulse pickup

1. Tightly secure the cuff on the subject's upper right arm. Attach the tubing from the cuff to the Statham transducer.
2. Test the cuff by pumping some air into it and noting if the pressure holds when the metering valve on the bulb is closed. Release the pressure on the subject's arm while you make other adjustments.

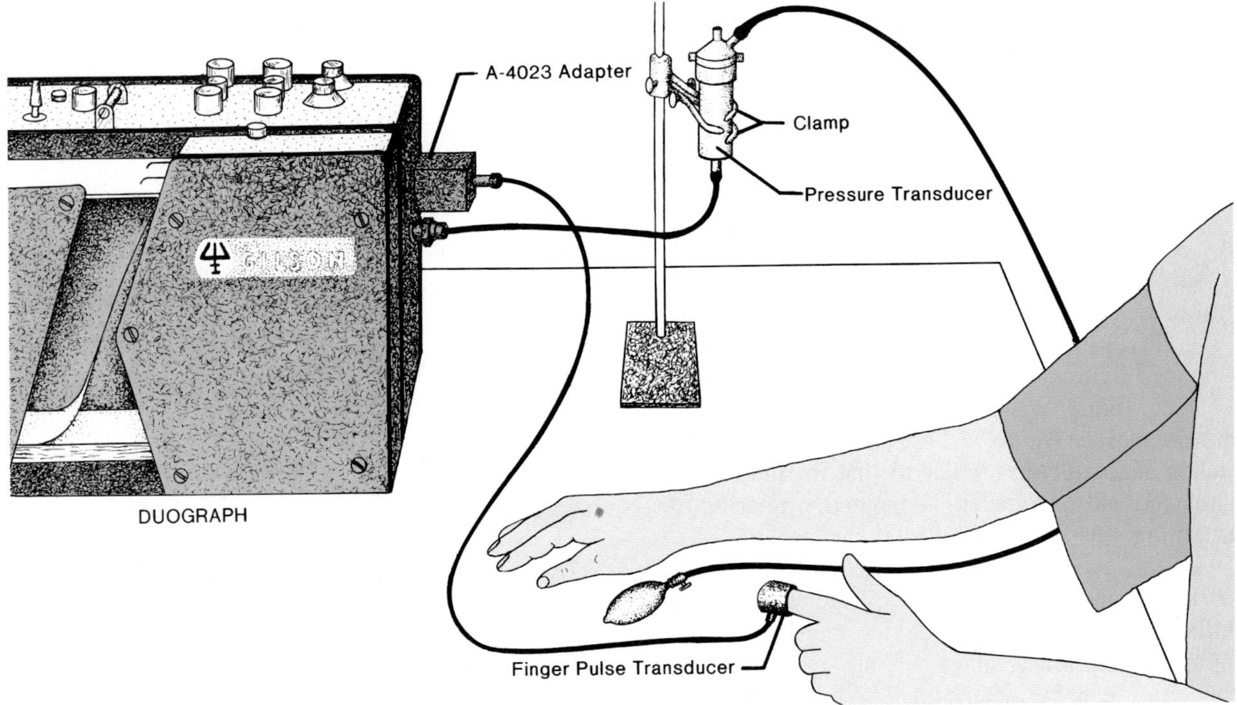

Figure 74.3 Duograph setup for monitoring the effects of the Valsalva maneuver.

3. Insert the subject's left index finger into the finger pulse pickup.
4. Plug in the Duograph, and turn on the power switch.
5. Set the speed control lever in the slow position and the stylus heat control knobs of both channels in the two o'clock position.
6. Set the gain control for channel 1 at 2 mV/cm, and for channel 2 at 1 mV/cm.
7. Turn the sensitivity control knobs of both channels completely counterclockwise (lowest values).
8. Set the mode control for channel 1 at DC and for channel 2 at TRANS.
9. Calibrate the Statham transducer on channel 2 by depressing the TRANS button and simultaneously adjusting the sensitivity knobs to get 2 cm deflection. The calibration line represents 100 mm Hg pressure.
10. Adjust the sensitivity on channel 1 so that you get 1.5 to 2 cm deflection on the pulse.

Experiment Procedure

Now that the equipment is properly adjusted and activated, proceed as follows:

1. Pump air into the cuff, and establish a baseline for 1 minute.
2. Have the subject take a deep breath and perform a Valsalva maneuver, exerting as much internal pressure as possible.
3. Mark the chart at the instant the subject begins the maneuver. On the Duograph, hold down the event marker button until the maneuver ends.
4. Repeat the maneuver several times to provide enough chart material for each team member.
5. Repeat the experiment, using other team members as subjects.

Assignment:
Complete the Laboratory Report for this exercise.

75

Spirometry: Lung Capacities

A **spirometer** measures the volume of air that moves in and out of the lungs during breathing. Several types are available, including the handheld type (see figure 75.2), the tank recording spirometer (see figure 76.1), and the Phipps and Bird spirometer (see figure 77.1) A recording spirometer makes a **spirogram** similar to that in figure 75.1. The handheld spirometer is designed, primarily, for screening measurements of vital capacity.

In this exercise, you use a handheld spirometer to determine individual respiratory volumes. Although this convenient device cannot measure inhalation volumes, most of the essential lung capacities can be determined. While most class members are working with handheld spirometers, some students can be working with a recording spirometer, as directed in Exercise 76.

Respiratory Volumes

Proceed as follows to measure or calculate the various respiratory volumes using a handheld spirometer. Each student should use a new mouthpiece for these experiments, and the stem of the spirometer should be swabbed with 70% alcohol or packaged alcohol swabs in between students. Disposable mouthpieces should be disposed of appropriately.

Materials:
handheld spirometer
disposable mouthpieces
70% alcohol and cotton swabs

 **Lab #6:** Pulmonary Function

Tidal Volume

The *tidal volume* (TV) is the amount of air that moves in and out of the lungs during a normal respiratory cycle (see figure 75.1). Although sex, age, and weight determine the TV and other capacities, the average normal TV is around 500 ml. Proceed as follows to determine your TV:

1. Swab the stem of the spirometer with 70% alcohol, and place a disposable mouthpiece over the stem.
2. Rotate the dial of the spirometer to zero, as figure 75.2 shows.
3. After three normal breaths, expire three times into the spirometer while inhaling through your nose. Do not exhale forcibly. **Note: Do not inhale through the spirometer, and always hold the spirometer with the dial upward.**

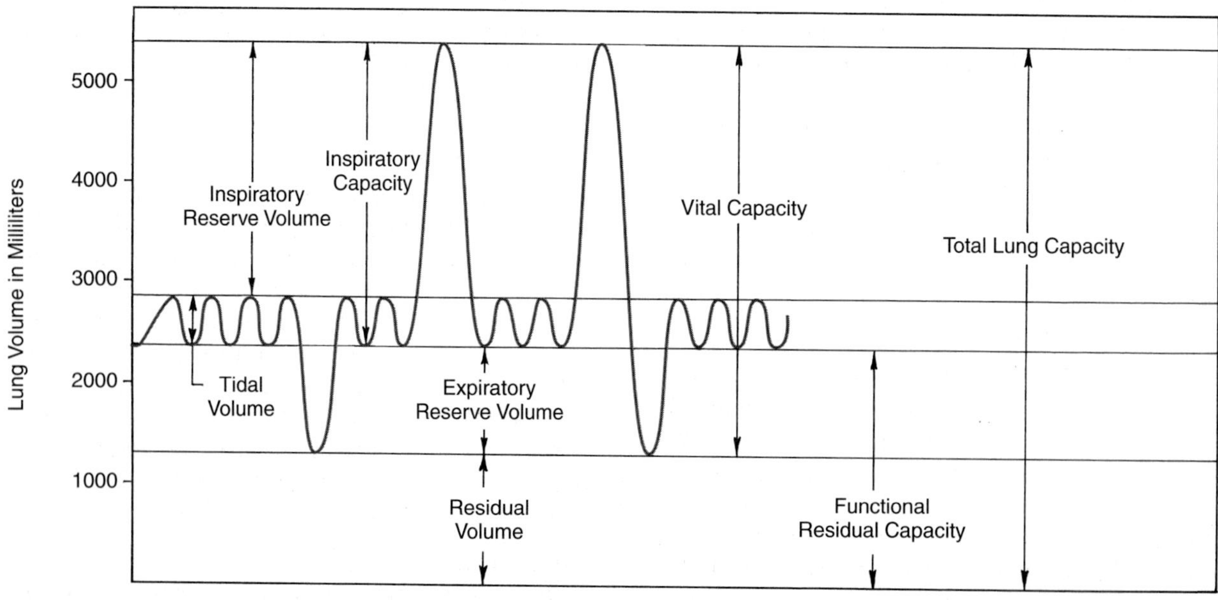

Figure 75.1 Spirogram of lung capacities

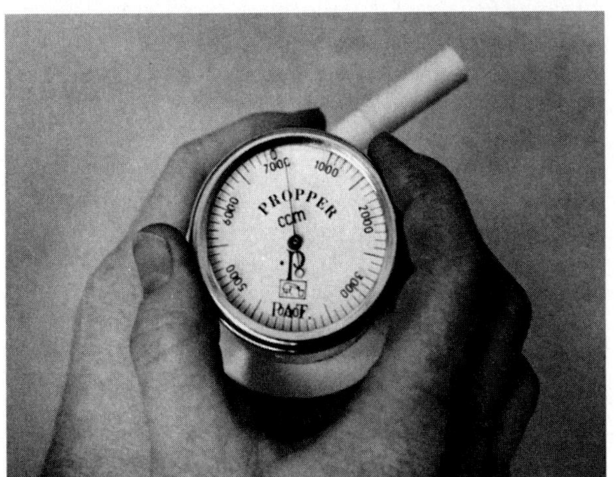

Figure 75.2 **Rotate the dial of the spirometer to zero before measuring exhalations.**

4. Divide the total volume of the three breaths by 3. This is your TV. Record this value on the Laboratory Report.

Minute Respiratory Volume

Your *minute respiratory volume* (MRV) is the amount of tidal air that passes in and out of your lungs in 1 minute. To determine this value, count your respirations for 1 minute, and multiply this number by your TV. Record this volume on the Laboratory Report.

Expiratory Reserve Volume

The *expiratory reserve volume* (ERV) is the amount of air that you can expire beyond the TV (see figure 75.1). It is usually around 1100 ml. To determine your ERV, proceed as follows:

1. Set the spirometer dial on 1000.
2. After making three normal expirations, expel all the air you can from your lungs through the spirometer.
3. Subtract 1000 from the reading on the dial to determine the exact volume.
4. Repeat steps 1 to 3 two more times to get an average ERV.
5. Record the average ERV on the Laboratory Report.

Vital Capacity

Note in figure 75.1 that the sum of the tidal, expiratory reserve, and inspiratory reserve volumes is the total functional or *vital capacity* (VC) of the lungs. The VC is determined by taking as deep a breath as possible and exhaling all the air possible. Although the average VC for men and women is around 4500 ml, age, height, and sex affect this volume appreciably. Even the established norms can vary as much as 20% and still be considered normal. Appendix A provides tables of normal VC values for men and women. Proceed as follows to determine your VC:

1. Set the spirometer dial on zero.
2. After taking two or three deep breaths and exhaling completely after each, take one final deep breath, and exhale all the air through the spirometer. A slow, even, forced exhalation is best.
3. Repeat this two or more times to verify your readings. Your VC should be within 100 ml each time.
4. Record your average VC on the Laboratory Report.
5. Consult tables V and VI in Appendix A for predicted (normal) VC values.

Inspiratory Capacity

If you take a deep breath to your maximum capacity after emptying your lungs of tidal air, you have reached your maximum *inspiratory capacity* (IC) (see figure 75.1). This volume is usually around 3000 ml. Note in the figure 75.1 spirogram that the IC is the sum of the tidal and inspiratory reserve volumes.

Since a handheld spirometer cannot record inhalations, you must calculate this volume with the following formula:

$$IC = VC - ERV$$

Inspiratory Reserve Volume

The *inspiratory reserve volume* (IRV) is the amount of air that can be drawn into the lungs in a maximal inspiration after filling the lungs with tidal air (see figure 75.1). Since the IRV is the IC less the TV, make this subtraction and record your results on the Laboratory Report.

Residual Volume

The *residual volume* (RV) is the volume of air in the lungs that cannot be forcibly expelled (see figure 75.1). No matter how hard you try to empty your lungs, a certain amount of air, usually around 1200 ml, remains trapped in the alveoli. The magnitude of the RV is often significant in the diagnosis of pulmonary impairment disorders.

The RV is determined by washing all the nitrogen from the lungs with pure oxygen and measuring the volume of nitrogen expelled. Obviously, you do not make this measurement here.

Assignment:
After recording your lung capacities on the Laboratory Report for this exercise, evaluate your results.

76 Spirometry: The FEV_T Test

A spirometer can detect impairment of pulmonary function in the form of asthma, emphysema, and cardiac insufficiency (left-sided heart failure). The symptom common to all these conditions is shortness of breath, or *dyspnea.* The pertinent lung capacity in these conditions is the vital capacity.

In several types of severe pulmonary impairment, the vital capacities of patients may be nearly normal. However, if the *rate of expiration* is recorded on a kymograph and timed, the extent of pulmonary damage becomes quite apparent. A kymograph is an instrument that, in conjunction with a spirometer, can record not only the respiratory volumes described in Exercise 75 but also the rate of expiration.

A timed expiratory test is called the *timed vital capacity* or *forced expiratory volume* (FEV_T). Figure 76.1 shows the spirometer (Collins Recording Vitalometer) used in this exercise. Figure 76.2 shows the sample record produced on such a kymograph. A record of this type is an *expirogram.*

A timed vital capacity test requires making a maximum inspiration and then expelling all the air from the lungs into the spirometer *as fast as possible.* The chart paper on the moving kymograph drum is calibrated vertically in liters and horizontally in seconds. A pen poised on the drum records how much air is expelled in a given period.

A normal person can expel approximately 95% of vital capacity within the first 3 seconds. The *percentage of vital capacity expelled during the first second* is of paramount diagnostic importance, however. An individual with no pulmonary impairment can usually expel 75% of the lungs' total capacity during the first second. Individuals with emphysema and asthma, however, expel a much lower percentage due to "air entrapment." Given sufficient time to exhale, they might be able to do much better.

In this exercise, you determine your FEV_T. Each student should use a new mouthpiece for this experiment.

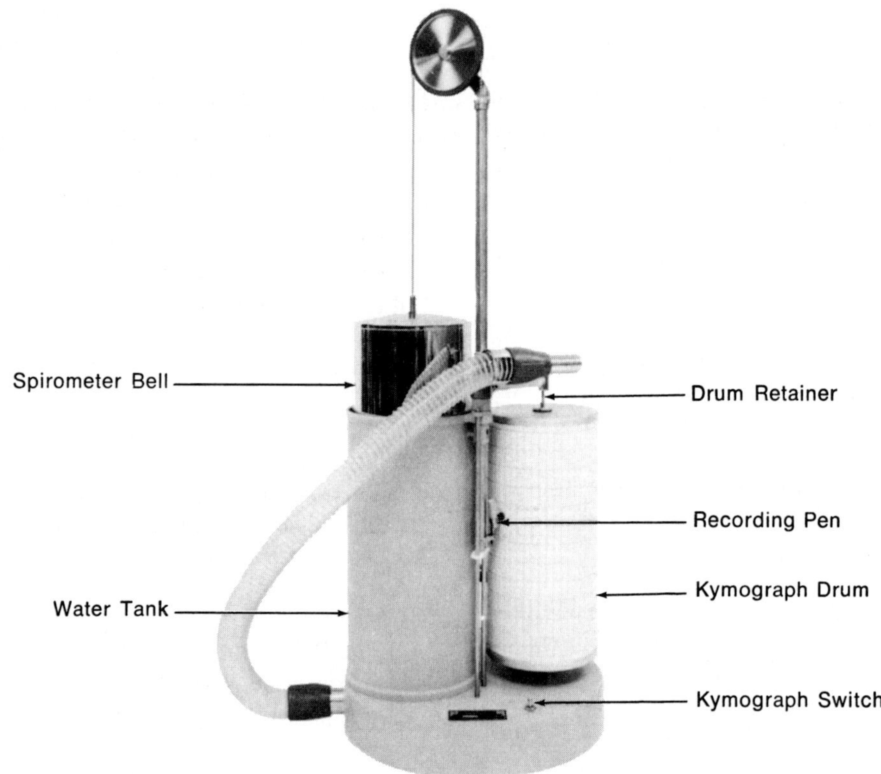

Figure 76.1 A Collins recording spirometer.

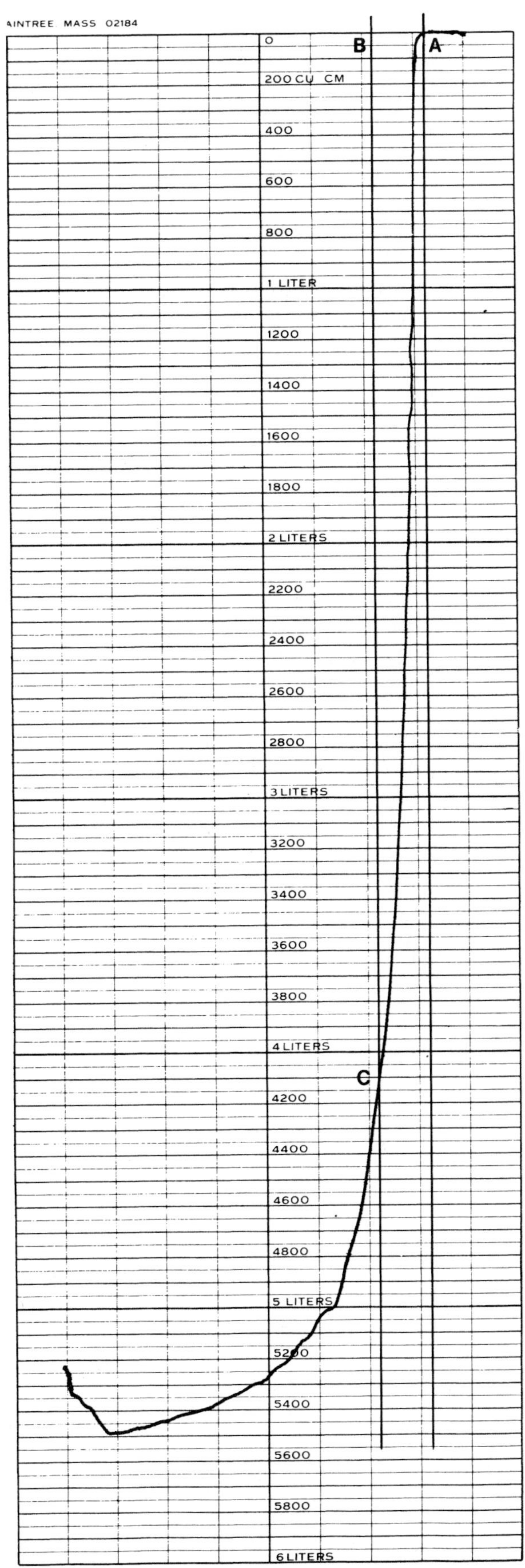

Figure 76.2 An expirogram.

Materials:
spirometer/kymograph (Collins Recording Vitalometer)
disposable mouthpieces (Collins #P612)
kymograph paper (Collins #P629)
noseclip
Scotch tape
dividers
ruler

Equipment Preparation

Prior to performing this test, prepare the spirometer:

1. Remove the spirometer bell, and fill the water tank to within 1.5 inches of the top (see figure 76.1). *Before filling, be sure to close the drain petcock.*
2. Replace the bell, making certain that the bead chain rests in the pulley groove.
3. Remove the kymograph drum by lifting the drum retainer first, and wrap a piece of chart paper around the drum. Attach paper to the drum with Scotch tape. Verify that the right edge overlaps the left edge.
4. Replace the kymograph drum, checking that the bottom spindle is in the hole in the bottom of the drum. Lower the kymograph drum retainer into the hole in the top of the drum (see figure 76.1).
5. Remove the protective cap from the recording pen, and determine if the pen is moist with ink. If the pen is dry, replace it.
6. Verify that the pen is oriented properly with respect to the drum. You can rotate the pen to the correct marking position.
7. Check the vertical position of the pen to see that it is recording on the zero line. Adjust the pen vertically within its socket by simply pushing or pulling on it. Loosen the set screw if major adjustments are necessary.
8. Plug the cord into an electrical outlet. *Do not use this apparatus in outlets that lack a ground circuit.* The three-pronged plug must fit into a three-holed receptacle. If you use an adapter, make the necessary ground hookup. Electricity and water can be a lethal combination!

Experiment Procedure

Before having a subject exhale into the mouthpiece, verify that everything is ready:

- The spirometer bell should be in the lowered position so that the recording pen is exactly on zero.
- A clean sheet of paper should cover the kymograph drum.

- A sterile mouthpiece should be on the end of the breathing tube.

Proceed as follows:

1. Apply a noseclip to the subject to prevent air leakage. Allow the subject to hold the breathing tube.
2. Before turning on the instrument, tell the subject:
   - to take as deep a breath as possible,
   - to expel as much air as possible into the mouthpiece, *as rapidly and completely as possible,* and
   - to inhale *before* the mouthpiece is placed in his or her mouth.
3. Turn on the kymograph, and tell the subject to perform the test as described in step 2.
4. As the subject blows into the tube, encourage him or her to *push, push, push* to get all the air out.
5. Turn off the kymograph.
6. After recording as many tracings as desired, remove the paper from the kymograph drum.

Analysis of Tracing

To determine the FEV_T from the expirogram, proceed as follows:

1. Draw a vertical line through the starting point of exhalation (see point A, figure 76.2).
2. Set a pair of dividers to the distance between two vertical lines. This distance represents 1 second.
3. Place one point of the dividers on point A, and establish point B to the left of point A. Draw a parallel vertical line through point B. This line intersects the expiratory curve at point C (see figure 76.2). The distance B–C represents the 1-second volume (FEV_1), or 1-second timed vital capacity. Read this volume directly from the volume markings. In figure 76.2, it is 4100 ml. The total vital capacity is also read directly from the volume markings. In figure 76.2, it is 5500 ml.
4. Next, correct the recorded total vital capacity for body temperature, ambient pressure, and water saturation. To do this, consult table IV in Appendix A for the conversion factor, and multiply the recorded vital capacity by this factor. For example, if the temperature in the spirometer is 25°C, the conversion factor is 1.075. Thus, for a total vital capacity of 5500 ml (as in figure 76.2):

   5500 ml × 1.075 = 5913 ml

5. Also convert the 1-second timed vital capacity (FEV_1) in the same manner. Thus, for an FEV_1 of 4100 ml (as in figure 76.2):

   4100 ml × 1.075 = 4408 ml

6. Divide the FEV_1 by the total vital capacity. As per the previous two cxamples:

   4408/5913 = 75%

7. Determine how the subject's vital capacity compares with the predicted (normal) values in tables V and VI of Appendix A. For example, the individual who produced the expirogram in figure 76.2 was a 26-year-old male, 6 feet (184 cm) tall. From table VI, his predicted vital capacity is 4545 ml.

   (Actual VC/Predicted VC) × 100 =
   (5913/4545) × 100 = 130%

Assignment:
Complete the Laboratory Report for this exercise.

Spirometry: Using Computerized Hardware

77

In this exercise, you use the Intelitool Spirocomp system to perform all the spirometry experiments in Exercises 75 and 76. Figure 77.1 illustrates the setup. The Spirocomp system allows you to measure and graph the following:

- tidal volume (TV)
- expiratory reserve volume (ERV)
- vital capacity (VC)

The Spirocomp system also computes and graphs the inspiratory reserve volume (IRV) and the predicted vital capacity (pVC). Forced expiratory volumes (FEV_T) for 1, 2, and 3 seconds are also determined with the Spirocomp system.

The principal advantage over the methods used in Exercises 75 and 76 is that this computerized setup makes all of the calculations *automatically* from data fed into the program. In addition, the program computes and averages the values of up to 30 subjects, allowing respiratory efficiency comparisons of smokers with nonsmokers, athletes with nonathletes, and males with females.

In this exercise, you work with a laboratory partner. While one person acts as the subject, breathing into the spirometer, the other person runs the computer and prompts the subject as to the procedures.

Note in the "Materials" list that two Spirocomp manuals, the *Spirocomp User Manual* (SUM) and the *Spirocomp Lab Manual* (SLM), are available for reference purposes. Page-number references to these manuals assist you in obtaining more information about certain procedures.

Materials:
computer with monitor and printer
Phipps and Bird wet spirometer
Spirocomp scale arm and interface box
computer game port to transducer cable
Spirocomp program diskette
Spirocomp User Manual
Spirocomp Lab Manual
blank diskette for saving data
continued on next page

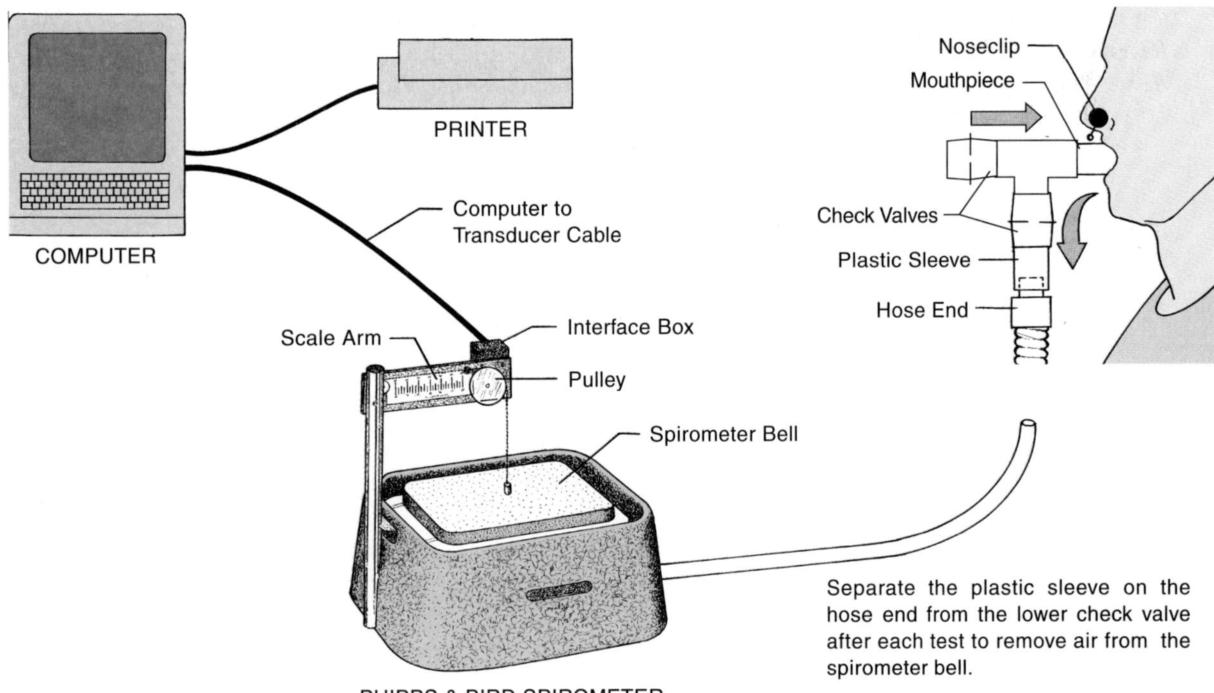

Figure 77.1 The Spirocomp setup for monitoring respiratory volumes.

valve assembly (Spirocomp)
four reusable mouthpieces
noseclip
stopwatch or watch with second hand
Lysol disinfectant (or Clorox bleach)
beaker of 70% alcohol
container of mild soap solution large enough to immerse one-way valve assembly

Equipment Hookup

The Spirocomp scale arm and interface box are already attached to the spirometer post (see figure 77.1), so all you have to do is hook up the components, fill the spirometer with water, and calibrate the spirometer. Proceed as follows:

1. Fill the spirometer to a little over the fill line with fresh tap water, and add Lysol or Clorox to the water for disinfection.
2. Plug in the computer and printer power cords to *grounded* 110 V electrical outlets.
3. *With the computer turned off,* connect the computer to transducer cable to the interface box and the game port of the computer.
4. Insert the program diskette, and turn on the computer. Strike any key, and the program will take over. If you are using an Apple IIe, IIc, or IIGS, *be sure to depress the Caps-Lock key.*
5. After you have described the system configuration by answering the appropriate questions (Reference: SUM, pp. 12–13), the Spirocomp Menu appears on the screen:

SPIROCOMP MENU

Experiment Menu
Review/Analyze Menu
Data File/Disk Commands
Preferences
Quit to DOS

6. Place all four mouthpieces into a beaker of 70% alcohol.
7. Select "Experiment Menu" from the Spirocomp Menu. Then select either "New Group File" or "Add New Record to Group File" from the Experiment Menu, as your instructor directs. If you are the first subject, you have to calibrate the Spirocomp transducer.
8. Calibrate the Spirocomp transducer (Reference: SUM, pp. 22–23) (see figure 77.1):
   a. Lift the chain of the pulley so that you raise the weight side, not the bell float side.
   b. While holding up the chain, turn the pulley until the computer reads 0 (± 100).
   c. Lay the chain back down over the pulley, and align the plastic pointer (with the rubber pointer guide right behind it) to 0 on the printed scale. Check the computer to verify that it still reads 0 (±100). Once zeroed in, press the space bar. *Note:* If the chain slips at any time during data acquisition, recalibrate the transducer before continuing.

Experiment Procedure

Now that all components are active, proceed with the experiments under the "Experiment Menu." Figure 77.2 illustrates the various respiration capacities to measure.

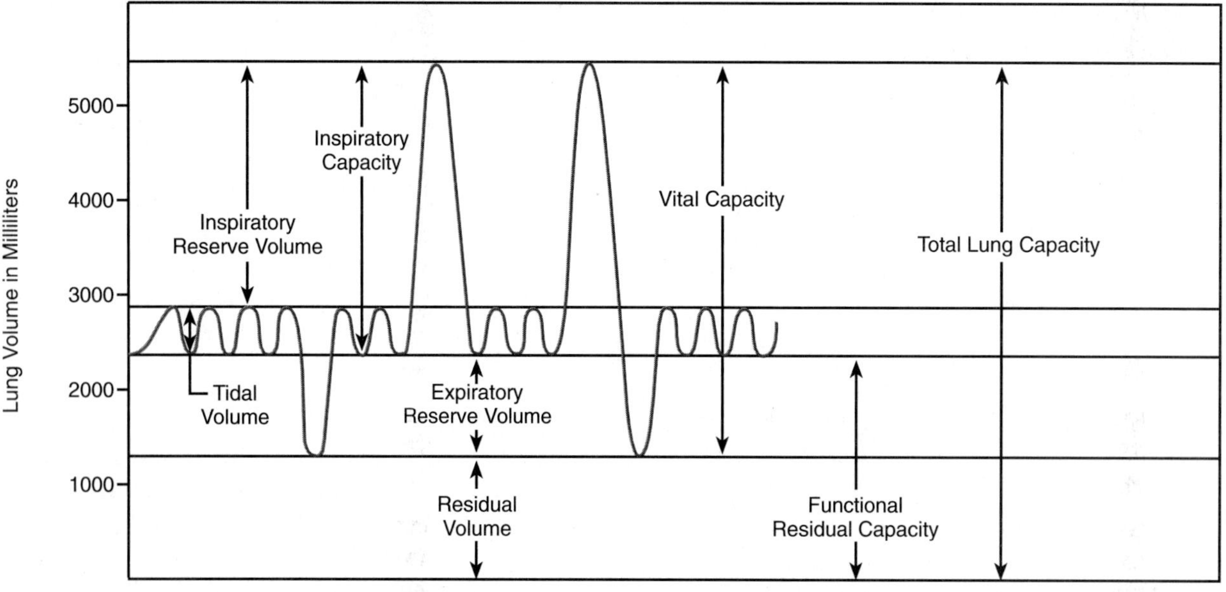

Figure 77.2 Spirogram of lung capacities.

Tidal Volume

The *tidal volume* (TV) is the amount of air that moves in and out of the lungs during a normal respiratory cycle (see figure 77.2). Although sex, age, and weight determine this value, the average normal TV is around 500 ml. For reliable results, the subject must be rested, calm, and breathing normally. Proceed as follows:

1. Seat the subject in a straight, high-backed chair.
2. Swish the valve assembly in 70% alcohol, wipe it and a mouthpiece off, and insert the mouthpiece into the valve assembly.
3. Dip the noseclip in 70% alcohol, wipe it off, and place it on the subject's nose.
4. Enter the requested subject data. The software will use this data to calculate the predicted vital capacity for a subject of a given gender, age, and height. *Note:* If you do not know the height in centimeters, enter it in inches, followed by an inch symbol (″). The software makes the conversion automatically. (Example: 69″ is converted to 175 cm.)
5. Disconnect the valve assembly from the hose (Reference: SLM, p. 26). Then empty the bell of the spirometer. *Gently* pressing down on the bell float hastens this process.
6. Have the subject breathe normally into the spirometer for one or two cycles. The valve system should cause the spirometer to fill on exhalation. The subject should draw in fresh air from the room during inhalation. The valves work best if they are wet. *Note:* If the valves are not working correctly, refer to SUM, pages 9 to11.
7. *During an inhalation,* press the **T** key, and have the subject follow the instructions on the screen, which read:

   BREATHE NORMAL TIDAL CYCLES

   The software reads the changes in the spirometer for three cycles, automatically stops, and graphs the average TV. If you feel that your results are not correct, repeat the reading of the TV before continuing (Reference: SLM, pp. 23–24).

Minute Respiratory Volume

The *minute respiratory volume* (MRV) is the amount of tidal air that passes in and out of the lungs in 1 minute. To determine the MRV, count your respirations for 1 minute, using a stopwatch or any watch with a second hand, and multiply this number by your TV. A normal respiratory rate is usually 12 to 15 cycles per minute. Example:

$$MRV = 12 \times 500 \text{ ml} = 6000 \text{ ml per minute}$$

Expiratory Reserve Volume

The *expiratory reserve volume* (ERV) is the amount of air that you can expire beyond the TV. It is usually around 1100 ml. Note its position on the spirogram in figure 77.2. To measure the ERV, proceed as follows:

1. Empty the spirometer bell float (step 5 from the TV procedure).
2. Have the subject breathe normally into the spirometer for one or two cycles.
3. *During an inhalation,* press the **E** key, and have the subject follow the instructions on the screen, which read:

   BREATHE NORMAL TIDAL CYCLES

4. *After two normal cycles,* the software prompts the subject to:

   STOP AFTER NEXT NORMAL EXHALE

   This means that at the end of the next (third) exhale of a normal tidal cycle, the subject should stop all air movement and wait for the next prompt, which is:

   EXHALE FULLY
   (the computer beeps here)

   This means that the subject should empty the lungs as forcefully and completely as possible. **It is also very important that the subject not inhale at this point.** As the subject holds his or her breath for a few seconds, the software calculates the ERV and graphs it on the screen. If you feel that your results are not correct, repeat the reading of the ERV before continuing (Reference: SLM, pp. 24–25).

Vital Capacity and FEV_T

The *vital capacity* (VC) is the sum of the tidal, expiratory reserve, and inspiratory reserve volumes (see figure 77.2). The VC is determined by taking as deep a breath as possible and exhaling all the air possible. The average VC for men and women is around 4500 ml.

While measuring the VC, you also can determine *timed vital capacity,* or the *forced expiratory volume* (FEV_T). The FEV_T is determined by taking a deep breath and expelling it as fast as possible. An individual with no respiratory impairment should be able to expire 95% of his or her VC within 3 seconds; however, the percentage of VC expelled within the first second is most important. **An individual with no pulmonary impairment should be able to expel 75% of his or her VC within 1 second.** Individuals with emphysema and asthma have a much lower percentage due to air entrapment.

Proceed as follows to determine the VC and FEV_T:

1. Empty the spirometer bell float (step 5 from the TV procedure).
2. Press the **V** key, and have the subject follow the instructions on the screen, which read:

   INHALE FULLY
   then
   EXHALE as FAST and COMPLETELY as you can.

   This means that the subject should first inhale maximally and then exhale as forcefully, rapidly, and completely as possible into the spirometer. The subject must keep trying to squeeze out every little bit of air until the software stops and graphs the VC on the screen. If you feel that your results are not satisfactory, repeat the reading of the VC before you continue (Reference: SLM, p. 25).
3. Read the VC and FEV_T on the screen (Reference: SLM, p. 27).

Evaluation and Next Subject

The subject's record is now complete. The software has calculated and graphed the subject's inspiratory reserve volume (IRV = VC − [TV + ERV]). The software has also calculated the predicted VC and displayed the FEV_T.

To print out the results, press the **P** key. If you wish to add the next subject to the group record, press the **N** key, and repeat the experiment procedure for that individual.

If you are finished with the group record, press the **ESC** key. If you pressed the ESC key and you still want to add to the group record, simply select "Run Spirocomp," and press the **A** key for ADD to the current group data. Remember to wash the valve assembly, mouthpiece, and nose clip in 70% alcohol prior to testing each new subject.

Compare your volumes and capacities to the following approximate norms:

	TV	ERV	IRV	VC
Males	500 ml	1000 ml	3000 ml	4500 ml
Females	400 ml	750 ml	2250 ml	3400 ml

Compare your VC to the VC predicted for your age, height, and gender (see tables V and VI in Appendix A).

Compare different groups to each other (smoker versus nonsmoker, etc.). Are there differences? Explain.

Assignment:
Since this exercise does not have a Laboratory Report, your instructor will indicate how to report this experiment.

PART 13 Autonomic Responses to Physical and Psychological Factors

When performing "piecemeal" physiological experiments, you sometimes tend to forget that the body is an integrated whole that responds in many ways to various stimuli, as some of the experiments in Part 8 revealed. The three exercises in this unit are based on this premise.

The three experiments utilize the Duograph or polygraph to monitor two or more autonomic responses in stressful situations. In Exercise 78, you monitor blood pressure and heart rate to determine the effectiveness of physical exercise. This experiment is not intended for all students, but for those who routinely exercise by jogging, bicycling, or some other similar activity. It is a lengthy, though simple, experiment to pursue over a 3- or 4-month period. It enables you to scientifically measure the effectiveness of a physical training program.

While Exercise 78 attempts to measure the effects of physical stress on certain autonomic responses, Exercise 79 demonstrates that emotions can also profoundly change autonomic responses. In this exercise, the polygraph monitors respiration, heart rate, muscle tension, and galvanic skin response in the detection of attempted deception. The experiment is simple but effective.

Exercise 80 is an experiment in biofeedback that requires a trained subject. If a suitable subject is not available, you may wish to experiment with the procedure as a special project to develop tracings that exhibit the existence of biofeedback.

78 Response to Physical Stress: Monitoring Physical Fitness Training

Awareness of the importance of physical fitness has been increasing for some time. More and more individuals have taken up weight lifting, jogging, bicycling, tennis, and other activities to increase physical fitness. Although these activities benefit the individual, scientific evaluation of the accompanying improvement in physical well-being is generally lacking.

In many instances, the only barometer for improvement in a physical fitness program is a sense of well-being. The individual feels good, and therefore, has an incentive to continue the program. What is needed is a convenient, inexpensive means of scientifically monitoring physiological changes that occur in such a program. This exercise provides a relatively simple means of evaluating an exercise program for beneficial physiological changes.

In general, the two categories of exercise testing are *recovery tests* and *effort tests.* Recovery tests immediately follow exercise. Effort tests are performed during exercise. This exercise is based on recovery tests, operating on the assumption that *quicker recovery signifies improvement in physical fitness.*

The program outlined here is intended to last from 10 to 16 weeks—normally, the length of an entire quarter or semester. The procedure is as follows: First, certain tests and a questionnaire establish the subject's qualifications for the experiment. Second, anthropomorphic data, such as height, weight, limb dimensions, and so on, are recorded. Finally, the subject's blood pressure and heart rate are measured while the subject is at rest and after a controlled exercise session in the laboratory.

These preliminaries are followed by weekly monitoring of the pulse rate and blood pressure. Careful records are kept during the entire period, and all data are plotted on graphs to note progress.

The results of this experiment depend on the length of the training period, the extent of the training, the level of fitness at the start, whether or not weight reduction is included, the presence or absence of smoking, and many other factors. Since physical training tends to *reduce the heart rate* and to *promote faster recovery to resting levels after exercise,* these are the two changes to look for.

Test Parameters

Physical fitness is probably best assessed by measuring the metabolic rate under a variety of environmental conditions. This type of testing, however, requires a well-equipped research facility and extensive technical training.

Fortunately, numerous studies reveal that **heart rate** bears a linear relationship to oxygen consumption. Since the correlation between heart rate and aerobic capacity is in the range of 50 to 90%, heart rate is a simple test parameter that the average person can use to monitor the degree of physical fitness.

Closely associated with heart rate is **blood pressure,** another easily monitored parameter. The tests in this exercise use a finger pulse pickup and Statham pressure transducer with a Duograph, as figure 78.1 shows. A Physiograph requires one ESG (electrosphygmomanometer) coupler with arm cuff and one transducer coupler with pulse pickup. These setups were used in Exercise 74 in the Valsalva maneuver experiment.

Meaningful measurements with this setup require careful control of the following six variables:

1. **Time of day.** Humans exhibit circadian rhythms to varying degrees that can influence pulse rate and body temperature. To minimize this variable, testing should be at about the same time each day, preferably before meals.
2. **Ambient temperature.** Increased temperature requires increased cardiac output to dissipate the heat load, resulting in higher readings. Testing in an air-conditioned building is helpful. Light clothing and comfortable shoes are essential.
3. **Diet and mealtimes.** Increased metabolism after a heavy meal elevates heart rate and ventilation for over an hour. Conversely, fasting diminishes blood sugar levels, possibly reducing performance. Testing 1.0 to1.5 hours after breakfast may be a suitable compromise. Stimulants such as tea, coffee, and chocolate should be avoided.
4. **Medications.** Drugs such as antihistamines or decongestants influence cardiovascular function and should be avoided for at least 4 hours before testing. All required medications should be entered into the records.
5. **Rest.** Unusually strenuous workouts should be avoided on the day before testing. The reasons for this precaution are related to peripheral sequestration leading to reduced central blood

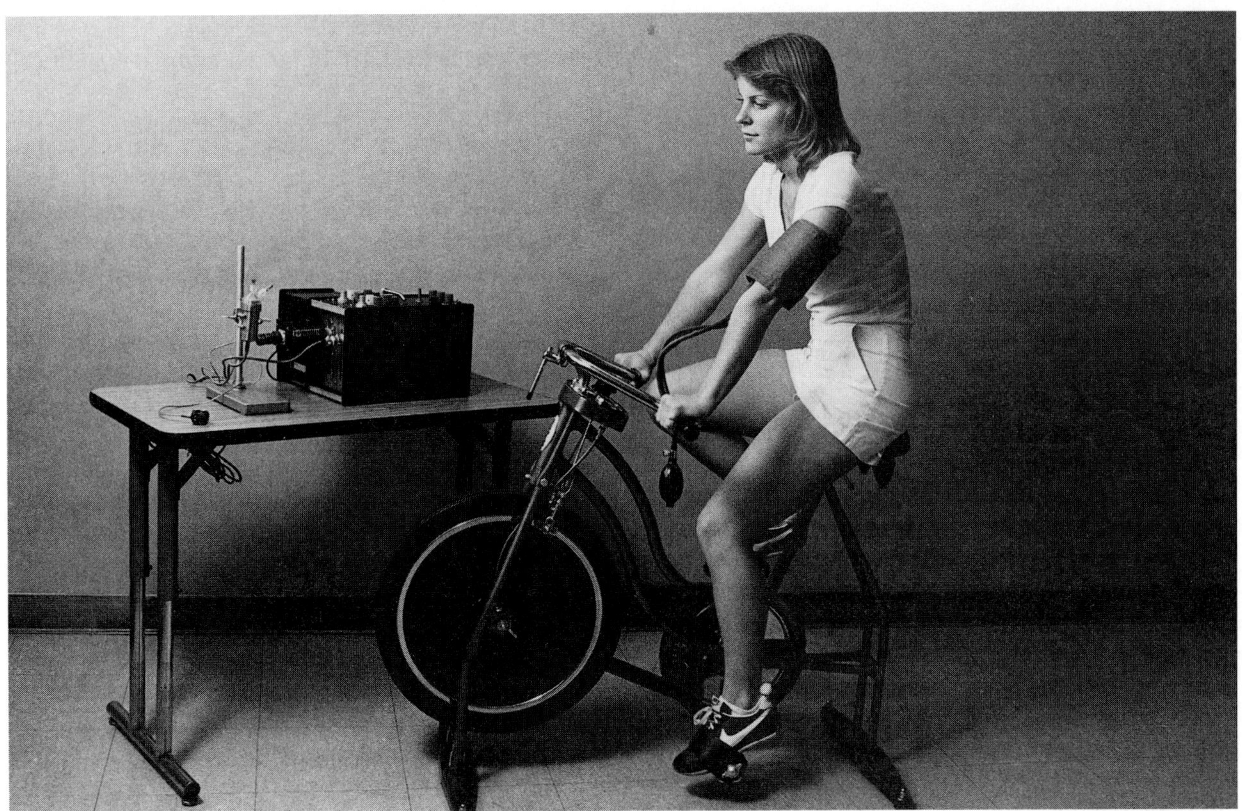

Figure 78.1 A subject in a physical fitness program can periodically monitor her recovery rates of blood pressure and heart rate on a Duograph (or Physiograph) after exercising on an ergometer such as a stationary bicycle.

volume and stiffness of the affected musculature. Furthermore, depletion of muscle glycogen stores reduces the capacity for sustained muscular effort.

6. **Anxiety.** Anxiety is a natural, subtle complication that frequently accompanies testing because the subject may be self-conscious about poor results. Apprehension is often responsible for seriously high readings, even in individuals who are outwardly calm.

   To minimize the effects of anxiety, the subject should undertake several practice trials. Single readings should be interpreted cautiously, because even normal, healthy individuals sometimes have erratic ECGs, and abnormal heart rate and blood pressure values.

Ergometric Standardization

The type of exercise employed in this physical fitness program will vary considerably from student to student. The one requirement is that the activity is enjoyable. People who jog, lift weights, or play tennis persist in these activities because they derive great personal satisfaction from them.

All subjects will use ergometers in the laboratory to make standardized recovery measurements. Whether the ergometer is a stationary bicycle or a specific flight of stairs, the standardization is based on the premise that the subject will achieve heart rates of 120 ±5 beats per minute following 2 to 5 minutes of exertion. An exception is the exceptionally physically fit student whose heart rate is already very low. In such a case, pulse rates of 100 ±5 beats per minute will suffice.

Experiment Procedures

This lengthy experiment is divided into four phases: (1) screening, (2) anthropomorphic measurements, (3) recovery tests, and (4) final evaluation. Your instructor may decide to modify several aspects of each portion of the experiment. Time and the availability of equipment will determine exactly how the experiment proceeds.

Materials:
tape measure
scale for body weight
for ECG, see Exercise 62
for heart rate and blood pressure, see Exercise 74
for urine analysis, see Exercise 86

Screening (First Laboratory Period)

This experiment is voluntary and is performed on your own time. Since the details of the exercise program are completely student-oriented without instructor control or responsibility, all training is done outside the jurisdiction of this course. However, if you have any physiological problem that could be life-threatening, you must not perform the experiment. It is your responsibility to obtain medical clearance to enter any type of physical fitness training. As a screening tool, perform the tests that follow in the first laboratory period. Since many of these tests are difficult to perform without help, work with a laboratory partner where necessary.

Electrocardiogram Work with your laboratory partner to record each other's ECG, following the instructions in Exercise 62. Make two recordings, one before exercise and one right after exercise. Attach both readings to Laboratory Report 78S.

Heart Rate and Blood Pressure With a finger pulse pickup and arm cuff, make recordings of the heart rate and blood pressure before and after exercise. See Exercise 74.

Urine Analysis Do a urine analysis according to instructions in Exercise 86, testing for bilirubin, urobilinogen, blood, ketones, glucose, and protein. Check the pH also. Record the tests in table II of Laboratory Report 78S.

After completing the previous tests, complete parts C and D on Laboratory Report 78S. Read the "Release" paragraph, and sign the report. Turn it in to your instructor for approval.

Anthropomorphic Measurements

Anthropomorphic measurements before and after the experimental period may reveal physical change improvements in addition to recovery improvement. Record these measurements in part B of the Laboratory Report. The following suggestions concerning each measurement provide for standardization:

Height Stand barefoot with back to wall, and look straight ahead. Measure to the closest 0.5 cm.

Weight Remove shoes and wear light clothing. Weigh to the nearest 0.1 kg.

Chest Measure chest circumference at the end of normal expiration. Apply tape to the bare chest at nipple level in males, to the highest point under the breasts in females.

Abdomen Measure the *maximum* circumference of the bare abdomen at the end of normal expiration.

Thigh Measure the *maximum* circumference of the bare thigh while standing.

Calf Measure the *maximum* circumference of the bare right calf while standing.

Arm Measure the *maximum* circumference of the upper arm while extended and relaxed.

Vital Capacity Determine with handheld or wet spirometers your vital capacity. See Exercise 75 or 77.

Recovery Tests

Whether you perform recovery tests once a week, every 2 weeks, or once a month depends on how you view the overall experiment. An advantage of frequent testing is that recovery progress can be used to increase the levels of exertion. For example, if you feel that recovery is slower than desired, you can experiment with a variety of stress-increasing activities to improve recovery rates.

The type of ergometer you use depends on the equipment available. Steps, treadmills, arm cranks, and stationary bicycles are most frequently used. Each type of ergometer has its advantages and disadvantages. An elaborate setup is not required here, since all you are concerned with is exerting enough effort to get the pulse rate up to 120 ±5 beats per minute (100 ±5 for conditioned athletes). Experimentation will reveal how much exercise (usually 3–5 minutes) is enough to produce this heart rate.

Before exercising, record your blood pressure and pulse rate on the chart recorder (slow speed), taking into consideration the six controlling factors mentioned previously (time of day, temperature, relation to mealtime, medications, rest, and anxiety). Record your results in the table in part C of the Laboratory Report.

Before exercising, disconnect the tubing to the arm cuff, and remove the finger pulse pickup. Leave the cuff on the arm, however. Exercise for 3 to 5 minutes, using a stationary bicycle, flight of stairs, or whatever is available to get the desired pulse rate. Hook up to the Duograph or Physiograph again as soon as possible, and record for 2 to 3 minutes. Record the results in the table in part C of the Laboratory Report.

Final Evaluation

After performing the last recovery tests and recording the data, proceed as follows:

1. **Anthropomorphic measurements.** Repeat all the anthropomorphic measurements made during the first week of the experiment, recording them in the table on the Laboratory Report in part B. Compare the data, and record any significant changes.
2. **Graphs.** Plot all the data from the tables onto the appropriate graphs in the Laboratory Report.
3. **Final statement.** Briefly summarize the results of the experiment on the Laboratory Report. If results seem negative, try to interpret what went wrong.

Assignment:

Complete the Laboratory Report for this exercise.

79 Response to Psychological Stress: Lie Detection

All of us are familiar with the physiological effects of attempting to conceal the truth. Surely, everyone has experienced such sensations as an increased heart rate, a rush of blood to the face, and an uncontrollable impulse to swallow during attempted deception. When we feel that the deception has been successful, we often sigh in relief and resume normal breathing.

Some individuals, of course, are better at deception than others. They are more successful either because lying does not disturb them or because they have better cortical control over the responses of the autonomic nervous system. Nevertheless, even the best deceivers have autonomic responses that cannot be masked. These physiological aberrations form the basis for the success of the lie detector (polygraph) in criminal investigation.

This exercise demonstrates the mechanics of lie detection with a subject from the class by monitoring five physiological parameters, as figure 79.1 shows. When the experiment has been completed, you will work in small groups to repeat the process, trying various techniques.

Considerable cooperation is required of all observing class members, or the experiment will not succeed. Ideally, the experiment should be performed in a quiet room, completely lacking in distraction. For obvious reasons, the presence of observers does not add to test reliability. Extraneous noises, such as subdued conversation or telephone ringing, can induce distorted recordings. Even seemingly insignificant objects, such as pictures on the wall or room ornaments, can either distract the attention of the subject or permit him or her to use them as a focus for practicing evasion. Therefore,

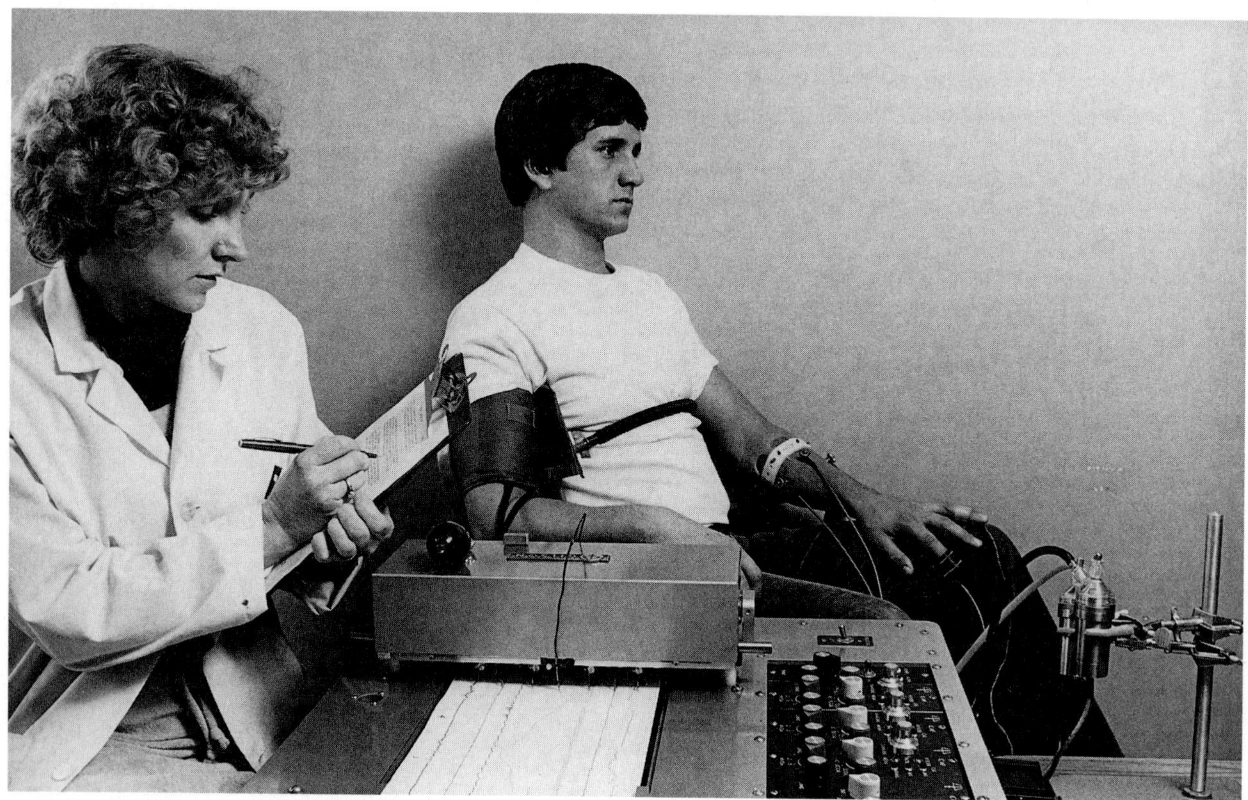

Figure 79.1 The five physiological responses monitored in lie detection are (1) blood pressure (cuff on right arm), (2) respiration (chest bellows), (3) heart rate (finger pulse pickup on index finger of left hand), (4) galvanic skin response (GSR) (fourth finger of left hand), and (5) electromyography (EMG) (left forearm).

the setting for this experiment must be carefully controlled.

In addition to minimal audience distraction, the success of this experiment depends on (1) skilled interrogation, (2) proper subject preparation, and (3) correct equipment hookup.

Role of the Interrogator

The interrogator should maintain a noncommittal expression and ask questions in an even, well-modulated tone, free of insinuation, incredulity, and provocation. Distracting activity, such as making notations, organizing materials, and operating the polygraph, should be minimized.

During the test, the interrogator should sit to one side of the subject—not in front of him or her. If the subject is not a class member and is not acquainted with the polygraph and the procedure, the interrogator should answer any of the subject's questions to satisfy curiosity or allay anxiety.

The Subject

Position, comfort, and lack of movement are the prime considerations for the subject. The subject should be seated in a comfortable armchair, facing neither the audience nor the interrogator. Ideally, the subject should face a blank wall of neutral color and moderate lighting. He or she should not face a window.

Binding straps and wire leads should be judiciously placed to permit maximum comfort. Since the pressure cuff on the arm causes the greatest discomfort, it should be inflated only while recording and always deflated during resting periods between tests.

The subject should refrain from coughing, sniffing, or clearing the throat to prevent distortion of breathing traces. He or she should look straight ahead, never turning the head, moving the arms, or shifting body position. He or she should remain still, but not rigid; any movement causes artifacts while recording.

Performing the Test

Since this lie detection demonstration does not pertain to the actual commission of a crime, the heightened anxiety expected with an actual criminal suspect is absent. Nevertheless, the subject in this demonstration is inclined to *beat the machine,* and the challenge for observers is to detect lies by skillfully formulating questions and interpreting results.

The protocol that follows utilizes regular playing cards. The subject selects a card from a group of eight cards and shows the card to two witnesses, but not to the interrogator. One of the witnesses then mixes the cards and turns up the top card for the subject and interrogator to see. Each time the witness turns over a card, the interrogator asks the subject to deny that the card is the one originally picked. As the subject makes the denials, the interrogator monitors the five parameters. The interrogator then tells the subject and the class which card the subject lied about. Class members and the subject can study the tracings to see what polygraph evidence supported the interrogator's decision.

The five-channel polygraph records respiration, blood pressure, heart rate, galvanic skin response (GSR), and muscle tension (electromyography [EMG]). You have made all of these measurements, except for the GSR, in previous experiments. Note the separate "Materials" lists for the Gilson and Narco polygraphs.

The procedure involves (1) prequestioning, (2) subject preparation, and (3) monitoring. Proceed as follows:

Materials:

armchair
electrode paste
felt-tip pen for marking chart
Scotchbrite pad
self-adhering wrap
eight nonface cards from a deck of playing cards

for Gilson Polygraph (five-channel) setup:

two Statham TP213D pressure transducers
pneumograph
blood pressure cuff
finger pulse pickup
Gilson A4023 adapter for pulse pickup
Gilson E4041 GSR electrodes
Gilson A4042 adapter for GSR electrodes
three ECG plate electrodes
cable for muscle (EMG) electrodes
two clamps for Statham transducers
stand for mounting Statham transducers

for Narco Physiograph setup:

two transducer couplers on Physiograph
EEG/EMG coupler on Physiograph
GSR coupler on Physiograph
cardiac coupler on Physiograph (optional)
pneumograph transducer (Narco 705–0190)
cable for pneumograph transducer
finger pulse pickup (Narco 705–0050)
cable for EEG/EMG coupler
GSR electrode kit (Narco PN 710–0008)
three ECG plate electrodes and straps

Prequestioning the Subject

Before the subject is connected to the polygraph, the interrogator talks privately with the subject to get personal information for 10 simple questions for establishing a control baseline. The questions are phrased for a "yes" or "no" response. A typical list of questions follows:

1. Is your full name *Mary Ann Jones?**
2. Do your friends call you *Peggy?*
3. Did you have a cup of coffee this morning?
4. Do you drive a *Toyota* to school?
5. Your real name is *Hillary Rodham Clinton,* isn't it?
6. Do you smoke?
7. Did you *play tennis* yesterday?
8. Isn't your favorite dessert a *hot fudge sundae?*
9. You're wearing a *red pair of shoes,* aren't you?
10. Today is *Thursday,* isn't it?

*Italicized words or phrases can be changed to match the subject.

Subject Preparation

Seat the subject in the armchair, and attach the pneumograph, finger pulse pickup, EMG plate electrodes, and blood pressure cuff to the appropriate parts of the body. For details on the various hookups, consult the following exercises:

Pneumograph—Exercise 72
Finger pulse pickup—Exercise 64
EMG hookup—Exercise 32
ECG hookup—Exercise 62

Modify the setups in the following ways:

Pneumograph Note in figure 79.1 that two Statham transducers are mounted on a single stand. One of these transducers is for the pneumograph, and the other is for the arm cuff on Gilson setups.

Blood Pressure Adjust cuff pressure to maintain the dicrotic notch in the midposition on the downslope of the pulse. To achieve this, you will need about 90 mm Hg. If a lower pressure still produces a high-amplitude tracing, reduce the cuff pressure even further. **Keep the cuff deflated until just before recording, and release the cuff pressure as soon as recording is completed.** This parameter causes the most subject discomfort.

Heart Rate The finger pulse pickup monitors the heart rate. If the Gilson polygraph has a cardiotachometer module (IC-CT), the A4023 adapter is not necessary. The adapter is used only with IC-MP modules.

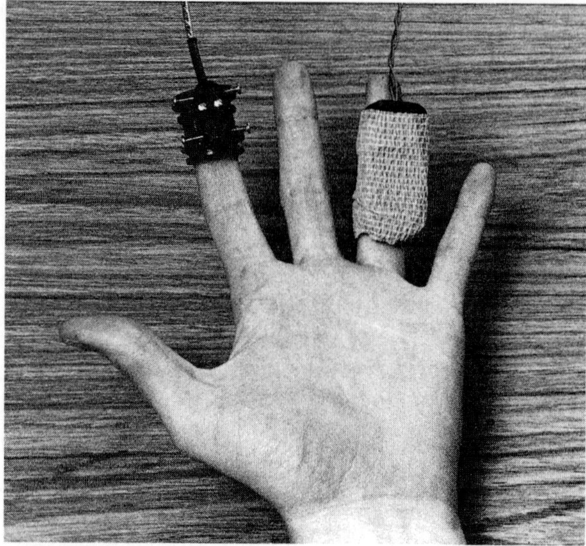

Figure 79.2 GSR electrodes and finger pulse pickup on the left hand.

Muscle Tension Note in figure 79.1 that three plate electrodes on the subject's left forearm monitor muscle tension. The electrodes can also be attached to the calf of either leg. On the Gilson polygraph, you can use either the IC-EMG or the IC-MP module for this parameter. On the Physiograph, you should use the EEG/EMG coupler.

GSR Electrodes Attach the GSR electrode to the fourth finger of the subject's left hand, as figure 79.2 shows. Before attaching it, scrub the skin with a Scotchbrite pad, and apply a light film of electrode paste to the surface. Secure the electrode to the finger with self-adhering wrap or Scotch tape.

Since the GSR is a comparatively slow response, connect it directly to the Servo channel. On a Gilson polygraph, a 4042 adapter is needed.

The GSR measures changes in conductivity of a low DC voltage on the skin. The autonomic nervous system regulates blood flow and perspiration, both of which affect skin conductivity. Emotional stress shows up clearly with this parameter.

Establishing a Control Baseline

1. Pump air into the arm cuff, and turn on the polygraph. Set the speed at 2.5 mm/sec.
2. Check the trace on all channels to verify that all are adequate.
3. To establish the appearance of artifacts, ask the subject to do the following as you record on the chart each activity:
   - Cough.
   - Move the arm with the pressure cuff.
   - Move the other arm.
   - Shift position in the chair.

4. To establish a baseline for truthful answers, ask the subject the 10 questions that must be answered *truthfully*. Record on the chart not only the number of the control question, but also a plus (+) for "yes" and a minus (−) for "no."
5. Deflate the cuff, and allow the subject to rest for 5 minutes.

Deception Test No. 1

You are now ready to test for deception. Proceed as follows with eight nonface cards of different suits from a deck of playing cards. Two witnesses are used. One handles the cards and reveals them to the subject and the other witness.

1. *Interrogator to subject:* "A witness will select one of the playing cards and show it to you and the other witness. Don't show it to me or the audience."
2. Witness A shuffles the cards and fans them out in front of the subject, *facedown.* Witness B picks one card. All three look at it. Witness A returns it to the other seven cards, mixes them up, and places all eight cards in a pile in front of the subject.
3. Witness A turns the top card faceup for all four to see.
4. *Interrogator to subject:* "Did you select the ____?" (Card is identified by number and suit, such as *three of diamonds.*) The subject answers "No" as the card is shown.
5. The interrogator marks the card number and suit on the chart.
6. After 20 seconds, Witness A turns over the second card.
7. The interrogator asks the subject the same question about the second card. The subject answers "No."
8. Every 20 seconds, a card is turned over, a question is asked, and the chart is marked until all eight cards have been exposed.
9. Now Witness A shuffles the cards again and exposes one card at a time again, as before. This time, however, the subject answers "Yes" to each question. The chart is marked accordingly.
10. Finally, the cards are shuffled and exposed a third time in the same manner, except that the *subject remains silent* when asked a question.
11. The polygraph chart is stopped, the arm cuff pressure is released, and the chart is studied. On the basis of tracing analysis, the interrogator tells the subject which card had been picked. The entire class examines the chart.

Deception Test No. 2

Another way the same playing cards can be used to detect lying is with a *peak of tension test.* The rationale behind this stratagem is to lead the subject to a climax of anxiety in which the greatest autonomic response presumably betrays the truth. In this case, the subject answers "No" to all questions asked. The procedure is as follows:

1. At 20-second intervals, ask the subject:
   "Did you select a card of spades?"
   "Did you select a card of diamonds?"
   "Did you select a card of hearts?"
   "Did you select a card of clubs?"
2. During step 1, one of the witnesses arranges the eight cards in numerical order, with the lowest number on top. The interrogator then looks at each card and asks the subject questions like:
   "Was your card a 2?"
   "Was your card a 3?"
3. At the end of the test, release the pressure cuff, stop the paper movement, and evaluate the chart.

Student Group Projects

The card test protocol can be adapted to finding out concealed information, such as a person's weight or IQ. If time is available, the class can divide into several groups of four or five students to design a test and formulate questions to disclose information. Test subjects should be volunteers from another team. The object is to devise a test that creates emotional arousal without angering or embarrassing the subject. In addition to weight or IQ, other concealed information could be the subject's birthday, age, GPA, and so on.

80 A Biofeedback Demonstration

The experiments in Exercise 79 confirm that cerebral activity greatly influences autonomic reflexes. Deliberate efforts to conceal the truth result in changes in heart rate, blood pressure, respiration, and galvanic skin response. These physiological changes are solid evidence that higher brain centers *unconsciously* influence visceral reflexes. The whole concept of psychosomatic pathologies is based on this premise. Even such ailments as peptic ulcers, heart palpitation, and heart attacks have been induced psychosomatically.

Much evidence also supports the concept that cerebral activity can *consciously* regulate certain autonomic reflexes. The ability of the mind to deliberately affect autonomic reflexes to improve well-being is **biofeedback.** In recent years, biofeedback research has focused on the development of techniques. Individuals have used biofeedback techniques to gain relief from certain types of headaches, muscle tension, tachycardia, and hypertension. Although biofeedback probably cannot completely eliminate pain or greatly intensify pleasure, some people have benefited considerably from biofeedback techniques.

This exercise attempts to demonstrate the existence of biofeedback, using four channels on a polygraph. Since biofeedback is essentially a learned skill, a trained biofeedback subject will achieve the most success.

To observe cerebral activity, you use one channel to monitor brain waves with EEG (electroencephalogram) electrodes, as in Exercise 47. The other three parameters are the galvanic skin response (GSR), heart rate, and skin temperature.

Since the last experiment used two of these parameters, this exercise is a logical extension of Exercise 79, and it is plagued by the same problems. Observers distracting the subject is a paramount problem. The subject may ask to be blindfolded and to use earplugs. Other problems pertain primarily to maintaining good electrode contact on the scalp and skin. Review the procedures in Exercise 47 that relate to EEG monitoring. The thermistor probe, used for detecting skin temperature, must be in close contact with the skin at all times; momentary separation from the skin causes air cooling and artifacts on the chart.

Although this experiment is set up primarily for the Gilson polygraph, it can be easily adapted to the Physiograph. Proceed as follows:

Materials:
Gilson polygraph
Gilson T4060 thermistor probe
Gilson A4062 thermistor adapter
Gilson T4020 finger pulse pickup
Gilson A4023 adapter for pulse pickup
Gilson E4041 GSR electrodes
electrode paste
Scotchbrite pad
Scotch tape
felt-tip pen for marking chart
patient cable for EEG (three leads)
three EEG disk-type electrodes
three circular Band-Aids
alcohol swabs
elastic wrapping for EEG
earplugs or cotton
blindfold
elastic banding for thermistor probe

Preparation of Subject

Attach three disk electrodes to the scalp, according to the procedures outlined in Exercise 47. As figure 80.1 illustrates, the finger pulse pickup attaches to the index finger of the right hand.

To the middle and index fingers of the left hand, attach the GSR electrodes and thermistor, respectively. Figure 80.2 shows how to slip the thermistor under the wrapping to maintain good skin contact.

After all attachments are secure, allow the subject to assume a comfortable horizontal position. If the subject wants a blindfold and earplugs, apply them before the subject lies down.

Equipment Setup

Connect all cables to the appropriate adapters and modules. The finger pulse pickup and GSR should be connected to the same modules as in Exercise 79. An IC-MP or IC-UM module may be used for the three-lead cable from the EEG electrodes. The thermistor probe requires the servo, IC-MP, or IC-UM module.

Monitoring

With the chart paper moving at a speed of 25 mm/sec, establish a baseline for about 3 minutes.

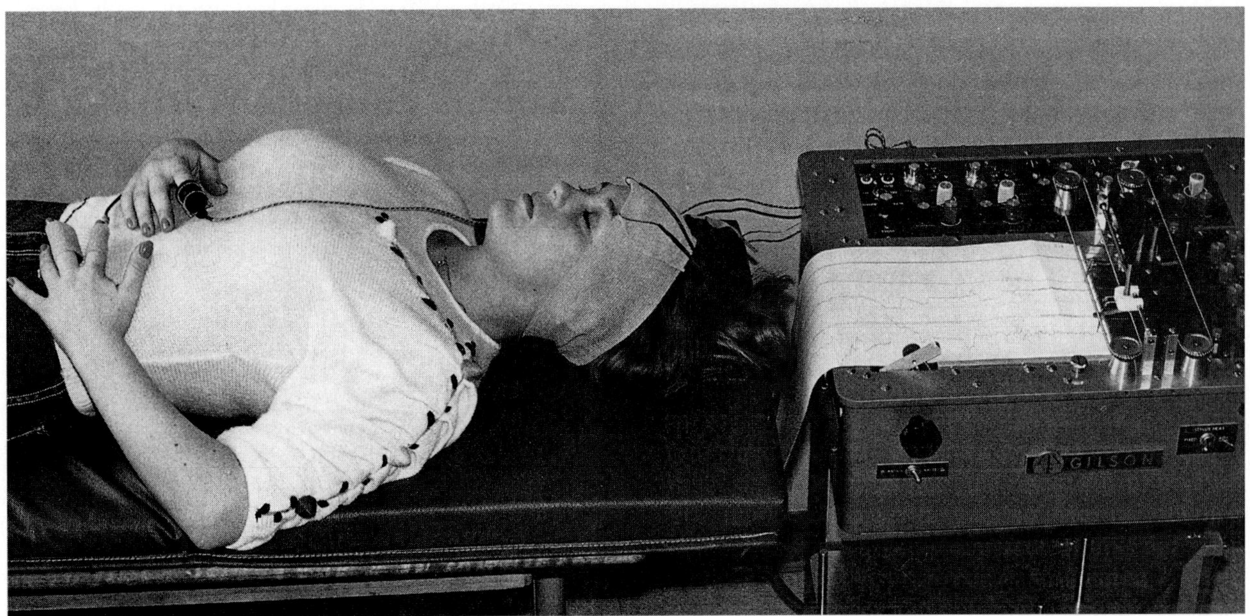

Figure 80.1 In this four-channel biofeedback demonstration, EEG electrodes attach to the scalp, a pulse pickup is on the right index finger, a GSR electrode is on the left middle finger, and a thermistor is on the left index finger.

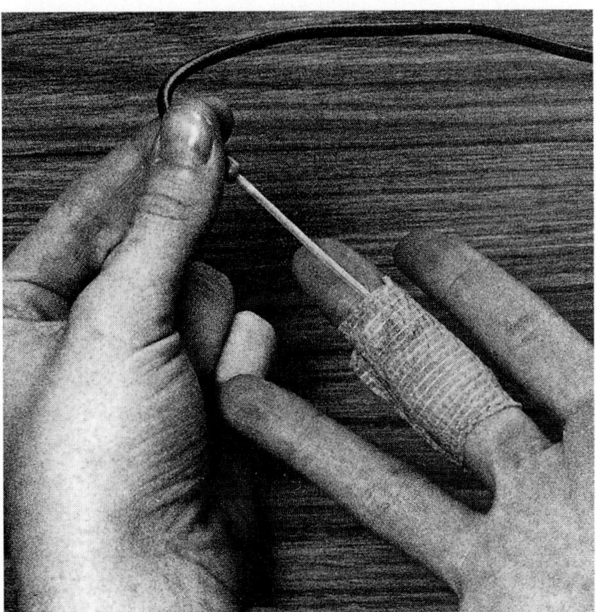

Figure 80.2 To get good thermistor-to-skin contact, wrap the finger with an adhesive wrapping, and then slip the thermistor between the wrapping and the finger.

Adjust sensitivities where necessary to get the desired stylus deflection.

Now ask the subject to attempt the type of biofeedback he or she has been trained to perform. Mark the chart at the point of beginning. Watch the brain-wave pattern to note any changes. If the subject is blindfolded, you may wish to quietly inform him or her of any changes. Look for alpha waves and changes in the other three parameters. Record continuously as long as necessary to get results.

Part 14 The Digestive System

Although this part contains only four exercises, one of them—Exercise 81—is lengthy and covers considerable material. You should study the material on human anatomy before coming to the laboratory to dissect the cat and to study the teeth of skulls.

Exercise 82 examines factors that influence intestinal motility, such as smooth muscle fibers and the autonomic nervous system. This exercise is also equally applicable to studies of skeletal muscle physiology and the nervous system.

Exercise 83 is an in-depth study of the three basic hydrolytic reactions that occur during digestion: carbohydrate, protein, and fat hydrolysis. After removing the pancreas from a rat, you extract and analyze the enzymes. Exercise 84 focuses on how environmental factors affect digestive enzymes.

81 Anatomy of the Digestive System

This exercise focuses on the alimentary canal, liver, and pancreas. Cat dissection provides an in-depth study of the various components.

The Alimentary Canal

Figure 81.1 is a simplified illustration of the digestive system. The alimentary canal, which is about 30 feet long, has been foreshortened in the intestinal area for clarity. Refer to this figure as you read how food passes through the alimentary canal's entire length.

Food taken into the oral cavity is chewed and mixed with saliva secreted by many salivary glands. The most prominent of these glands are the parotid, sublingual, and submandibular glands. The **parotid glands** are in the cheeks in front of the ears, one on each side of the head. The **sublingual glands** are

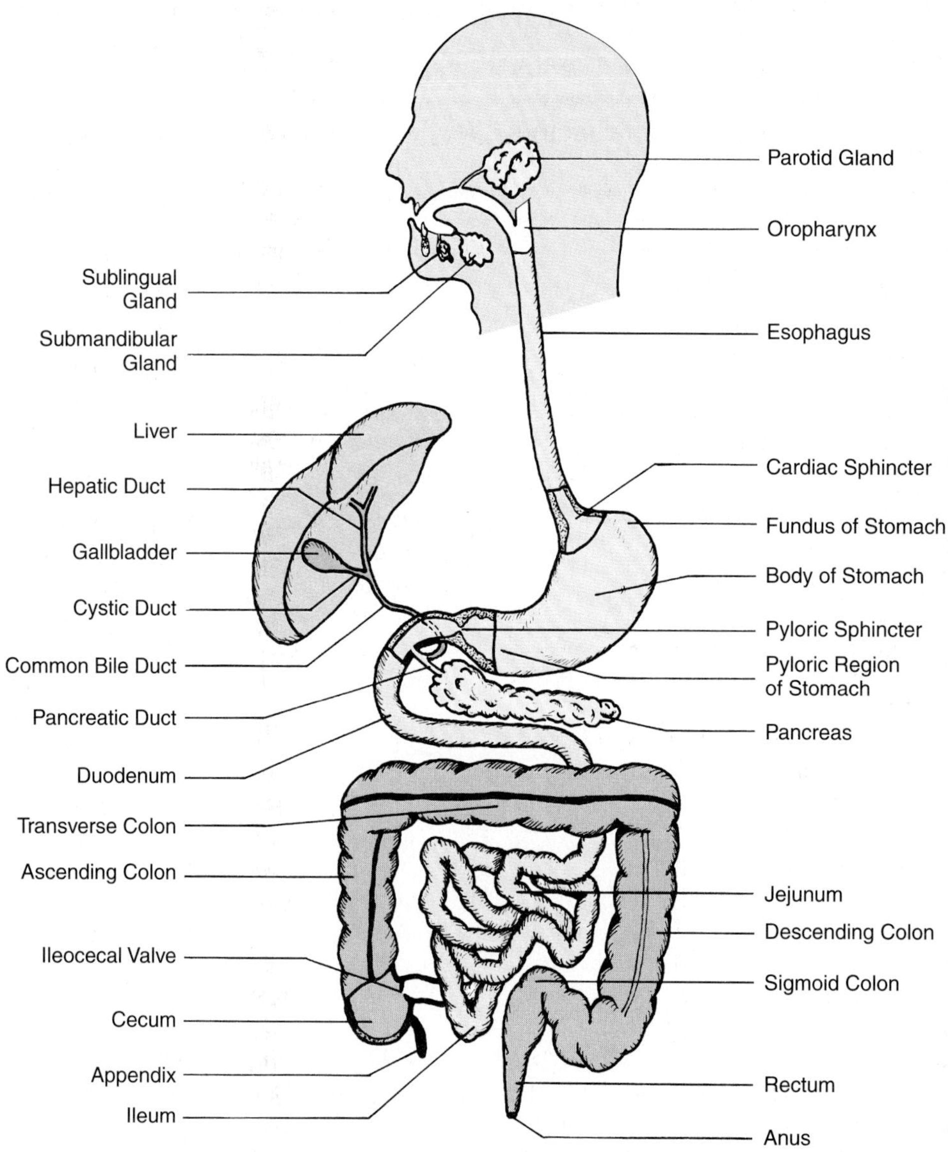

Figure 81.1 The alimentary canal.

under the tongue and are the most anterior of the three pairs of glands. The **submandibular glands** are posterior to the sublinguals, just inside the body of the mandible. (Figure 81.4 shows the position of these glands more precisely.)

After the food mixes with saliva, it passes to the **stomach** by way of a long tube, the **esophagus.** Wavelike constrictions called *peristaltic waves* move the food through the esophagus. Peristaltic waves originate in the **oropharynx,** which is the cavity at the top of the esophagus, posterior to the tongue. The upper opening of the stomach through which the food enters is the **cardiac sphincter.** The upper rounded portion, or **fundus,** of the stomach holds the bulk of the food to be digested. The lower portion, or **pyloric region,** is smaller in diameter, more active, and accomplishes most of the digestion that occurs in the stomach. That part of the stomach between the fundus and the pyloric region is the **body.** After various gastric enzymes have acted on the food, it enters the small intestine through the **pyloric sphincter** of the stomach.

The *small intestine* is approximately 23 feet long and consists of three parts: the duodenum, jejunum, and ileum. The first 10 to 12 inches is the **duodenum.** The **jejunum** comprises the next 7 or 8 feet. The last coiled portion is the **ileum.** Food is digested and absorbed in the small intestine.

Indigestible food and water pass from the ileum into the large intestine, or **colon,** through the **ileocecal valve.** Figure 81.1 shows this valve in a cutaway section. The large intestine has four sections: the ascending, transverse, descending, and sigmoid colons. The ileum empties into the **ascending colon.** At its lower end, the ascending colon has an enlarged compartment, or pouch, called the **cecum.** A narrow tube, the **appendix,** extends downward from the cecum.

The ascending colon rises on the right side of the abdomen until it reaches the undersurface of the liver, where it bends abruptly to the left, becoming the **transverse colon.** The **descending colon** passes down the left side of the abdomen and then changes direction to become the **sigmoid colon.** The last 5 inches or so of the alimentary canal is the **rectum,** which terminates in an opening, the **anus.**

Leading downward from the inferior surface of the liver is the **hepatic duct.** This duct joins the **cystic duct,** which connects with the round, saclike **gallbladder.** Bile, a secretion of the liver, passes down the hepatic duct and up the cystic duct to the gallbladder, which stores the bile until it is needed. Gallbladder contractions force bile down the cystic duct and into the **common bile duct** that extends from the juncture of the cystic and hepatic ducts to the intestine.

Between the duodenum and the stomach lies another gland, the **pancreas.** Its duct, the **pancreatic duct,** joins the common bile duct and empties into the duodenum.

Assignment:
Disassemble a laboratory manikin, and identify all of the structures discussed in the preceding section. Then complete part A of the Laboratory Report for this exercise.

Oral Anatomy

Figure 81.2 shows a typical normal mouth. To maximally expose the oral structures, the lips (*labia*) have been retracted away from the teeth, and the cheeks (*buccae*) have been cut.

The lips are flexible folds that meet laterally at the angle of the mouth, where they are continuous with the cheeks. Mucous membrane, the *mucosa,* lines the lips, cheeks, and other oral surfaces. Near the median line of the mouth on the inner surface of the lips, the mucosa thickens to form folds, the **labial frenula,** or *frena.* Of the two frenula, the upper one is usually stronger. The delicate mucosa that covers the neck of each tooth is the **gingiva.**

Figure 81.2 shows the hard and soft palates of the mouth. The **hard palate** is adjacent to the teeth. The **soft palate** is posterior to the hard palate. The **uvula,** a fingerlike projection above the back of the tongue, is a part of the soft palate. It varies considerably in size and shape in different individuals.

The **palatine tonsils** are on each side of the tongue at the back of the mouth. The recess in which each tonsil lies is bounded by two membranes: the **glossopalatine arch** anteriorly and the **pharyngopalatine arch** posteriorly.

The tonsils consist of lymphoid tissue covered by epithelium. The epithelial covering dips into the lymphoid tissue, forming glandlike pits called **tonsillar crypts** (see figure 81.3). These crypts connect with channels that course through the tonsil's lymphoid tissue. Infection impairs tonsils' protective function, and tonsils may actually become foci of infection. Inflammation of the palatine tonsils is *tonsillitis.* Enlargement of the tonsils obstructs the throat cavity and interferes with air passage to the lungs.

The Tongue

Figure 81.3 shows the tongue and adjacent structures. The tongue is a mobile mass of striated muscle completely covered with mucous membrane. It has three parts: apex, body, and root. The **apex** of the tongue is the most anterior tip that rests against

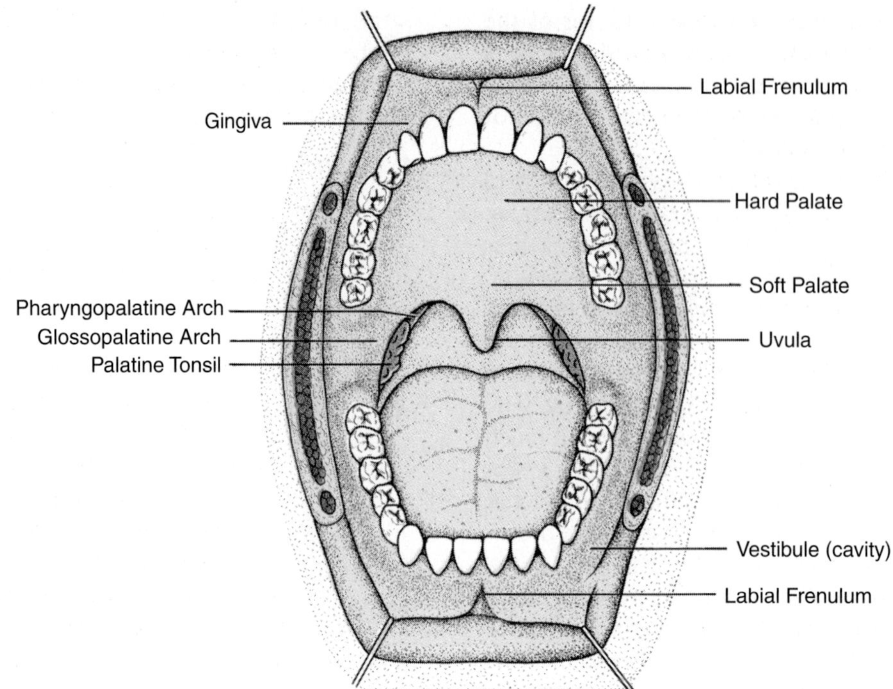

Figure 81.2 The oral cavity.

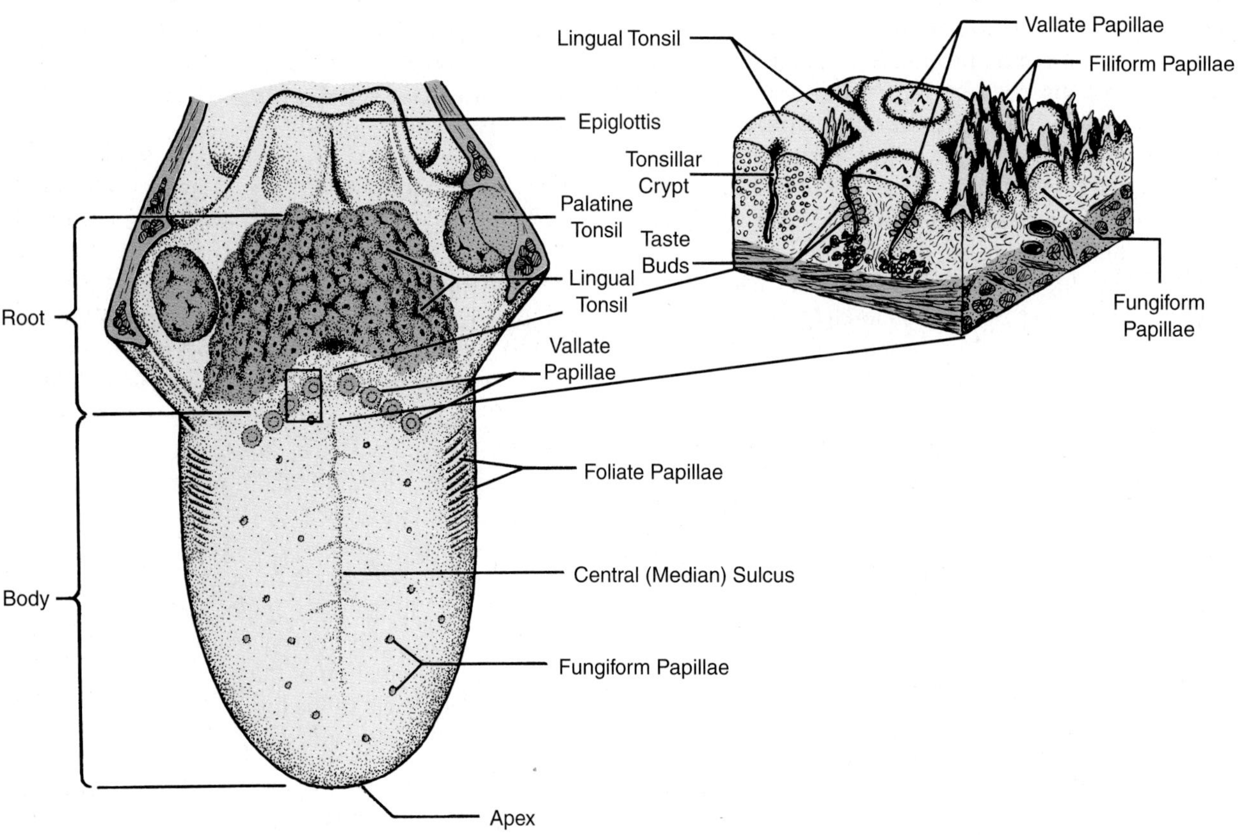

Figure 81.3 Tongue anatomy.

the inside surfaces of the anterior teeth. The **body** (*corpus*) is the bulk of the tongue that extends posteriorly from the apex to the root and is visible by simple inspection—that is, without the aid of a mirror. The posterior border of the body is arbitrarily located somewhere anterior to the tonsillar material of the organ. A groove, the **central** (*median*) **sulcus,** extends down the median line of the body. The **root** of the tongue is the most posterior portion. The **lingual tonsil** covers most of the root's surface.

The **epiglottis** extends upward from the posterior margin of the root of the tongue. The oval **palatine tonsils** are on each side of the tongue.

Several kinds of projections called *papillae* cover the dorsum of the tongue. The cutout section in figure 81.3 is an enlarged portion of the tongue's surface that shows the anatomical differences among these papillae. Most of the projections have tapered points and are called **filiform papillae.** These projections are sensitive to touch and give the dorsum a rough texture that provides friction for food handling. Scattered among the filiform papillae are the larger, rounded **fungiform papillae.** A third type of papilla is the large, donut-shaped **vallate** (*circumvallate*) **papilla.** These papillae are arranged in a "V" near the posterior margin of the dorsum.

The fungiform and vallate papillae contain **taste buds,** the taste receptors. Taste buds cluster on the walls of the circular furrow surrounding each vallate papilla. The vallate papillae contain many more taste buds than the fungiform papillae.

A fourth type of papilla is the **foliate papilla.** These projections form vertical rows of folds of mucosa on each side of the tongue, posteriorly. A few taste buds are also scattered among these papillae.

The Salivary Glands

The salivary glands are grouped according to size. The three pairs of large glands, called the *major salivary glands,* are the parotids, submandibulars, and sublinguals. The smaller glands, which average only 2 to 5 mm in diameter, are the *minor salivary glands.*

Major Salivary Glands

Figure 81.4 illustrates the relative positions of the three major salivary glands on the left side of the face. All of these glands are paired: the right side of the face has a corresponding set.

The **parotid glands** are the largest major salivary glands. They lie between the skin layer of the cheek in front of the ear and the *masseter muscle.* The **parotid duct** passes over the masseter and through the *buccinator muscle* into the oral cavity. This duct usually drains near the upper second molar.

The parotid glands secrete a clear, watery fluid that has a cleansing action in the mouth. It contains the digestive enzyme *salivary amylase,* which splits starch molecules into disaccharides (double sugars). Secretions from this gland increase when sour (acid) substances are in the mouth.

Inside the arch of the mandible lie the **submandibular** (*submaxillary*) **glands.** In figure 81.4, the mandible has been cut away to reveal two cut surfaces. Note that the lower margin of the left submandibular gland extends down somewhat below the inferior border of the mandible. Also note that a flat muscle, the *mylohyoid,* extends somewhat into the gland. The submandibular glands empty into the oral cavity through the **submandibular duct.** The **opening of the submandibular duct** is under the tongue near the lingual frenulum. The **lingual frenulum** is a mucosal fold on the median line between the tongue and the floor of the mouth. Figure 81.4 shows it as a triangular membrane.

The submandibulars secrete a fluid that is similar in consistency to that of the parotid glands but slightly more viscous due to the presence of *mucin.* Mucin helps hold food together in a bolus. Bland substances such as bread and milk stimulate the submandibular glands.

The **sublingual glands** are the smallest of the three pairs of major salivary glands. They are encased in a fold of mucosa, the **sublingual fold,** under the tongue in the floor of the mouth (see figure 81.5). The sublingual glands drain through several **lesser sublingual ducts**. The number of openings varies in different people. The sublingual glands differ from the other two pairs of major salivary glands in that they primarily secrete mucus. Its high mucin content makes it somewhat ropy.

Minor Salivary Glands

The four principal groups of minor salivary glands in the mouth are the palatine, lingual, buccal, and labial glands. In figure 81.5A, the palatine mucosa has been partially removed to reveal the closely packed nature of the **palatine glands.** These glands, which are 2 to 4 mm in diameter, almost completely cover the roof of the oral cavity. Removal of the mucosa near the lower third molar in figure 81.5A shows that the glands even occur next to the teeth. The glands in this particular area are called **retromolar glands.**

Figure 81.5B shows the oral cavity with the tongue pointed upward and the inferior mucosal surface of the tongue partially removed. The gland

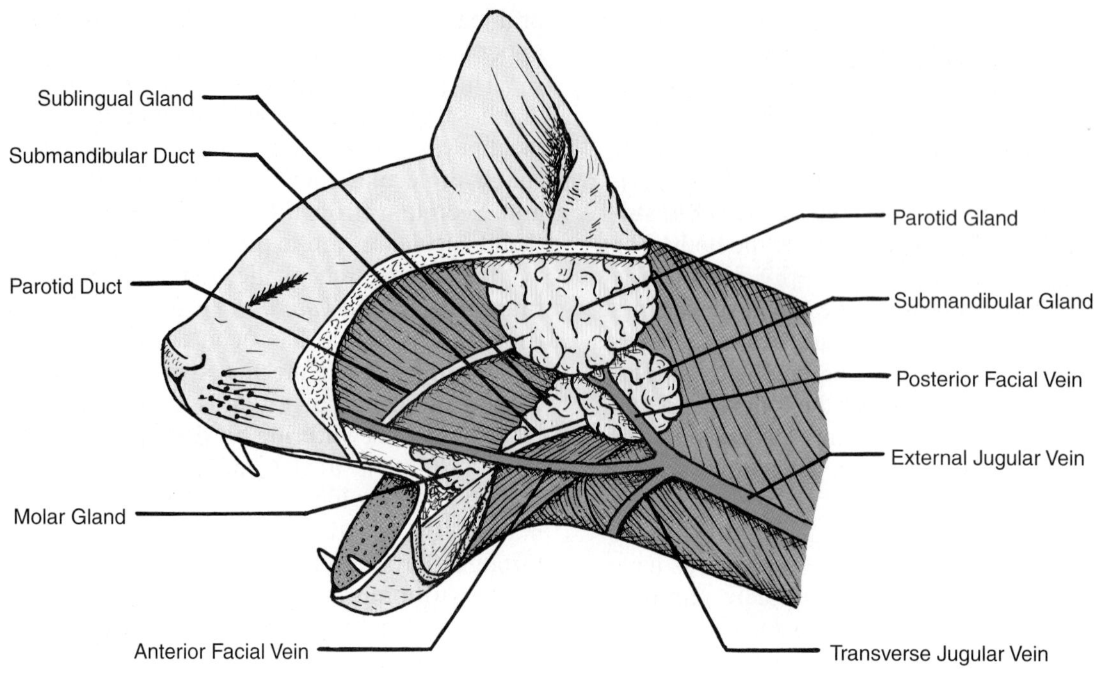

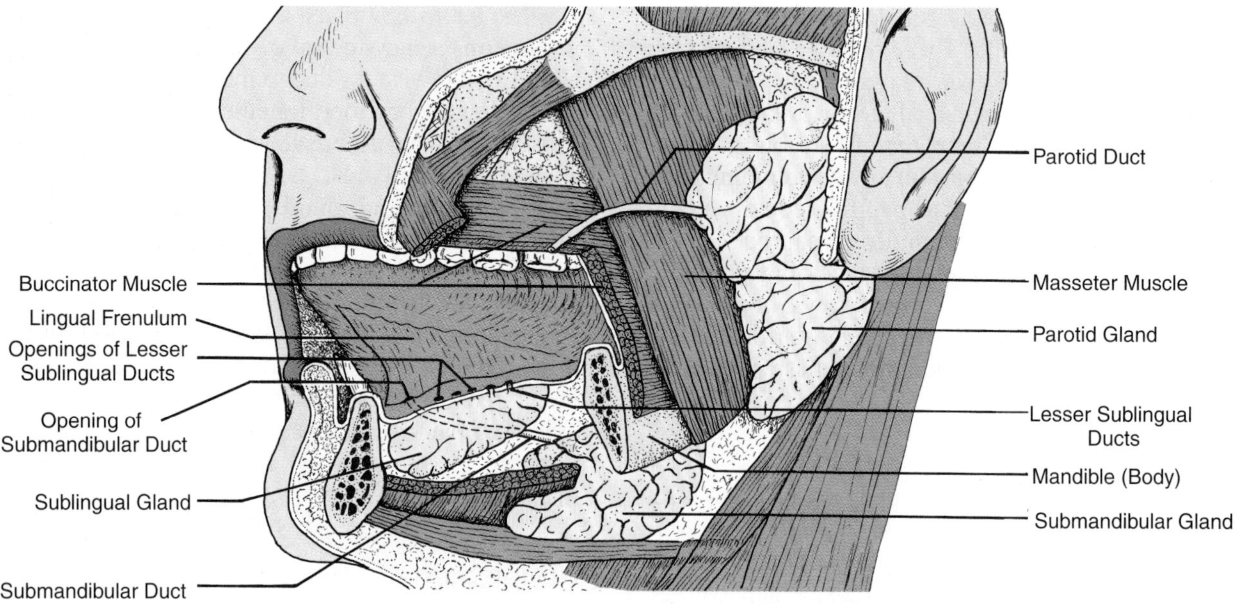

Figure 81.4 Major salivary glands, cat and human.

exposed in this area of the tongue is the **anterior lingual gland.**

The **buccal glands** cover most of the inner surface of the cheeks, and the **labial glands** are under the inner surface of the lips. All of the minor glands except for the palatine glands produce salivary amylase.

Assignment:
Complete part B of the Laboratory Report for this exercise.

The Teeth

Everyone develops two sets of teeth during the first 21 years of life. The first set, which begins to appear at approximately 6 months of age, are the *deciduous teeth,* also known as *primary, baby,* or *milk teeth.* The second set of teeth, known as *permanent* or *secondary teeth,* begins to appear when a child is about 6 years old. As a result of normal growth from the sixth year on, the jaws enlarge, the permanent teeth begin to exert pressure on the

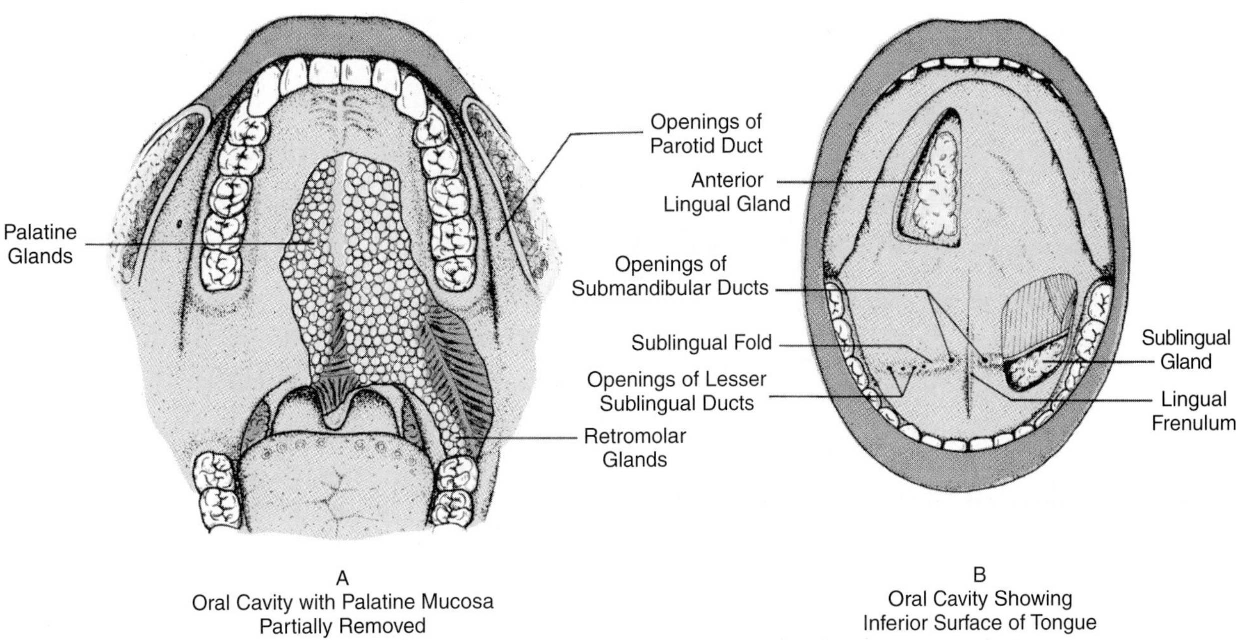

Figure 81.5 Minor salivary glands.

deciduous teeth, and the deciduous teeth are *exfoliated,* or shed.

The Deciduous Teeth

The deciduous teeth number 20 in all—5 in each quadrant of the jaws. Normally, all 20 have erupted by age two. Starting with the first tooth (either upper or lower) at the median line, they are named as follows: **central incisor, lateral incisor, cuspid** (*canine*), **first molar,** and **second molar.** The deciduous teeth do not visibly change until around age six. At this time, the first permanent teeth begin to erupt.

The Permanent Teeth

From approximately ages seven to twelve, the child has a *mixed dentition,* consisting of both deciduous and permanent teeth. As the submerged permanent teeth enlarge in the tissues, the roots of the deciduous teeth are resorbed. This removal of the underpinnings of the deciduous teeth results eventually in exfoliation.

Figure 81.6 illustrates the permanent dentition. Note that each half of the mouth has 16 teeth, for a total of 32 teeth. In sequence from the median line of the mouth, they are the **central incisor, lateral incisor, cuspid** (*canine*), **first bicuspid** (premolar), **second bicuspid** (premolar), **first molar, second molar,** and **third molar.** The third molar is also called the *wisdom tooth.*

Comparisons of the teeth in figure 81.6 reveal that the anterior teeth have single roots and the posterior ones have several roots. Of the bicuspids, only the maxillary first bicuspids have two roots (*bifurcated*). Although figure 81.6 shows all of the maxillary molars with three roots (*trifurcated*), the number of roots can vary considerably, particularly with respect to the third molars. The roots of the mandibular molars are generally bifurcated. The maxillary cuspids have the longest roots.

Assignment:
Identify the teeth of skulls that are available in the laboratory.

Tooth Anatomy

Figure 81.7 shows the anatomy of an individual tooth. Longitudinally, it divides into two portions, the **crown** and the **root.** The **cervical line,** or *cemento-enamel juncture,* is where these two parts meet.

A dentist sees the crown from two aspects: the anatomical and clinical crowns. The **anatomical crown** is the portion of the tooth that is covered with enamel. The **clinical crown,** is that portion of the crown that is exposed in the mouth. The structure and physical condition of the soft tissues around the neck of the tooth determine the size of the clinical crown.

The tooth is composed of four tissues: enamel, dentin, cementum, and pulp. **Enamel** is the most densely mineralized and hardest material in the body. Ninety-six percent of enamel is mineral. The remaining 4% is a carbohydrate-protein complex. Calcium and phosphorous make up

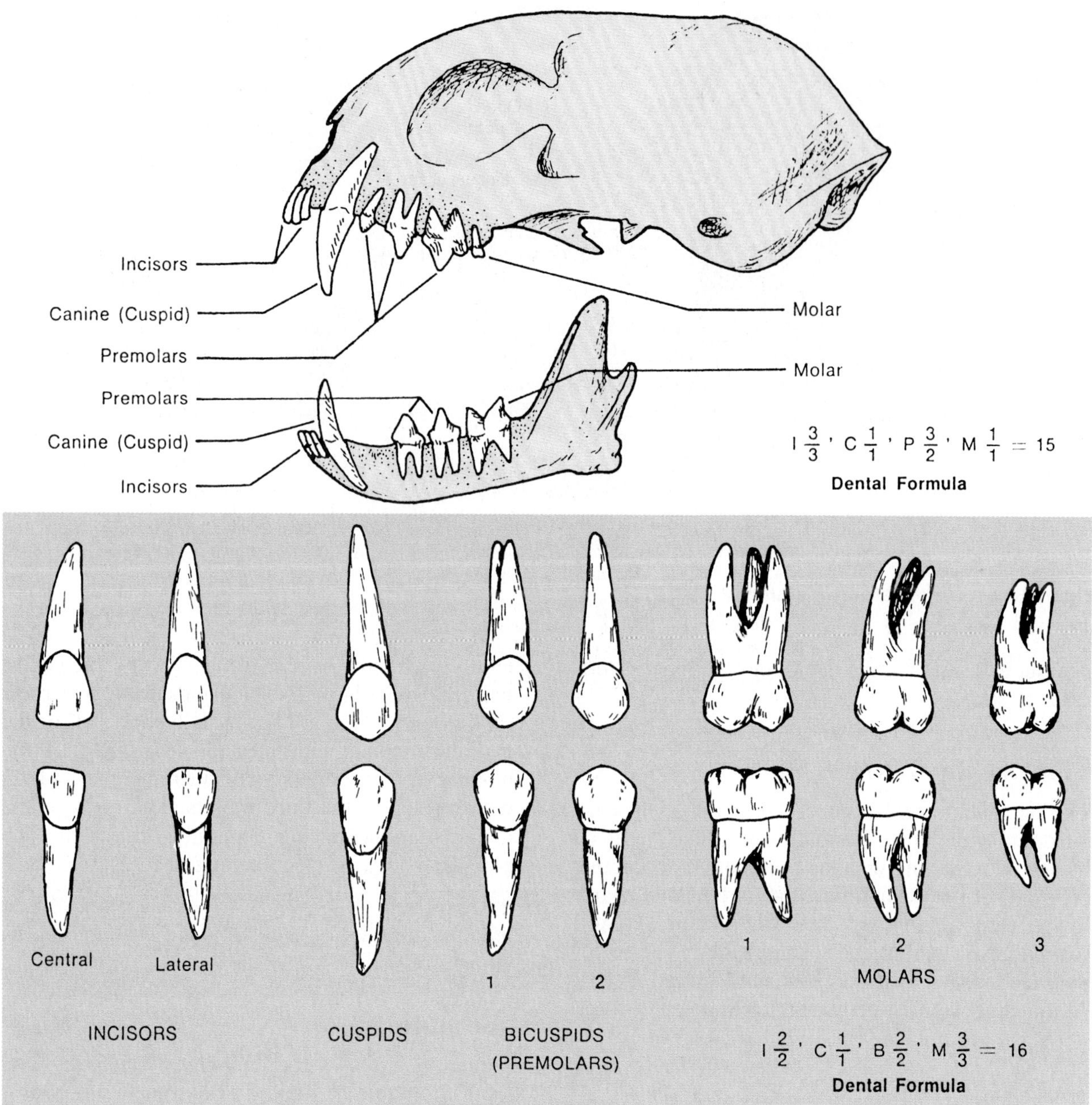

Figure 81.6 The permanent teeth, cat (*top*) and human (*bottom*).

over 50% of enamel's chemical structure. Microscopically, enamel consists of very fine rods or prisms that lie approximately perpendicular to the crown's outer surface. The upper molars may have as many as 12 million of these small prisms per tooth. The hardness of enamel enables the tooth to withstand the abrasive action of one tooth against the other.

Dentin lies beneath the enamel and makes up the bulk of the tooth. It is not as hard or brittle as enamel, and it resembles bone in composition and hardness. A layer of **odontoblasts** in the outer margin of the pulp cavity produces dentin.

Cementum is the hard dental tissue that covers the tooth's anatomical root. It is a modified bone tissue, somewhat resembling osseous tissue. Cells called *cementocytes,* which are very similar to osteocytes, produce cementum. Cementum attaches the tooth to surrounding tissues in the alveolus.

The cavity in the center of the tooth is called the **pulp cavity.** Where it extends down into the roots, this cavity becomes the **root canals.** Where it extends up into the cusps of the crown, it forms the **pulpal horns.**

The **pulp** is essentially loose connective tissue, consisting of fibroblasts, intercellular ground sub-

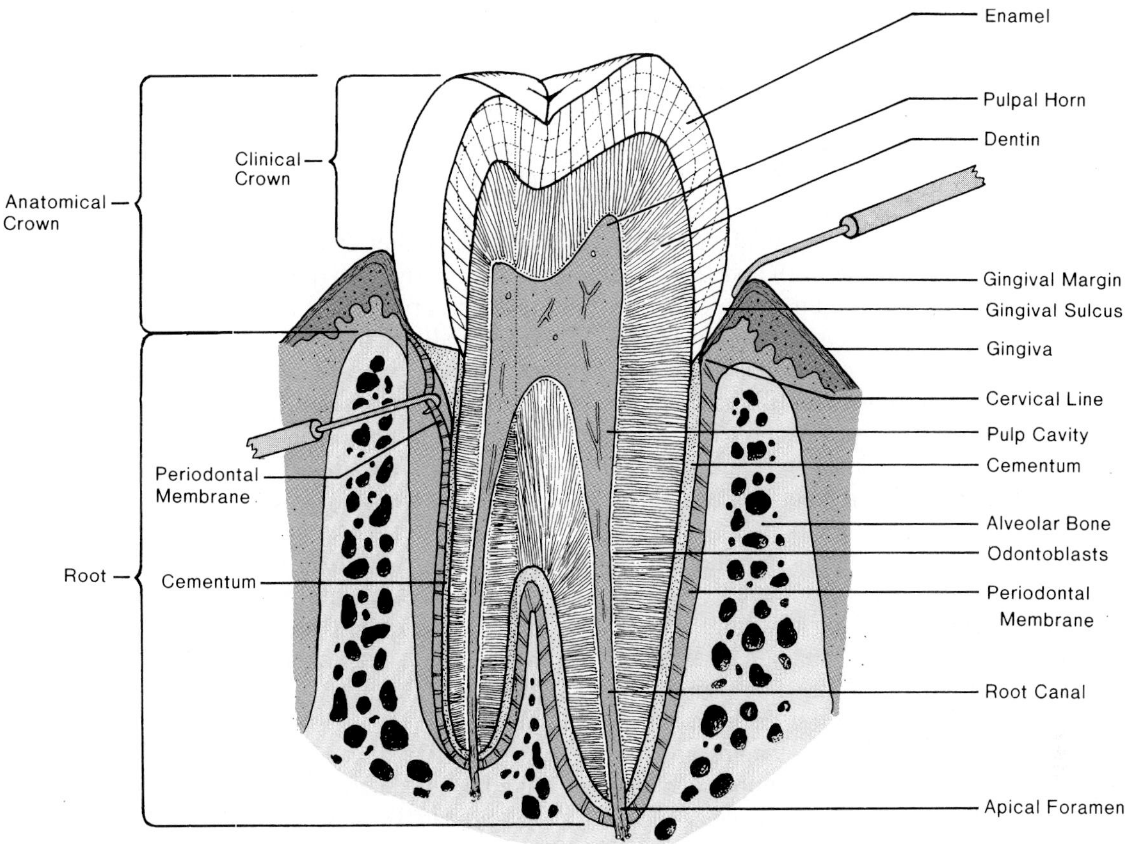

Figure 81.7 Tooth anatomy.

stance, and white fibers. Permeating this tissue are blood vessels, lymphatic vessels, and nerve fibers.

The pulp provides nourishment for the living cells of the tooth, supplies some sensation to the tooth, and produces dentin. Blood vessels and nerve fibers enter the pulp cavity through openings, the **apical foramina,** in the tips of the roots. Inflammation of the pulp, or *pulpitis,* may destroy the blood vessels and nerves, producing a dead or *devitalized tooth.*

The **periodontal membrane** is a vascular layer of connective tissue that consists of bundles of fibers that extend from the cementum to the alveolar bone. The left side of figure 81.7 shows it pulled away from the root. The periodontal membrane firmly holds the tooth in place. It also acts as a cushion to reduce the trauma of occlusal action. The membrane provides sensitivity to touch, due to the presence of free nerve ending receptors. The slightest touch at the surface of the tooth is transmitted to these nerve endings through the periodontal membrane. Even removing apical portions of the membrane, as in root tip resection, does not impair the sense of touch.

A probe on the right side of figure 81.7 demonstrates the depth of the **gingival sulcus.** This crevice is frequently a site for bacterial putrefaction, which is the origin of *gingivitis.* The upper edge of the gingiva is the **gingival margin.** The portion of the gingiva that lies over the alveolar bone is the *alveolar mucosa.*

Assignment:
Complete part C of the Laboratory Report for this exercise.

Cat Dissection

Since the anatomy of the circulatory, respiratory, and digestive systems of the cat cannot be studied independently of each other, this dissection will likely be performed simultaneously with the study of the other two systems. Because of these interrelationships, some of the discussion of digestive anatomical structures may be repetitive.

Oral Cavity

If you have already done Exercise 71, you have identified the *oral cavity, nasal cavity, nasopharynx, oropharynx, soft palate, hard palate, palatine tonsils, auditory tube aperture, esophagus, larynx,* and *trachea.* If you have not already studied these organs, refer to page 355 to identify them. Some specimens need to be cut along the median line with a bone saw, as described on page 356, to reveal some of these structures.

In addition to the structures just mentioned, identify the transverse ridges, or **rugae,** of the hard palate that assist in holding food in the mouth during swallowing. Also identify the **oral vestibule,** which is the space between the teeth and lips.

Lift up the tongue, and examine its inferior surface. Do you see any evidence of a **lingual frenulum?** The cat tongue's dorsal surface has distinct **filiform** and **fungiform papillae.** The filiform papillae are spinelike and more numerous than the fungiform papillae. In addition to being tactile receptors, the filiform papillae also act as scrapers for removing meat from bones and as brushes for grooming. Can you locate the **vallate** (*circumvallate*) **papillae** at the back of the tongue? How many are there?

Examine the teeth, referring to figure 81.6 to identify them. How does the number of **incisors** compare with the human? Note how small the incisors are. The important anterior teeth on the cat are the **canines.** They kill small prey and tear away portions of food. The cat has three upper and two lower **premolars.** Observe the insignificance of the upper **molar,** compared to the lower molar. Since cats are carnivores and swallow chunks of meat without chewing, the posterior teeth act as shears rather than grinders. The cat's posterior teeth do not have any flat occlusal surfaces.

Salivary Glands

Remove the skin and platysma muscle from the left side of the head if this has not already been done. Remove any lymph nodes that obscure the anterior facial vein.

By referring to figure 81.4, identify the **parotid, submandibular,** and **sublingual glands.** What gland do cats have that humans do not? Can you locate the **parotid** and **submandibular ducts?** A fifth gland, the **infraorbital gland,** is in the floor of each eye orbit but cannot be seen unless the eye is removed.

Esophagus

Force a probe into the upper end of the esophagus, and stretch its walls laterally. Note how much it can be distended. Trace the esophagus down to where it penetrates the diaphragm and liver to enter the stomach.

Abdominal Organs

If you have already studied the circulatory system, the abdominal cavity will be open for the remainder of this study. If the abdomen has not been opened yet, make an incision with scissors along the median line from the thoracic region to the symphysis pubis. Also, make additional cuts laterally from the median line in the pubic region so that you can pull the muscle of the abdominal wall aside to expose all the organs.

Removing the abdominal wall reveals the liver and the greater omentum. The **greater omentum** is a double sheet of peritoneum that attaches to the greater curvature of the stomach and to the dorsal body wall. Figure 81.8 only shows a small segment of it because most has been cut away to expose the intestines underneath.

Lift up the greater omentum, and examine both surfaces. Note that fat fills the potential space (*omental bursa*) between its two layers of peritoneum. Cut a slit into the surface of the greater omentum to confirm its structure. That portion of the greater omentum between the stomach and spleen is called the **gastrosplenic ligament.** Identify this ligament, as well as the **spleen** and **stomach.** Cut away and discard most of the greater omentum.

Identify the right medial, right lateral, left medial, left lateral, and cavdate **lobes of the liver.** Also locate the soft, greenish **gallbladder.** Refer to figure 81.9 to identify the falciform and round ligaments. The **falciform ligament** is a sickle-shaped (Latin: *falx, falcis,* sickle), double-layered peritoneal fold that lies in the notch between the right and left medial lobes of the liver. Spread these lobes, and lift out the falciform ligament with forceps. Note that its left margin attaches to the diaphragm. Its free margin contains a thickened structure, the **round ligament.** The round ligament is a vestige of the umbilical vein that functioned during prenatal existence.

Identify the **lesser omentum** that extends from the liver to the lesser curvature of the stomach and a portion of the duodenum (see figure 81.8). The common bile duct, hepatic artery, and portal vein lie within the lateral border of the lesser omentum. Dissect this region to locate these structures.

Probe into the space under the right medial lobe of the liver, and locate the **cystic duct** that leads from the gallbladder into the **common bile duct.**

Raise the liver sufficiently to see where the esophagus enters the stomach. Identify the **fundus, body,** and **pylorus** of the stomach (see figure 81.8). The **cardiac portion** is where the esophagus joins the stomach. Open the stomach by making an incision along its long axis. Observe that the stomach lining has long folds called **rugae.**

Identify the **duodenum** (see figure 81.8). Note that it passes caudally about 3 inches and then doubles back on itself for another 3 to 4 inches. Within this duodenal loop lies the duodenal portion of the

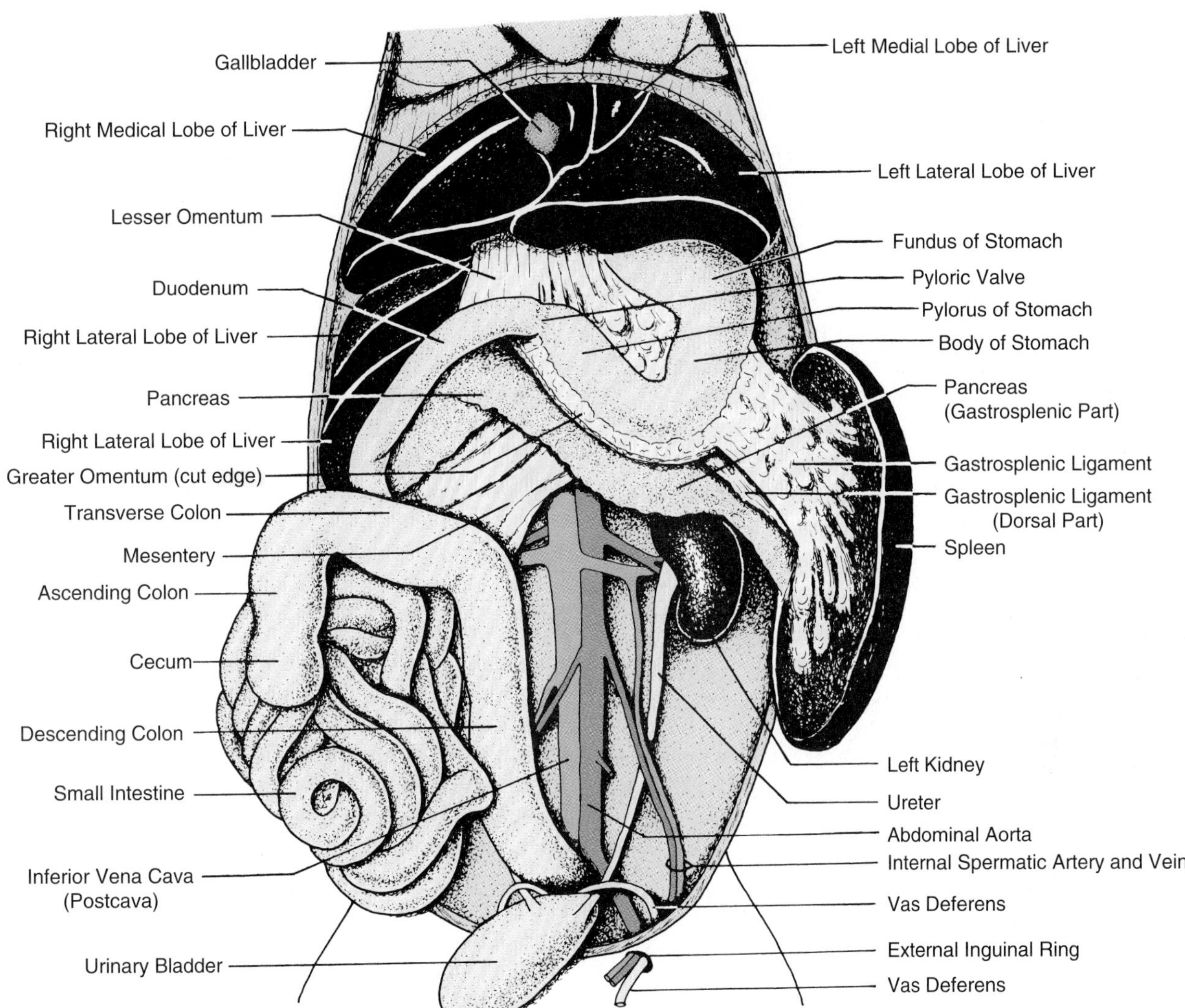

Figure 81.8 Abdominal viscera of the cat.

pancreas. The pancreas is quite long in that it extends from the duodenum to the spleen. The mesentery that supports the duodenum and pancreas in this region is called the *mesoduodenum.*

Expose the **pancreatic duct** by dissecting away some of the pancreatic tissue. This duct lies embedded in the gland. It joins the common bile duct to form a short duct, the **ampulla of Vater,** that empties into the duodenum. A small **accessory pancreatic duct** that enters the duodenum about 2 cm caudad to the ampulla of Vater is often confused with small arteries in this region.

The remainder of the small intestine is divided into a proximal half, the **jejunum,** and a distal half, the **ileum.** Trace the ileum to where it enters the cecum.

Remove a 2-inch section of the ileum or jejunum, slit it open longitudinally, and wash out its interior. Examine its inner lining with a dissecting microscope or hand lens, looking for the minute tubular projections called **villi.**

The large intestine consists of the **cecum, ascending colon, transverse colon,** and **descending colon** (see figure 81.8). The descending colon empties into the **rectum** and exits through the **anus.** Identify all of these structures. Make an incision through the distal end of the ileum and the cecum to reveal the **ileocecal valve.**

Histological Studies

Make a systematic histological study of the organs of the digestive system, using the Histology Atlas for reference.

Materials:
Prepared slides of:
salivary glands
taste buds
tooth embryology
esophagus
continued on next page

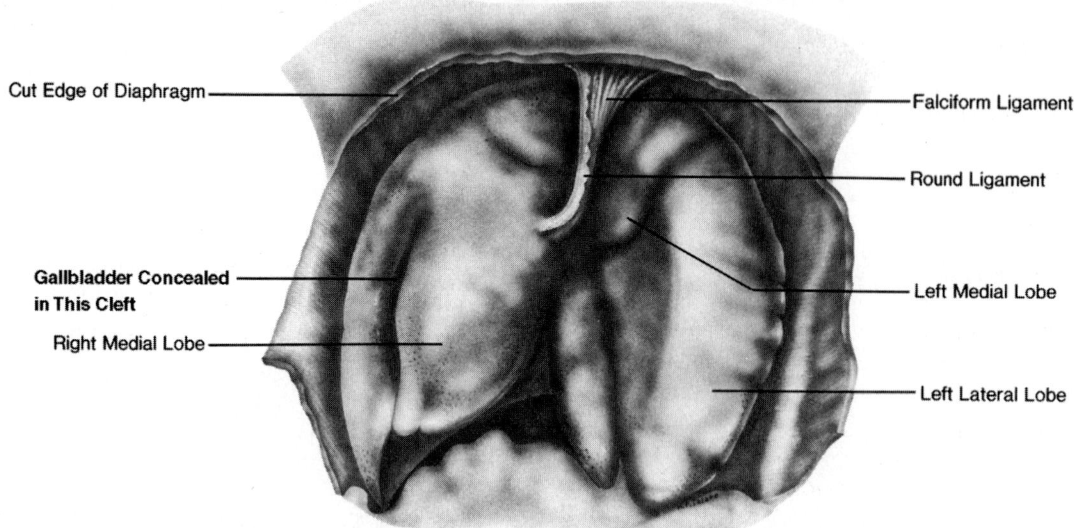

Figure 81.9 Anatomy of the cat liver, anterior view.

Materials (continued)
stomach
duodenum, jejunum, and ileum
colon
appendix
anal canal

Salivary Glands Examine a slide of the three major salivary glands, using figure HA–18 in the Histology Atlas to note the histological differences between them.

Taste Buds Examine slides of vallate or foliate papillae. Identify the structures shown on figure HA–19 in the Histology Atlas.

Tooth Embryology Read the legend for figure HA–20 in the Histology Atlas, and examine slides to identify the structures shown in figure HA–20. In particular, note what cells give rise to dentin and enamel.

Esophagus Identify the four coats that make up the wall of the esophagus. In particular, note the type of epithelial tissue that lines the esophagus. Use figure HA–21 in the Histology Atlas for reference.

Stomach Wall Examine a slide of the fundus of the stomach, and compare it to figures HA–21C and D in the Histology Atlas. Study the deep cells within the pits, and differentiate the **parietal** and **chief cells.** How do these cells differ in function? How does the mucosa of the stomach differ from that of the esophagus?

Small Intestine Study a slide that has all three sections of the small intestine. Use figures HA–22 and HA–23 in the Histology Atlas for reference. Note the differences between the duodenal and intestinal glands. How do these glands differ in function? What is the function of the **lacteal** in a villus? Look for lymphoidal tissue (**Peyer's patches**) in the ileum (see figure HA–27D in the Histology Atlas). This tissue constitutes the gut-associated lymphoid tissue (GALT) of the digestive tract, an important part of the immune system.

Large Intestine Examine a slide of the colon, and refer to figures HA–24A and B in the Histology Atlas. What is lacking on the mucosa of the colon that is seen in the small intestine? Identify all structures.

Appendix Refer to figure HA–24C in the Histology Atlas while studying a slide of the appendix. Compare the structure of the appendix with that of the colon. What function would lymph nodules perform here?

Liver and Pancreas While examining slides of the liver and pancreas, read the legend for figure HA–25 in the Histology Atlas, which explains the significance of the labeled structures.

Gallbladder Refer to figures HA–26A and B in the Histology Atlas while studying a slide of the gallbladder. How does the mucosa of the gallbladder resemble the small intestine? How does it differ?

Anal Canal Determine what kind of tissue makes up the mucosa in the anal canal. Consult figure HA–24D in the Histology Atlas.

Assignment:
Complete the Laboratory Report for this exercise by drawing your microscopic observations in the space provided.

Intestinal Motility

82

The smooth muscle fibers of the intestinal walls cause a variety of movements that greatly facilitate food digestion and absorption. The principal movements are *peristaltic, segmenting,* and *pendular.* In addition, the muscularis mucosa exhibits *rippling movements,* and the microscopic villi have *pumping action.*

Segmenting contractions occur rhythmically at approximately 17 per minute in the duodenal area and 10 per minute in the ileal region. Intestinal motility appears to be related to the metabolic rate, since experimentally induced temperature variations greatly affect motility. Neural elements in the wall of the digestive tract are primarily responsible for regulation, however.

In this exercise, a small segment of rat intestine is mounted on a Combelectrode. Intestinal movements are monitored as the ambient temperature is changed, as the tissue is electrically stimulated, and as the hormones epinephrine and acetylcholine are administered. Figure 82.1 illustrates the setup.

You will work in four-member teams. While two team members help with dissection and tissue preparation, the other two balance the transducer and set up the equipment. Procedures are provided for both Gilson and Narco equipment.

Equipment Setup

While other team members are dissecting the rat and preparing the tissue, set up the equipment as follows:

Materials:

for Unigraph setup:
Unigraph
electronic stimulator (Grass SD–9)
force transducer (Trans-Med Biocom #1030)
event synchronization cable

for Physiograph setup:
Physiograph with transducer coupler
stimulator (Narco SM–1)
event marker cable
myograph (Narco)
steel rod for myograph

for both setups:
ring stand
ring and wire gauze
continued on next column

Materials, continued
alcohol lamp, or small Bunsen burner
Harvard isometric clamp
two double clamps
oxygen supply, or air pump

1. Set up a ring stand, isometric clamp, and transducer as figure 82.1 shows. To mount a Narco myograph to the Harvard clamp, refer to figure 31.3.
2. Place a wire gauze on the ring, and adjust the ring to the height that accommodates an alcohol burner or small Bunsen burner.
3. Plug the transducer into the recorder.
4. Plug in the power cords of the stimulator and recorder.

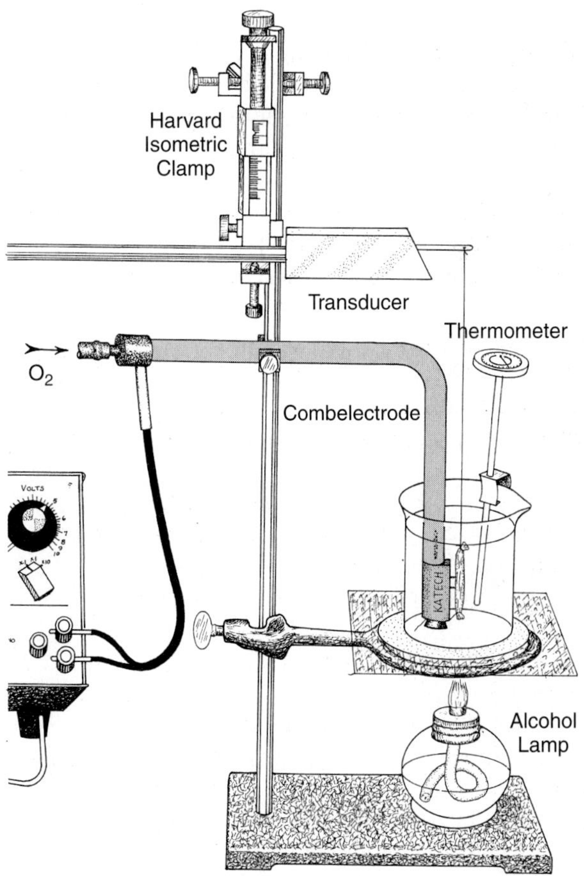

Figure 82.1 Instrumentation setup.

5. Connect the event marker (synchronization) cable between the stimulator and recorder.
6. Balance the transducer (myograph) according to the instructions in Appendix C.
7. Calibrate the recorder according to the instructions in Appendix C.

You are now ready to accept the Combelectrode with mounted rat intestine tissue that other team members have prepared.

Anesthetization of Rat

One rat will provide enough tissue for six or seven setups. To anesthetize the rat, one person must hold the animal while another person injects Nembutal into the abdominal cavity.

Materials:
rat, 200 g weight
Nembutal
1 cc syringe and needle

1. Draw 0.25 ml of Nembutal into the syringe.
2. Use your right hand to pull the rat from the cage by its tail. When you have the rat near the door of the cage, grasp the skin of the back of the neck firmly with your left hand.
3. Have your assistant inject the Nembutal into the peritoneal cavity about 0.5 cm below the sternum. The rat will become unconscious within 30 seconds.

Dissection and Tissue Preparation

As soon as the rat is unconscious, the abdomen must be opened to expose the intestines. Before sections of the intestines are excised, motility is observed. Sections of the intestine are then removed, placed in individual dishes of Tyrode's solution, and dispensed to individual teams for attachment to Combelectrodes. Proceed as follows:

Materials:
dissecting pan
dissecting instruments
Petri dishes
Tyrode's solution

1. Fill a beaker with 200 ml of Tyrode's solution. In addition, pour about 1/4 inch of Tyrode's solution into as many Petri dishes as needed to supply tissue to each team.
2. Open up the abdominal wall of the rat with scissors, exposing the intestines. Examine the intestines before disturbing them to observe the degree of motility.
3. Before cutting off lengths of the intestine, cut away the mesentery from that portion to be used.
4. Cut as many 2 cm lengths as needed. Place each section into a separate Petri dish of Tyrode's solution.

Final Hookup

Attach the intestinal tissue to a Combelectrode, and complete the hookup as figure 82.1 shows, using the following steps:

Materials:
thermometer
Combelectrode
medicine droppers
250 ml graduated beaker
thread

1. While the tissue is still in the Petri dish, flush out the lumen of the intestine with a medicine dropper and Tyrode's solution.
2. Tie an 18-inch length of thread to each end of the tissue section.
3. Attach one thread to the lower end of the Combelectrode, securing it with the stopper. Be sure the tissue completely covers the electrical contacts. Cut off excess thread.
4. Insert the end of the Combelectrode into a beaker of Tyrode's solution on the ring stand. Allow the free thread to hang over the edge of the beaker.
5. Secure the Combelectrode to the clamp on the ring stand.
6. Attach the free thread to the first two leaves of the transducer. Cut off excess thread. Take up any slack in the thread with the isometric clamp.
7. Plug the Combelectrode into the stimulator posts.
8. Connect the oxygen supply hose to the Combelectrode, and adjust the oxygen metering valve to produce 5 to 10 bubbles per second.
9. Attach the thermometer to the side of the beaker. You are now ready to begin the experiment.

Experiment Procedure

You will monitor intestinal contractions as temperature is gradually increased from 22°C (73°F), as the tissue is electrically stimulated, and as the tissue is exposed to epinephrine and acetylcholine.

Materials:
epinephrine (1:20,000)
acetylcholine (1:20,000)
1 cc syringe and needle

Effect of Temperature

Adjust the temperature of the Tyrode's solution in the beaker to 22°C (73°F). Turn on the recorder, and proceed as follows:

1. Set the speed at 2.5 mm/sec (0.25 cm/sec), and start the chart moving. Observe the trace.
2. If no contractions are evident, increase the gain by one step (2 mV/cm to 1 mV/cm) without altering the sensitivity knob on the Unigraph or the inner knob on the Physiograph. (*Changing the fine adjustment alters the calibration.*)

   If no contractions are evident at this setting, increase the gain to 0.5 mV/cm, which should produce contractions that are weak and quite slow. Record for about 2 minutes at this temperature. Mark the temperature on the paper.
3. Place an alcohol lamp or very small Bunsen burner under the beaker to warm the Tyrode's solution. An alcohol lamp can usually be left under the beaker without overheating; Bunsen burners require more attention.
4. Control the heat so that the temperature increases **at a rate of about 2°C per minute.** Record the temperature every minute on the paper.
5. Continue heating to 41°C. The normal body temperature of the rat is 38.1°C (100.5°F). A temperature of 42.5°C is lethal to the animal.
6. Note how the regularity and frequency of contractions increase with rising temperature. Do contractions maintain constancy of contraction, or do they cyclically strengthen and weaken? At what temperature is the frequency of contraction the greatest?
7. Record your results in part A of the Laboratory Report.

Effect of Electrical Stimulation

Allow the solution in the beaker to cool to around 30°C (90°F), and then stimulate the tissue with single pulses as follows:

1. Set the stimulator controls: duration (or width) control at 15 msec on the Grass SD9, or 2 msec on the Narco SM–1; voltage at 10 V; and the mode control at Off.
2. Turn on the power switch of the stimulator.
3. Depress the mode control to Single, and record "10 V" on the chart. Does the muscle respond? Repeat several times to see if the muscle responds at any particular point on the contraction cycle.
4. If you get no response at 10 V, increase the voltage by 10 V increments to see if the muscle responds.
5. Record your results in part B of the Laboratory Report.

Effects of Epinephrine and Acetylcholine

Nerve endings of the autonomic nervous system produce hormones that stimulate and inhibit contractions of smooth muscle in the intestinal wall. To simulate these conditions, you will expose the tissue to acetylcholine and epinephrine. The nerve endings of parasympathetic fibers produce acetylcholine. Epinephrine is similar, if not identical, to the action of norepinephrine of the sympathetic nervous system. Proceed as follows:

1. With the temperature in the beaker adjusted to 38°C, observe the contractions for about 1 minute, and then add 0.5 cc of epinephrine to the solution with a hypodermic syringe. On the chart, record the temperature and "epinephrine." Watch for recovery, noting how long it takes. Is the muscle stimulated or inhibited?
2. Lower the ring on the ring stand so that you can lift off the beaker. *Do not disturb the Combelectrode.* Replace the Tyrode's solution in the beaker with fresh solution at 38°C.
3. After contractions have stabilized, add 0.25 cc of acetylcholine to the beaker, and observe the effects.

Assignment:
Complete the Laboratory Report for this exercise.

83 The Chemistry of Hydrolysis

The basic nutrients of the body consist of small quantities of vitamins and minerals, plus large amounts of carbohydrates, fats, and proteins. While vitamins and minerals play a vital role in all physiological activities, the carbohydrates, fats, and proteins provide the energy and raw materials for growth and tissue repair.

Food consists of complex molecules of carbohydrates, fats, and proteins that are too large to pass through the intestinal wall. Digestive enzymes in the stomach and intestines split these large molecules into smaller ones by hydrolysis.

Hydrolysis is the process of splitting larger molecules into smaller molecules by combining with water. Although hydrolysis occurs automatically at body temperature, the process is extremely slow; the catalytic action of digestive enzymes greatly hastens the process.

Hydrolysis reduces carbohydrates to monosaccharides, fats to fatty acids and glycerol, and proteins to amino acids. Some food molecules are so large that different enzymes must work at different levels to produce intermediate-size molecules first.

The salivary glands, the stomach wall, the intestinal wall, and the pancreas produce digestive enzymes. Since the pancreas produces all three of the basic types of enzymes (proteases, lipases, and carbohydrases), the focus here is on this gland.

In this exercise, you analyze an enzyme extract from the pancreas of a freshly killed rat. Figure 83.1 shows the steps for extracting the pancreatic juice. Since enzymes are relatively unstable and easily denatured by certain adverse environmental conditions, chemical purity, cleanliness, and temperature control are critical.

Once the pancreatic juice has been extracted, you will test for evidence of carbohydrase, lipase, and protease activity. Small amounts of pancreatic extract will be added to substrates of starch, fat, and protein to observe the hydrolytic action of the various enzymes. Color changes in each test indicate the breaking of bonds within the large molecules to produce smaller molecules. Positive and negative test controls will be compared with the actual tests.

You will work in four-member teams. While two team members dissect the rat, the other two set up the necessary supplies and equipment for the extraction and assay procedures.

Extraction Procedure

Remove the pancreas from a freshly killed rat, and produce the pancreatic extract, using the following procedures:

Materials:
freshly killed rat
refrigerated centrifuge
centrifuge tubes
balance
Sorvall blender
beaker, 30 ml size
Erlenmeyer flask, 125 ml size
graduated cylinder, 100 ml size
crushed ice
cold mammalian Ringer's solution
dissecting pan (wax bottom)
dissecting instruments
dissecting pins
Parafilm

1. Follow the steps in figures 83.2 through 83.5 to open the abdominal cavity of a freshly killed rat and remove the pancreas.
2. Place the gland into a beaker containing a small amount (about 20 ml) of cold mammalian Ringer's solution (see figure 83.1).
3. Snip off a small piece of the intestine, and add it to the beaker.
4. After rinsing the tissues in the Ringer's solution, macerate and homogenize them in the blender (see figure 83.1). Blend the tissues of several teams together to get ample bulk.
5. Distribute the blended mixture into evenly balanced centrifuge tubes. Weigh the tubes carefully, and adjust the contents until the tubes are equal. Spin the tubes in a refrigerated centrifuge (Beckman or other) for 7 minutes at 10,000 rpm.
6. Decant the supernatant from the tubes into Erlenmeyer flasks—one for each team. The supernatant will be a milky pink mixture of digestive enzymes. Cap each flask with Parafilm.
7. Store the flasks of pancreatic extract in a container of ice to prevent enzyme deterioration.

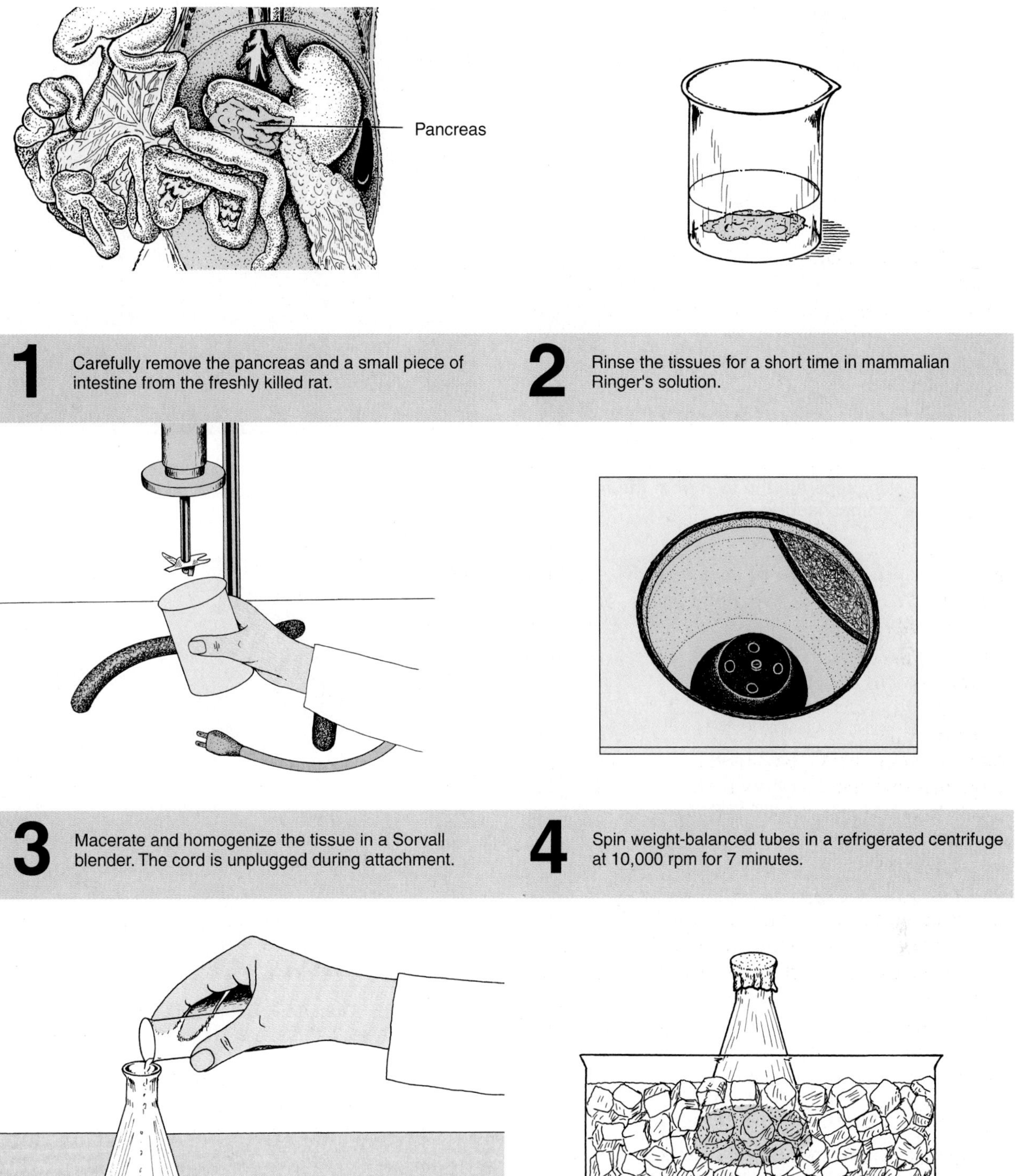

5 Dispense the supernatant in centrifuge tubes into Erlenmeyer flasks, and cap with Parafilm.

6 Keep the flask of pancreatic extract on ice until the extract is dispensed into assay tubes.

Figure 83.1 Procedure for extracting pancreatic enzymes.

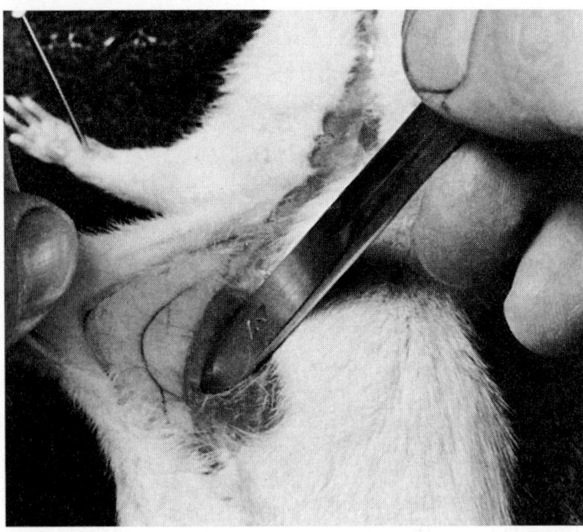

Figure 83.2 After pinning down the feet and cutting through the skin, separate the skin from the musculature with the scalpel handle.

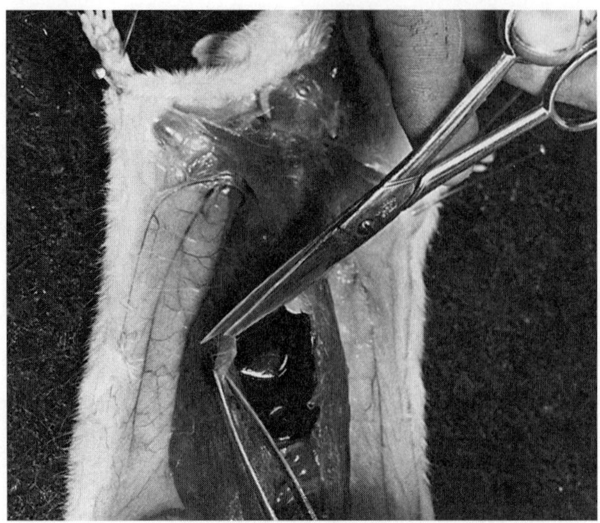

Figure 83.4 As the muscular body wall is opened, pull the flaps of muscle wall laterally, and pin them down to keep the cavity open.

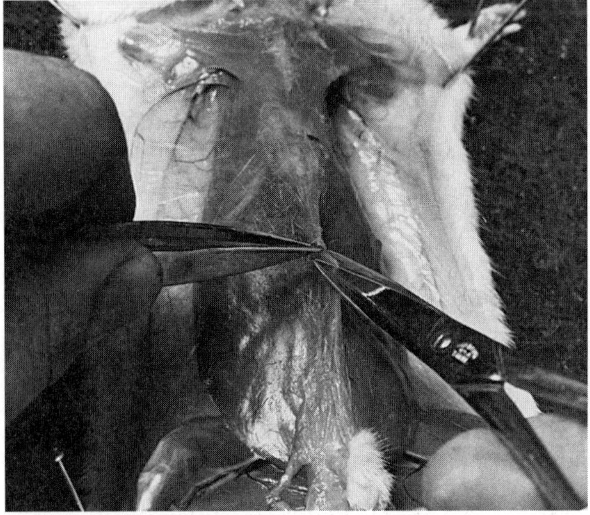

Figure 83.3 Hold the musculature up with forceps while cutting through the wall. Be careful not to damage any organs.

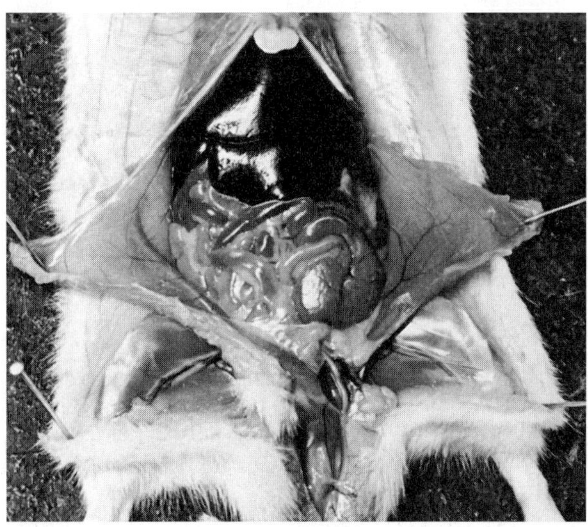

Figure 83.5 Once the abdominal organs are exposed, as shown here, left the liver with forceps to expose the pancreas for removal.

Tube Preparation

Figure 83.6 shows the test tube setup to assay the pancreatic extract for carbohydrase, protease, and lipase. The detection of each enzyme depends on a specific color test after the tubes have been incubated in a 38°C water bath for 1 hour.

Tube 1 in each series contains pancreatic extract, a substrate, and necessary buffering agents. The last tube in each series is a **positive test control** that contains the enzyme being tested for in tube 1.

All other tubes in each series should give negative results because of certain deficiencies. Note that tube 2 in each series contains the same ingredients as tube 1. Two significant variables, however, alter the outcome: (1) the tube is placed on ice while the others are incubated, and (2) the pancreatic extract is not added until after 1 hour on ice. This tube is designated the **zero time control.**

All four team members work cooperatively to set up these 16 tubes. After the tubes have been in the water baths for 1 hour, 1 ml of pancreatic extract is added to the three zero time control tubes. Various tests are then performed to detect hydrolysis.

Materials:
water baths (38°C)
Vortex mixer
pipetting devices
continued on next page

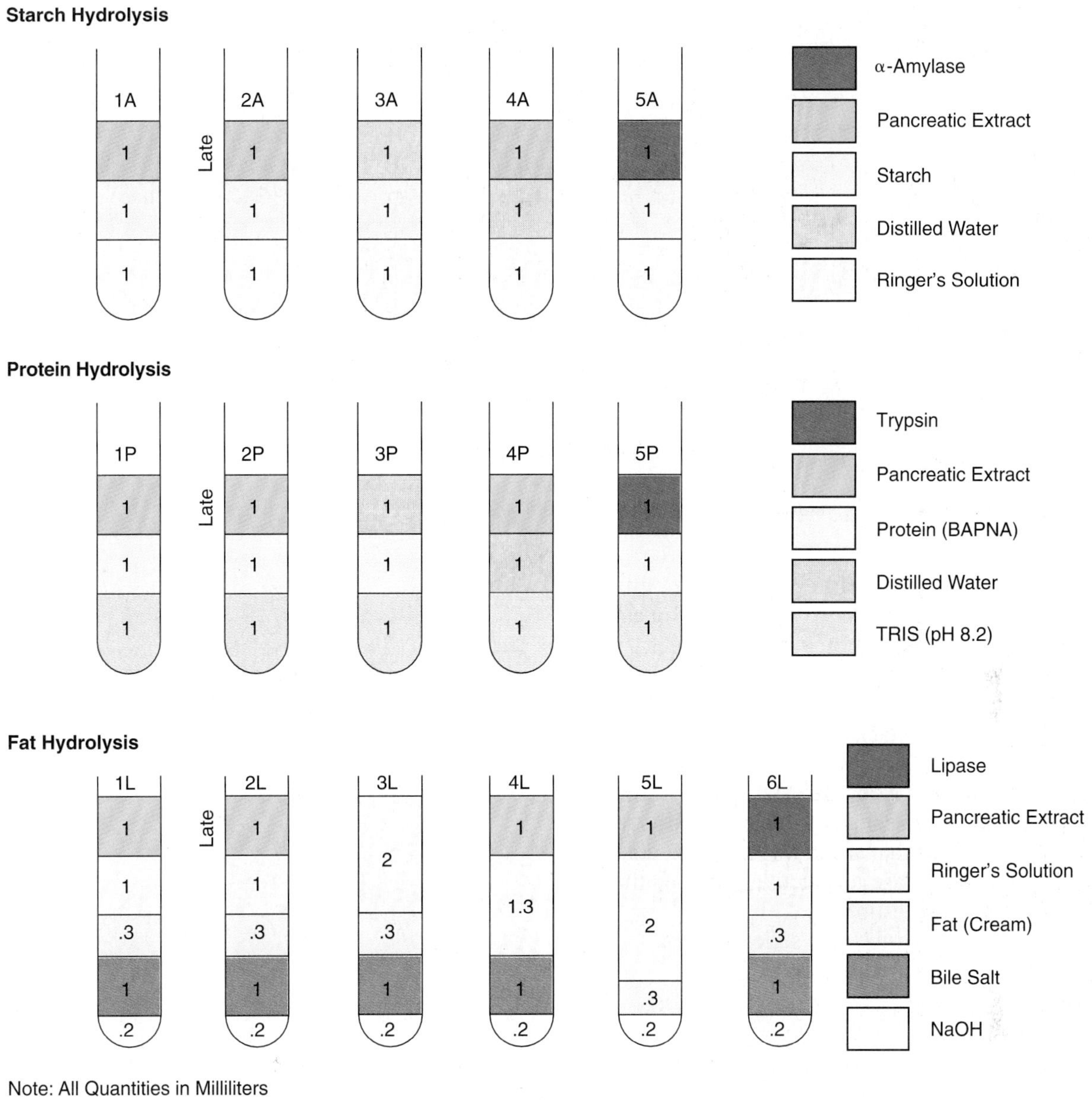

Starch Hydrolysis

1A	2A (Late)	3A	4A	5A
1	1	1	1	1
1	1	1	1	1
1	1	1	1	1

α-Amylase
Pancreatic Extract
Starch
Distilled Water
Ringer's Solution

Protein Hydrolysis

1P	2P (Late)	3P	4P	5P
1	1	1	1	1
1	1	1	1	1
1	1	1	1	1

Trypsin
Pancreatic Extract
Protein (BAPNA)
Distilled Water
TRIS (pH 8.2)

Fat Hydrolysis

1L	2L (Late)	3L	4L	5L	6L
1	1	2	1	1	1
1	1	.3	1.3	2	1
.3	.3	1	1	.3	.3
1	1	.2	.2	.2	1
.2	.2				.2

Lipase
Pancreatic Extract
Ringer's Solution
Fat (Cream)
Bile Salt
NaOH

Note: All Quantities in Milliliters

Figure 83.6 Protocol for enzyme assay of pancreatic extract.

Materials, continued

test tube rack
test tubes (16 mm X 150 mm), 16 per team
hot plates
pipettes, 1 ml, 5 ml, and 10 ml sizes
graduated cylinder (10 ml size)
bromthymol blue pH color standards
china marking pencil
photocolorimeter (Spectronic 20)
spot plates

solutions in dropping bottles

IKI solution
Benedict's solution
soluble starch, 0.1%
cream, unhomogenized
2% sodium taurocholate (a bile salt)
Barfoed's solution
bromthymol blue
albumin, 0.1%
NaCl, 0.2%
NaOH, 0.1N
BAPNA solution
amylase, 1%
trypsin, 1%
lipase, 1%
TRIS (pH 8.2)

Lab #8: Digestion of Fat
Lab #10: Enzyme Characteristics

Carbohydrase

Plants store carbohydrate primarily in the form of starch. Starch is composed of 98% amylose and 2% amylopectin. Amylose is a straight chain polysaccharide of glucose units attached by **α-1,4-glucosidic bonds.** Amylopectin is a

AMYLOPECTIN

AMYLOSE

α - 1,6 - Glucosidic Bond

α - 1,4 - Glucosidic Bond

Figure 83.7 Carbohydrate linkage where hydrolysis occurs.

branching polysaccharide with side chains attached to the main chain by **α-1,6-glucosidic bonds.** Figure 83.7 illustrates these linkages.

The enzyme α-amylase is present in saliva and pancreatic juice, and catalyzes starch digestion. This enzyme randomly attacks the α-1,4 bonds. An α-1,6-glucosidase is necessary to break the α-1,6 bond. The action of α-amylase on starch results in the formation of molecules of maltose and isomaltose. The enzymes maltase and isomaltase convert maltose and isomaltase to glucose in the epithelial cells of the small intestine.

Set up the five-tube test series for α-amylase according to the following protocol:

1. With a china marking pencil, label the test tubes 1A through 5A ("A" for amylase).
2. With a 5 ml or 10 ml pipette, deliver 1 ml of **mammalian Ringer's solution** to each of the five tubes. Use a mechanical pipetting device such as the one figure 83.8 shows.
3. Flush the pipette with water, and deliver 1 ml of **distilled water** to tubes 3 and 4.
4. With the same pipette, deliver 1 ml of **starch solution** to tubes 1, 2, 3, and 5. Note that tube 4 does not get any starch.
5. With a fresh 5 ml pipette, deliver 1 ml of **pancreatic extract** to tubes 1 and 4.
6. With a fresh 1 ml pipette, deliver 1 ml of **α-amylase** to tube 5.
7. Mix each tube on a Vortex mixer (see figure 83.9) for several seconds.
8. Remove tube 2, cover it with Parafilm, and place it in a rack immersed in ice water.
9. Cover the other four tubes with Parafilm, and place the test tube rack with these tubes in a 38°C water bath for **1 hour.**

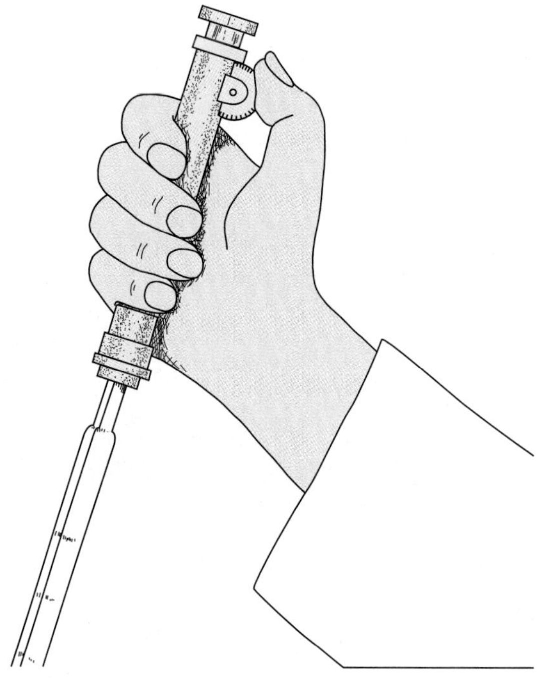

Figure 83.8 The thumb controls the volume of deliveries with this pipetting device.

Protease

Meats and vegetables provide most dietary proteins. These proteins consist of long chains of amino acids connected by **peptide bonds.** Note in figure 83.10 that when a peptide bond is hydrolyzed, the carbon of one amino acid takes on OH^- to form a carboxyl group (COOH), and H^+ is added to the other amino acid to form an amine (NH_2) group.

Protein digestion begins in the stomach with the action of pepsin and is completed in the small

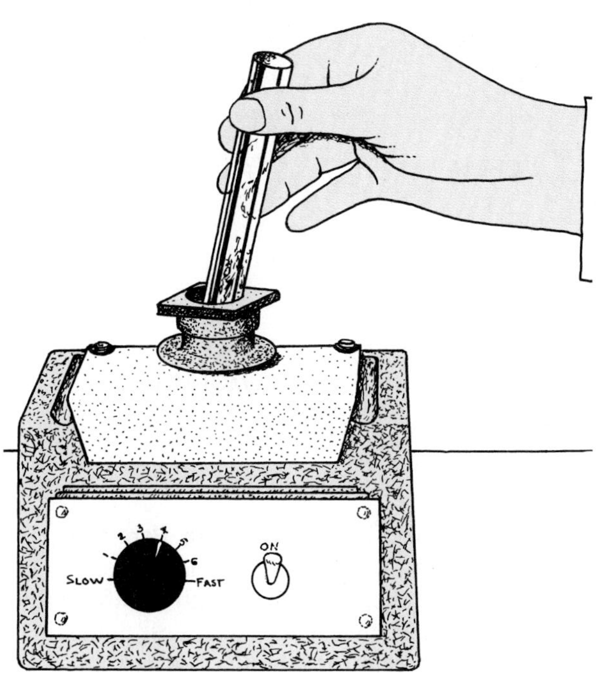

Figure 83.9 Mix each tube in the assay procedure on the Vortex mixer.

intestine. Pancreatic juice contains trypsin, chymotrypsin, and carboxypolypeptidase, which convert proteoses, peptones, and polypeptides to polypeptides and amino acids. Amino polypeptidase and dipeptidase within the epithelial cells of the small intestine hydrolyze the final peptide bonds, and amino acids pass through the intestinal mucosa into the portal blood.

The assay of pancreatic extract uses a synthetic peptide substrate designated BAPNA (benzoyl D-L p-arginine nitroaniline). If the pancreatic extract contains trypsin, BAPNA will be hydrolyzed to release a yellow aniline dye, giving visual evidence of hydrolysis. Trypsin is used in the positive test control.

Set up the five-tube test series for trypsin according to the following protocol:

1. With a china marking pencil, label the test tubes 1P through 5P ("P" for protease).
2. With a 5 ml or 10 ml pipette, deliver 1 ml of **pH 8.2 buffered solution** to each of the five tubes. Use a mechanical pipetting device such as the one figure 83.8 shows.
3. Flush the pipette with water, and deliver 1 ml of **distilled water** to tubes 3 and 4.
4. With the same pipette, deliver 1 ml of **BAPNA solution** to tubes 1, 2, 3, and 5. Note that tube 4 does not get any BAPNA; it is a negative test control.
5. With a fresh 5 ml pipette, deliver 1 ml of **pancreatic extract** to tubes 1 and 4.
6. With a fresh 1 ml pipette, deliver 1 ml of **trypsin** to tube 5.
7. Mix each tube on a Vortex mixer (see figure 83.9) for several seconds.
8. Remove tube 2, cover it with Parafilm, and place it in a rack immersed in ice water.
9. Cover the other four tubes with Parafilm, and place the test tube rack with these tubes in a 38°C water bath for **1 hour.**

Lipase

Lipase catalyzes the hydrolysis of fats to glycerol and fatty acids. This enzyme is present in pancreatic secretions, and fat digestion occurs in the small intestine.

Lipase hydrolysis differs from carbohydrase and protease hydrolysis in that it requires an emulsifying agent to hasten the process. Bile salts, produced by the liver, act like detergents to break large fat globules into smaller globules, greatly increasing the surface area of fat particles. Since lipases, like all enzymes, can only work on the surfaces of their substrate molecules, the increased surface area greatly enhances the speed of fat hydrolysis.

The most common fats in food are neutral fats, or **triglycerides.** As figure 83.11 shows, each triglyceride molecule is composed of a glycerol portion and three fatty acids. Although the final end

Peptide Bond

$$NH_2-C(R)(H)-\underbrace{C(=O)-N(H)}-C(R)(H)-COOH + H_2O \xrightarrow{\text{Peptidase}} NH_2-C(R)(H)-COOH + H-C(R)(NH_2)-COOH$$

PEPTIDE AMINO ACIDS

Figure 83.10 Protein hydrolysis.

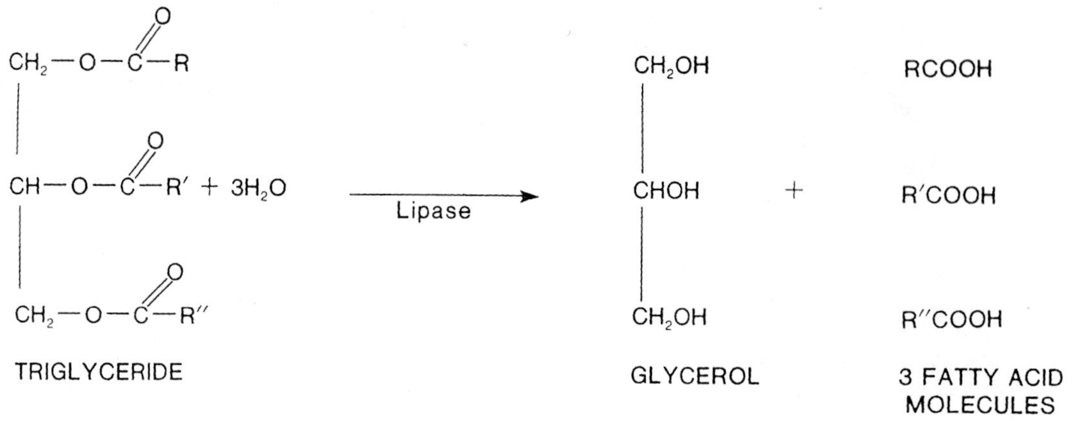

$$\begin{array}{l} CH_2-O-\overset{O}{\overset{\|}{C}}-R \\ | \\ CH-O-\overset{O}{\overset{\|}{C}}-R' + 3H_2O \\ | \\ CH_2-O-\overset{O}{\overset{\|}{C}}-R'' \end{array} \xrightarrow[\text{Lipase}]{} \begin{array}{l} CH_2OH \\ | \\ CHOH \\ | \\ CH_2OH \end{array} + \begin{array}{l} RCOOH \\ \\ R'COOH \\ \\ R''COOH \end{array}$$

TRIGLYCERIDE — GLYCEROL — 3 FATTY ACID MOLECULES

Figure 83.11 Fat hydrolysis.

products of triglyceride hydrolysis are fatty acids and glycerol, monoglycerides and diglycerides are intermediate products. Diagrammatically, the process looks like this:

E → F → Γ → I + Ξ

TRIGLYCERIDE DIGLYCERIDE MONOGLYCERIDE GLYCEROL FATTY ACIDS

Set up *six* test tubes to assay pancreatic extract for the presence of lipase, as follows:

1. With a china marking pencil, label the test tubes 1L through 6L ("L" for lipase).
2. With a 1 ml pipette, deliver 0.2 ml of **0.1N NaOH** to each of the six tubes. Use a mechanical pipetting device such as the one figure 83.8 shows.
3. With a 5 ml pipette, deliver 1.0 ml of **mammalian Ringer's solution** to tubes 1, 2, and 6; 2 ml to tubes 3 and 5; and 1.3 ml to tube 4.
4. Flush the pipette with water, and deliver 1 ml of **bile salt solution** to each of the tubes except tube 5.
5. Flush the pipette with water again, and deliver 0.3 ml of **cream** to each of the tubes except tube 4.
6. With a fresh 5 ml pipette, deliver 1 ml of **pancreatic extract** to tubes 1, 4, and 5.
7. With a fresh 1 ml pipette, deliver 1 ml of **lipase** to tube 6.
8. Mix each tube on a Vortex mixer (see figure 83.9) for several seconds.
9. Add two drops of bromthymol blue to each tube. Check the color against a bromthymol blue standard of pH 7.2. If the correct color does not develop, slowly add NaOH, drop by drop, until each tube reaches the correct color of pH 7.2.
10. Remove tube 2, cover it with Parafilm, and place it in a rack immersed in ice water.
11. Cover the other five tubes with Parafilm, and place the test tube rack with these tubes in a 38°C water bath for **1 hour.**

Evaluation of Tests

Remove the tubes from both water baths after 1 hour, and proceed as follows to complete the experiment:

1. To tube 2 of each series, add 1 ml of pancreatic extract, and mix for several seconds on a Vortex mixer. Put each tube 2 in place within its series.
2. **Carbohydrase test:** Place a sample of test solution from each of the five tubes (1A through 5A) in separate depressions of a spot plate. Add two drops of IKI solution to each of the solutions on the plate. If starch is present due to lack of hydrolysis, the spot solution turns blue or black. If hydrolysis has occurred, test the remaining solution in the tube with Barfoed's and Benedict's tests to determine the degree of digestion. (See Appendix C for these tests.) Record your results in part A of the Laboratory Report.
3. **Protease test:** Since hydrolysis produces a yellow color in the test tube, record the color of each tube in part B of the Laboratory Report.
4. **Lipase test:** Since fat hydrolysis lowers pH due to fatty acid formation, the bromthymol blue in the tubes changes from blue to yellow. Compare the six tubes to a bromthymol blue color standard to determine the pH changes.

Assignment:
Complete the Laboratory Report for this exercise.

Factors Affecting Hydrolysis

84

Enzymes are highly complex protein molecules and somewhat unstable due to weak hydrogen bonds. Of the many factors that influence hydrolysis rates, temperature and hydrogen ion concentration are probably the most important. This exercise examines the influence of these two conditions on the enzyme salivary amylase.

Salivary amylase hydrolyzes the starch amylose to produce soluble starch, maltose, dextrins, achrodextrins, and erythrodextrin. When iodine is used as an indicator of amylase activity, starch and soluble starch produce a blue color, and erythrodextrin yields a red color. Benedict's solution detects the presence of maltose. With heat, maltose reduces soluble cupric oxide to insoluble red cuprous oxide in Benedict's solution.

This exercise uses the saliva of one individual, diluted to 10% with distilled water. To determine the effect of temperature on the rate of hydrolysis, iodine solution (IKI) is added to a series of 10 tubes containing starch and saliva. Adding the iodine at 1-minute intervals to successive tubes allows you to determine, by color, when the hydrolysis of starch is complete. Two water baths—one at 20°C and the other at 37°C—allow temperature comparisons. You also investigate the effect of boiling amylase and use buffered solutions of pH 5, 7, and 9 to determine the effects of hydrogen ion concentration on the action of amylase. Solutions in this exercise can be dispensed with medicine droppers, graduates, or serological pipettes.

If laboratory time is limited, your instructor may divide the class into thirds to each perform only a portion of the entire experiment. Group results can be tabulated on the chalkboard so that all students can record results for the complete experiment. The best arrangement, however, is for students to work in pairs. The assignments that follow work well for a 2-hour laboratory period:

Saliva Preparation

Prior to this experiment, the instructor prepares a 10% saliva solution for the entire class as follows:

Materials:
50 ml graduate
250 ml graduated beaker
paraffin
12 dropping bottles with labels

Place a small piece (1/4-inch cube) of paraffin under your tongue, and allow it to soften for several minutes before chewing it. As you chew, expectorate all saliva into a 50 ml graduate.

When you have collected 15 to 20 ml of saliva, add enough distilled water to the saliva to make a 10% solution. Dispense the diluted saliva into labeled ("10% saliva") dropping bottles. For a class of 24 students, you will need 12 bottles. All students should wear disposable latex gloves during this experiment.

Controls

A set of five control tubes is needed for color comparisons with test results. Tubes 1 and 2 are for starch detection. Tubes 3, 4, and 5 are for maltose detection. Table 84.1 shows the ingredient in each tube and the significance of test results. Proceed as follows to prepare your set of controls.

Materials:
10% saliva solution (in dropping bottle)
0.1% starch solution
1% maltose solution
Benedict's solution (in dropping bottle)
IKI solution (in dropping bottle)
test tubes (13 mm × 100 mm)
test tube rack (Wassermann type)
medicine droppers
250 ml beaker
electric hot plate
1 ml sterile pipettes
pipetting device

1. Label five clean test tubes 1 through 5.
2. Fill the five tubes with the following reagents, using a different pipette for each reagent:

   Tube 1: 1 ml starch solution and two drops IKI solution
   Tube 2: 1 ml distilled water and two drops IKI solution
   Tube 3: 1 ml starch solution, two drops saliva solution, and five drops Benedict's solution

Table 84.1 Control Tube Contents and Functions

TUBE NUMBER	CARBOHYDRATE	SALIVA	REAGENT	TEST RESULTS
1	Starch	—	IKI	Positive for starch
2	—	—	IKI	Negative for starch
3	Starch	2 drops	Benedict's	Positive for sugar Positive for hydrolysis
4	—	2 drops	Benedict's	Negative for sugar
5	Maltose	—	Benedict's	Positive for sugar

Tube 4: 1 ml distilled water, two drops saliva solution, and five drops Benedict's solution

Tube 5: 1 ml maltose solution and five drops Benedict's solution

3. Place tubes 3, 4, and 5 into a beaker of warm water, and boil on an electric hot plate for **5 minutes** to bring about the desired color changes in tubes 3 and 5.
4. Set the five tubes in a test tube rack for color comparisons. Tube 1 is your positive starch control. Tubes 3 and 5 arc your positive sugar controls. Tubes 2 and 4 are your negative controls for starch and sugar.

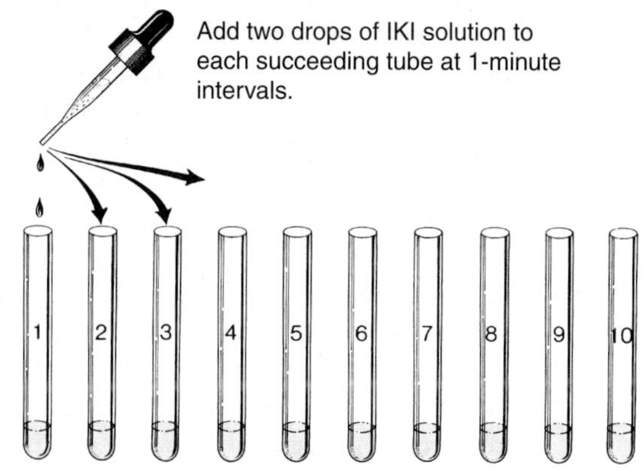

Figure 84.1 Amylase activity at 20°C and 37°C.

Effect of Temperature (20°C and 37°C)

In this portion of the experiment, you determine how room temperature (20°C) and body temperature (37°C) affect the rate at which amylase acts. Note in figure 84.1 that tubes of starch and saliva receive two drops of IKI solution at 1-minute intervals to observe how long it takes for hydrolysis to occur at a given temperature.

When IKI is added to the first tube, the solution is blue, indicating no hydrolysis. At some point, however, the blue color disappears, indicating complete hydrolysis. Note that the "Materials" list is for performing the test at only one temperature.

Materials:
10 serological test tubes (13 mm × 100 mm)
test tube rack (Wassermann type)
water bath (20°C or 37°C)
10% saliva solution
0.1% starch solution
china marking pencil
thermometer (centigrade scale)
1 ml serological pipettes or 10 ml graduate
mechanical pipetting device
IKI solution

1. Label 10 test tubes 1 through 10 with a china marking pencil, and arrange them sequentially in the front row of the test tube rack.
2. With a 1 ml pipette, dispense 1 ml of starch solution to each tube.
3. Place the rack of tubes and bottle of saliva in the water bath (20°C or 37°C).
4. Insert a clean thermometer into the bottle of saliva. When the temperature of the saliva reaches the temperature of the water bath (3–5 minutes), proceed to the next step.
5. Lift the bottle of saliva from the water bath, and **record the time.** Leave the rack of tubes in the water bath. Put **one drop** of saliva into each test tube, starting with tube 1. Be sure that the drop falls directly into the starch solution without touching the sides of the tube.
6. Shake the rack of tubes to mix their contents.
7. Exactly **1 minute** after tube 1 has received saliva, add 2 drops of IKI solution to it, and mix again. Note the color.
8. **One minute later,** add two drops of IKI solution to tube 2. Repeat this process every minute until all 10 tubes have received IKI solution.
9. Compare the colors of the tubes with the control tubes, and determine the time required to digest the starch.

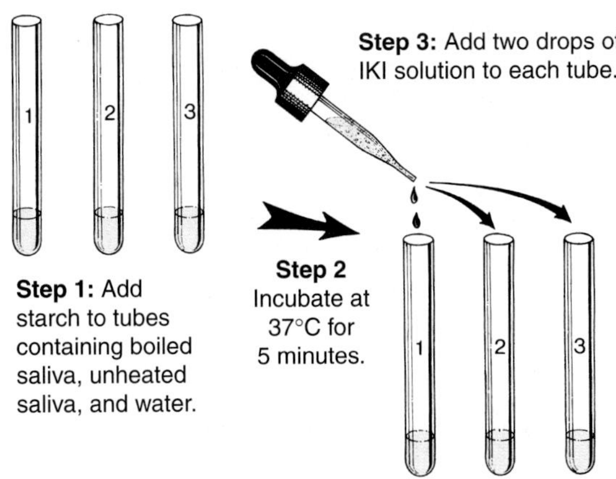

Figure 84.2 Effect of boiling on amylase activity.

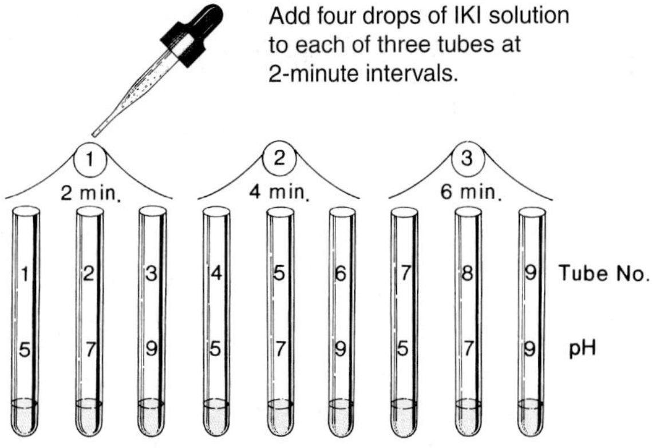

Figure 84.3 Effect of pH on amylase activity.

10. Record your results in table I on the Laboratory Report.
11. Wash the insides of all glassware with bleach water, and rinse thoroughly.

Effect of Boiling

In this portion of the experiment, you compare the action of boiled amylase with that of unheated amylase. Figure 84.2 illustrates the procedure.

Materials:
three test tubes
china marking pencil
test tube holder
Bunsen burner
10% saliva solution
0.1 starch solution
IKI solution
distilled water
water bath (37°C)

1. Label the test tubes 1 to 3 with a china marking pencil.
2. Dispense 1 ml of saliva solution to tubes 1 and 2, and 1 ml of distilled water to tube 3.
3. With a test tube holder, hold tube 1 over a Bunsen burner flame so that the saliva solution boils for several seconds.
4. Add 1 ml of starch solution to each of the three tubes, and place them in a 37°C water bath for 5 minutes.
5. Remove the tubes from the water bath, and test for the presence of starch by adding 2 drops of IKI solution to each tube. Record the results in table II on the Laboratory Report.
6. Wash and rinse the glassware in bleach water.

Effect of pH on Enzyme Activity

In this portion of the experiment, you determine the effect of different hydrogen ion concentrations on amylase activity. Amylase, starch, and buffered solutions are incubated at 37°C (see figure 84.3).

Materials:
test tube rack (Wassermann type)
12 test tubes (13 mm × 100 mm)
china marking pencil
1 ml pipettes and pipetting device
10 ml graduate
water bath (37°C)
thermometer
10% saliva solution
0.1% starch solution
IKI solution
pH 5 buffer solution
pH 7 buffer solution
pH 9 buffer solution

1. Label nine test tubes as figure 84.3 shows. Note that the tubes are arranged in three groups and that each group has three tubes of pH 5, 7, and 9 in sequence.
2. With a 1 ml pipette, deliver 1 ml of pH 5 buffer solution to tubes 1, 4, and 7.
3. With a fresh pipette, deliver 1 ml of pH 7 solution to tubes 2, 5, and 8.
4. With another fresh pipette, deliver 1 ml of pH 9 solution to tubes 3, 6, and 9.
5. Add two drops of saliva solution to each of the nine tubes.

6. Fill three empty test tubes about two-thirds full of starch solution (total of approximately 15 ml).
7. Place a clean thermometer in one of the three starch tubes, and insert the whole rack of 12 tubes in the 37°C water bath.
8. When the thermometer in the starch tube registers 37°C, use a pipette to deliver 1 ml of warm starch solution to each of the nine numbered tubes. Use the starch from the three tubes in the rack.
9. **Record the time** that you add starch to tube 1.
10. **Two minutes** after this recorded time, add four drops of IKI solution to tubes 1, 2, and 3.
11. **Four minutes** after the recorded time, add four drops of IKI solution to tubes 4, 5, and 6.
12. **Six minutes** after the recorded time, add four drops of IKI solution to tubes 7, 8, and 9.
13. Remove the tubes from the water bath, and compare them with the controls.
14. Record your observations on the Laboratory Report.
15. Wash all glassware with bleach water. Wipe down the lab table with a 10% bleach solution.

Assignment:

Complete the remainder of the Laboratory Report for this exercise.

Part 15 The Excretory System

This part includes only two exercises. Exercise 85 is a microscopic and macroscopic exploration of the urinary system. You dissect the urinary system of the cat and examine the detailed structure of a sheep kidney. As in previous exercises of this type, you should study the material on human anatomy prior to coming to the laboratory.

In Exercise 86, a routine urine analysis is demonstrated or performed individually. Reagent test strip methods as well as classic methods are examined.

85 Anatomy of the Urinary System

This exercise first provides a general description of the human urinary system. The laboratory activities consist of three parts: (1) histological studies, (2) sheep kidney dissection, and (3) cat dissection. Although the cat kidney is dissected, a sheep kidney is also included here because of its greater size.

Organs of the Urinary System

Figure 85.1 illustrates the components of the urinary system and its principal blood supply. The urinary system consists of two kidneys, two ureters, the urinary bladder, and a urethra. The dark brown, bean-shaped **kidneys** are behind the peritoneum (a condition referred to as *retroperitoneal*). The right kidney is usually somewhat lower than the left one, probably because of its displacement by the liver. Each kidney receives blood through a **renal artery** that branches off the **abdominal aorta.** Blood leaving each kidney drains via a **renal vein** into the **inferior vena cava.**

Urine passes from each kidney to the **urinary bladder** through a **ureter** that enters the bladder on its posterior surface. A short tube, the **urethra,** leads from the urinary bladder to the exterior of the body. The male urethra is about 20 cm long; the female urethra is approximately 4 cm long.

Two sphincter muscles—the sphincter vesicae and the sphincter urethrae—control the exit of urine from the bladder (*micturition*). The **sphincter vesicae** is a smooth muscle sphincter near the bladder exit. When approximately 300 ml of urine has accumulated in the bladder, the muscular walls of the bladder are stretched sufficiently to initiate a parasympathetic reflex that contracts the bladder wall. These contractions force urine past the sphincter vesicae into the urethra above the **sphincter urethrae.** This second sphincter, approximately 1 to 3 cm below the sphincter vesicae, consists of skeletal muscle fibers under voluntary control. The presence of urine in the urethra above this sphincter creates a desire to micturate; however, since the valve is under voluntary control, micturition can be inhibited. When both sphincters relax, urine passes from the body.

Kidney Anatomy

Figure 85.2 shows a frontal section of a human kidney. A thin, fibrous **renal capsule** covers the kidney's outer surface. In addition, a fatty capsule (not shown in figure 85.2) completely encases the kidney and provides support and protection.

Immediately under the renal capsule is the kidney **cortex,** which is reddish brown due to its exten-

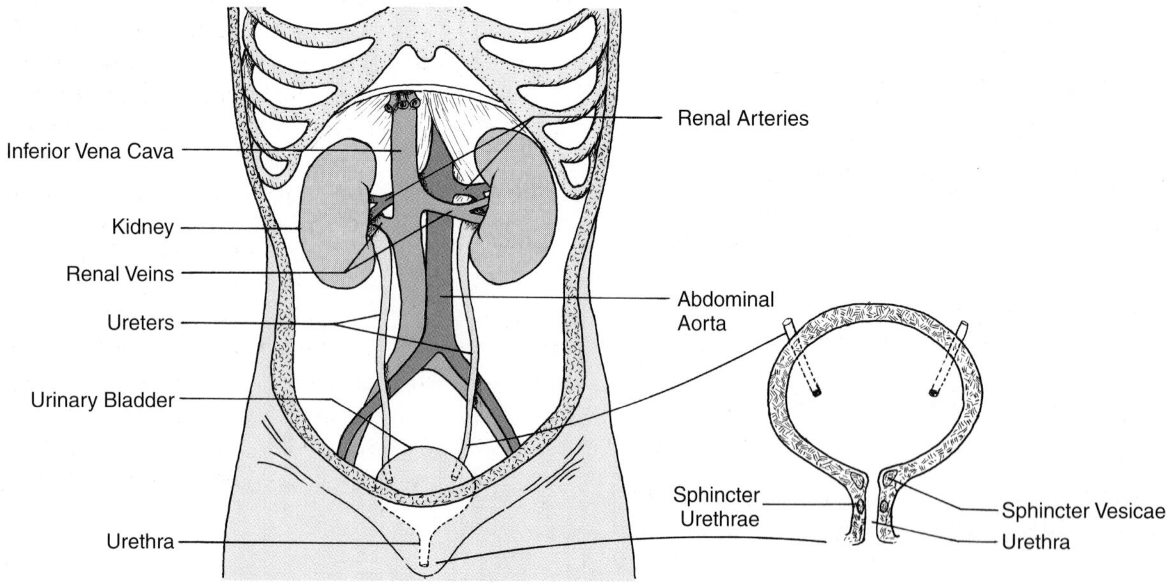

Figure 85.1 The urinary system.

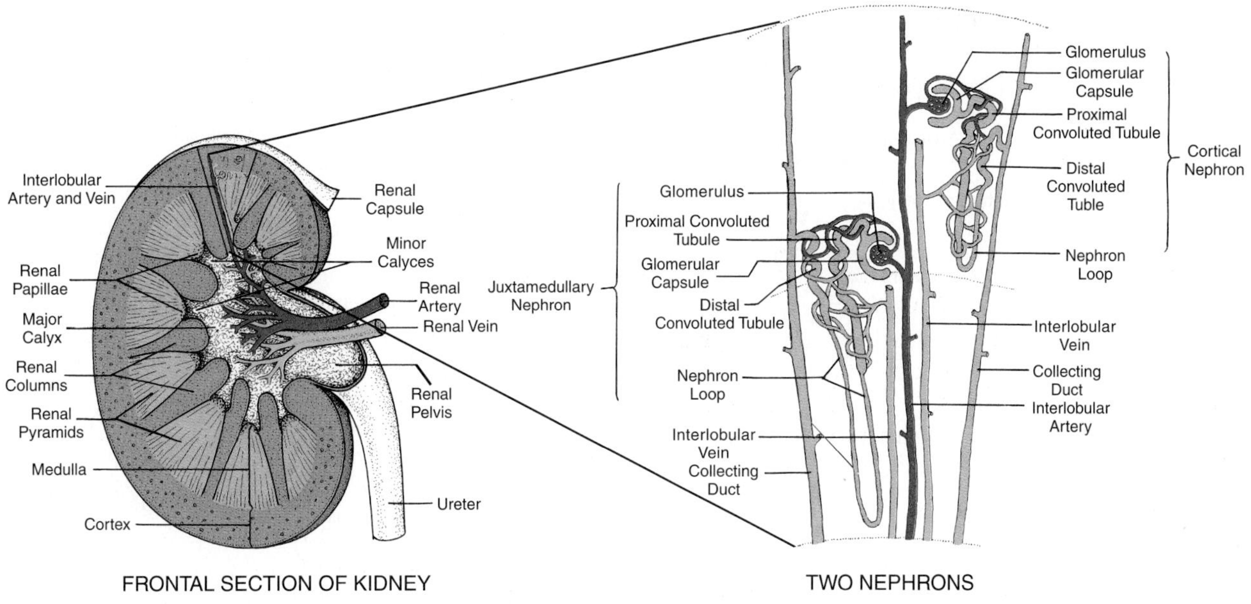

Figure 85.2 Anatomy of the kidney.

sive blood supply. The lighter inner portion of the kidney is the **medulla,** which divides into cone-shaped **renal pyramids.** Cortical tissue, in the form of **renal columns,** extends down between the pyramids. Each renal pyramid terminates as a **renal papilla** that projects into a minor calyx. The **minor calyces** are short tubes that receive urine from the papillae and empty into a central **major calyx** of the kidney. The funnel-like **renal pelvis** receives urine from the major calyx and empties it into the **ureter.**

The Nephrons

The basic functioning unit of the kidney is the *nephron.* The enlarged section in figure 85.2 shows two nephrons. Figure 85.3 illustrates a single nephron in greater detail. Each kidney has approximately 1 million nephrons. Eighty percent of the nephrons are in the cortex (**cortical nephrons**). The remaining 20% are partially in the cortex and partially in the medulla (**juxtamedullary nephrons**).

The nephron forms urine through three physiological activities: (1) filtration, (2) reabsorption, and (3) secretion. Because of the differing functions of various regions of the nephron, the fluid that first forms at the beginning of the nephron is quite unlike the urine that finally enters the calyces of the kidney.

Each nephron consists of an enlarged end, the **renal corpuscle,** and a long tubule that empties eventually into the calyx of the kidney. Each renal corpuscle has two parts: (1) an inner tuft of capillaries, the **glomerulus,** and (2) an outer, double-walled, cuplike structure, the **glomerular** *(Bowman's)* **capsule.** Blood that enters the kidney through the **renal artery** reaches each nephron through an **interlobular artery.** A short **afferent arteriole** conveys blood from the interlobular artery into the glomerulus. Blood exits the glomerulus through the **efferent arteriole,** which has a much smaller diameter than the afferent arteriole.

The high intraglomerular blood pressure forces a highly dilute fluid, the glomerular filtrate, to pass into the glomerular capsule. This fluid consists of glucose, amino acids, urea, salts, and a great deal of water.

The glomerular filtrate passes next through the **proximal convoluted tubule,** the **descending limb of the nephron loop,** the **ascending limb of the nephron loop,** and the **distal convoluted tubule** before emptying into the large **collecting duct.** Several nephrons may empty into a single collecting duct.

The **peritubular capillaries** that enmesh the glomerular filtrate's entire route reabsorb water, glucose, amino acids, and other substances in the filtrate back into the blood. The walls of the proximal convoluted tubules reabsorb 80% of the water. The antidiuretic hormone of the posterior pituitary and aldosterone of the adrenal cortex facilitate water reabsorption. Cells lining the collecting ducts alter urine composition by secreting ammonia, uric acid, and other substances into the lumen of the duct.

Assignment:
Complete parts A and B of the Laboratory Report for this exercise.

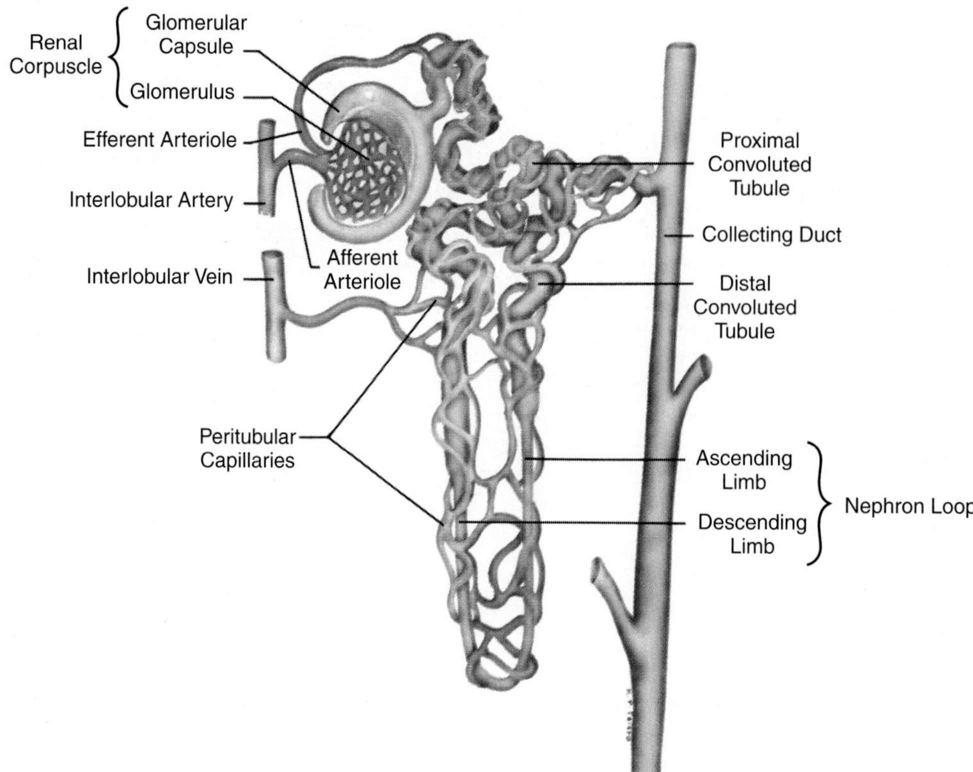

Figure 85.3 The nephron.

Histological Study

Materials:
prepared slides of:
kidney tissue
ureter
urinary bladder

Examine slides of the **kidney cortex, kidney medulla, urinary bladder,** and **ureter.** Compare the slides with figures HA-26C and D, HA-29, HA-30, and HA-33 A and B in the Histology Atlas. Note the different kinds of tissue in the tubules, calyx, urinary bladder, and ureters.

Assignment:
Draw your microscopic observations in part C of the Laboratory Report for this exercise.

Sheep Kidney Dissection

Combining the sheep kidney dissection with the cat dissection provides greater anatomical clarity. Although the injected cat kidney better illustrates the blood supply, the fresh sheep kidney more vividly shows some of the organ's lifelike characteristics.

Materials:
dissecting kit, tray, long knife
fresh sheep kidneys

1. If the kidneys are still encased in fat, peel the fat off carefully. As you do so, look carefully for the **adrenal gland,** which should be embedded in the fat near one end of the kidney.
2. Remove the adrenal gland from the fat, and cut it in half. Note that the gland has a distinct outer **cortex** and an inner **medulla.**
3. Probe the surface of the kidney with a sharp dissecting needle to see if you can differentiate the **renal capsule** from the underlying tissue.
4. With a long knife, slice the kidney longitudinally to produce a frontal section similar to figure 85.2. Wash out the cut halves with running water.
5. Identify all the structures you see in the kidney section of figure 85.2.

Cat Dissection

Because the urinary and reproductive organs are closely associated, you examine these two systems together. Most emphasis here, however, is on the urinary organs.

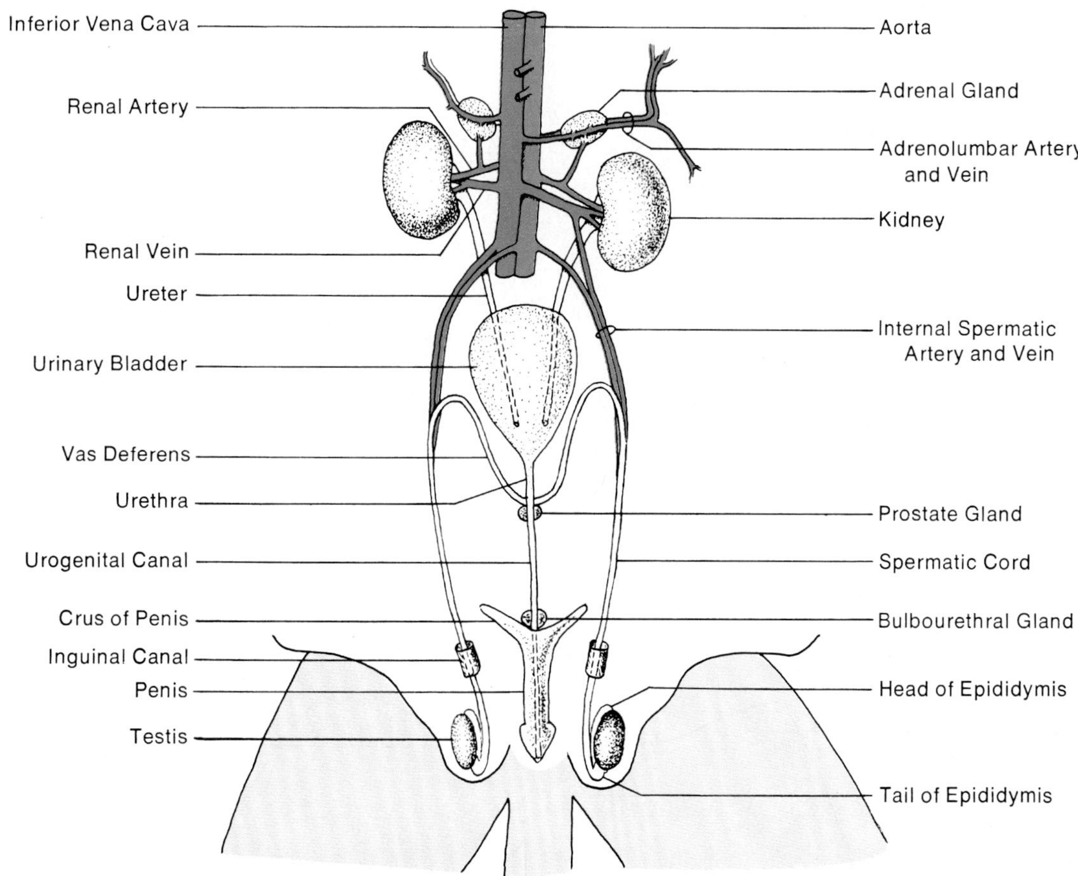

Figure 85.4 Urogenital system of the male cat.

As you proceed through the dissection, refer to figures 85.4 and 85.5 to identify the various organs. Exercise 88 provides a more detailed study of the reproductive organs.

1. To expose the urogenital system, remove the liver, spleen, and stomach.
2. With forceps and scalpel, clear away excess fat, the peritoneum, and other connective tissue in the area. *Take care not to injure the internal spermatic (or ovarian) blood vessels and the ureters that are embedded in fat.*
3. Note that the **kidneys** are retroperitoneal in that they lie between the peritoneum and the body wall. Note also that the left kidney is somewhat caudad of the right one.
4. Identify the two **adrenal glands** embedded in connective tissue between the anterior ends of the kidneys and the large abdominal blood vessels (aorta and inferior vena cava).
5. Remove both kidneys. With a razor blade or sharp scalpel, slice one kidney longitudinally to produce a series of thin frontal sections that reveal the kidney's internal structure.
6. Slice the other kidney to produce a series of thin transverse sections.
7. Identify the **cortex, medulla,** and **renal papilla** on the sections. Only one renal papilla in the cat kidney empties into the pelvis of the ureter. Humans have approximately 12 papillae.
8. Trace the **ureter** of each kidney to the urinary bladder, where the ureter enters the bladder's dorsal surface.
9. Note that the **urinary bladder** is also retroperitoneal. It is a musculomembranous sac that ligaments formed from folds of peritoneum attach to the abdominopelvic walls.

   The **medial ligament** attaches the ventral side of the bladder to the linea alba. Two **lateral ligaments,** with sizable fat deposits, connect the sides of the bladder to the dorsal body wall.
10. Make a linear ventral incision in the bladder with a scalpel, and examine the inside wall. Locate the openings of the two ureters and the urethra. Also, cut into the urethra where it joins the bladder. Do you see any evidence of **sphincter muscles** in this region?
11. Observe that the urethra of the female cat empties into a **urogenital sinus.** In the male, the urethra passes through the penis.
12. Trade specimens with students who have the opposite sex of your specimen to become familiar with the anatomy of both sexes.

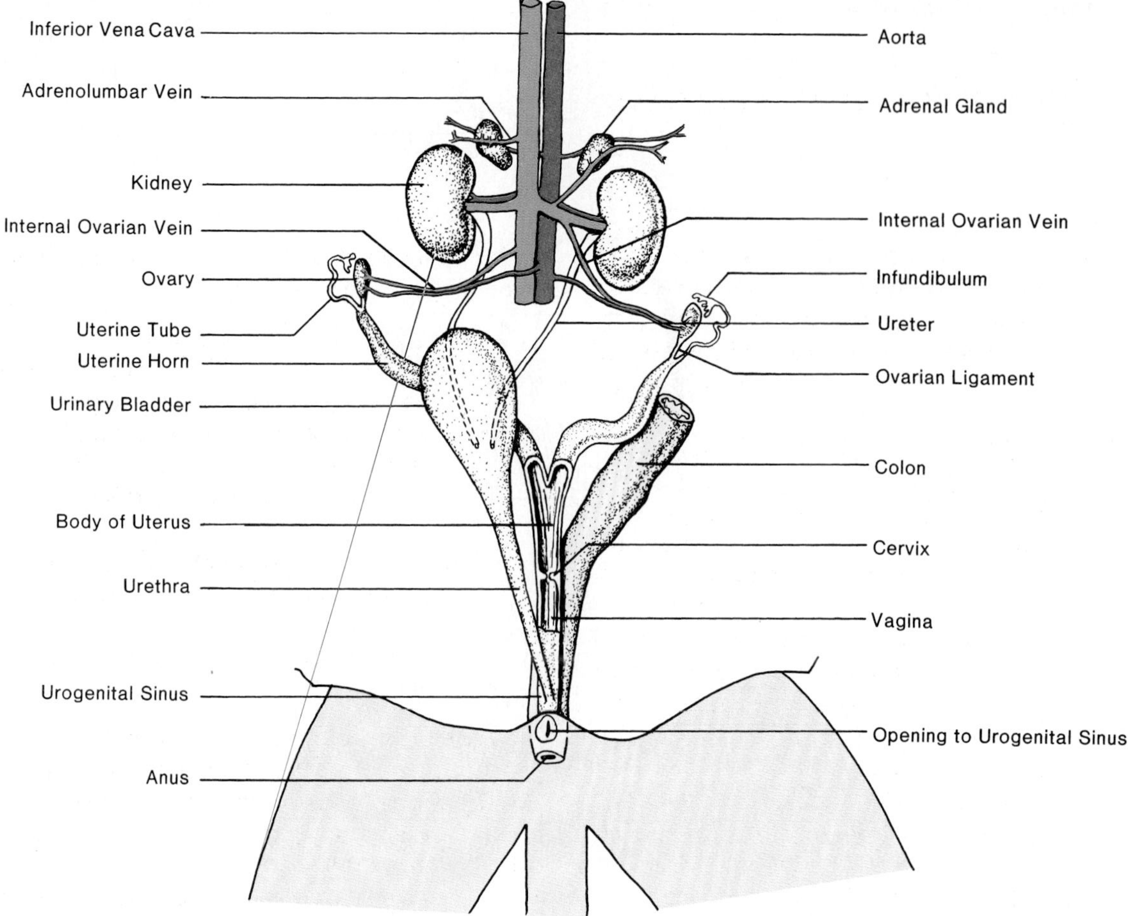

Figure 85.5 Urogenital system of the female cat.

Urine: Composition and Tests

86

Urine is a highly complex aqueous solution of organic and inorganic substances. Urea, uric acid, creatinine, sodium chloride, ammonia, and water are the principal ingredients. Most substances are either waste products of cellular metabolism or are derived directly from certain foods.

This exercise focuses on detecting abnormal substances in urine that may have pathological significance. Of the 10 specific tests performed, some are for physical properties, but most are for chemical substances. Your instructor may provide artificial urine samples for urinalysis or you may be required to collect urine samples as described below. If you will be working with an actual urine sample, be sure to wear disposable latex gloves during each exercise. Dispose of all gloves and test strips in a biohazard/infectious materials container. All contaminated glassware should be washed in a basin of bleach water. At the end of the lab period, wipe down the table with a 10% bleach solution.

The chemical tests are of two types: (1) classical in vitro methods, and (2) test-strip methods. In most laboratories, the test-strip method is preferred because of time saved and test reliability.

Specimen Collection

Collect urine in a clean container, and store it in a cool place until tested. If you are to collect the sample just prior to lab testing, plastic disposable containers will be on the supply table. The best time for collecting a urine sample is 3 hours after a meal. Urine samples collected first thing in the morning are least likely to reveal abnormal substances.

Appearance of Test Specimen

The color and transparency of a urine sample can provide evidence of pathology. Evaluate the sample's color and transparency according to the following standards:

Color

Normal urine varies from light straw to amber colored. Urine color is due to the pigment urochrome, which is the end product of hemoglobin breakdown:

Hemoglobin → Hematin → Bilirubin → Urochromogen → Urochrome

The following deviations from normal color have pathological implications:

Milky: pus, bacteria, fat, or chyle.
Reddish amber: urobilinogen or porphyrin. The action of bacteria on bile pigment produces urobilinogen in the intestine. Porphyrin may be evidence of liver cirrhosis, jaundice, Addison's disease, and other conditions.
Brownish yellow or green: bile pigments. Yellow foam is definite evidence of bile pigments.
Red to smoky brown: blood and blood pigments.

Carrots, beets, rhubarb, and certain drugs and vitamins may color the urine, yet have no pathological significance. Carrots may cause increased yellow color due to carotene, beets cause reddening, and rhubarb may turn urine brown.

Evaluate your urine sample according to the previous criteria, and record the information on the Laboratory Report.

Transparency (Cloudiness)

A fresh sample of normal urine should be clear but may become cloudy after standing for a while. Cloudy urine may be evidence of phosphates, urates, pus, mucus, bacteria, epithelial cells, fat, or chyle. Phosphates disappear with the addition of dilute acetic acid, and urates dissipate with heat. Other causes of turbidity can be analyzed by microscopic examination.

After shaking your sample, determine the degree of cloudiness, and record it on the Laboratory Report.

Specific Gravity of Specimen

The specific gravity of a 24-hour specimen of normal urine is between 1.015 and 1.025. Single urine specimens may range from 1.002 to 1.030. The more solids in solution, the higher the specific gravity. The greater the volume of urine in a 24-hour specimen, the lower its specific gravity. A low specific gravity may signal chronic nephritis or diabetes insipidus. A high specific gravity may indicate diabetes mellitus, fever, or acute nephritis.

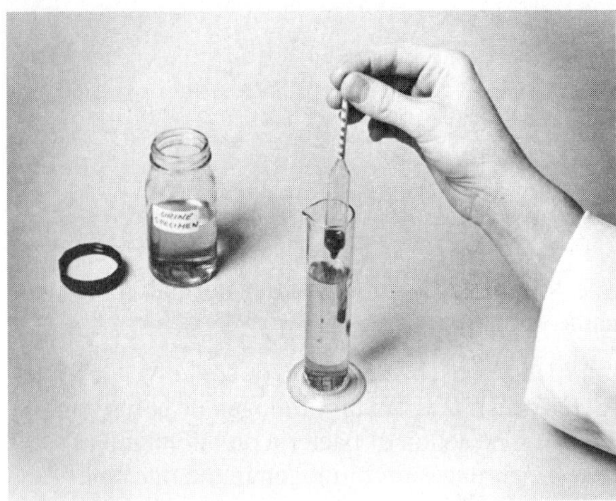

Figure 86.1 Specific gravity. Lower the hydrometer into the urine after removing the foam with filter paper. Take the reading from the bottom of the meniscus.

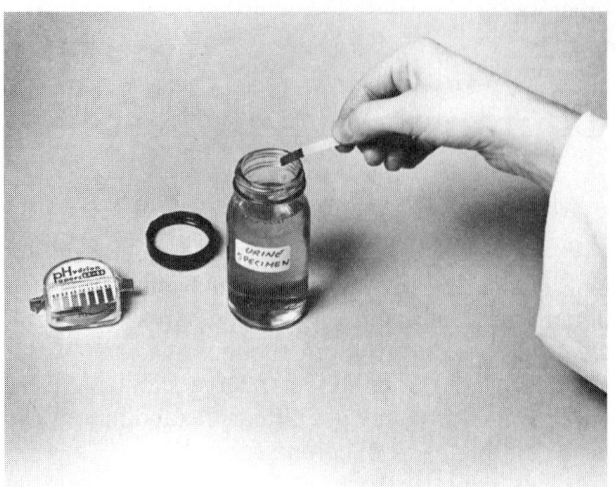

Figure 86.2 Determination of pH. Test only fresh urine samples because pH tends to rise with aging of urine. Read the pH 1 minute after dipping the paper.

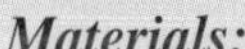

Materials:
urinometer (cylinder and hydrometer)
thermometer (centigrade scale)
filter paper

1. Fill the urinometer cylinder three-fourths full of well-mixed urine (see figure 86.1). Remove any foam on the surface with filter paper.
2. Insert the hydrometer into the urine, and read the graduation on the stem at the level of the bottom of the meniscus.
3. Measure the temperature of the urine, and compensate for the temperature by adding 0.001 to the specific gravity for each 3°C above 25°C, and subtracting the same amount for each degree below 25°C. This value is the adjusted specific gravity.
4. Record the adjusted specific gravity on the Laboratory Report.
5. Wash the cylinder and hydrometer with bleach water after the test.

Hydrogen Ion Concentration of Specimen

Although freshly voided urine is usually acidic (around pH 6), the normal range is between 4.8 and 7.5. The pH varies with the time of day and diet. Twenty-four-hour specimens are less acidic than fresh specimens and may become alkaline after standing due to bacterial decomposition of urea.

High acidity is present in acidosis, fevers, and high-protein diets. Excess alkalinity may be due to urine retention in the bladder, chronic cystitis, anemia, obstructive gastric ulcers, and alkaline therapy. The simplest way to determine pH is to use pH indicator paper strips, as figure 86.2 shows.

Materials:
pH indicator paper strip (pHydrion or nitrazine papers)

1. Dip a strip of pH paper into the urine sample three consecutive times, and shake off excess liquid.
2. After 1 minute, compare the color with the color chart. Record your observations on the Laboratory Report. Discard pH paper strips in the appropriate container.

Protein Detection

Although the large size of protein molecules prohibits their appearance in normal urine, certain conditions allow them to filter through. Excessive muscular exertion, prolonged cold baths, and excessive protein ingestion may result in what is termed physiologic albuminuria.

Pathologic albuminuria, on the other hand, exists when albumin in the urine results from kidney congestion, toxemia of pregnancy, febrile diseases, and anemias.

Exton's Method

Albumin can be detected by precipitation with chemicals, heat, or both. This test uses Exton's method, which employs sulfasalicylic acid and heat.

Materials:
Exton's reagent
test tubes, one for each urine sample
continued on next page

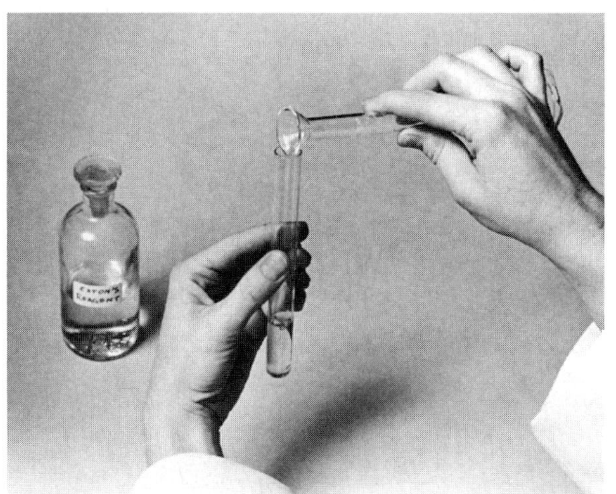

Figure 86.3 **Albumin test.** If albumin is present in the urine, adding Exton's reagent to the urine forms a precipitate.

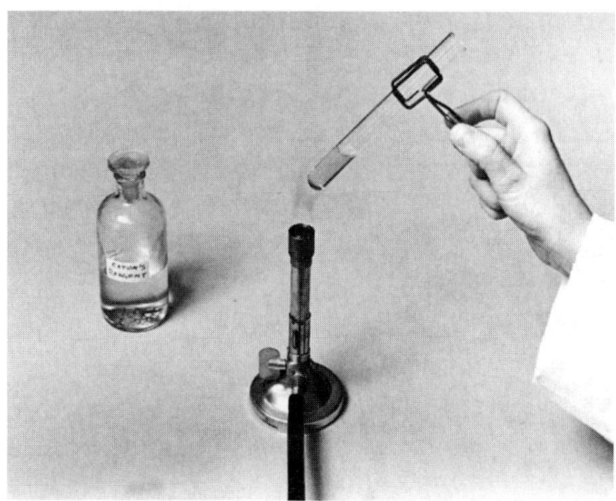

Figure 86.4 **Albumin confirmation.** If the precipitate formed from the addition of Exton's regent persists after heating, the presence of albumin is confirmed.

Materials (continued)
test tube holder
Bunsen burner
pipettes (5 ml size) or graduates

1. Pour or pipette approximately 3 ml of both urine and Exton's reagent into a test tube (see figure 86.3).
2. Shake the tube from side to side, or roll it between the palms of your hands to mix thoroughly. If precipitate forms, albumin is present.
3. Heat the tube in a Bunsen flame (see figure 86.4). If the precipitate persists, albumin is definitely present in the urine.
4. Record your results on the Laboratory Report.

Test-Strip Method

Materials:
Albustix test strip

Shake the urine sample, and dip the test portion of an Albustix test strip into the urine. Touch the tip of the strip against the edge of the urine container to remove excess urine. Immediately compare the test area with the color chart on the bottle label. Note that the color scale runs from yellow (negative) to turquoise (++++). Record your results on the Laboratory Report.

Detection of Mucin
(Using Glacial Acetic Acid)

Inflammation of the mucous membranes of the urinary tract and vagina may result in large amounts of mucin in urine. Since mucin can be confused with albumin, its presence should be confirmed. Glacial acetic acid is used here to detect mucin.

Materials:
test tubes and test tube holder
Bunsen burner
acetic acid (glacial)
sodium hydroxide (10%)
pipettes (5 ml size)
filter paper
funnel
ring stand
small graduate

1. If the urine sample was positive for albumin, remove the albumin by boiling 5 ml of urine for several minutes and then filtering it while hot (see figure 86.5).
2. After the urine has cooled, add 6 ml of distilled water to 2 ml of the urine to prevent precipitation of urates.
3. Add a few drops of glacial acetic acid. If mucin is present, the urine will become turbid.
4. To further prove that the precipitate is mucin, add a few drops of sodium hydroxide. The precipitate will disappear if it is mucin.
5. Record the results of this test on your Laboratory Report.

Detection of Glucose

Normally, urine has no more than 0.03 g of glucose per 100 ml of urine. Excess glucose in urine is glycosuria , which usually indicates diabetes mellitus.

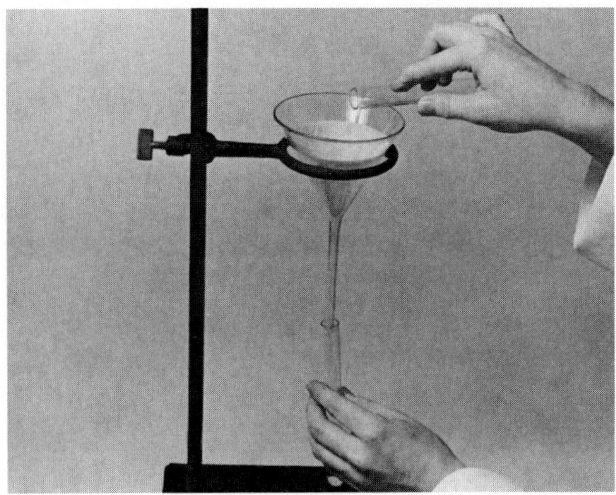

Figure 86.5 Mucin test. Before testing for mucin, remove albumin by filtering boiled urine. Acetic acid causes mucin to precipitate.

Figure 86.6 Glucose test. To detect glucose in urine, boil 0.5 ml of urine in 5 ml of Benedict's regent for 5 minutes to get a color change.

The renal threshold of glucose is around 160 mg per 100 ml. Glycosuria indicates that blood levels of glucose exceed this amount and that the kidneys are unable to reabsorb all of the carbohydrate.

Benedict's Reagent

Glucose in urine is usually detected with Benedict's reagent. In the presence of glucose, a precipitate ranging from yellowish green to red forms. The color of the precipitate indicates the approximate percentage of glucose present.

Materials:
two test tubes
electric hot plate
beaker (250 ml size)
Benedict's qualitative reagent
pipettes (5 ml size) or graduates
urine sample to be tested
urine sample (positive for glucose)

1. Label one test tube for each sample to be tested. One tube should be for a known positive sample, if available.
2. Pour 5 ml of Benedict's reagent into each tube.
3. Add 8 drops (or 0.5 ml) of urine to each tube of Benedict's reagent. Use separate clean pipettes for each urine sample.
4. Place the tubes in a beaker of warm water, and boil the contents of the tubes for 5 minutes (see figure 86.6).
5. Determine the amount of glucose present by color:

Negative = clear blue to cloudy green
\+ = yellowish green (0.5 to 1.0 g %)
++ = greenish yellow (1.0 to 1.5 g %)
+++ = yellow (1.5 to 2.5 g %)
++++ = orange (2.5 to 4 g %)
+++++ = red (4 g % and over)

Test-Strip Method

Materials:
Clinistix test strip

Shake the urine sample, and dip the test portion of a Clinistix test strip into the urine. Ten seconds after wetting, compare the color of the test area with the color chart on the bottle label. Note the three degrees of positivity.

The light intensity generally indicates 0.25% or less glucose. The dark intensity indicates 0.5% or more glucose. The medium intensity has no quantitative significance. Record your results on the Laboratory Report.

Detection of Ketones

Normal catabolism of fats produces carbon dioxide and water as final end products. When carbohydrate in the diet is inadequate, or when carbohydrate metabolism is defective, the body begins to utilize an increasing amount of fatty acids. When this increased fat metabolism reaches a certain point, fatty acid utilization becomes incomplete, and intermediary products of fat metabolism appear in the blood and urine. These intermediary substances are the ketones acetoacetic acid, acetone, and beta hydroxybutyric acid. The presence of these substances in urine is called ketonuria.

Ketonuria may occur in diabetes mellitus. Progressive diabetic ketosis causes diabetic acidosis, which can eventually lead to coma or death. For this reason, ketonuria detection in diabetics is very important.

Rothera's Test

The test method used here is Rothera's test. The active ingredients in Rothera's reagent are sodium nitroprusside and ammonium sulfate. In the presence of ketones and ammonium hydroxide, this reagent produces a pinkish purple color.

Materials:
Rothera's reagent
test tubes
ammonium hydroxide (concentrated)

1. Add about 1 g of Rothera's reagent to 5 ml of urine in a test tube.
2. Allow 1 to 2 ml of concentrated ammonium hydroxide to flow gently down the side of the inclined test tube and layer over the urine.
3. If a pink-purple ring develops at the interface, ketones are present. No ring, or a brown ring, is negative.
4. Record your results on the Laboratory Report.

Test-Strip Method

Materials:
Ketostix test strip

Shake the urine sample, and dip the test portion of a Ketostix test strip into the urine. Tap it dry on the edge of the urine container. Fifteen seconds after wetting, compare the color of the test strip with the color chart on the bottle label. Record your results on the Laboratory Report.

Detection of Hemoglobin (Using Hemastix Test Strips)

Red blood cells that disintegrate (hemolyze) in the body release hemoglobin into the surrounding fluid. If hemolysis occurs in the blood vessels, the hemoglobin becomes a constituent of the plasma, and the kidneys excrete some of it into the urine. If red blood cells enter the urinary tract due to disease or trauma to parts of the urinary system, the cells hemolyze in the urine. Regardless of the cause of hemoglobin in urine, the condition is called hemoglobinuria.

Hemoglobinuria may be evidence of a number of conditions, including hemolytic anemia, transfusion reactions, yellow fever, smallpox, malaria, hepatitis, mushroom poisoning, renal infarction, and burns. The simplest way to test for hemoglobin is to use Hemastix test strips as follows:

Materials:
Hemastix test strip

1. Shake the urine sample, and dip the test portion of a Hemastix test strip into the urine.
2. Tap the edge of the strip against the edge of the urine container, and let it dry for 30 seconds.
3. Compare the color of the test strip with the color chart on the bottle label.
4. Record your results on the Laboratory Report.

Detection of Bilirubin (Using Bili-Labstix)

Bilirubin forms in the second stage of hemoglobin degradation from hematin. It is normally present in urine in very small quantities. When present in large amounts, however, it usually indicates that hepatotoxic agents (chemical or biological) have damaged the liver. The presence of significant amounts of bilirubin is called bilirubinuria. The simplest way to test for this constituent in urine is to use Bili-Labstix. The procedure is as follows:

Materials:
Bili-Labstix test strip

1. Shake the urine sample, and dip the test portion of a Bili-Labstix test strip into the urine.
2. Tap the edge of the strip against the edge of the urine container, and let it dry for 20 seconds.
3. Compare the color of the test strip with the color chart on the bottle label. The results are interpreted as negative, small (+), moderate (++), or large (+++) amounts of bilirubin.
4. Record your results on the Laboratory Report.

Microscopic Study

Microscopic examination of elements in urine is the most important phase of urinalysis. However, the scope of this phase goes considerably beyond the

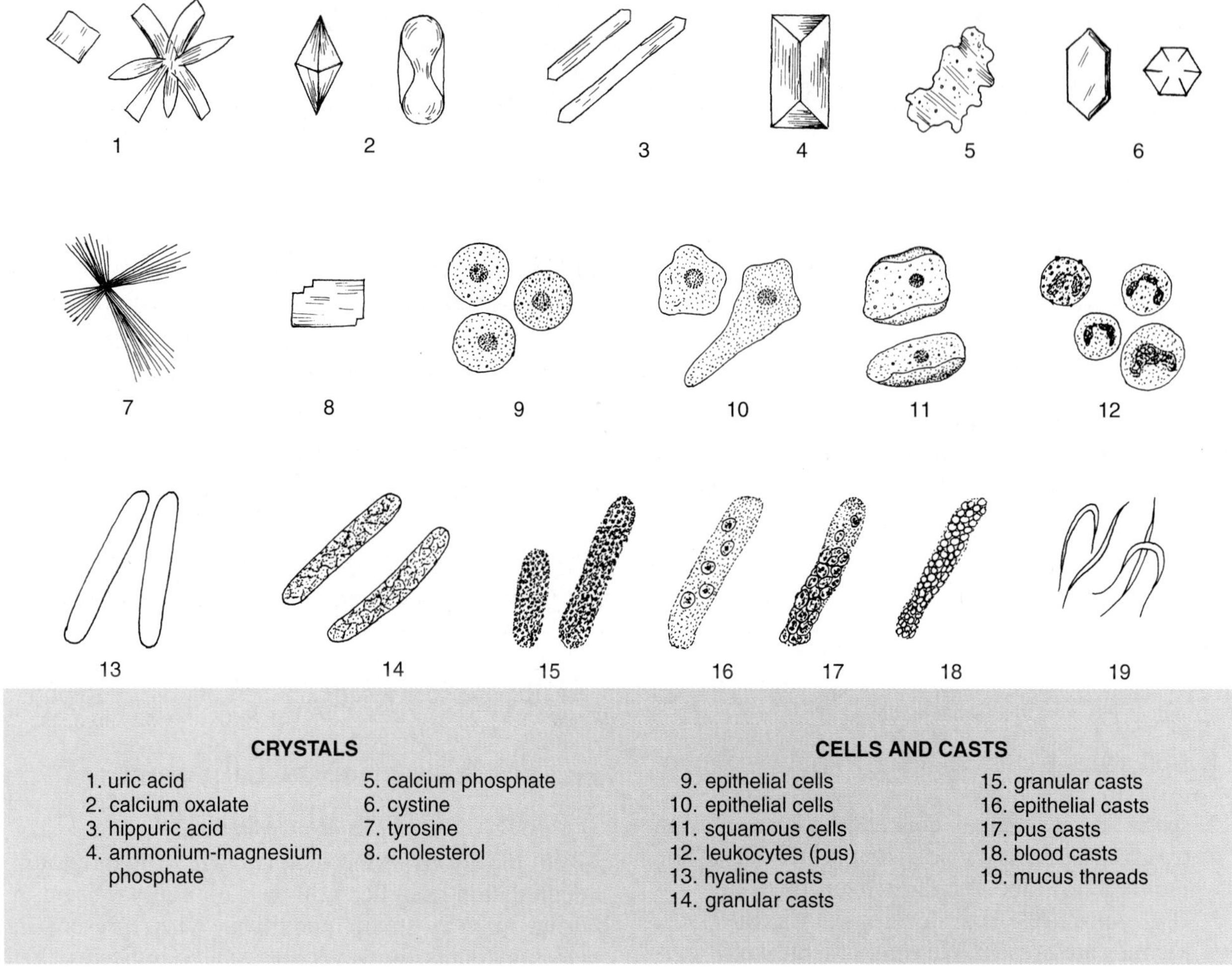

Figure 86.7 Microscopic elements in urine.

limitations of this course. A complete microscopic examination includes not only visual analysis of the sediment, but also bacteriological differentiation.

Normal urine contains an occasional leukocyte, some epithelial cells, mucin, bacteria, and crystals of various kinds. The experienced technologist has to determine when these substances are in excess and be able to differentiate the various types of casts, cells, and crystals that predominate. Figure 86.7 illustrates only a few of the elements that might be found in urine. Do not attempt in this exercise to identify all particulate matter in urine.

Materials:
microscope slides and cover glasses
Pasteur pipettes (optional)
centrifuge
centrifuge tubes (conical tipped)
pipettes (5 ml size) or graduates
mechanical pipetting device
wire loop

1. Shake the urine sample to resuspend the sediment. Use a mechanical pipetting device or a graduate to measure 5 ml of urine into a centrifuge tube. Balance the centrifuge with an even number of loaded tubes.
2. Centrifuge the tubes for 5 minutes at a slow speed (1500 rpm).
3. Pour off all urine, and allow the remaining sediment in the tube to settle to the bottom.
4. With a Pasteur pipette or flamed wire loop, transfer a small amount of the sediment to a microscope slide, and cover with a cover glass.
5. Examine under the microscope with low- and high-power objectives. Adjust the diaphragm to reduce the lighting. Refer to figure 86.7 to identify structures.

Assignment:
Complete the Laboratory Report for this exercise.

PART 16 The Endocrine and Reproductive Systems

The endocrine and reproductive systems are logically studied together because the endocrine glands have a regulatory role in reproduction. Exercise 87 focuses on the histological nature of the endocrine glands and the specific hormones that each gland produces. Microscope slides of the various glands show the specific cells that produce the hormones. Since most of the glands have already been observed in previous dissections, no cat dissection is performed here.

Exercise 88 examines the anatomy of the male and female reproductive organs, as well as spermatogenesis and oogenesis. The cat is used for dissection.

Exercise 89 studies the fertilization of an egg and the resultant stages in early cleavage, using sea urchin eggs and spermatozoa.

87 The Endocrine Glands

In the cat dissections of previous exercises, you encountered and briefly studied various endocrine glands. In this exercise, you study all of the endocrine glands (see figure 87.1) to review what has been stated previously and to explore in greater detail the histology of each gland. By examining stained microscope slides of the various glands, and by comparing the slides with photomicrographs, you should be able to identify the specific cells in most glands that produce the various hormones. This laboratory experience should help to dispel some of the abstractness usually encountered when trying to assimilate a mass of endocrinological facts.

The Thyroid Gland

The thyroid gland consists of two lobes joined by a connecting isthmus (see figure 87.1). Figure 87.2 shows a posterior view of this gland. Note in figure 87.3 that, microscopically, the thyroid gland consists of large numbers of spherical sacs called **follicles.** A colloidal suspension of a glycoprotein, **thyroglobulin,** fills these follicles. The principal hormones of the thyroid gland are *thyroid hormone* and *calcitonin.*

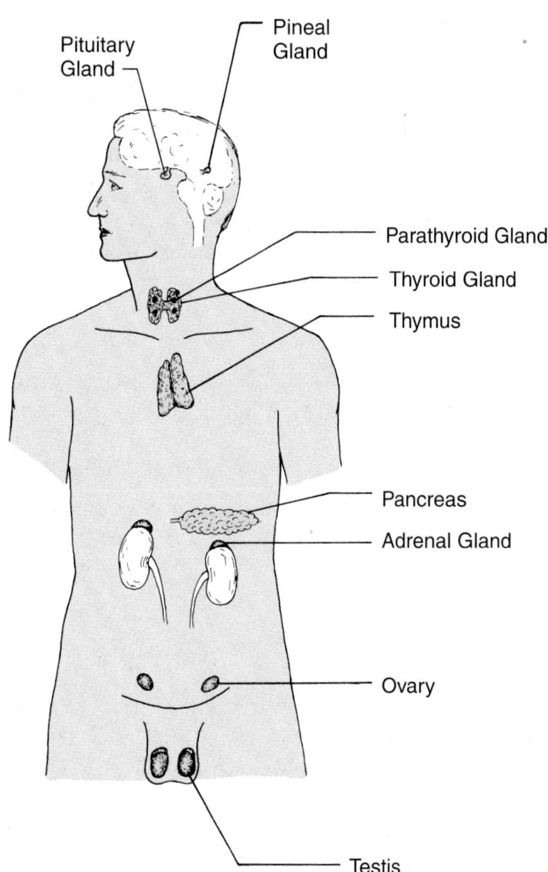

Figure 87.1 Secretory components of the endocrine system.

Thyroid Hormone

Thyroid hormone consists of thyroxine, triiodothyronine, and a small quantity of closely related iodinated hormones. They form within the thyroglobulin, emerge from the follicles, and enter blood vessels in the gland for transport to all body tissues. Thyroglobulin stores excess hormones.

Action Together, thyroxine and triiodothyronine increase the metabolic activity of most body tissues: Bone growth in children accelerates; carbohydrate metabolism is enhanced; and cardiac output and heart rate, respiratory rate, and mental activity increases.

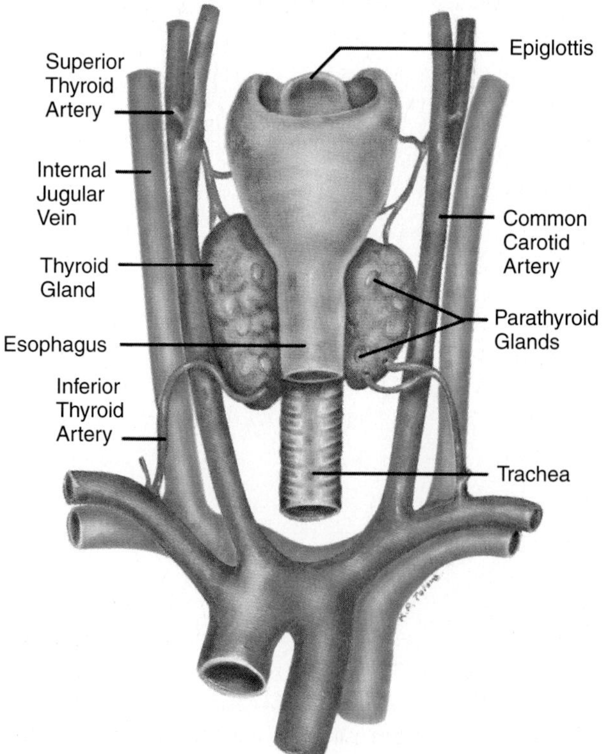

Figure 87.2 The thyroid and parathyroid glands.

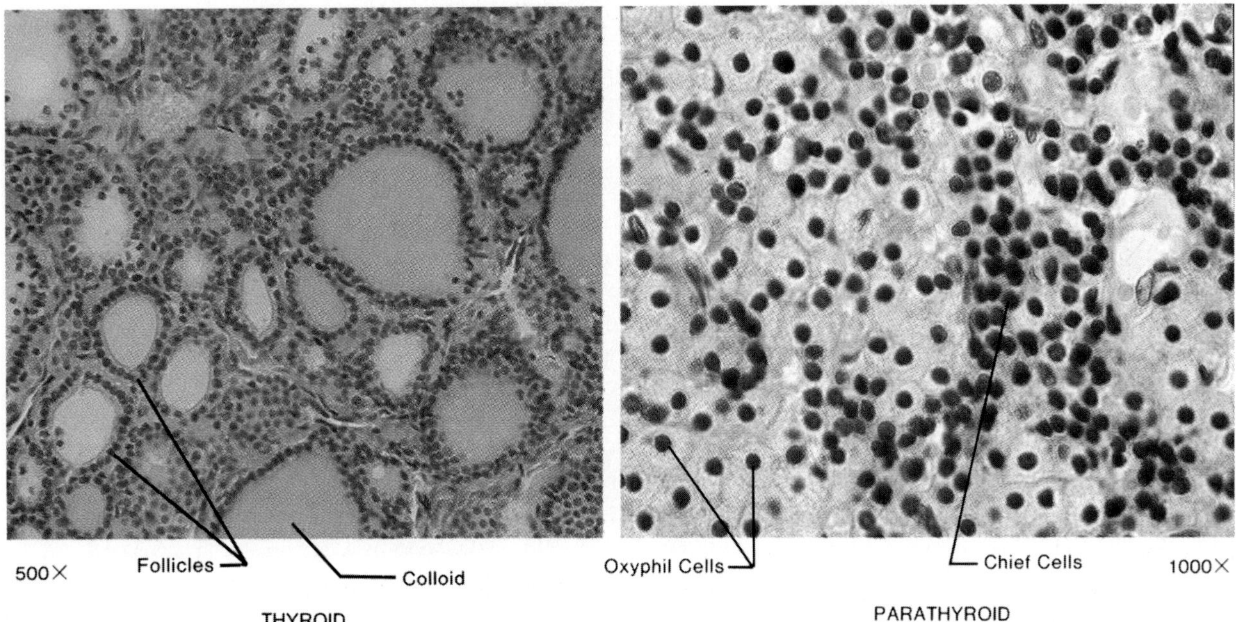

Figure 87.3 The thyroid and parathyroid glands.

Deficiency Symptoms In children, hypothyroidism may result in *cretinism,* characterized by physical and mental retardation. In adults, severe, long-term deficiency causes *myxedema,* characterized by obesity, slow pulse, lack of energy, and mental depression.

Oversecretion Symptoms Weight loss, rapid pulse, intolerance to heat, nervousness, and inability to sleep characterize hyperthyroidism. *Exophthalmia,* or protrusion of the eyes, is often present.

Goiters Enlarged thyroid glands ("goiters") may be present in both hypo- and hyperthyroidism. Hypothyroid goiters may be due to iodine deficiency (*endemic goiter*) or some other unknown cause (*idiopathic nontoxic goiter*). Hyperthyroid goiters may be due to malignancy of the gland.

Regulation The anterior pituitary gland produces thyroid-stimulating hormone (TSH), or thyrotropin, which stimulates the thyroid to produce thyroid hormone.

Calcitonin

The parafollicular cells in the interstitium of the thyroid gland produce **calcitonin.** When injected experimentally, calcitonin causes a decrease in blood calcium, due, presumably, to increased osteoblastic and decreased osteoclastic activity. Calcitonin is important in bone remodeling during growth in children, but in adults, it has very little, if any, effect on long-term blood calcium levels.

The Parathyroid Glands

Four small parathyroid glands are embedded in the posterior surface of the thyroid gland (see figures 87.1 and 87.2). Occasionally, parathyroid tissue is outside of the thyroid gland in the neck region.

Parathyroid glands have two kinds of cells: **chief** and **oxyphil cells** (see figure 87.3). Chief cells are smaller, more numerous, and arranged in cords. They produce the only hormone of the parathyroid glands—**parathyroid hormone (PTH)**. The function of the larger oxyphil cells is unknown at this time.

Parathyroid Hormone

Action PTH raises blood calcium levels and lowers blood phosphorus levels by activating osteoclasts in bone tissue. To prevent calcium loss in the kidneys, PTH also promotes calcium reabsorption in the renal tubules. Removal of all parathyroid tissue results in *tetany* and death.

Regulation Human blood calcium levels are remarkably stable—about 10 mg/100 ml—and appear to regulate PTH production. Excess blood calcium levels inhibit PTH production, while decreased blood calcium levels increase PTH production.

The Pancreas

Figure 87.4 illustrates the histology of the pancreas (see also figure 87.1). The endocrine-secreting

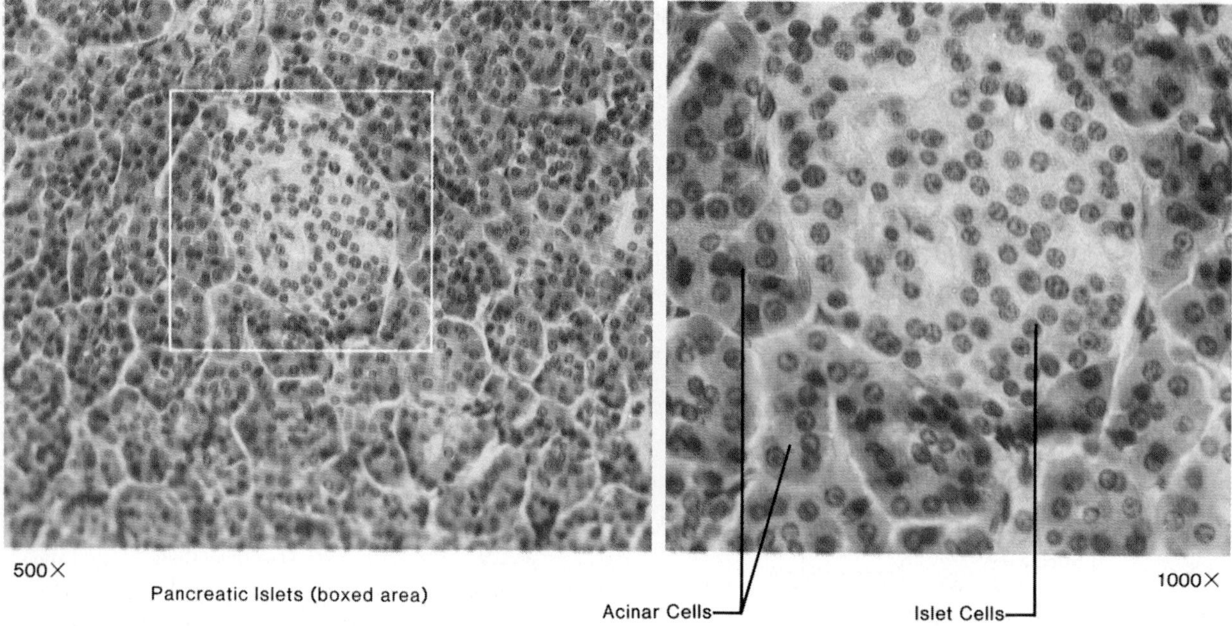

Figure 87.4 Pancreas histology.

portions of the pancreas are clusters of cells called **pancreatic islets** (islets of Langerhans). These cells are located between the saclike glands (*acini*) that produce the pancreatic digestive enzymes. Different types of cells in the pancreatic islets produce insulin, glucagon, and somatostatin.

Insulin

Insulin is an anabolic hormone that the **beta cells** of the pancreatic islets produce. Beta cells have many granules in the cytoplasm, which differentiates beta cells from other islet cells.

Action Insulin promotes glucose storage by converting glucose to glycogen. It also promotes the storage of fatty acids and amino acids. Insulin deficiency results in *diabetes mellitus,* characterized by *hyperglycemia* (high blood sugar).

Regulation Although a variety of stimulatory and inhibitory factors affect insulin production, blood glucose level is the major controlling factor. Elevated blood glucose levels increase insulin production; normal or low blood glucose levels decrease insulin production.

Glucagon

Glucagon is a catabolic hormone that the **alpha cells** of the pancreatic islets produce. Its action is the opposite of insulin in that it mobilizes glucose, fatty acids, and amino acids in tissues. Glucagon deficiency results in hypoglycemia. Glucagon also stimulates the production of growth hormone, insulin, and pancreatic somatostatin.

Somatostatin

The **delta cells** of the pancreatic islets produce **somatostatin,** a growth-inhibiting hormone. With Mallory's stain, the cytoplasm of these cells stains blue.

Somatostatin is growth inhibiting in that it inhibits the anterior pituitary gland's release of growth hormone. In addition, it inhibits insulin and glucagon production. Tumors involving the delta cells result in hyperglycemia and other diabetes-like symptoms. Removal of the tumors eradicates the symptoms.

The Adrenal Glands

Each of the two adrenal glands (see figure 87.1) has an outer *cortex* and an inner *medulla.* A capsule of fibrous connective tissue surrounds each gland.

The embryological origins of the adrenal cortex and medulla explain their differences in function. The cells of the adrenal medulla originate in the neural crest of the embryo. This embryonic tissue also gives rise to ganglionic cells of the sympathetic nervous system. Thus, the adrenal medulla and the sympathetic nervous system have a close kinship in function. Cells of the adrenal cortex, on the other hand, arise from embryonic tissue associated with the gonads and therefore produce hormones that affect the reproductive organs.

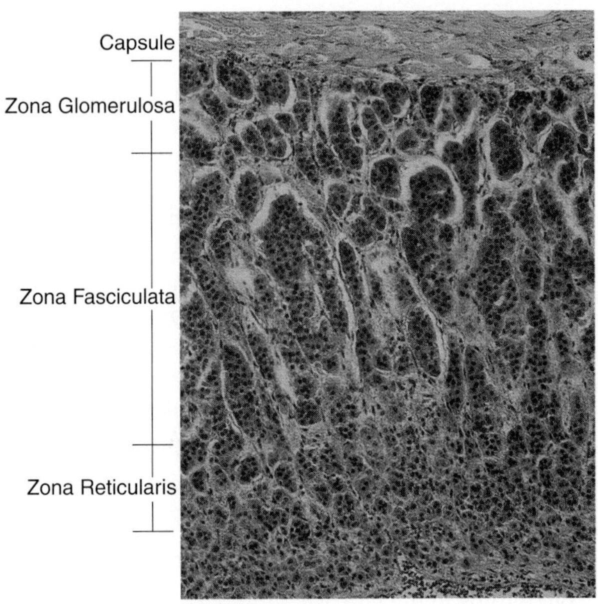

Figure 87.5 The adrenal cortex.

The Adrenal Cortex

Figures 87.5 and 87.6 show the three layers of the cortex. The **zona glomerulosa** is the layer immediately under the capsule. The innermost layer of cells, which interfaces with the medulla, is the **zona reticularis.** The thick middle layer is the **zona fasciculata.**

All three layers have vascular channels called **sinusoids.** Sinusoids are larger in diameter than capillaries, but they resemble capillaries in that their walls are one cell thick. Hormones collect in sinusoids before passing directly into the general circulation.

The adrenal cortex secretes many different hormones, all of which are *steroids*—or more specifically, **corticosteroids.** Corticosteroids are synthesized from cholesterol and have similar molecular structures. Corticosteroids fall into three groups: mineralocorticoids, glucocorticoids, and androgenic hormones. *Mineralocorticoids* control the excretion of sodium and potassium ions that affects electrolyte balance. *Glucocorticoids* primarily affect the metabolism of glucose and protein. *Androgenic hormones* have some of the same effects on the body as the male sex hormone (testosterone). Of the 30 or more corticosteroids that the adrenal cortex produces, aldosterone and cortisol are the most important.

Aldosterone **Aldosterone** is a mineralocorticoid that cells of the zona glomerulosa produce. Its most important function is to increase the rate of renal tubular absorption of sodium.

Any condition that impairs aldosterone production results in death within 2 weeks if salt replacement or mineralocorticoid therapy is not provided. In the absence of aldosterone, the potassium ion concentration in extracellular fluids rises, sodium and chloride ion concentrations decrease, and the total volume of body fluids becomes greatly reduced. These conditions cause reduced cardiac output, shock, and death.

Factors that affect aldosterone production are adrenocorticotropic hormone (ACTH), potassium ion concentration, sodium ion concentration, and the renin-angiotensin system.

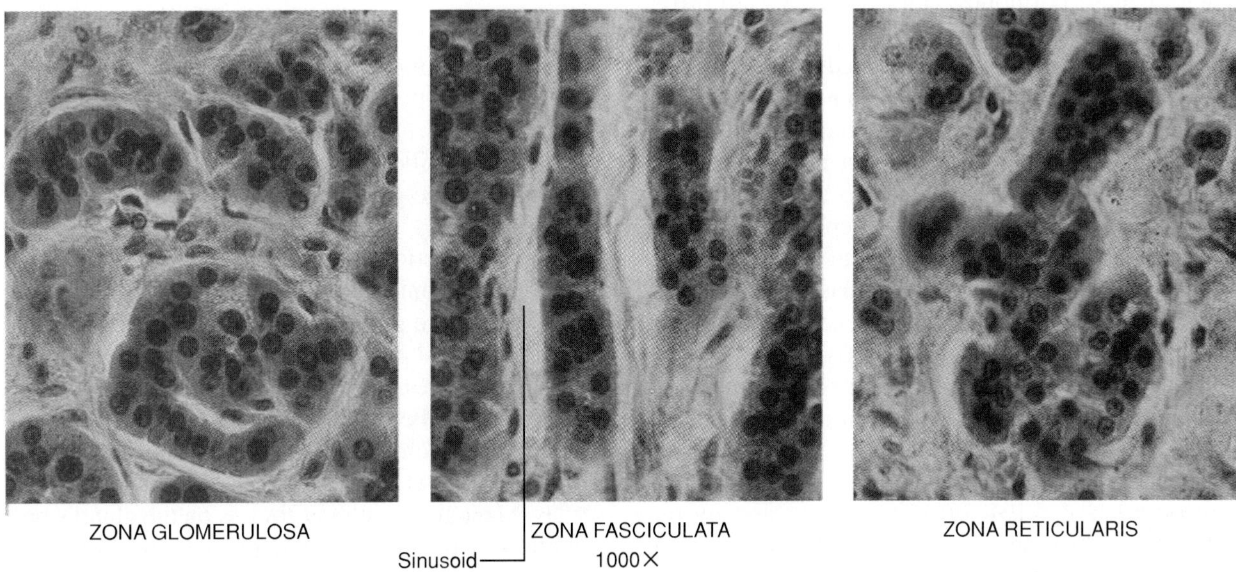

Figure 87.6 Histology of the three layers of the adrenal cortex.

Cortisol (*Hydrocortisone, Compound F*) **Cortisol** is responsible for approximately 95% of glucocorticoid activity. Cells primarily in the zona fasciculata, but also in the zona reticularis, produce cortisol.

Excess cortisol results in: (1) an increased rate of glucogenesis, (2) decreased cell utilization of glucose, (3) increased protein catabolism in all cells except the liver, (4) decreased amino acid transport to muscle cells, and (5) mobilization of fatty acids from adipose tissue. Many of these reactions characterize *Cushing's syndrome,* in which a pituitary hormone induces an *excess of glucocorticoids.* Protein depletion in individuals with Cushing's syndrome causes poorly developed muscles, weak bones (due to bone dissolution), slow wound healing, hyperglycemia, and thin, scraggly hair.

Total glucocorticoid insufficiency characterizes *Addison's disease.* The destruction of the adrenal cortex by cancer, tuberculosis, or some other infectious agent causes Addison's disease.

Although several substances, such as vasopressin, serotonin, and angiotensin II, stimulate the adrenal cortex, ACTH is primarily responsible for inducing the cells of the zona fasciculata and zona reticularis to produce cortisol. The anterior pituitary's production of ACTH is partially initiated by stress through the hypothalamus. Stress induces the hypothalamus to produce corticotropin-releasing factor (CRF), which passes to the anterior pituitary and causes it to produce ACTH.

The Adrenal Medulla

The adrenal medulla produces two **catecholamines: norepinephrine** and **epinephrine.** Although different cells in the medulla produce each of these hormones, cell differences are difficult to detect microscopically.

Approximately 80% of the medullary secretion is epinephrine, with norepinephrine making up the remaining 20%. Certain nerve endings of the autonomic nervous system also produce norepinephrine.

Action The effects of norepinephrine and epinephrine in different tissues depend on the types of receptors in the tissues. The two classes of adrenergic receptors are alpha and beta, each of which has two types (α_1 and α_2 and β_1 and β_2).

Both norepinephrine and epinephrine increase the force and rate of heart contraction. The β_1 receptors mediate these reactions. Norepinephrine causes vasoconstriction in all organs through α_1 receptors.

Epinephrine causes vasoconstriction everywhere except in the muscles and liver, where beta receptors bring about vasodilation. Both catecholamines also increase mental alertness, possibly because of increased blood pressure.

Both epinephrine and norepinephrine increase blood glucose levels by glycogenolysis (hydrolysis of glycogen to glucose) in the liver. The β_2 receptors account for this reaction.

Norepinephrine and epinephrine are equally potent in mobilizing free fatty acids through beta receptors. They also increase the metabolic rate, although how this takes place is not yet precisely understood.

Regulation Sympathetic nerve fibers stimulate the adrenal medulla to produce epinephrine and norepinephrine.

The Thymus Gland

The thymus gland consists of two long tubes and lies in the upper chest region above the heart (see figure 87.1). It is most highly developed before birth and during the growing years. After puberty, it begins to regress and continues to do so throughout life.

Histologically, the thymus gland consists of a cortex and medulla (see figure 87.7). The **cortex** consists of lymphoidlike tissue filled with a large number of lymphocytes. The **medulla** contains a smaller number of lymphocytes, as well as **thymic corpuscles**, whose function is unknown.

The principal role of the thymus gland is processing lymphocytes into T-cells, as described on page 351. The thymus gland may produce one or more hormones that function in T-cell formation. Researchers are studying thymosin and several other extracts from this gland.

The Pineal Gland

The pineal gland is on the roof of the third ventricle under the posterior end of the corpus callosum (see figure 87.1). A stalk of postganglionic sympathetic nerve fibers supports but does not seem to extend into the gland. The pineal gland consists of neuroglial and parenchymal cells that suggest a secretory function.

In young animals and infants, the pineal gland is large and more glandlike; cells tend to be arranged in alveoli. Just before puberty, however, the gland begins to regress, and small concretions of calcium carbonate, called **pineal sand,** form (see figure 87.7, *right*). Some evidence, though not conclusive, indicates that the gland contains gonadotropin peptides. Most attention, however, is on the gland's production of melatonin. **Melatonin** is an indole synthesized from serotonin. Daylight (circadian rhythm) appears to regulate its production.

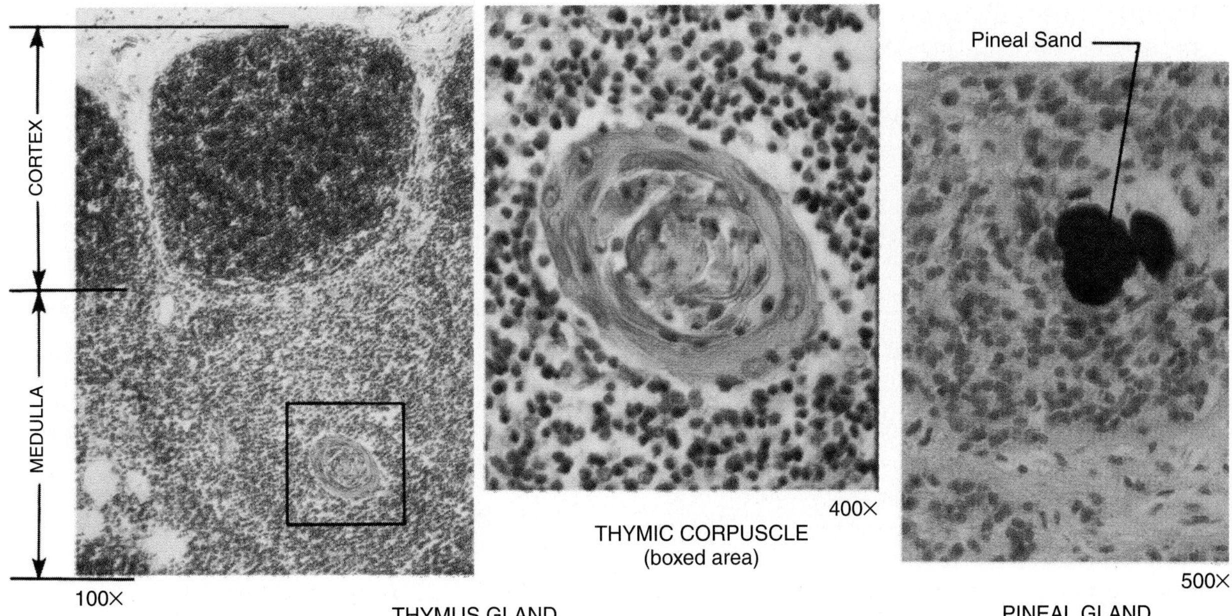

Figure 87.7 The thymus and pineal glands.

During daylight, melatonin production is suppressed. At night, the pineal gland becomes active and produces considerable melatonin. Regulation occurs through the eyes. Light striking the retina sends messages to the pineal gland through a retina-hypothalamic pathway. Norepinephrine reaches the cells from postganglionic sympathetic nerve endings in the pineal stalk. Beta adrenergic receptors to the cells are mediators in the inhibition of melatonin production.

Many researchers speculate that melatonin inhibits the estrus (menstrual) cycle in humans, as it probably does in lower animals. This has not yet been proven, however.

The Testes

The testes are the primary male reproductive organs (see figure 87.1). A section through a testis shows coils of **seminiferous tubules** that give rise to spermatozoa, and **interstitial cells** that secrete the male sex hormone **testosterone,** (see figure 87.8). Note that interstitial cells lie in the spaces between seminiferous tubules.

The testes produce large amounts of testosterone during embryological development, but very little of the hormone from early childhood to puberty. With the onset of puberty at age 10 or 11, the testes begin to produce large quantities of testosterone as sexual maturity develops.

During embryological development, testosterone is responsible for the development of the male sex organs. If the testes are removed from a fetus at an early stage, the fetus develops a clitoris and vagina instead of a penis and scrotum, even though the fetus is a male. Testosterone also controls the development of the prostate gland, seminal vesicles, and male ducts, while suppressing the formation of female genitals.

The embryological testes develop within the body cavity. During the last two months of development the testes descend into the scrotum through the inguinal canals. Testosterone controls the descent of the testes.

During puberty, the testes enlarge and produce a great deal of testosterone, which considerably enlarges the penis and scrotum. At the same time, the male secondary sexual characteristics develop: (1) Hair appears on the face, axillae, chest, and in the pubic region; (2) the larynx enlarges, and the voice deepens; (3) the skin thickens; (4) muscular development increases; and (5) bones thicken and strengthen.

During embryological development, **chorionic gonadotropin,** a hormone that the placenta produces, regulates testosterone production. At the onset of puberty, the anterior lobe of the pituitary gland produces luteinizing hormone (LH), which stimulates the interstitial cells of the testes to produce testosterone. The anterior pituitary gland also produces follicle-stimulating hormone (FSH), which controls spermatozoa maturation in the testes. Testosterone assists FSH in this process.

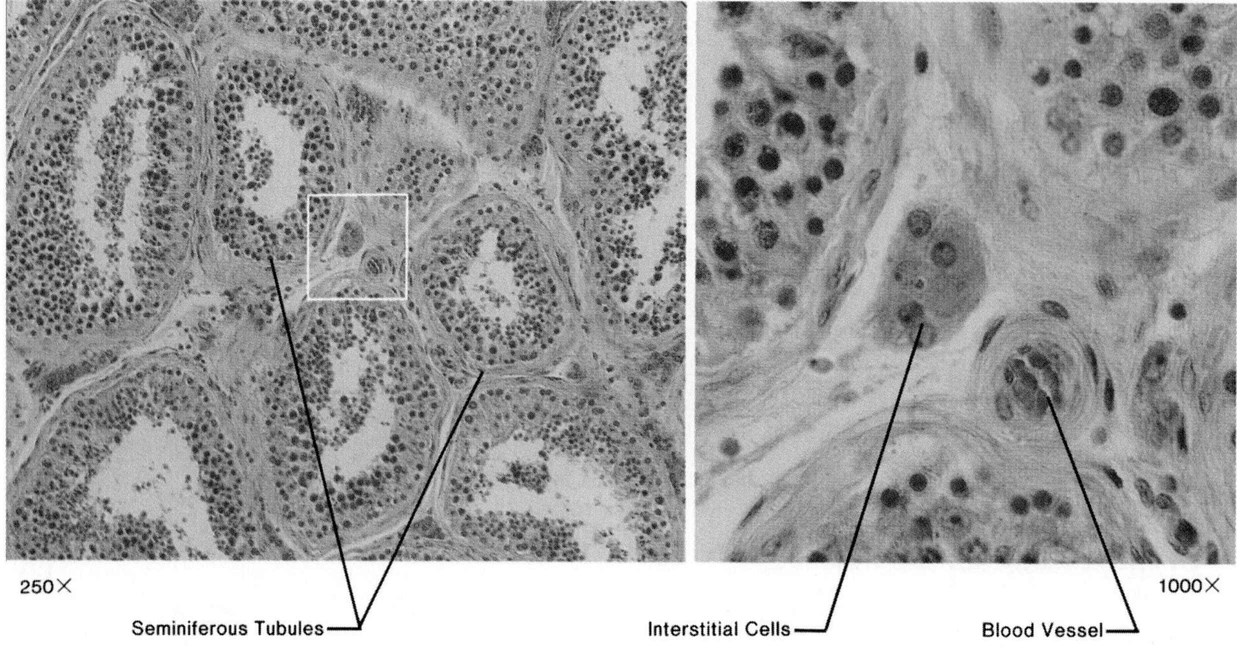

Figure 87.8 Testicular tissue.

The Ovaries

The ovaries are the primary female reproductive organs (see figure 87.1). During fetal development, small groups of cells move inward from the germinal epithelium of each ovary and develop into primordial follicles. As figure 87.9 shows, the **germinal epithelium** is a layer of cuboidal cells near the ovarian surface, and the **primordial follicles** are the small, round bodies that contain **ova.**

At the onset of puberty, the ovaries of a young female contain 100,000 to 400,000 immature follicles. During a female's reproductive years, only about 400 of these follicles will mature and expel their ova.

During the reproductive years, selected primordial follicles enlarge to form **Graafian follicles,** and every 28 days or so, a maturing Graafian follicle expels an ovum (ovulation). The development of these follicles and the subsequent corpus luteum results in the production of **estrogens** and **progesterone,** the principal female sex hormones.

Estrogens

Estrogens promote the growth of specific cells in the body and control the development of female secondary sexual characteristics. Graafian follicles produce estrogens, as do the corpus luteum, placenta, adrenal cortex, and testes. FSH secretion by the anterior lobe of the pituitary gland initiates estrogen production at puberty.

Of the six or seven estrogens that have been isolated from female plasma, **β-estradiol, estrone,** and **estriol** are the most abundant. The most potent of all the estrogens is β-estradiol.

During puberty, when the estrogens are produced in large quantities, female secondary sexual characteristics develop: (1) female reproductive organs develop fully; (2) the vaginal epithelium changes from cuboidal to stratified epithelium; (3) the uterine lining (endometrium) becomes more glandular in preparation for implantation of the fertilized ovum; (4) the breasts form; (5) osteoblastic activity increases with more rapid bone growth; (6) calcification of the epiphyses in long bones is hastened; (7) pelvic bones enlarge and change shape, increasing the size of the pelvic outlet; (8) fat deposition under the skin, and in the hips, buttocks, and thighs increases; and (9) skin vascularization increases.

Progesterone

After ovulation, the **corpus luteum** replaces the Graafian follicle and produces progesterone. Progesterone promotes secretory changes in the endometrium in preparation for implantation of the fertilized ovum. It also causes mucosal changes in the uterine tubes and promotes the proliferation of alveolar cells in the breasts in preparation for milk production.

If the plasma level of progesterone falls, due to regression of the corpus luteum, the endometrium deteriorates, and menstruation occurs. If pregnancy occurs, however, the corpus luteum enlarges and produces additional progesterone. The conversion

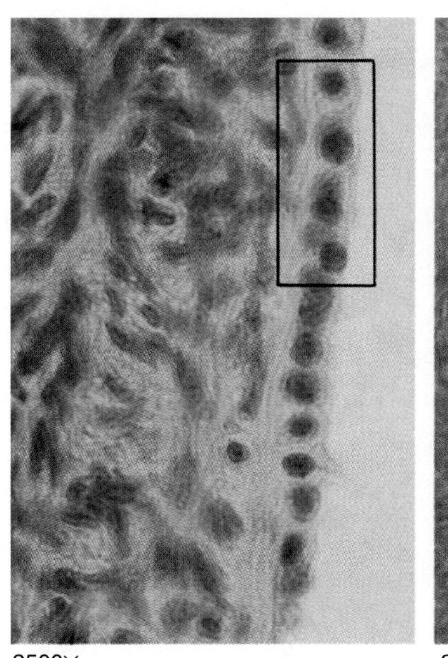

GERMINAL EPITHELIUM (boxed cells)

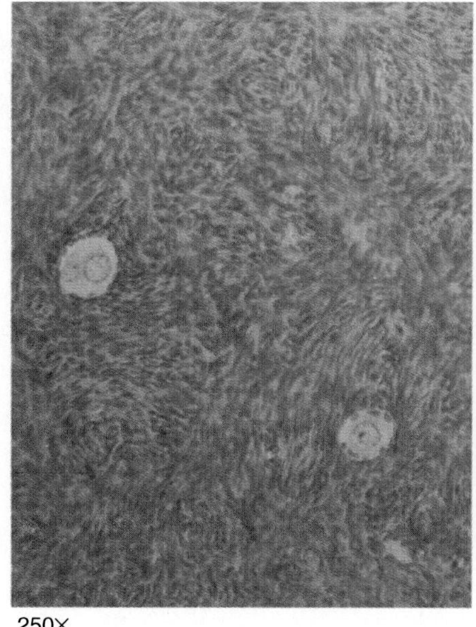

PRIMORDIAL FOLLICLES

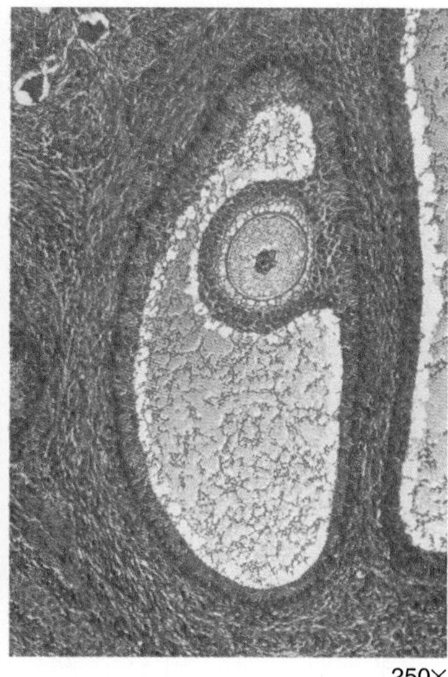

MATURE GRAAFIAN FOLLICLE

Figure 87.9 Ovarian tissue.

of a Graafian follicle into a corpus luteum depends completely on LH production.

The Pituitary Gland

The pituitary gland, or hypophysis, consists of two lobes: anterior and posterior. The anterior lobe, or **adenohypophysis,** develops from the roof of the oral cavity of the embryo and consists of glandular cells. The posterior lobe, or **neurohypophysis,** develops as an outgrowth of the floor of the embryonic brain. Its cells are nonsecretory and resemble neuroglial tissue. An extension of the dura mater invests the entire pituitary gland.

Figure 87.10 shows most of the hormones produced by the pituitary gland that regulate growth and glands of the body. The figure also shows many of the hormones produced by glands that the pituitary gland regulates.

Figure 87.11 shows portions of the two pituitary lobes. Note that the neurohypophysis in this region consists of two parts: the **pars intermedia** and the **pars nervosa.**

The Adenohypophysis

The adenohypophysis has two basic types of glandular cells: chromophobes and chromophils (see figure 87.12). **Chromophobes** lack an affinity for routine dyes used in staining tissues. **Chromophils** stain readily and are of two types: acidophils and basophils. **Acidophils** take on the pink stain of eosin. **Basophils** stain readily with basic stains, such as methylene blue and crystal violet.

These various cells in the adenohypophysis produce six hormones. Except for somatotropin, all of these hormones are targeted specifically for glands. The hypothalamus regulates this hormone production. Between the hypothalamus and adenohypophysis is a vascular connection, the *hypophyseal portal system,* that transports releasing factors from the hypothalamus to the secretory cells. Cells in the hypothalamus secrete the releasing factors.

The six hormones of the adenohypophysis are as follows:

Somatotropin **Somatotrophs,** a type of acidophil, produce **somatotropin (SH),** also called **growth hormone (GH).** In males and nonpregnant females most acidophils are somatotrophs.

SH increases the growth rate of all body cells by enhancing amino acid uptake and protein synthesis. Excess SH production during the growing years causes **gigantism,** due to the stimulation of growth in the epiphyses of the long bones. An adult with excess SH production develops **acromegaly,** characterized by enlargement of the mandible and forehead, as well as the small bones of the hands and feet. SH deficiency can cause **dwarfism.**

Prolactin Another type of acidophil, called a **mammotroph,** produces **prolactin** (also known as luteotropic hormone [LTH]). With suitable staining

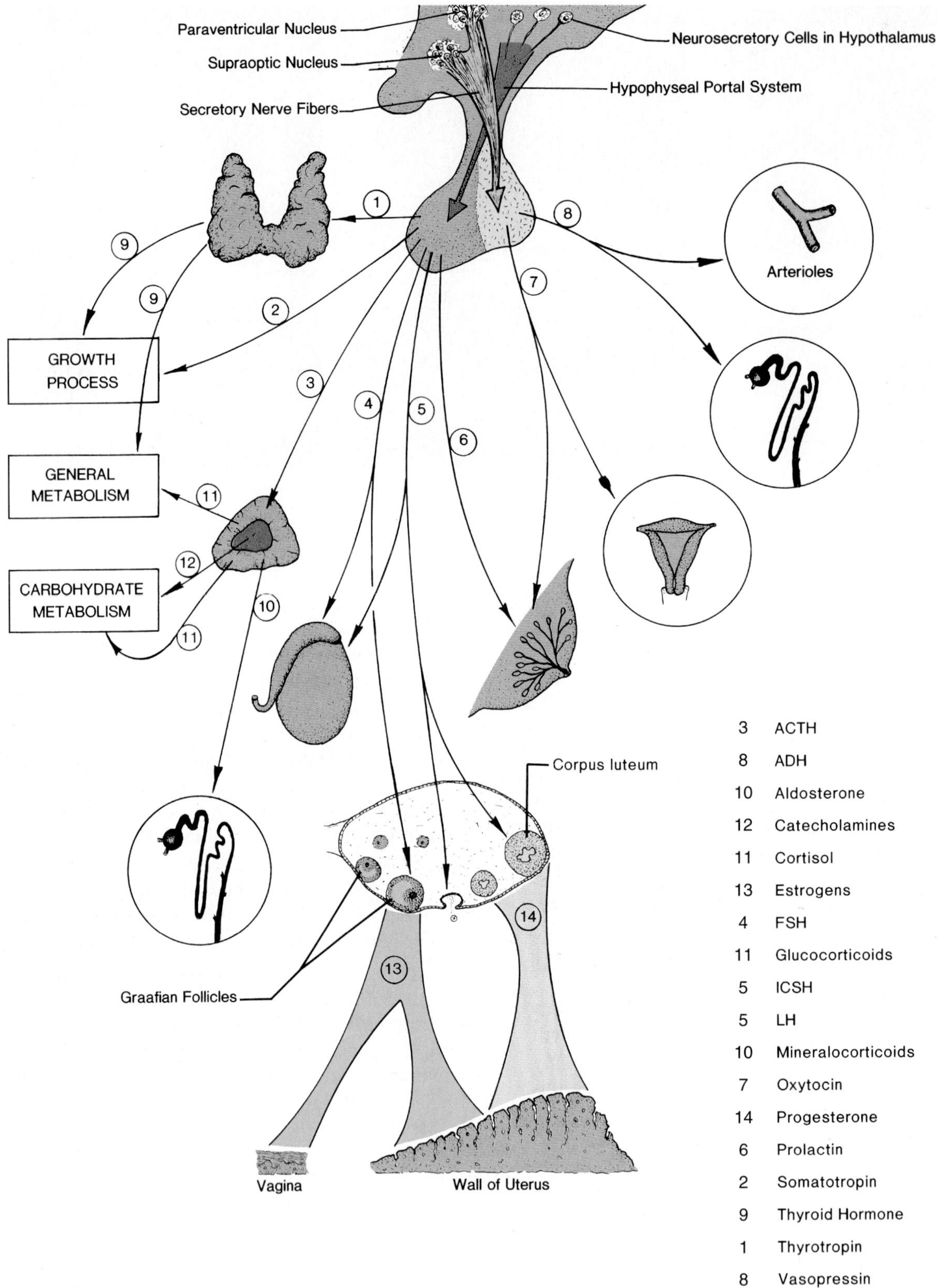

3	ACTH
8	ADH
10	Aldosterone
12	Catecholamines
11	Cortisol
13	Estrogens
4	FSH
11	Glucocorticoids
5	ICSH
5	LH
10	Mineralocorticoids
7	Oxytocin
14	Progesterone
6	Prolactin
2	Somatotropin
9	Thyroid Hormone
1	Thyrotropin
8	Vasopressin

Figure 87.10 The pituitary regulatory mechanism.

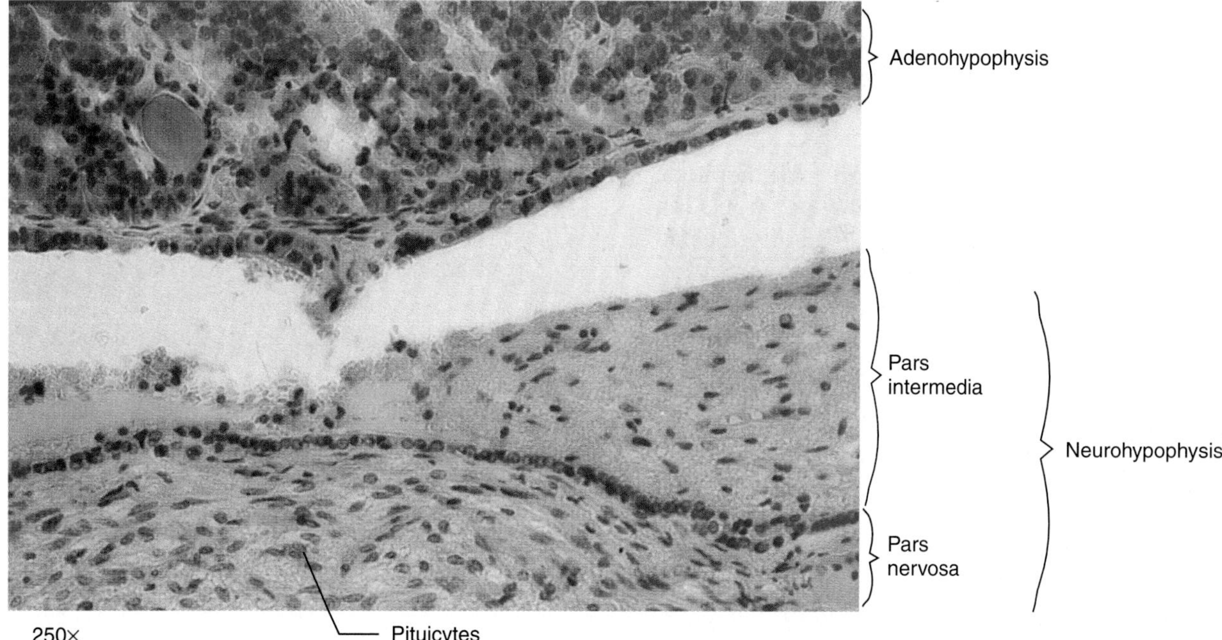

Figure 87.11 The pituitary gland.

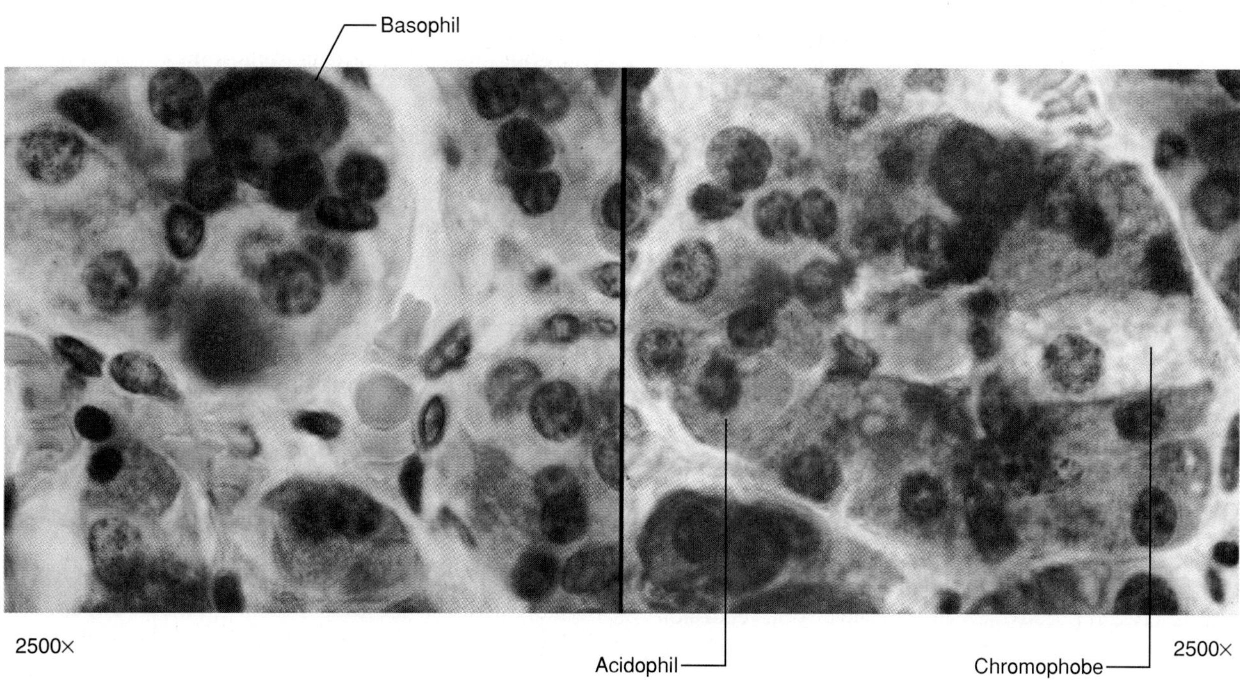

Figure 87.12 Cells of the adenohypophysis.

methods, mammotrophs can be differentiated from somatotrophs.

Prolactin promotes the milk production in the breasts after childbirth. Prior to childbirth, the hypothalamus produces *PIF (prolactin-inhibiting factor),* which inhibits prolactin release. Large quantities of estrogens and progesterone prior to childbirth cause the hypothalamus to produce PIF.

Thyrotropin Large, irregular basophilic cells, called **thyrotrophs** produce **thyrotropin** (thyroid-stimulating hormone [TSH]). Thyrotropin regulates the rate of iodine uptake and the thyroid gland's synthesis of thyroid hormone. Excess thyrotropin causes hyperthyroidism. A thyrotropin deficiency causes hypothyroidism.

Hypothalamic *thyrotropin-releasing factor (TRF)* regulates thyrotropin production. Nerve endings in the hypothalamus secrete TRF, which then passes to the adenohypophysis via the hypophyseal portal system. A thyroid hormone feedback mechanism also regulates thyrotropin production.

Adrenocorticotropin A type of chromophobe probably produces **adrenocorticotropin (ACTH).** The suspected cell is a large, peripherally located chromophobe with large cytoplasmic granules. ACTH controls glucocorticoid production in the adrenal cortex.

As stated previously, CRF (corticotropin-releasing factor) initiates ACTH production. Stress causes cells in the hypothalamus to release CRF into the hypophyseal portal system, activating the adenohypophysis to produce ACTH. In addition to CRF, a glucocorticoid feedback mechanism regulates the adenohypophysis.

Follicle-Stimulating Hormone Basophilic **gonadotrophs** produce **follicle-stimulating hormone (FSH).** FSH promotes the development of Graafian follicles in the ovaries and the maturation of spermatozoa in the testes. Estrogen and testosterone feedback mechanisms regulate FSH production in the adenohypophysis.

Luteinizing Hormone Basophilic **gonadotrophs** that are somewhat larger than the FSH gonadotrophs produce **luteinizing hormone (LH).** While FSH-producing cells tend to be near the periphery of the adenohypophysis, LH-producing cells are more generally distributed.

Maturation of the Graafian follicles and ovulation during the menstrual cycle depend on LH. Although FSH contributes much to early follicle development, ovulation depends entirely on LH. Estrogen and progesterone production also depend on LH.

In addition, LH stimulates the interstitial cells of the testes to produce testosterone; for this reason, it is also called **interstitial cell–stimulating hormone (ICSH).** Estrogen and progesterone feedback mechanisms regulate LH production by the adenohypophysis.

The Neurohypophysis

The neuroglial-like cells in the neurohypophysis are **pituicytes.** Pituicytes are small, have numerous processes, and unlike neuroglia, contain fat and pigment granules in their cytoplasm (see figure 87.13).

Pituicytes are nonsecretory. They provide support for nerve fibers from tracts that originate in the hypothalamus. Two hormones—**antidiuretic hormone** and **oxytocin,**—are liberated by the ends of these nerve tracts. The hormones are actually synthesized in the cell bodies of **neurosecretory neurons** of the hypothalamus and pass down the fibers into the neurohypophysis, where they are released.

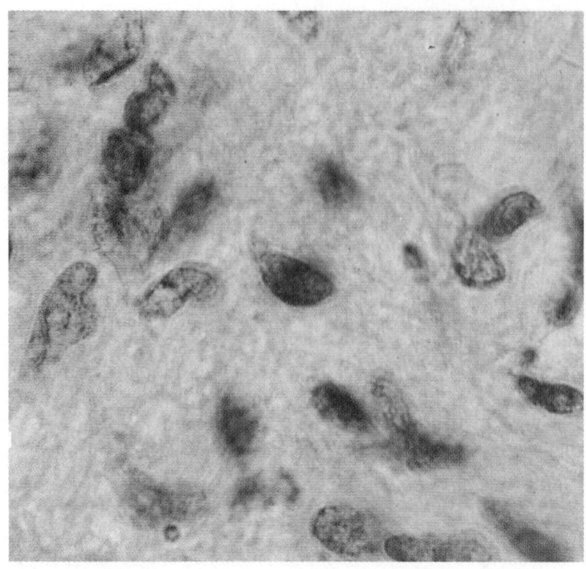

Figure 87.13 Pituicytes of the neurohypophysis (2500×).

Antidiuretic Hormone Antidiuretic hormone (ADH) controls the permeability of collecting tubules in nephrons to water absorption. When ADH is present in even minute amounts, the walls of the collecting tubules easily reabsorb water back into the blood. When ADH is lacking, water is rapidly lost through the kidneys. In *diabetes insipidus,* the neurohypophysis fails to provide enough ADH. Without this hormone, the osmotic balance of body fluids destabilizes.

Another function of ADH is to increase arterial blood pressure when a severe blood loss has reduced blood volume. A 25% loss of blood increases ADH production 25 to 50 times. ADH increases blood pressure by constricting the arterioles. Because of this potent pressor capability, ADH is also called **vasopressin.** The mechanism for increased ADH production during blood loss lies in baroreceptors in the walls of the carotid, aortic, and pulmonary arteries.

Oxytocin Any substance that causes uterine contractions during pregnancy is designated "oxytocic." During childbirth, the pressure of the unborn child on the uterine cervix causes a neurogenic reflex to the neurohypophysis, stimulating it to produce the hormone oxytocin. This hormone, being oxytocic, increases the strength of uterine contractions to facilitate childbirth. Oxytocin also activates the myoepithelial cells of the mammary gland, resulting in milk flow.

Microscopic Studies

Do a systematic study of the various endocrine glands using the slides available. Make drawings, if required.

Materials:
prepared slides of the following glands:
thyroid
parathyroid
thymus
adrenal
ovary
testis
pituitary
pineal

While examining these slides, refer to the photomicrographs in this exercise to identify the various cells discussed. Use low power for scanning and oil immersion wherever necessary.

Assignment:
Complete the Laboratory Report for this exercise.

88 The Reproductive System

This exercise examines the human reproductive system, using cat dissection for anatomical studies. You explore the histology of the various organs, as well as engage in microscopic studies of spermatogenesis and oogenesis.

The Male Organs

Figure 88.1 is a sagittal section of the male reproductive system. Models and/or wall charts are helpful in identifying all the structures. The primary sex organs are the paired oval **testes** (testicles), which lie enclosed in a sac, the **scrotum.** The external location of the testes provides a slightly lower localized body temperature (94–95°F) that favors the development of mature spermatozoa.

Lying over the superior and posterior surfaces of each testis is an elongated, flattened body, the **epididymis,** which stores immature sperm after they leave the testis.

The cutaway section of the testis in figure 88.1 shows the relationship of the epididymis to the testis. Partitions, or **septa,** that contain coiled-up **seminiferous tubules** divide each testis into several chambers. The figure shows a single seminiferous tubule held out of one of the compartments. All of the seminiferous tubules anastomose to form a network of tubules called the **rete testis.** Cilia within the rete testis move the immature spermatozoa through this maze of tubules into 10 or 15 **vasa efferentia** that lead directly into the epididymis. The outer wall, or **tunica albuginea,** that surrounds each testis consists of fibrous connective tissue.

During ejaculation, the spermatozoa leave the epididymis by way of the **vas deferens.** This duct passes over the pubic bone and bladder into the pelvic cavity. The terminus of the vas deferens enlarges to form the **ampulla of the vas deferens.** A pair of glands—the **seminal vesicles**—and the ampulla empty into the **common ejaculatory duct,** which passes through the prostate gland to join the **prostatic urethra.** From here, sperm pass through the **penile urethra** out of the body.

The prostate gland, seminal vesicles, and bulbourethral glands contribute alkaline secretions to the seminal fluid. This alkalinity stimulates sperm motility.

The **prostate gland** is the largest of the secondary sex glands, while the **bulbourethral glands** are the smallest. The pea-sized bulbourethral glands have 1-inch ducts that empty into the urethra at the base of the penis. The bulbourethral glands secrete a clear, mucoid fluid that lubricates the end of the penis and prepares the urethra for seminal fluid.

The **penis** consists of three cylinders of erectile tissue: two corpora cavernosa and one corpus spongiosum. The **corpus spongiosum** is a cylinder of tissue that surrounds the penile urethra. A septum separates the two **corpora cavernosa** on the midline. The distal end of the corpus spongiosum enlarges to form a cone-shaped **glans penis.** The enlarged portion of the urethra within the glans is the **navicular fossa.** Penile erection occurs when the spongelike tissue of the three corpora fills with blood.

Over the end of the glans penis lies a circular fold of skin, the **prepuce.** Around the neck of the glans and on the inner surface of the prepuce are small *preputial glands.* These sebaceous glands produce a secretion that readily decomposes to form a whitish substance called *smegma. Circumcision* is the surgical removal of the prepuce to facilitate sanitation.

Assignment:
Complete part A of the Laboratory Report for this exercise.

Spermatogenesis

Spermatogenesis is the process whereby spermatozoa are produced in the testes. This function of the testes begins at puberty and continues without interruption throughout life.

Figure 88.2 shows a section of a seminiferous tubule, along with a diagram of the various stages in the development of mature spermatozoa. These germ cells form as a result of both mitosis and meiosis.

In this study of spermatogenesis, you examine slides of human or animal testes under the microscope. Before examining the slides, however, familiarize yourself with the characteristic differences between meiosis and mitosis.

Materials:
prepared stained slides of:
human testis
rat testis
rabbit testis

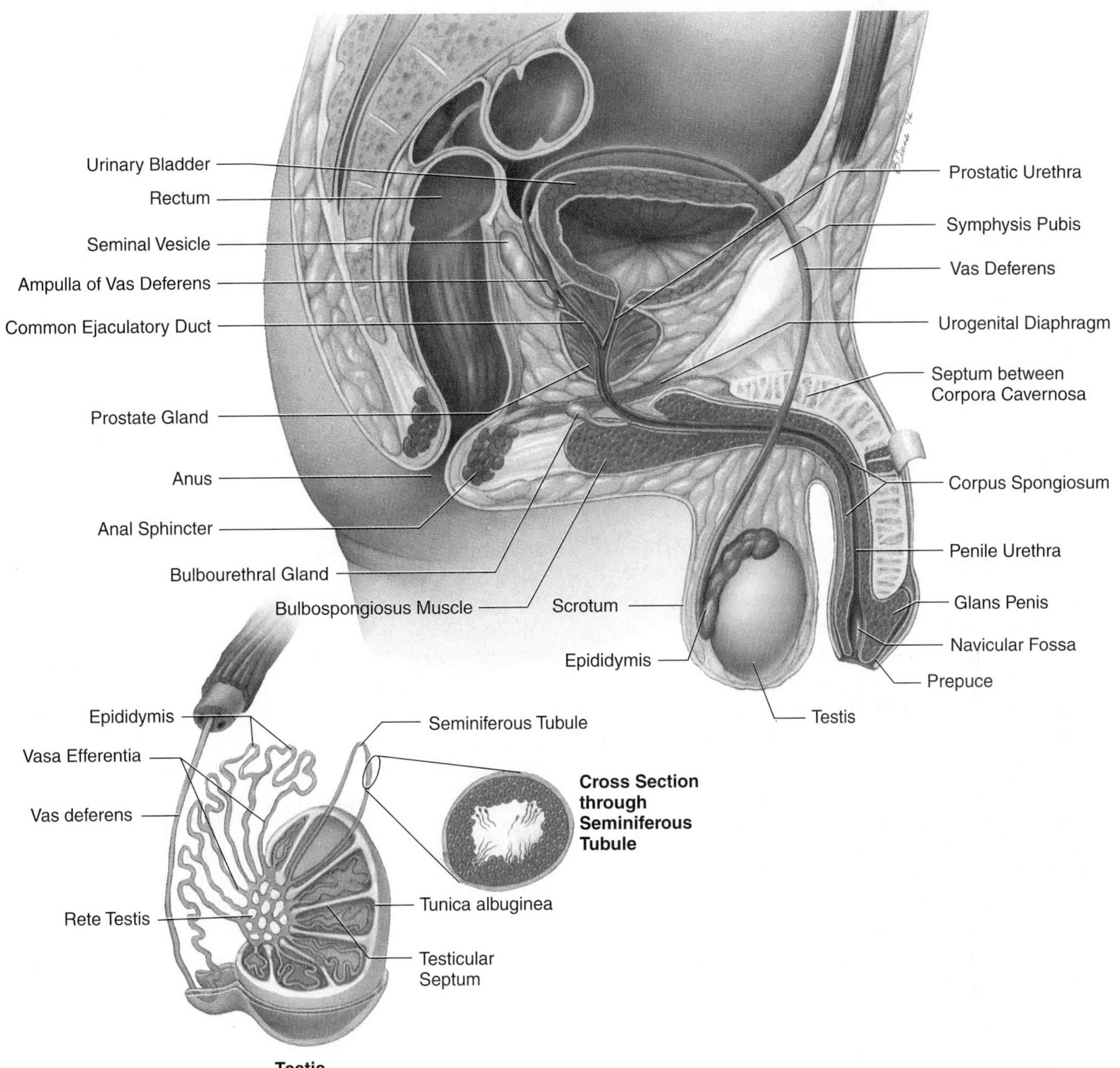

Figure 88.1 Male reproductive organs.

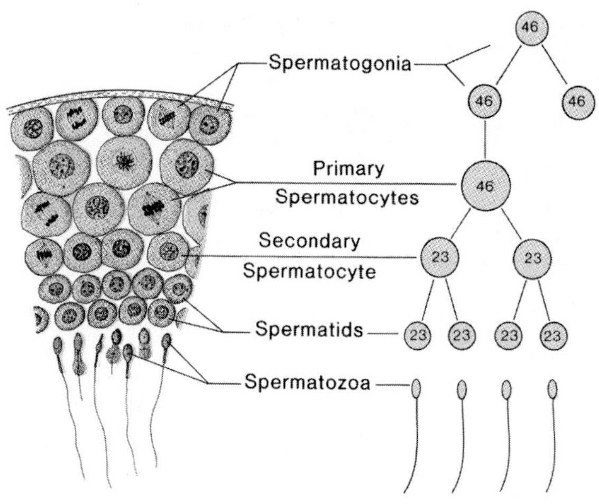

Figure 88.2 Spermatogenesis.

Mitosis

All spermatozoa originate from **spermatogonia** of the **primary germinal epithelium,** a layer of cells at the periphery of the seminiferous tubule. These cells contain the same number of chromosomes as other body cells (i.e., 23 pairs, or 46 chromosomes) and thus are *diploid.* These cells constantly divide mitotically to produce other spermatogonia.

As spermatogonia move toward the center of the seminiferous tubule, they enlarge to become **primary spermatocytes.** In this stage, the homologous chromosomes unite to form chromosomal units called *tetrads.* This union of homologous chromosomes is *synapsis.*

Meiosis

Each primary spermatocyte divides by meiosis to produce two **secondary spermatocytes.** *Meiosis,* or reduction division, results in each secondary spermatocyte receiving a haploid number of chromosomes—that is, 23 chromosomes instead of 46. The individual chromosomes of the secondary spermatocytes form when the chromosomes of the tetrads separate to form chromosomes called *dyads.*

Each secondary spermatocyte divides by meiosis to produce two haploid **spermatids.** In this division, the dyads split to form single-stranded chromosomes, or *monads.* Each spermatid metamorphoses directly into a mature sperm cell. Note that four **spermatozoa** form from each spermatogonium.

Assignment:
Examine slides of human, rat, or rabbit testicular tissue to identify the various stages of spermatogenesis. Of the three listed in the "Materials" list, the rabbit may be best to use. Refer to figure HA-36 of the Histology Atlas to identify the various types of cells. Then draw a representative section of a seminiferous tubule, labeling all cell types in part B of the Laboratory Report.

The Female Organs

Figures 88.3, 88.4, and 88.5 illustrate the female reproductive organs. Models and/or wall charts are helpful in identifying all the structures.

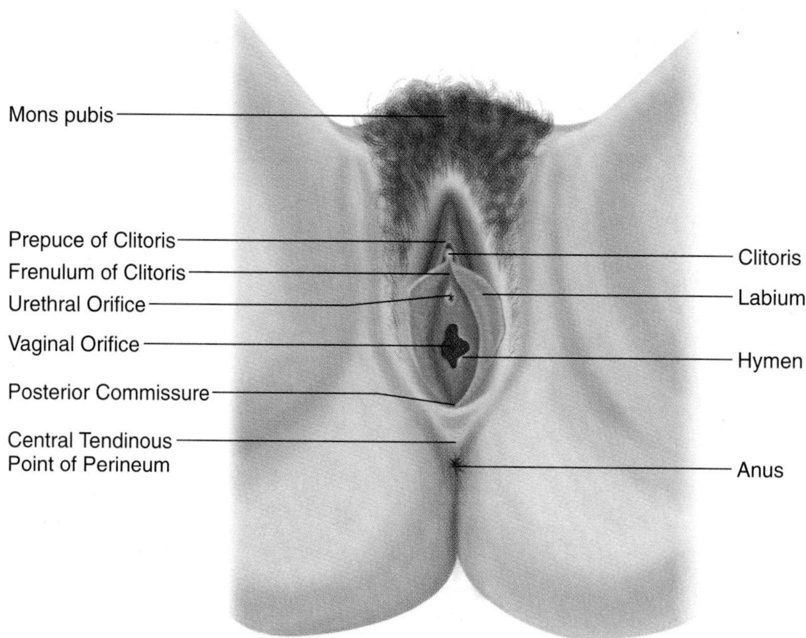

Figure 88.3 Female genitalia.

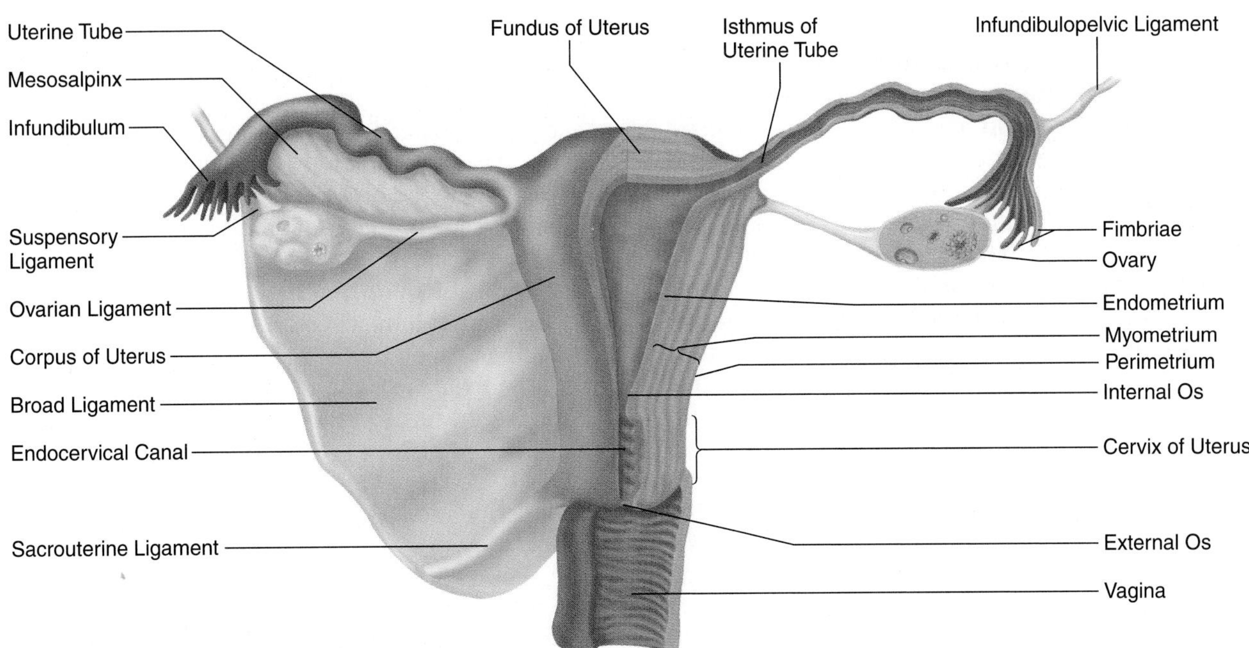

Figure 88.4 Posterior view of the female reproductive organs.

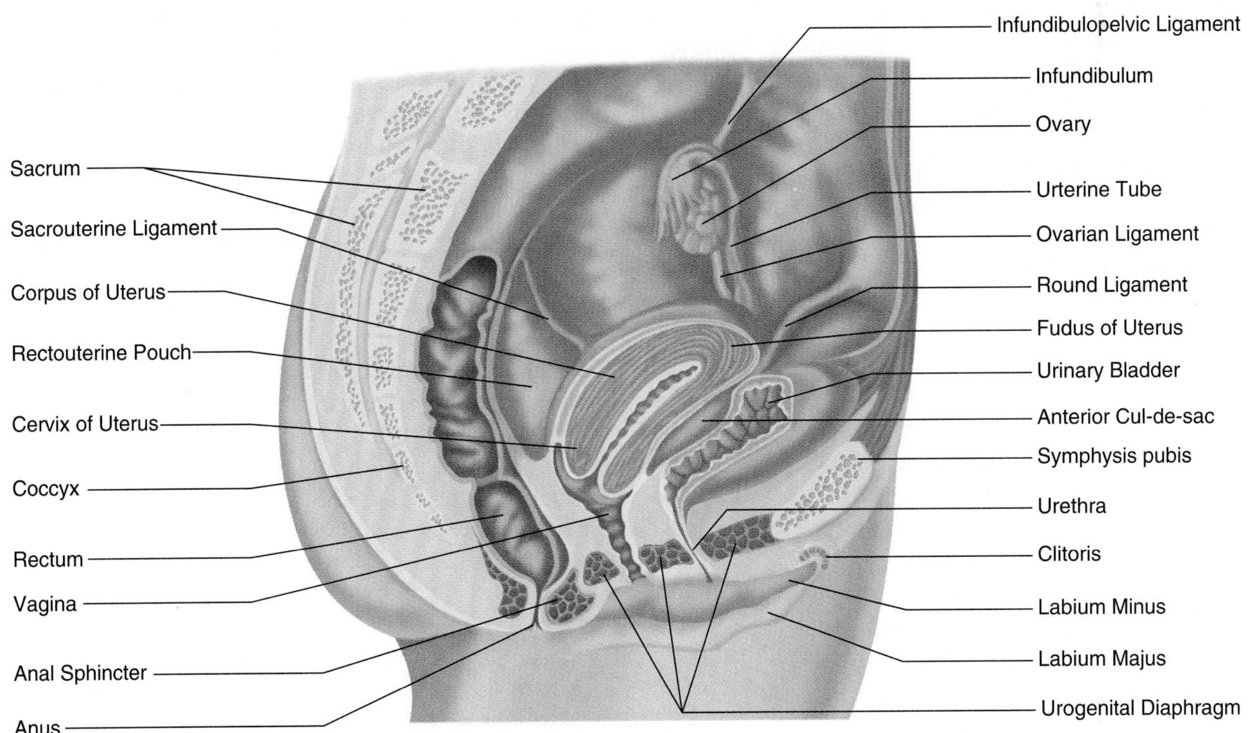

Figure 88.5 Midsagittal section of female reproductive organs.

Female Genitalia

The **vulva** of the external female reproductive organs includes the mons pubis, labia majora, labia minora, and hymen (see figure 88.3). The **mons pubis** (*mons veneris*) is the most anterior portion and consists of a firm, cushionlike elevation over the symphysis pubis. Hair covers it.

Two folds of skin on each side of the **vaginal orifice** lie over the opening. The larger exterior folds are the **labia majora** (*labium majus,* singular). Hair covers their exterior surfaces; their inner surfaces are smooth and moist. The labia majora are homologous (of similar embryological origin) to the male scrotum.

Medial to the labia majora are the smaller **labia minora** (*labium minus,* singular). These folds meet anteriorly on the median line to form a fold of skin, the **prepuce of the clitoris.** The **clitoris** is a small protuberance of erectile tissue under the prepuce that is homologous to the male penis. It is highly sensitive to stimulation. The fold of skin that extends from the clitoris to each labium minus is the **frenulum of the clitoris.** Posteriorly, the labia minora join to form a transverse fold of skin, the **posterior commissure,** or **fourchette.** Between the anus and the posterior commissure is the **central tendinous point of the perineum.** The **perineum** corresponds to the outlet of the pelvis.

The *vestibule* is the area between the labia minora that extends from the clitoris to the fourchette. Within the vestibule are the vaginal orifice, hymen, urethral orifice, and openings of the vestibular glands.

The **urethral orifice** lies 2 to 3 cm posterior to the clitoris. Many small **paraurethral glands** surround this opening. These glands are homologous to the male prostate gland.

On either side of the vaginal orifice are openings from the two **greater vestibular glands.** These glands are homologous to the male bulbourethral glands. They secrete mucus for vaginal lubrication.

The **hymen** is a thin fold of mucous membrane that separates the vagina from the vestibule. It may be completely absent or cover the vaginal orifice partially or completely. Its condition or absence is not a determinant of virginity.

Internal Female Organs

Figure 88.4 is a posterior view of the female reproductive organs that shows the relationship of the vagina, uterus, ovaries, and uterine tubes. It also reveals many of the principal supporting ligaments.

The Uterus The **uterus** is a pear-shaped, thick-walled, hollow organ that consists of a fundus, corpus, and cervix. The **fundus** is the dome-shaped portion at the large end. The body, or **corpus,** extends from the fundus to the cervix. The **cervix,** or neck, of the uterus is the narrowest portion—about 1 inch long. Within the cervix is an **endocervical canal** that has an outer opening, the **external os,** that opens into the **vagina.** An inner opening, the **internal os,** opens into the cavity of the corpus.

The thick, muscular portion of the uterine wall is the **myometrium.** The inner lining, or **endometrium,** consists of mucosal tissue. **Peritoneum** covers the fundus and the anterior and posterior surfaces of the corpus.

The Uterine (Fallopian) **Tubes** A **uterine tube** extends laterally from each side of the uterus. The distal end of each tube enlarges to form a funnel-shaped, fimbriated **infundibulum** that surrounds an **ovary.** The infundibulum does not usually touch the ovary, but one or more of the fingerlike **fimbriae** on the edge of the infundibulum usually do. The infundibula and fimbriae receive the oocytes the ovaries release during ovulation.

Ciliated columnar cells line the uterine tubes and help transport egg cells and developing zygotes from the ovary into the uterus. The muscular wall of the uterine tubes produces peristaltic movements that help propel egg cells. Fertilization usually occurs somewhere along the uterine tube. Note that each uterine tube narrows to form an **isthmus** near the uterus.

The Urinary Bladder Figure 88.5 shows the relationship of the urinary bladder to the reproductive organs. Note that the bladder lies between the uterus and the **symphysis pubis.** The base of the bladder is in direct contact with the anterior vaginal wall. The **urethra** is between the vagina and the symphysis pubis. The neck of the bladder lies on the superior surface of the **urogenital diaphragm.**

Peritoneum covers the superior surface of the bladder and is continuous with the peritoneum on the anterior face of the uterus. The small cavity lined with peritoneum between the bladder and uterus is the **anterior cul-de-sac.** The space between the uterus and rectum is the **rectouterine pouch**.

Ligaments Various ligaments suspend and support the uterus, ovaries, and uterine tubes. These ligaments are formed from muscle and connective tissue encased in peritoneum, or simply, from folds of peritoneum. Figure 88.4 shows most of the ligaments; figure 88.5 shows a few of them.

The largest supporting structure is the **broad ligament.** It is an extension of the peritoneal layer that covers the fundus and corpus of the uterus. Extending up over the uterine tubes, this ligament forms a mesentery, the **mesosalpinx,** on each side of the uterus. The mesosalpinx contains blood vessels that supply nutrients to the uterine tube.

Attached to the cervical region of the uterus are two **sacrouterine ligaments** that anchor the uterus to the sacral wall of the pelvic cavity. A **round ligament** extends from each side of the uterus to the body wall. Figure 88.4 does not show the round ligament because of the ligament's anterior placement.

Ovarian and suspensory ligaments hold the ovaries in place. The **ovarian ligament** extends

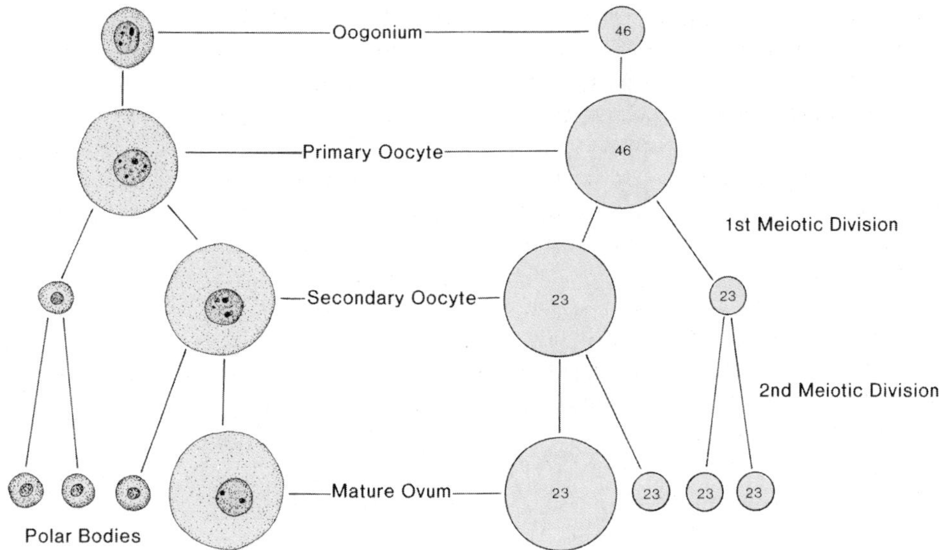

Figure 88.6 Oogenesis.

from the medial surface of the ovary to the uterus. The **suspensory ligament** is a peritoneal fold on the other side of the ovary that attaches the ovary to the uterine tube.

An **infundibulopelvic ligament** attached to the posterior surface of the infundibulum holds each uterine tube in place.

Assignment:
Complete parts C and D of the Laboratory Report for this exercise.

Oogenesis

Oogenesis is the development of egg cells in the ovary. As mentioned earlier, as many as 400,000 immature follicles develop in the ovaries from the germinal epithelium. At the onset of puberty, all potential ova in these follicles contain 46 chromosomes and are **primary oocytes.** Only about 400 of these oocytes mature during a female's reproductive lifetime.

Note in figure 88.6 that the primary oocyte, which forms from an oogonium of the germinal epithelium during embryological development, has 46 chromosomes. In the prophase stage of a dividing oocyte, double-stranded chromosomes (*dyads*) unite to form 23 homologous pairs of chromosomes during synapsis. Each pair consists of four chromatids; thus, it is called a *tetrad.* When the primary oocyte divides in the first meiotic division, each dyad member of each homologous pair enters a different cell. The first meiotic division produces a **secondary oocyte** with all the yolk of the primary oocyte and a **polar body** with no yolk. Both the secondary oocyte and polar body contain 23 chromosomes.

The second meiotic division produces a **mature ovum** with 23 chromosomes that are *monads;* that is, each chromosome consists of a single chromatid. Another polar body also forms in this division. The first polar body also divides to produce two new polar bodies with monads. The end result is that one primary oocyte produces one mature ovum and three polar bodies. The polar bodies never serve any direct function in the fertilization process.

At ovulation, the "ovum" is a secondary oocyte. Usually, a sperm must penetrate the cell to trigger the second meiotic division. The 23 chromosomes in the sperm unite with the 23 chromosomes of the mature ovum to form a zygote with 46 chromosomes.

Assignment:
Complete part E of the Laboratory Report for this exercise.

Cat Dissection

In Exercise 85, the study of the urinary system required a cursory examination of the reproductive organs in the cat. You now study these organs in greater detail. Refer to figures 85.4 and 85.5, as well as to figures 88.7 and 88.8 to identify the various organs in this dissection.

The Male Reproductive System

This dissection begins with an examination of the penis, which necessitates the eventual removal of the right hind leg to reveal the inner structures. Proceed as follows:

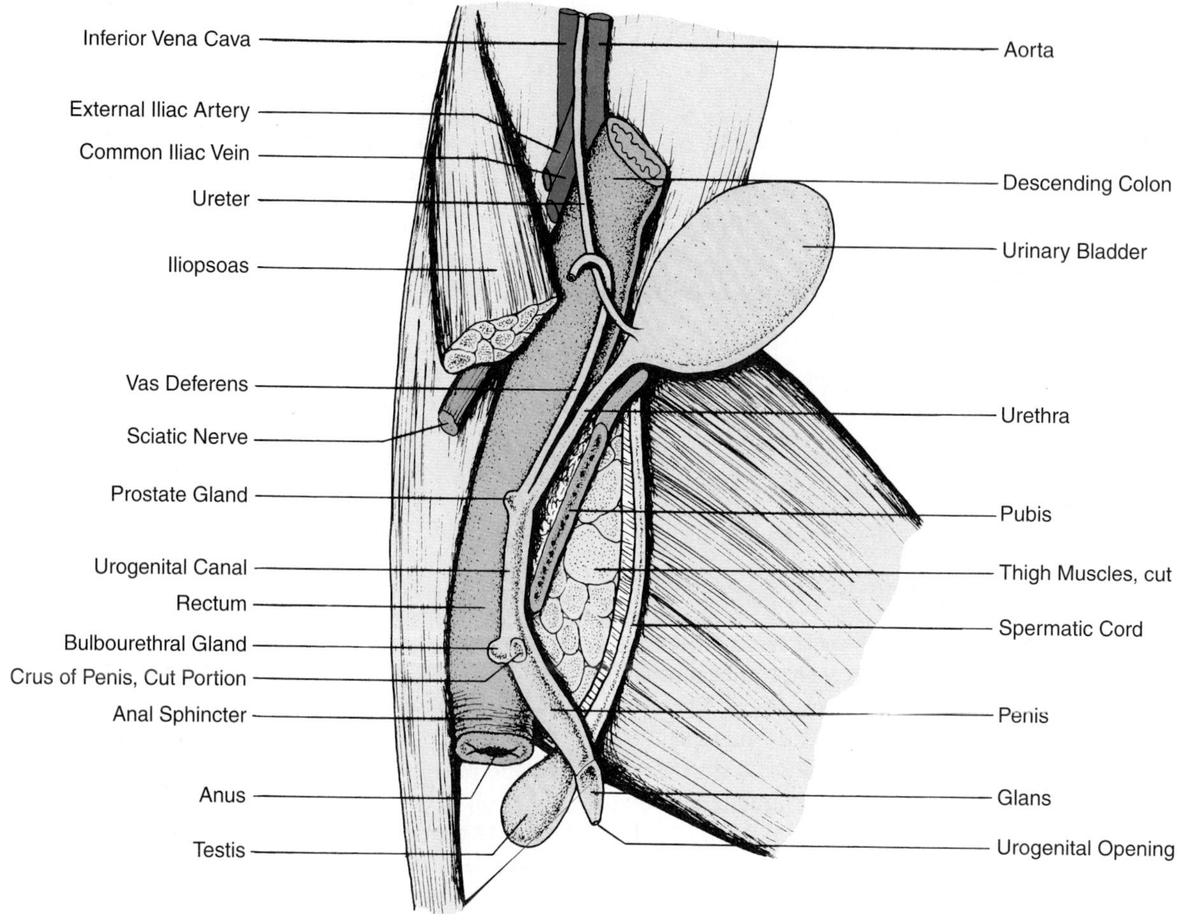

Figure 88.7 Lateral view of the urogenital system of the male cat.

1. Locate the **penis.** The skin that forms the sheath around the organ is the **prepuce.** Pull back the prepuce to expose the enlarged end, the **glans.** The glans in the cat is covered with minute horny papillae.
2. Remove the **scrotum,** the skin that encloses the two **testes.** Note that a *fascial sac* covers each testis. Do not open this sac at this time.
3. Trace the **spermatic cord** from one of the testes to the body wall. The cord is composed of the **vas deferens** (*ductus deferens*), blood vessels, nerves, and lymphatic vessels that supply and drain the testis. Note that the covering of the spermatic cord is continuous with the fascial sac of the testis.
4. The vas deferens and blood vessels pass through the abdominal wall via the *inguinal canal.* From the inguinal canal, the vas deferens passes up over the ureter and enters the penis near the prostate gland.

   To study the structures shown in figure 88.7, remove the right leg near the hip joint. You must cut through the symphysis pubis and also through the hipbone above the acetabulum. Clear away the pelvic muscles to expose the structures figure 88.7 shows.
5. Locate the **prostate gland** where the vas deferens joins the urethra. The tube leading from the prostate through the penis is the **urogenital canal** (penile urethra). At the base of the penis is a pair of **bulbourethral glands.**
6. Cut through the penis to produce a transverse section. Identify the two **corpora cavernosa** that cause erection. Near the symphysis pubis, the two corpora diverge to form the **right** and **left crura** (*crus,* singular). Each crus attaches to a part of the ischium. Figure 85.4 shows both crura.
7. Cut through the fascial sac, exposing the testis. The space between this sac and the testis is the **vaginal sac.** This cavity is a downward extension of the peritoneal cavity.
8. Trim away the fascial sac, and trace the vas deferens toward the testis, noting its convoluted nature. Identify the **epididymis,** which lies on the dorsal side of the testis and is continuous with the vas deferens.

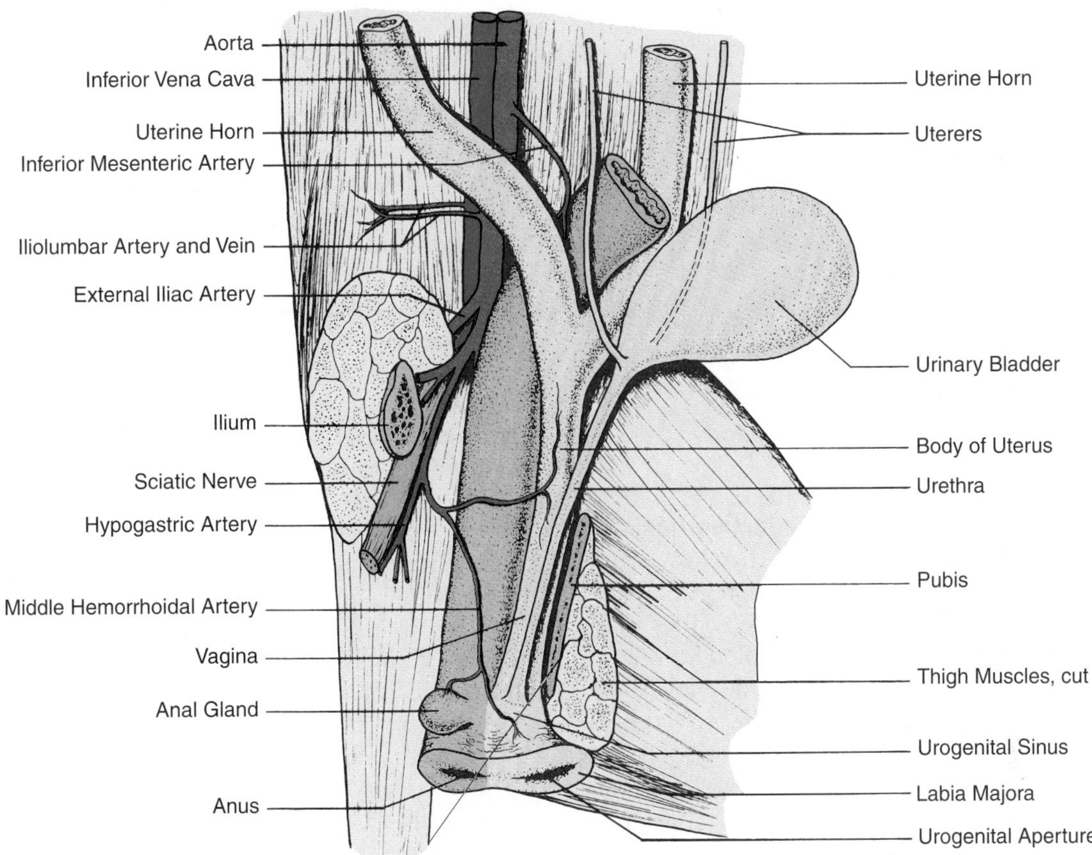

Figure 88.8 Lateral view of the urogenital system of the female cat.

9. With a sharp scalpel or razor blade, cut through the testis and epididymis to produce a section similar to that in figure 88.1. Identify the structures labeled in this diagram.

The Female Reproductive System

Examination of the female reproductive organs begins with the ovaries. Figure 85.5 is useful for this portion of the study. Examination of the lower portion of the reproductive tract, however, requires removal of the right leg. Proceed as follows:

1. Identify the small, light-colored, oval ovaries posterior to the kidneys. Note that a fold of tissue holds them to the dorsal body wall. This peritoneal fold is the **mesovarium.** Note also that the **ovarian ligament** attaches the ovary to the uterine horn.
2. Identify the **uterine tube** (oviduct) with its enlarged, funnel-like terminus, the **infundibulum** (abdominal ostium). Note that the perimeter of the infundibulum has many irregular projections called **fimbriae.** When ova leave the ovary, ciliated epithelium on the tissues of the uterine tube draw the ova into the uterine tube through the infundibulum.
3. Observe that the uterus is Y-shaped, having two horns and a body. The **uterine horns** lie along each side of the abdominal cavity. The **uterine body** lies between the urinary bladder and the rectum (see figure 88.8).
4. To observe the relationship of the vagina to other structures, as figure 88.8 shows, remove the right leg near the hip joint. To do this, cut through the hipbone above the acetabulum. Remove the cut portion of the hipbone, and clear away the pelvic muscles to expose the structures shown in figure 88.8. Preserve the blood vessels as much as possible.
5. Identify the general areas of the uterine body, vagina, and urogenital sinus.
6. Remove the entire female reproductive tract, and open it to reveal the inner structures shown in figure 85.5, including the **cervix, vagina,** and **urogenital sinus.**

Histological Study

The focus here is primarily on a histological study of the various reproductive organs of both sexes. Refer to the appropriate pages of the Histology

Atlas and to some of the illustrations in Exercise 87 ("The Endocrine Glands") as necessary.

Materials:
prepared slides:
ovary
uterus
uterine tube
vagina
vas deferens
penis
seminal vesicle
prostate gland

Ovary Study various slides of the ovary to identify **primordial follicles,** the **germinal epithelium,** and **Graafian follicles.** Since the Histology Atlas does not show the ovaries, refer to figure 87.9. Note that an incomplete layer of low cuboidal or flattened epithelium surrounds the developing ova in primordial follicles. Some slides do not show mature Graafian follicles. If your slide is sectioned properly, you should see a large space surrounding the oocyte and containing large quantities of estrogen that the follicular cells produced.

Uterus When studying a slide of the uterus, take into consideration the phase of the menstrual cycle represented on the slide. The photomicrographs in figure HA-31 in the Histology Atlas were made from Turtox slide H9.331, which represents about the third week in the cycle. Both the endometrium and myometrium look much different in earlier and later phases.

Differentiate the **endometrium** from the **myometrium.** Identify the **uterine glands** and **endometrial epithelium.**

Uterine Tube Scan a slide of the uterine tube with the low-power objective to find the lumen of the tube. Refer to figures HA-31C and D in the Histology Atlas. Note the extremely irregular outline of the mucosa, which consists of a mixture of ciliated and nonciliated columnar epithelium. How many layers of smooth muscle tissue do you find in the muscularis?

Vagina Examine a slide of the mucosa of the vagina, and compare it to figure HA-32C in the Histology Atlas. How does the vaginal mucosa differ from the skin? What function do the lymphocytes perform in the lamina propria?

Uterine Cervix Study a longitudinal section through the cervical area of the uterus that shows where the vaginal lining meets the cervical region. Consult figure HA-32A in the Histology Atlas. Note the large number of spaces, or **cervical crypts,** that lie deep in the cervical tissue. Read about the significance of these crypts in the legend for figure HA-32.

Vas Deferens Use minimal magnification to examine a cross-sectional slide of the vas deferens. Compare your slide with figure HA-34 in the Histology Atlas. Note that between the inner **epithelium** and outer **adventitia** are three layers of smooth muscle tissue: an inner longitudinal layer, an outer longitudinal layer, and a middle layer of circular fibers. Read what the legend for figure HA-34 states about the **stereocilia** on the epithelium.

Penis Examine a cross-sectional preparation of the penis. Refer to the photomicrographs in figures HA-33C and D in the Histology Atlas.

Examine the epithelium under high-dry magnification. Does the epithelium on your slide look like figure HA-33D? If not, why not? Read the comments in the legend for figure HA-33 that pertain to epithelial differences.

Seminal Vesicle Figures HA-35A and B of the Histology Atlas show a cross section of a portion of a seminal vesicle. Examine a slide first with low-power magnification, and note the striking way in which the mucosa folds upon itself to produce a maze of pockets.

Examine the epithelium under high-dry or oil immersion. What is the nature and function of seminal fluid?

Prostate Gland Examine a slide of the prostate gland under low power first to identify the structures shown in figure HA-35C of the Histology Atlas. Note the large number of **tubuloalveolar glands** that secrete the prostatic fluid. Why does the prostate gland tend to harden and to restrict micturition in older men?

Assignment:
Complete the remainder of the Laboratory Report for this exercise.

Fertilization and Early Embryology (Sea Urchin)

89

The human ovum is usually fertilized in the distal third of the uterine tubes. Although many spermatozoa may be on the surface of a fertilized ovum, the nucleus of only one sperm cell migrates through the ovum to unite with the ovum nucleus. The union of the two nuclei (**fertilization**) produces a **zygote** with 46 chromosomes.

The sperm cell's penetration of the ovum is called **activation.** As a result of this process, fluid rapidly accumulates between the inner and outer membranes around the ovum. This outer **fertilization membrane** prevents other spermatozoa from entering the cell.

Because fertilization occurs *internally* in mammals, observation of the process is not a simple procedure. Fortunately, however, you can easily observe activation and early embryology in sea urchins because these steps occur *externally* in seawater. By chemical or electrical stimulation, sea urchins can be induced to secrete their gametes.

In this exercise, you observe activation and some of the early stages of sea urchin embryology. You will not be able to observe the actual union of sperm and egg nuclei in fertilization, however. You will study living embryos like those in figure 89.1, as well as prepared slides. The early stages of development in the sea urchin and human are quite similar.

Procurement of Gametes

Since sexual differentiation of sea urchins by external examination is very difficult, you should have several animals to increase your chances of having at least one male and one female. You only need one specimen of each sex to provide all the gametes for an entire class. The following procedure should yield gametes:

Materials:
hypodermic syringe and needle
0.5 M potassium chloride solution
empty dropping bottles and pipettes
beaker (size depends on sea urchin size)
graduate
seawater
four or five sea urchins
paper toweling

1. With a hypodermic syringe, inject each animal with 1 ml of potassium chloride solution. Make the injection through the membranous region around the mouth into the coelom, or body cavity (see figure 89.2).
2. Turn the animals over, and place them, mouth downward, on paper toweling (see figure 89.3).

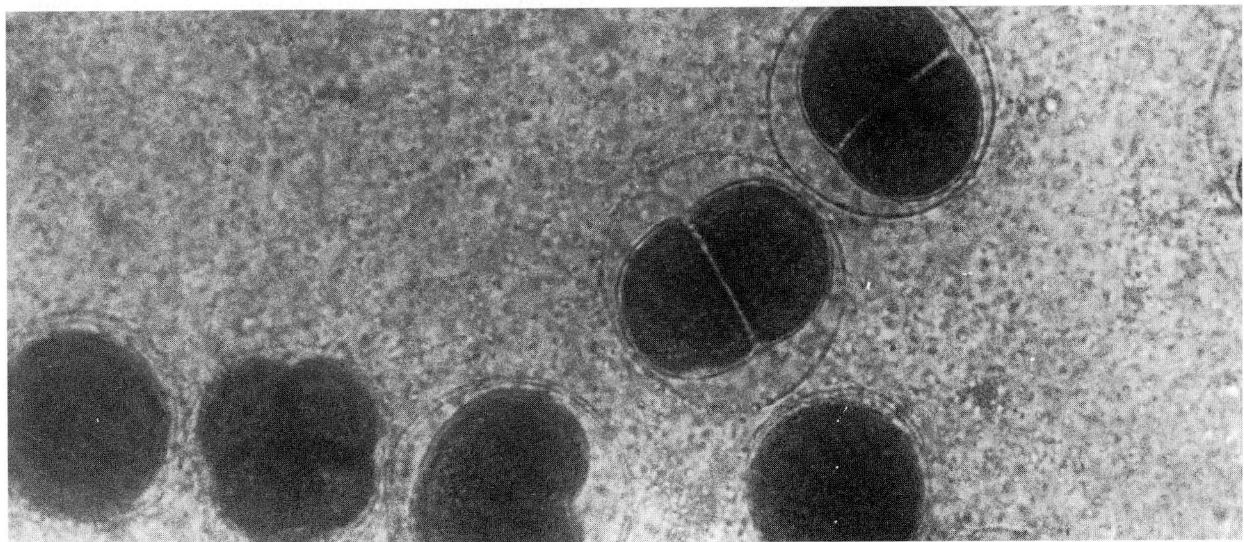

Figure 89.1 Early cleavage stages of the sea urchin.

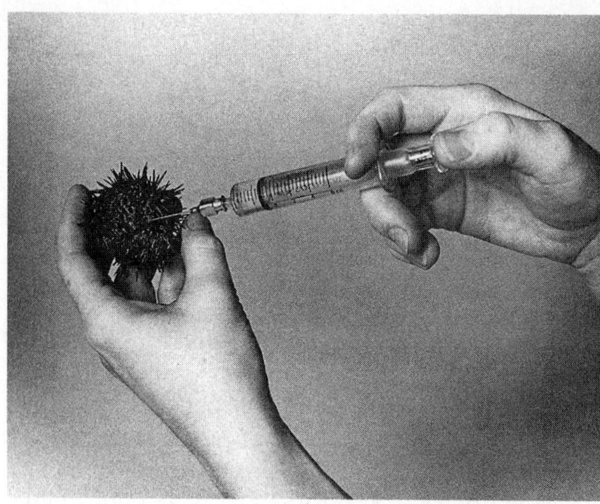

Figure 89.2 Inject sea urchins in the soft tissue near the mouth with 0.5M potassium chloride solution to stimulate gamete secretion.

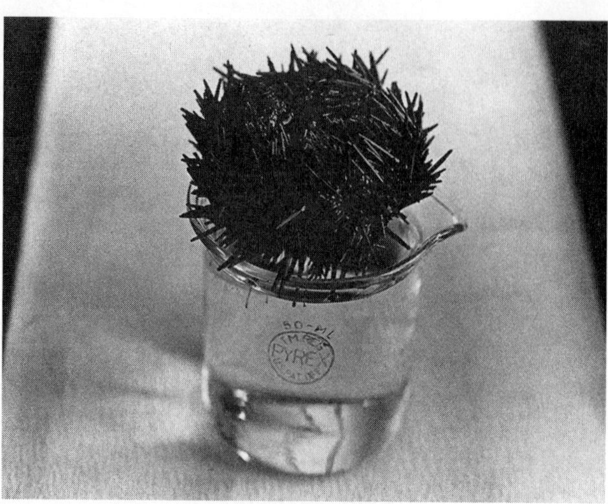

Figure 89.4 Place females over a beaker of seawater so that the aboral surface touches the water. Eggs drift to the bottom of the beaker.

Figure 89.3 Place injected animals mouth down on paper toweling. Gametes are secreted on the aboral surface.

Figure 89.5 Remove sperm secretions from the aboral surfaces of males with a pipette, and place them in a dropping bottle.

Within a few minutes, reproductive secretions should appear on the aboral surfaces. Sperm secretions are white, and eggs are yellowish brown.

3. Fill a beaker with fresh seawater.
4. Collect the egg secretions by placing the female over the beaker of seawater so that her aboral surface touches the water (see figure 89.4).
5. When all the eggs have been shed into the water, remove the female, swirl the eggs and water, and let the eggs settle to the bottom of the beaker. Pour off the seawater and add fresh seawater. Washing the eggs in this manner facilitates union of the gametes. These eggs will remain viable for up to 3 days if stored in the refrigerator.
6. Use a pipette to remove sperm secretions from male sea urchins (see figure 89.5). If the sperm are to be used immediately, dilute by adding one or two drops of sperm solution to 25 ml of seawater in a clean dropping bottle. If the sperm are to be used later, do not dilute at this time. The sperm will remain viable for 2 to 3 days in the refrigerator if undiluted.

Union of Gametes

To ensure fertilization, water temperature must be around 20°C. Temperatures in excess of 22°C are deleterious. If the room temperature of the laboratory is high (hot summer day without air-conditioning),

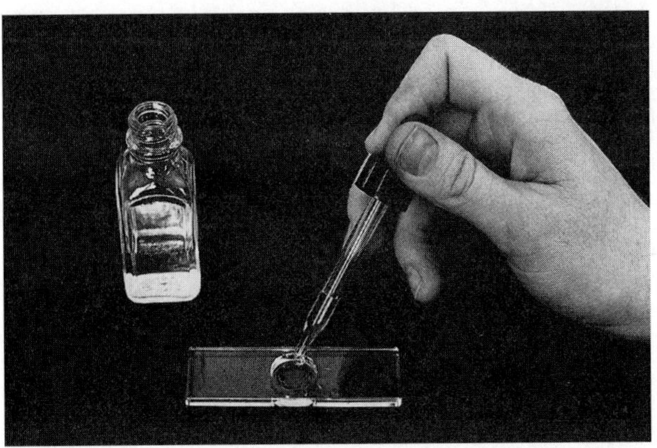

Figure 89.6 Add a drop of diluted sperm to the eggs in the depression of a slide. Then study activation under a microscope.

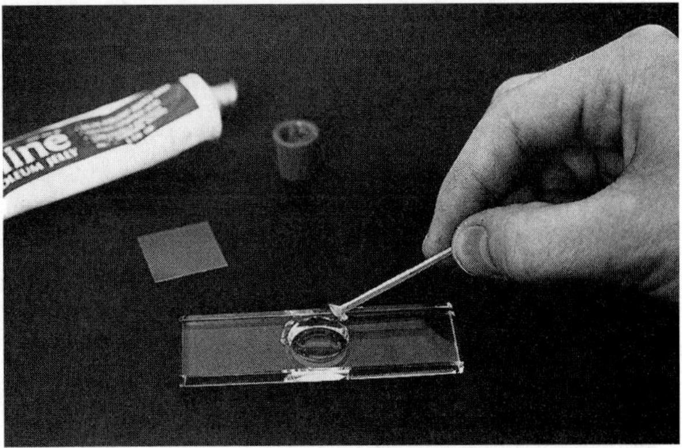

Figure 89.7 Add Vaseline and a cover glass to the slide to prevent drying. The temperature of the slide during the study should not exceed 22°C.

keep the slide cool by using a refrigerator (or low-temperature incubator) set at 20°C.

Materials:
depression slides and cover glasses
Vaseline
medicine droppers
diluted sperm solution (two drops of sperm solution in 25 ml of seawater)
sea urchin egg solution

1. Transfer one or two drops of seawater, containing 10 to 30 eggs, to a depression slide.
2. Examine the eggs with the 10× objective of your microscope. Look for the **egg nucleus.**
3. Add a drop of diluted sperm solution to the eggs (see figure 89.6). Watch the cells carefully through your 10X objective. Note the activity of the sperm and how they cluster around the eggs. This attraction results from the interaction of chemicals that both eggs and spermatozoa produce. Mammals exhibit a similar phenomenon.
4. Observe the rapid formation of the **fertilization membrane.** Do not expect to see sperm nuclei actually penetrate the eggs.
5. Place a little Vaseline around the edge of the depression, and place a clean cover glass over the depression. The Vaseline seals the chamber so that it does not dry out (see figure 89.7). You can observe such a slide for several days if the temperature does not exceed 22°C.

Early Embryology

The first cleavage division should occur 45 to 60 minutes after activation. Succeeding divisions occur at 30-minute intervals. Proceed as follows in your study of the early cleavage stages for as long as possible (depends on lab time available).

Materials:
depression slides from previous procedure
developing embryos (6, 12, 24, 48, and 96 hours old)
stained slides of starfish blastula and gastrula stages

1. Examine your slide every 15 to 20 minutes to note the mitotic divisions as they occur. While waiting for the cells to divide, examine the living embryos resulting from gamete union 6, 12, 24, 48, and 96 hours earlier. Note that cleavage divisions produce progressively smaller cells.
2. Draw 2-, 4-, 8-, and 16-cell stages as they appear in the depression slide.
3. Examine stained slides of starfish blastula and gastrula stages. (Starfish embryos are essentially identical to sea urchins and are more easily procured.) Make drawings on the Laboratory Report.

Assignment:
Complete the Laboratory Report for this exercise.

LABORATORY REPORT 1

Student: ____________________

Section: ____________________

Anatomical Terminology

A. *Terminology*

From the lists of positions and sections, select those that apply to the statements that follow. In some cases, more than one answer may apply.

Relative Positions		*Sections*
anterior—1	lateral—7	frontal—13
caudal—2	medial—8	midsagittal—14
cranial—3	posterior—9	sagittal—15
distal—4	proximal—10	transverse—16
dorsal—5	superior—11	
inferior—6	ventral—12	

Sections:

1. Divides body into front and back portions
2. Divides body into unequal right and left sides
3. Three sections that are longitudinal
4. Divides body into equal right and left sides
5. Two sections that expose both lungs and heart in each section

Positions:

6. The tongue is ______________ to the palate.
7. The shoulder of a dog is ______________ to its hip.
8. A hat is worn on the ______________ surface of the head.
9. The cheeks are ______________ to the tongue.
10. The fingertips are ______________ to all structures of the hand.
11. The human shoulder is ______________ to the hip.
12. The shoulder is the ______________ (*proximal, distal*) portion of the arm.

Select the surfaces on which the following are located:

13. Ear
14. Adam's apple
15. Kneecap
16. Palm of hand

Answers

Terminology

1. 13
2. 14
3. 13, 14, 15
4. 14
5. 13, 16
6. 6
7. 3
8. 11
9. 7
10. 4
11. 11
12. 10
13. ______
14. ______
15. ______
16. ______

B. Localized Areas

Identify the specific areas of the body that the statements describe.

antebrachium—1	calf—6	gluteal—11	lumbar—16
antecubital—2	costal—7	groin—12	pectoral—17
axilla—3	cubital—8	hypochondriac—13	plantar—18
brachium—4	epigastric—9	hypogastric—14	popliteal—19
buttocks—5	flank—10	iliac—15	umbilical—20

1. Elbow area
2. "Rump" area
3. Underarm area
4. Sole of foot
5. Upper arm
6. Forearm
7. Upper chest region
8. Depression on back of leg behind knee
9. Side of abdomen between lower edge of rib cage and upper edge of hipbone
10. Back portion of abdominal body wall that extends from lower edge of rib cage to hipbone
11. Area over ribs on dorsum
12. Posterior portion of lower leg
13. Anterior surface of elbow
14. Abdominal area that surrounds navel
15. Abdominal area that is lateral to pubic region
16. Area of abdominal wall that covers stomach
17. Abdominal area that is lateral to epigastric area
18. Abdominal area that is lateral to umbilical area
19. Area where the thigh meets the abdomen
20. Abdominal areas between the transpyloric and transtubercular planes

Answers

Localized Areas

1. 8
2.
3.
4.
5.
6.
7.
8. 19
9.
10.
11.
12.
13.
14.
15.
16.
17.
18.
19.
20.

LABORATORY REPORT 2

Student: ______________________

Section: ______________________

Body Cavities and Membranes

A. Body Cavities

From the list of cavities, select those that apply to the statements.

abdominal—1
abdominopelvic—2
cranial—3
dorsal—4
pelvic—5
pericardial—6
pleural—7
spinal—8
thoracic—9
ventral—10

1. Cavity that consists of cranial and spinal cavities
2. Most inferior portion of abdominopelvic cavity
3. Cavity that the diaphragm separates into two major divisions
4. Cavity inferior to diaphragm that consists of two parts

Select the most specific cavity in which the following organs are located:

5. Uterus
6. Duodenum
7. Brain
8. Heart
9. Rectum
10. Pancreas
11. Lungs and heart
12. Liver
13. Spleen
14. Spinal cord
15. Urinary bladder
16. Stomach

B. Membranes

From the list of membranes, select those that apply to the statements that follow.

mesentery—1
pericardial sac—2
parietal peritoneum—3
parietal pleura—4
visceral pericardium—5
visceral peritoneum—6
visceral pleura—7

1. Serous membrane attached to lung surface
2. Serous membrane attached to thoracic wall
3. Double-layered membrane that holds abdominal organs in place
4. Double-layered membrane that surrounds heart

Answers

Body Cavities	Membranes
1. 4	1.
2. 5	2.
3. 2	3.
4.	4.
5. 5	
6. 1	
7. 3	
8. 6	
9. 1	
10. 1	
11. 9	
12.	
13.	
14.	
15.	
16.	

LABORATORY REPORT 3

Student: ______________________

Section: ______________________

Organ Systems: Rat Dissection

A. System Functions

Select the system or systems that perform the body functions described. More than one answer may apply.

circulatory—1	lymphatic—5	respiratory—9
digestive—2	muscular—6	reticuloendothelial—10
endocrine—3	nervous—7	skeletal—11
integumentary—4	reproductive—8	urinary—12

1. Removes carbon dioxide from blood
2. Shields vital organs such as the brain and heart
3. Movement of ribs in breathing
4. Carries heat from muscles to surface of body for dissipation
5. Provides rigid support for muscle attachment
6. Transports food from intestines to all parts of body
7. Disposes of urea, uric acid, and other cellular wastes
8. Moves arms and legs
9. Helps to maintain normal body temperature by dissipating excess heat
10. Destroys microorganisms once they break through skin
11. Phagocytic cells of this system remove bacteria from lymph
12. Carries messages from receptors to centers of interpretation
13. Controls rate of growth
14. Returns tissue fluid to blood
15. Produces various kinds of blood cells
16. Enables body to adjust to changing internal and external environmental conditions
17. Ensures continuity of species
18. Eliminates excess water from body
19. Carries hormones from glands to all tissues
20. Shields against bacterial invasion
21. Carries fats from intestines to blood
22. Coordinates all body movements
23. Breaks food particles into small molecules
24. Controls development of the ovaries and testes
25. Produces spermatozoa and ova

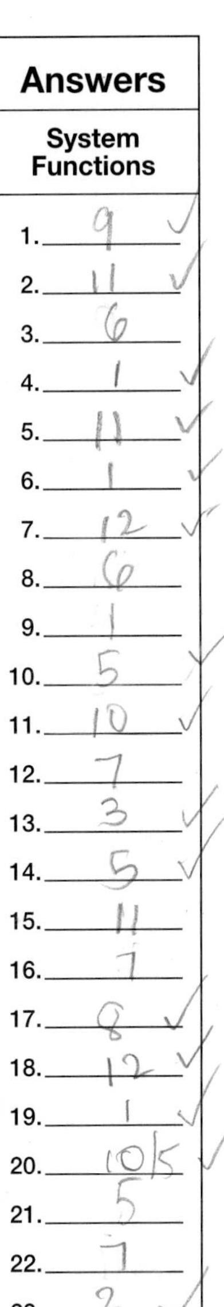

B. Organ Placement

Select the system on the right that includes the organs that follow.

1. Bones
2. Brain
3. Bronchi
4. Dermis
5. Esophagus
6. Hair
7. Heart
8. Kidneys
9. Lungs
10. Pancreas
11. Spleen
12. Stomach
13. Teeth
14. Trachea
15. Toenails
16. Tonsils
17. Thyroid
18. Ureters
19. Uterus
20. Veins

circulatory—1
digestive—2
endocrine—3
integumentary—4
lymphatic—5
muscular—6
nervous—7
reproductive—8
respiratory—9
skeletal—10
urinary—11

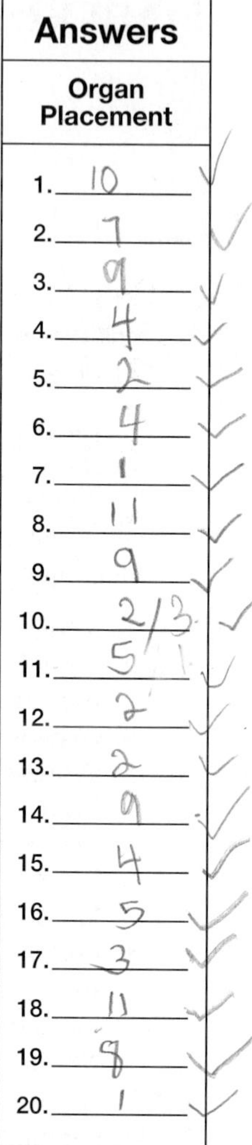

LABORATORY REPORT 4

Student: ____________________

Section: ____________________

Microscopy

A. Questions

1. List the three lens systems in a compound light microscope.
2. How can you greatly increase the bulb life on a microscope lamp if the voltage is variable?
3. What characteristic of a microscope enables you to switch from one objective to another without altering the focus?
4. What effect (*increase* or *decrease*) does closing the diaphragm have on
   a. image brightness?
   b. image contrast?
   c. resolution?
5. In general, at what position should the condenser be kept?
6. Express the maximum resolution of the compound microscope in micrometers (μm).
7. If you are getting 225× magnification with a 45× high-dry objective, what is the power of the eyepiece?
8. What is the magnification of objects observed through a 100× oil immersion objective with a 7.5× eyepiece?
9. Immersion oil must have the same refractive index as _____ to be of any value.
10. Substage filters should be of a _____ color to get the maximum resolution of the optical system.

B. True–False

Indicate whether the statements are true or false.

1. Only the fine adjustment knob should be used when viewing a specimen with the high-dry or oil immersion objectives.
2. Lenses can be safely cleaned with almost any kind of tissue or cloth.
3. When swinging the oil immersion objective into position after using high-dry, always increase the distance between the lens and slide to prevent damaging the oil immersion lens.
4. Instead of starting with the oil immersion lens, it is best to use one of the lower magnifications first and then swing the oil immersion into position.
5. The 45× and 100× objectives have shorter working distances than the 10× objective.

Answers

Questions

1a. ____________________
b. ____________________
c. ____________________
2. ____________________
3. ____________________
4a. ____________________
b. ____________________
c. ____________________
5. ____________________
6. ____________________
7. ____________________
8. ____________________
9. ____________________
10. ____________________

True–False

1. ____________________
2. ____________________
3. ____________________
4. ____________________
5. ____________________

C. *Multiple Choice*

Select the answer that best completes each statement.

Answers
Multiple Choice
1. ________
2. ________
3. ________
4. ________
5. ________

1. The resolution of a microscope is increased by
   a. using blue light.
   b. stopping down the diaphragm.
   c. lowering the condenser.
   d. raising the condenser to its highest point.
   e. Both *a* and *d* are correct.

2. The magnification of an object seen through the 10× objective with a 10× ocular is
   a. 10 times.
   b. 20 times.
   c. 1000 times.
   d. None of these is correct.

3. The most commonly used ocular is
   a. 5×.
   b. 10×.
   c. 15×.
   d. 20×.

4. Microscope lenses can be cleaned with
   a. lens paper.
   b. a soft linen handkerchief.
   c. paper towels.
   d. Both *a* and *b* are correct.
   e. *a, b,* and *c* are correct.

5. When changing from low power to high power, it is generally necessary to
   a. lower the condenser.
   b. open the diaphragm.
   c. close the diaphragm.
   d. Both *a* and *b* are correct.
   e. Both *a* and *c* are correct.

Laboratory Report 5

Student: ______________________

Section: ______________________

Basic Cell Structure

A. *Cell Drawings*

In the space provided, sketch one or two epithelial cells as seen under high-dry magnification. Label the **nucleus, cytoplasm,** and **plasma membrane.**

B. *Cell Structure*

From the list of structures, select those that apply to the statements. More than one structure may apply to some statements.

centrosome—1	Golgi apparatus—6	mitochondria—11	RER—16
chromatin—2	lysosome—7	nuclear envelope—12	ribosomes—17
cilia—3	microvilli—8	nucleolus—13	SER—18
cytoplasm—4	microfilaments—9	nucleus—14	transport vesicles—19
flagella—5	microtubules—10	plasma membrane—15	

1. Prominent body in nucleus that produces ribosomes
2. Cell membrane
3. Nonmembranous organelle consisting of two bundles of microtubules
4. Chromosomal material within nucleus
5. ER that lacks ribosomes
6. Small bodies of RNA and protein attached to surfaces of RER
7. Extensive system of membranous tubules and sacs within cytoplasm
8. All cellular material between plasma membrane and nucleus
9. Sacs in cytoplasm that contain digestive enzymes
10. Organelle that contains cristae
11. Short, hairlike appendages on surface of some cells
12. ER covered by ribosomes
13. Extranuclear bodies that possess DNA and can self-replicate
14. Layered concave structure in cytoplasm that is similar to basic structure of SER
15. Surface protuberances that form delicate brush border
16. Bodies produced by ER

Answers
Cell Structure
1. ________
2. ________
3. ________
4. ________
5. ________
6. ________
7. ________
8. ________
9. ________
10. ________
11. ________
12. ________
13. ________
14. ________
15. ________
16. ________

C. *Organelle Functions*

Select the organelles that perform the functions that follow. More than one structure may apply to some statements.

centrosome—1	microvilli—5	plasma membrane—9
flagellum—2	mitochondrion—6	ribosomes—10
Golgi apparatus—3	nucleolus—7	RER—11
lysosome—4	nucleus—8	SER—12

1. Associate with spindle fibers during cell division
2. Produces ribosomes
3. Microcirculatory system of cell
4. Contains genetic code of cell
5. Place where oxidative respiration occurs
6. Synthesize protein for use within cell
7. Brings about rapid hydrolysis (digestion) of cell after death
8. Place where ATP is synthesized
9. Regulates flow of materials into and out of cell
10. Synthesizes protein for extracellular distribution
11. Disposal organelles in cytoplasm that digest protein
12. Increases absorptive area of cell
13. Synthesis, packaging, and transportation of substances
14. Plays a role in flagellar development
15. Provides some motile function for some cells

D. *True–False*

Indicate whether the statements are true or false.

1. Secretory cells have well-developed Golgi apparatuses.
2. Mitochondria are able to replicate themselves.
3. The plasma membrane is permeable to practically all molecules.
4. Ribosomes are found in the cytoplasm, mitochondria, RER, and SER.
5. The inner and outer layers of the plasma membrane are essentially composed of lipids, with scattered molecules of protein and glycoprotein.
6. A "brush border" consists of microvilli.
7. Cilia propel spermatozoa.
8. Ribosomes are primarily involved in oxidative respiration.
9. Cilia form from basal bodies derived from centrioles.
10. Spindle fibers develop from mitochondria.

Answers

Organelle Functions

1. ________
2. ________
3. ________
4. ________
5. ________
6. ________
7. ________
8. ________
9. ________
10. ________
11. ________
12. ________
13. ________
14. ________
15. ________

True-False

1. ________
2. ________
3. ________
4. ________
5. ________
6. ________
7. ________
8. ________
9. ________
10. ________

Laboratory Report 6

Student: ______________________

Section: ______________________

Mitosis

A. Phases of Mitosis

Select the phase in which the events occur.

1. Nuclear membrane disappears.
2. Cleavage furrow forms.
3. Spindle fibers begin to form.
4. New nuclear membrane forms.
5. Asters form from centrioles.
6. Cytokinesis begins.
7. DNA replicates.
8. Sister chromatids separate and pass to opposite poles.
9. These form after telophase is complete.
10. Chromosomes become oriented on equatorial plane.

anaphase—1
daughter cells—2
interphase—3
metaphase—4
prophase—5
telophase—6

Answers
Phases of Mitosis
1.________
2.________
3.________
4.________
5.________
6.________
7.________
8.________
9.________
10.________

B. Diagram

Diagram the four stages of mitosis observed in the laboratory in the space provided. Label all identifiable structures.

LABORATORY REPORT 7

Student: ______________________

Section: ______________________

Osmosis and Plasma Membrane Integrity

A. *Molecular Movement*

1. Brownian Movement

a. Were you able to see any movement of carbon particles under the microscope? ______________

b. What forces propel these particles? ______________________________

____

____

____

2. Diffusion. Plot the distance of diffusion of molecules for the two crystals on the graphs.

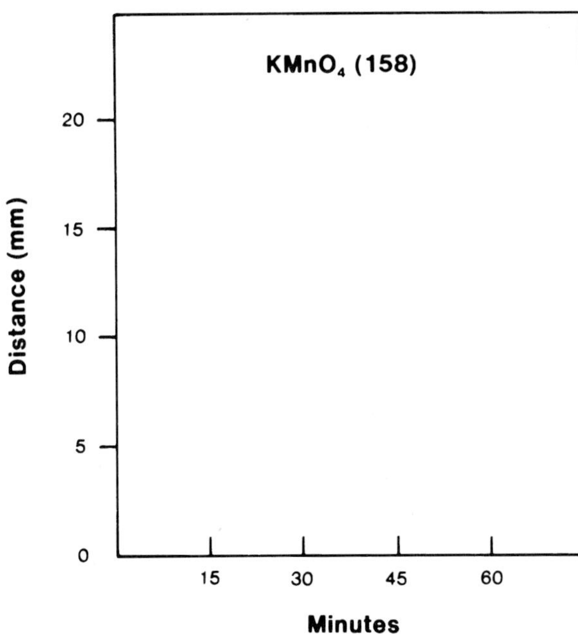

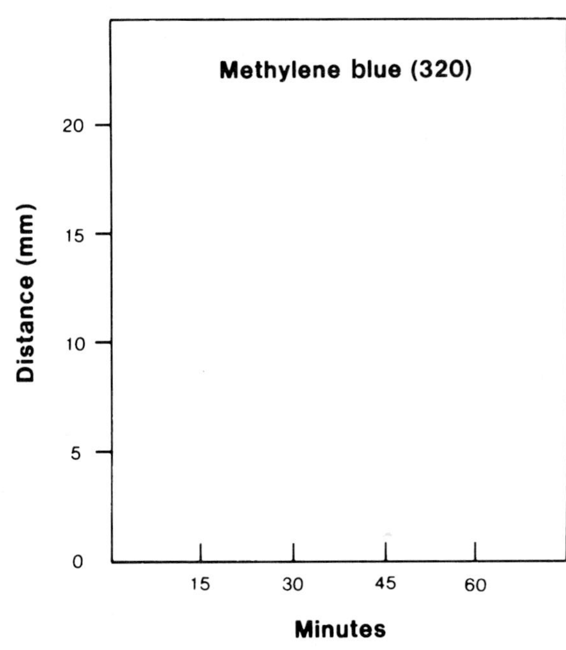

3. Conclusion. What is the relationship of molecular weight to the rate of diffusion? ______________

↑ Weight , ↓ diffusion

____

____

____

B. Osmotic Effects

As you examine each tube macroscopically and microscopically, record your results in the table.

SOLUTION	MACROSCOPIC APPEARANCE (lysis or no lysis)	MICROSCOPIC APPEARANCE (crenation, lysis, or isotonic)
.15M Sodium chloride		
.30M Sodium chloride		
.28M Glucose $C_6H_{12}O_6$		
.30M Glycerine $C_3H_5(OH)_3$		
.30M Urea $CO(NH_2)_2$		

1. Are any of the solutions isotonic for red blood cells? ______________________

   If so, which ones? ______________________

2. With the aid of the atomic weight table in Appendix A, calculate the percentages of solutes in those solutions that proved to be isotonic. ______________________

   ____

   ____

   ____

LABORATORY REPORT 8

Student: ______________________________

Section: ______________________________

Hydrogen Ion (H^+) Concentration and pH

A. *Comparison Testing*

As each solution is tested colorimetrically with pH paper and electronically with a pH meter, record your results in the table. To promote objectivity, your instructor will not furnish the known (actual) pH values until the end of the period. At that time, add these values to the table.

SOLUTION	pH VALUES			SOLUTION	pH VALUES		
	Paper Method	Meter Method	Actual		Paper Method	Meter Method	Actual
A				G			
B				H			
C				I			
D				J			
E				K			
F				L			

Compare the merits and disadvantages of each method. ______________________________

____

____

____

B. *Body Fluids*

Record the pH of the various body fluids. Obtain several values for each fluid.

Urine ______________________________

Blood ______________________________

Saliva ______________________________

Tears ______________________________

C. *pH Computations*

With the aid of the log tables in Appendix A, make the following pH conversions:

pH	gms H^+/liter
5.0	
5.8	
7.9	
	4.8×10^9
	8.3×10^3

LABORATORY REPORT 9

Student: ____________________

Section: ____________________

Buffering Systems

Record the number of drops of 0.0001N HC1 added to each solution.

Test Solutions		Drops of 0.0001N HCl
A	Distilled H_2O	
B	Bicarbonate	
C	Phosphate	
D	Albumin	
E	Blood	
F	Plasma	

Which mixture exhibits the greatest buffering capacity? ____________________

The least? ____________________

LABORATORY REPORT 10

Student: ____________________

Section: ____________________

Epithelial Tissues

A. *Tissue Characteristics*

From the list of tissues, select those that apply to the statements. More than one tissue may apply.

1. Elongated cells, single layer
2. Cells with blocklike cross section
3. Tissue with goblet cells
4. Some binucleate cells
5. Have basal lamina
6. Lack extracellular matrix
7. Flattened cells, single layer
8. Flattened cells, several layers
9. Have extensive vascularization
10. Have apical and basal surfaces
11. Dome-shaped cells near surface
12. Produce mucus on free surface
13. Elongated cells, with some cells not extending up to free surface
14. Cells with microvilli on free surface

columnar,
- pseudostratified—1
- simple—2
- stratified—3

cuboidal—4
squamous,
- simple—5
- stratified—6

transitional—7
all of the above—8
none of the above—9

B. *Tissue Locations*

Identify the epithelial tissues found in the structures.

columnar, ciliated—1
columnar, plain—2
columnar, stratified—3
cuboidal—4
pseudostratified, ciliated—5
pseudostratified, plain—6
squamous, simple—7
squamous, stratified—8
transitional—9

1. Epidermis of skin
2. Lining of kidney calyces
3. Peritoneum
4. Auditory tube lining
5. Cornea of the eye
6. Capillary walls
7. Capsule of lens of the eye
8. Lining of the pharynx
9. Lining of the mouth
10. Pleural membranes
11. Intestinal lining
12. Lining of trachea
13. Lining of blood vessels
14. Lining of urinary bladder
15. Lining of stomach
16. Lining of male urethra
17. The conjunctiva
18. Lining of vagina
19. Anal lining
20. Salivary glands

Answers

Tissue Characteristics

1. ____________
2. ____________
3. ____________
4. ____________
5. ____________
6. ____________
7. ____________
8. ____________
9. ____________
10. ____________
11. ____________
12. ____________
13. ____________
14. ____________

Tissue Locations

1. ________	11. ________
2. ________	12. ________
3. ________	13. ________
4. ________	14. ________
5. ________	15. ________
6. ________	16. ________
7. ________	17. ________
8. ________	18. ________
9. ________	19. ________
10. ________	20. ________

LABORATORY REPORT 11

Student: ______________________

Section: ______________________

Connective Tissues

A. *Tissue Characteristics*

From the list of tissues, select those that apply to the statements. More than one tissue may apply.

1. Contains all three types of fibers
2. Reinforced mostly with collagenous fibers
3. Cells contained in lacunae
4. Cells with large fat vacuoles
5. Contain mast cells
6. Nonyielding support tissue
7. Devoid of blood supply
8. Contain macrophages
9. Most flexible of support tissues
10. Converted to bone during growing years
11. Matrix impregnated with calcium and phosphorous salts
12. Toughest ordinary tissue (for stretching)
13. Cells with appearance of a signet ring
14. Contains elastic fibers woven into flat sheets
15. Looks like white plastic

adipose—1
areolar—2
bone—3
cartilage,
elastic—4
fibrocartilage—5
hyaline—6
dense irregular—7
dense regular—8
reticular—9

B. *Tissue Locations*

From the list of tissues in part A, select the connective tissue found in each of the following.

1. Fetal skeleton
2. Tendon sheaths
3. Intervertebral disks
4. Superficial fasciae
5. Epiglottis
6. Beneath epithelia
7. Lymph nodes
8. Aponeuroses
9. Nerve sheaths
10. Symphysis pubis
11. Auditory tube
12. Skeletal components
13. Periosteum (outer layer)
14. Sharpey's fiber
15. Ligaments and tendons
16. Hemopoietic tissue
17. Tracheal rings
18. Pinna of outer ear

Answers	
Tissue Characteristics	**Tissue Locations**
1.________	1.________
2.________	2.________
3.________	3.________
4.________	4.________
5.________	5.________
6.________	6.________
7.________	7.________
8.________	8.________
9.________	9.________
10.________	10.________
11.________	11.________
12.________	12.________
13.________	13.________
14.________	14.________
15.________	15.________
	16.________
	17.________
	18.________

C. *Terminology*

Select the terms on the right that apply to the statements.

1. Protein in white fibers
2. Hemoposetic tissue
3. Material between fibroblasts
4. Cartilage-producing cells
5. Inner layer of periosteum
6. Cell that produces fibers
7. Synonym for Haversian system
8. Protein in reticular fibers
9. Bony plates in spongy bones
10. Anchors periosteum to bone tissue
11. Layers of bony matrix around central (Haversian) canal
12. Layer of reticular fibers at base of epithelial tissues
13. Cell found in loose fibrous tissue
14. Minute canals radiating out from lacunae in bone tissue
15. Canal in bone that contains capillaries
16. Horizontal canals that connect adjacent central canals

basement lamina—1
canaliculi—2
central canal—3
chondrocytes—4
collagen—5
fibroblast—6
goblet cell—7
lamellae—8
macrophage—9
mast cell—10
matrix—11
osteocyte—12
osteogenic layer—13
osteon—14
perforating canal—15
Red bone marrow—16
Sharpey's fibers—17
trabeculae—18

Answers
Terminology
1.________
2.________
3.________
4.________
5.________
6.________
7.________
8.________
9.________
10.________
11.________
12.________
13.________
14.________
15.________
16.________

D. *Microscopic Study*

If any drawings of tissue are to be made, make them on separate sheets of paper, and label all identifiable structures.

LABORATORY REPORT 12

Student: ____________________

Section: ____________________

The Integument

A. *Multiple Choice*

Select the answer that best completes each statement.

1. Granules of the stratum granulosum consist of
   a. melanin.
   b. keratin.
   c. eleidin.
2. Pacinian corpuscles are sensitive to
   a. temperature.
   b. pressure.
   c. touch.
3. Meissner's corpuscles are sensitive to
   a. temperature.
   b. pressure.
   c. touch.
4. The epidermis has _____ distinct layers.
   a. three
   b. four
   c. five
   d. six
5. The papillary layer is a part of the
   a. dermis.
   b. epidermis.
   c. stratum spinosum.
6. Melanin granules are produced by the
   a. stratum granulosum.
   b. melanocytes.
   c. corium.
   d. stratum corneum.
7. New cells of the epidermis originate in the
   a. dermis.
   b. stratum basale.
   c. stratum corneum.
   d. corium.
8. The outermost layer of the epidermis is the
   a. stratum corneum.
   b. stratum basale.
   c. stratum lucidum.
   d. stratum spinosum.

Answers
Multiple Choice
1.________
2.________
3.________
4.________
5.________
6.________
7.________
8.________

9. The secretions of the apocrine glands usually empty
   a. into a hair follicle.
   b. through the skin surface.
   c. Both *a* and *b* are correct.
   d. None of these is correct.
10. Sebum is secreted by
   a. sweat glands.
   b. eccrine glands.
   c. apocrine glands.
   d. None of these is correct.
11. Keratin's function is primarily to
   a. destroy bacteria.
   b. provide waterproofing.
   c. cool the body.
12. Meissner's corpuscles are in the
   a. papillary layer.
   b. reticular layer.
   c. subcutaneous layer.
13. Arrector pili assist in
   a. maintaining skin tonus.
   b. limiting excessive sweating.
   c. forcing out sebum.
   d. None of these is correct.
14. The hair follicle is
   a. a shaft of hair.
   b. the root of a hair.
   c. a tube in the skin.
   d. None of these is correct.
15. Pacinian corpuscles are in the
   a. papillary layer.
   b. reticular layer.
   c. subcutaneous layer.
16. Sweat is produced by
   a. sebaceous glands.
   b. eccrine glands.
   c. ceruminous glands.
   d. None of these is correct.
17. The coiled-up portion of an apocrine gland is in the
   a. hair follicle.
   b. epidermis.
   c. dermis.
   d. subcutaneous layer.
18. Apocrine glands are
   a. in the axillae.
   b. on the scrotum.
   c. in the ear canal.
   d. All of these are correct.

Answers
Multiple Choice
9. ________
10. ________
11. ________
12. ________
13. ________
14. ________
15. ________
16. ________
17. ________
18. ________

B. Microscopic Study

Make your histology drawings on a separate sheet of paper.

LABORATORY REPORT 13

Student: ____________________

Section: ____________________

The Skeletal Plan

A. *Long Bone Structure*

Match the terms on the right with their descriptions.

1. Shaft portion of bone
2. Hollow chamber in bone shaft
3. Type of marrow in medullary cavity
4. Enlarged end of a bone
5. Type of bone in diaphysis
6. Fibrous covering of bone shaft
7. Linear growth area of long bone
8. Smooth gristle covering bone end
9. Type of bone marrow in bone ends
10. Lining of medullary cavity
11. Type of bone tissue in bone ends
12. Lining of central canals

articular cartilage—1
compact bone—2
diaphysis—3
endosteum—4
epiphyseal disk—5
epiphysis—6
medullary cavity—7
periosteum—8
red marrow—9
spongy bone—10
yellow marrow—11

B. *Bone Identification*

Match the structures on the right with their descriptions.

1. Shoulder blade
2. Collarbone
3. Breastbone
4. Shinbone
5. Kneecap
6. Upper arm bone
7. Bones of spine
8. Thighbone
9. Lateral bone of forearm
10. Joint between ossa coxae
11. Horseshoe-shaped bone
12. Bones of shoulder girdle
13. One half of pelvic girdle
14. Medial bone of forearm
15. Thin bone paralleling tibia (calf bone)

clavicle—1
femur—2
fibula—3
humerus—4
hyoid—5
os coxa—6
patella—7
radius—8
scapula—9
sternum—10
symphysis pubis—11
tibia—12
ulna—13
vertebrae—14

Answers	
Long Bone Structure	Bone Identification
1.______	1.______
2.______	2.______
3.______	3.______
4.______	4.______
5.______	5.______
6.______	6.______
7.______	7.______
8.______	8.______
9.______	9.______
10.______	10.______
11.______	11.______
12.______	12.______
	13.______
	14.______
	15.______

C. *Medical*

Select the condition that each statement describes. Since not all of these conditions are described in this manual, consult your lecture text or medical dictionary for some of the terminology.

comminuted fracture—1	fissured fracture—5	simple fracture—9
compacted fracture—2	greenstick fracture—6	spiral fracture—10
compound fracture—3	oblique fracture—7	transverse fracture—11
displaced fracture—4	segmental fracture—8	

1. Fracture at right angle to a bone's longitudinal axis
2. Fracture in which skin is not broken
3. Fracture caused by torsional forces
4. Fracture in which bone fragments move out of alignment
5. Fracture caused by severe vertical forces
6. Fracture in which piece of bone breaks out of shaft
7. Fracture characterized by two or more fragments
8. Fracture that extends only partially through bone; an incomplete fracture
9. Fracture that occurs at an angle to bone's longitudinal axis

D. *Terminology*

Identify the term on the right that each statement describes. More than one term may apply.

1. Narrow slit	condyle—1
2. Cavity in a bone	fissure—2
3. Very large process to which muscle attaches	foramen—3
4. Small process for muscle attachment	fossa—4
5. Large roughened process for muscle attachment	meatus—5
6. Rounded, knucklelike process that articulates with another bone	sinus—6
7. Depression in a bone	spine—7
8. Tubelike passageway	trochanter—8
9. Hole through which nerves and blood vessels pass	tubercle—9
10. Sharp or long, slender process	tuberosity—10

Answers

Medical

1. ________
2. ________
3. ________
4. ________
5. ________
6. ________
7. ________
8. ________
9. ________

Terminology

1. ________
2. ________
3. ________
4. ________
5. ________
6. ________
7. ________
8. ________
9. ________
10. ________

Laboratory Report 14

Student: ______________________

Section: ______________________

The Skull

A. Bone Names

Select the bones that the statements describe. More than one answer may apply.

1. Forehead bone
2. Upper jaw
3. Cheekbone
4. Lower jaw
5. Sides of skull
6. On walls of nasal fossa
7. Unpaired bones (usually)
8. Bones of hard palate (two pair)
9. Back and bottom of skull
10. Contain sinuses (four bones)
11. External bridge of nose
12. On median line in nasal cavity

ethmoid—1
frontal—2
lacrimal—3
mandible—4
maxilla—5
nasal—6
nasal conchae—7
occipital—8
palatine—9
parietal—10
sphenoid—11
temporal—12
vomer—13
zygomatic—14

B. Terminology

From the list of bones in part A, select those that have the structures that follow.

1. Alveolar process
2. Angle
3. Antrum of Highmore
4. Chiasmatic groove
5. Coronoid process
6. Cribriform plate
7. Crista galli
8. External acoustic meatus
9. Incisive fossa
10. Inferior temporal line
11. Internal acoustic meatus
12. Mandibular condyle
13. Mandibular fossa
14. Mastoid process
15. Mylohyoid line
16. Occipital condyles
17. Ramus
18. Sella turcica
19. Styloid process
20. Superior nasal concha
21. Superior temporal line
22. Symphysis
23. Tear duct
24. Zygomatic arch

C. Sutures

Identify the sutures that the statements describe.

1. Between parietal and occipital
2. Between two maxillae on median line
3. Between two parietals on median line
4. Between frontal and parietal
5. Between parietal and temporal
6. Between maxillary and palatine

coronal—1
lambdoidal—2
median palatine—3
sagittal—4
squamosal—5
transverse palatine—6

Answers

Bones

1. ________	7. ________
2. ________	8. ________
3. ________	9. ________
4. ________	10. ________
5. ________	11. ________
6. ________	12. ________

Terminology

1. ________	13. ________
2. ________	14. ________
3. ________	15. ________
4. ________	16. ________
5. ________	17. ________
6. ________	18. ________
7. ________	19. ________
8. ________	20. ________
9. ________	21. ________
10. ________	22. ________
11. ________	23. ________
12. ________	24. ________

Sutures

1. ________	4. ________
2. ________	5. ________
3. ________	6. ________

D. Foramina Location

Refer to the illustrations in Exercise 14 to identify the bones on which the foramina or canals are located.

1. Anterior palatine foramen
2. Carotid canal
3. Condyloid canal
4. Foramen rotundum
5. Foramen cecum
6. Foramen ovale
7. Foramen spinosum
8. Infraorbital foramen
9. Foramen magnum
10. Mental foramen
11. Hypoglossal canal
12. Supraorbital foramen
13. Internal acoustic meatus
14. Mandibular foramen
15. Optic canal
16. Zygomaticofacial foramen

frontal—1
mandible—2
maxilla—3
occipital—4
palatine—5
parietal—6
sphenoid—7
temporal—8
zygomatic—9

E. Foramina Function

Select the cranial nerves and other structures that pass through the following foramina, canals, and fissures. Note that the number following each cranial nerve corresponds to the number designation of that particular cranial nerve. (Example: The abducens nerve is also known as the sixth cranial nerve.) Also, note that the fifth cranial nerve (trigeminal) has three divisions.

1. Carotid canal
2. Cribriform plate foramina
3. Foramen magnum
4. Foramen rotundum
5. Foramen ovale
6. Internal acoustic meatus
7. Optic canal
8. Hypoglossal canal
9. Superior orbital fissure

Cranial nerves
- olfactory—1
- optic—2
- oculomotor—3
- trochlear—4
- trigeminal branches
  - ophthalmic—5.1
  - maxillary—5.2
  - mandibular—5.3
- abducens—6
- facial—7
- vestibulocochlear—8
- spinal accessory—11
- hypoglossal—12

brain stem—13
internal carotid artery—14
internal jugular vein—15

F. Completion

Name the structures that the statements describe.

1. Lighten skull
2. Provides attachment site for falx cerebri
3. Allow compression of skull bones during birth
4. Anchors sternocleidomastoid muscle
5. Point of attachment for some tongue muscles
6. Allow sound waves to enter skull
7. Point of articulation between mandible and skull
8. Points of articulation between skull and first vertebra

Answers

Foramina Location

1. ________
2. ________
3. ________
4. ________
5. ________
6. ________
7. ________
8. ________
9. ________
10. ________
11. ________
12. ________
13. ________
14. ________
15. ________
16. ________

Foramina Function

1. ________
2. ________
3. ________
4. ________
5. ________
6. ________
7. ________
8. ________
9. ________

Completion

1. ________
2. ________
3. ________
4. ________
5. ________
6. ________
7. ________
8. ________

LABORATORY REPORT 15

Student: ____________________

Section: ____________________

The Vertebral Column and Thorax

A. *Bone Names*

Select the proper anatomical terminology for each bone.

1. Tailbone	atlas—1
2. Breastbone	axis—2
3. False rib	body—3
4. Floating rib	coccyx—4
5. True rib	manubrium—5
6. Second cervical vertebra	sacrum—6
7. First cervical vertebra	sternum—7
8. Inferior portion of breastbone	vertebral ribs—8
9. Midportion of breastbone	vertebrochondral ribs—9
10. Superior portion of breastbone	vertebrosternal ribs—10
	xiphoid—11

B. *Structures*

Select the structures or foramina with the functions described.

body—1	spinal curves—6
intervertebral disks—2	spinous process—7
intervertebral foramina—3	sternum—8
odontoid process—4	transverse process—9
rib cage—5	transverse foramina—10

1. Provides anterior attachment for vertebrosternal ribs
2. Openings through which spinal nerves exit
3. Portion of vertebra that supports body weight
4. Protects vital thoracic organs
5. Impart springiness to vertebral column (two things)
6. Part of thoracic vertebra that provides attachment for rib
7. Acts as pivot for atlas to move around
8. Points of attachment for muscles of the neck and back

Answers
Bone Names
1. ________
2. ________
3. ________
4. ________
5. ________
6. ________
7. ________
8. ________
9. ________
10. ________
Structures
1. ________
2. ________
3. ________
4. ________
5. ________
6. ________
7. ________
8. ________

C. *Vertebral Differences*

Select the vertebrae from the column on the right that are unique for the characteristics that follow.

1. Transverse foramina
2. Massive body
3. Odontoid process
4. Pelvic curvature
5. Rib facets on transverse process

atlas—1
axis—2
cervical vertebrae—3
thoracic vertebrae—4
lumbar vertebrae—5
sacral vertebrae—6

D. *Numbers*

Indicate the number of components that constitute each of the following.

1. Cervical vertebrae
2. Thoracic vertebrae
3. Lumbar vertebrae
4. Sacrum (fused vertebrae)
5. Coccyx (rudimentary vertebrae)
6. Pairs of ribs
7. Spinal curvatures
8. Pairs of vertebrochondral ribs
9. Pairs of vertebrosternal ribs
10. Pairs of vertebral ribs

Answers
Vertebral Differences
1.__________
2.__________
3.__________
4.__________
5.__________
Numbers
1.__________
2.__________
3.__________
4.__________
5.__________
6.__________
7.__________
8.__________
9.__________
10.__________

LABORATORY REPORT 16

Student: ______________________

Section: ______________________

The Appendicular Skeleton

A. Shoulder

Identify the parts of the scapula and humerus that the statements describe. Refer to figures 16.1 and 16.2 for assistance.

acromion process—1	glenoid cavity—5	lesser tubercle—9
axillary margin—2	greater tubercle—6	scapular spine—10
coracoid process—3	head of humerus—7	vertebral margin—11
deltoid tuberosity—4	inferior angle—8	

1. Part of humerus that articulates with scapula
2. Depression on scapula that articulates with humerus
3. Process on humerus to which deltoid muscle attaches
4. Process on scapula that is anterior to glenoid cavity
5. Most inferior tip of scapula
6. Large lateral process near head of humerus
7. Edge of scapula nearest to vertebral column
8. Process on scapula that articulates with clavicle
9. Lateral margin of scapula that extends from glenoid cavity to inferior angle
10. Small process just above surgical neck of humerus
11. Long oblique ridge on posterior surface of scapula that extends from the lateral border to the acromion

B. Arm

Refer to figures 16.1 and 16.3 to identify the parts of the arm described.

capitulum—1	medial epicondyle—5	radial tuberosity—9
coronoid process—2	olecranon fossa—6	semilunar notch—10
head of radius—3	olecranon process—7	styloid process—11
lateral epicondyle—4	radial notch—8	trochlea—12

1. Distal medial condyle of humerus
2. Distal lateral condyle of humerus
3. Part of radius that articulates with humerus
4. Small distal medial process of ulna
5. Process on radius for biceps brachii attachment
6. Depression on proximal end of ulna that articulates with trochlea of humerus
7. Depression on ulna that is in contact with head of radius
8. Condyle of humerus that articulates with ulna
9. Fossa on distal posterior surface of humerus
10. Distal lateral prominence of radius
11. Condyle of humerus that articulates with radius
12. Process on anterior surface of ulna just below semilunar notch
13. Epicondyle adjacent to capitulum
14. Proximal posterior process of ulna, sometimes called the "funny bone"
15. Epicondyle of humerus adjacent to trochlea

Answers
Shoulder
1. __________
2. __________
3. __________
4. __________
5. __________
6. __________
7. __________
8. __________
9. __________
10. __________
11. __________
Arm
1. __________
2. __________
3. __________
4. __________
5. __________
6. __________
7. __________
8. __________
9. __________
10. __________
11. __________
12. __________
13. __________
14. __________
15. __________

C. *Hand*

Refer to figure 16.3 to identify the parts of the wrist and hand described.

1. Part of ulna that articulates with fibrocartilaginous disk of the wrist
2. Two carpal bones that articulate with radius
3. Bone that articulates with the fourth and fifth metacarpals
4. Two carpal bones that articulate with thumb
5. Bones of fingers
6. Four distal carpal bones, from lateral to medial
7. Bones that make up the palm of the hand
8. Four proximal carpal bones, from lateral to medial

capitate—1
hamate—2
head—3
lunate—4
metacarpals—5
phalanges—6
pisiform—7
scaphoid—8
trapezium—9
trapezoid—10
triquetral—11

D. *Pelvis*

Refer to figures 16.4 and 16.5 to identify the parts of the pelvis described.

1. Three bones of os coxa
2. Upper bone of os coxa
3. Anterior bone of os coxa
4. Part of os coxa that one sits on
5. Uppermost process on posterior edge of ischium
6. Superior ridge of hipbone
7. Atriculation site for the head of the femur
8. Large foramen on os coxa
9. Joint between os coxa and sacrum
10. Large prominent process of ischium
11. Joint between two ossa coxae on the median line
12. Lower posterior bone of os coxa

acetabulum—1
iliac crest—2
ilium—3
ischial spine—4
ischium—5
obturator foramen—6
posterior inferior spine—7
posterior superior spine—8
pubis—9
sacroiliac joint—10
symphysis pubis—11
tuberosity of ischium—12

E. *Pelvis Comparisons*

Compare a male pelvis to a female pelvis, and then determine whether these statements are true or false.

1. The aperture of the female pelvis is heart-shaped.
2. The male greater pelvis (concavity of the expanded iliac bones above the pelvic brim) is narrower.
3. The male sacrum and coccyx are more curved.
4. The obturator foramina of the female pelvis tend to be triangular and smaller.
5. The aperture of the female pelvis is larger and more oval.
6. The muscular impressions on the female pelvis are less distinct (bone is smoother).
7. The iliac crests of the female pelvis protrude more.
8. The acetabula of the female face more anteriorly.
9. The sacrum of the female pelvis is shorter and wider.
10. The obturator foramina of the female pelvis are rounder and larger.

Answers
Hand
1.________
2.________
3.________
4.________
5.________
6.________
7.________
8.________
Pelvis
1.________
2.________
3.________
4.________
5.________
6.________
7.________
8.________
9.________
10.________
11.________
12.________
Pelvis Comparisons
1.________
2.________
3.________
4.________
5.________
6.________
7.________
8.________
9.________
10.________

F. *Leg*

Refer to figures 16.5 and 16.6 to identify the parts of the leg described.

femur—1	lateral condyle—9
fibula—2	lateral malleolus—10
gluteal tuberosity—3	lesser trochanter—11
greater trochanter—4	linea aspera—12
head of femur—5	medial condyle—13
head of fibula—6	medial malleolus—14
intertrochanteric crest—7	tibia—15
intertrochanteric line—8	tibial tuberosity—16

1. Distal lateral process of femur
2. Distal medial process of femur
3. Distal lateral process of fibula that forms outer anklebone
4. Distal medial process of tibia that forms inner prominence of ankle
5. Proximal lateral process of tibia that articulates with femur
6. Proximal anterior process of tibia that protrudes below the kneecap
7. Proximal medial process of tibia that articulates with femur
8. Ridge on posterior surface of femur (upper portion)
9. Ridge on posterior surface of femur (lower portion)
10. Thin lateral bone of lower leg
11. Strongest bone of lower leg
12. Process just inferior to neck of femur on medial surface
13. Oblique ridge between trochanters on posterior surface of femur
14. Part of femur that fits into acetabulum
15. Longest, heaviest bone of leg
16. End of fibula that articulates with upper end of tibia
17. Large process lateral to neck of femur
18. Oblique ridge between trochanters on anterior surface of femur

G. *Foot*

Refer to figure 16.6 to identify the parts of the foot described.

calcaneous—1	navicular—5
cuboid—2	phalanges—6
cuneiform—3	talus—7
metatarsals—4	

1. Heelbone
2. Largest tarsal bone
3. Tarsal bone that articulates with tibia
4. Tarsal bone on side of foot in front of calcaneous
5. Five elongated bones of foot that form instep
6. Bones of toes
7. Three similarly named tarsal bones anterior to navicular
8. Tarsal bone anterior to talus and on medial side of foot

Answers
Leg
1.________
2.________
3.________
4.________
5.________
6.________
7.________
8.________
9.________
10.________
11.________
12.________
13.________
14.________
15.________
16.________
17.________
18.________
Foot
1.________
2.________
3.________
4.________
5.________
6.________
7.________
8.________

H. Numbers

Supply the correct numbers for the statements that follow.

1. Number of phalanges in large toe
2. Total number of carpal bones in each wrist
3. Total number of tarsal bones in each foot
4. Total number of metatarsals in each foot
5. Number of phalanges in thumb
6. Number of phalanges on each of the other four fingers
7. Number of phalanges in the second, third, fourth, and fifth toes
8. Number designation for thumb metacarpal
9. Number designation for small toe metatarsal
10. Number designation for little finger metacarpal

Answers
Numbers
1.________
2.________
3.________
4.________
5.________
6.________
7.________
8.________
9.________
10.________

LABORATORY REPORT 17

Student: ______________________

Section: ______________________

Articulations

A. Characteristics

Select the joints from the list on the right that have the characteristics described. More than one answer may apply in some cases.

1. Movement in all directions
2. Slightly movable
3. Immovable
4. Freely movable in one plane only
5. Freely movable in two planes
6. Rotational movement only
7. Condyle end in elliptical cavity
8. Each bone end concave in one direction and convex in another direction
9. Held together with interosseous ligament
10. Fibrous connective tissue joins bone ends
11. Flat or slightly convex bone ends
12. Fused joint of hyaline cartilage
13. Contains fibrocartilaginous pad

ball-and-socket—1
condyloid—2
gliding—3
hinge—4
pivot—5
saddle—6
sutures—7
symphysis—8
synchondrosis—9
syndesmosis—10

B. Joint Classification

By examining and manipulating the bones of a skeleton, classify the joints that follow. Two terms apply to each joint.

1. Elbow joint
2. Knee joint
3. Phalanges to metacarpals
4. Phalanges to metatarsals
5. Hip joint
6. Shoulder joint
7. Intervertebral joint (between vertebral bodies)
8. Metaphysis of long bone of child
9. Interlocking joint between cranial bones
10. Symphysis pubis
11. Articulation between tibia and fibula at distal ends
12. Atlas to skull
13. Toe joints (between phalanges)
14. Finger joint (between phalanges)
15. Wrist (carpals to carpals)
16. Ankle (talus to tibia)
17. Ankle (tarsals to tarsals)
18. Axis-atlas rotation

Basic Types

freely movable—1
immovable—2
slightly movable—3

Subtypes

ball-and-socket—4
condyloid—5
gliding—6
hinge—7
pivot—8
saddle—9
suture—10
symphysis—11
synchondrosis—12
syndesmosis—13

Answers

Characteristics

1. ________
2. ________
3. ________
4. ________
5. ________
6. ________
7. ________
8. ________
9. ________
10. ________
11. ________
12. ________
13. ________

Joint Classification

1. ________
2. ________
3. ________
4. ________
5. ________
6. ________
7. ________
8. ________
9. ________
10. ________
11. ________
12. ________
13. ________
14. ________
15. ________
16. ________
17. ________
18. ________

C. *True–False*

Indicate whether the statements are true or false.

1. Bursae are sacs of synovial membrane.
2. Cartilage that is surgically removed from a knee joint is usually the hyaline type.
3. The ligamentum teres femoris is primarily nutritional in function.
4. The acetabular labrum increases the depth of the acetabulum.
5. Freely movable joints are lubricated with serum.
6. The capsule around a joint consists of a ligament lined with synovial membrane.
7. Hyaline cartilage covers the ends of amphiarthrotic and diarthrotic joints.
8. "Water on the knee" involves the suprapatellar bursa.
9. The knee joint is substantially reinforced to withstand lateral forces.
10. The most highly stressed joint in the body is the hip joint.
11. Articular capsules consist of ligaments and tendons.
12. The strongest ligament holding the hip together is the ligamentum teres femoris.

D. *Medical*

Consult your lecture text, medical dictionary, or other source to match the conditions with their descriptions.

ankylosis—1	rheumatism—7
bursitis—2	rheumatoid arthritis—8
chronic fibrositis—3	sprain—9
dislocation—4	strain—10
gouty arthritis—5	subluxation—11
osteoarthritis—6	tenosynovitis—12

1. Partial or incomplete dislocation
2. Displacement of bones from normal orientation in joint
3. Inflammation of synovial bursa
4. Inflammation of fibrous connective tissue, causing tenderness and stiffness
5. Injury to joint by displacement, resulting in swelling and discoloration
6. Inflammation of tendon and its sheath
7. Injury to joint in which ligaments stretch without swelling or discoloration
8. Arthritis of confusing etiology, characterized by deformity and immobility of joints; often treated with cortisone
9. Arthritis due to excess levels of uric acid in blood
10. Arthritis resulting from degenerative changes in joint (wear and tear)
11. Fixation or fusion of a joint; usually preceded by infection
12. General term applied to soreness and stiffness in muscles and joints

Answers
True–False
1.________
2.________
3.________
4.________
5.________
6.________
7.________
8.________
9.________
10.________
11.________
12.________
Medical
1.________
2.________
3.________
4.________
5.________
6.________
7.________
8.________
9.________
10.________
11.________
12.________

LABORATORY REPORT 18

Student: ______________________

Section: ______________________

The Nerve Cells

A. *Nerve Cell Types*

Select the cell type that applies to each description. More than one cell type may apply in some cases.

1. Afferent neurons
2. Efferent neurons
3. Association cells for pyramidal cells and Purkinje cells
4. Neurons in the molecular layer of the cerebellum
5. Linkage cells between sensory and motor neurons
6. Predominant neuron type in cerebral cortex
7. Small cells covering perikaryon of sensory neuron

basket cells—1
interneurons—2
motor neuron—3
Purkinje cells—4
pyramidal cells—5
satellite cells—6
sensory neuron—7
stellate cells—8

B. *True–False*

Indicate whether the statements concerning nerve cells are true or false.

1. Neuroglia cells do not carry nerve impulses.
2. Perikarya of afferent neurons are in the spinal ganglia.
3. Most neurons have many dendrites with only a single axon.
4. Perikarya of motor neurons are in the white matter of the spinal cord.
5. Myelinated fibers carry nerve impulses at a slower speed than unmyelinated ones.
6. Schwann cells are always present along myelinated fibers.
7. Part of the axon of a sensory neuron functions like a dendrite.
8. Some neurons lack axons.
9. Neurolemmocytes form both the myelin sheath and the neurolemma.
10. Motor neurons carry nerve impulses away from the CNS.

Answers

Nerve Cell Types

1. ________
2. ________
3. ________
4. ________
5. ________
6. ________
7. ________

True–False

1. ________
2. ________
3. ________
4. ________
5. ________
6. ________
7. ________
8. ________
9. ________
10. ________

C. *Neuron Structure*

Select the neuron structure that each statement describes. More than one structure may apply in some cases.

1. Neuron cell body
2. Side branch of axon
3. Long process on motor neuron
4. Short process on motor neuron
5. Lacking in unmyelinated nerve fibers
6. Process that receives nerve impulses
7. Slender filaments in perikaryon
8. Formed by neurolemmocytes
9. Synonym for neurolemmal sheath
10. Interruptions in neurolemma
11. Outer membrane covering myelin sheath
12. Bodies on ER that synthesize protein
13. Promotes regeneration of damaged nerve fiber
14. Enables nerve fiber to carry impulses by saltatory conduction
15. Bodies in perikaryon that stain well with basic aniline dyes
16. Process that carries nerve impulses away from perikaryon

axon—1
collateral—2
dendrite—3
myelin sheath—4
neurofibrils—5
neurolemma—6
neurolemmocytes—7
Nissl bodies—8
nodes of Ranvier—9
perikaryon—10

Answers
Neuron Structure
1.________
2.________
3.________
4.________
5.________
6.________
7.________
8.________
9.________
10.________
11.________
12.________
13.________
14.________
15.________
16.________

D. *Cell Drawings*

On a separate sheet of paper, sketch representative cells of the various types of neurons from available slides. Attach the drawings to this report.

LABORATORY REPORT 19

Student: ____________________

Section: ____________________

The Muscle Cells

A. *Muscle Cell Characteristics*

Identify the types of muscle fibers with the characteristics described.

cardiac—1
skeletal—2
smooth—3

1. Attached to bones
2. Spindle-shaped cells
3. Nonsyncytial
4. Syncytial structure
5. Voluntarily controlled (generally)
6. Lack cross-striations
7. Long, multinucleated cells
8. In walls of venae cavae where they enter right atrium
9. In walls of other veins and arteries
10. Involuntarily controlled (generally)
11. Branched cells
12. Nuclei near surface of cell

B. *Muscle Cell Structures*

Identify the muscle cell structures that the statements describe.

endomysium—1
fascicle—2
intercalated disks—3
perimysium—4
sarcolemma—5
sarcoplasm—6

1. Plasma membrane of skeletal muscle fibers
2. Plasma membranes between individual cardiac muscle fibers
3. Connective tissue between individual skeletal muscle fibers
4. Cytoplasmic material of skeletal muscle fibers
5. Bundle of skeletal muscle fibers

C. *Cell Drawings*

On a separate sheet of paper, sketch representative muscle cells from available slides. Attach the drawings to this report.

Answers
Muscle Cell Characteristics
1. ________
2. ________
3. ________
4. ________
5. ________
6. ________
7. ________
8. ________
9. ________
10. ________
11. ________
12. ________
Muscle Cell Structures
1. ________
2. ________
3. ________
4. ________
5. ________

LABORATORY REPORT 20

Student: ____________________

Section: ____________________

The Neuromuscular Junction

Summary

Complete these paragraphs about the neuromuscular junction.

The unmyelinated ending of a motor neuron consists of several knoblike structures called ___1___ . Between these knobs and the sarcolemma is a ___2___ cleft filled with a gelatinous substance. When a nerve impulse reaches the end of the neuron, a neurotransmitter, called ___3___ , bridges the gap to initiate an action potential in the muscle fiber.

The neurotransmitter can stimulate the muscle fiber because of the presence of receptors in the ___4___ membrane, which the neurotransmitter activates. The synaptic vesicles of a motor neuron contain large quantities of the neurotransmitter, which are released into the synaptic cleft via ___5___. The neurotransmitters then ___6___ across the synaptic cleft and bind with membrane receptors.

Depolarization of the neurolemma results in ___7___ ions flowing into the neuron and ___8___ ions flowing outward. Depolarization in the muscle fiber progresses from the sarcolemma inward to the ___9___ system, and the release of ___10___ ions by the ___11___ .

Within 1/500 of a second after its passage across the gap, the neurotransmitter is inactivated by the enzyme ___12___ . The neurotransmitter is inactivated by being split into its two components: ___14___ and ___13___ . Resynthesis of the neurotransmitter from these two components is accomplished by the enzyme ___15___ , which is produced by the ___16___ of the neuron, and stored in the cytoplasm of the knoblike structures.

Answers

Summary

1. ____________________
2. ____________________
3. ____________________
4. ____________________
5. ____________________
6. ____________________
7. ____________________
8. ____________________
9. ____________________
10. ____________________
11. ____________________
12. ____________________
13. ____________________
14. ____________________
15. ____________________
16. ____________________

Laboratory Report 21–26

Student: ______________________________

Section: ______________________________

Electronic Instrumentation

A. Transducers and Electrodes

1. What determines whether you use an electrode or a transducer for monitoring a particular biological phenomenon? ______________________________

____

____

2. Differentiate between:

Input transducer ______________________________

____

Output transducer ______________________________

____

3. Indicate what kind of pickup device (*electrode* or *transducer*) to use for the following. If a transducer is to be used, indicate the specific type.

Sound ______________________________

Light ______________________________

Muscle pull ______________________________

Temperature ______________________________

Skin conductivity ______________________________

Brain waves ______________________________

B. Unigraph Functions

Indicate after each of the following the specific control you would use to produce the desired effect.

1. Turn on the unit ______________________________
2. Change the paper speed ______________________________
3. Increase the intensity of the tracing ______________________________
4. Record on the chart the exact time that a new event occurs ______________________________
5. Increase the sensitivity ______________________________
6. Balance the transducer ______________________________
7. Start and stop the paper movement ______________________________

C. Physiograph Functions

1. How do you increase or decrease ink flow to the pens? ______________________________

____

2. Into what unit do you insert the cable from electrodes or transducers? ______________________

____

3. At what sensitivity setting (*low* or *high*) do you usually start an experiment? ______________
4. At what speeds can the paper move on a Physiograph? ______________________________
5. What button do you press to start the chart moving? ______________________________
6. How do you remove ink from the pens on the instrument at the end of the laboratory period? ______

____

D. Electronic Stimulator Functions

Indicate how you would set the controls on the Grass and Narco stimulators to get the outputs that follow.

Grass SD9		***Narco SM–1***	
4.5 V:			
Round control knob	________	Voltage range control	________
Decade switch	________	Variable control knob	________
80 impulses per second:			
Round control knob	________	Round control knob (Hz) ...	________
Decade switch	________	Rate switch	________
Impulse duration (width) of 1.5 msec:			
Round control knob	________	Round control knob	________
Decade switch	________		

E. The Intelitool System

1. List several advantages that computerized hardware has over chart recorders.

____

____

2. What advantage does a Duograph or polygraph have over Intelitool computerized hardware?

____

3. When can you *not* select "Review/Analyze Menu" from the Main Menu?

____

LABORATORY REPORT 27

Student: ______________________

Section: ______________________

The Physicochemical Properties of Muscle Contraction

A. *Length Changes*

Record muscle fiber lengths before and after adding activants.

Original length __________ mm

Length after adding ATP, K^+, and Mg^{++} __________ mm

B. *Drawings*

Sketch the muscle fibers as seen under high-dry or oil immersion before and during contraction.

Before contraction	During contraction

C. *Structure*

Select the muscle fiber component that each statement describes. More than one answer is required in some cases.

actin—1
F-actin protein—2
heavy meromyosin—3
light meromyosin—4
myosin—5
sarcomere—6
tropomyosin—7
troponin—8

1. Thick myofilament
2. Principal constituent of A band
3. Two types of myofilaments that make up a myofibril
4. Three components of actin
5. Type of myofilament that makes up I band
6. Protein of a myosin molecule that forms a head
7. Component of a myofibril that extends between two Z lines

Answers

Structure

1. __________
2. __________
3. __________
4. __________
5. __________
6. __________
7. __________

D. *Physiology*

Supply the terms to complete the explanation of muscle contraction.

actin—1	heads—6	sarcoplasmic reticulum—10
ATP—2	magnesium—7	tropomyosin—11
calcium—3	myosin—8	troponin—12
cross-bridges—4	potassium—9	
F-actin protein—5		

The interaction of actin and myosin during muscle contraction depends on ample amounts of ATP, calcium, and __1__ ions. The calcium ions are released by the __2__ during depolarization. When these ions are released, __3__ myofilaments are pulled toward each other by the ratcheting action of __4__ that are present on the __5__ myofilaments. The calcium ions combine with __6__ , causing the __7__ molecule to be drawn deeper, which, in turn, exposes active sites on the __8__ filament. When active sites are exposed, __9__ form between the myofilaments. The energy for muscle contraction is derived from __10__ .

Answers

Physiology

1. ____________
2. ____________
3. ____________
4. ____________
5. ____________
6. ____________
7. ____________
8. ____________
9. ____________
10. ____________

LABORATORY REPORT 28

Student: ______________________

Section: ______________________

Mapping of Motor Points

A. Minimum Voltages

Record in the spaces under each number the minimum voltage that stimulates each motor point. If no reaction was induced at a particular spot, write *none.*

1	2	3	4	5	6
12	11	10	9	8	7
13	14	15	16	17	18

B. Minimum Duration

Record the minimum duration at each motor point that could induce a twitch.

1	2	3	4	5	6
12	11	10	9	8	7
13	14	15	16	17	18

C. *Reverse Polarity*

Did you note any differences when the polarity was reversed? ______________________________

____

____

D. *Questions*

1. How do you account for the differences in the threshold levels of different motor points? ___________

   ____

   ____

   ____

   ____

2. Were there differences in the duration required for stimulating motor points? ___________

Laboratory Report 29

Student: ______________________

Section: ______________________

Muscle Contraction Experiments: Using Chart Recorders

A. *Visible Twitches (Unrecorded)*

Record the **threshold stimuli** that cause visible twitches without using the recorder. Note that the muscle is stimulated directly and through the nerve.

A visible twitch with probe on **muscle** . . . . ____________ V

____________ msec

Extension of foot with probe on **muscle** . . . . ____________ V

____________ msec

A visible twitch with probe on **nerve** . . . . ____________ V

____________ msec

Extension of foot with probe on **nerve** . . . . ____________ V

____________ msec

Comparisons

Determine how much greater the stimulus requirement is for direct muscle stimulation by dividing one voltage into the other as follows:

$$\frac{\text{Voltage by direct muscle stimulation}}{\text{Voltage by nerve stimulation}} =$$

for visible twitch . . . . ____________

for foot extension . . . . ____________

B. *Recorded Twitches*

Attach a segment of the chart in the space provided.

C. Evaluation of Recorded Twitches

From the chart record on the previous page, determine the following:

Latent period (A) . . . . ____________ msec

Contraction phase (B) . . . . ____________ msec

Relaxation phase (C) . . . . ____________ msec

Entire duration (A+ B + C) . . . . ____________ msec

D. Spatial Summation

Attach a segment of the chart that illustrates spatial summation. Be sure to indicate the voltages used for each stimulation.

E. Temporal (Wave) Summation

Attach a segment of the chart that illustrates temporal (wave) summation and tetanization.

F. Definitions

Define the terms.

Motor unit ____

Threshold stimulus ____

Spatial summation ____

Tetany ____

LABORATORY REPORT 31

Student: ____________________

Section: ____________________

The Effects of Drugs on Muscle Contraction

A. Tabulation of Results

Indicate each drug's effect on muscle contraction. Add any optional drugs you tested in the blank spaces.

DRUG	EFFECT
Valium	
Epinephrine	
Acetylcholine	
EDTA	

B. Conclusions

Can you draw any conclusions from this experiment? Did you encounter any problems that would cause you to do this experiment differently if it were repeated? ____________________

____

____

____

Laboratory Report 32

Student: ____________________

Section: ____________________

Electromyography

A. *Recordings*

Attach your sample myograms to this page.

B. *Applications*

List several useful applications of electromyography. ____________________

____

____

____

____

LABORATORY REPORT 33

Student: ____________________

Section: ____________________

Muscle Structure

A. *Muscle Terminology*

Match the terms at right with the statements.

1. Bundle of muscle fibers
2. Sheath of fibrous connective tissue that surrounds each fasciculus
3. Movable end of muscle
4. Immovable end of muscle
5. Thin layer of connective tissue surrounding each muscle fiber
6. Layer of connective tissue between skin and deep fascia
7. Flat sheet of connective tissue that attaches muscle to skeleton or another muscle
8. Layer of connective tissue that surrounds a group of fasciculi
9. Outer connective tissue that covers entire muscle
10. Band of fibrous connective tissue that connects muscle to bone or to another muscle

aponeurosis—1
deep fascia—2
endomysium—3
epimysium—4
fasciculus—5
insertion—6
ligament—7
origin—8
perimysium—9
superficial fascia—10
tendon—11

Answers
Muscle Terminology
1. __________
2. __________
3. __________
4. __________
5. __________
6. __________
7. __________
8. __________
9. __________
10. __________

LABORATORY REPORT 34

Student: ____________________

Section: ____________________

Body Movements

A. Muscle Types

Match the types of muscles on the right with the statements.

1. Prime movers	agonists—1
2. Opposing muscles	antagonists—2
3. Muscles that assist prime movers	fixation muscles—3
4. Muscles that hold structures steady to allow agonists to act smoothly	synergists—4

B. Movement Descriptions

Match the types of movements on the right with their descriptions.

1. Movement of limb away from median line of body	abduction—1
2. Movement of limb toward median line of body	adduction—2
3. Movement around central axis	circumduction—3
4. Turning of sole of foot inward	dorsiflexion—4
5. Turning of sole of foot outward	eversion—5
6. Flexion of toes	extension—6
7. Backward movement of head	flexion—7
8. Extension of foot at ankle	hyperextension—8
9. Increased angle between two bones	inversion—9
10. Decreased angle between two bones	plantar flexion—10
11. Movement of hand from palm-down to palm-up position	pronation—11
12. Conelike rotational movement	rotation—12
13. Movement of hand from palm-up to palm-down position	supination—13

Answers

Muscle Types

1. ________
2. ________
3. ________
4. ________

Movement Descriptions

1. ________
2. ________
3. ________
4. ________
5. ________
6. ________
7. ________
8. ________
9. ________
10. ________
11. ________
12. ________
13. ________

C. *Movement Visualizations*

Identify the type of movement that each drawing shows.

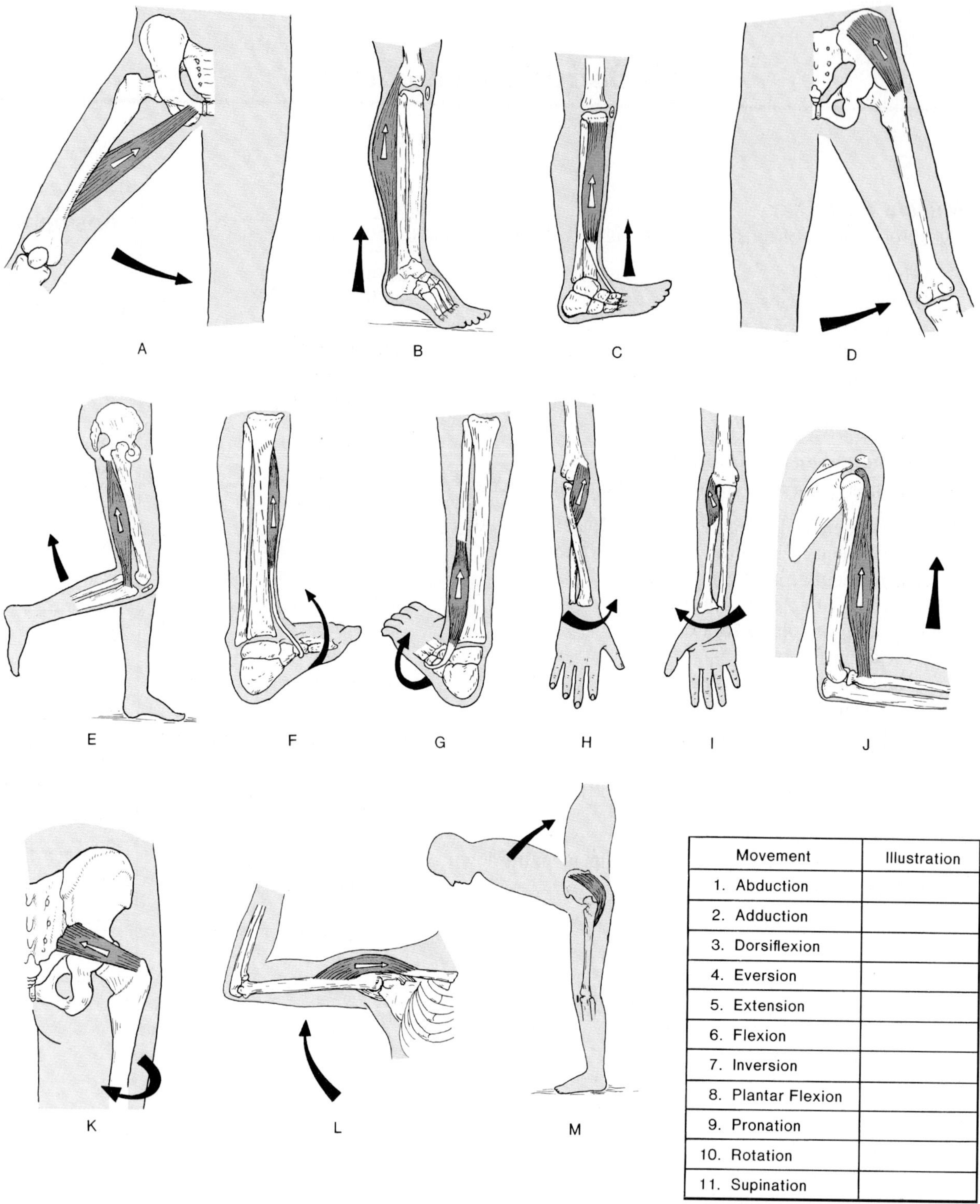

Movement	Illustration
1. Abduction	
2. Adduction	
3. Dorsiflexion	
4. Eversion	
5. Extension	
6. Flexion	
7. Inversion	
8. Plantar Flexion	
9. Pronation	
10. Rotation	
11. Supination	

LABORATORY REPORT 36

Student: ____________________

Section: ____________________

Head and Neck Muscles

A. *Facial Expressions*

Consult figure 36.1 and the text to match the facial muscles on the right with the descriptions.

1. Smiling
2. Horror
3. Sadness
4. Contempt or disdain
5. Pouting lips
6. Blinking or squinting
7. Horizontal wrinkling of forehead

Origins

8. On frontal bone
9. On zygomatic bone

Insertions

10. On lips
11. On eyebrows
12. On eyelid

frontalis—1
orbicularis oculi—2
orbicularis oris—3
platysma—4
quadratus labii inferioris—5
quadratus labii superioris—6
triangularis—7
zygomaticus—8

B. *Mastication*

Consult figure 36.1 and the text to match the muscles on the right with the masticatory movements.

1. Lowers the mandible
2. Raises the mandible
3. Retracts the mandible
4. Protrudes the mandible
5. Holds food in place

Origins

6. On zygomatic arch
7. On sphenoid only
8. On frontal, parietal, and temporal bones
9. On maxilla, mandible, and pteromandibular raphe
10. On maxilla, sphenoid, and palatine bones

Insertions

11. On orbicularis oris
12. On coronoid process of mandible
13. On medial surface of body of mandible
14. On lateral surface of ramus and angle of mandible
15. On neck of mandibular condyle and articular disk

buccinator—1
external pterygoid—2
internal pterygoid—3
masseter—4
temporalis—5

Answers

Facial Expressions

1. ________
2. ________
3. ________
4. ________
5. ________
6. ________
7. ________
8. ________
9. ________
10. ________
11. ________
12. ________

Mastication

1. ________
2. ________
3. ________
4. ________
5. ________
6. ________
7. ________
8. ________
9. ________
10. ________
11. ________
12. ________
13. ________
14. ________
15. ________

C. Neck Muscles

Consult figure 36.2 and the text to identify the neck muscles from their descriptions.

digastric—1
longissimus capitis—2
mylohyoid—3
omohyoid—4
platysma—5
semispinalis capitis—6
splenius capitis—7
sternocleidomastoid—8
sternohyoid—9
sternothyroid—10
trapezius—11

1. Broad, sheetlike muscle on side of neck just under skin
2. Broad, triangular muscle of back that functions only partially as a neck muscle
3. Provides support for floor of mouth
4. V-shaped muscle attached to hyoid and mandibular bones
5. Posterior neck muscles that insert on mastoid process
6. Anterior neck muscle that attaches to thyroid cartilage and part of sternum
7. Anterior neck muscle that extends from hyoid bone to clavicle and manubrium
8. Long muscle of neck that extends from mastoid process to clavicle and sternum
9. Muscles that pull head backward (hyperextension)
10. Muscles that pull head forward

Answers
Neck Muscles
1.________
2.________
3.________
4.________
5.________
6.________
7.________
8.________
9.________
10.________

Laboratory Report 37

Student: ______________________

Section: ______________________

Trunk and Shoulder Muscles

A. Anterior Trunk Muscles

Refer to figure 37.1 and the text to match the muscles with the descriptions.

deltoid—1
external intercostal—2
internal intercostal—3
pectoralis major—4
pectoralis minor—5
serratus anterior—6
subscapularis—7

Origins

1. On upper eight or nine ribs
2. On scapular spine and clavicle
3. On third, fourth, and fifth ribs
4. On clavicle, sternum, and costal cartilages

Insertions

5. On coracoid process
6. On deltoid tuberosity of humerus
7. On anterior surface of scapula
8. On upper anterior end of humerus

Actions

9. Pulls scapula forward, downward, and inward
10. Rotates arm medially
11. Raises ribs (inhaling)
12. Lowers ribs (exhaling)
13. Adducts arm; also, rotates it medially
14. Abducts arm

B. Posterior Trunk Muscles

Refer to figure 37.2 and the text to match the muscles with the descriptions.

infraspinatus—1
latissimus dorsi—2
levator scapulae—3
rhomboideus major—4
rhomboideus minor—5
sacrospinalis—6
supraspinatus—7
teres major—8
teres minor—9
trapezius—10

Origins

1. Near inferior angle of scapula
2. On lateral (axillary) margin of scapula
3. On fossa of scapula above spine
4. On infraspinous fossa of scapula
5. On thoracic and lumbar vertebrae, sacrum, lower ribs, and iliac crest
6. On lower part of ligamentum nuchae and first thoracic vertebra
7. On transverse processes of first four cervical vertebrae

Answers

Anterior Trunk Muscles

1. ________
2. ________
3. ________
4. ________
5. ________
6. ________
7. ________
8. ________
9. ________
10. ________
11. ________
12. ________
13. ________
14. ________

Posterior Trunk Muscles

1. ________
2. ________
3. ________
4. ________
5. ________
6. ________
7. ________

B. Posterior Trunk Muscles (continued)

infraspinatus—1
latissimus dorsi—2
levator scapulae—3
rhomboideus major—4
rhomboideus minor—5
sacrospinalis—6
supraspinatus—7
teres major—8
teres minor—9
trapezius—10

Insertions

8. On lower third vertebral border of scapula
9. On vertebral border of scapula near origin of scapular spine
10. On crest of lesser tubercle of humerus
11. On intertubercular groove of humerus
12. On middle facet of greater tubercle of humerus
13. On ribs and vertebrae

Actions

14. Abducts arm
15. Raises scapula and draws it medially
16. Extends spine to maintain erectness
17. Abducts and rotates arm medially
18. Abducts and rotates arm laterally

Answers

Trunk Muscles Posterior

8. ________
9. ________
10. ________
11. ________
12. ________
13. ________
14. ________
15. ________
16. ________
17. ________
18. ________

Laboratory Report 38

Student: ______________________

Section: ______________________

Upper Extremity Muscles

A. Arm Movements

Consult figure 38.1 to match the muscles with the descriptions. Verify your answers with the text.

1. Extends forearm
2. Adducts arm
3. Flexes forearm
4. Carries arm forward in flexion
5. Flexes forearm and rotates radius outward to supinate hand

biceps brachii—1
brachialis—2
brachioradialis—3
coracobrachialis—4
triceps brachii—5

Origins

6. On lower half of humerus
7. On coracoid process of scapula
8. Above the lateral epicondyle of the humerus
9. Three heads: on scapula, on posterior surface of humerus, and just below the radial groove of humerus
10. Two heads: on coracoid process and on supraglenoid tubercle of humerus

Insertions

11. On olecranon process
12. On radial tuberosity
13. Near the middle, medial surface of humerus
14. On front surface of coronoid process of ulna
15. On lateral surface of radius just above styloid process

B. Hand Movements

Consult figures 38.2 and 38.3 to match the muscles with the descriptions. Verify your answers with the text.

1. Supinates hand
2. Pronates hand
3. Flexes and abducts hand
4. Extends wrist and hand
5. Flexes thumb
6. Extends all fingers except thumb
7. Abducts thumb
8. Flexes all distal phalanges except thumb
9. Flexes all fingers except thumb
10. Extends thumb

abductor pollicis—1
extensor carpi radialis brevis—2
extensor carpi radialis longus—3
extensor carpi ulnaris—4
extensor digitorum communis—5
extensor pollicis longus—6
flexor carpi radialis—7
flexor carpi ulnaris—8
flexor digitorum profundus—9
flexor digitorum superficialis—10
flexor pollicis longus—11
pronator quadratus—12
pronator teres—13
supinator—14

Answers
Arm Movements
1.________
2.________
3.________
4.________
5.________
6.________
7.________
8.________
9.________
10.________
11.________
12.________
13.________
14.________
15.________
Hand Movements
1.________
2.________
3.________
4.________
5.________
6.________
7.________
8.________
9.________
10.________

B. Hand Movements (continued)

abductor pollicis—1
extensor carpi radialis brevis—2
extensor carpi radialis longus—3
extensor carpi ulnaris—4
extensor digitorum communis—5
extensor pollicis longus—6
flexor carpi radialis—7
flexor carpi ulnaris—8
flexor digitorum profundus—9
flexor digitorum superficialis—10
flexor pollicis longus—11
pronator quadratus—12
pronator teres—13
supinator—14

Origins

11. On interosseous membrane between radius and ulna
12. On radius, ulna, and interosseous membrane
13. On lateral epicondyle of humerus and part of ulna
14. On humerus, ulna, and radius
15. On medial epicondyle of humerus

Insertions

16. On distal phalanges of second, third, fourth, and fifth fingers
17. On distal phalanx of thumb
18. On middle phalanges of second, third, fourth, and fifth fingers
19. On radial tuberosity and oblique line of radius
20. On upper lateral surface of radius
21. On first metacarpal and trapezium
22. On second metacarpal
23. On proximal portion of second and third metacarpals
24. On middle metacarpal
25. On fifth metacarpal

Answers
Hand Movements
11.________
12.________
13.________
14.________
15.________
16.________
17.________
18.________
19.________
20.________
21.________
22.________
23.________
24.________
25.________

Laboratory Report 39

Student: ______________________

Section: ______________________

Abdominal, Pelvic, and Intercostal Muscles

A. *Abdominal Muscles*

Consult figure 39.1 to match the muscles with the descriptions. Verify your answers with the text.

external oblique—1	rectus abdominis—3
internal oblique—2	transversus abdominis—4

1. Flexion of spine in lumbar region
2. Maintain intraabdominal pressure
3. Antagonists of diaphragm

Origins

4. On pubic bone
5. On external surface of lower eight ribs
6. On lateral half of inguinal ligament and anterior two-thirds of iliac crest
7. On inguinal ligament, iliac crest, and costal cartilages of lower six ribs

Insertions

8. On linea alba and crest of pubis
9. On cartilages of fifth, sixth, and seventh ribs
10. On the linea alba where fibers of aponeurosis interlace
11. On costal cartilages of lower three ribs, linea alba, and crest of pubis

B. *Pelvic Muscles*

Consult figure 39.3 to match the muscles with the descriptions. Verify your answers with the text.

iliacus—1
psoas major—2
quadratus lumborum—3

1. Flexes femur on trunk
2. Flexes lumbar region of vertebral column
3. Extension of spine at lumbar vertebrae

Origins

4. On lumbar vertebrae
5. On iliac crest, iliolumbar ligament, and transverse processes of lower four lumbar vertebrae
6. On iliac fossa

Insertions

7. On inferior margin of last rib and transverse processes of upper four lumbar vertebrae
8. On lesser trochanter of femur

Answers

Abdominal Muscles

1. ________
2. ________
3. ________
4. ________
5. ________
6. ________
7. ________
8. ________
9. ________
10. ________
11. ________

Pelvic Muscles

1. ________
2. ________
3. ________
4. ________
5. ________
6. ________
7. ________
8. ________

Laboratory Report 40

Student: ______________________

Section: ______________________

Lower Extremity Muscles

A. Thigh Movements

Consult figure 40.1 to match the muscles with the descriptions. In some cases, more than one muscle may apply.

adductor brevis—1
adductor longus—2
adductor magnus—3
gluteus maximus—4
gluteus medius—5
gluteus minimus—6
piriformis—7

1. Gluteus muscle that extends femur
2. Gluteus muscles that abduct femur
3. Three muscles that pull femur toward median line
4. Gluteus muscle that rotates femur outward
5. Three muscles attached to linea aspera that flex femur
6. Small muscle that abducts, extends, and rotates femur

Origins

7. On anterior surface of sacrum
8. On pubis
9. On external surface of ilium
10. On ilium, sacrum, and coccyx
11. On inferior surface of ischium and portion of pubis

Insertions

12. On linea aspera of femur
13. On anterior border of greater trochanter
14. On upper border of greater trochanter of femur
15. On lateral part of greater trochanter of femur
16. On iliotibial tract and posterior part of femur

B. Thigh Muscles

Match the thigh muscles with the descriptions.

biceps femoris—1
gracilis—2
rectus femoris—3
sartorius—4
semimembranosus—5
semitendinosus—6
tensor fasciae latae—7
vastus intermedius—8
vastus lateralis—9
vastus medialis—10

1. Largest quadriceps muscle; on side of thigh
2. Portion of quadriceps beneath the rectus femoris
3. Portion of quadriceps on medial surface of thigh
4. Hamstring muscle on lateral surface of thigh
5. Longest muscle of thigh
6. Most medial component of hamstring muscles
7. Smallest hamstring muscle
8. Superficial muscle on medial surface of thigh that inserts with sartorius on tibia
9. Four muscles of a group that extend leg
10. Flexes thigh upon pelvis
11. Two muscles other than parts of quadriceps that rotate thigh medially (inward)
12. Rotates thigh laterally (outward)
13. Adducts thigh

Answers

Thigh Movements

1. ________
2. ________
3. ________
4. ________
5. ________
6. ________
7. ________
8. ________
9. ________
10. ________
11. ________
12. ________
13. ________
14. ________
15. ________
16. ________

Thigh Muscles

1. ________
2. ________
3. ________
4. ________
5. ________
6. ________
7. ________
8. ________
9. ________
10. ________
11. ________
12. ________
13. ________

C. *Lower Leg and Foot Muscles*

Match the lower leg and foot muscles with their descriptions. In some cases, several answers apply.

extensor digitorum longus—1
extensor hallucis longus—2
flexor digitorum longus—3
flexor hallucis longus—4
gastrocnemius—5
peroneus brevis—6
peroneus longus—7
peroneus tertius—8
soleus—9
tibialis anterior—10
tibialis posterior—11

1. Located on front of lower leg; causes dorsiflexion and inversion of foot
2. Deep portion of triceps surae
3. Superficial portion of triceps surae
4. Three muscles that originate on fibula and cause various movements of foot
5. Two muscles that join to form Achilles tendon
6. Part of triceps surae that originates on femur
7. Inserts on proximal superior portion of fifth metatarsal bone of foot
8. Inserts on proximal inferior portion of fifth metatarsal bone of foot
9. Flexes great toe
10. On posterior surfaces of tibia and fibula; causes plantar flexion and inversion of foot
11. Originates on tibia and fascia of tibialis posterior; flexes second, third, fourth, and fifth toes
12. Flexes calf of thigh
13. Originates on tibia, fibula, and interosseous membrane; extends toes (dorsiflexion) and inverts foot
14. Two muscles that insert on calcaneous
15. Peroneus muscle that causes dorsiflexion and eversion of foot
16. Two peroneus muscles that cause plantar flexion
17. Muscle on tibia that supports foot arches

D. *Questions*

1. List the four muscles that are collectively referred to as the quadriceps femoris.
2. List two muscles associated with the iliotibial tract.
3. List three muscles that constitute the hamstrings.
4. What two muscles are associated with the Achilles tendon?
5. What two muscles constitute the triceps surae?

E. *Surface Muscles Review*

Label the muscles in the figure.

Answers

Lower Leg and Foot Muscles

1.______	10.______
2.______	11.______
3.______	12.______
4.______	13.______
5.______	14.______
6.______	15.______
7.______	16.______
8.______	17.______
9.______	

Questions

1a.______
b.______
c.______
d.______
2a.______
b.______
3a.______
b.______
c.______
4a.______
b.______
5a.______
b.______

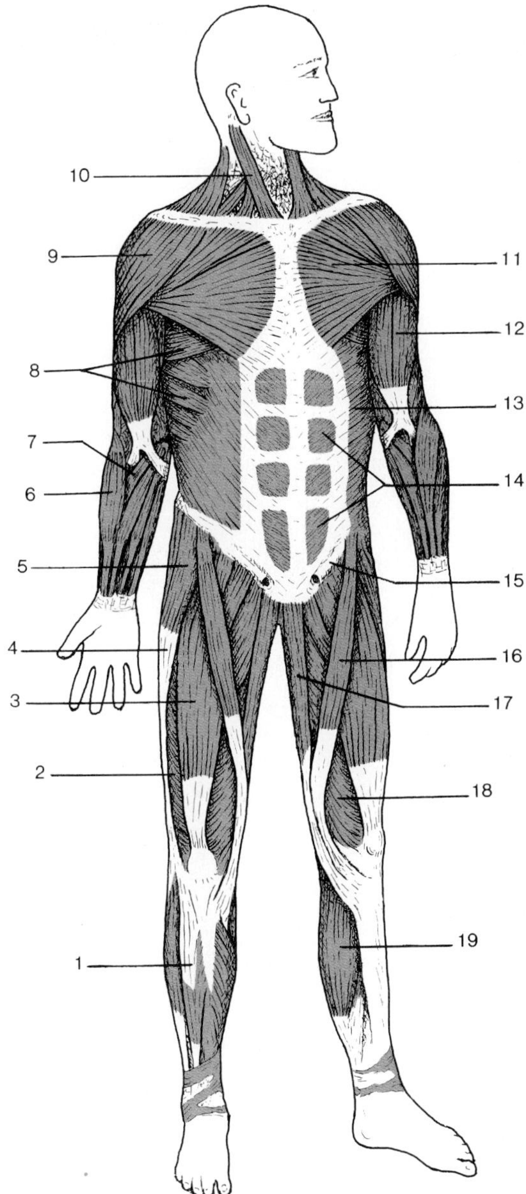

ANTERIOR SURFACE MUSCLES

_____ Biceps brachii
_____ Brachioradialis
_____ Deltoid
_____ External Oblique
_____ Gastrocnemius
_____ Gracilis
_____ Iliotibial Tract
_____ Inguinal Ligament
_____ Pectoralis major
_____ Pronator teres
_____ Rectus abdominis
_____ Rectus femoris
_____ Sartorius
_____ Serratus anterior
_____ Sternocleidomastoid
_____ Tensor fasciae latae
_____ Tibialis anterior
_____ Vastus lateralis
_____ Vastus medialis

POSTERIOR SURFACE MUSCLES

_____ Achilles Tendon
_____ Biceps femoris
_____ Extensor carpi ulnaris
_____ Extensor digitorum communis
_____ Gastrocnemius
_____ Gluteus maximus
_____ Gluteus medius
_____ Hamstring Muscles
_____ Iliotibial Tract
_____ Infraspinatus
_____ Latissimus dorsi
_____ Semimembranosus
_____ Semitendinosus
_____ Teres major
_____ Trapezius
_____ Triceps brachii

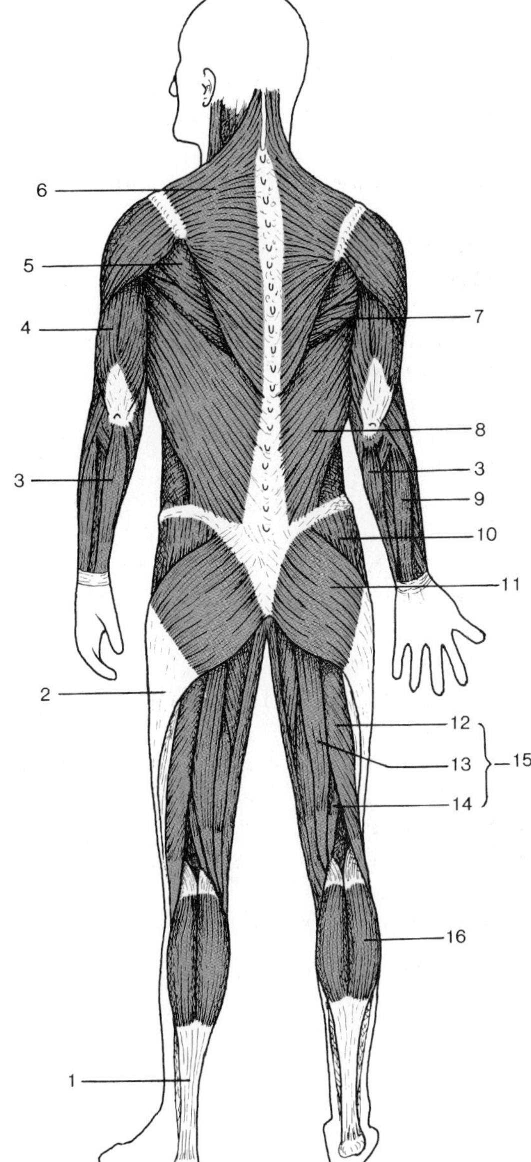

Laboratory Report 41

Student: ______________________

Section: ______________________

The Spinal Cord, Spinal Nerves, and Reflex Arcs

A. Completion

1. The dural sac of the spinal cord is continuous with the ____________ mater that surrounds the brain.
2. The caudal end of the spinal cord is the ____________ .
3. The caudal end of the spinal cord terminates at the lower border of the ____________ lumbar vertebra.
4. The cluster of nerves that extends downward from the end of the spinal cord is the ____________ .
5. The coccygeal nerve emerges from the ____________ plexus.
6. The femoral nerve forms by the union of the following three lumbar nerves: ____________ , ____________ , and ____________ .
7. The largest nerve emerging from the sacral plexus is the ____________ nerve.
8. The iliohypogastric and ilioinguinal nerves are branches of the first ____________ nerve.
9. The innermost meninx surrounding the spinal cord is the ____________ mater.
10. The meninx that lies just within the dura mater is the ____________ .
11. The spinal ganglion is an enlargement of the ____________ root.
12. The central canal of the spinal cord is lined with ____________ ependymal cells.
13. The cavity within the spinal cord that is continuous with the ventricles of the brain is the ____________ canal.
14. The inner portion of the spinal cord, forming the pattern of a butterfly, consists of ____________ matter.
15. The space between the dura mater and vertebral bone is the ____________ space.
16. ____________ (*vertebral* or *collateral*) ganglia are usually located close to the organ that is innervated.
17. A somatic reflex arc with one or more interneurons is said to be ____________ .
18. Cell bodies of afferent neurons of visceral reflexes are in ____________ ganglia.
19. If a somatic reflex lacks an interneuron, it is said to be ____________ .
20. The two divisions of the autonomic nervous system are ____________ and ____________ .

Answers
Completion
1. ____________
2. ____________
3. ____________
4. ____________
5. ____________
6a. ____________
b. ____________
c. ____________
7. ____________
8. ____________
9. ____________
10. ____________
11. ____________
12. ____________
13. ____________
14. ____________
15. ____________
16. ____________
17. ____________
18. ____________
19. ____________
20a. ____________
b. ____________

B. *True–False*

Indicate whether the statements are true or false.

1. The fetal spinal cord fills the entire spinal cavity.
2. The lumbar plexus forms by the union of L_1, L_2, L_3, and most of L_4 nerves.
3. The brachial plexus forms by the union of the lower four cervical nerves and the first thoracic nerve.
4. The cervical plexus forms by the union of the first three pairs of cervical nerves.
5. The filum terminale externa is a downward extension of the filum terminale interna.
6. The innermost fiber of the cauda equina (on the median line) is the filum terminale interna.
7. Cerebrospinal fluid fills the subarachnoid space.
8. The white matter of the spinal cord consists of myelinated fibers of nerve cells.
9. The posterior cutaneous nerve emerges from the lumbar plexus.
10. The sacral plexus forms by the union of the femoral and sciatic nerves.
11. The effector in a somatic reflex is always skeletal muscle.
12. Fibers of the sympathetic and parasympathetic divisions of the autonomic nervous system innervate most organs of the body.
13. Collateral ganglia of the autonomic nervous system form a chain along the vertebral column.
14. All visceral reflex arcs have two afferent neurons.
15. The central canal of the spinal cord is continuous with the ventricles of the brain.
16. The dura mater of the spinal cord extends peripherally to form a covering for spinal nerves.
17. Ciliated epithelium lines the central canal of the spinal cord.
18. All somatic reflex arcs have at least one interneuron.
19. All visceral reflex arcs have two efferent neurons.
20. Spinal ganglia are in the anterior roots of spinal nerves.

C. *Numbers of Spinal Nerves*

Indicate the number of pairs normally present.

1. Cervical nerves
2. Thoracic nerves
3. Lumbar nerves
4. Sacral nerves
5. Coccygeal nerves

Answers

True–False

1. ________
2. ________
3. ________
4. ________
5. ________
6. ________
7. ________
8. ________
9. ________
10. ________
11. ________
12. ________
13. ________
14. ________
15. ________
16. ________
17. ________
18. ________
19. ________
20. ________

Numbers of Spinal Nerves

1. ________
2. ________
3. ________
4. ________
5. ________

Laboratory Report 42

Student: ____________________

Section: ____________________

Somatic Reflexes

A. Functional Nature

Unsuspended Frog: Does the spinal frog attempt to control the position of its hind legs? ___1___ Are leg muscles of the spinal frog *flaccid* or *firm?* ___2___ Does tonus exist in these muscles ___3___? Does the frog jump when prodded? ___4___ Can the frog swim? ___5___ Does the frog float or sink? ___6___

Suspended Frog: Does the frog attempt to remove the acid? ___7___ Do you think that the frog experiences a burning sensation? ___8___ Explain. ____________________

____

Does applying acid to the abdomen or other leg cause a response in the frog? ___9___ If you try to restrain the frog's attempt to remove the acid, does it try some other maneuver? ___10___

Comment on the apparent functional nature of reflexes.

____

____

____

____

Answers

Functional Nature

1. ____________________
2. ____________________
3. ____________________
4. ____________________
5. ____________________
6. ____________________
7. ____________________
8. ____________________
9. ____________________
10. ____________________

Reflex Inhibition

1. ____________________
2. ____________________
3. ____________________
4. ____________________
5. ____________________

B. Reaction Time

Record the reaction time (seconds) for each concentration in the table. Plot these times on the graph.

HCL (percentage)	Time (seconds)
0.05	
0.1	
0.2	
0.3	
0.4	
0.5	
1.0	

Time (seconds)

0.05 0.1 0.2 0.3 0.4 0.5 1.0

Percentage of HCL

Conclusion ____________________

____

C. *Reflex Radiation*

Describe the response of the spinal frog to increased electrical stimulation of the foot. ______________

____

____

____

____

D. *Reflex Inhibition*

Was it possible to inhibit right foot withdrawal from the acid by electrically stimulating the left foot? ___1___ If so, how much voltage was required? ___2___ At what pH? ___3___ At what acid concentration was reflex action inhibited? ___4___ At what voltage? ___5___

E. *Synaptic Fatigue*

Where do you believe fatigue occurred in this experiment? ______________________________

____

F. *Diagnostic Reflex Tests*

Record the responses for the first five reflexes with a "0" for *no response,* a "+" for *diminished response,* and "++" for *good response.* For Hoffmann's reflex and the plantar flexion reflex, check the appropriate box. In the last column, list the nerves affected in abnormal responses.

REFLEX	RESPONSES		NERVES AFFECTED WHEN RESPONSE IS ABNORMAL
	Right Side	Left Side	
1. Biceps			
2. Triceps			
3. Brachioradialis			
4. Patellar			
5. Achilles			
6. Hoffmann's	Absent (normal)	Present (abnormal)	
7. Plantar flexion	Present (normal)	Babinski's Sign (abnormal)	

LABORATORY REPORT 44

Student: ______________________

Section: ______________________

Action Potential Velocities

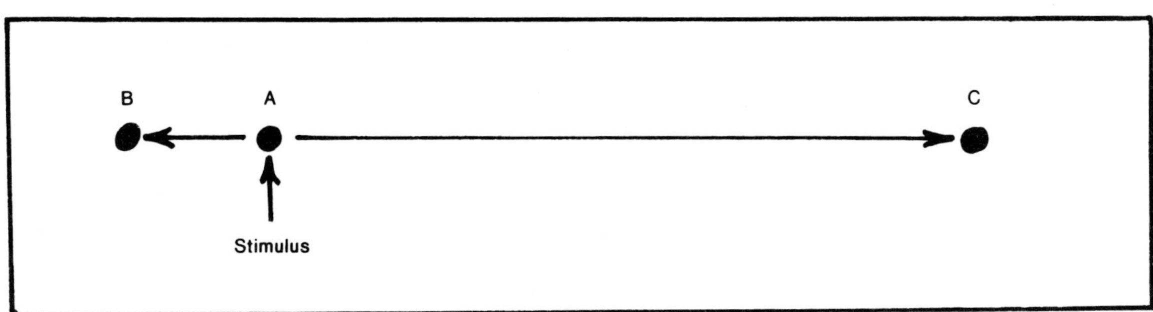

A. Calculations

With a meter stick, measure the distances on the arm between points A and B, and points A and C in millimeters, and convert to meters.

Distances
AB = _______ m
AC = _______ m

Given that each millimeter on the Unigraph chart represents 1/25 second, or 0.04 second, determine how long it took the action potentials to move from A to B, and from A to C.

Times
A to B = _______ sec
A to C = _______ sec

To determine the velocity of the action potentials, divide the distances by the times.

$$\text{Velocity} = \frac{\text{Distance (meters)}}{\text{Time (seconds)}}$$

Velocities
A to B = _______ m/sec
A to C = _______ m/sec

B. Questions

1. What is the threshold for muscle stimulation in this experiment? ______________________
2. What is the threshold for nerve stimulation in this experiment? ______________________
3. The motor fibers of the ulnar nerve have a known velocity of 46 to 73 m per second. How do your experimental results compare to this? ______________________
4. Is the velocity of the nerve impulse in motor fibers distinguishable from the velocity of the impulse in sensory fibers? ______________ Explain. ______________________

____

5. Which would be most susceptible to conduction blockage due to pressure: motor, sensory, or autonomic fibers? ________________ Explain. ____

____

6. Can you identify the reflex potential? ________________ What is its latency from the onset of stimulation? ____

7. What factors account for the latency of the reflex potential as the impulse completes the reflex arc?

____

____

____

8. Describe the sensations the subject perceived in this experiment. ________________________

____

____

To what areas did the sensations seem to be projected? ________________________

____

C. Chart

Attach the Duograph chart in the space provided.

LABORATORY REPORT 45

Student: ______________________

Section: ______________________

Brain Anatomy: External

A. Meninges

Select the meninx or meningeal space that each statement describes.

1. Fibrous outer meninx
2. Middle meninx
3. Meninx that surrounds sagittal sinus
4. Meninx that forms falx cerebri
5. Meninx attached to brain surface
6. Space that contains cerebrospinal fluid

arachnoid mater—1
dura mater—2
epidural space—3
pia mater—4
subarachnoid space—5
subdural space—6

Answers	
Meninges	
1.	2
2.	1
3.	2
4.	2
5.	4
6.	5

B. Brain Dissections

Answer the questions regarding your observations of the sheep brain dissections.

1. Does the arachnoid mater appear to be attached to the dura mater? No

   ________ To the brain? No.

2. Describe the appearance of the dura mater.

   ____

3. Within what structure is the sagittal sinus enclosed?

   dura mater

4. Differentiate:

   Gyrus ridges

   Sulcus Grooves

5. Of what value are the sulci and gyri of the cerebrum?

   increase surface area

6. List five structures that you identified on the dorsal surface of the midbrain. ______________________

7. Is the cerebellum of the sheep brain divided on the median line like the human brain? ______________________

8. What is the significance of the size differences of the olfactory bulbs on the sheep and human brains?

   ____

9. Which cranial nerve is the largest in diameter? ______________________

C. *Fissures and Sulci*

Select the fissure or sulcus that each statement describes.

1. Between frontal and parietal lobes
2. On median line between cerebral hemispheres
3. Between occipital and parietal lobes
4. Between temporal and parietal lobes

anterior central sulcus—1
central sulcus—2
lateral cerebral fissure—3
longitudinal cerebral fissure—4
parieto-occipital fissure—5

D. *Brain Functions*

Select the part of the brain responsible for the activities described.

1. Maintenance of posture
2. Cardiac control
3. Respiratory control
4. Vasomotor control
5. Consciousness
6. Voluntary muscular movements
7. Contains centers for some visual reflexes
8. Contains fibers that connect cerebral hemispheres
9. Contains nuclei of fifth, sixth, seventh, and eighth cranial nerves
10. Coordinates complex muscular movements
11. Function unknown in humans

cerebellum—1
cerebrum—2
corpora quadrigemina—3
corpus callosum—4
medulla oblongata—5
midbrain—6
pineal gland—7
pons—8

E. *Cerebral Functional Localization*

Select the area of the cerebrum that each statement describes.

Location

1. On postcentral gyrus
2. On parietal lobe
3. On superior temporal gyrus
4. On temporal lobe
5. On frontal lobe
6. On occipital lobe
7. On precentral gyrus
8. On angular gyrus
9. In area anterior to precentral gyrus

association area—1
auditory area—2
common integrative area—3
motor speech area—4
olfactory area—5
premotor area—6
somatomotor area—7
somatosensory area—8
visual area—9

Answers

Fissures and Sulci

1. ________
2. ________
3. ________
4. ________

Brain Functions

1. ________
2. ________
3. ________
4. ________
5. ________
6. ________
7. ________
8. ________
9. ________
10. ________
11. ________

Cerebral Functional Localization

1. ________
2. ________
3. ________
4. ________
5. ________
6. ________
7. ________
8. ________
9. ________

E. Cerebral Functional Localization (continued)

Function

10. Speech production
11. Speech understanding
12. Cutaneous sensibility
13. Sight
14. Sense of smell
15. Visual interpretation
16. Voluntary muscular movement
17. Integration of sensory association areas
18. Influences motor area function

association area—1
auditory area—2
common integrative area—3
motor speech area—4
olfactory area—5
premotor area—6
somatomotor area—7
somatosensory area—8
visual area—9

F. Cranial Nerves

Record the number of the cranial nerve that applies to each statement. More than one nerve may apply to some statements.

1. Sensory nerves
2. Mixed nerves
3. Emerges from midbrain
4. Emerges from pons
5. Emerges from medulla oblongata

Structures Innervated

6. Cochlea of ear
7. Heart
8. Salivary glands
9. Abdominal viscera
10. Retina of eye
11. Receptors in nasal membranes
12. Taste buds at back of tongue
13. Lateral rectus muscle of eye
14. Taste buds of anterior two-thirds of tongue
15. Semicircular canals of ear
16. Thoracic viscera
17. Three extrinsic eye muscles (superior rectus, medial rectus, inferior oblique) and levator palpebrae
18. Superior oblique muscle of eye
19. Pharynx, upper larynx, uvula, and palate

1. Olfactory
2. Optic
3. Oculomotor
4. Trochlear
5. Trigeminal
6. Abducens
7. Facial
8. Vestibulocochlear
9. Glossopharyngeal
10. Vagus
11. Accessory
12. Hypoglossal

Answers

Cerebral Functional Localization

10. ______
11. ______
12. ______
13. ______
14. ______
15. ______
16. ______
17. ______
18. ______

Cranial Nerves

1. ______
2. ______
3. ______
4. ______
5. ______
6. ______
7. ______
8. ______
9. ______
10. ______
11. ______
12. ______
13. ______
14. ______
15. ______
16. ______
17. ______
18. ______
19. ______

G. *Trigeminal Nerve*

Select the branch of the trigeminal nerve that innervates the structures that follow.

1. All lower teeth
2. All upper teeth
3. Tongue
4. Lacrimal gland
5. Outer surface of nose
6. Buccal gum tissues of mandible
7. Lower teeth, tongue, muscles of mastication, and gum surfaces

incisive—1
inferior alveolar—2
infraorbital—3
lingual—4
long buccal—5
mandibular—6
maxillary—7
ophthalmic—8

H. *Multiple Choice*

Select the answer that best completes each statement.

1. Shallow furrows on the surface of the cerebrum are
   a. sulci.
   b. fissures.
   c. gyri.
2. Deep furrows on the surface of the cerebrum are
   a. sulci.
   b. fissures.
   c. gyri.
3. The infundibulum supports the
   a. mammillary body.
   b. hypophysis.
   c. hypothalamus.
4. The mammillary bodies are a part of the
   a. medulla.
   b. midbrain.
   c. hypothalamus.
5. The corpora quadrigemina are on the
   a. cerebrum.
   b. medulla.
   c. midbrain.
6. The major ganglion of the trigeminal nerve is the
   a. Gasserian.
   b. sphenopalatine.
   c. ciliary.
7. Convolutions on the surface of the cerebrum are
   a. sulci.
   b. fissures.
   c. gyri.
8. The hypophysis is on the inferior surface of the
   a. medulla.
   b. midbrain.
   c. hypothalamus.
9. The spinal bulb is the
   a. medulla oblongata.
   b. pons.
   c. midbrain.
10. The pineal gland is on the
    a. medulla.
    b. pons.
    c. midbrain.

Answers
Trigeminal Nerve
1.________
2.________
3.________
4.________
5.________
6.________
7.________
Multiple Choice
1.________
2.________
3.________
4.________
5.________
6.________
7.________
8.________
9.________
10.________

LABORATORY REPORT 46

Student: ____________________

Section: ____________________

Brain Anatomy: Internal

A. Locations

Identify the part of the brain in which each structure is located.

cerebellum—1	diencephalon—3	midbrain—5
cerebrum—2	medulla oblongata—4	pons—6

1. Caudata nuclei
2. Cerebral aqueduct
3. Cerebral peduncles
4. Corpus callosum
5. Fornix
6. Globus pallidus
7. Hypothalamus
8. Lateral ventricles
9. Mammillary bodies
10. Putamen
11. Rhinencephalon
12. Thalamus
13. Third ventricle

B. Functions

Identify the structures that perform the functions described.

arachnoid granulations—1	infundibulum—9
caudate nuclei—2	intermediate mass—10
cerebral aqueduct—3	interventricular foramen—11
cerebral peduncles—4	lateral aperture—12
choroid plexus—5	lentiform nucleus—13
corpus callosum—6	median aperture—14
fornix—7	rhinencephalon—15
hypothalamus—8	thalamus—16

1. Relay station for all messages to cerebrum
2. Temperature regulation
3. Entire olfactory mechanism of cerebrum
4. Assists in muscular coordination
5. Fiber tracts of olfactory mechanism
6. Connects halves of thalamus
7. Exerts steadying effect on voluntary movements
8. Allows cerebrospinal fluid to pass from third to fourth ventricle
9. Consists of ascending and descending tracts
10. Supporting stalk of hypophysis
11. Allows cerebrospinal fluid to return to blood
12. Commissure that unites cerebral hemispheres
13. Secretes cerebrospinal fluid into ventricle
14. Coordinates autonomic nervous system

Answers
Locations
1. ________
2. ________
3. ________
4. ________
5. ________
6. ________
7. ________
8. ________
9. ________
10. ________
11. ________
12. ________
13. ________
Functions
1. ________
2. ________
3. ________
4. ________
5. ________
6. ________
7. ________
8. ________
9. ________
10. ________
11. ________
12. ________
13. ________
14. ________

C. *Multiple Choice*

Select the answer that best completes each statement.

1. The brain stem consists of the
   a. cerebrum, pons, midbrain, and medulla.
   b. cerebellum, medulla, and pons.
   c. pons, medulla, and midbrain.
2. The lateral ventricles are separated by the
   a. thalamus.
   b. fornix.
   c. septum pellucidum.
3. Pain is perceived in the
   a. somatomotor area.
   b. thalamus.
   c. pons.
4. The hypothalamus regulates
   a. the hypophysis, appetite, and wakefulness.
   b. body temperature, hypophysis, and vision.
   c. reproductive functions, body temperature, and voluntary movements.
5. Feelings of pleasantness and unpleasantness appear to be associated with the
   a. somatomotor area.
   b. hypothalamus.
   c. thalamus.
   d. medulla oblongata.
6. The reticular formation consists of
   a. gray matter.
   b. white matter.
   c. an interlacement of white and gray matter.
7. Alert consciousness is partially regulated by
   a. caudate nuclei.
   b. lentiform nuclei.
   c. nuclei of the reticular formation.
8. The intermediate mass passes through the
   a. midbrain.
   b. third ventricle.
   c. hypothalamus.
9. The reticular formation is in the
   a. spinal cord.
   b. cerebrum.
   c. brain stem.
   d. spinal cord, brain stem, and diencephalon.
10. Centers for vomiting, coughing, swallowing, and sneezing are in the
    a. pons.
    b. medulla.
    c. midbrain.
    d. thalamus.

Answers
Multiple Choice
1. ________
2. ________
3. ________
4. ________
5. ________
6. ________
7. ________
8. ________
9. ________
10. ________

D. *Cerebrospinal Fluid*

You should be able to trace the path of the cerebrospinal fluid from its point of origin to where it is reabsorbed into the blood. If you can complete the following paragraph without referring to the text, you know the sequence fairly well. If you can state the sequence from memory, better yet.

Ventricles
fourth—1
lateral—2
third—3

Passageways
acoustic meatus—4
cerebral aqueduct—5
foramen magnum—6
interventricular foramen—7
lateral aperture—8
median aperture—9

Structures
arachnoid granulations—10
cerebellum—11
cerebrum—12
choroid plexus—13
cisterna cerebellomedullaris—14
cisterna superior—15
dura mater—16
sagittal sinus—17
septum pellucidum—18
subarachnoid space—19

Cerebrospinal fluid is secreted into each ventricle by a __1__ . From the __2__ ventricles, which are in the cerebral hemispheres, the fluid passes to the __3__ ventricle through an opening called the __4__ . A canal, called the __5__ , allows the fluid to pass from the latter ventricle to the __6__ ventricle . From this last ventricle, the cerebrospinal fluid passes into a subarachnoid space called the __7__ through three foramina: one __8__ and two __9__ . From this cavity, the fluid passes over the cerebellum into another subarachnoid space called the 10____ . It also passes __11__ (*up, down*) the posterior side of the spinal cord and __12__ (*up, down*) the anterior side of the spinal cord. From the cisterna superior, the cerebrospinal fluid passes to the subarachnoid space around the __13__ . This fluid is reabsorbed back into the blood through delicate structures called the __14__ . The blood vessel that receives the cerebrospinal fluid is the __15__ .

Answers
Cerebrospinal Fluid
1.________
2.________
3.________
4.________
5.________
6.________
7.________
8.________
9.________
10.________
11.________
12.________
13.________
14.________
15.________

Laboratory Report 47

Student: ______________________

Section: ______________________

Electroencephalography

A. Records

Attach EEG samples in the spaces provided. Trim excess paper from the chart.

Calibration

Alpha rhythm

Alpha block

Hyperventilation (alkalosis)

B. Results

1. Were you able to demonstrate changes in the EEG pattern? ______________________

   If not, provide a possible explanation. ______________________

2. What was the maximum amplitude (in millivolts) in a brain wave? ______________________

3. What is the significance of the following in a normal person? ______________________

   Alpha rhythm ______________________

   Beta rhythm ______________________

   Delta rhythm ______________________

LABORATORY REPORT 48

Student: ______________________

Section: ______________________

Anatomy of the Eye

A. *Structures*

Identify the structures that the statements describe.

1. Small, nonphotosensitive area on retina
2. Small pit in retina of eye
3. Outer layer of wall of eye
4. Fluid between lens and retina
5. Fluid between lens and cornea
6. Inner, light-sensitive layer
7. Round, yellow spot on retina
8. Delicate membrane that lines eyelids
9. Middle vascular layer of wall of eyeball
10. Drainage tubes for tears in eyelids
11. Clear, transparent portion of front of eyeball
12. Tube that drains tears from lacrimal sac
13. Chamber between iris and cornea
14. Conical body in medial corner of eye
15. Chamber between iris and lens
16. Circular color band between lens and cornea
17. Circular band of smooth muscle tissue surrounding lens
18. Connective tissue between lens perimeter and surrounding muscle
19. Semicircular fold of conjunctiva in medial canthus of eye

anterior chamber—1
aqueous humor—2
blind spot/optic disk—3
caruncula—4
choroid coat—5
ciliary body—6
conjunctiva—7
cornea—8
fovea centralis—9
iris—10
lacrimal ducts—11
lacrimal sac—12
macula lutea—13
medial canthus—14
nasolacrimal duct—15
plica semilunaris—16
posterior chamber—17
pupil—18
retina—19
sclera—20
suspensory ligament—21
vitreous body—22

B. *Extrinsic Eye Muscles*

Identify the muscles and structures that the statements describe.

1. Inserted on top of eyeball
2. Inserted on side of eyeball
3. Inserted on medial surface
4. Inserted on bottom of eyeball
5. Inserted on eyelid
6. Raises eyelid
7. Rotates eyeball inward
8. Rotates eyeball downward
9. Rotates eyeball outward
10. Rotates eyeball upward
11. Cartilaginous loop through which superior oblique muscle acts

inferior oblique—1
inferior rectus—2
lateral rectus—3
medial rectus—4
superior levator palpebrae—5
superior oblique—6
superior rectus—7
trochlea—8

Answers

Structures

1. 3
2. 9
3. 20
4. 22
5. 2
6. 19
7. ______
8. ______
9. ______
10. ______
11. 8
12. 15
13. ______
14. ______
15. ______
16. 10
17. 6
18. ______
19. ______

Extrinsic Eye Muscles

1. 7
2. 3
3. 4
4. 2
5. ______
6. 1
7. 4
8. 2
9. 3
10. 7
11. 8

C. Functions

Select the part of the eye that performs the function described.

aqueous humor—1
blind spot/optic disk—2
choroid coat—3
ciliary body—4
conjunctiva—5
iris—6
lacrimal ducts—7
lacrimal puncta—8
lacrimal sac—9
macula lutea—10
nasolacrimal duct—11
pupil—12
retina—13
sclera—14
scleral venous sinus—15
suspensory ligament—16
trabeculae—17
trochlea—18
vitreous body—19

1. Where nerve fibers of retina leave eyeball
2. Furnishes blood supply to retina and sclera
3. Large vessel in wall of eye that collects aqueous humor from trabeculae
4. Exerts force on lens, changing its contour
5. Controls amount of light that enters eye
6. Maintains firmness and roundness of eyeball
7. Part of retina where critical vision occurs
8. Provides most of the strength to wall of eyeball

Answers
Functions
1. ________
2. ________
3. ________
4. 4
5. 12
6. 16
7. 10
8. ________

D. Cow Eye Dissection Questions

1. What is the shape of the pupil? ____________________
2. Why do you suppose it is so difficult to penetrate the sclera with a sharp scalpel? ____________________
3. What is the function of the black pigment in the eye? ____________________
4. Compare the consistency of the two fluids in the eye.
   Aqueous humor ____________________
   Vitreous humor ____________________
5. When you hold up the lens and look through it, what is unusual about the image? ____________________
6. Does the lens magnify printed matter when placed directly on it? ____________________
7. Compare the consistencies of the following portions of the lens:
   Center ____________________
   Edge ____________________
8. What is the reflective portion of the choroid coat called? ____________________
9. Is there a macula lutea on the retina of the cow eye? ____________________

E. *Ophthalmoscopy*

Record your ophthalmoscope measurements here, and answer the questions.

1. Diopter Measurements

   The diopter value (D) of a lens is the reciprocal of its focal length (f) in meters, or $D = \frac{1}{f}$.

   A lens of one diopter (1D) has a focal length of 1 m, $(D = \frac{1}{f} = \frac{1}{1} = 1)$; a 2D lens has a focal length of 0.5 meter, or $\frac{1}{f} = \frac{1}{0.5} = 2$; etc.

   Record your measurements for: 5D ________; 10D ________; 20D ________; and 40D ________.

   Calculated diopter distances: 5D ________; 10D ________; 20D ________; and 40D ________.

   Do your calculations match the measured distances? ______________________________

2. If you were able to examine your laboratory partner's eye with "0" in the diopter window of the ophthalmoscope, what would this indicate about the curvature of the lens of your eye? ______________

   About your laboratory partner's eye? ______________________________

Laboratory Report 49

Student: ____________________

Section: ____________________

Visual Tests

A. The Purkinje Tree

Describe in a few sentences the image you observed. ____________________

____

____

____

____

____

B. Tabulations

Record your results for the five eye tests.

TEST	RIGHT EYE	LEFT EYE
Blind spot (inches)		
Near point (inches)		
Scheiner's experiment (inches)		
Visual acuity (X/20)		
Astigmatism (present or absent)		

C. *Color Blindness Test*

Have your laboratory partner record your responses to each test plate. An *x* indicates inability to read correctly.

Plate Number	Subject's Response	Normal Response	Response If Red-Green Deficiency				Response If Totally Color-Blind
1		12	12				12
2		8	3				x
3		5	2				x
4		29	70				x
5		74	21				x
6		7	x				x
7		45	x				x
8		2	x				x
9		x	2				x
10		16	x				x
11		Traceable	x				x
			Protan		Deutan		
			Strong	Mild	Strong	Mild	
12		35	5	(3)5	3	3(5)	
13		96	6	(9)6	9	9(6)	
14		Can trace two lines	Purple	Purple (red)	Red	Red purple	x

x—Indicates inability of subject to respond in any way to test.

Conclusion ____

D. *Pupillary Reflexes*

1. Accommodation to Distance

a. How did pupillary size change in this experiment? ______________________________

____

b. Can you suggest a benefit that might result from this happening? ______________________________

____

2. Accommodation to Light Intensity

a. Did the pupil of the unexposed eye become smaller when the right eye was exposed to light?

____

b. Trace the pathways of the nerve impulses in effecting the response. ______________________________

____

____

c. Can you suggest a possible benefit that might result from this phenomenon? ______________________________

____

LABORATORY REPORT 50

Student: ______________________

Section: ______________________

The Ear: Its Role in Hearing

A. Structure

Identify the structures that the statements describe.

basilar membrane—1
cochlear duct—2
endolymph—3
hair cells—4
helicotrema—5
incus—6
malleus—7
perilymph—8
reticular lamina—9
rods of Corti—10
scala tympani—11
scala vestibuli—12
stapes—13
tectorial membrane—14
tympanic membrane—15
vestibular membrane—16

1. Ossicle that fits in oval window
2. Ossicle activated by tympanic membrane
3. Chamber of cochlea into which round window opens
4. Chamber of cochlea into which oval window opens
5. Membrane set in vibration by sound waves in the air
6. Membrane that contains 20,000 fibers of varying lengths
7. Membrane in which hair tips of hair cells are embedded
8. Two fluids found in cochlea
9. Fluid within scala tympani
10. Fluid within cochlear duct
11. Fluid within scala vestibuli
12. Parts of organ of Corti
13. Transfers vibrations from basilar membrane to reticular lamina
14. Receptors that initiate action potentials in cochlear nerve
15. Opening between scala vestibuli and scala tympani

Answers

Structure

1. 13
2. 7
3. ________
4. ________
5. ________
6. ________
7. ________
8. ________
9. ________
10. ________
11. ________
12. ________
13. ________
14. ________
15. ________

B. Watch-Tick Method (Screening Test)

Record the distances at which you and three other persons could hear the watch tick.

SUBJECT	INCHES FROM EAR			
	Right Ear		Left Ear	
	Approaching	Receding	Approaching	Receding

C. *Physiology of Hearing*

Indicate whether the statements are true or false.

1. The loudness of sound is directly proportional to the square of the amplitude.
2. The amplitude of the sound wave determines the pitch of a sound.
3. Quality and pitch are two different characteristics of sound.
4. Some hearing takes place through the bones of the skull rather than through the auditory ossicles.
5. While there is usually no cure for nerve deafness, conduction deafness can often be corrected.
6. Endolymph in the scala vestibuli flows into the scala tympani through the helicotrema.
7. Nerve deafness can be detected by placing a tuning fork on the mastoid process.
8. The frequency range used in speech is from 30 to 20,000 cps.
9. The bending of hairs of hair cells causes an alternating electrical charge known as the endocochlear potential.
10. Before using an audiometer, the operator must calibrate it.

Answers

Physiology of Hearing

1. ________
2. ________
3. ________
4. ________
5. ________
6. ________
7. ________
8. ________
9. ________
10. ________

D. *Tuning Fork Methods*

Record the results of the two tuning fork methods.

Rinne Test

Right ear ________________

Left ear ________________

Weber Test

Right ear ________________

Left ear ________________

E. *Audiometry*

Record the threshold levels in the right ear with a red "O" and in the left ear with a blue "X." Connect the "O's" with red lines and the "X's" with blue lines.

Frequency (cycles per second)

Hearing Threshold Level in dB	125	250	500	750	1000	1500	2000	3000	4000	6000	8000	
−10												−10
0												0
10												10
20												20
30												30
40												40
50												50
60												60
70												70
80												80
90												90
100												100
110												110

(Shaded area indicates critical area of speech interpretation)

LABORATORY REPORT 51

Student: ____________________

Section: ____________________

The Ear: Its Role in Equilibrium

A. *Anatomy of Vestibular Apparatus*

1. What are the sensory receptors for static equilibrium called?
2. Where are the static equilibrium sensory receptors located? (two places)
3. What are the small calcium carbonate crystals of the macula called?
4. Where are the sensory receptors of dynamic equilibrium located?
5. What nerve branch of the eighth cranial nerve supplies the vestibular apparatus?
6. What part of the vestibular apparatus connects directly to the cochlear duct?
7. What is the name of the fluid within the vestibular apparatus?
8. What structures in joints prevent a sense of malequilibrium when you tilt your head to one side?
9. What fluid do the semicircular canals contain?
10. Within how many planes do the semicircular canals lie?

Answers

Anatomy of Vestibular Apparatus

1. ____________
2. ____________
3. ____________
4. ____________
5. ____________
6. ____________
7. ____________
8. ____________
9. ____________
10. ____________

B. *Nystagmus*

After having observed nystagmus, answer the questions.

1. What direction do the eyes move in relation to the direction of rotation? ____________
2. What seems to be the function of slow movement of the eye? ____________
3. What seems to be the function of fast movement of the eye? ____________
4. What part of the vestibular apparatus triggers the reflexes that control the eye muscles? ____________

C. Proprioceptive Influences

1. At-Rest Reactions

a. Was the subject able to place the heel of the foot on the toes of the other foot with eyes closed?

____

b. Was the subject able to touch the nose with the eyes closed? ______________________________

c. Explain what mechanisms in the body make this possible. ______________________________

d. Describe the subject's ability to touch the pencil eraser with eyes closed after practicing with eyes open. ______________________________

2. Effects of Rotation

a. Was the subject, with eyes open, able to point directly to the pencil eraser with the finger after rotation? ______________________________

b. When the blindfolded subject tried to locate the pencil eraser after rotation, what was the result?

____

LABORATORY REPORT

52-58

Student: ______________________

Section: ______________________

Hematological Tests

Test Results

Except for blood typing, record all test results in table II on page 329. Follow the directions given here for calculating blood cell counts.

A. *Differential White Blood Cell Count* *(Exercise 52)*

Move the slide in the pattern indicated in figure 52.4, and record all the different types of cells in table I. Refer to figures 52.1 and 52.2 for cell identification. Use this method of tabulation: ~~1111~~ ~~1111~~ 1 1. Identify and tabulate 100 leukocytes. Divide the total of each kind of cell by 100 to determine percentages.

Table I Leukocyte Tabulation

NEUTROPHILS	LYMPHOCYTES	MONOCYTES	EOSINOPHILS	BASOPHILS
Totals				
Percent				

B. *Calculations for Blood Cell Counts* *(Exercise 53)*

1. Total White Blood Cell Count
   Total WBCs counted in 4 “W” areas × 50 = __________ WBCs per cu mm
2. Red Blood Cell Count
   Total RBCs counted in 5 “R” areas × 10,000 = __________ RBCs per cu mm

C. *Blood Typing* *(Exercise 58)*

Record your blood type here: __________

Refer to your text, or lecture notes, in order to answer the following questions concerning blood transfusions.

1. If you needed a blood transfusion, what types of blood could you receive? ____________________
   ____
2. If your blood were to be given to someone in need, what type should that person have?
   ____

Questions

Although most of the answers to these questions can be derived from this manual, consult your lecture text or medical dictionary for some of the answers.

D. Blood Components

Identify the blood components that the statements describe. More than one answer may apply.

antibodies—1	fibrin—5	neutrophils—10
basophils—2	fibrinogen—6	platelet factor—11
eosinophils—3	lymphocytes—7	platelets—12
erythrocytes—4	lysozyme—8	prothrombin—13
	monocytes—9	thromboplastin—14

1. Cells that transport O_2 and CO_2
2. Phagocytic granulocyte
3. Phagocytic agranulocyte cell concerned, primarily, with generalized infections
4. Protein substances in blood that bind to foreign antigens
5. Noncellular formed elements essential for blood clotting
6. Substances necessary for blood clotting
7. Gelatinous material formed in blood clotting
8. Substance that injured cells release to initiate blood clotting
9. Enzyme in blood plasma that destroys some kinds of bacteria
10. Leukocyte with distinctive red-stained granules
11. Leukocyte that probably produces heparin
12. Most numerous type of leukocyte
13. Contains hemoglobin
14. Most numerous type of blood cell

E. Blood Diseases

Identify the pathological conditions that the statements characterize. More than one condition may apply to some statements.

anemia—1	leukemia—5	neutropenia—9
eosinophilia—2	leukocytosis—6	neutrophilia—10
erythroblastosis fetalis—3	leukopenia—7	pernicious anemia—11
hemophilia—4	lymphocytosis—8	polycythemia—12
		sickle-cell anemia—13

1. Too few red blood cells
2. Too many red blood cells
3. Too few leukocytes
4. Too many neutrophils
5. Too many lymphocytes
6. Too little hemoglobin
7. Defective RBCs due to stomach enzyme deficiency
8. Rh-factor incompatibility present at birth
9. Hereditary blood disease of African Americans; RBCs of abnormal shape
10. Heritable bleeder's disease
11. Cancerlike condition in which leukocytes are too numerous

Answers

Blood Components

1. 4
2. 10
3. 9
4. 1
5. 12
6. 12, 5, 11, 13, 14, 6
7. 5
8. 14
9. 8
10. 3
11. 2
12. 10
13. 4
14. 4

Blood Diseases

1. 1
2. 12
3. 7
4. 10
5. 8
6. 1, 11, 13
7. 11
8. 3
9. 13
10. 4
11. 5, 10

F. *Terminology*

Differentiate between the pairs of related terms.

Plasma ____________________

Lymph ____________________

Coagulation ____________________

Agglutination ____________________

Antibody ____________________

Antigen ____________________

G. *True-False*

Indicate whether the statements are true or false.

1. The most common blood types are O and A.
2. A universal donor has O/Rh-negative blood.
3. Anemia is usually due to iodine deficiency. iron
4. Hemoglobin combines with both oxygen and carbon dioxide.
5. The maximum life span of an erythrocyte is about 30 days. 120 days
6. White blood cells live much longer than red blood cells. F
7. Worm infestations may cause eosinophilia.
8. Viral infections may cause an increase in the number of lymphocytes or monocytes.
9. Megakaryocytes of bone marrow produce blood platelets.
10. Bloods that are matched for ABO and Rh factor can be mixed with complete assurance of no incompatibility.
11. The rarest type of blood is AB negative.
12. Heparin is essential to fibrin formation.
13. Blood typing can prove that an individual is not the father in a paternity suit.
14. Blood typing can prove that an individual is the father in a paternity suit.
15. Approximately 87% of the population is Rh negative.
16. Fibrin forms when prothrombin reacts with fibrinogen.

Answers
True-False
1. T
2. T
3. F
4. T
5. F
6. F
7. T
8. T
9. T
10. F
11. T
12. F
13. T
14. F
15. T
16. T

H. Summarization of Results

Record blood cell counts and other test results in table II.

Table II

TEST	NORMAL VALUES	TEST RESULTS	EVALUATION (over, under, normal)
Differential WBC count	Neutrophils: 50 - 70%		
	Lymphocytes: 20 - 30%		
	Monocytes: 2 - 6%		
	Eosinophils: 1 - 5%		
	Basophils: 0.5 - 1%		
Total WBC count	5000–9000 per cu mm		
RBC count	Males: 4.8–6.0 million/cu mm		
	Females: 4.1–5.1 million/cu mm		
Hemoglobin percentage	Males: 13.4–16.4 g/100ml		
	Females: 12.2–15.2 g/100 ml		
Hematocrit	Males: 40–54% (Av. 47%)		
	Females: 37–47% (Av. 42%)		
Coagulation time	2–6 minutes		
Sedimentation rate	Males 1–15 min/hr		
	Females 1–20 min/hr		

I. Materials

Identify the various blood tests that require the supplies listed.

1. Capillary tubes
2. Centrifuge
3. Hemacytometer
4. Hemoglobinometer
5. Hemolysis applicators
6. Landau rack
7. Microscope slides
8. Sealing clay
9. Sodium citrate
10. Unopette
11. Wright's stain

clotting time—1
differential WBC count—2
hematocrit—3
hemoglobin determination—4
RBC and WBC counts—5
sedimentation rate—6

Answers
Materials
1.________
2.________
3.________
4.________
5.________
6.________
7.________
8.________
9.________
10.________
11.________

LABORATORY REPORT 59

Student: ______________________

Section: ______________________

Anatomy of the Heart

A. *Structures*

Identify the structures of the heart that the statements describe.

aorta—1	mitral valve—13
aortic semilunar valve—2	myocardium—14
bicuspid valve—3	papillary muscles—15
chordae tendineae—4	parietal pericardium—16
endocardium—5	pulmonary semilunar valve—17
epicardium—6	pulmonary trunk—18
inferior vena cava—7	right atrium—19
interventricular septum—8	right pulmonary artery—20
left atrium—9	right ventricle—21
left pulmonary artery—10	superior vena cava—22
left ventricle—11	tricuspid valve—23
ligamentum arteriosum—12	visceral pericardium—24

1. Lining of heart
2. Partition between right and left ventricles
3. Fibroserous saclike structure surrounding heart
4. Two chambers of heart that contain deoxygenated blood
5. Large artery that carries blood from right ventricle
6. Blood vessel that returns blood to heart from head and arms
7. Remnant of a functional prenatal vessel between the pulmonary trunk and the aorta
8. Structure formed from ductus arteriosus
9. Valve at base of pulmonary trunk
10. Two arteries that are branches of pulmonary trunk
11. Synonym for bicuspid valve
12. Structures on cardiac wall to which chordae tendineae attach
13. Muscular portion of cardiac wall
14. Thin covering on surface of heart (two names)
15. Large vein that empties blood into top of right atrium
16. Two chambers of heart that contain oxygenated blood
17. Chamber of heart that receives blood from lungs
18. Large artery that carries blood out from left ventricle
19. Large vein that empties blood into lower part of right atrium
20. Atrioventricular valves
21. Valve at base of aorta
22. Blood vessel in which openings to coronary arteries are located
23. Atrium into which coronary sinus empties
24. Valvular restraints that prevent atrioventricular valve cusps from being forced back into atria

Answers
Structures
1. ________
2. ________
3. ________
4. ________
5. ________
6. ________
7. ________
8. ________
9. ________
10. ________
11. ________
12. ________
13. ________
14. ________
15. ________
16. ________
17. ________
18. ________
19. ________
20. ________
21. ________
22. ________
23. ________
24. ________

B. *Coronary Circulation*

Identify the arteries and veins of the coronary circulatory system that the statements describe.

anterior interventricular artery—1	middle cardiac vein—6
circumflex artery—2	posterior descending right coronary artery—7
coronary sinus—3	posterior interventricular vein—8
great cardiac vein—4	right coronary artery—9
left coronary artery—5	small cardiac vein—10

1. Two principal branches of left coronary artery
2. Coronary vein that lies in anterior interventricular sulcus
3. Coronary artery that lies in right atrioventricular sulcus
4. Vein that receives blood from great cardiac vein
5. Vein that lies in posterior interventricular sulcus
6. Two major coronary arteries that take their origins in wall of aorta
7. Large coronary vessel that empties into right atrium
8. Coronary vein that parallels right coronary artery in right atrioventricular sulcus
9. Branch of right coronary artery on posterior surface of heart
10. Vein that lies alongside of posterior descending right coronary artery

Answers

Coronary Circulation

1. ________
2. ________
3. ________
4. ________
5. ________
6. ________
7. ________
8. ________
9. ________
10. ________

C. *Sheep Heart Dissection*

After completing the sheep heart dissection, answer these questions. You encountered some of these questions during the dissection; others pertain to structures not shown in figures 59.1 and 59.2.

1. Identify the location of and descibe the following structures:

   Pectinate muscle ________________________________

   ____

   Moderator band ________________________________

   ____

   Ligamentum arteriosum ________________________________

   ____

2. How many papillary muscles did you find in the right ventricle? ____________

   left ventricle? ____________

3. How many pouches are present in the pulmonary semilunar valve? ____________

   aortic semilunar valve? ____________

4. Where does blood enter the myocardium? ________________________________

   ____

5. Where does blood leave the myocardium and return to the circulatory system?

   ____

LABORATORY REPORT 60

Student: ____________________

Section: ____________________

Cardiovascular Sounds

A. *Phonocardiogram*

Attach a segment of a phonocardiogram that you made on the recorder. Identify on the chart those segments that represent first and second sounds.

B. *Effect of Exercise*

How do heart sounds differ before and after exercise? ____________________

____

____

C. *Causes*

Explain what causes:

1. First sound ____________________

   ____

   ____

2. Second sound ____________________

   ____

   ____

3. Third sound ____________________

   ____

4. Murmur ____

____

____

D. Questions

1. Why does the aortic valve close sooner than the pulmonary valve during inspiration? ___________

____

____

2. Explain why pulmonary stenosis is associated with greater splitting of the second heart sound.

____

____

____

LABORATORY REPORT 61

Student: ____________________

Section: ____________________

Heart Rate Control

A. *Tracings*

Attach six tracings of your experiment in the spaces provided. Trim excess paper from the tracings to fit them into the allotted spaces. Also, record the contractions per minute in the space provided.

No.	Tracings	Contractions Per Minute
1	Normal contractions: Ringer's solution	
2	0.05 ml epinephrine	
3	0.05 ml acetylcholine	
4	0.05 ml epinephrine	
5	0.25 ml epinephrine	
6	0.25 ml acetylcholine	

Since the paper moves at 2.5 mm per second (slow speed), 5 mm on the paper represents 2 seconds. Also, note that the white margin on the Gilson paper has a line every 7.5 centimeters. The distance between these margin lines represents 30 seconds.

B. Results

1. How did the first injection of epinephrine affect the

   heart rate? ____________________

   strength of contraction? ____________________

2. How did the heart rate change after the first injection of acetylcholine?

   Previous rate __________ Acetylcholine rate ____________________

   Previous amplitude __________ (mm) A/C amplitude ____________________ (mm)

3. Did any injection stop the heart rate? ____________________

   If so, which one? ____________________

   Did the heart ever recover? ____________________

C. Questions

1. Does acetylcholine affect both the rate and strength of contraction? ____________________
2. Does epinephrine affect both the rate and strength of contraction? ____________________

LABORATORY REPORT 62

Student: ______________________

Section: ______________________

Electrocardiogram Monitoring: Using Chart Recorders

A. Tracings

Attach three tracings of your experiment in the spaces provided. Trim excess paper from the tracings to fit them into the allotted spaces.

TRACINGS	EVALUATION
Calibration	
Subject at rest	*Heart rate: _____ per min QR potential: _____ millivolts **Duration of cycle: _____ msec
After exercise	*Heart rate: _____ per min QR potential: _____ millivolts **Duration of cycle: _____ msec

*Space between two margin lines is 3 seconds.

**Time from beginning of P wave to end of T wave.

B. Questions

1. During what part of ECG wave pattern does atrial depolarization and contraction occur? __________

2. During what part of ECG wave pattern does ventricular depolarization and contraction occur? __________

3. Would a heart murmur necessarily show up on an ECG? __________

   Explain. __________

Laboratory Report 64

Student: ____________________

Section: ____________________

Pulse Monitoring

A. *Tracings*

Attach samples of tracings made in this experiment in the spaces provided here and on the next page. Use the right-hand column for relevant comments.

TRACINGS	EVALUATION
Transducer at heart level	Pulse rate: _____
Arm slowly raised above heart level	Pulse rate: _____
Arm lowered toward floor	Pulse rate: _____
Holding breath for 15 seconds	Pulse rate: _____

TRACINGS	EVALUATION
Brachial artery occluded for 15–20 seconds	
Chart speed increased to 25mm per second	
	Pulse rate: _____
Subject startled	

B. Pulse Tabulations

Record your resting pulse rate on the chalkboard. Once the pulse rates of all students are recorded, record the highest, lowest, and median pulse rates:

Highest __________ Lowest __________ Median __________

LABORATORY REPORT 65

Student: ____________________

Section: ____________________

Blood Pressure Monitoring

A. *Stethoscope-sphygmomanometer Method*

Record the blood pressures of the test subject.

	BEFORE EXERCISE	AFTER EXERCISE
Systolic		
Diastolic		
*Pulse Pressure		
**Mean Arterial Pressure		

*Pulse pressure = systolic pressure minus diastolic pressure.

**Mean arterial pressure ≃ diastolic pressure + 1/3 pulse pressure.

B. *Chart Recorder Method*

In the spaces provided here and on the next page, attach chart recordings for the various portions of the experiments.

Calibration of recorder
Subject in chair

Subject standing erect
Subject reclining
After exercise

C. Questions

You may need to refer to your text or lecture notes to answer the following questions.

1. What are the systolic and diastolic pressures for normal blood pressure? ______________________
2. What systolic pressure indicates hypertension? ______________________
3. Why is a low diastolic pressure considered harmful? ______________________
4. Why are diuretics helpful in treating hypertension? ______________________

   ____
5. Why is reduced salt intake helpful in regulating blood pressure? ______________________

   ____

   ____

Laboratory Report 66

Student: ____________________

Section: ____________________

A Polygraph Study of the Cardiac Cycle

A. Tracing

Attach a portion of the tracing in the space provided.

B. Events

Identify the stages in which each of the events occurs. More than one answer may apply in some cases.

early atrial systole—1
late atrial systole—2
isometric ventricular systole—3
midventricular ejection—4
late ventricular ejection—5
early diastole—6
late diastole—7

1. First sound
2. Second sound
3. Third sound
4. Dicrotic notch
5. T wave
6. P wave
7. U wave
8. A-V valves open
9. Aortic and pulmonary valves open
10. All four valves momentarily closed
11. Arterial pressure highest
12. Arterial pressure lowest
13. Ventricular depolarization
14. Ventricular repolarization
15. Atrial depolarization
16. Atrial repolarization

Answers
Events
1.________
2.________
3.________
4.________
5.________
6.________
7.________
8.________
9.________
10.________
11.________
12.________
13.________
14.________
15.________
16.________

Laboratory Report 67

Student: ______________________

Section: ______________________

Peripheral Circulation Control (Frog)

A. Results

1. What change did you observe in the frog's foot when histamine was applied? ______________________

2. How did histamine affect the arterioles (*vasodilation* or *vasoconstriction*)? ______________________

3. What change did you observe in the frog's foot when epinephrine was applied? ______________________

4. How did epinephrine affect the arterioles (*vasodilation* or *vasoconstriction*)? ______________________

B. Questions

It may be necessary to consult your lecture text or medical dictionary for some of the answers to the following questions.

1. Does epinephrine have the same effect on arterioles of skeletal muscles? ______________________

2. Provide a theoretical explanation of how epinephrine can cause vasodilation in one part of the body and vasoconstriction in another region. ______________________

3. What are baroreceptors, and what role do they play in regulating blood flow? ______________________

Laboratory Report 68

Student: ______________________

Section: ______________________

The Arteries and Veins

A. The Circulatory Plan

Complete this diagram by filling in the proper blood vessels and providing correct labels. Color the oxygenated vessels red and the deoxygenated ones blue, and provide arrows to show direction of blood flow.

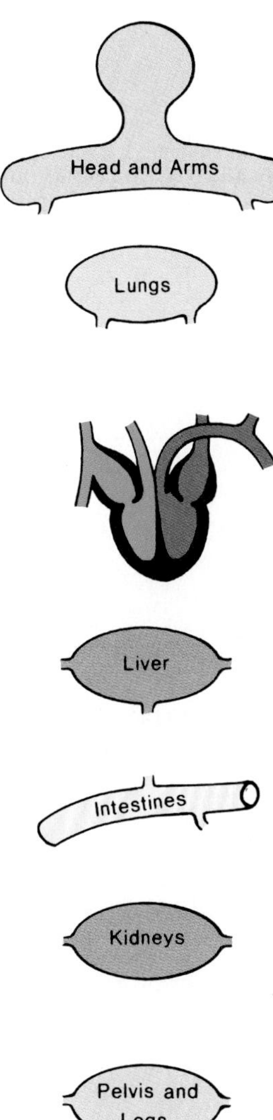

B. *Arteries*

Refer to figure 68.1 to identify the arteries the statements describe.

anterior tibial—1
aorta—2
aortic arch—3
axillary—4
brachial—5
brachiocephalic—6
celiac—7
common iliac—8
deep femoral—9
external iliac—10
femoral—11
inferior mesenteric—12
internal iliac—13
left common carotid—14
left subclavian—15
popliteal—16
posterior tibial—17
radial—18
renal—19
right common carotid—20
subclavian—21
superior mesenteric—22
ulnar—23

1. Artery of shoulder
2. Artery of upper arm
3. Artery of armpit
4. Gives rise to right common carotid and right subclavian
5. Lateral artery of forearm
6. Medial artery of forearm
7. Three branches of aortic arch
8. Gives rise to femoral artery
9. Major artery of thigh
10. Artery of knee region
11. Artery of calf region
12. Large branch of common iliac
13. Small branch of common iliac
14. Supplies stomach, spleen, and liver
15. Major artery of chest and abdomen
16. Supplies kidney
17. Also known as hypogastric artery
18. Supplies blood to most of small intestines and part of colon
19. Artery in interior portion of lower leg
20. Branch of femoral that parallels medial surface of femur
21. Curved vessel that receives blood from left ventricle
22. Supplies large intestine and rectum

Answers
Arteries
1.________
2.________
3.________
4.________
5.________
6.________
7.________
8.________
9.________
10.________
11.________
12.________
13.________
14.________
15.________
16.________
17.________
18.________
19.________
20.________
21.________
22.________

C. *Veins*

Refer to figures 68.2 and 68.3 to identify the veins the statements describe.

accessory cephalic—1	hepatic—14
axillary—2	hepatic portal—15
basilic—3	inferior mesenteric—16
brachial—4	inferior vena cava—17
brachiocephalic—5	internal iliac—18
cephalic—6	internal jugular—19
common iliac—7	median cubital—20
coronary—8	popliteal—21
dorsal venous arch—9	posterior tibial—22
external iliac—10	pyloric—23
external jugular—11	subclavian—24
femoral—12	superior mesenteric—25
great saphenous—13	superior vena cava—26

1. Largest vein in neck
2. Small vein in neck
3. Vein of armpit
4. Empty into brachiocephalic veins
5. Collects blood from veins of head and arms
6. Short vein between basilic and cephalic veins
7. Collects blood from veins of chest, abdomen, and legs
8. On posterior surface of humerus
9. Large veins that unite to form inferior vena cava
10. On medial surface of upper arm
11. On lateral portion of forearm
12. On lateral portion of upper arm
13. Vein from liver to inferior vena cava
14. Receives blood from descending colon and rectum
15. Empties into popliteal
16. Collects blood from two brachiocephalic veins
17. Collects blood from top of foot
18. Superficial vein on medial surface of leg
19. Vein of knee region
20. Collects blood from intestines, stomach, and colon
21. Vein that receives blood from posterior tibial
22. Receives blood from ascending colon and part of ileum
23. Empties into great saphenous
24. Small vein that empties into common iliac at juncture of external iliac
25. Vein into which great saphenous empties
26. Two veins that drain blood from stomach into hepatic portal vein

Answers
Veins
1.________
2.________
3.________
4.________
5.________
6.________
7.________
8.________
9.________
10.________
11.________
12.________
13.________
14.________
15.________
16.________
17.________
18.________
19.________
20.________
21.________
22.________
23.________
24.________
25.________
26.________

LABORATORY REPORT 69

Student: ______________________________

Section: ______________________________

Fetal Circulation

Questions

1. What blood vessel in the umbilical cord supplies the fetus with nutrients?
2. What blood vessels in the umbilical cord return blood to the placenta from the fetus?
3. What blood vessel shunts blood from the pulmonary trunk to the aorta?
4. What vessel in the liver carries blood from the umbilical vein to the inferior vena cava?
5. What structure in the liver forms from the ductus venosus?
6. What structure forms from the ductus arteriosus?
7. What structure forms from the umbilical vein?
8. What structure enables blood to flow from the right atrium to the left atrium before birth?
9. Give two occurrences at childbirth that prevent the infant from losing excessive blood through the cut umbilical cord.

Answers
Questions
1. ______________
2. ______________
3. ______________
4. ______________
5. ______________
6. ______________
7. ______________
8. ______________
9a. ______________
b. ______________

LABORATORY REPORT 70

Student: ______________________

Section: ______________________

The Lymphatic System and the Immune Response

A. *Components*

Refer to figures 70.1 and 70.2 to identify the lymphatic system structures described.

1. Large vessel in thorax and abdomen that collects lymph from lower extremities
2. Short vessel that collects lymph from right arm and right side of head
3. Microscopic lymphatic vessels situated among cells of tissues
4. Saclike structure that receives chyle from intestine
5. Blood vessel that receives fluid from thoracic duct
6. Filters of lymphatic system
7. Vessels of arms and legs that convey lymph to collecting ducts
8. Depression in lymph node through which blood vessels enter node
9. Outer covering of lymph node
10. Structural units of lymph node that are packed with lymphocytes
11. Activated T-cells
12. Cells that produce antibodies

capsule—1
cisterna chyli—2
germinal centers—3
hilum—4
left subclavian vein—5
lymphatic vessels—6
lymph capillaries—7
lymph nodes—8
lymphoblasts—9
plasma cells—10
right lymphatic duct—11
thoracic duct—12

Answers
Components
1.________
2.________
3.________
4.________
5.________
6.________
7.________
8.________
9.________
10.________
11.________
12.________

B. *Questions*

1. How does lymph in the lymphatic vessels of the legs differ in composition from lymph in the thoracic duct? ______________________

____

____

2. Where do stem cells for all lymphocytes originate? ______________________

____

____

3. Give the cell type that accounts for each type of immunity:

   humoral immunity ______________________

   cell-mediated immunity ______________________

4. What tissues program stem cells for

   humoral immunity? ______________________

   cell-mediated immunity? ______________________

Laboratory Report 71

Student: ______________________

Section: ______________________

The Respiratory Organs

A. *Cat Dissection*

After completing the cat dissection, answer the questions.

1. Are the cartilaginous rings of the trachea continuous around the organ? ______________
2. Which lung has four lobes? ______________

   Which lung has three lobes? ______________

B. *Sheep Pluck Dissection*

After completing the sheep pluck dissection, answer the questions.

1. Are the cartilaginous rings of the trachea continuous around the organ? ______________
2. Describe the texture of the surface of the lung. ______________
3. How many lobes make up the right lung? ______________
4. How many lobes make up the left lung? ______________
5. Why does lung tissue collapse so readily when you quit blowing into it with a straw? ______________
6. Identify the membrane on the surface of the lung. ______________

C. *Histological Study*

On a separate sheet of paper, make drawings, as required by your instructor, of nasal, tracheal, and lung tissues.

D. *Organ Identification*

Identify the respiratory structures that the statements describe.

alveoli—1	nasopharynx—11
bronchi—2	oral cavity proper—12
bronchioles—3	oral vestibule—13
cricoid cartilage—4	oropharynx—14
epiglottis—5	palatine tonsils—15
hard palate—6	pharyngeal tonsils—16
larynx—7	pleural cavity—17
lingual tonsils—8	soft palate—18
nasal cavity—9	thyroid cartilage—19
nasal conchae—10	trachea—20

1. Partition between nasal and oral cavities
2. Cartilage of Adam's apple
3. Small air sacs of lung tissue
4. Cavity above soft palate
5. Cavity above hard palate
6. Cavity that contains the tongue
7. Cavity near palatine tonsils
8. Voice box
9. Small tubes leading into alveoli
10. Another name for adenoids
11. Tube between larynx and bronchi
12. Fleshy lobes in nasal cavity
13. Cavity between lips and teeth
14. Tonsils on sides of pharynx
15. Tonsils attached to base of tongue
16. Flexible, flaplike cartilage over larynx
17. Most inferior cartilaginous ring of larynx
18. Tubes formed by bifurcation of trachea
19. Potential cavity between lung and thoracic wall

E. *Organ Function*

Select the respiratory structures in part D that perform the functions described.

1. Initiates swallowing
2. Prevents food from entering larynx when swallowing
3. Contains the vocal folds
4. Warms air as it passes through nasal cavity
5. Provides surface for gas exchange in lungs
6. Essential for speech

Answers
Organ Identification
1.__________
2.__________
3.__________
4.__________
5.__________
6.__________
7.__________
8.__________
9.__________
10.__________
11.__________
12.__________
13.__________
14.__________
15.__________
16.__________
17.__________
18.__________
19.__________
Organ Function
1. __________
2. __________
3. __________
4. __________
5. __________
6. __________

LABORATORY REPORT 72

Student: ____________________

Section: ____________________

Hyperventilation and Rebreathing

A. *Questions for Nonrecording Portion of Experiment*

1. Why does breathing into a bag after hyperventilating restore normal breathing more rapidly than breathing into the air? ____________________

2. How long were you able to hold your breath after only one deep inspiration? ____________________
3. How long were you able to hold your breath after hyperventilating? ____________________
4. After exercising, how long were you able to hold your breath? ____________________
5. Explain your observations from questions 2 through 4. ____________________

B. *Tracings*

In the spaces provided here and on the next page, attach chart recordings for the various portions of this experiment.

Normal respiration (slow speed)	Speed: _____ cm/sec

Normal respiration (fast speed) Speed: _____ cm/sec

Rebreathing and recovery

Hyperventilation

C. *Generalizations*

What generalizations can you make from the tracings?

____

____

____

____

____

____

____

Laboratory Report 73, 74

Student: ______________________

Section: ______________________

The Diving Reflex (Ex. 73)

A. *Tracings*

Attach chart recordings for the various portions of this experiment in the spaces provided here and on the next page.

Tidal breathing before immersion
Holding breath after tidal inspiration
Immersion in water at 70°F
Immersion in water at 60°F

Hyperventilation followed by holding breath (no immersion)

Hyperventilation followed by immersion Total time immersion: _____

B. Generalizations

____

____

____

____

____

____

The Valsalva Maneuver (Ex. 74)

Tracing

Attach a chart recording made during a Valsalva maneuver in the space provided.

LABORATORY REPORT 75

Student: ______________________________

Section: ______________________________

Spirometry: Lung Capacities

Tabulation

In the following table, record the results of your spirometer readings and calculations.

LUNG CAPACITIES	NORMAL (ml)	YOUR CAPACITIES
Tidal volume (TV)	500	
Minute respiratory volume (MRV) (MRV = TV × resp. rate)	6000	
Expiratory reserve volume (ERV)	1100	
Vital capacity (VC) (VC = TV + ERV + IRV)	See Appendix A (Table V and VI)	
Inspiratory capacity (IC) (IC = VC − ERV)	3000	
Inspiratory reserve volume (IRV) (IRV = IC − TV)	2500	

LABORATORY REPORT 76

Student: ______________________

Section: ______________________

Spirometry: The FEV_T Test

A. *Expirogram*

Trim the tracing of your expirogram to as small a piece as you can without removing volume values, and tape it in the space shown. Do not tape sides and bottom.

Tape Here

B. *Calculations*

Record your computations for determining the various percentages.

1. **Vital Capacity.** From the expirogram, determine the total volume of air expired (vital capacity) . . . . ________

2. **Corrected Vital Capacity.** By consulting table IV, Appendix A, determine the corrected vital capacity (step 4 on page 374).

   VC × Conversion factor = ... ________

   Math:

3. **One-Second Timed Vital Capacity (FEV_1).** After determining the FEV_1 (steps 1, 2, and 3 on page 374) record your results here . . . . ________

4. **Corrected FEV_1.** Correct the FEV_1 for the temperature of the spirometer (step 5 on page 374).

   FEV_1 × Conversion factor = ... ________

5. **FEV_1** Percentage. Divide the corrected FEV_1 by the corrected vital capacity to determine the percentage expired in the first second (step 6 on page 374).

   $$\frac{\text{Corrected FEV}_1}{\text{Corrected vital capacity}} =$$ . . . . ________

   Math:

6. **Percent of Predicted VC.** Determine what percentage your vital capacity is of the predicted vital capacity (Step 7 on page 374).

   $$\frac{\text{Your corrected VC}}{\text{Predicted VC}} =$$ . . . . ________

   Math:

C. *Questions*

1. Why isn't a measure of your vital capacity as significant as the FEV_T? ________________________

   ____

   ____

   ____

2. What respiratory diseases does this test most readily detect? ________________________

   ____

   ____

LABORATORY REPORT 78S

Student: ______________________

Section: ______________________

Physical Fitness Clearance

A. *Cardiovascular (Table 1)*

Report blood pressure, heart rates, and ECG results here. Attach your ECG tracing in the space provided.

	ECB			OTHER	
	S-T Segment Depression	PVCs	Other Arrhythmia	Heart Rates	Blood Pressure
Before Exercise					
After Exercise					

ECG Tracing

B. *Urine Analysis (Table II)*

Record results of urine analysis here.

Urobilinogen	Blood	Bilirubin	Ketones	Glucose	Protein	pH

C. *Adverse Postexercise Signs*

Record *present* or *absent* for each of the following:

Pallor	Cyanosis	Cold, Moist Skin	Staggering	Head Nodding	Confusion

D. *Questionnaire*

After moderate to strenuous exercise, do you experience any of the following?

Yes	No	Symptoms
		Severe shortness of breath
		Pains in the chest
		Severe fatigue
		Fainting spells
		Tendency to limp

E. *Release*

I voluntarily participate in this fitness program to monitor my physical progress while engaged in an exercise regimen of my choice. Although my instructor has taken all reasonable precautions to prevent injuries, I reserve the right to withdraw from further testing if I feel endangered in any way. Furthermore, I do not hold my instructor or the college liable for any accidents arising from circumstances beyond their control.

Signature ______________________ Date ____________ Age ____________

Additional Tracings

Laboratory Report 78

Student: ____________________

Section: ____________________

Response to Physical Stress: Monitoring Physical Fitness Training

A. Data

Record the following pertinent information:

Sex: ____________ Age: ____________

Nature of exercise: ____________ Starting date: ____________ Completion date: ____________

B. Anthropomorphic Measurements

Record all dimensions here. First line is for beginning of experiment; second line is for completion date.

Date	Height (cm)	Weight (kg)	Chest (cm)	Abdomen (cm)	Thigh (cm)	Calf (cm)	Arm (cm)	VC

C. Cardiovascular Measurements

Record the blood pressures and heart rates before and after exercise in the tables.

BLOOD PRESSURE		
Date	At Rest	After Exercise
1		
2		
3		
4		
5		

HEART RATE												
Date	At Rest	Minutes after Exercising										
		0	½	1	1½	2	2½	3	3½	4	4½	5
1												
2												
3												
4												
5												

D. Recordings

Attach tracings of the various recovery tests.

E. Graphs

Plot the data from the various tables on the graphs.

Blood Pressure at Rest

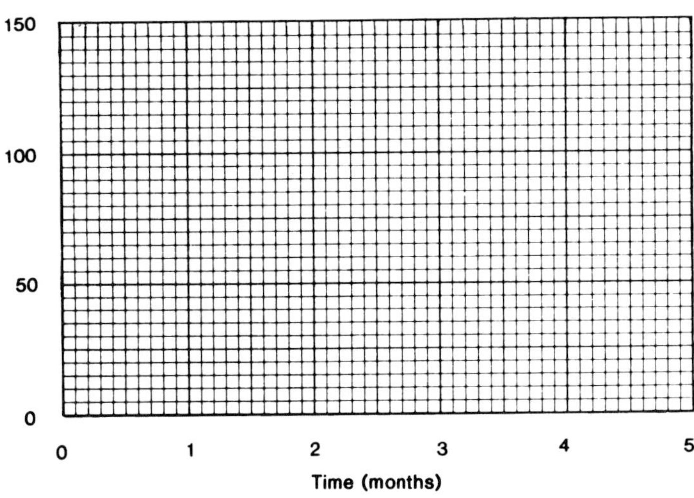

Blood Pressure After Exercise

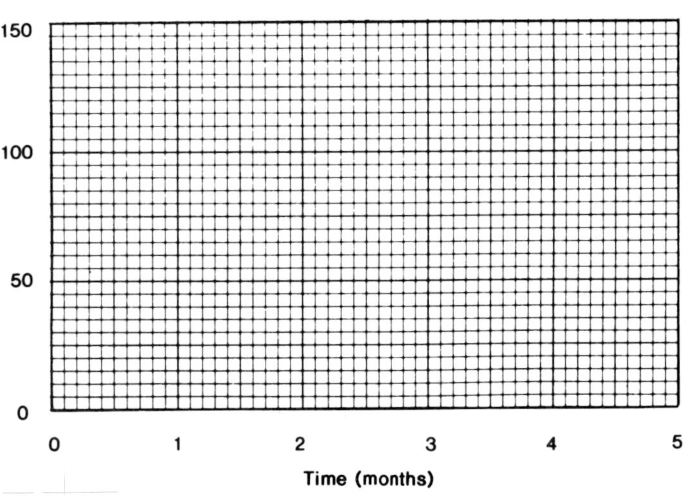

Heart Rate

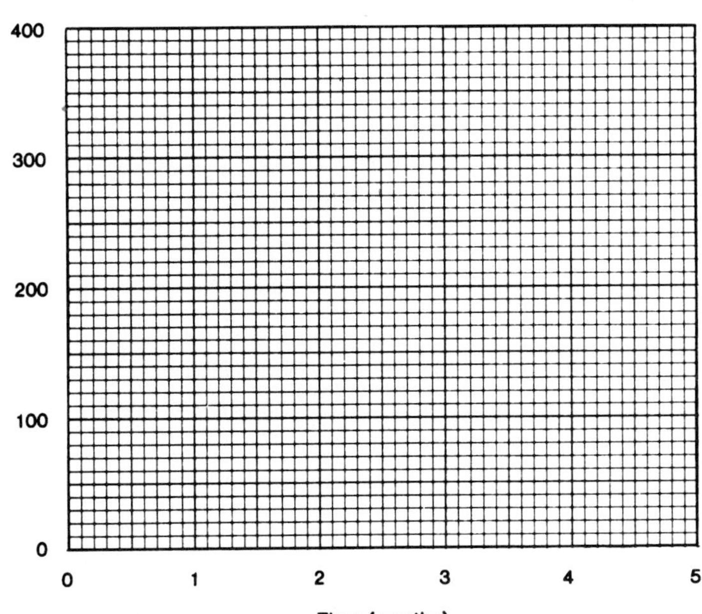

F. Statement

Briefly summarize the effectiveness (or ineffectiveness) of your exercise regimen.

LABORATORY REPORT 81

Student: ____________________

Section: ____________________

Anatomy of the Digestive System

A. *Alimentary Canal*

Identify the parts of the digestive system that the statements describe.

anus—1	hepatic duct—10
cardiac sphincter—2	ileocecal valve—11
cecum—3	ileum—12
colon—4	pancreatic duct—13
common bile duct—5	pharynx—14
cystic duct—6	pyloric portion of stomach—15
duodenum—7	pyloric sphincter—16
esophagus—8	rectum—17
fundus of stomach—9	sigmoid colon—18

1. Where swallowing (peristalsis) begins
2. Entrance opening of stomach
3. Exit opening of stomach
4. Tube between mouth and stomach
5. Distal coiled portion of small intestine
6. First 12 inches of small intestine
7. Proximal pouch or compartment of large intestine
8. Section of large intestine between descending colon and rectum
9. Most active portion of stomach
10. Duct that drains liver
11. Valve between small and large intestines
12. Duct that drains gallbladder
13. Duct that joins common bile duct before entering intestine
14. Structure on which appendix is located
15. Duct that conveys bile to intestine
16. Exit of alimentary canal
17. Part of digestive tract where most water absorption (conservation) occurs
18. Last 6 inches of alimentary canal

Answers
Alimentary Canal
1. ________
2. ________
3. ________
4. ________
5. ________
6. ________
7. ________
8. ________
9. ________
10. ________
11. ________
12. ________
13. ________
14. ________
15. ________
16. ________
17. ________
18. ________

B. *Oral Cavity*

Identify the structures of the mouth that the statements describe.

buccae—1
filiform papillae—2
foliate papillae—3
fungiform papillae—4
gingiva—5
glossopalatine arch—6
labial frenulum—7
lesser sublingual ducts—8
lingual frenulum—9
lingual tonsil—10
mucosa—11
palatine tonsil—12
parotid duct—13
parotid glands—14
pharyngeal tonsil—15
pharyngopalatine arch—16
sublingual glands—17
submandibular duct—18
submandibular glands—19
uvula—20
vallate papillae—21

1. Lining of mouth
2. Name for cheeks
3. Duct that drains parotid gland
4. Fold of skin between lip and gums
5. Duct that drains submandibular gland
6. Fold of skin between tongue and floor of mouth
7. Portion of mucosa around teeth
8. Tonsils at back of mouth
9. Tonsils at root of tongue
10. Fingerlike projection at end of soft palate
11. Vertical ridges on side of tongue
12. Rounded papillae on dorsum of tongue
13. Small, tactile papillae on surface of tongue
14. Large papillae at back of tongue
15. Ducts that drain sublingual gland
16. Where taste buds are located
17. Salivary glands located inside and below mandible
18. Membrane in front of palatine tonsil
19. Membrane in back of palatine tonsil
20. Salivary glands under tongue
21. Salivary glands in cheeks

C. *The Teeth*

Select the answer that best completes each statement.

1. All deciduous teeth usually erupt by the time a child is
   a. 1 year old.
   b. 2 years old.
   c. 4 years old.
2. A complete set of primary teeth consists of
   a. 10 teeth.
   b. 20 teeth.
   c. 32 teeth.
3. The permanent teeth with the longest roots are the
   a. incisors.
   b. cuspids.
   c. molars.
4. The smallest permanent molars are the
   a. first molars.
   b. second molars.
   c. third molars.

Answers
Oral Cavity
1.________
2.________
3.________
4.________
5.________
6.________
7.________
8.________
9.________
10.________
11.________
12.________
13.________
14.________
15.________
16.________
17.________
18.________
19.________
20.________
21.________
Teeth
1.________
2.________
3.________
4.________

C. The Teeth (continued)

5. A complete set of permanent teeth consists of
   a. 24 teeth.
   b. 28 teeth.
   c. 32 teeth.
6. The primary dentition lacks
   a. molars.
   b. incisors.
   c. bicuspids.
7. A child's first permanent tooth usually erupts during the
   a. second year.
   b. third year.
   c. fifth year.
   d. sixth year.
8. Trifurcated roots exist on
   a. upper cuspids.
   b. upper molars.
   c. lower molars.
9. Bifurcated roots exist on
   a. upper first bicuspids and lower molars.
   b. cuspids and lower bicuspids.
   c. all molars.
10. A tooth is considered "dead" or devitalized if
    a. the enamel is destroyed.
    b. the pulp is destroyed.
    c. the periodontal membrane is infected.
11. Dentin is produced by cells called
    a. odontoblasts.
    b. ameloblasts.
    c. cementocytes.
12. Cementum is secreted by cells called
    a. odontoblasts.
    b. ameloblasts.
    c. cementocytes.
13. The wisdom tooth is
    a. a supernumerary tooth.
    b. a succedaneous tooth.
    c. the third molar.
14. Caries (cavities) are caused primarily by
    a. using the wrong toothpaste.
    b. acid production by bacteria.
    c. fluorides in water.
15. Enamel covers the
    a. entire tooth.
    b. clinical crown only.
    c. anatomical crown only.
16. The tooth receives nourishment through
    a. the apical foramen.
    b. the periodontal membrane.
    c. both the apical foramen and the periodontal membrane.

Answers
Teeth
5. ________
6. ________
7. ________
8. ________
9. ________
10. ________
11. ________
12. ________
13. ________
14. ________
15. ________
16. ________

D. *Microscopic Studies*

Draw your microscopic observations in the space provided. If this space is insufficient for all required drawings, use a separate sheet of paper.

LABORATORY REPORT 82

Student: ____________________

Section: ____________________

Intestinal Motility

A. *Temperature*

Record your observations of intestinal motility changes that occurred as the temperature of Tyrode's solution was gradually increased.

1. Frequency

Plot the frequency of contraction as related to temperature on the graph.

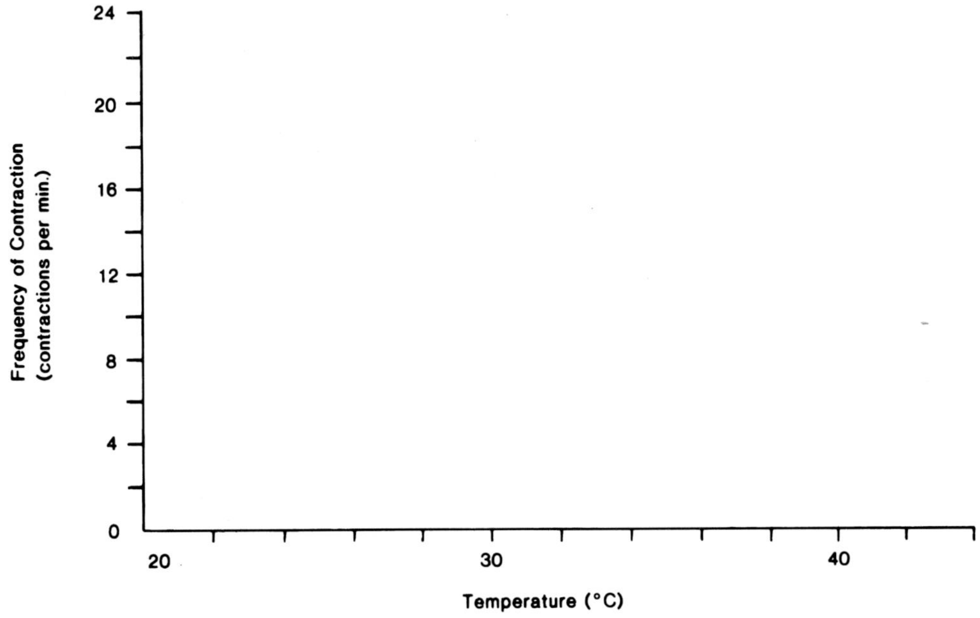

2. Strength of Contraction

a. Did the strength of contraction change with the change in temperature? ____________________

____

b. If the answer to the previous question is yes, what temperature was optimum for strength of contraction? ____________________

B. *Electrical Stimulation*

Describe in detail the results of electrical stimulation. ____________________

____

____

____

____

____

C. Neurohumoral Control

1. What effect does epinephrine have on muscular contraction? ______________________________

____

____

2. What effect does acetylcholine have on muscular contraction? ______________________________

____

____

3. Since the parasympathetic nerves produce acetylcholine, and the sympathetic nerves produce norepinephrine, what generalizations can you make concerning nervous control of intestinal motility?

____

____

____

____

D. Tracings

Attach samples of tracings with labels in the space provided.

LABORATORY REPORT 83

Student: ______________________

Section: ______________________

The Chemistry of Hydrolysis

A. Carbohydrate Digestion

1. **IKI Test.** Record here the presence (+) or absence (−) of starch as revealed by the IKI test on spot plates.

1A	2A	3A	4A	5A

a. Did the pancreatic extract hydrolyze starch? ______________________

b. Which tube is your evidence for this conclusion? ______________________

c. How do tubes 1A and 5A compare in color? ______________________

d. What do you conclude from question *c?* ______________________

e. What is the value of tube 4A? ______________________

f. What is the value of tube 2A? ______________________

2. **Barfoed's and Benedict's Tests.** After performing Barfoed's and Benedict's tests on tubes 1A and 5A, what are your conclusions about the degree of digestion of starch by amylase? ______________________

____

____

B. Protein Digestion

Record the color and optical densities (OD) of each tube in the chart.

	1P	2P	3P	4P	5P
Color					
OD					

1. Did pancreatic juice hydrolyze the BAPNA? ______________________
2. Which tube is your evidence for this conclusion? ______________________

3. What other tube shows protein hydrolysis? ____________________
4. What can you conclude from the observation in question 3? ____________________

____

5. Why is there no hydrolysis in tube 4P? ____________________
6. Why is there no hydrolysis in tube 3P? ____________________
7. Why didn't tube 2P show hydrolysis? ____________________

____

C. *Fat Digestion*

Determine the pH of each tube by comparing with a bromthymol blue color standard. Record these values in the table.

1L	2L	3L	4L	5L	6L

1. Did the pancreatic extract hydrolyze fat in the presence of bile? ____________________
2. Did the pancreatic extract hydrolyze fat in the absence of bile? ____________________
3. How do tubes 1L and 6L compare with regard to degree of hydrolysis? ____________________

____

4. What can you conclude from the observation in question 3? ____________________

____

5. What is the value of tube 2L? ____________________

____

D. *Enzyme Review*

Consult your text and lecture notes to answer these questions concerning digestive enzymes.

Digestive Juices

Select the enzymes listed on the right that are present in the digestive juices that follow.

1. Saliva
2. Gastric juice
3. Pancreatic fluid
4. Intestinal juice (succus entericus)

Substrates

Select the enzymes that act on the substrates that follow.

1. Fats
2. Lactose
3. Maltose
4. Peptides
5. Proteins
6. Starch
7. Sucrose

Carbohydrases
lactase—1
maltase—2
pancreatic amylase—3
salivary amylase—4
sucrase—5

Proteases
carboxypeptidase—6
chymotrypsin—7
erepsin (peptidase)—8
pepsin—9
rennin—10
trypsin—11

Lipases
pancreatic lipase—12

End Products

Select the enzymes from the previous list that produce the following end products in digestion.

1. Amino acids
2. Fatty acids
3. Fructose
4. Galactose
5. Glucose
6. Glycerol
7. Maltose
8. Polypeptides

Answers
Digestive Juices
1. ________
2. ________
3. ________
4. ________
Substrates
1. ________
2. ________
3. ________
4. ________
5. ________
6. ________
7. ________
End Products
1. ________
2. ________
3. ________
4. ________
5. ________
6. ________
7. ________
8. ________

LABORATORY REPORT 84

Student: ______________________

Section: ______________________

Factors Affecting Hydrolysis

A. *Tabulations*

If the class has been subdivided into separate groups to perform different parts of this experiment, these tabulations should be recorded on the chalkboard so that all students can copy the results onto their Laboratory Reports.

1. **Temperature.** After IKI solution has been added to all 10 tubes, record the color and degree of amylase activity in each tube in the tables that follow. Refer to the carbohydrate differentiation separation outline pertaining to the IKI test in Appendix C that follow.

Table 1 Amylase Action at 20°C

TUBE NO.	1	2	3	4	5	6	7	8	9	10
Color										
Starch Hydrolysis										

Table II Amylase Action at 37°C

TUBE NO.	1	2	3	4	5	6	7	8	9	10
Color										
Starch Hydrolysis										

2. **Hydrogen Ion Concentration.** After IKI solution has been added to all nine tubes, record, as before, the color and degree of amylase action for each tube in the table that follows.

Table III pH and Amylase action at 37°C

TIME	2 Minutes			4 Minutes			6 Minutes		
Tube No.	1	2	3	4	5	6	7	8	9
pH	5	7	9	5	7	9	5	7	9
Color									
Starch Hydrolysis									

B. Conclusions

1. How long did it take for complete starch hydrolysis at 20°C? __________ at 37°C? __________
2. Why would you expect these results? ______________________________

____

____

3. What effect does boiling have on amylase? ______________________________
4. At which pH was starch hydrolysis most rapid? ______________________________

   How does this pH compare with the pH of normal saliva? ______________________________

   ____

5. What is the pH of gastric juice? ______________________________

   What do you suppose happens to the action of amylase in the stomach? ______________________________

   ____

LABORATORY REPORT 85

Student: ______________________

Section: ______________________

Anatomy of the Urinary System

A. *Anatomy*

Identify the urinary system structures that the statements describe.

calyces—1	medulla—6	renal pelvis—11
collecting duct—2	nephron—7	renal pyramids—12
cortex—3	renal capsule—8	ureters—13
glomerular capsule—4	renal column—9	urethra—14
glomerulus—5	renal papilla—10	

1. Tubes that drain kidneys
2. Tube that drains urinary bladder
3. Portion of kidney that contains renal corpuscles
4. Cone-shaped areas of medulla
5. Portion of kidney that consists primarily of collecting tubules
6. Basic functioning unit of kidney
7. Two portions of renal corpuscle
8. Distal tip of renal pyramid
9. Short tubes that receive urine from renal papillae
10. Funnel-like structure that collects urine from calyces of each kidney
11. Structure that receives urine from several nephrons
12. Cup-shaped membranous structure that surrounds glomerulus
13. Thin, fibrous outer covering of kidney
14. Cortical tissue between renal pyramids
15. Tuft of capillaries within the glomerular capsule

B. *Physiology*

Select the answer that best completes each statement about the physiology of urine production. For definitions of medical terms, consult a medical dictionary or your lecture text.

1. Blood enters the glomerulus through the
   a. efferent arteriole.
   b. arcuate artery.
   c. afferent arteriole.
   d. None of these is correct.
2. Surgical removal of a kidney is called
   a. nephrectomy.
   b. nephrotomy.
   c. nephrolithotomy.
3. Water reabsorption from the glomerular filtrate into the peritubular blood is facilitated by
   a. antidiuretic hormone.
   b. renin.
   c. aldosterone.
   d. Both *a* and *c* are correct.

Answers

Anatomy

1. ____
2. ____
3. ____
4. ____
5. ____
6. ____
7. ____
8. ____
9. ____
10. ____
11. ____
12. ____
13. ____
14. ____
15. ____

Physiology

1. ____
2. ____
3. ____

4. The following substances are reabsorbed through the walls of the nephron into the peritubular blood:
   a. glucose and water.
   b. urea and water.
   c. glucose, amino acids, salts, and water.
   d. glucose, amino acids, urea, salts, and water.
5. The amount of urine produced is affected by
   a. blood pressure.
   b. environmental temperature.
   c. amount of solute in glomerular filtrate.
   d. *a, b, c,* and additional factors are correct.
6. The amount of urine normally produced in 24 hours is about
   a. 100 ml.
   b. 500 ml.
   c. 1.5 l.
   d. 4.5 l.
7. The reabsorption of sodium ions from the glomerular filtrate into the peritubular blood draws the following back into the blood:
   a. potassium ions.
   b. chloride ions.
   c. water.
   d. chloride ions and water.
8. Cells of the walls of collecting tubules secrete the following substances into urine:
   a. amino acids.
   b. uric acid.
   c. ammonia.
   d. Both *b* and *c* are correct.
9. Renal diabetes is due to
   a. a lack of ADH.
   b. a lack of insulin.
   c. faulty reabsorption of glucose in the nephron.
   d. Both *a* and *b* are correct.
10. Glucose in the urine and a low level of insulin in the blood is diagnosed as
    a. diabetes insipidus.
    b. diabetes mellitus.
    c. renal diabetes.
    d. None of these is correct.
11. Approximately 80% of water in the glomerular filtrate is reabsorbed in the
    a. proximal convoluted tubule.
    b. nephron loop.
    c. distal convoluted tubule.
    d. None of these is correct.
12. The desire to micturate normally occurs when the following amount of urine is present in the bladder:
    a. 100 ml.
    b. 200 ml.
    c. 300 ml.
    d. 400 ml.
13. Removal of calculus (kidney stones) is called
    a. nephrectomy.
    b. nephrotomy.
    c. nephrolithotomy.
    d. Both *b* and *c* are correct.

Answers
Physiology
4.________
5.________
6.________
7.________
8.________
9.________
10.________
11.________
12.________
13.________

C. *Microscopy*

Use the space provided for your histology drawings, or put them on a separate sheet of paper.

Laboratory Report 86

Student: ______________________

Section: ______________________

Urine: Composition and Tests

A. Test Results

Record in the chart the results of all urine tests performed. If the tests are done as a demonstration, the results are tabulated in columns A and B. If students perform the tests on their own urine, the last column is for their results.

TEST	NORMAL VALUES	ABNORMAL VALUES	A Positive Test Control	B Unknown Sample	C Student's Urine
Color	Colorless Pale straw Straw Amber	Milky Reddish amber Brownish yellow Green Smoky brown			
Cloudiness	Clear	+ Slight + + Moderate + + + Cloudy + + + + Very cloudy			
Specific Gravity	1.001–1.060	Above 1.060			
pH	4.8–7.5	Below 4.8 Above 7.5			
Albumin (protein)	None	+ Barely visible + + Granular + + + Flocculent + + + + Large flocculent			
Mucin	None	Visible amounts			
Glucose	None	+ Yellow green + + Greenish yellow + + + Yellow + + + + Orange			
Ketones	None	+ Pink-purple ring			
Hemoglobin	None	See color chart			
Bilirubin					

B. *Microscopy*

Record here by sketch and word any structures seen on microscopic examination.

C. *Interpretation*

Indicate the probable significance of each of the following substances appearing in urine. More than one condition may apply in some instances.

bladder infection—1	diabetes mellitus—4	gonorrhea—7
cirrhosis of liver—2	exercise (extreme)—5	hepatitis—8
diabetes insipidus—3	glomerulonephritis—6	kidney tumor—9
		normally present—10

1. Albumin (constantly)
2. Albumin (periodic)
3. Bacteria
4. Bile pigments
5. Blood
6. Creatinine
7. Glucose
8. Mucin
9. Porphyrin
10. Pus cells (neutrophils)
11. Sodium chloride
12. Specific gravity (high)
13. Specific gravity (low)
14. Urea
15. Uric acid

D. *Terminology*

Select the terms that the statements describe. Use your text or medical dictionary.

albuminuria—1	cystitis—4	glycosuria—7
anuria—2	diuretic—5	nephritis—8
catheter—3	enuresis—6	pyelitis—9
		uremia—10

1. Absence of urine production
2. High urea level in blood
3. Sugar in urine
4. Albumin in urine
5. Inflammation of nephrons
6. Inflammation of pelvis of kidney
7. Bladder infection
8. Device for draining bladder
9. Involuntary bed-wetting during sleep
10. Chemical that stimulates urine production

Answers

Interpretation

1. ________
2. ________
3. ________
4. ________
5. ________
6. ________
7. ________
8. ________
9. ________
10. ________
11. ________
12. ________
13. ________
14. ________
15. ________

Terminology

1. ________
2. ________
3. ________
4. ________
5. ________
6. ________
7. ________
8. ________
9. ________
10. ________

Laboratory Report 87

Student: ____________________

Section: ____________________

The Endocrine Glands

A. Microscopy

Make the drawings of the various microscopic studies on a separate sheet of paper, and attach the paper to the Laboratory Report.

B. Sources

Select the glandular tissues from the list on the right that produce the hormones.

1. ACTH
2. ADH
3. Aldosterone
4. Calcitonin
5. Chorionic gonadotropin
6. Cortisol
7. Epinephrine
8. Estrogens
9. FSH
10. Glucagon
11. ICSH
12. Insulin
13. LH
14. Melatonin
15. Norepinephrine
16. Oxytocin
17. Parathyroid hormone
18. PIF
19. Progesterone
20. Prolactin
21. Somatostatin
22. Testosterone
23. Thymosin
24. Thyrotropin
25. Thyroxine
26. Triiodothyronine
27. Vasopressin

adrenal cortex
- zona fasciculata—1
- zona glomerulosa—2
- zona reticularis—3

adrenal medulla—4
hypothalamus—5
ovary
- corpus luteum—6
- Graafian follicle—7

pancreas
- acini cells—8
- alpha cells—9
- beta cells—10
- delta cells—11

parathyroid gland—12
pineal gland—13
pituitary gland
- adenohypophysis—14
- neurohypophysis—15

placenta—16
testis—17
thymus gland—18
thyroid gland
- follicular cells—19
- parafollicular cells—20

C. Physiology

Select the hormones that produce the physiological effects described. Note that a separate list of hormones is provided for each group.

Group I

1. Increases metabolism
2. Increases blood pressure
3. Increases strength of heartbeat
4. Promotes sodium absorption in nephron
5. Promotes gluconeogenesis
6. Causes vasoconstriction in all organs
7. Causes vasoconstriction in all organs except muscles and liver
8. Causes vasodilation in skeletal muscles
9. Inhibits estrus cycle in lower animals

aldosterone—1
epinephrine—2
glucocorticoids—3
melatonin—4
norepinephrine—5
none of these—6

Answers

Sources

1. ________
2. ________
3. ________
4. ________
5. ________
6. ________
7. ________
8. ________
9. ________
10. ________
11. ________
12. ________
13. ________
14. ________
15. ________
16. ________
17. ________
18. ________
19. ________
20. ________
21. ________
22. ________
23. ________
24. ________
25. ________
26. ________
27. ________

Physiology Group I

1. ________
2. ________
3. ________
4. ________
5. ________
6. ________
7. ________
8. ________
9. ________

Select the hormones that produce the physiological effects described.

Group II

1. Inhibits insulin production
2. Inhibits production of SH
3. Stimulates osteoclasts
4. Mobilizes fatty acids
5. Inhibits glucagon production
6. Raises blood calcium level
7. Increases cardiac output and respiratory rate
8. Impedes phosphorus reabsorption in renal tubules
9. Promotes storage of glucose, fatty acids, and amino acids
10. Mobilizes glucose, fatty acids, and amino acids
11. Stimulates metabolism of all cells in body
12. Lowers blood phosphorus levels
13. Decreases blood calcium when injected intravenously

calcitonin—1
glucagon—2
insulin—3
parathyroid hormone—4
somatostatin—5
thyroid hormone—6
none of these—7

Group III

1. Promotes breast enlargement
2. Promotes rapid bone growth
3. Stimulates testicular descent
4. Causes milk ejection from breasts
5. Promotes maturation of spermatozoa
6. Constricts arterioles
7. Stimulates production of thyroxine
8. Stimulates glucocorticoid production
9. Inhibits prolactin production
10. Causes thickening and vascularization of endometrium
11. Causes interstitial cells to produce testosterone
12. Causes myometrium contraction during childbirth
13. Promotes milk production after childbirth
14. Promotes stratification of vaginal epithelium
15. Suppresses development of female genitalia in male
16. Prepares endometrium for egg cell implantation
17. Promotes growth of all cells in body

ACTH—1
ADH—2
chorionic gonadotropin—3
estrogens—4
FSH—5
LH—6
oxytocin—7
PIF—8
progesterone—9
somatotropin—10
testosterone—11
thyrotropin—12
none of these—13

D. *Hormonal Imbalance*

Select the hormones responsible for the disorders described. After each hormone number, indicate with a (+) or (−) whether the condition is due to *excess* (+) or *deficiency* (−). More than one hormone may apply in some cases.

1. Acromegaly
2. Addison's disease
3. Cretinism
4. Cushing's syndrome
5. Diabetes insipidus
6. Diabetes mellitus
7. Dwarfism
8. Exophthalmia
9. Gigantism
10. Hyperglycemia
11. Hypothyroidism
12. Myxedema
13. Tetany

ADH—1
glucagon—2
glucocorticoids—3
insulin—4
mineralocorticoids—5
parathyroid hormone—6
somatotropin—7
thyroid hormone—8
none of these—9

Answers

Physiology Group II

1._____	8._____
2._____	9._____
3._____	10._____
4._____	11._____
5._____	12._____
6._____	13._____
7._____	

Physiology Group III

1._____	10._____
2._____	11._____
3._____	12._____
4._____	13._____
5._____	14._____
6._____	15._____
7._____	16._____
8._____	17._____
9._____	

Hormonal Imbalance

1._____	8._____
2._____	9._____
3._____	10._____
4._____	11._____
5._____	12._____
6._____	13._____
7._____	

LABORATORY REPORT 88

Student: ____________________

Section: ____________________

The Reproductive System

A. Male Organs

Identify the structures that the statements describe.

bulbourethral gland—1
common ejaculatory duct—2
corpora cavernosa—3
corpus spongiosum—4
epididymis—5
glans penis—6
penis—7
prepuce—8
prostate gland—9
prostatic urethra—10
seminal vesicle—11
seminiferous tubule—12
testes—13
vas deferens—14

1. Copulatory organ of male
2. Source of spermatozoa
3. Erectile tissue of penis (three answers)
4. Fold of skin over end of penis
5. Structure that stores spermatozoa
6. Contribute fluids to semen (three answers)
7. Duct that passes from urinary bladder through prostate gland
8. Coiled-up structures in testes where spermatozoa originate
9. Tube that carries sperm from epididymis to ejaculatory duct
10. Distal end of penis

B. Microscopy

Draw the various stages of spermatogenesis on a separate piece of paper to be included with this Laboratory Report.

C. Female Organs

Identify the structures that the statements describe.

cervix—1
clitoris—2
external os—3
fimbriae—4
greater vestibular glands—5
hymen—6
infundibulum—7
internal os—8
labia majora—9
labia minora—10
mons pubis—11
myometrium—12
ovaries—13
paraurethral glands—14
uterine tubes—15
vagina—16

1. Copulatory organ of female
2. Source of ova in female
3. Muscular wall of uterus
4. Neck of uterus
5. Provides vaginal lubrication during coitus
6. Fingerlike projections around edge of infundibulum
7. Protuberance of erectile tissue sensitive to sexual excitation
8. Entrance to uterus at vaginal end
9. Homologous to bulbourethral glands of male
10. Homologous to penis of male
11. Homologous to scrotum of male
12. Homologous to prostate gland of male
13. Ducts that convey ovum to uterus
14. Membranous fold of tissue surrounding entrance to vagina
15. Open, funnel-like end of uterine tube

Answers
Male Organs
1.________
2.________
3.________
4.________
5.________
6.________
7.________
8.________
9.________
10.________
Female Organs
1.________
2.________
3.________
4.________
5.________
6.________
7.________
8.________
9.________
10.________
11.________
12.________
13.________
14.________
15.________

D. *Ligaments*

Identify the supporting structures that hold the ovaries, uterus, and uterine tubes in place.

broad ligament—1
infundibulopelvic ligament—2
mesosalpinx—3
ovarian ligament—4
round ligament—5
sacrouterine ligament—6
suspensory ligament—7
none of these—8

1. Ligament that extends from uterus to ovary
2. Ligament between cervix and sacral part of pelvic wall
3. Ligament that extends from infundibulum to wall of pelvic cavity
4. Ligament that extends from corpus of uterus to body wall
5. Large, flat ligament of peritoneal tissue that extends from uterine tube to cervical part of uterus
6. Mesentery between ovary and uterine tubes that contains blood vessels leading to uterine tube
7. Fold of peritoneum that attaches ovary to uterine tube

E. *Germ Cells*

Identify the types of cells of oogenesis and spermatogenesis that the statements describe. More than one answer may apply.

Spermatogenesis
primary spermatocyte—1
secondary spermatocyte—2
spermatids—3
spermatogonia—4
spermatozoa—5

Oogenesis
oogonia—6
polar bodies—7
primary oocyte—8
secondary oocyte—9

1. Haploid cells
2. Cells that contain tetrads
3. Cells that contain monads
4. Cells that contain dyads
5. Diploid cells
6. Cells at periphery of ovary that produce all ova
7. Cells that undergo mitosis
8. Cells in which first meiotic division occurs
9. Cells at periphery of seminiferous tubule that give rise to all spermatozoa
10. Cells that develop directly into mature spermatozoa
11. Cells in which second meiotic division occurs

F. *Physiology of Reproduction and Development*

Select the answer that best completes each statement. You may need to consult your textbook or a medical dictionary for help in answering these questions.

1. The ovum gets from the ovary to the uterus by
   a. amoeboid movement.
   b. ciliary action.
   c. peristalsis and ciliary action of the uterine tube.
2. The vaginal lining is normally
   a. alkaline.
   b. acidic.
   c. neutral.
3. The seminal vesicle secretion is
   a. alkaline.
   b. acidic.
   c. neutral.

Answers

Ligaments

1. ________
2. ________
3. ________
4. ________
5. ________
6. ________
7. ________

Germ Cells

1. ________
2. ________
3. ________
4. ________
5. ________
6. ________
7. ________
8. ________
9. ________
10. ________
11. ________

Physiology of Reproduction and Development

1. ________
2. ________
3. ________

F. *Physiology of Reproduction and Development (continued)*

4. The birth canal consists of the
   a. vagina.
   b. uterus.
   c. vagina and uterus.
5. Spermatozoan viability is enhanced by a temperature that is
   a. 98.6°F.
   b. above 98.6°F.
   c. below 98.6°F.
6. The prostatic secretion is
   a. alkaline.
   b. acidic.
   c. neutral.
7. Circumcision involves excisement of the
   a. prepuce.
   b. perineum.
   c. glans penis.
8. The human ovum is usually fertilized in the
   a. uterus.
   b. vagina.
   c. uterine tube.
9. Uterine muscle consists of
   a. smooth muscle.
   b. striated muscle.
   c. both smooth and striated muscle.
10. The lining of the uterus is the
   a. myometrium.
   b. endometrium.
   c. epimetrium.

Answers
Physiology of Reproduction and Development
4.________
5.________
6.________
7.________
8.________
9.________
10.________

G. *Menstrual Cycle*

With the help of your textbook or a medical dictionary, select the answer that best completes each statement.

Answers
Menstrual Cycle
1.________
2.________
3.________
4.________
5.________
6.________
7.________
8.________

1. Cessation of menstrual flow in a woman in her late forties is
   a. menarche.
   b. menopause.
   c. amenorrhea.
2. The first menstrual flow is known as
   a. menarche.
   b. climacteric.
   c. menopause.
3. A distinct rise in the basal body temperature usually occurs
   a. at ovulation.
   b. during the proliferative phase.
   c. during the quiescent period.
4. Menstrual flow is the result of
   a. a deficiency of progesterone and estrogen.
   b. a deficiency of estrogen.
   c. an excess of progesterone and estrogen.
5. Ovulation usually (not always) occurs the following number of days before the next menstrual period:
   a. three.
   b. seven.
   c. fourteen.
   d. eighteen.
6. Pain during ovulation, as experienced by some women, is known as
   a. dysmenorrhea.
   b. oligomenorrhea.
   c. mittelschmerz.
7. The proliferative phase in the menstrual cycle begins on about the
   a. second day of the menstrual cycle.
   b. fifth day of the menstrual cycle.
   c. seventh day of the menstrual cycle.
8. Repair of the endometrium after menstruation is due to
   a. estrogen.
   b. progesterone.
   c. luteinizing hormone.

9. Progesterone secretion ceases entirely within the following number of days after the onset of menses:
   a. 10.
   b. 14.
   c. 26.
   d. 28.
10. Excessive blood flow during menstruation is known as
    a. oligomenorrhea.
    b. dysmenorrhea.
    c. menorrhagia.
11. Excessive discomfort and pain during menstruation is known as
    a. dysmenorrhea.
    b. oligomenorrhea.
    c. menorrhagia.
12. Occasional or irregular menses is known as
    a. amenorrhea.
    b. oligomenorrhea.
    c. dysmenorrhea.
13. Absence of menstruation is called
    a. dysmenorrhea.
    b. oligomenorrhea.
    c. amenorrhea.

Answers

Menstrual Cycle

9. ________
10. ________
11. ________
12. ________
13. ________

Medical

1. ________
2. ________
3. ________
4. ________
5. ________
6. ________
7. ________
8. ________
9. ________
10. ________
11. ________
12. ________
13. ________
14. ________
15. ________

H. Medical

Use a medical dictionary to select the type of surgery or condition that the statements describe.

anteflexion—1	hysterectomy—6	retroflexion—11
cryptorchidism—2	mastectomy—7	salpingectomy—12
endometritis—3	oophorectomy—8	salpingitis—13
episiotomy—4	oophorhysterectomy—9	syphilis—14
gonorrhea—5	oophoroma—10	vasectomy—15

1. Surgical removal of breast
2. Failure of testes to descend into scrotum
3. Surgical removal of uterus
4. Spirochaetal sexually transmitted disease
5. Sterilization procedure in males
6. Type of incision made in perineum at childbirth to prevent excessive damage to anal sphincter
7. Surgical removal of one or both ovaries
8. Ovarian malignancy
9. Sexually transmitted disease that affects mucous membranes rather than blood
10. Surgical removal or sectioning of uterine tubes
11. Most common sexually transmitted disease
12. Inflammation of uterine wall
13. Malpositioned uterus (two types)
14. Sexually transmitted disease caused by a coccoidal (spherical) organism
15. Surgical removal of ovaries and uterus

LABORATORY REPORT 89

Student: ______________________

Section: ______________________

Fertilization and Early Embryology (Sea Urchin)

A. *Drawings*

Put all drawings of cleavage stages on a separate sheet of paper. These should include the zygote (1 cell), 2-, 4-, 8-, and 16-cell stages. If prepared slides showing blastula and gastrula stages are available, make drawings of them, also.

B. *Questions*

1. Differentiate between

   Activation ______________________

   ____

   Fertilization ______________________

   ____

2. What germ layer forms as a result of gastrula formation? ______________________
3. Where does mesoderm form? ______________________
4. Indicate the germ layer that gives rise to the structures that follow. You may need to consult your textbook for assistance.

   Epidermis ______________________

   Nervous system ______________________

   Muscles ______________________

   Skeleton ______________________

   Digestive tract ______________________

   Hair ______________________

   Kidneys ______________________

Appendix Tables

Table I International atomic weights.

Element	Symbol	Atomic Number	Atomic Weight
Aluminum	Al	13	26.97
Antimony	Sb	51	121.76
Arsenic	As	33	74.91
Barium	Ba	56	137.36
Beryllium	Be	4	9.013
Bismuth	Bi	83	209.00
Boron	B	5	10.82
Bromine	Br	35	79.916
Cadmium	Cd	48	112.41
Calcium	Ca	20	40.08
Carbon	C	6	12.010
Chlorine	Cl	17	35.457
Chromium	Cr	24	52.01
Cobalt	Co	27	58.94
Copper	Cu	29	63.54
Fluorine	F	9	19.00
Gold	Au	79	197.2
Hydrogen	H	1	1.0080
Iodine	I	53	126.92
Iron	Fe	26	55.85
Lead	Pb	82	207.21
Magnesium	Mg	12	24.32
Manganese	Mn	25	54.93
Mercury	Hg	80	200.61
Nickel	Ni	28	58.69
Nitrogen	N	7	14.008
Oxygen	O	8	16.0000
Palladium	Pd	46	106.7
Phosphorus	P	15	30.98
Platinum	Pt	78	195.23
Potassium	K	19	39.096
Radium	Ra	88	226.05
Selenium	Se	34	78.96
Silicon	Si	14	28.06
Silver	Ag	47	107.880
Sodium	Na	11	22.997
Strontium	Sr	38	87.63
Sulfur	S	16	32.066
Tin	Sn	50	118.70
Titanium	Ti	22	47.90
Tungsten	W	74	183.92
Uranium	U	92	238.07
Vanadium	V	23	50.95
Zinc	Zn	30	65.38
Zirconium	Zr	40	91.22

Table II Four place logarithms.

N	0	1	2	3	4	5	6	7	8	9
10	0000	0043	0086	0128	0170	0212	0253	0294	0334	0374
11	0414	0453	0492	0531	0569	0607	0645	0682	0719	0755
12	0792	0828	0864	0899	0934	0969	1004	1038	1072	1106
13	1139	1173	1206	1239	1271	1303	1335	1367	1399	1430
14	1461	1492	1523	1553	1584	1614	1644	1673	1703	1732
15	1761	1790	1818	1847	1875	1903	1931	1959	1987	2014
16	2041	2068	2095	2122	2148	2175	2201	2227	2253	2279
17	2304	2330	2355	2380	2405	2430	2455	2480	2504	2529
18	2553	2577	2601	2625	2648	2672	2695	2718	2742	2765
19	2788	2810	2833	2856	2878	2900	2923	2945	2967	2989
20	3010	3032	3054	3075	3096	3118	3139	3160	3181	3201
21	3222	3243	3263	3284	3304	3324	3345	3365	3385	3404
22	3424	3444	3464	3483	3502	3522	3541	3560	3579	3598
23	3617	3636	3655	3674	3692	3711	3729	3747	3766	3784
24	3802	3820	3838	3856	3874	3892	3909	3927	3945	3962
25	3979	3997	4014	4031	4048	4065	4082	4099	4116	4133
26	4150	4166	4183	4200	4216	4232	4249	4265	4281	4298
27	4314	4330	4346	4362	4378	4393	4409	4425	4440	4456
28	4472	4487	4502	4518	4533	4548	4564	4579	4594	4609
29	4624	4639	4654	4669	4683	4698	4713	4728	4742	4757
30	4771	4786	4800	4814	4829	4843	4857	4871	4886	4900
31	4914	4928	4942	4955	4969	4983	4997	5011	5024	5038
32	5051	5065	5079	5092	5105	5119	5132	5145	5159	5172
33	5185	5198	5211	5224	5237	5250	5263	5276	5289	5302
34	5315	5328	5340	5353	5366	5378	5391	5403	5416	5428
35	5441	5453	5465	5478	5490	5502	5514	5527	5539	5551
36	5563	5575	5587	5599	5611	5623	5635	5647	5658	5670
37	5682	5694	5705	5717	5729	5740	5752	5763	5775	5786
38	5798	5809	5821	5832	5843	5855	5866	5877	5888	5899
39	5911	5922	5933	5944	5955	5966	5977	5988	5999	6010
40	6021	6031	6042	6053	6064	6075	6085	6096	6107	6117
41	6128	6138	6149	6160	6170	6180	6191	6201	6212	6222
42	6232	6243	6253	6263	6274	6284	6294	6304	6314	6325
43	6335	6345	6355	6365	6375	6385	6395	6405	6415	6425
44	6435	6444	6454	6464	6474	6484	6493	6503	6513	6522
45	6532	6542	6551	6561	6571	6580	6590	6599	6609	6618
46	6628	6637	6646	6656	6665	6675	6684	6693	6702	6712
47	6721	6730	6739	6749	6758	6767	6776	6785	6794	6803
48	6812	6821	6830	6839	6848	6857	6866	6875	6884	6893
49	6902	6911	6920	6928	6937	6946	6955	6964	6972	6981
50	6990	6998	7007	7016	7024	7033	7042	7050	7059	7067
51	7076	7084	7093	7101	7110	7118	7126	7135	7143	7152
52	7160	7168	7177	7185	7193	7202	7210	7218	7226	7235
53	7243	7251	7259	7267	7275	7284	7292	7300	7308	7316
54	7324	7332	7340	7348	7356	7364	7372	7380	7388	7396
N	**0**	**1**	**2**	**3**	**4**	**5**	**6**	**7**	**8**	**9**

Table II Four place logarithms—continued.

N	0	1	2	3	4	5	6	7	8	9
55	7404	7412	7419	7427	7435	7443	7451	7459	7466	7474
56	7482	7490	7497	7505	7513	7520	7528	7536	7543	7551
57	7559	7566	7574	7582	7589	7597	7604	7612	7619	7627
58	7634	7642	7649	7657	7664	7672	7679	7686	7694	7701
59	7709	7716	7723	7731	7738	7745	7752	7760	7767	7774
60	7782	7789	7796	7803	7810	7818	7825	7832	7839	7846
61	7853	7860	7868	7875	7882	7889	7896	7903	7910	7917
62	7924	7931	7938	7945	7952	7959	7966	7973	7980	7987
63	7993	8000	8007	8014	8021	8028	8035	8041	8048	8055
64	8062	8069	8075	8082	8089	8096	8102	8109	8116	8122
65	8129	8136	8142	8149	8156	8162	8169	8176	8182	8189
66	8195	8202	8209	8215	8222	8228	8235	8241	8248	8254
67	8261	8267	8274	8280	8287	8293	8299	8306	8312	8319
68	8325	8331	8338	8344	8351	8357	8363	8370	8376	8382
69	8388	8395	8401	8407	8414	8420	8426	8432	8439	8445
70	8451	8457	8463	8470	8476	8482	8488	8494	8500	8506
71	8513	8519	8525	8531	8537	8543	8549	8555	8561	8567
72	8573	8579	8585	8591	8597	8603	8609	8615	8621	8627
73	8633	8639	8645	8651	8657	8663	8669	8675	8681	8686
74	8692	8698	8704	8710	8716	8722	8727	8733	8739	8745
75	8751	8756	8762	8768	8774	8779	8785	8791	8797	8802
76	8808	8814	8820	8825	8831	8837	8842	8848	8854	8859
77	8865	8871	8876	8882	8887	8893	8899	8904	8910	8915
78	8921	8927	8932	8938	8943	8949	8954	8960	8965	8971
79	8976	8982	8987	8993	8998	9004	9009	9015	9020	9025
80	9031	9036	9042	9047	9053	9058	9063	9069	9074	9079
81	9085	9090	9096	9101	9106	9112	9117	9122	9128	9133
82	9138	9143	9149	9154	9159	9165	9170	9175	9180	9186
83	9191	9196	9201	9206	9212	9217	9222	9227	9232	9238
84	9243	9248	9253	9258	9263	9269	9274	9279	9284	9289
85	9294	9299	9304	9309	9315	9320	9325	9330	9335	9340
86	9345	9350	9355	9360	9365	9370	9375	9380	9385	9390
87	9395	9400	9405	9410	9415	9420	9425	9430	9435	9440
88	9445	9450	9455	9460	9465	9469	9474	9479	9484	9489
89	9494	9499	9504	9509	9513	9518	9523	9528	9533	9538
90	9542	9547	9552	9557	9562	9566	9571	9576	9581	9586
91	9590	9595	9600	9605	9609	9614	9619	9624	9628	9633
92	9638	9643	9647	9652	9657	9661	9666	9671	9675	9680
93	9685	9689	9694	9699	9703	9708	9713	9717	9722	9727
94	9731	9736	9741	9745	9750	9754	9759	9763	9768	9773
95	9777	9782	9786	9791	9795	9800	9805	9809	9814	9818
96	9823	9827	9832	9836	9841	9845	9850	9854	9859	9863
97	9868	9872	9877	9881	9886	9890	9894	9899	9903	9908
98	9912	9917	9921	9926	9930	9934	9939	9943	9948	9952
99	9956	9961	9965	9969	9974	9978	9983	9987	9991	9996
100	0000	0004	0009	0013	0017	0022	0026	0030	0035	0039
N	**0**	**1**	**2**	**3**	**4**	**5**	**6**	**7**	**8**	**9**

Table III Temperature conversion table—Centigrade to Fahrenheit.

°C	0	1	2	3	4	5	6	7	8	9
−50	**−58.0**	**−59.8**	**−61.6**	**−63.4**	**−65.2**	**−67.0**	**−68.8**	**−70.6**	**−72.4**	**−74.2**
−40	−40.0	−41.8	−43.6	−45.4	−47.2	−49.0	−50.8	−52.6	−54.4	−56.2
−30	−22.0	−23.8	−25.6	−27.4	−29.2	−31.0	−32.8	−34.6	−36.4	−38.2
−20	− 4.0	− 5.8	− 7.6	− 9.4	−11.2.	−13.0	−14.8	−16.6	−18.4	−20.2
−10	+14.0	+12.2	+10.4	+ 8.6	+ 6.8	+ 5.0	+ 3.2	+ 1.4	− 0.4	− 2.2
− 0	+32.0	+30.2	+28.4	+26.6	+24.8	+23.0	+21.2	+19.4	+17.6	+15.8
0	**32.0**	**33.8**	**35.6**	**37.4**	**39.2**	**41.0**	**42.8**	**44.6**	**46.4**	**48.2**
10	50.0	51.8	53.6	55.4	57.2	59.0	60.8	62.6	64.4	66.2
20	68.0	69.8	71.6	73.4	75.2	77.0	78.8	80.6	82.4	84.2
30	86.0	87.8	89.6	91.4	93.2	95.0	96.8	98.6	100.4	102.2
40	104.0	105.8	107.6	109.4	111.2	113.0	114.8	116.6	118.4	120.2
50	122.0	123.8	125.6	127.4	129.2	131.0	132.8	134.6	136.4	138.2
60	**140.0**	**141.8**	**143.6**	**145.4**	**147.2**	**149.0**	**150.8**	**152.6**	**154.4**	**156.2**
70	158.0	159.8	161.6	163.4	165.2	167.0	168.8	170.6	172.4	174.2
80	176.0	177.8	179.6	181.4	183.2	185.0	186.8	188.6	190.4	192.2
90	194.0	195.8	197.6	199.4	201.2	203.0	204.8	206.6	208.4	210.2
100	212.0	213.8	215.6	217.4	219.2	221.0	222.8	224.6	226.4	228.2
110	**230.0**	**231.8**	**233.6**	**235.4**	**237.2**	**239.0**	**240.8**	**242.6**	**244.4**	**246.2**
120	248.0	249.8	251.6	253.4	255.2	257.0	258.8	260.6	262.4	264.2
130	266.0	267.8	269.6	271.4	273.2	275.0	276.8	278.6	280.4	282.2
140	284.0	285.8	287.6	289.4	291.2	293.0	294.8	296.6	298.4	300.2
150	302.0	303.8	305.6	307.4	309.2	311.0	312.8	314.6	316.4	318.2
160	**320.0**	**321.8**	**323.6**	**325.4**	**327.2**	**329.0**	**330.8**	**332.6**	**334.4**	**336.2**
170	338.0	339.8	341.6	343.4	345.2	347.0	348.8	350.6	352.4	354.2
180	356.0	357.8	359.6	361.4	363.2	365.0	366.8	368.6	370.4	372.2
190	374.0	375.8	377.6	379.4	381.2	383.0	384.8	386.6	388.4	390.2
200	392.0	393.8	395.6	397.4	399.2	401.0	402.8	404.6	406.4	408.2
210	**410.0**	**411.8**	**413.6**	**415.4**	**417.2**	**419.0**	**420.8**	**422.6**	**424.4**	**426.2**
220	428.0	429.8	431.6	433.4	435.2	437.0	438.8	440.6	442.4	444.2
230	446.0	447.8	449.6	451.4	453.2	455.0	456.8	458.6	460.4	462.2
240	464.0	465.8	467.6	469.4	471.2	473.0	474.8	476.6	478.4	480.2
250	482.0	483.8	485.6	487.4	489.2	491.0	492.8	494.6	496.4	498.2

$°F = °C \times 9/5 + 32$ $°C = °F - 32 \times 5/9$

Table IV Conversion factors for temperature differentials (spirometry).

°C	°F	Conversion Factor
20	68.0	1.102
21	69.8	1.096
22	71.6	1.091
23	73.4	1.085
24	75.2	1.080
25	77.0	1.075
26	78.8	1.068
27	80.6	1.063
28	82.4	1.057
29	84.2	1.051
30	86.0	1.045
31	87.8	1.039
32	89.6	1.032
33	91.4	1.026
34	93.2	1.020
35	95.0	1.014
36	96.8	1.007
37	98.6	1.000

Table V Predicted vital capacities for females.

AGE	HEIGHT IN CENTIMETERS AND INCHES																		
CM	152	154	156	158	160	162	164	166	168	170	172	174	176	178	180	182	184	186	188
IN	59.8	60.6	61.4	62.2	63.0	63.7	64.6	65.4	66.1	66.9	67.7	68.5	69.3	70.1	70.9	71.7	72.4	73.2	74.0
16	3,070	3,110	3,150	3,190	3,230	3,270	3,310	3,350	3,390	3,430	3,470	3,510	3,550	3,590	3,630	3,670	3,715	3,755	3,800
17	3,055	3,095	3,135	3,175	3,215	3,255	3,295	3,335	3,375	3,415	3,455	3,495	3,535	3,575	3,615	3,655	3,695	3,740	3,780
18	3,040	3,080	3,120	3,160	3,200	3,240	3,280	3,320	3,360	3,400	3,440	3,480	3,520	3,560	3,600	3,640	3,680	3,720	3,760
20	3,010	3,050	3,090	3,130	3,170	3,210	3,250	3,290	3,330	3,370	3,410	3,450	3,490	3,525	3,565	3,605	3,645	3,695	3,720
22	2,980	3,020	3,060	3,095	3,135	3,175	3,215	3,255	3,290	3,330	3,370	3,410	3,450	3,490	3,530	3,570	3,610	3,650	3,685
24	2,950	2,985	3,025	3,065	3,100	3,140	3,180	3,220	3,260	3,300	3,335	3,375	3,415	3,455	3,490	3,530	3,570	3,610	3,650
26	2,920	2,960	3,000	3,035	3,070	3,110	3,150	3,190	3,230	3,265	3,300	3,340	3,380	3,420	3,455	3,495	3,530	3,570	3,610
28	2,890	2,930	2,965	3,000	3,040	3,070	3,115	3,155	3,190	3,230	3,270	3,305	3,345	3,380	3,420	3,460	3,495	3,535	3,570
30	2,860	2,895	2,935	2,970	3,010	3,045	3,085	3,120	3,160	3,195	3,235	3,270	3,310	3,345	3,385	3,420	3,460	3,495	3,535
32	2,825	2,865	2,900	2,940	2,975	3,015	3,050	3,090	3,125	3,160	3,200	3,235	3,275	3,310	3,350	3,385	3,425	3,460	3,495
34	2,795	2,835	2,870	2,910	2,945	2,980	3,020	3,055	3,090	3,130	3,165	3,200	3,240	3,275	3,310	3,350	3,385	3,425	3,460
36	2,765	2,805	2,840	2,875	2,910	2,950	2,985	3,020	3,060	3,095	3,130	3,165	3,205	3,240	3,275	3,310	3,350	3,385	3,420
38	2,735	2,770	2,810	2,845	2,880	2,915	2,950	2,990	3,025	3,060	3,095	3,130	3,170	3,205	3,240	3,275	3,310	3,350	3,385
40	2,705	2,740	2,775	2,810	2,850	2,885	2,920	2,955	2,990	3,025	3,060	3,095	3,135	3,170	3,205	3,240	3,275	3,310	3,345
42	2,675	2,710	2,745	2,780	2,815	2,850	2,885	2,920	2,955	2,990	3,025	3,060	3,100	3,135	3,170	3,205	3,240	3,275	3,310
44	2,645	2,680	2,715	2,750	2,785	2,820	2,855	2,890	2,925	2,960	2,995	3,030	3,060	3,095	3,130	3,165	3,200	3,235	3,270
46	2,615	2,650	2,685	2,715	2,750	2,785	2,820	2,855	2,890	2,925	2,960	2,995	3,030	3,060	3,095	3,130	3,165	3,200	3,235
48	2,585	2,620	2,650	2,685	2,715	2,750	2,785	2,820	2,855	2,890	2,925	2,960	2,995	3,030	3,060	3,095	3,130	3,160	3,195
50	2,555	2,590	2,625	2,655	2,690	2,720	2,755	2,785	2,820	2,855	2,890	2,925	2,955	2,990	3,025	3,060	3,090	3,125	3,155
52	2,525	2,555	2,590	2,625	2,655	2,690	2,720	2,755	2,790	2,820	2,855	2,890	2,925	2,955	2,990	3,020	3,055	3,090	3,125
54	2,495	2,530	2,560	2,590	2,625	2,655	2,690	2,720	2,755	2,790	2,820	2,855	2,885	2,920	2,950	2,985	3,020	3,050	3,085
56	2,460	2,495	2,525	2,560	2,590	2,625	2,655	2,690	2,720	2,755	2,790	2,820	2,855	2,885	2,920	2,950	2,980	3,015	3,045
58	2,430	2,460	2,495	2,525	2,560	2,590	2,625	2,655	2,690	2,720	2,750	2,785	2,815	2,850	2,880	2,920	2,945	2,975	3,010
60	2,400	2,430	2,460	2,495	2,525	2,560	2,590	2,625	2,655	2,685	2,720	2,750	2,780	2,810	2,845	2,875	2,915	2,940	2,970
62	2,370	2,405	2,435	2,465	2,495	2,525	2,560	2,590	2,620	2,655	2,685	2,715	2,745	2,775	2,810	2,840	2,870	2,900	2,935
64	2,340	2,370	2,400	2,430	2,465	2,495	2,525	2,555	2,585	2,620	2,650	2,680	2,710	2,740	2,770	2,805	2,835	2,865	2,895
66	2,310	2,340	2,370	2,400	2,430	2,460	2,495	2,525	2,555	2,585	2,615	2,645	2,675	2,705	2,735	2,765	2,800	2,825	2,860
68	2,280	2,310	2,340	2,370	2,400	2,430	2,460	2,490	2,520	2,550	2,580	2,610	2,640	2,670	2,700	2,730	2,760	2,795	2,820
70	2,250	2,280	2,310	2,340	2,370	2,400	2,425	2,455	2,485	2,515	2,545	2,575	2,605	2,635	2,665	2,695	2,725	2,755	2,780
72	2,220	2,250	2,280	2,310	2,335	2,365	2,395	2,425	2,455	2,480	2,510	2,540	2,570	2,600	2,630	2,660	2,685	2,715	2,745
74	2,190	2,220	2,245	2,275	2,305	2,335	2,360	2,390	2,420	2,450	2,475	2,505	2,535	2,565	2,590	2,620	2,650	2,680	2,710

From: Archives of Environmental Health
February 1966, Vol. 12, pp. 146–189
E. A. Gaensler, MD and G. W. Wright, MD

Table VI Predicted vital capacities for males.

AGE	HEIGHT IN CENTIMETERS AND INCHES																		
CM	152	154	156	158	160	162	164	166	168	170	172	174	176	178	180	182	184	186	188
IN	59.8	60.6	61.4	62.2	63.0	63.7	64.6	65.4	66.1	66.9	67.7	68.5	69.3	70.1	70.9	71.7	72.4	73.2	74.0
16	3,920	3,975	4,025	4,075	4,130	4,180	4,230	4,285	4,335	4,385	4,440	4,490	4,540	4,590	4,645	4,695	4,745	4,800	4,850
18	3,890	3,940	3,995	4,045	4,095	4,145	4,200	4,250	4,300	4,350	4,405	4,455	4,505	4,555	4,610	4,660	4,710	4,760	4,815
20	3,860	3,910	3,960	4,015	4,065	4,115	4,165	4,215	4,265	4,320	4,370	4,420	4,470	4,520	4,570	4,625	4,675	4,725	4,775
22	3,830	3,880	3,930	3,980	4,030	4,080	4,135	4,185	4,235	4,285	4,335	4,385	4,435	4,485	4,535	4,585	4,635	4,685	4,735
24	3,785	3,835	3,885	3,935	3,985	4,035	4,085	4,135	4,185	4,235	4,285	4,330	4,380	4,430	4,480	4,530	4,580	4,630	4,680
26	3,755	3,805	3,855	3,905	3,955	4,000	4,050	4,100	4,150	4,200	4,250	4,300	4,350	4,395	4,445	4,495	4,545	4,595	4,645
28	3,725	3,775	3,820	3,870	3,920	3,970	4,020	4,070	4,115	4,165	4,215	4,265	4,310	4,360	4,410	4,460	4,510	4,555	4,605
30	3,695	3,740	3,790	3,840	3,890	3,935	3,985	4,035	4,080	4,130	4,180	4,230	4,275	4,325	4,375	4,425	4,470	4,520	4,570
32	3,665	3,710	3,760	3,810	3,855	3,905	3,950	4,000	4,050	4,095	4,145	4,195	4,240	4,290	4,340	4,385	4,435	4,485	4,530
34	3,620	3,665	3,715	3,760	3,810	3,855	3,905	3,950	4,000	4,045	4,095	4,140	4,190	4,225	4,285	4,330	4,380	4,425	4,475
36	3,585	3,635	3,680	3,730	3,775	3,825	3,870	3,920	3,965	4,010	4,060	4,105	4,155	4,200	4,250	4,295	4,340	4,390	4,435
38	3,555	3,605	3,650	3,695	3,745	3,790	3,840	3,885	3,930	3,980	4,025	4,070	4,120	4,165	4,210	4,260	4,305	4,350	4,400
40	3,525	3,575	3,620	3,665	3,710	3,760	3,805	3,850	3,900	3,945	3,990	4,035	4,085	4,130	4,175	4,220	4,270	4,315	4,360
42	3,495	3,540	3,590	3,635	3,680	3,725	3,770	3,820	3,865	3,910	3,955	4,000	4,050	4,095	4,140	4,185	4,230	4,280	4,325
44	3,450	3,495	3,540	3,585	3,630	3,675	3,725	3,770	3,815	3,860	3,905	3,950	3,995	4,040	4,085	4,130	4,175	4,220	4,270
46	3,420	3,465	3,510	3,555	3,600	3,645	3,690	3,735	3,780	3,825	3,870	3,915	3,960	4,005	4,050	4,095	4,140	4,185	4,230
48	3,390	3,435	3,480	3,525	3,570	3,615	3,655	3,700	3,745	3,790	3,835	3,880	3,925	3,970	4,015	4,060	4,105	4,150	4,190
50	3,345	3,390	3,430	3,475	3,520	3,565	3,610	3,650	3,695	3,740	3,785	3,830	3,870	3,915	3,960	4,005	4,050	4,090	4,135
52	3,315	3,353	3,400	3,445	3,490	3,530	3,575	3,620	3,660	3,705	3,750	3,795	3,835	3,880	3,925	3,970	4,010	4,055	4,100
54	3,285	3,325	3,370	3,415	3,455	3,500	3,540	3,585	3,630	3,670	3,715	3,760	3,800	3,845	3,890	3,930	3,975	4,020	4,060
56	3,255	3,295	3,340	3,380	3,425	3,465	3,510	3,550	3,595	3,640	3,680	3,725	3,765	3,810	3,850	3,895	3,940	3,980	4,025
58	3,210	3,250	3,290	3,335	3,375	3,420	3,460	3,500	3,545	3,585	3,630	3,670	3,715	3,755	3,800	3,840	3,880	3,925	3,965
60	3,175	3,220	3,260	3,300	3,345	3,385	3,430	3,470	3,500	3,555	3,595	3,635	3,680	3,720	3,760	3,805	3,845	3,885	3,930
62	3,150	3,190	3,230	3,270	3,310	3,350	3,390	3,440	3,480	3,520	3,560	3,600	3,640	3,680	3,730	3,770	3,810	3,850	3,890
64	3,120	3,160	3,200	3,240	3,280	3,320	3,360	3,400	3,440	3,490	3,530	3,570	3,610	3,650	3,690	3,730	3,770	3,810	3,850
66	3,070	3,110	3,150	3,190	3,230	3,270	3,310	3,350	3,390	3,430	3,470	3,510	3,550	3,600	3,640	3,680	3,720	3,760	3,800
68	3,040	3,080	3,120	3,160	3,200	3,240	3,280	3,320	3,360	3,400	3,440	3,480	3,520	3,560	3,600	3,640	3,680	3,720	3,760
70	3,010	3,050	3,090	3,130	3,170	3,210	3,250	3,290	3,330	3,370	3,410	3,450	3,480	3,520	3,560	3,600	3,640	3,680	3,720
72	2,980	3,020	3,060	3,100	3,140	3,180	3,210	3,250	3,290	3,330	3,370	3,410	3,450	3,490	3,530	3,570	3,610	3,650	3,680
74	2,930	2,970	3,010	3,050	3,090	3,130	3,170	3,200	3,240	3,280	3,320	3,360	3,400	3,440	3,470	3,510	3,550	3,590	3,630

From: Archives of Environmental Health
February 1966, Vol. 12, pp. 146–189
E. A. Gaensler, MD and G. W. Wright, MD

Appendix B
Solutions and Reagents

Physiological Solutions

Working with freshly dissected or excised vertebrate tissues requires that they be perfused or immersed in an environment that approximates as nearly as possible the ionic, osmotic, and pH qualities of their own tissue fluids. Doing so prevents dysfunction and allows for maintenance and observation of the tissue over longer intervals. A partial list of physiological solutions follows.

Physiological Saline: A solution of sodium chloride in water that can allow for moisture maintenance and tonicity of most tissues for short periods. Consists of 0.9% for mammals and 0.7% for amphibians.

Frog Ringer's: An all-purpose solution for amphibian tissues (nerve, muscle, skin, etc.).

Turtle Ringer's: Replaces the tissue fluids of most reptiles.

Mammalian Ringer's: An all-purpose solution for general applications in mammalian tissue studies, both short and long term.

Locke's Solution: Devised for use primarily with isolated smooth muscle and cardiac muscle of mammals.

Tyrode's Solution: Designed for use with mammalian smooth muscle preparations.

Krebs-Henseleit Solution: May be used for work with mammalian nerves and for metabolic measurements of other mammalian tissues.

In this appendix, the individual recipes for some of the aforementioned solutions will be given in alphabetical order; however, where large quantities of solutions are required during the semester, it will be more convenient to make up stock solutions from which the various solutions can be quickly made. The longer shelf life of stock solutions enables one to be able to make up fresh physiological solutions daily as needed.

Stock Solutions

Make up five flasks in which the amounts listed are *grams per liter of distilled water.* The molarity of each solution is also given.

Sol'n	Compound	Amount	Molarity
A	NaCl	58.54 gm	1M
B	KCl	7.45 gm	0.1M
C	$CaCl_2$	11.1 gm	0.1M
D	NaH_2PO_4	12.0 gm	0.1M
E	$MgSO_4$	12.0 gm	0.1M

In addition to the above solutions, $NaHCO_3$ in dry form is used. Its instability in solution precludes using a stock solution of this ingredient. Glucose, in dry form, is also added for certain applications. *Refrigeration of all glucose solutions is necessary to inhibit bacterial action.*

Table I Physiological solutions from stock solutions.

Stock	Frog Ringer's	Turtle Ringer's	Mammalian Ringer's	Tyrode's	Locke's	Krebs-Henseleit
A NaCl	103 ml	117 ml	155 ml	137 ml	155 ml	111 ml
B KCl	30 ml	40 ml	65 ml	27 ml	56 ml	47 ml
C $CaCl_2$	20 ml	20 ml	22 ml	18 ml	22 ml	25 ml
$NaHCO_3$ (dry)	.2 gm	.2 gm	.2 gm	.1 gm	.2 gm	.2 gm
D NaH_2PO_4	---	---	8.3 ml	2.5 ml	---	---
E $MgSO_4$	---	---	---	---	---	12.5 ml
*Glucose (dry)	1.0 gm	1.0 gm	1.0 gm	1.0 gm	1.0 gm	1.0 gm

*Addition of glucose is optional. Keep refrigerated. Lasts only 1 week in refrigerator.

Dilutions

To prepare a liter of any of the physiological solutions, fill a volumetric flask half full of distilled water and add the amounts given in Table I. Each compound is added in sequence and mixed thoroughly before adding succeeding compounds. After all ingredients are added, the flask is filled to the one liter mark.

Caution If excess dissolved CO_2 is present in the water, it will tend to precipitate out the calcium in the form of $CaCO_3$. This may be prevented by (1) pre-aerating the water with an oil-free oxygen or air line for 5–10 minutes or (2) quickly adding solution C ($CaCl_2$) with rapid agitation at the very last. If a precipitate does occur, the solution is not usable.

Other Solutions and Reagents

BAPNA (Ex. 83)

BAPNA is benzoyl DL arginine p-nitroaniline HCl. To prepare this substrate for Ex. 83, dissolve 43 mg of BAPNA in 1 ml of dimethyl sulfoxide (caution) and make up to a volume of 100 ml with TRIS buffer solution.

Barfoed's Reagent

Dissolve 12 gm copper acetate in 450 ml of boiling distilled water. Do not filter if precipitate forms. To this hot solution, quickly add 13 ml of 8.5% lactic acid. Most of precipitate will dissolve. Cool the mixture and dilute to 500 ml. Remove any final precipitate by filtration.

Benedict's Solution (Qualitative)

Sodium citrate	173.0 gm
Sodium carbonate, anhydrous	100.0 gm
Copper sulfate, pure crystalline CuSO4 • 5H2O	17.3 gm

Dissolve the sodium citrate and sodium carbonate in 700 ml distilled water with aid of heat and filter.

Then, dissolve the copper sulfate in 100 ml of distilled water with aid of heat and pour this solution slowly into the first solution, stirring constantly. Make up to 1 liter volume with distilled water.

Buffer Solution, pH 5 (Ex. 84)

Make up 1 liter each of the following solutions:

M/10 potassium acid phthalate
20.418 gm $KHC_8H_4O_4$ to distilled water to make one liter.

M/10 sodium hydroxide
4.0 gm sodium hydroxide to distilled water to make one liter.

Use 50 ml of the first solution and 23.9 ml of the second solution to make 73.9 ml of pH 5 buffer solution.

Buffer Solution, pH 7 (Ex. 84)

Make up 1 liter each of the following solutions:

M/15 potassium acid phosphate
9.08 gm KH_2PO_4 to distilled water to make one liter.

M/15 disodium phosphate
9.47 gm Na_2HPO_4 (anhydrous) to distilled water to make one liter.

Use 38.9 ml of the first solution and 61.1 ml of the second solution to make 100 ml of pH 7 buffer solution.

Buffer Solution, pH 9 (Ex. 84)

Make up 1 liter each of the following solutions:

.2M boric acid
5.96 gm H_3BO_3 to distilled water to make one liter.

.2M potassium chloride
14.89 gm KCl to distilled water to make one liter.

.2M sodium hydroxide
8 gm NaOH to distilled water to make one liter.

Use 50 ml each of the first two solutions, 21.5 ml of the third solution and dilute to 200 ml of distilled water.

EDTA (Ex. 31)

Ethylenediamine tetracetate is a chelating agent which binds with calcium ions, removing them from the solution. For this experiment, a 2% solution is used.

Exton's Reagent (for albumin test)

Sodium sulfate (anhydrous)	88 gm
Sulfosalicylic acid	50 gm
Distilled water to make one liter.	

Dissolve the sodium sulfate in 80 ml of water with heat. Cool, add the sulfosalicylic acid, and make up to volume with water.

Frog Ringer's Solution

This solution is used with most amphibian muscle and nerve preparations. Frog skin and toad bladder preparations can also be handled with this saline solution.

NaCl	6.02 gm
KCl	0.22 gm
$CaCl_2$	0.22 gm
$NaHCO_3$	2.00 gm
Glucose	1.00 gm

Distilled water to make 1 liter.

Iodine (IKI) Solution

Potassium iodide	20 gm
Iodine crystals	4 gm
Distilled water	1 liter

Dissolve the potassium iodide in 1 liter of distilled water and add the iodine crystals, stirring to dissolve. Store in dark bottles.

Locke's Solution

NaCl	9.06 gm
KCl	0.42 gm
$CaCl_2$	0.24 gm
$NaHCO_3$	2.00 gm
Glucose	1.00 gm

Distilled water to make 1 liter.

Mammalian Ringer's Solution

NaCl	9.00 gm
KCl	0.48 gm
$CaCl_2$	0.24 gm
$NaHCO_3$	0.20 gm
NaH_2PO_4	0.10 gm
Glucose	1.00 gm

Distilled water to make one liter.

Methylene Blue (Loeffler's)

Solution A: Dissolve 0.3 gm of methylene blue (90% dye content) in 30.0 ml ethyl alcohol (95%).

Solution B: Dissolve 0.01 gm potassium hydroxide in 100.0 ml distilled water. Mix solutions A and B.

Potassium Chloride, .5M

(Sea Urchin Spawning)

37.22 gm KCl to distilled water to make one liter.

Red Blood Cell (RBC) Diluting Fluid (Hayem's)

Mercuric chloride	1.0 gm
Sodium sulfate (anhydrous)	4.4 gm
Sodium chloride	2.0 gm
Distilled water	400.0 ml

Rothera's Reagent

The addition of 1 gm of this reagent to 5 ml of urine with ammonium hydroxide is used to detect ketones.

Sodium nitroprusside	7.5 gm
Ammonium sulfate	700.0 gm

Mix and pulverize in mortar with pestle.

Starch Solution (1%)

Add 10 grams of cornstarch to 1 liter of distilled water. Bring to boil, cool, and filter. Keep refrigerated.

TRIS, pH 8.2 (Ex. 83)

The molecular weight of 2-amino-2(hydroxymethyl)-1,3-propanediol, or TRIS, is 121.14. Its formula is $(HOCH_2)_3CNH_2$. To make up this solution for Ex. 83, make up a .05M solution first by dissolving 6.066 gms in a liter of distilled water. To 500 ml of the 0.5M TRIS buffer, add 130 ml of .05N HCl.

Tyrode's Solution

This solution is used with various smooth muscle preparations of mammals.

NaCl	8.00 gm
KCl	0.20 gm
$CaCl_2$	0.20 gm
$NaHCO_3$	1.00 gm
NaH_2PO_4	0.03 gm

Distilled water to make one liter.

White Blood Cell (WBC) Diluting Fluid

Hydrochloric acid	5 ml
Distilled water	495 ml

Add 2 small crystals of thymol as a preservative.

Wright's Stain

In most instances, it is best to purchase this stain already prepared from biological supply houses. If the powder is available, however, it can be prepared as follows:

Wright's stain powder	0.3 gm
Glycerine	3.0 cc
Methyl alcohol (acetone-free)	100.0 cc

Combine the powder with glycerine by grinding with mortar and pestle. Add the alcohol gradually to bring to solution. Store in brown bottle for one week and filter before using. During first week, agitate bottle occasionally.

Appendix C
Tests and Methods

Carbohydrate Differentiation

Several experiments in this book require that sugars and starches be differentiated. An understanding of the application of these tests can be derived from the separation outline at the bottom of this page.

Barfoed's Test To 5 ml of reagent in test tube, add 1 ml of unknown. Place in boiling water bath for 5 minutes. Interpretation is based on separation outline.

Benedict's Test This is a semiquantitative test for amounts of reducing sugar present. To 5 ml of reagent in test tube, add 8 drops of unknown. Place in boiling water bath for 5 minutes. A green, yellow, or orange-red precipitate determines the amount of reducing sugar present. See Ex. 86.

Electronic Equipment Adjustments

Lengthy instructions pertaining to recorder and transducer manipulations that are used in several experiments are outlined here for reference.

Balancing Transducer on Unigraph (Ex. 29 and 87)

The Unigraph is designed to function in a sensitivity range of .1 MV/CM to 2 MV/CM. When an experiment is begun, the sensitivity control is set at 2 MV/CM, its least sensitive position. As the experiment progresses it may become necessary to shift to a more sensitive setting, such as 1 MV/CM or .5 MV/CM. If the Wheatstone bridge in the transducer is not balanced, the shifting from one sensitivity position to another will cause the stylus to change position. This produces an unsatisfactory record. Thus, it is essential that one go through the following steps to balance the transducer.

1. Connect the transducer jack to the transducer socket in the end of the Unigraph. The small upper socket is the one to use.
2. Before plugging in the Unigraph, make sure that the power switch is off, the chart control switch is on STBY, the gain selector knob is on 2 MV/CM, the sensitivity knob is turned

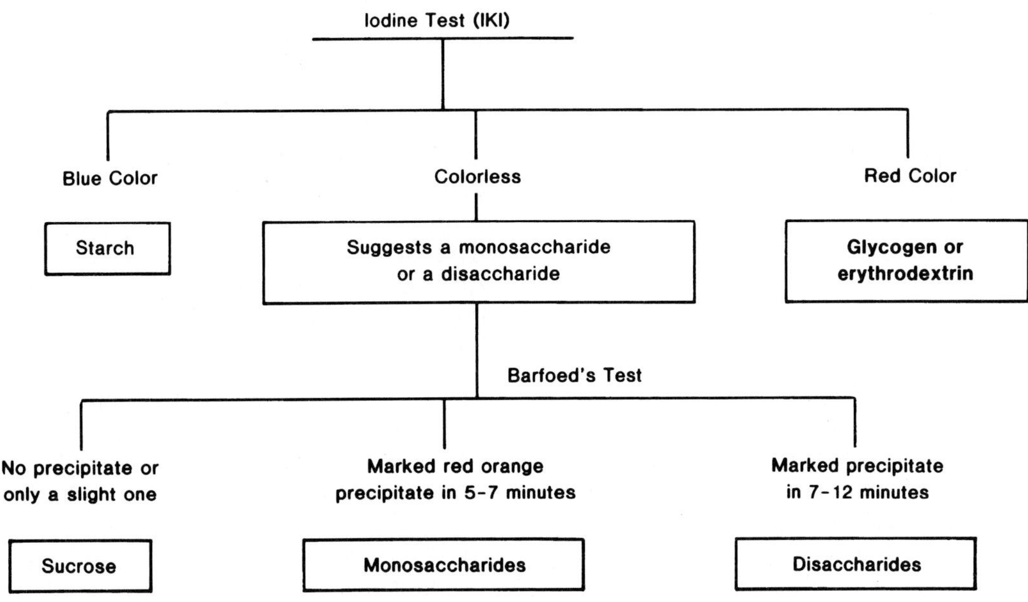

completely counterclockwise, the mode selector control is on TRANS, and the Hi Filter and Mean switches are positioned toward Normal. Now plug in the Unigraph.

3. Turn on the Unigraph power switch and set the heat control knob at the two o'clock position.
4. Place the speed control lever at the slow position and the c.c. lever at Chart On.
5. As the chart moves along, bring the stylus to the center of the paper by turning the centering knob.
6. Turn the sensitivity control completely clockwise to maximum sensitivity. If the bridge is unbalanced, the stylus will move away from the center. To return the stylus to the center, unlock the TRANS-BAL control by pushing the small lever on the control and turning the knob in either direction to return the stylus to the center of the chart.
7. Set the gain control to 1. If the bridge is unbalanced, the stylus will not be centered. Center with the TRANS-BAL control.
8. Set the gain control to .5 and re-center the stylus with the TRANS-BAL control. Repeat this same procedure for .2 and .1 settings of the gain control. The bridge is now completely balanced at its highest gain and sensitivity. Lock the TRANS-BAL control with the lock lever.

   Note: Although we have gone through this lengthy process to achieve a balanced transducer, we may find that we will still get baseline shifts when going from one setting to another. However, they will be minor compared to one that has not been through this procedure. Any minor deviations should be corrected with the centering control.
9. Return the gain control to 2 and the chart control switch to STBY.

Calibration of Unigraph with Strain Gauge (Ex. 29 and 82)

When the transducer has been balanced, it is necessary to calibrate the tracing on the Unigraph to a known force on the transducer. Calibration establishes a linear relationship between the magnitude of deflection and the degree of strain (extent of bending) of the transducer leaf before the muscle is attached to the transducer. Calibrations must be made for each gain at maximum sensitivity that one expects to use in the experiment (your muscle preparations will probably require gain settings of 2, 1, or .5). After calibration, it is possible to directly determine the force of each contraction by simply reading the maximum height of the tracing.

Materials:
small paper clip weighing around 0.5 gram

1. Weigh a convenient object, such as a small paper clip, to the nearest 0.1 mg.
2. With the chart control switch on STBY, suspend the paper clip on to the two smallest leaves of the transducer.
3. Put the c.c. switch to Chart On and note the extent of deflection on the chart. Record about 1 centimeter on the chart and place the c.c. switch on STBY. Label this deflection in mm/mg.
4. Set the gain control at 1, c.c. to Chart On, and record for another centimeter. Place the c.c. switch on STBY. Label the deflection.
5. Repeat the above procedure for the other two gain settings (.5, .2, and .1). Note how the stylus vibrates as you increase the sensitivity of the Unigraph. This is normal.
6. Return the c.c. switch to STBY, gain to 2, and remove the paper clip. The Unigraph is now calibrated and the muscle can now be attached to it.

Calibration for EMG on Unigraph (Ex. 32)

Calibration of the Unigraph for EMG measurements is necessary to establish that one centimeter deflection of the stylus equals 0.1 millivolt. When calibration is accomplished, the recording is made with the EEG mode since there is no EMG mode. An EMG can be made very well in this mode at lower sensitivities than would be used when making an EEG. Proceed as follows to calibrate:

1. Turn the mode selector control to CC/Cal. (Capacitor Coupled Calibrate.)
2. Turn the small main switch to ON.
3. Turn the stylus heat control to the two o'clock position.
4. Check the speed selector lever to see that it is in the slow position. The free end of the lever should be pointed toward the styluses.
5. Move the chart-stylus switch to Chart On and observe the width of the tracing that appears on

the paper. Adjust the stylus heat control to produce the desired width of tracing.

6. Set the gain knob to .1 MV/CM.
7. Push down the .1 MV button to determine the amount of deflection. Hold the button down for 2 seconds before releasing. If it is released too quickly the tracing will not be perfect. Adjust the deflection so that it deflects exactly 1 centimeter in each direction by turning the sensitivity knob. The unit is now calibrated so that you get 1 cm deflection for 0.1 millivolt.
8. Reset the gain knob to .5 MV/CM. This reduces the sensitivity of the instrument.
9. Return the mode selector control to EEG.
10. Return the chart-stylus switch to STBY. Place the speed selector in the fast position. The instrument is now ready for use.

Balancing the Myograph to Physiograph® Transducer Coupler

(Ex. 29)

Balancing (matching) of the transducer signal with the amplifier is necessary to get the recording pen to stay on a preset baseline when the RECORD button is in either the OFF or ON position. Proceed as follows:

1. Before starting, make sure that the RECORD switch is OFF and that the myograph jack is inserted into the transducer coupler.
2. Set the sensitivity control (outer knob) to its lowest numbered setting (highest sensitivity).
3. Set the paper speed at 0.5 cm/sec and lower the pen lifter.
4. Adjust the pen position so that the pen is writing *exactly* on the center line.
5. Place the RECORD switch in the ON position. The pen will probably be moving a large distance either up or down.
6. With the BALANCE control, adjust the pen so that it is again writing *exactly* on the center line.
7. Check your match, or balance, by placing the RECORD button in the OFF position. **Your system is balanced if the pen remains on the center line.**
8. If the pen does not remain on the center line, relocate the pen to the center line again with the

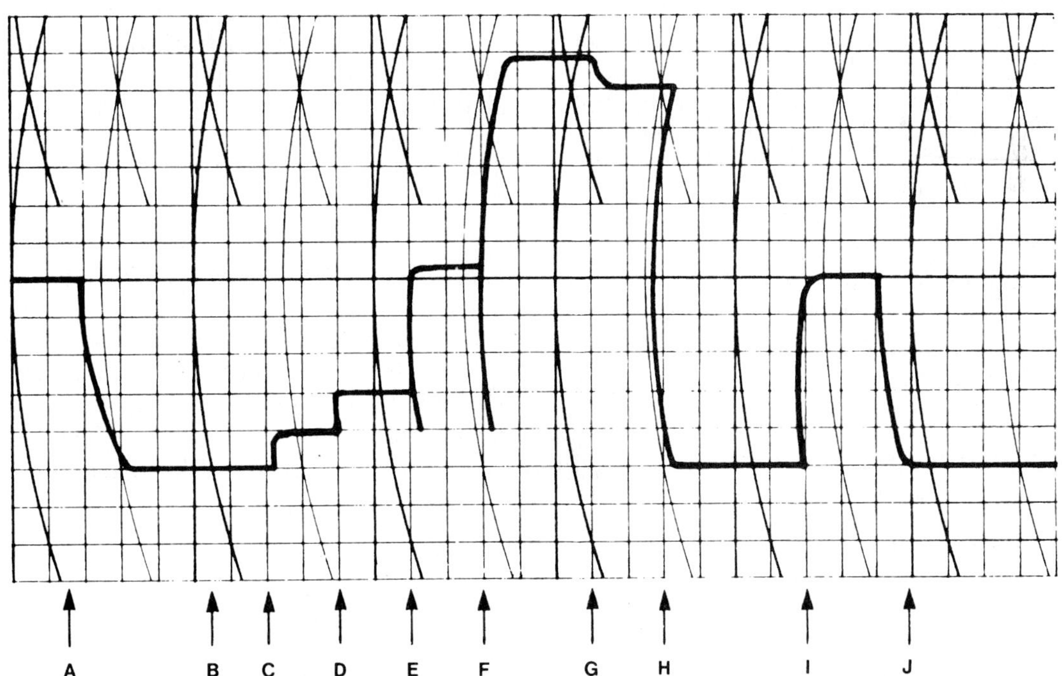

A. Baseline set 2.5 cm below center line. B. Record button turned on. C. 100 gram weight applied. D., E., F. Outer sensitivity knob rotated to lower numbers (higher sensitivity). G. Center sensitivity knob rotated counterclockwise to attain exactly 5 cm deflection. H. Weight removed. I., J. 50 gram weight applied and removed.

Figure C-1 Sample record of calibration at 100 grams per 5 centimeters

position control, place the RECORD button in the ON position, and repeat the balancing procedure. You have a balanced system *only* when the pen remains on a preset baseline with the RECORD button in both OFF and ON positions.

Calibration of Myograph Transducer on Physiograph® (Ex. 29)

Calibration of the transducer is necessary so that we know exactly how much tension, in grams, is being exerted by the muscle during contraction. The channel amplifier must be balanced first. Our calibration here will be to get 5 cm pen deflection with 100 grams. *These instructions apply to couplers that lack built-in calibration buttons.* Figure C-1 on the previous page illustrates the various steps.

Materials:
100 gram weight
50 gram weight

1. Start the paper drive at 0.1 cm/sec and lower the pen lifter so that we are recording on the desired channel.
2. Check to see that the channel amplifier is balanced.
3. With the pen position control, set the baseline exactly 2.5 cm (5 blocks) below the channel center line.
4. Rotate the outer knob of the sensitivity control fully counterclockwise to the lowest sensitivity level (i.e., 1000 setting). Be sure that the inner knob of the sensitivity control is in its fully clockwise "clicked" position.
5. Place the RECORD button in the ON position.
6. Suspend a 100 gram weight to the actuator of the myograph. Note that the addition of the weight has caused the baseline to move upward approximately one block (0.5 cm). See step C, Figure C–1.
7. Rotate the *outer* knob of the sensitivity control clockwise until the pen exceeds 5 cm (10 blocks) of deflection from the original baseline. See step F, Figure C–1.
8. Rotate the *inner* knob of the sensitivity control counterclockwise to bring the pen back downward until it is *exactly* 5 cm (10 blocks) of deflection from the original baseline.
9. Remove the weight. The pen should return to the original baseline. If it does not return exactly to the baseline, reset the baseline with the position control, reapply the weight, and rotate the inner knob until the pen is writing exactly 5 cm above the original baseline.
10. Attach a 50 gram weight to the myograph to see if you get 5 blocks of pen deflection, as shown in the last deflection in Figure C–1. The unit is now properly calibrated.

Calibration of Myograph Transducer with Calibration Button (Ex. 29)

When the coupler has a calibration button, the use of a 100 gram weight is not necessary. To calibrate, all that is necessary is to press the 100 gram button, and follow the steps above to produce the 5 cm pen deflection.

Appendix Physiograph® Sample Records

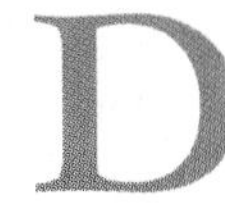

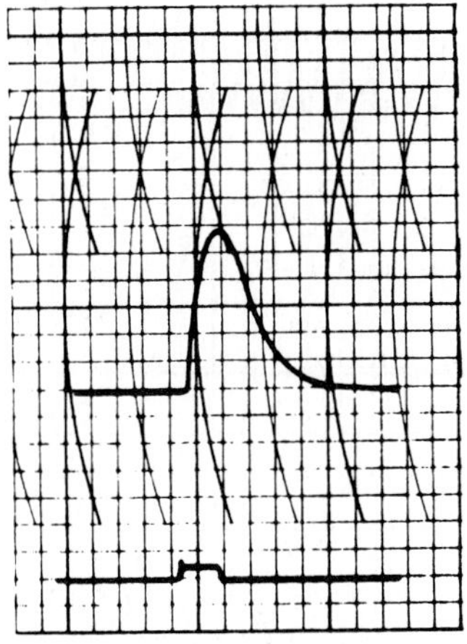

A. Isotonic Muscle Twitch

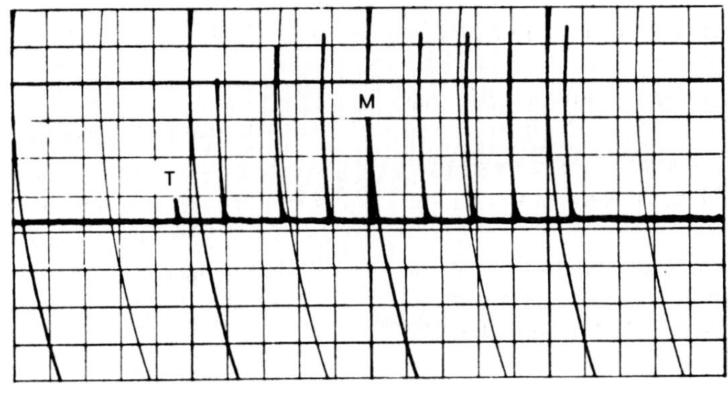

B. Threshold (T) and Maximal Response (M)

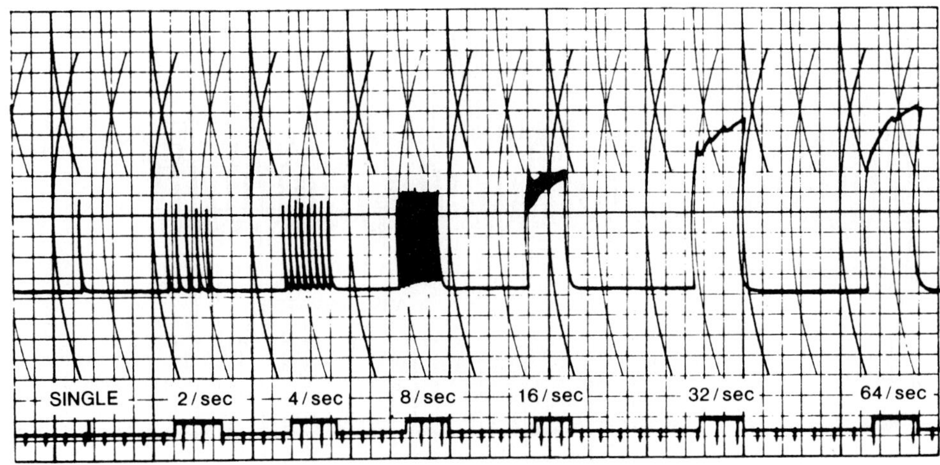

C. Effect of Increasing Frequency with Voltage Constant

Figure SR-1 Physiograph® sample records of frog muscle contraction (Ex. 29).

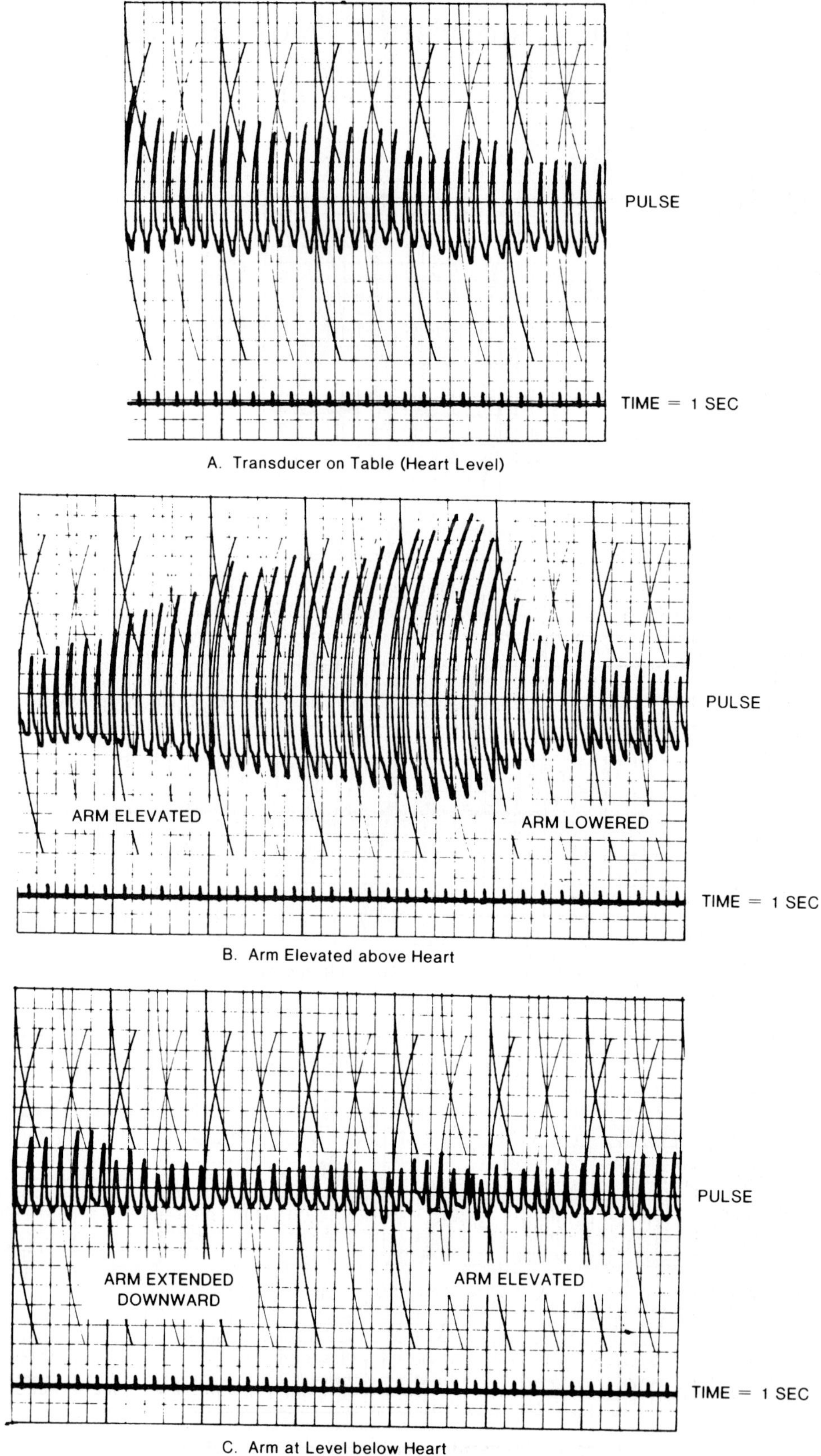

Figure SR-2 Effects of hand position on pulse (Ex. 64).

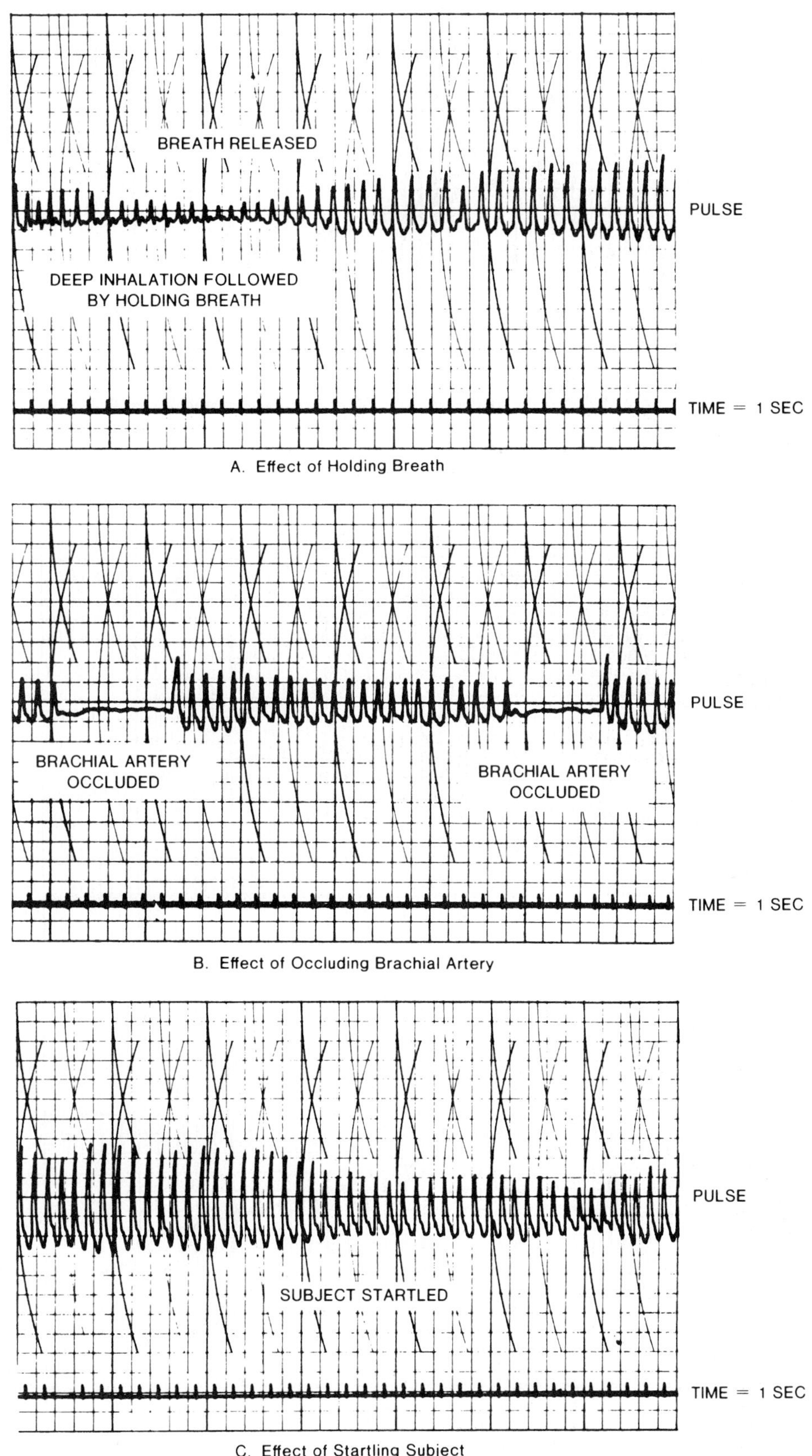

Figure SR-3 Pulse variations due to various factors (Ex. 64).

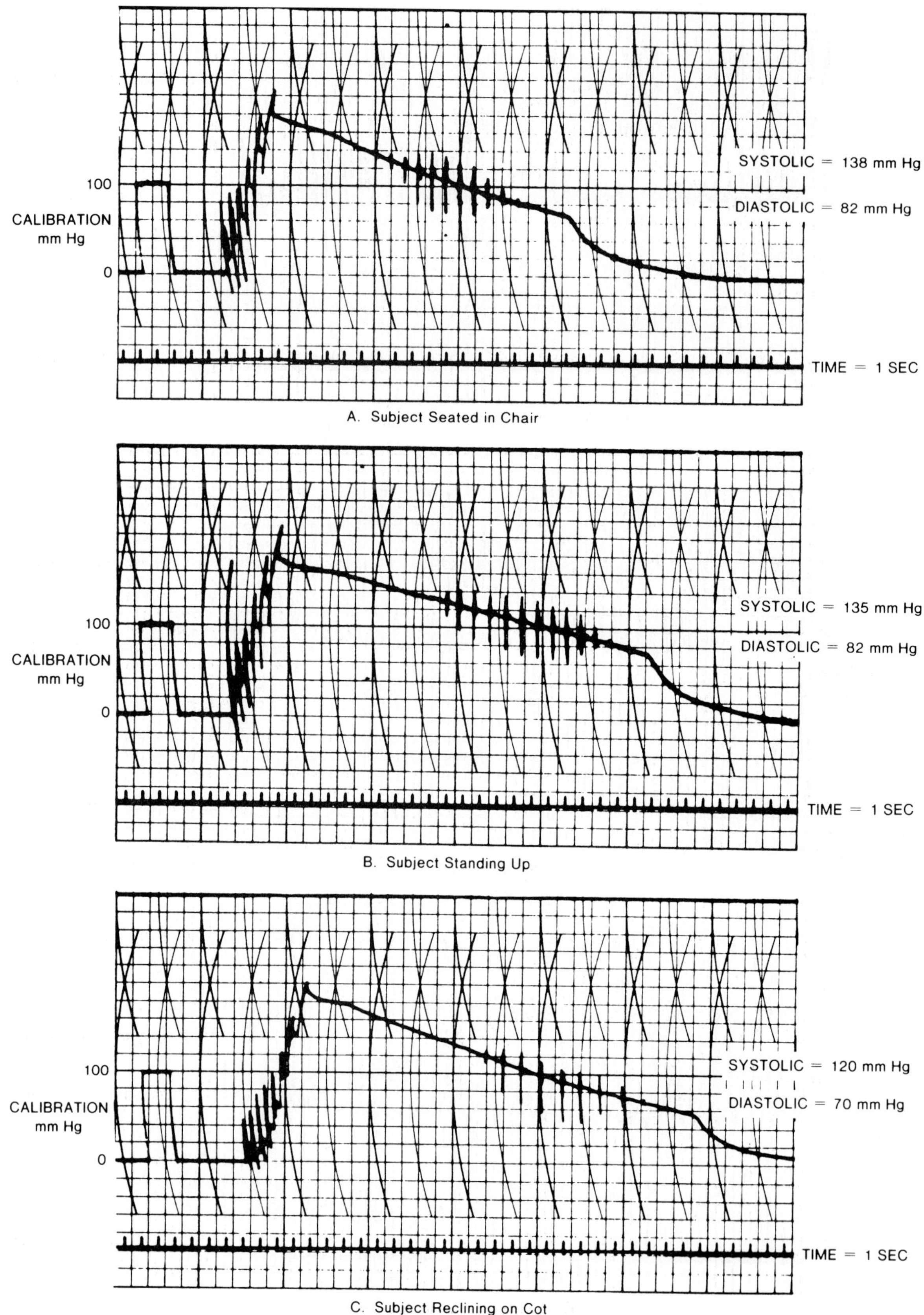

Figure SR-4 Effects of posture on blood pressure (Ex. 65).

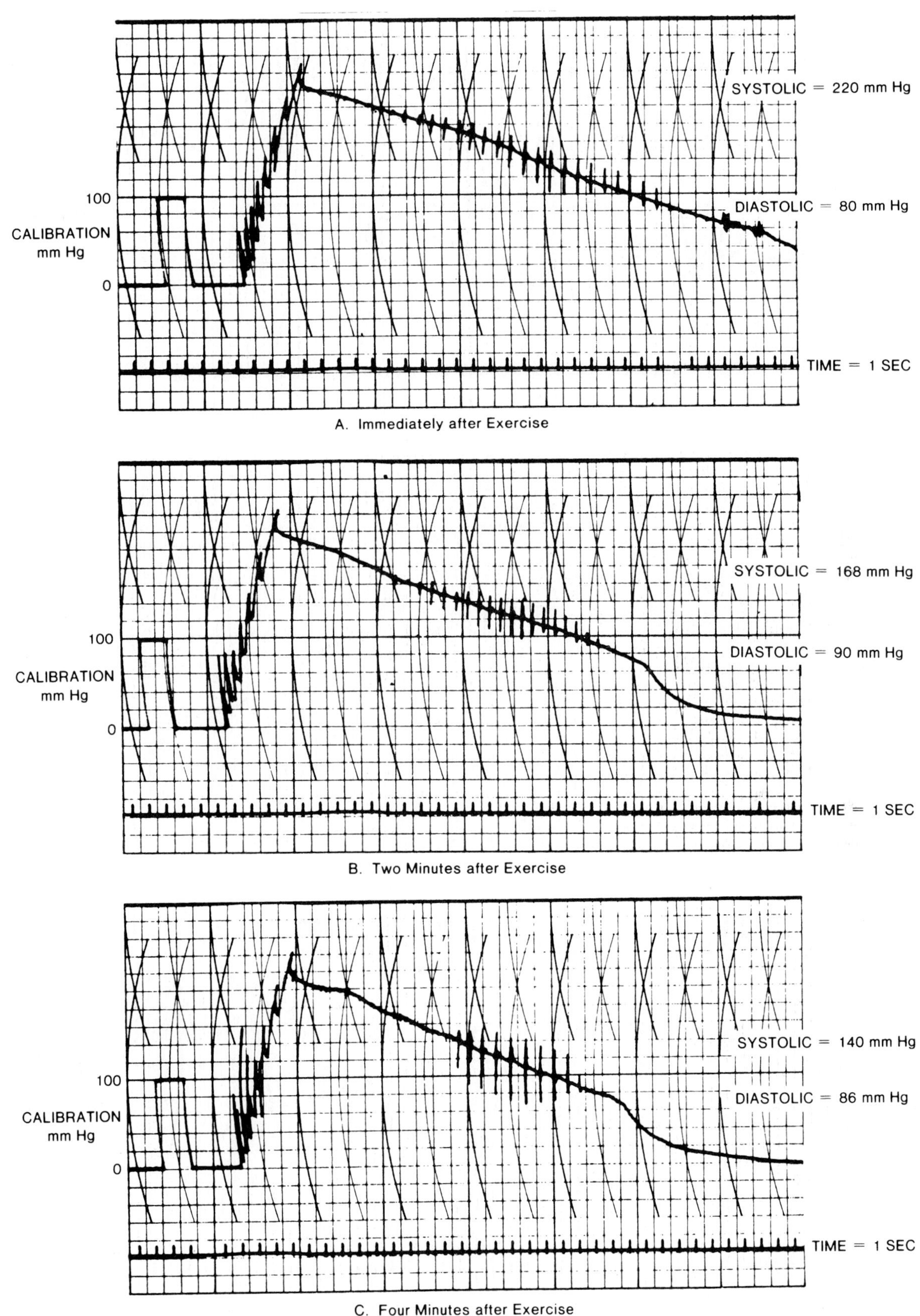

Figure SR-5 Sample records of blood pressure monitoring after exercise (Ex. 65).

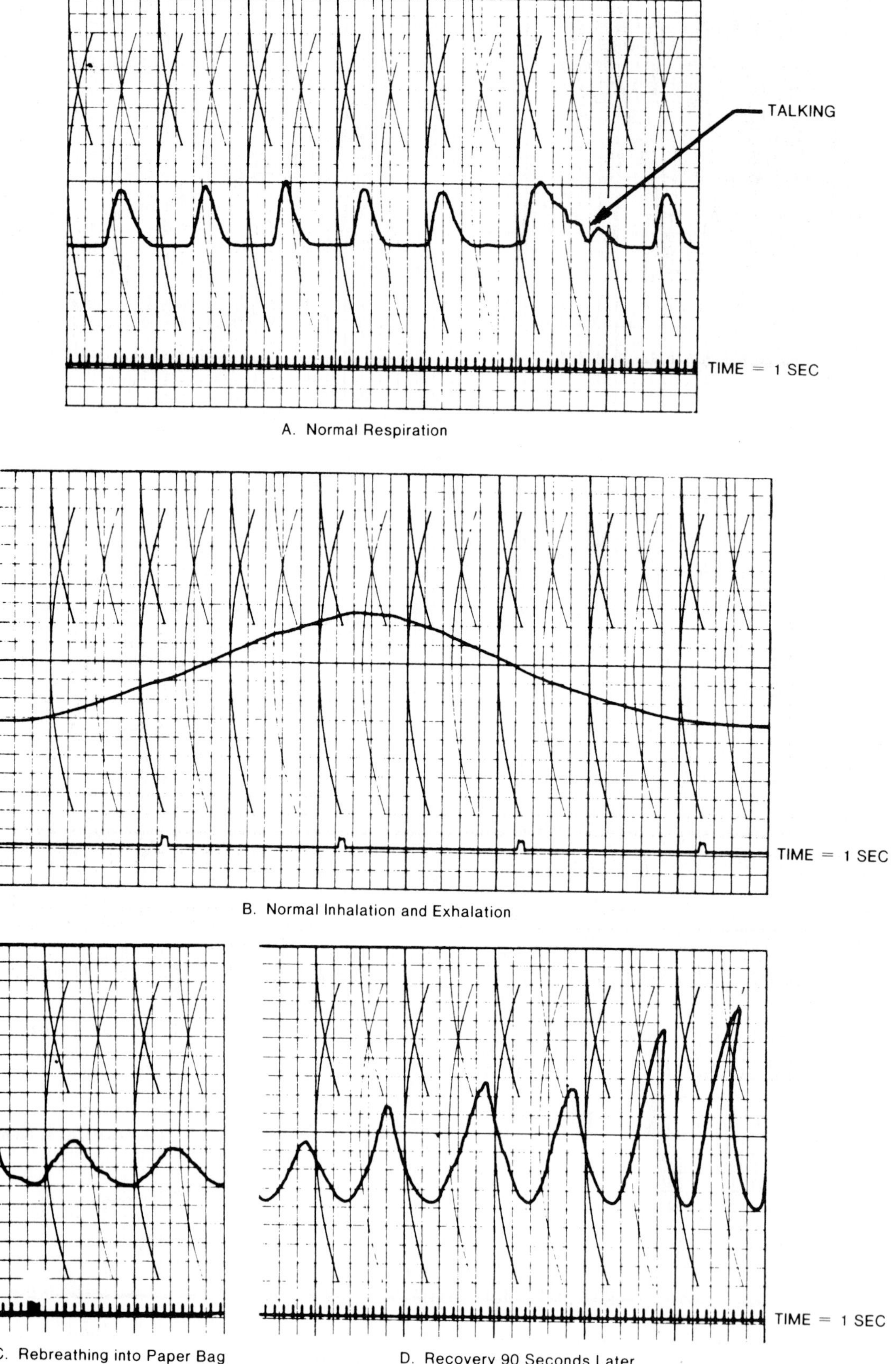

Figure SR-6 Pneumograph recording records (Ex. 72).

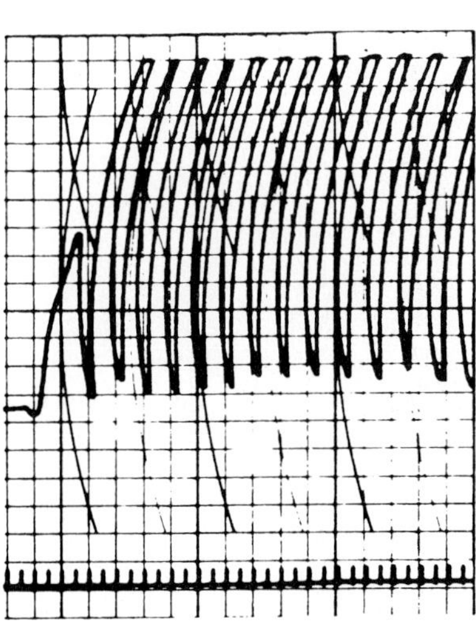

A. Hyperventilation Begun

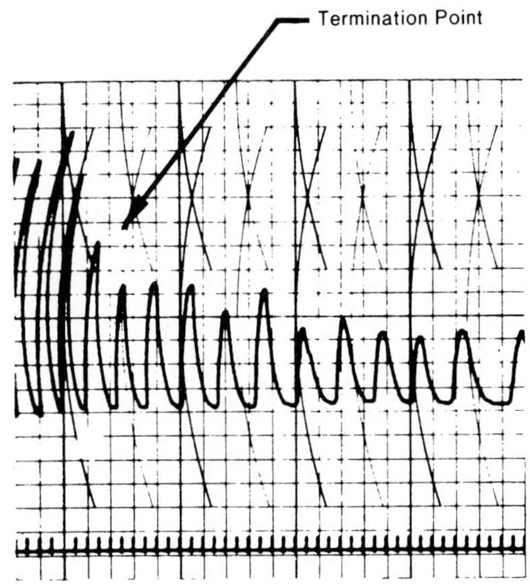

B. Hyperventilation Terminated

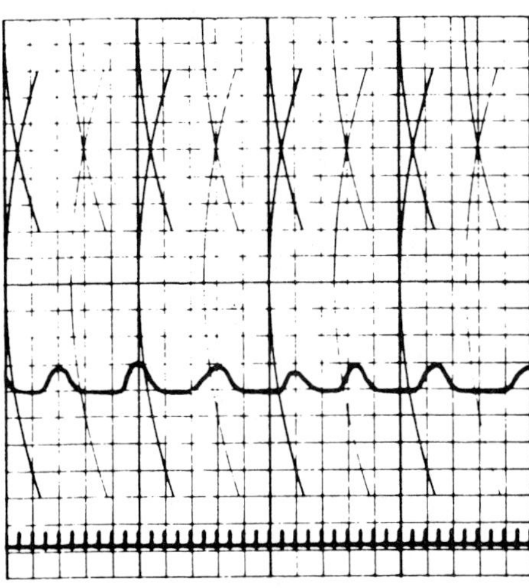

C. Recovery One Minute Later

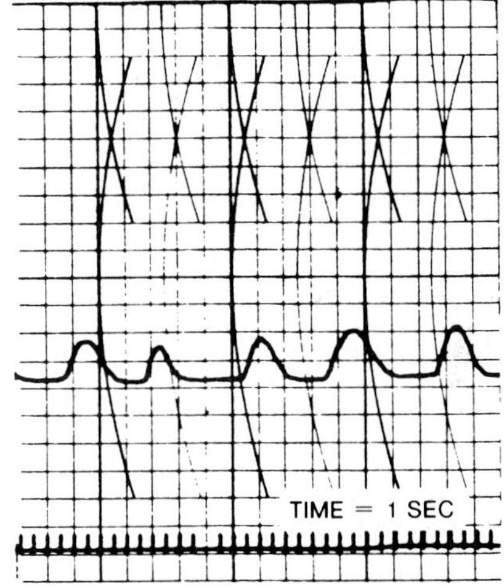

D. Recovery Three Minutes Later

Figure SR-7 Hyperventilation sample record (Ex. 72).

Appendix Microorganisms

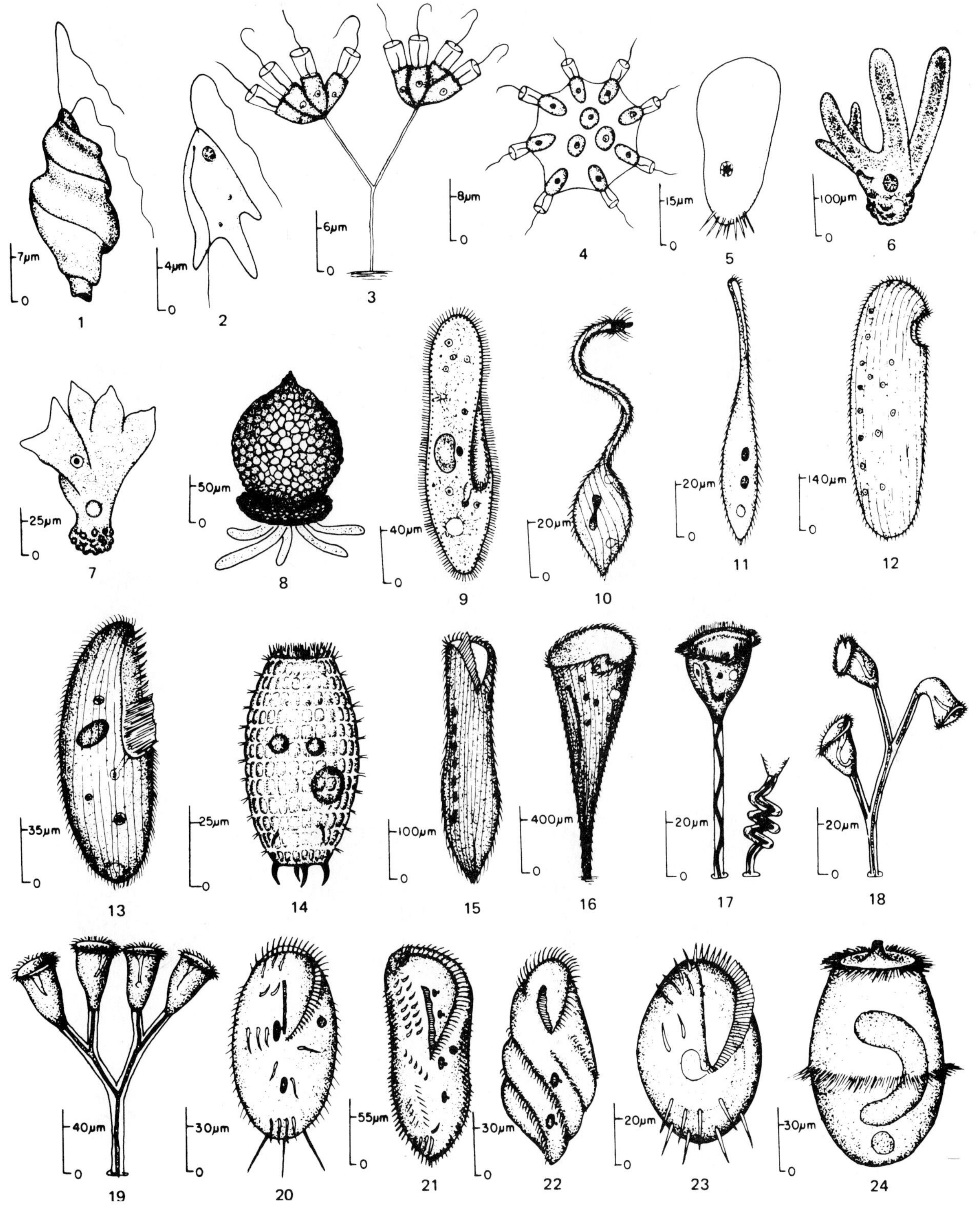

1. *Heteronema*
2. *Cercomonas*
3. *Codosiga*
4. *Protospongia*
5. *Trichamoeba*
6. *Amoeba*
7. *Mayorella*
8. *Diffugia*
9. *Paramecium*
10. *Lacrymaria*
11. *Lionotus*
12. *Loxodes*
13. *Blepharisma*
14. *Coleps*
15. *Condylostoma*
16. *Stentor*
17. *Vorticella*
18. *Carchesium*
19. *Zoothamnium*
20. *Stylonychia*
21. *Onycnodromos*
22. *Hypotrichidium*
23. *Euplotes*
24. *Didinium*

Figure MO-1 Protozoans.

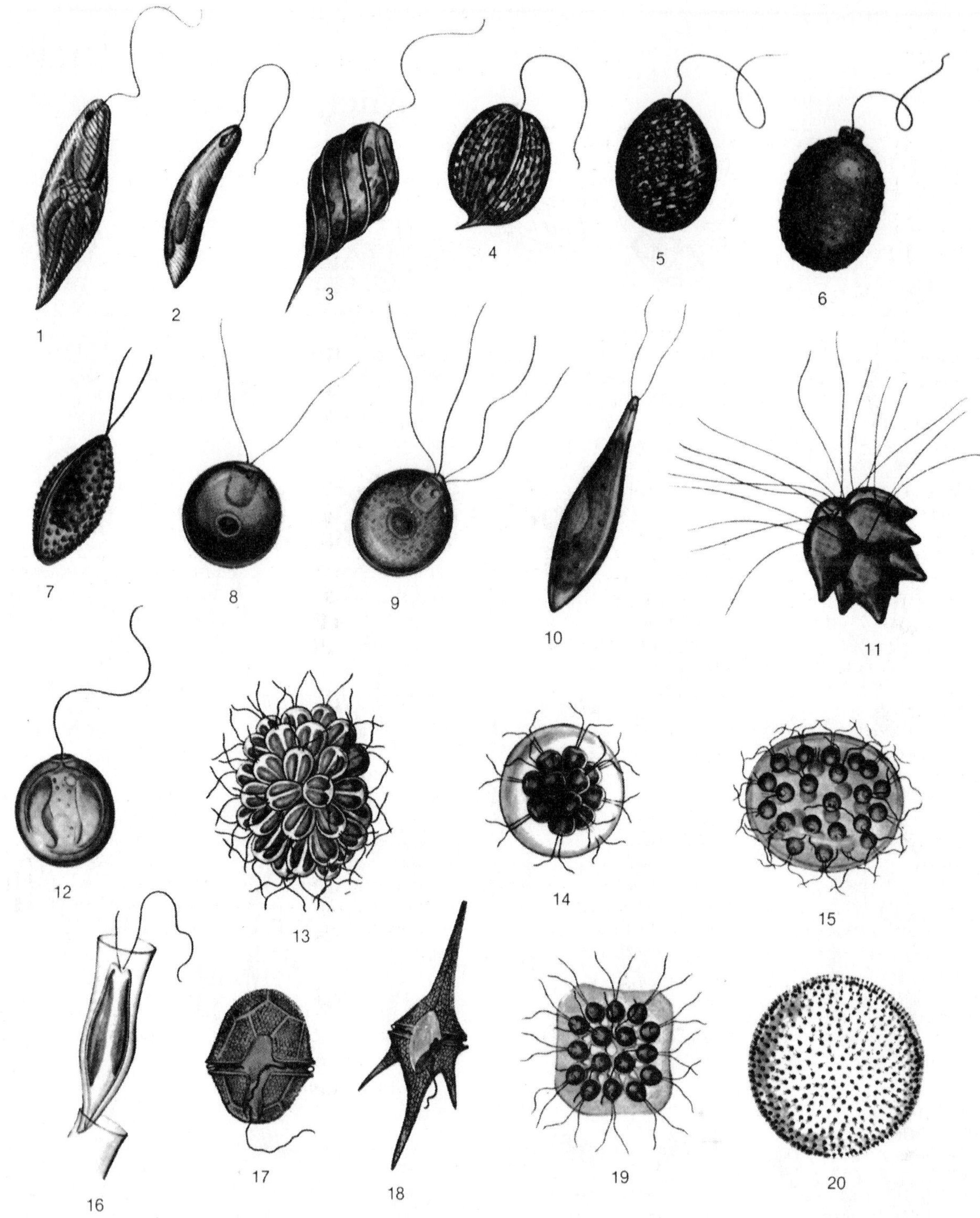

Courtesy of the U.S. Environmental Protection Agency, Office of Research & Development, Cincinnati, Ohio 45268.

1. *Euglena* (700X)
2. *Euglena* (700X)
3. *Phacus* (1000X)
4. *Phacus* (350X)
5. *Lepocinclis* (350X)
6. *Trachelomonas* (1000X)
7. *Phacotus* (1500X)
8. *Chlamydomonas* (1000X)
9. *Carteria* (1500X)
10. *Chlorogonium* (1000X)
11. *Pyrobotrys* (1000X)
12. *Chrysococcus* (3000X)
13. *Synura* (350X)
14. *Pandorina* (350X)
15. *Eudorina* (175X)
16. *Dinobyron* (1000X
17. *Peridinium* (350X)
18. *Ceratium* (175X)
19. *Gonium* (350X)
20. *Volvox* (100X)

Figure MO-2 Flagellated algae.

Index

Page numbers in *italic* indicate illustrations.